GAMMA-RAY BURSTS
SECOND WORKSHOP

AIP CONFERENCE PROCEEDINGS 307

GAMMA-RAY BURSTS

SECOND WORKSHOP
HUNTSVILLE, AL OCTOBER 1993

EDITORS:
GERALD J. FISHMAN
NASA/MARSHALL SPACE FLIGHT CENTER
JEROME J. BRAINERD
UNIVERSITY OF ALABAMA IN HUNTSVILLE
KEVIN HURLEY
UNIVERSITY OF CALIFORNIA, BERKELEY

American Institute of Physics New York

Authorization to photocopy items for internal or personal use, beyond the free copying permitted under the 1978 U.S. Copyright Law (see statement below), is granted by the American Institute of Physics for users registered with the Copyright Clearance Center (CCC) Transactional Reporting Service, provided that the base fee of $2.00 per copy is paid directly to CCC, 27 Congress St., Salem, MA 01970. For those organizations that have been granted a photocopy license by CCC, a separate system of payment has been arranged. The fee code for users of the Transactional Reporting Service is: 0094-243X/87 $2.00.

© 1994 American Institute of Physics.

Individual readers of this volume and nonprofit libraries, acting for them, are permitted to make fair use of the material in it, such as copying an article for use in teaching or research. Permission is granted to quote from this volume in scientific work with the customary acknowledgment of the source. To reprint a figure, table, or other excerpt requires the consent of one of the original authors and notification to AIP. Republication or systematic or multiple reproduction of any material in this volume is permitted only under license from AIP. Address inquiries to Series Editor, AIP Conference Proceedings, AIP Press, American Institute of Physics, 500 Sunnyside Boulevard, Woodbury, NY 11797-2999.

L.C. Catalog Card No. 94-71317
ISBN 1-56396-336-1
DOE CONF-9310252

Printed in the United States of America.

CONTENTS*

Preface xvii

GENERAL OBSERVATIONS AND GLOBAL STUDIES

General Burst Observations

Two and a Half Years of BATSE Burst Observations 3
    C. Meegan, G. Fishman, R. Wilson, M. Brock, J. Horack, W. Paciesas,
    G. Pendleton, M. Briggs, T. Koshut, C. Kouveliotou, B. Teegarden, J. Matteson,
    and J. Hakkila

Gamma Ray Bursts Observed with WATCH-EURECA 13
    S. Brandt, N. Lund, and A. J. Castro-Tirado

WATCH Observations of Gamma-Ray Bursts During 1990–1992 17
    A. J. Castro-Tirado, S. Brandt, N. Lund, I. Y. Lapshov, O. Terekhov, and
    R. A. Sunyaev

EGRET Observations of Three Gamma-Ray Bursts at Energies >30 MeV 22
    B. L. Dingus, P. Sreekumar, E. J. Schneid, D. L. Bertsch, C. E. Fichtel,
    A. K. Harding, R. C. Hartman, S. D. Hunter, D. J. Thompson, J. R. Mattox,
    K. Hurley, D. A. Kniffen, G. Kanbach, H. A. Mayer-Hasselwander, M. Sommer,
    C. von Montigny, Y. C. Lin, P. F. Michelson, and P. L. Nolan

The Ulysses Supplement to the First BATSE Catalog of Gamma-Ray
Bursts 27
    K. Hurley, M. Sommer, C. Kouveliotou, G. Fishman, C. Meegan, T. Cline,
    M. Boër, and M. Niel

Status of Burst Detection by the Mars Observer GRS Experiment 32
    J. G. Laros, W. Boynton, R. McCloskey, A. Metzger, R. Starr, and J. Trombka

DMSP Satellites as Gamma-Ray Burst Detectors 34
    J. Terrell, P. Lee, R. W. Klebesadel, and J. W. Griffee

Intensity Distributions and Isotropy Measures

The Effect of Repeating Burst Sources on $\langle V/V_{max} \rangle$ 39
    D. L. Band

The Isotropy of Gamma-Ray Bursts: Dipole and Quadrupole Tests 44
    M. S. Briggs, W. S. Paciesas, G. N. Pendleton, G. J. Fishman, C. A. Meegan,
    R. B. Wilson, M. N. Brock, and C. Kouveliotou

Model Independent Constraints on Heliocentric Gamma-Ray Burst
Models 49
    L. E. Brown, D. H. Hartmann, and L.-S. The

*NOTE: Authors should be contacted directly for preprints.

Are Gamma Ray Bursts Nearby? 54
    D. Eichler
Additional Constraints on Galactic Coronal/Extended Halo Source Distributions from BATSE Observations 59
    J. Hakkila, C. A. Meegan, G. J. Fishman, R. B. Wilson, M. N. Brock,
    J. M. Horack, G. N. Pendleton, W. S. Paciesas, and M. S. Briggs
Integral Moment Analysis of the BATSE Gamma-Ray Burst Intensity Distribution 64
    J. M. Horack and A. G. Emslie
Implications of the BATSE Data for a Heliocentric Origin of Gamma-Ray Bursts 69
    J. M. Horack, S. D. Storey, T. M. Koshut, R. S. Mallozzi, and A. G. Emslie
Evidence for the Galactic Origin of Gamma-Ray Bursts 74
    D. Q. Lamb and J. M. Quashnock
Comments on Galactic Anisotropy of Gamma-Ray Bursts 79
    H. Li, I. A. Smith, and R. C. Duncan
The Slope of the Log N-Log S Distribution for BATSE Gamma-Ray Bursts ... 84
    V. Petrosian, W. J. Azzam, and C. A. Meegan
Comparing and Combining the Gamma-Ray Bursts from SMM, SIGNE and BATSE 88
    V. Petrosian, W. J. Azzam, K. Hurley, and G. Share
Distribution of Peak Counts and Duration of Gamma-Ray Bursts 93
    V. Petrosian, T. T. Lee, and W. J. Azzam
On the Galactic Distribution of Gamma-Ray Bursts 98
    R. Rutledge and W. H. G. Lewin
Comments on C_{max}/C_{min} Distributions 103
    I. A. Smith

Studies of and Searches for Burst Repetition and Clustering

Evidence that Gamma-Ray Burst Sources Repeat 107
    J. M. Quashnock and D. Q. Lamb
The Angular Correlation Function of Gamma-Ray Bursts 117
    G. R. Blumenthal, D. H. Hartmann, and E. V. Linder
The Nearest Neighbors of BATSE Gamma-Ray Bursts: Narrowing the Possibilities 122
    J. J. Brainerd, W. S. Paciesas, C. A. Meegan, and G. J. Fishman
Do Classical Gamma-Ray Bursts Repeat? 127
    D. H. Hartmann, G. R. Blumenthal, K. Hurley, E. V. Linder, J. Hakkila,
    G. J. Fishman, C. A. Meegan, R. B. Wilson, M. Brock, J. M. Horack,
    C. Kouveliotou, M. S. Briggs, W. S. Paciesas, and G. N. Pendleton
Do Gamma-Ray Burst Sources Repeat? 132
    R. Narayan and T. Piran

Correlations with Gamma-Ray Bursts 137
    R. J. Nemiroff, G. F. Marani, J. R. Cebral, and J. P. Norris
Analysis of the Density Fluctuations in the Gamma-Ray Bursts
Distribution 141
    G. F. Zharkov, V. G. Kurt, and V. G. Zharkov
Search for Periodically-Repeating Gamma-Ray Bursters from BATSE 145
    K. W. Chuang
Searching Gamma-Ray Bursts for Gravitational Lensing Echoes 150
    R. J. Nemiroff, W. A. D. T. Wickramasinghe, J. P. Norris, C. Kouveliotou,
    G. J. Fishman, C. A. Meegan, and W. S. Paciesas
BATSE Burst Location Accuracy and Constraints on the Fraction of
Repeating GRB Sources 155
    T. E. Strohmayer, E. E. Fenimore, and J. A. Miralles
Repeating Sources of Classical Gamma-Ray Bursts 160
    V. C. Wang and R. E. Lingenfelter

TEMPORAL STUDIES

Two Classes of Gamma-Ray Bursts 167
    C. Kouveliotou, C. A. Meegan, G. J. Fishman, P. N. Bhat, M. S. Briggs,
    T. M. Koshut, W. S. Paciesas, and G. N. Pendleton
Exploration of Bi-Modality in Gamma-Ray Burst Duration and Hardness
Distributions 172
    J. P. Norris, R. J. Nemiroff, S. P. Davis, C. Kouveliotou, G. J. Fishman,
    C. A. Meegan, and W. S. Paciesas
Measuring the Distribution of Temporal Structure in Gamma-Ray Bursts ... 177
    J. P. Norris
Pulse Width Distributions and Total Counts as Indicators of Cosmological
Time Dilation in Gamma-Ray Bursts 182
    S. P. Davis, J. P. Norris, C. Kouveliotou, G. J. Fishman, C. A. Meegan,
    and W. S. Paciesas
The Average Temporal Profile of BATSE Gamma-Ray Bursts: Comparison
Between Strong and Weak Events 187
    I. G. Mitrofanov, A. M. Chernenko, A. S. Pozanenko, W. S. Paciesas,
    C. Kouveliotou, C. A. Meegan, G. J. Fishman, and R. Z. Sagdeev
Study of the Temporal Behavior of Fourteen Fast Gamma-Ray Bursts
of the SIGNE Experiment 192
    B. M. Belli
Morphological Study of Short Gamma Ray Bursts 197
    P. N. Bhat, G. J. Fishman, C. A. Meegan, R. B. Wilson, and W. S. Paciesas
Frequency-Dependent Energy Phase Lags in Gamma-Ray Bursts 202
    E. Chipman

**Duration Versus Brightness of Gamma-Ray Bursts: Comparisons
Between SIGNE and BATSE** 207
    V. E. Kargatis, H. Li, E. P. Liang, I. A. Smith, K. Hurley, C. Barat,
    and M. Niel
**Gamma-Ray Burst Variability: A Search for Correlations with Other
Parameters.** 212
    J. P. Lestrade, M. Briggs, W. B. Paciesas, G. J. Fishman, C. A. Meegan,
    and R. B. Wilson
**Effects of Pulse Shape on the Details of the "Fluence Edge" for Short
Bursts** 217
    H. Li, E. Liang, I. A. Smith, and V. Kargatis
Gamma-Ray Burst Repetition and the Definition of Bursts. 222
    R. E. Lingenfelter, V. C. Wang, and J. C. Higdon
Evidence for Two Distinct Morphological Classes of Gamma-Ray Bursts. 227
    C. Graziani and D. Q. Lamb
On Burst Classification Using Peak Rate Ratios 232
    C. Meegan and C. Kouveliotou
Time Asymmetry in Gamma-Ray Burst Light Curves 237
    R. J. Nemiroff, J. P. Norris, C. Kouveliotou, G. J. Fishman, C. A. Meegan,
    and W. S. Paciesas
On the Usefulness of "Variability" in Gamma-Ray Bursts 242
    R. Rutledge and W. H. G. Lewin

SPECTRAL STUDIES

**The Search for Gamma-Ray Burst Spectral Features in the Compton
GRO BATSE Data** 247
    D. M. Palmer, B. J. Teegarden, B. E. Schaefer, T. L. Cline, D. L. Band,
    L. A. Ford, J. L. Matteson, R. D. Preece, W. S. Paciesas, G. N. Pendleton,
    M. S. Briggs, G. J. Fishman, C. A. Meegan, R. B. Wilson, and J. P. Lestrade
Cyclotron Line Search: BATSE-Ginga Consistency 256
    D. Band, L. Ford J. Matteson, D. Palmer, B. Teegarden, B. Schaefer,
    M. Briggs, W. Paciesas, G. Pendleton, and R. Preece
**A Candidate Absorption Feature from a Gamma Ray Burst Observed
by BATSE** 261
    L. Ford, D. Band, J. Matteson, D. Palmer, B. Schaefer, B. Teegarden,
    R. Preece, M. Briggs, W. Paciesas, and G. Pendleton
Consistency Analysis of a Candidate Line Feature in GRB930506 266
    R. Preece, M. Briggs, W. Paciesas, G. Pendleton, L. Ford, D. Band,
    J. Matteson, D. Palmer, B. Teegarden, B. Schaefer, and M. Brock
BATSE Cyclotron Line Search Protocol 271
    B. E. Schaefer, B. J. Teegarden, D. M. Palmer, T. L. Cline, S. Mitruka,
    L. A. Ford, D. L. Band, J. L. Matteson, M. S. Briggs, W. S. Paciesas,
    G. N. Pendleton, and R. D. Preece

Spectral Properties of Gamma-Ray Bursts Observed by COMPTEL 275
  L. O. Hanlon, K. Bennett, O. R. Williams, C. Winkler, W. Collmar, R. Diehl,
  J. Greiner, V. Schöfelder, H. Steinle, A. Strong, M. Varendorff, R. van Dijk,
  J. W. den Herder, W. Hermsen, L. Kuiper, A. Connors, R. M. Kippen,
  M. McConnell, and J. Ryan

Cross Calibration of Burst Spectra with BATSE, EGRET, and COMPTEL for GRB910503 280
  B. E. Schaefer, B. Teegarden, T. Cline, B. L. Dingus, G. J. Fishman,
  C. A. Meegan, R. B. Wilson, W. S. Paciesas, G. N. Pendleton, D. L. Band,
  E. J. Schneid, P. W. Kwok, V. Schönfelder, C. Winkler, W. Hermsen,
  and R. M. Kippen

Comparison of BATSE, COMPTEL, EGRET, and OSSE Spectra of GRB910601 283
  G. H. Share, W. N. Johnson, J. D. Kurfess, R. J. Murphy, A. Connors,
  B. L. Dingus, B. E. Schaefer, D. Band, J. Matteson, W. Collmar,
  V. Schönfelder, C. E. Fichtel, P. W. Kwok, B. J. Teegarden, G. Fishman,
  L. Kuiper, G. V. Jung, S. M. Matz, P. L. Nolan, E. J. Schneid, and C. Winkler

Spectral Evolution of a Sub-Class of Gamma Ray Bursts 288
  P. N. Bhat, G. J. Fishman, C. A. Meegan, R. B. Wilson, C. Kouveliotou,
  W. S. Paciesas, G. N. Pendleton, and B. E. Schaefer

Decomposition of a Cosmic Gamma-Ray Burst into Two Non-Correlated Radiation Components 293
  A. M. Chernenko and I. G. Mitrofanov

Continuum Evolution of Bright Gamma Ray Bursts Observed by BATSE 298
  L. Ford, D. Band, J. Matteson, B. Teegarden, and W. Paciesas

A Study of Continuum Spectra of Short-Duration Gamma-Ray Bursts Observed by BATSE 303
  T. M. Koshut, G. N. Pendleton, R. S. Mallozzi, W. S. Paciesas,
  and M. S. Briggs

The Energy Emission of Gamma-Ray Bursts and Solar Flares 308
  R. S. Mallozzi, G. N. Pendleton, T. M. Koshut, W. S. Paciesas,
  and M. S. Briggs

Continuum Spectral Characteristics of Bursts Measured with the BATSE Large Area Detectors 313
  G. N. Pendleton, W. S. Paciesas, M. S. Briggs, R. S. Mallozzi, T. M. Koshut,
  G. J. Fishman, R. B. Wilson, C. A. Meegan, and C. Kouveliotou

Spectral Curvature in High-Energy Gamma Ray Bursts Observed by the BATSE Large Area Detectors 318
  R. D. Preece, M. S. Briggs, W. S. Paciesas, G. N. Pendleton, C. Kouveliotou,
  and M. N. Brock

The Color of Gamma-Ray Bursts 323
  F. De Paolis, S. Pezzuto, and M. Tavani

Color Diagrams of Gamma-Ray Bursts 328
  W. A. D. T. Wickramasinghe, R. J. Nemiroff, J. P. Norris, and C. Kouveliotou

A Spectral Study of an "X-Ray Rich" Gamma-Ray Burst 333
    A. Yoshida and T. Murakami
The Effect of an Intrinsic Column on GRB Spectra........................ 336
    A. Owens
Distance to Gamma Ray Bursts from Their Soft X-Ray Spectra 341
    B. E. Schaefer
The Compton Attenuation Model of Cosmological Gamma-Ray Bursts 346
    J. J. Brainerd
A Cold Absorption Model of Gamma Ray Burst Spectra 351
    E. P. Liang

BURST LOCALIZATIONS AND SEARCHES FOR COUNTERPARTS

Burst Localization

Precise Localizations and Counterpart Searches of GRBs from the 2nd Interplanetary Network... 359
    F. Hack, K. Hurley, J.-L. Atteia, C. Barat, M. Niel, T. Cline, B. Dennis,
    C. Kouveliotou, R. Klebesadel, J. Laros, V. Kurt, A. Kuznetsov,
    and V. Zenchenko
Comparison of WATCH and IPN Locations of Gamma-Ray Bursts 364
    K. Hurley, N. Lund, S. Brandt, A. J. Castro-Tirado, M. Sommer, I. Lapshov,
    J. Laros, R. Klebesadel, G. Fishman, C. Kouveliotou, C. Meegan, T. Cline,
    M. Boër, and M. Niel
Precise Triangulation of the Jan 31 1993 ("Superbowl") Burst 369
    K. Hurley, M. Sommer, G. Fishman, C. Kouveliotou, C. Meegan, T. Cline,
    M. Boër, and M. Niel
Study of the Precision of the Gamma-Ray Burst Source Locations Obtained with the Ulysses/PVO/CGRO Network 373
    T. L. Cline, K. C. Hurley, M. Sommer, M. Boër, M. Niel, G. J. Fishman,
    C. Kouveliotou, C. A. Meegan, W. S. Paciesas, R. B. Wilson, J. G. Laros,
    and R. W. Klebesadel
A Search for High Energy Gamma-Ray Bursts in the EGRET Data Utilizing Space-Time Correlation.. 377
    R. Buccheri, M. C. Maccarone, J. R. Mattox, D. J. Thompson, G. Kanbach,
    U. Camerini, and W. F. Fry

Counterparts—General

Search for Gamma Ray Burst Counterparts 382
    B. E. Schaefer

Optical Counterparts

Rapid Optical Follow-Up Observations of Three Recent Gamma Ray Bursts 392
    S. D. Barthelmy, D. M. Palmer, and B. E. Schaefer

The ESO-Schmidt Survey of Gamma-Ray Bursts 396
    M. Boër, H. Pederson, A. Smette, J. Fishman, C. Kouveliotou, and K. Hurley

A Real-Time Search for Optical Counterparts to Gamma-Ray Bursts 401
    J. T. Bonnell, J. P. Norris, S. D. Barthelmy, T. L. Cline, N. Gehrels, G. J. Fishman, C. Kouveliotou, and C. A. Meegan

Optical Follow-Up of Gamma-Ray Bursts Observed by WATCH 404
    A. J. Castro-Tirado, S. Brandt, N. Lund, and S. S. Guziy

Simultaneous Optical/Gamma-Ray Observations of GRBs 408
    J. Greiner, W. Wenzel, R. Hudec, E. I. Moskalenko, V. Metlov, N. S. Chernych, V. S. Getman, R. Ziener, K. Birkle, N. Bade, S. B. Tritton, G. J. Fishman, C. Kouveliotou, C. A. Meegan, W. S. Paciesas, and R. B. Wilson

Gamma Ray Bursts at Optical Wavelengths: First Optically Identified GRB? 413
    R. Hudec and J. Soldán

First Results of the BATSE/COMPTEL/NMSU Rapid Burst Response Campaign. 418
    R. M. Kippen, A. Connors, J. Macri, M. McConnell, J. Ryan, W. Collmar, J. Greiner, V. Schönfelder, M. Varendorff, G. J. Fishman, C. Meegan, C. Kouveliotou, B. McNamara, T. Harrison, W. Hermsen, L. Kuiper, K. Bennett, L. Hanlon, and C. Winkler

Searches for Optical Counterparts of BATSE Gamma-Ray Bursts 423
    H. A. Krimm, R. K. Vanderspek, and G. R. Ricker

Stellar Flares and Gamma-Ray Bursts 428
    P. Li, K. Hurley, G. J. Fishman, and C. Kouveliotou

X-Ray and Optical Observations of the COMPTEL Error Box for GRB910601 433
    B. McNamara, T. Harrison, and C. Williams

The Search for Optical Transients with the Explosive Transient Camera 438
    R. Vanderspek, H. A. Krimm, and G. R. Ricker

Results from the USNO Gamma-Ray Burst Optical Counterpart Search 443
    F. J. Vrba, D. H. Hartmann, and M. C. Jennings

Is a QSO the Source of OT050510, and is GRB910219 Related? 448
    F. J. Vrba, C. B. Luginbuhl, D. H. Hartmann, R. Hudec, F. H. Chaffee, C. B. Foltz, and K. C. Hurley

X-Ray Counterparts

The X-Ray Survey of the Second Catalog Gamma-Ray Burst Error Boxes 453
    M. Boër, J. Greiner, P. Kahabka, C. Motch, and W. Voges

Recent Small Gamma-Ray Burst Error Boxes in the ROSAT All-Sky Survey 458
    M. Boër, J. Greiner, P. Kahabka, C. Motch, W. Voges, M. Sommer,
    K. Hurley, M. Niel, J. Laros, R. Klebesadel, C. Kouveliotou,
    G. Fishman, and T. Cline

X-Ray Burst Rates from ROSAT 463
    A. Kahn and H. Ögelman

An ASCA Attempt of GRB Observation in X-Ray Range 466
    A. Yoshida, Y. Ogasaka, and T. Murakami

Very High Energy Counterpart Searches

Searches for Bursts of TeV Gamma Rays on Time-Scales of Seconds 470
    V. Connaughton, M. Chantell, A. C. Rovero, T. Whitaker, T. C. Weekes,
    C. W. Akerlof, D. I. Meyer, M. S. Schubnell, D. J. Fegan, S. Fennell,
    J. Hagan, N. A. Porter, M. Punch, J. Gaidos, G. Sembroski, C. Wilson,
    A. M. Hillas, J. Rose, M. West, A. D. Kerrick, P. Kwok, D. A. Lewis,
    R. C. Lamb, and G. Mohanty

Soudan 2 Muons in Coincidence with BATSE Bursts 475
    D. M. DeMuth, M. L. Marshak, and G. L. Wagner

Search for Ultra High Energy Radiation from Gamma-Ray Bursts 481
    R. Schnee (Representing the CYGNUS Collaboration)

Soft Gamma-Ray Repeaters—Counterparts

The Identification of a Supernova Remnant with a Soft Gamma Ray Repeater 486
    D. A. Frail and S. R. Kulkarni

X-Ray Identification of SGR1806-20 489
    T. Murakami, T. Sonobe, Y. Ogasaka, T. Aoki, A. Yoshida,
    and S. R. Kulkarni

BURST ORIGINS AND EMISSION PROCESSES (THEORY)

Cosmological Models

Fireballs 495
    T. Piran

Shock Models and O, X, γ Signatures of Gamma-Ray Burst Sources 505
    P. Mészáros and M. J. Rees

Gamma-Ray Bursts and Gamma-Ray Blazars 510
    C. D. Dermer and R. Schlickeiser

Focusing of Alfvénic Power in Neutron Star Magnetospheres 515
    M. Fatuzzo and F. Melia
Escape of High-Energy Photons from Relativistically Expanding Gamma-Ray Burst Sources 520
    A. K. Harding and M. G. Baring
Gamma-Ray Bursts from Black Hole–Neutron Star Mergers 525
    J. Isern, M. Hernanz, R. Mochkovitch, and X. Martin
Radio and Optical Emission, Spectral Shapes and Breaks in GRB 529
    J. I. Katz
Axion Bursts from Supernovae at Cosmological Distances 533
    A. Loeb
Gamma-Ray Bursts from Relativistic Beams in Neutron Star Mergers 537
    R. Mochkovitch, S. Loiseau, M. Hernanz, and J. Isern
Radio and Neutrino Emission from Theoretical Gamma-Ray Bursters 542
    B. Paczyński, J. Rhoads, and G. Xu
Gamma-Ray Bursts from Neutron Star Mergers 543
    T. Piran
GRBs from Compton Drag of Relativistic Flows 547
    A. Shemi
Cosmic Fireballs and Gamma Ray Bursts 548
    A. Shemi
On the Nature of Nonthermal Radiation from Cosmological Gamma Ray Bursters 552
    V. V. Usov
Relativistic Bulk Motion and Statistics of Beamed Gamma-Ray Bursts 557
    I. Yi

Galactic Models

Searching for a Galactic Origin of Gamma-Ray Bursts 562
    D. H. Hartmann
Gamma-Ray Burst Continuum Spectra from Magnetic Inverse Compton Scattering 572
    M. G. Baring
The Observation of Unusual Gamma Ray Bursts from Primordial Black Hole Evaporation 577
    D. B. Cline and W. Hong
Gamma-Ray Bursts from the Accretion of Solid Bodies onto High-Velocity Galactic Neutron Stars 581
    S. A. Colgate and P. J. T. Leonard
Dual Population, Galactic Neutron Star Models of Gamma-Ray Bursts Revisited 586
    J. C. Higdon and R. E. Lingenfelter

Halo Population of Quark Nuggets as Gamma-Bursters: The Glow of Dark Matter? 591
    J. E. Horvath

Nearby Neutron Stars as the Sources of the Gamma-Ray Bursts 596
    W. Kundt and H.-K. Chang

Beamed Gamma-Ray Bursts from the Galactic Halo: Model Comparisons with BATSE Data 600
    H. Li, R. Duncan, and C. Thompson

Why 'Galactic' Gamma-Ray Bursts Might Depend on Environment: Blast Waves Around Neutron Stars 605
    M. J. Rees, P. Mészáros, and M. C. Begelman

Galactic Arm and Disk Plus Halo Models of Gamma-Ray Burst Sources 610
    I. A. Smith

A Possible Cyclotron Line Signature from Quiescent Gamma-Ray Burst Counterparts 615
    J. C. L. Wang and R. W. Nelson

BATSE Requirements for a Colliding Comet Source of Gamma-Ray Bursts 620
    R. S. White

Soft Gamma-Ray Repeaters—Theory

Astrophysics of Very Strongly Magnetized Neutron Stars: A Model for the Soft Gamma Repeaters 625
    R. C. Duncan and C. Thompson

INSTRUMENTATION AND NEW ANALYSIS TECHNIQUES

Gamma-Ray Optical Counterpart Search Experiment (GROCSE) 633
    C. Akerlof, M. Fatuzzo, B. Lee, R. Bionta, A. Ledebuhr, H.-S. Park,
    S. Barthelmy, T. Cline, and N. Gehrels

Albedo Effect on the Expected In-Flight Performance of the Gamma-Ray Burst Monitor on Board the SAX Satellite 638
    F. Alberghini, D. Dal Fiume, F. Frontera, and G. Pizzichini

BACODINE: The Real-Time BATSE Gamma-Ray Burst Coordinates Distribution Network 643
    S. D. Barthelmy, T. L. Cline, N. Gehrels, T. G. Bialas, M. A. Robbins,
    J. R. Kuyper, G. J. Fishman, C. Kouveliotou, and C. A. Meegan

BATSE: Burst Performance and Experiment Status 648
    G. J. Fishman, C. A. Meegan, C. Kouveliotou, R. Mallozzi, J. Horack,
    T. Koshut, G. Pendleton, W. S. Paciesas, R. B. Wilson, and M. N. Brock

The Fourth Interplanetary Network: Arcsecond Localizations from Spacecraft at 100 AU 653
        K. Hurley and T. Cline

X-Ray Telescope Array for Gamma-Ray Burst Localization 658
        N. Kawai

X-Rays from Gamma Ray Bursts 662
        E. P. Liang

RULER: An Instrument to Measure Gamma-Ray Burster Distances 665
        A. Owens, J. Greiner, T. Mineo, K. Pounds, M. Rees, B. Sacco, L. Scarsi,
        B. Schaefer, S. Sembay, O. Terekhov, and A. Wells

Ideas for a Large Detector for Optical Transients 670
        H. Pedersen, M. Andersen, and M. Boër

The Kapteyn (Lognormal) Distribution as a Model of GRB Spectra and Pulse Profiles 672
        M. Brock, C. Meegan, R. Wilson, C. Kouveliotou, W. Paciesas, G. Pendleton,
        M. Briggs, and R. Preece

New Techniques in the Fitting of Gamma-Ray Burst Cyclotron Lines 677
        P. E. Freeman, C. Graziani, D. Q. Lamb, and T. J. Loredo

Timing Accuracy of the Ulysses GRB Experiment 682
        K. Hurley and M. Sommer

Cross-Correlating Gamma-Ray Burst Time Histories 687
        K. Hurley

Trigger Efficiencies of BATSE and PVO 692
        J. J. M. in 't Zand and E. E. Fenimore

An Evaluation of BATSE Burst Locations Computed with the MAXBC Datatype 697
        T. M. Koshut, W. S. Paciesas, G. N. Pendleton, M. N. Brock,
        G. J. Fishman, C. A. Meegan, and R. B. Wilson

Wavelet Analysis of Gamma Ray Bursts 701
        D. C. Meredith, J. M. Ryan, C. A. Young, and J. P. Lestrade

A Search Technique for Weak and Long-Duration Gamma-Ray Bursts from Background Model Residuals 706
        R. T. Skelton and W. A. Mahoney

Using Contour Maps to Search for Red-Shifted 511 keV Features in BATSE GRB Spectra 711
        P. G. Varmette, J. P. Lestrade, G. J. Fishman, C. A. Meegan, M. N. Brock,
        R. D. Preece, M. S. Briggs, G. N. Pendleton, and W. S. Paciesas

MISCELLANEOUS

Bursters and the Quest for "Cosmological Effects" 717
        V. Trimble

The Angular Size of Cosmic Gamma-Ray Bursts Sources 722
        I. G. Mitrofanov

A Gamma-Ray Burst Bibliography, 1973–1993 726
    K. Hurley
A Century of Gamma-Ray Burst Models 730
    R. J. Nemiroff
Gamma-Ray Burst Workshop Attendees 735
Author Index 741

PREFACE

It is now three years after the launch of the Compton Gamma-Ray Observatory (CGRO), and the enigma of gamma-ray bursts is as great as ever. The discovery by the Burst and Transient Source Experiment (BATSE) on CGRO that bursts are distributed isotropically and inhomogeneously has altered the thinking of many theorists. Burst models once again cover all distance scales. The once dominant galactic neutron star model of gamma-ray bursts, while still prominent, now occupies a minority position. Numerous new cosmological models have appeared, and old ones have been revived. These models associate gamma-ray bursts with merging compact stars, massive black holes, or other, often more exotic, sources. Even solar system models have their advocates. Despite this diversity, a definite shift to cosmological models has occurred over the past two years, reinforced by the growing BATSE data base.

The Second Huntsville Gamma-Ray Burst Workshop was held in Huntsville, Alabama, USA, from the 20th to the 22nd of October, 1993. It was the largest conference on gamma-ray bursts held to date. While theorists attended in large numbers, the principle focus of the workshop was the presentation and interpretation of recent observational results. Many contentious issues were presented and debated at length.

In these proceedings an attempt has been made to group together similar observations and issues. However, the reader is cautioned that many papers could have fit equally well in more than one section. It is suggested that one should at least superficially peruse all of the titles and abstracts for papers of interest on a particular topic.

Many attempts to solve the mystery of burst origins fall into two categories: statistical tests on ensembles of bursts and observational attempts to locate burst counterparts. In the first category are attempts to find evidence of burst repetitions, of deviations from isotropy, of time dilation, and of morphological differentiation. In the second category are the searches of burst locations for quiescent sources at other wavelengths, the searches of archives for transient sources, and the simultaneous observation of bursts at other wavelengths. Neither category of test has yet provided unambiguous answers. Although detailed spectral and temporal analyses of large numbers of bursts have been made, the results have been either difficult to characterize in a systematic way or subject to controversy.

In spite of this, there have been two possibly significant breakthroughs: one is that the distribution of burst durations is double peaked; the other is evidence of a correlation of burst duration with burst intensity in the manner expected for a cosmological time dilation. There has also been at least one very significant development in the spectral observations of individual bursts: emission into the GeV energy range is clearly seen by EGRET on CGRO in several bursts. Simultaneous observations of gamma-ray burst spectra by all four experiments on CGRO were presented at the workshop, showing the ability of CGRO to derive an accurate spectrum over five decades of photon energy for the strongest bursts. The issue of whether the continued absence of spectral lines in the bursts observed by BATSE is inconsistent with the observations of KONUS, Ginga, and HEAO was discussed and is represented in this volume by several papers.

The most contentious topic at the workshop was the claim that some burst sources produce more than one gamma-ray burst. The impetus of this claim were analyses by several different groups of the First BATSE Gamma-Ray Burst Catalog, which was released early in 1993 and sparked a rapid flurry of new work. In this volume 11 papers address this topic, and at the workshop two hours of open debate were devoted to this topic. The importance of the outcome is clear: many burst models are unable to produce more than one burst per source.

Observing an optical transient simultaneously with a gamma-ray burst, or finding a counterpart at any wavelength within hours of a burst, is believed by many to be the best hope of solving the burst mystery. For this reason a number of new optical and radio search programs have been implemented or planned. Recent results from some of these programs are presented.

The mystery shrouding soft-gamma repeaters was largely resolved by multiwavelength counterpart observations. Shortly before this workshop, ASCA and BATSE simultaneously observed an outburst from SGR1806-20, placing the source within a galactic radio supernova remnant recently identified by Kulkarni and Frail. We were fortunate to have these results presented at a special session at the workshop and in these proceedings.

Resolution of the mystery surrounding classical gamma-ray bursts may require observations by new experiments or new techniques of analyzing current burst data. Among the ideas discussed at the workshop are improved methods of burst localization, which would enable deeper counterpart searches, and experiments that determine a burster's distance scale by measuring the photoelectric absorption of soft gamma rays by the galactic interstellar medium.

Many persons helped to make the workshop a success. We are grateful to the following persons for assisting with the organization of the workshop program: C. Kouveliotou, D. Q. Lamb, O. Blaes, and T. L. Cline. Those that served as session chairs did an outstanding job of staying within difficult time limitations for talks, while allowing for stimulating and beneficial discussions. They included E. Fenimore and D. Hartmann, as well as the above named. J. P. Lestrade provided the manuscript TeX macros. M. Rees, on short notice, prepared and gave and enlightening summary of the workshop.

Finally, we thank Dannah McCauley, Paula Cushman, and other staff members of the Universities Space Research Association for the excellent logistical support they provided for the workshop.

G. J. Fishman
J. J. Brainerd
K. Hurley

Gamma-Ray Burst Workshop

Huntsville, October 20–22, 1993

GENERAL OBSERVATIONS AND GLOBAL STUDIES

General Burst Observations
Intensity Distributions and Isotropy Measures
Studies of and Searches for Burst Repetition and Clustering

TWO AND A HALF YEARS OF BATSE BURST OBSERVATIONS

Charles Meegan, Gerald Fishman, Robert Wilson, Martin Brock, John Horack
NASA/Marshall Space Flight Center

William Paciesas, Geoffrey Pendleton, Michael Briggs, Thomas Koshut
University of Alabama, Huntsville

Chryssa Kouveliotou
USRA/Marshall Space Flight Center

Bonnard Teegarden
Goddard Space Flight Center

James Matteson
University of California, San Diego

Jon Hakkila
Mankato State University

ABSTRACT

We summarize the global properties of 743 gamma-ray bursts observed by BATSE since launch in April of 1991. We emphasize those observations relevant to the central problem of the spatial distribution of the burst sources. The high degree of isotropy, combined with an intensity distribution inconsistent with homogeneity, severely constrains geometrical models for source distributions. Various representations of the intensity distribution are discussed. Evidence for repeating sources in the first BATSE catalog do not appear to be confirmed in later bursts. The distribution of burst durations shows a bimodality with peaks at about 0.3 seconds and 30 seconds. Spectral characteristics are summarized.

INTRODUCTION

The Burst and Transient Source Experiment (BATSE) on the Compton Gamma Ray Observatory has detected 743 cosmic gamma-ray bursts between activation of the instrument on April 19, 1991 and August 17, 1993. BATSE consists of eight detector modules, situated at the corners of the GRO spacecraft. Each module contains a Large Area Detector (LAD) comprising a 50.8 cm diameter by 1.27 cm thick NaI(Tl) scintillator, and a Spectroscopy Detector (SD) comprising a 12.7 cm diameter by 7.62 cm thick NaI(Tl) scintillator. The instrument recognizes bursts by testing for statistically significant increases in the background count rate of the LADs on time scales of 64, 256, and 1024 ms. The trigger threshold corresponds to a flux of about 0.2 photons cm^{-2} s^{-1}. Details of the instrument are described elsewhere[1].

Early results from BATSE showed that the angular distribution of gamma-ray bursts is isotropic, while the brightness distribution implied a decrease in the density of sources with distance[2]. Galactic disk components do not exhibit these features, leading to models in which the bursts sources are thought to reside either in an extended galactic halo or at cosmological distances.

The First BATSE Burst Catalog contains data on 260 bursts observed up to March 5, 1992. Information available includes locations, peak count rates and thresholds, fluxes, fluences, and durations. The data are available from the GRO Science Support Center. A catalog that also includes plots of time histories is in press[3].

This review will emphasize those aspects of the BATSE observations directly relevant to the question "Where are the burst sources?" Specifically, the angular and brightness distributions are discussed in detail. We also describe the duration distribution and attempts to classify bursts. The important topic of spectral analysis, including the search for cyclotron lines, is briefly summarized.

ANGULAR DISTRIBUTION

The direction to each burst is determined using the relative count rates on the eight LADs. The locations have a statistical error that is about 13 degrees for the weakest events. There is also a systematic error that is about 4 degrees for the 260 bursts in the first catalog and currently about 6 or 7 degrees for subsequent bursts. We expect to reprocess all the subsequent bursts using the current best location algorithm, which should yield a 4 degree systematic error for all bursts. As systematic effects become better understood, further improvements in location accuracy are still possible. The accuracy of burst locations has been verified by locating solar flares, IPN-located bursts, Cygnus X-1 fluctuations, and by the observed cutoff in the geocenter angle distribution. The exposure as a function of celestial coordinates has been determined for the first 260 bursts. It exhibits a small quadrupole moment (due to earth blockage) and dipole moment (due to the South Atlantic Anomaly) in equatorial coordinates.

The locations of 743 bursts in galactic coordinates is shown in Figure 1. A variety of statistical tests for isotropy are available. The most directly relevant to galactic models are the dipole and quadrupole moments in galactic coordinates. A measure of the dipole moment is $\langle \cos \theta \rangle$, where θ is the angle between the burst location and the galactic center. A measure of the quadrupole moment is $\langle \sin^2 b - 1/3 \rangle$, where b is the galactic latitude. These measures are both zero for an isotropic distribution of bursts. The values derived from the BATSE data for 743 bursts, corrected for sky exposure, are $\langle \cos \theta \rangle = 0.031 \pm 0.021$ and $\langle \sin^2 b - 1/3 \rangle = -0.007 \pm 0.011$. The error bars are the statistical errors resulting from the finite sample size. Errors due to location measurement uncertainty and sky exposure uncertainty are negligible [4]. These data are plotted on Figure 2. The moments of the angular distribution, particularly the dipole moment, indicate the distance scale for extended galactic halo models. Hartmann et al.[5] pointed out that any spherically symmetric distribution of sources can be constructed by adding shells of sources, each of which contributes a dipole moment determined by its radius. Figure 2 includes a scale of shell radii in kpc. The extended galactic halo model requires most burst sources to be at least 100 kpc from the galactic center, regardless of the freedom to adjust the form of the radial distribution.

Additional details concerning the isotropy of the BATSE sample of bursts are presented by Briggs et al.[6], who analyze various subsets of the burst sample, and also consider coordinate system independent measures of anisotropy, as described by Briggs[7]. The sample of 743 bursts show no convincing evidence for statistically significant deviation from isotropy in either the dipole or quadrupole moments. There has been a report of significant dipole and quadrupole moments in a subset of the first 260 bursts[8], but the anisotropy is not present in the subset of subsequent bursts selected with the same criteria[6].

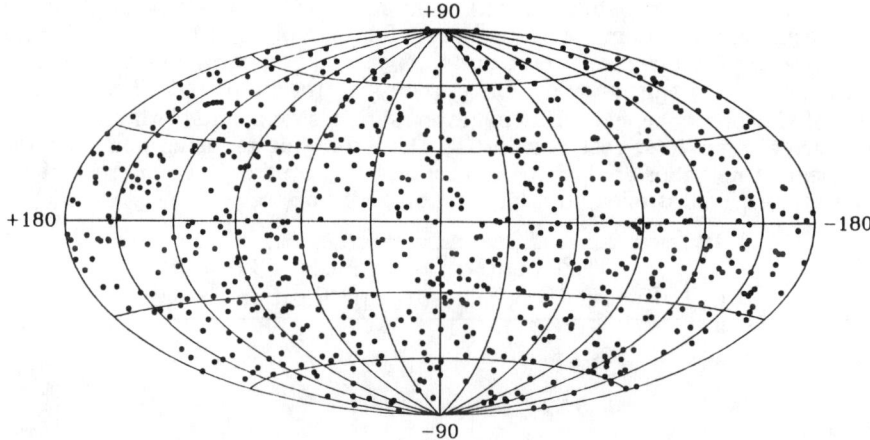

Figure 1. The distribution of 743 bursts in galactic coordinates.

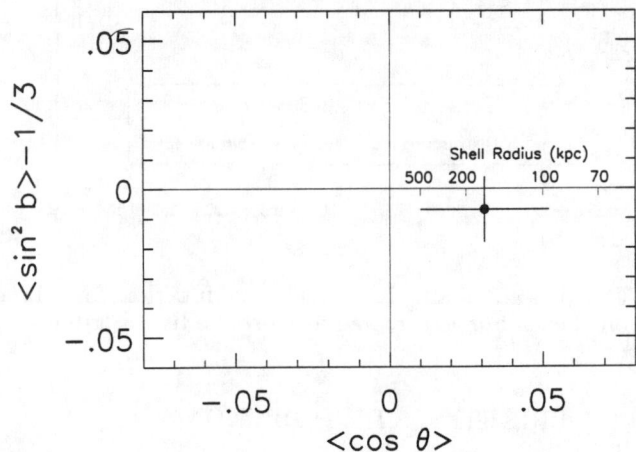

Figure 2. Dipole and quadrupole moments of the burst distribution.

The existence of anisotropy on smaller angular scales can be tested by the two-point angular correlation function (TPAC) or by nearest neighbor analysis. The first BATSE catalog exhibited a weak peak in the TPAC at small angles, as well as an excess in the nearest neighbor analysis. This was interpreted as evidence for either repeaters[9] or a systematic effect in the computed BATSE locations[10]. Brainerd et al.[11] reviewed the nearest neighbor analysis and concluded that the effect was of marginal statistical significance. The excess occurs on an angular scale of about 4 degrees, significantly less than the typical error in burst location. However, Quashnock and Lamb[12] argue that multiple repetitions will produce such an effect in the nearest neighbor analysis. The excess of nearby bursts is not evident in data subsequent to the first BATSE catalog. Figure 3 shows the TPAC for 743 bursts, using a technique that treats the location of each burst as a probability distribution given by the location error estimate. This technique results in correlated errors for the points in Figure 3. The point at zero angle gives a measure of burst repeaters. There is no evidence for repeating bursts in this sample. The crosses give the predicted TPAC assuming 10% of the bursts repeat. Hartmann et al.[13] also concluded that there is no evidence of repeaters, based on the TPAC of 743 bursts, using point locations, and nearest neighbor analysis. Resolution of this controversy may require more accurate burst locations and a larger sample of bursts.

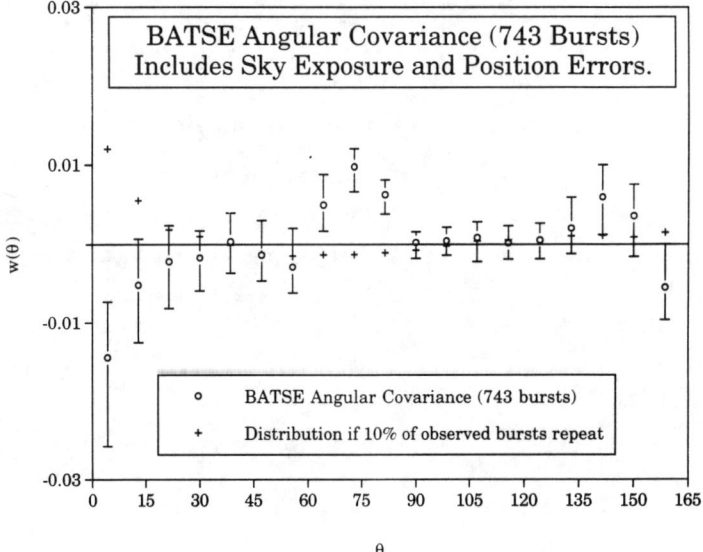

Figure 3. The two-point angular correlation function for 743 BATSE bursts. The position of each burst is treated as a probability distribution defined by its location error.

BRIGHTNESS DISTRIBUTION

There are several conventional ways to characterize the distribution of burst intensity, each with its own advantages and disadvantages. The best test for homogeneity is the now well-known $V/V_{\max}$ test[14]. For 520 BATSE bursts for which these data are available, $V/V_{\max} = 0.321 \pm 0.013$. This result provides the single important datum that the burst sources are not homogeneous within the volume of space sampled by BATSE. Although the measured value of $V/V_{\max}$ is occasionally used for model testing, it is important to remember that this is equivalent to characterizing the brightness distribution by a single instrument-specific number. Different experiments obtain different values of $V/V_{\max}$ because they sample different volumes according to their sensitivity. If BATSE sensitivity were reduced by a factor of 20, we would obtain $V/V_{\max} = 0.47 \pm 0.05$. Figure 4 shows the integral number of bursts as a function of $C_{\max}/C_{\min}$, where $C_{\max}$ is the maximum count rate during a burst and C_{min} is the threshold count rate for burst detection. This form of presentation shares with V/V_{max} the advantage that threshold effects are taken into account in a way that provides a valid test for homogeneity. This is achieved, however, at the cost of a measure of intensity that convolves detector and burst characteristics equally. It is not strictly valid to compare such plots from different instruments or to fit models to them. Additional complications specific to the BATSE triggering scheme are that the rates apply to the second most brightly illuminated detector (since two are required for a burst trigger), and that the rates apply to different time scales for different bursts (since the highest C_{max}/C_{min} value is used). Petrosian, et al.[15] describe a statistical technique to produce peak rate distributions that remove the bias caused by a varying threshold.

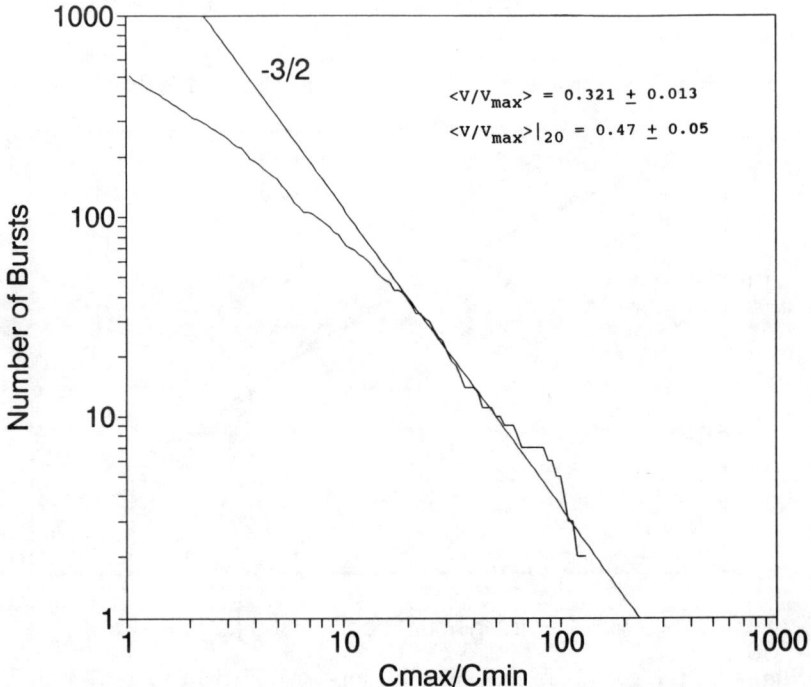

Figure 4. Intensity distribution for BATSE bursts. The measure of intensity is the maximum count rate divided by the threshold count rate.

A more physically meaningful measure of burst brightness is peak flux. An integration time scale and energy range must be specified to define the peak flux, and corrections for trigger efficiency must be applied to the number of bursts observed near the trigger threshold. The efficiency corrections are managable only if the time scale and energy range are those employed by the burst trigger criteria. For BATSE, the energy range is 50 to 300 keV, and the time scale can be 64, 256, or 1024 ms. The appropriate time scale to use depends on the purpose. When comparing results from different experiments, similar time scales should be used. Any time scale could be used for model fitting, but the 1024 ms threshold appears to sample the largest volume. It shows the greatest deviation from homogeneity and therefore provides the greatest constraints on models. Figure 5 shows the integral peak flux distribution for 455 BATSE bursts for the 1024 ms integration time. The solid area below 0.3 photons/cm^2-s represents the trigger efficiency correction. Its lower bound is the number of detected bursts detected, while its upper bound is defined by a correction factor that does not include atmospheric scattering. Since this correction is known to be too large at fluxes very close to the threshold, the true curve lies somewhere within the solid region. Also shown on Figure 5 is the $-3/2$ power law expected for homogeneity. The figure clearly shows that the deviation from homogeneity occurs at fluxes well above the instrument threshold. An important advantage of peak flux as a measure of burst intensity is that it is in principle possible to combine results for different experiments. Fenimore et al.[16] have combined PVO and BATSE data to produce an intensity distribution covering three decades in peak flux.

A disadvantage of the integral peak flux distribution is that the errors on the points are correlated, which can mislead the eye. Figure 6 shows the same data plotted as a differential distribution in equal logarithmic flux bins.

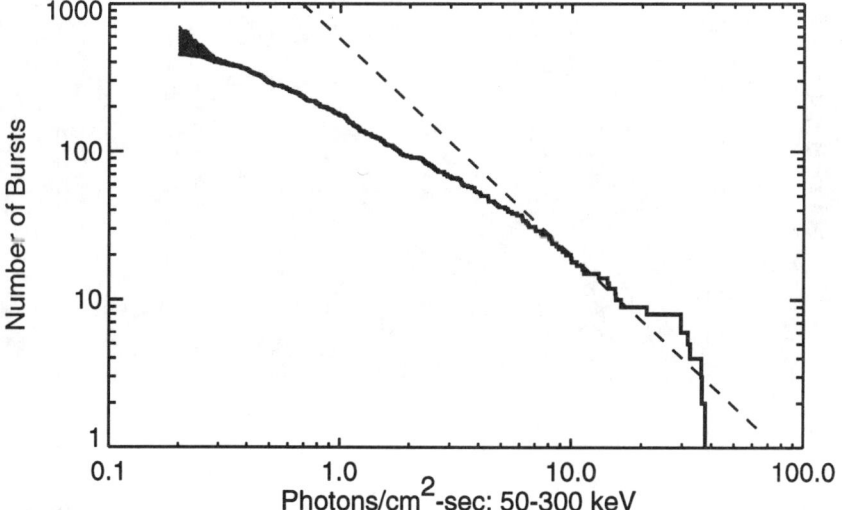

Figure 5. Integral peak flux distribution for 455 BATSE bursts that exceeded the 1024 ms trigger threshold, using a 1024 ms integration time for defining peak flux.

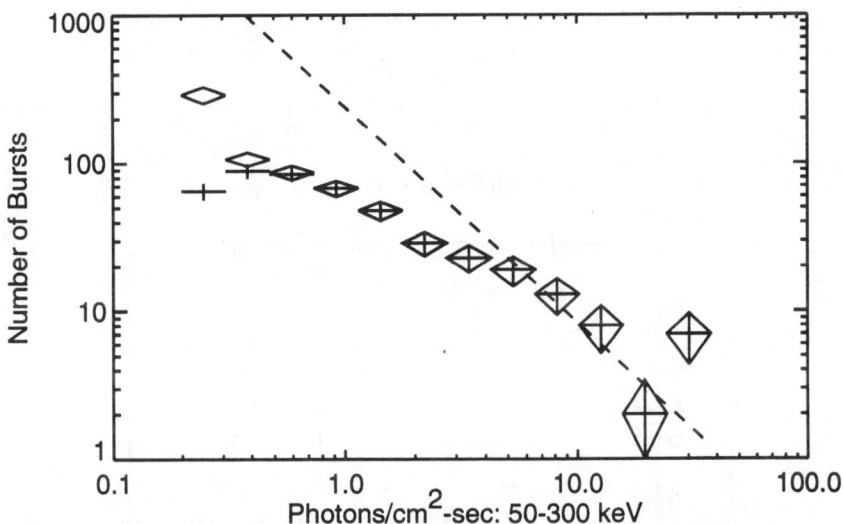

Figure 6. Differential peak flux distribution for 455 BATSE bursts that exceeded the 1024 ms trigger threshold, using a 1024 ms integration time for defining peak flux.

A number of standard statistical techniques are available for comparing the observed intensity distribution to theoretical predictions. Horack & Emslie[17] propose using the moments of the intensity distribution. This technique has the advantages of reducing the brightness distribution to just a few relevant numbers and nicely handles the convolution of the luminosity and distance distributions in the model.

Extensive modeling of geometrical distributions of galactic sources has been performed by Hakkila et al.[18,19] These studies demonstrate the importance of including more information on the intensity distribution than just the value of V/V_{max}, and the significance of the failure to observe bursts from M31 in constraining galactic models.

BURST CLASSES

The attempt to discover distinct classes of bursts has been largely futile. However, BATSE does confirm earlier reports[20,21] that the distribution of burst durations exhibits a bimodality. Figure 7 shows the number of bursts as a function of T_{90}, which is the time interval during which 90% of the total counts arrive. There is evidence for two peaks, with a deficit of bursts at around 2 seconds. Further details are provided by Kouveliotou et al.[22,23] The duration measures T_{90} and a similarly defined T_{50} are, to a large extent, unbiased by burst intensity. Instruments with very different sensitivities and trigger criteria should derive very similar T_{90} or T_{50} durations. An exception would be a burst that had a weak, well-separated precursor or postcursor that might not be recognized in a proportionally less intense burst, or by a less sensitive instrument.

Lamb et al.[24] proposed classifying bursts by a variability parameter V, which is defined as the ratio of peak counts on the 64 ms and 1024 ms time scales. This parameter is said to sort bursts into two classes: Type 1, which are variable, short, and generally weaker, and Type 2, which are smooth, long, and generally stronger. Critiques of this classification scheme argue that V simply distinguishes long bursts from short bursts by definition, that it does not measure variability, and that bursts observed by PVO do not verify the claimed correlation with intensity[25,26,27].

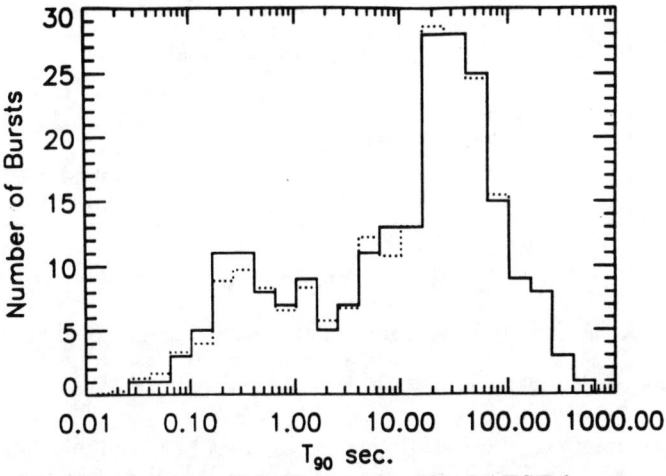

Figure 7. The duration distribution for 222 BATSE bursts, as measured by T_{90}. The solid histogram represents the raw data; the dashed histogram represents the data convolved with measurement errors.

SPECTRAL PROPERTIES

This section presents a brief summary of the spectral properties of bursts detected by BATSE. The search for spectral features has so far produced no convincing cyclotron line candidates. Details on this crucial program are presented by Palmer et al.[28].

Band et al.[29] found that BATSE burst spectra are well represented at low energies by a power law with an exponential cutoff, and a steeper power law at higher energies. Figure 8 shows a histogram of the number of bursts as a function of the energy at which the energy flux per logarithmic energy band peaks. Pendleton et al.[30] fit single power law spectra to 260 bursts in three energy intervals. The spectra were generally softer in higher energy bands. Figure 9 shows a distribution of spectral indices in the 50 to 300 keV band.

BATSE has good sensitivity for detecting rapid spectral evolution. Ford et al.[31] find that there is, in general, a softening of the spectra with time, both within individual pulses in a burst, and over the entire burst duration. Spectral evolution on a time scale of about 2 ms was seen in the very strong burst GB930131[32].

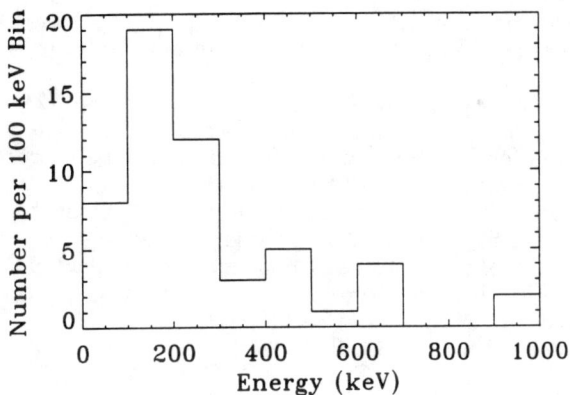

Figure 8. The distribution of the energy of the peak emission per unit logarithmic energy interval.

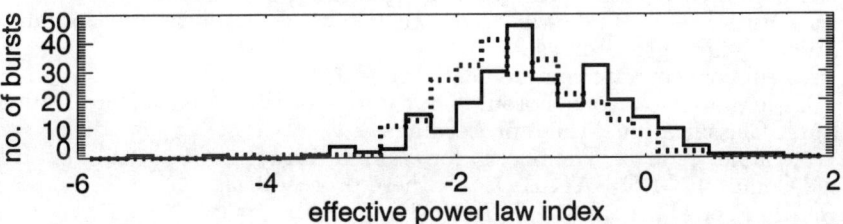

Figure 9. The distribution of burst spectral indices in the 50 to 300 keV energy band. The solid line represents the distribution for the peak rate spectrum and the dotted line represents the distribution for the total fluence spectrum.

REFERENCES

1. G. Fishman et al., Proceedings of the GRO Science Workshop, 2-39 (1989).
2. C. Meegan et al., Nature **355**, 143 (1992).
3. G. Fishman et al., Ap.J. (Supplements), in press (1994).
4. J. Horack et al., Ap.J **413**, 293 (1993).
5. D. Hartmann, E. Linder, & L-S. The, Compton Gamma Ray Observatory, 1003 (1993).
6. M. Briggs et al.,"The Isotropy of Gamma-Ray Bursts: Dipole and Quadrupole Tests", these proceedings.
7. M. Briggs, Ap.J **407**, 126 (1993).
8. J. Quashnock & D. Lamb, MNRAS, in press (1994).
9. J. Quashnock & D. Lamb, MNRAS, in press (1994).
10. Narayan and Piran, "Do Gamma-Ray Burst Sources Repeat?", these proceedings.
11. J. Brainerd et al., "The Nearest Neighbors of BATSE Gamma-Ray Bursts: Narrowing the Possibilities", these proceedings.
12. J. Quashnock & D. Lamb,"Evidence that Gamma-Ray Burst Sources Re-

peat", these proceedings.
13. D. Hartmann et al., "Do Classical Gamma-Ray Bursts Repeat?", these proceedings.
14. M. Schmidt, J. Higdon, & G. Heuter, Ap.J.(Letters) **329**, 85 (1988).
15. V. Petrosian, W. J. Azzam & C. A. Meegan, "The Slope of the Log N – Log S Distribution for BATSE Gamma-Ray Bursts", these proceedings.
16. E. Fenimore et al., Nature **366**, 40 (1993).
17. J. Horack & A. Emslie, "Integral Moment Analysis of the BATSE Gamma-Ray Burst Intensity Distribution", these proceedings.
18. J. Hakkila et al.,"Additional Constraints on Galactic Coronal/Extended Halo Source Distributions from BATSE Observations, these proceedings.
19. J. Hakkila et al., Ap.J. , in press (1994).
20. J. Norris et al., Nature **434**, 308 (1984).
21. K. Hurley, in Gamma-Ray Bursts, AIP Conf. Proc. 265, ed. W. S. Paciesas & G. J. Fishman (AIP:New York) p.3 , (1991).
22. C. Kouveliotou et al., Ap.J. **413**, L101 (1993).
23. C. Kouveliotou et al., "Two Classes of Gamma-Ray Bursts", these proceedings.
24. D. Lamb, C. Graziana, & I. Smith, Ap.J. **413**, L11 (1993).
25. R. Rutledge & W. Lewin, "On the Use of "Variability" in Gamma-Ray Bursts", these proceedings.
26. E. Fenimore, private communication , (1993).
27. C. Meegan & C. Kouveliotou, "On the Use of Peak Count Rate Ratios for Burst Classification", these proceedings.
28. D. Palmer et al., " The Search for Gamma-Ray Burst Spectral Features in the Compton GRO BATSE Data", these proceedings.
29. D. Band et al., Ap.J **413**, 281 (1993).
30. G. Pendleton et al., in preparation , (1994).
31. L. Ford et al., "Continuum Evolution of Bright Gamma-Ray Bursts Observed by BATSE", these proceedings.
32. C. Kouveliotou et al., Ap.J , in press (1994).

GAMMA RAY BURSTS OBSERVED WITH WATCH-EURECA

S. Brandt, N. Lund, & A.J. Castro-Tirado

Danish Space Research Institute, Gl. Lundtoftevej 7, DK-2800 Lyngby, Denmark

ABSTRACT

The WATCH wide field X-ray monitor has the capability of independently locating bright Gamma Ray Bursts to 1° accuracy. We report the preliminary positions of 12 Gamma Ray Bursts observed with the WATCH monitor flown on the ESA spacecraft EURECA during its 11 month mission. Also the recurrence of the Soft Gamma Repeater SGR 1900+14 in 1992 is verified.

INTRODUCTION

The EURECA (**EU**-ropean **RE**-trievable **CA**-rrier) platform was developed by the European Space Agency (ESA), primarily as a free flying reusable micro-gravity laboratory. In addition to the core of micro-gravity experiments, the EURECA-1 mission carried several experiments from other scientific fields. Among them was the WATCH wide field X-ray monitor.

The WATCH instrument was activated on August 8 1992 and operated almost continuously until the end of the EURECA mission in June 1993.

The WATCH instrument was developed at the Danish Space Research Institute, and also flown on the Russian GRANAT mission.[1,2] WATCH is a wide field X-ray monitor based on the Rotation Modulation Collimator (RMC) principle with a 65° radius field of view. The energy range is 6-150 keV, and the effective area of NaI and CsI scintillators is about 45 cm^2. A conservative error estimate of the position of a source in the preliminary analysis is within a 1° radius.

The effective observation time of WATCH during the 318 day EURECA mission was about 120 days, corresponding to about 30 days of continuous all-sky coverage.

GRBs OBSERVED BY WATCH-EURECA

WATCH on EURECA localized a total of 12 Gamma Ray Bursts (GRBs), summarized in Table I. Below follows a brief description of one of the individual, localized events and its light curve.

GRB 921022 was a peculiar and interesting event. It started with a very intense and sharp peak followed by a long rather structureless emission (fig. 1). The sharp peak in the beginning of the burst triggered the burst logic, and a burst modulation pattern of one rotation was collected. This pattern is unsuited for localization of the burst, since the peak is much shorter than one rotation of the modulation grid. But it gives an account of the count rate with 3.5 ms time resolution (fig. 2). In the 12-120 keV band the effect of modulation is taken out by adding the NaI and CsI patterns. We see, that the total duration of the first sharp spike is about 30 ms. The rise time to maximum intensity is less than 10 ms, and the peak is not resolved at 3.5 ms per bin. In the bin of maximum intensity, an average count rate of nearly 9000 counts per seconds was reached.

© 1994 American Institute of Physics

Table I. GRBs observed with WATCH–EURECA.					
Date	Time of peak	Dur.	α	δ	Conf.
GRB 92 08 14	06:10:31.2	20 s	$17^h 20^m 22^s$	$-46°10'$	B,U,WG
GRB 92 09 18	09:41:20	80 s	$16^h 23^m 30^s$	$-04°18'$	P,S,U
GRB 92 10 01	05:57:15.3	5 s	$14^h 58^m 35^s$	$+23°12'$	no
GRB 92 10 04	14:00:21.9	15 s	$14^h 37^m 30^s$	$+34°06'$	U
GRB 92 10 22	15:20:59.477	65 s	$16^h 56^m 09^s$	$-10°04'$	B U
GRB 92 11 18	22:12:12	140 s	$23^h 23^m 26^s$	$+49°34'$	B U
GRB 92 12 07	16:00:48.2	15 s	$20^h 36^m 05^s$	$-41°50'$	B U
GRB 93 03 18	12:31:30.5	30 s	$10^h 03^m 16^s$	$-20°02'$	B U
GRB 93 04 10	14:19:08.4	5 s	$04^h 44^m 41^s$	$-07°11'$	U
GRB 93 04 26	12:40:34.9	30 s	$04^h 49^m 52^s$	$+12°19'$	B U
GRB 92 06 12	06:45:02	50 s	$07^h 58^m 40^s$	$-02°39'$	no
GRB 92 06 14	03:40:32.0	8 s	$11^h 20^m 40^s$	$+23°04'$	U

Table 1. Localized GRBs observed with WATCH–EURECA. The preliminary positions given here have 99% confidence areas with a radius of 1° (epoch J2000). An exception is GRB 921001, where the position is less certain. For the confirmation U means Ulysses, P Phebus, S SIGMA, WG WATCH-GRANAT and B means BATSE.

We also note, that a statistically significant second peak of lower intensity appears in the 12-120 keV band about 130 ms after the first spike. The very sharp peak in this burst holds the possibility of a very accurate position through the combination of the interplanetary network timing and the WATCH position.[3]

The burst appears to be rather soft, with a considerable emission in the 6-16 keV band and no detection above 120 keV in the WATCH detector. The appearance of the burst is somewhat similar to the famous super burst on March 5 1979, GRB 790305b, with a sharp spike followed by much less intense emission. However, the March 5 event showed pulsations with a period of 8 seconds in the extended emission, often interpreted as the spin period of a neutron star. There is no indication of periodicity in the WATCH data on GRB 921022. We just note the 130 ms separation of the two peaks at the start of the burst. The Galactic longitude of the event is 9.6°, and the Galactic latitude is 20.0°. The event is not as close to the Galactic plane as the two SGRs believed to be located in the Galaxy, but the relative proximity of the Galactic Center is interesting. The evidence presented for GRB 921022 to be a Soft Gamma Repeater is rather weak, but it is suggested to look for short repeating events from this direction.

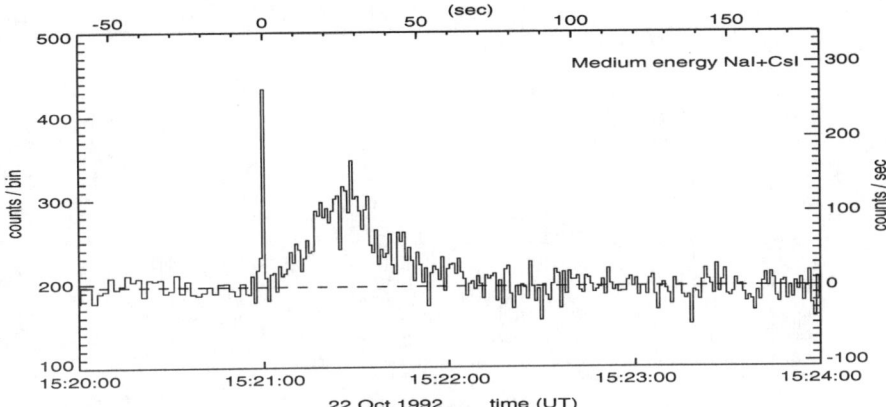

Figure 1. The 12-120 keV NaI+CsI count rate for GRB 921022. Note the sharp peak at the start of the burst, which is resolved on a finer time scale in fig.2. The time resolution is 0.892 sec per bin during the burst and twice that just before the burst.

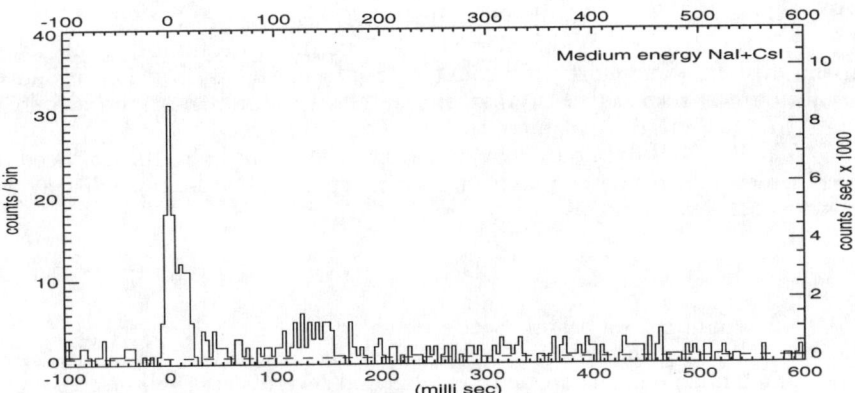

Figure 2. The short intense peak in GRB 921022 in the 12-120 keV band. The time resolution is 3.5 ms per bin.

WATCH DETECTION OF SGR 1900+14

Soft Gamma–Ray Repeaters (SGRs) are believed to constitute a separate class of objects consisting so far of only 3 objects.[4] In contrast to the classical GRBs, which never have been seen to repeat, the SGRs are recurrent sources. The SGRs are generally believed to be associated with neutron stars.

The soft Gamma–Ray repeater SGR 1900+14 was detected to show recurrent activity in 1992 by the BATSE experiment on the Compton Gamma Ray Observatory (GRO).[5] During the summer of 1992, BATSE triggered on 3 events believed to originate from SGR 1900+14.

One of the three events seen by BATSE was also detected by WATCH–EURECA on 19 August 1992 (fig. 3). The event seen by BATSE was of duration

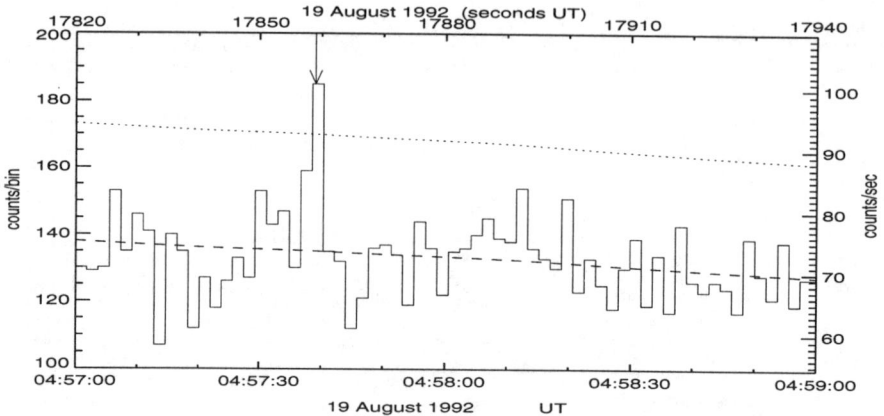

Figure 3. SGR 1900+14 detected in the CsI band (15-120 keV) of WATCH–EURECA. The dashed line shows the background level and the dotted line shows the 3 σ excess level. The arrow indicates the time of the BATSE trigger.

80 ms. This short event would normally have triggered the WATCH burst logic, but unfortunately it was disabled at the time due to software problems early in the mission. Therefore only count rate information with a 1.8 second time resolution was recorded. Anyway, such a short event is not localizeable with WATCH since the RMC rotates with a period of about 1 second.

The WATCH data base may hold other detections of SGRs, but need to be in coincidence with a detection in another instrument in order to be distinguished from particle events.

CONCLUSION

The small set of bright GRBs detected by WATCH on EURECA may not in itself bring the solution of the GRB mystery any closer. But the relatively small independent error circles, which in several cases will be considerably smaller than 1 square degree, combined with timing circles from the Inter Planetary Network can provide final error boxes suitable for deep searches at other wave lengths. The data base of WATCH GRB positions may also prove useful in connection with the recent suggestion, that classical GRBs may repeat.[6]

This work was supported by the Danish Space Board.

1. N. Lund, X-Ray Instrumentation in Astronomy (SPIE 597, 1985), p. 95.
2. S. Brandt, N. Lund, A.R. Rao, Adv. Space Res. **10(2)**, 239 (1990).
3. K. Hurley, N. Lund, S. Brandt, et al.,(These proceedings).
4. J.P. Norris et al., Astronomy & Astrophysics **366**, 240 (1991).
5. C. Kouveliotou, G.J Fishman, C.A. Meegan et al., Nature **362**, 728 (1993).
6. J.M. Quashnock, D.Q. Lamb, M.N.R.A.S. , in press (1993).

WATCH OBSERVATIONS OF GAMMA-RAY BURSTS DURING 1990–1992

Alberto J. Castro-Tirado,* Soren Brandt, Niels Lund
Danish Space Research Institute, Gl. Ludtoftevej 7,
DK-2800 Lyngby, Denmark

Igor Y. Lapshov, O. Terekhov, R.A. Sunyaev
Space Research Institute, Russia Academy of Sciences
Profsoyuznaya 84/32, Moscow

ABSTRACT

The First WATCH/GRANAT Gamma-Ray Burst Catalogue comprises 70 events which have been detected by WATCH during the period December 1989 - September 1992. 32 GRBs could be localized within a 3σ error radii of 1°. We have found a weak (2.2σ) clustering of these 32 bursts towards the Galactic Center. However we conclude that there is no strong evidence of concentration of the bursts towards the Galactic Center or Plane. Around ~10% of the 70 bursts showed X-ray precursor or/and X-ray tail. We discuss the possibility that two events, GRB 900126 and GRB 920311, would have been produced by the same source.

INTRODUCTION

The WATCH all-sky monitor on GRANAT has been observing the X-ray sky since December 1989. It consists of four identical units, based in the rotation modulation collimator principle[1]. Three of them are operational, covering $\sim 3\pi$ steradians of the sky. One of the main advantage of WATCH is the rapid localization of some Gamma-Ray Bursts (GRBs) with less than 1° error radius (at a 3σ confidence level). The useful period of WATCH, between 1989 December 18 and 1992 September 25, corresponds to a continuous whole-sky coverage of 230 days.

OBSERVATIONS

The First WATCH/GRANAT Catalogue includes 70 GRBs which have been detected[2] during the period mentioned above. This is ~ 110 events per year over the full sky. The cosmic origin of the events have been confirmed, in most cases via detections on other spacecrafts, and for the rest through the unambiguous localization to a unique position in the sky based on several independent data sets, ruling out a solar burst or particle origin. Thus, 32 events could be localized with less than 1° error radius (at a 3σ confidence level). We are currently working to reduce the systematic errors which for the time being

---

* current address: Laboratorio de Astrofisica Espacial y Fisica Fundamental; P.O. box 50727, 28080, Madrid, Spain

are dominating our error budget. A rapid localization of the bursts allowed in several cases to carry out optical searches on optical plates, however, no obvious optical counterpart has been found[3].

SOME IMPLICATIONS

- Dipole and quadrupole moments of the distribution.

It is possible to quantify the degree of isotropy of GRBs sources in the sky by means of the dipole and quadrupole moments with respect to the Galactic Center and the Galactic Plane. For the 32 GRBs which have been located by WATCH, the calculated measure of the dipole moment towards the Galactic Center is

$$< \cos \theta > = 0.22 \pm 0.10$$

This value is corrected for the non-uniform sky exposure of the WATCH data. We note that this value differs (although with marginal statistical significance) from the near zero value derived from the First BATSE Catalogue[4] (The 1B Catalogue). It is interesting to note that an homogeneus sample of 55 bright 1B GRBs exhibited a Galactic dipole moment[5,6] of $< \cos \theta > = 0.23\pm 0.08$, a value that it is not very different from the one derived from the WATCH sample. We note that only 2 of the 55 strong BATSE GRBs were included in our WATCH sample, so the two determinations of the dipole moment were essentially independent.

A second parameter that provides information on the distribution with respect to the Galactic Plane, is the quadrupole moment $< \sin^2 b >$ where b is the galactic latitude of the burst source. For the WATCH sample of 32 GRBs,

$$< \sin^2 b > = 0.29 \pm 0.05$$

consistent with the expected value for an isotropic distribution of 1/3. This value is also corrected for the non-uniform sky exposure. No concentration towards the galactic plane is evident . Our value differs slightly from the one calculated[5] for the 55 strong BATSE GRBs aboved mentioned above ($< \sin^2 b > = 0.21\pm 0.04$) and is closer to the value derived for all the GRBs in the 1B Catalogue (0.305±0.004).

In order to test the results of the WATCH and 1B distribution, we have calculated the moments of the distribution for the 32 GRBs with highest fluences ($S \geq 5.7 \ 10^{-6}$erg cm^{-2}) in the Konus Catalogue[7], and found $< \cos \theta > = -0.022\pm 0.10$ and $< \sin^2 b > = 0.25\pm 0.05$. This is compatible with isotropy in the distribution of bursts. We conclude that there is no strong evidence for concentration of the bursts towards the Galactic center or Plane.

- "X-ray rich" GRBs.

The observation of X-rays below 20 keV is important in order to understand the nature of the physical processes in the GRBs sources. Most of the experiments that have been designed for GRB observation had a lower cutoff energies of the order of 20~40 keV. However a few of them covered part of the soft X-rays (down to 1.5 keV in the case of Ginga[8]). In the case of the 70 GRBs observed by WATCH, it seems that a significant fraction (~10%) display X-ray

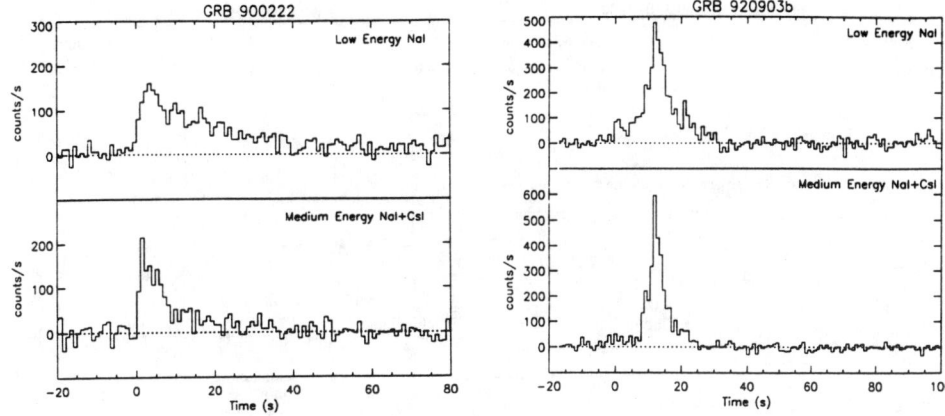

Fig.1. Two "X-ray rich" GRBs. GRB 900222 showed a long X-ray tail, and GRB 920903a had a precursor and extended X-ray activity during the whole event.

tails and/or X-ray precursor activity. No preferential concentration for these X-ray rich bursts (for instance, towards the Galactic Plane) is noticeable. Figure 1 display the time profile of two of these bursts.

Some models involving highly magnetized neutron stars predict high X-ray fluxes because of heating of the neutron star by the gamma-rays, assuming than the source is located in the outer astmosphere [9]. Soft X-ray tails may be interpreted as resulting from crustquakes on neutron stars[10]. And a mechanism in which a precursor arise is the reignition of a post-Vela Gamma-ray pulsar [11], but they are also expected by extragalactic models in which a fireball of electron-positron pairs becomes optically thin[12].

- Are some of the classical GRBs repeaters?

No classical sources of GRBs have been seen to repeat. However, in the recent analysis of data from the 1B Catalogue, evidences have been found[13] for a repeating source (GBS 0855-00) that produced 5 bursts over a $\sim$ 3 months period. The probability of such a random occurrence was estimated to be 2 10^{-6}. Furthermore, 201 bursts of the 1B Catalogue seemed to be clustered[14] on an angular scale of $\sim$ 4° and it was proposed that GRB sources repeat on a timescale of months. However, this result has been questioned when a larger sample of bursts of the future 2B and 3B Catalogues were considered[15].

In the First GRANAT-WATCH Catalogue there are two pairs of events for which the positions of the two bursts are within the WATCH systematic errors: GRB 900126 / GRB 920311 (separation 0.9°) and GRB 911209 / GRB 920814 (separation 2°). In the case of the 32 GRBs localized by WATCH, the Poisson probabilities of finding the pairs are 0.034 and 0.15 respectively. These

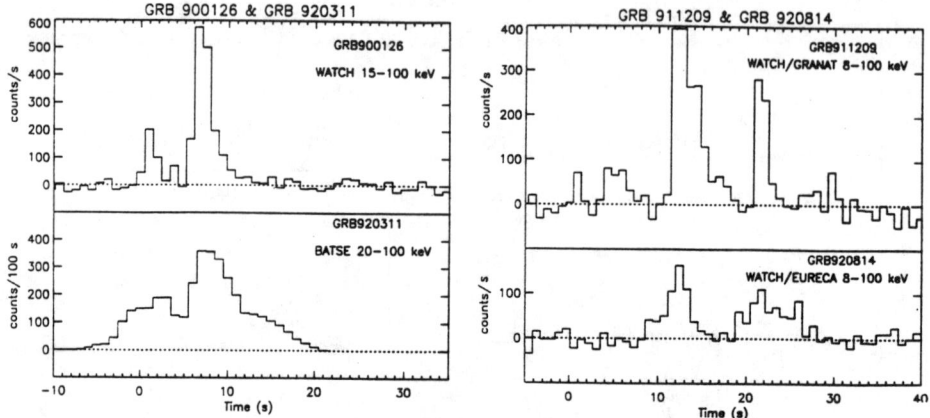

Fig.2. GRB 900126 and GRB 920311. For GRB 920311, and due to the fact that the burst did not trigger the WATCH logic, the BATSE light curve is shown (here reproduced by kindly permission of G. Fishman).It has been rebinned in order to get the 1 s time resolution of WATCH.

Fig.3. GRB 911209 and GRB 920814. It has been ruled out as a possible repeater by the Interplanetary Network.

probabilities would be slightly higher when the non uniformity in the WATCH sky exposure is taking into account. Could these pairs of bursts originate from repeating sources? Let us comment each pair separately:

i) GRB 900126/GRB 920311. The centroids of the two error boxes are separated by 0.9°. GRB 900126 is one of the brightest GRBs observed by WATCH, with a total fluence in the two WATCH energy bands of $\sim 6\ 10^{-6}$ erg cm^{-2}. Two peaks are present and separated by ~ 5.5 s in time. The second burst, GRB 920311 was observed to have a fluence of $\sim 4.4\ 10^{-6}$ erg cm^{-2} in the same WATCH energy band. Fig. 2 displays these two GRBs. This second burst is broader than the first, however a similar double peak structure is noticeable, but the separation is shorter (~ 4 s). If these events originated from the same source, they cannot be explained as a single burst originated at cosmological distances that has been gravitationally lensed by a foreground object. We have checked several GRBs catalogues and no bursts have been previously observed at this position.

ii) GRB 911209/GRB 920814. The centroids of the two error boxes are separated by 2°. The fluences of the two bursts are $\sim 1.4\ 10^{-6}$ erg cm^{-2} and 10^{-6} erg cm^{-2}, respectively. Fig. 3 displays these two bursts. It is striking the fact that the main two peaks in the bursts are separated by 10 s. We have checked for a possibility of a repetition around 1993 April 20, but no burst arising from that particular direction of the sky was seen (C. Kouveliotou, private communication). More precise positions from the 3rd. Interplanetary Network (K. Hurley, priv. comm.) have proved that indeed a separation of 2° is correct and certainly the two bursts did not come from the same position in the sky. Therefore it seems clear that two different sources produced the bursts, unless we consider the source to be a nearby object at a distance not larger than ~20000 AU, the beginning of the Oort Cloud of comets, and having a high transversal velocity v = 500 km/s.

CONCLUSIONS

In our analysis of the First WATCH Gamma-Ray Burst Catalogue, we have found a weak (2.2σ) clustering of our 32 bursts towards the Galactic Center. However we conclude that there is no strong evidence of concentration of the bursts towards the Galactic Center or Plane. Around ~10% of the 70 bursts showed X-ray precursor or/and X-ray tail. One of our candidates to a repeater source has been ruled out by the IPN Network. The other one, GRB 900126/GRB 920311, may originate from a single source. Studies of this position at other wavelengths are in progress.

Acknowledgements. We are very grateful to the BATSE team, in particular to C. Kouveliotou, G. Fishman W. Paciesas and S. Cole, for making avalaible GRB trigger times during March-September 1992. This made possible the inclusion of several bursts in the First WATCH/GRANAT Catalogue, specially GRB 920311 and GRB 920814, for which the light curves were obtained and compare with the ones obtained by WATCH. Also to O. Terekhov and F. Pelaez, for providing the PHEBUS and SIGMA lists of GRBs. Discussions with K. Hurley, were very valuable. Thanks to N. Khavenson at IKI, for providing with the instantaneous pointing of GRANAT for the trigger times of some GRBs. Conversations with J. Isern, and W.A.D.T. Wickramasinghe, were very useful.

REFERENCES

1. Lund, N., X-ray instrumentation in Astronomy, SPIE 597 (Culhane, J.L., 1985), p. 95.
2. Castro-Tirado, A.J., Ph.D. Thesis, University of Copenhagen (1993)
3. Castro-Tirado, A.J., Brandt, S., Lund, N. and Guziy, S., These Proceedings
4. Fishman, G. J. et al., ApJ Suppl. Ser., in press , (1994).
5. Quashnock. J. M., Lamb, D. Q., MNRAS, submitted , (1993).
6. Atteia, J. L. and Dezalay, J.-P., A&A, submitted , (1993).
7. Mazets, E. P. et al., Ap&SS **80**, 1 (1981).
8. Nishimura. J., Fujii, M. and Yamagami, T., A&SS **93**, 87 (1983).
9. Yoshida, A. et al., These proceedings .
10. Blaes, O. et al., ApJ **343**, 839 (1989).
11. Ruderman, M. and Cheng, K.S., ApJ **335**, 306 (1988).
12. Meszaros. P., Laguna, P. and Rees, M. J., ApJ letters, submitted , (1993).
13. Wang, V. C., and Lingenfelter, R. E., ApJ **416**, L13 (1993).
14. Quashnock, J. M., Lamb, D. Q., MNRAS, submitted , (1993).
15. Meegan, C. et al., These proceedings .

EGRET OBSERVATIONS OF THREE GAMMA-RAY BURSTS AT ENERGIES > 30 MEV

B. L. Dingus and P. Sreekumar
Universities Space Research Association, NASA/GSFC, Greenbelt, MD

E. J. Schneid
Grumman Aerospace Corporation, Mail Stop A01-26, Bethpage, NY

D. L. Bertsch, C. E. Fichtel, A. K. Harding,
R. C. Hartman, S. D. Hunter, and D. J. Thompson
NASA/Goddard Space Flight Center, Greenbelt, MD

J. R. Mattox
Compton Science Support Center, Computer Sciences Corp.,Greenbelt, MD

K. Hurley
University of California, Space Sciences Lab, Berkeley, CA

D. A. Kniffen
Hampden-Sydney College, P. O. Box 862, Hampden-Sydney, VA

G. Kanbach, H. A. Mayer-Hasselwander, M. Sommer, and C. von Montigny
Max Planck Institut fur Extraterrestrische Physik, 8046 Garching, GERMANY

Y. C. Lin, P. F. Michelson, and P. L. Nolan
Hansen Experimental Physics Laboratory, Stanford University, CA

ABSTRACT

EGRET has imaged three gamma-ray bursts at energies above 30 MeV. The bursts are GRB930131, GRB910503, and GRB910601 – listed in order of decreasing intensity. GeV emission has been detected in two bursts which increases by an order of magnitude the maximum energy gamma rays observed from a gamma-ray burst. The high energy emission is a significant contributor to the fluence of these burst, and the duration of the high energy emission is at least as long and in some cases longer than the emission observed by BATSE. The high energy gamma rays are also used to determine the most probable direction of the burst.

EGRET AS A GAMMA-RAY BURST DETECTOR

EGRET, the instrument on the Compton Observatory sensitive to the highest energies, has the capability of studying gamma rays from a burst in three overlapping energy bands. First, if the energy of the gamma-ray is greater than 30 MeV, it can pair produce in the tracking detector, and the resulting electron and positron are imaged. The effective area is about 1500 cm^2 from 200 to 1000 MeV decreasing at lower and higher energies and dropping by about one-half at

20° from the instrument axis and one-sixth at 30°. In this mode, only a few gamma-rays are seen, but their directions and energy can be measured. Also, due to the clear gamma-ray identification, the only significant radiation besides sources are the diffuse galactic and extragalactic radiation. Second, the energy spectrum between 1 and 200 MeV may be measured in the $\sim$ 6000 cm^2 NaI (Tl) crystals which have an independent, self-triggered low energy mode of analysis. The third manner in which a burst can be seen, albeit crudely, is in the count rate measured in the large anticoincidence scintillator dome. This plastic scintillator has a low energy threshold of about 50 keV, and a time resolution of 0.256 seconds. A description of the instrument[1], and the results of the instrument calibration, both before and after launch,[2] are described elsewhere.

EGRET has detected four gamma-ray bursts above an MeV in the NaI crystals and measured their spectra.[3,4,5] Tracking data was also available for three of these bursts: GRB930131, GRB910503, and GRB910601. For the other burst, GRB910814, the tracking was disabled because this burst occurred near the earth's horizon. The information available from the energy, time distribution, and direction of the few high energy gamma rays detected in these three gamma-ray bursts is presented below. These observations provide evidence that gamma-ray burst spectra can extend to GeV energies. Prior to the launch of the Compton Observatory, the highest energy observations were made with the Gamma Ray Spectrometer on the Solar Maximum Mission and only extended to 10 to 100 MeV.[6,7] Also, the sensitivity and low background of EGRET allows the time history to be examined. The high energy fluence is observed to be a significant part of the fluence even after the flux of lower energy gamma rays has dropped. Finally, the directions of the high energy gamma rays significantly constrains the burst direction.

GRB930131

On January 31, 1993 EGRET detected the most gamma rays above 30 MeV ever observed in a gamma-ray burst. A total of 16 gamma rays with direction consistent with the burst direction were imaged in the 25 second time interval after the BATSE burst trigger. Two of these gamma-rays had energies of approximately 1 GeV. The probability of one 1 GeV gamma ray, that is not associated with the burst, being detected from this direction in a 25 second time interval is 4×10^{-4}, and only 0.04 gamma rays of energy greater than 30 MeV are expected in this time interval, as derived from the flux measured by EGRET in the days prior to the burst. A crude spectrum of these 16 gamma rays shows a power law of spectral index -2.0 ± 0.4 and no evidence for a high energy cut off below a GeV.

The times and energies of all the gamma rays are listed in Figure 1. The figure also shows the light curve of > 50 keV gamma-rays as detected by the plastic anticoincidence dome of EGRET. Figure 1 is similar to the light curve of BATSE except that BATSE also detected low level emission extending for approximately 50 seconds.[8] The fluence detected by BATSE in this tail was over an order of magnitude less than the fluence of 10^{-5}ergs/cm^2 detected by EGRET at these later times. The high energy fluence during the intense spikes of emission is difficult to determine because of EGRET instrumental deadtime, but it is at least as large as 10^{-5}ergs/cm^2, which is comparable to the fluence measured by BATSE at lower energies during the first peak of emission. Thus, the high energy emission is the dominant part of the energy produced following the intense spike of intensity, and is a significant, if not dominant, part of the energy produced in the spike.

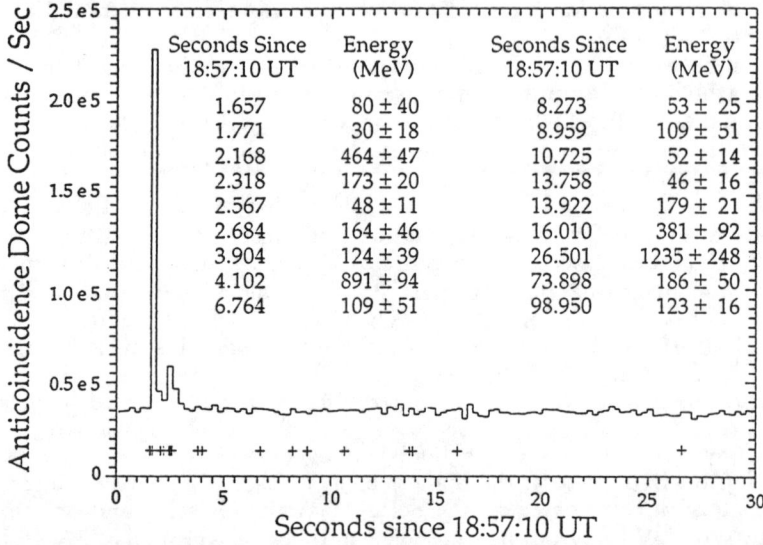

Figure 1. GRB930131 light curve of gamma-rays > 50 keV as detected by the EGRET plastic anticoincidence dome. The crosses indicate the gamma rays > 30 MeV which are consistent with the burst direction. The energy and time of each gamma ray is listed.

The higher the energy of the gamma ray the better the position information, so an energy dependent point spread function is used with a maximum likelihood technique to combine the directional information of the 16 gamma rays. The resulting confidence intervals are drawn in Figure 2 (a).

GRB910503

Another bright BATSE burst occured within EGRET's field of view on May 3, 1991. Again, the EGRET deadtime was high during the intense part of the burst, and 9 gamma rays were detected with the maximum energy of about 100 MeV.[4] However, 84 seconds after the BATSE trigger, EGRET detected a single gamma ray of energy 9.6 ± 1.2 GeV with a position consistent with the Interplanetary Network arc and the position determined from the gamma rays detected earlier. EGRET detects only a few 10 GeV gamma rays per day even from the direction of this burst which is near the galactic anticenter. The probability of detecting a gamma ray, not associated with the burst, of energy greater than 4 GeV in 100 sec within a square degree of the burst direction is 7×10^{-6}. The fluence associated with this single photon has a large uncertainty, but it is of the order of 10^{-3} ergs/cm^2 since the effective area at 10 GeV and 30° off axis is only about 20 cm^2. The fluence detected by BATSE was 5×10^{-5} ergs/cm^2.[9] Therefore, if this single late gamma-ray is associated with the burst, a large amount of the energy observed from the burst is delayed and is very high energy.

GRB910601

The least intense burst observed by EGRET was on June 1, 1991. No gamma rays were detected by EGRET until 33 seconds after the BATSE trigger. Ninety percent of the BATSE flux was contained within 28 seconds starting 12 seconds after the trigger.[9] This direction is occulted by the earth ~ 400 seconds later. Ten of the 19 gamma rays detected in this 400 seconds are consistent with the burst direction. The rate of gamma rays from this direction prior to the burst was used to determine that 3 gamma rays are expected in this time interval. Therefore, the probability of 10 gamma rays when 3 are expected is 10^{-3}. In addition to the increase in rate from this direction, a maximum likelihood analysis of all 19 gamma rays in this time interval indicate a point source of ~ 6 gamma rays with a confidence of 99.9%. The confidence intervals of the point source position are shown in Figure 2 (b). The Interplanetary Network arc crosses these intervals and COMPTEL's best position[10] along the arc is also consistent.

The energy of the gamma rays consistent with the burst direction varies from 50 to 300 MeV with an average energy of 200 MeV. The total fluence in these gamma rays is thus a few times 10^{-6}ergs/cm^2. The fluence detected by BATSE[9] is 5×10^{-5} ergs/cm^2.[9]

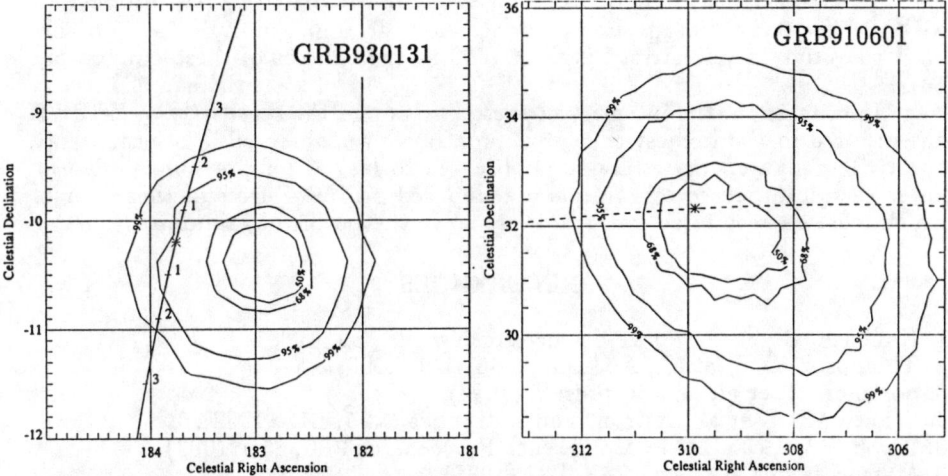

Figure 2. (a) EGRET determination of GRB 930131 position confidence intervals are the labeled contours. EGRET's most probable position on the Interplanetary Network arc is indicated with an asterisk and the 1, 2, and 3 sigma confidence intervals on the arc are marked. (b) EGRET determination of GRB 910601 position confidence intervals are the labeled contours. COMPTEL's most probable position[10] on the Interplanetary Network arc is indicated by the asterisk.

IMPLICATIONS

The presence of GeV gamma rays increases the highest energy at which gamma-ray bursts have been detected by an order of magnitude. While these bursts are unique in intensity at lower energies and in the maximum energy observed, many other bursts have been detected that show no evidence of a cut off above a few MeV.[6,7,11] The low energy properties of these bursts are in many ways similar to the other bursts detected by BATSE.[9] Therefore, the theoretical constraints imposed on the source of these three bursts may also be applicable to all gamma-ray bursts.

Both the energies of the observed gamma rays and their long duration after the start of the burst provide interesting constraints on the energy release and particle acceleration. Cosmic fireball models[12,13] which predict modified blackbody spectra at temperatures of a few MeV are excluded by these observations. Particles must be accelerated to GeV energies and either persist over the burst duration or the particle lifetime must be longer than the burst duration.

Also, higher energy gamma rays suffer photon-photon pair production attenuation in the source. BATSE observations of these bursts show rapid time variability implying a small emission region and high photon densities. If the high energy photons are produced in the same region as the low energy photons, then all the gamma rays must be highly collimated to allow the high energy gamma rays to escape. Photons are effectively collimated if the source is moving with a large bulk Lorentz factor. However, if the source distance is a Gpc, then bulk Lorentz factors of over 1000 are required.[14,15] A narrower beam also requires many more sources to explain the rate of bursts observed by BATSE.

The requirement that the source also be unattenuated by pair production with the microwave background photons[16] puts an upper limit of $Z < 80$. This limit rules out cosmic strings at $Z = 1000$ as the sources of these gamma-ray bursts.[17,18]

In conclusion, the intensity of these bursts and the sensitivity of EGRET has resulted in the highest energy gamma rays ever observed in a gamma-ray burst. The high energy emission is observed to last as long or longer than at lower energies. Severe constraints are imposed on the sources of these bursts which must be accounted for in order to solve the puzzle of gamma-ray bursts.

REFERENCES

1. Kanbach, G., et al., Space Science Reviews **46**, 69 (1988).
2. Thompson, D. J. et al., ApJ Supp. **86**, 629 (1993).
3. Sommer, M. et al, ApJ in press , (1994).
4. Schneid, E. J. et al., Astron. and Astrophys. **255**, L13 (1992).
5. Kwok, P.W., et al., AIP Conference Proceedings **280**, 855 (1993).
6. Matz, S.M., et al. , ApJ **288**, L37 (1985).
7. Share, G.H. et al., Adv. Space Res. **V.6, No.4**, 15 (1986).
8. Kouveliotou, C. et al., ApJ in press , (1994).
9. Fishman, G. J., et al., ApJ Supp. in press , (1994).
10. Hanlon, L. et al., A & A in press , (1994).
11. Winkler, C. et al., AIP Conference Proceedings **280**, 845 (1993).
12. Goodman, J., ApJ **308**, L47 (1986).
13. Paczynski, B., ApJ **308**, L43 (1986).
14. Harding, A. K. & Baring, M. G., These proceedings , (1994).
15. Baring, M. G., ApJ, in press , (1993).
16. Fazio, G. G. & Stecker, F. W., Nature **226**, 135 (1970).
17. Babul, A., Paczynski, B. & Spergel, D. , ApJ **316**, L49 (1987).
18. Paczynski, B., ApJ **335**, 525 (1988).

THE ULYSSES SUPPLEMENT TO THE FIRST BATSE CATALOG OF GAMMA-RAY BURSTS

K. Hurley
University of California, Space Sciences Laboratory
Berkeley, CA 94720

M. Sommer
Max-Planck Institut für Extraterrestrische Physik
D8046 Garching-bei-München, Germany

C. Kouveliotou, G. Fishman, C. Meegan
NASA - Marshall Space Flight Center
Huntsville, AL 35812

T. Cline
NASA-Goddard Space Flight Center
Greenbelt, MD 20771

M. Boer, M. Niel
Centre d'Etude Spatiale des Rayonnements
31029 Toulouse Cedex, France

Abstract

The first BATSE catalog contains localization information on 260 gamma-ray bursts from April 1991 to March 1992. Approximately 53 of these bursts were detected by one or more instruments in the third interplanetary network. Of this subset, 39 bursts were observed only by Ulysses and one or more near-Earth spacecraft (but always including BATSE). For these bursts, it is possible to reduce the size of the BATSE error region by a factor of several hundred, by deriving triangulation annuli whose widths are of the order of arcminutes and larger. Combined with the BATSE error circle, this leads to localization areas of the order of 1000 square arcminutes, which are useful for optical transient, soft X-ray, VHE, and UHE counterpart searches, as well as for searching for burster recurrence. We present data for these events in tabular form, display an example of an error box, and show the distribution of the minimum separation between the annuli and the centers of the BATSE error boxes.

Introduction

The Ulysses GRB experiment[1] has been in operation since November 1990. Its main features are a 20 cm^2 effective area from practically all angles of incidence, a 4π sr field of view, a duty cycle >95%, and sensitivity in the 25-150 keV range. Since the launch of Compton-GRO, we have systematically searched the Ulysses GRB data for responses to all BATSE events. The stronger bursts are readily identified in the triggered data. However, many of the weaker bursts, while below the trigger threshold, appear prominently in the real-time data, which include 25-150 keV count rates with a time resolution of 0.25 - 2 s, depending upon the telemetry rate. Since the Ulysses spacecraft reached a distance >6 AU from Earth, even these relatively low time resolution data are useful for triangulation. Here we present the preliminary triangulation annuli for 39 Ulysses/BATSE bursts which occurred during the first year of operation of CGRO and were not observed by PVO. Improvements in some of the locations, and particularly in the annulus widths, can be expected. The final data will be presented elsewhere[2].

Table 1 lists the date and BATSE number of the burst, the right ascension, declination, and radius of the trianulation annulus, the full width of the annulus, and the minimum separation S_{min} between the annulus and the center of the BATSE error circle. S_{min} is given by

$$S_{min}=|\cos^{-1}\{\sin\delta_T\sin\delta_B + \cos\delta_T\cos\delta_B\cos(\alpha_T-\alpha_B)\}-R|$$

where α_T,δ_T is the center of the triangulation annulus, α_B,δ_B is the center of the BATSE error circle, and R is the triangulation annulus radius. The narrowest annulus has a width <1', the widest has a width ~16', and the average width is 4.3'. If the radius of the BATSE error circle is taken to be

$$\sigma_B = \sqrt{\sigma_{stat}^2 + \sigma_{sys}^2}$$

where σ_{stat} is the statistical error and σ_{sys} is the systematic error (4°), then the smallest BATSE error circle has radius 4.01°, the largest 8.15°, and the average is 4.6°. (This is significantly less than the overall BATSE catalog average of 7.5°, since this subset comprises only the brighter bursts.) Thus the average BATSE/Ulysses error region has an area of ~1300 square arcminutes, or .007 times the area of the BATSE error circle. Figure 1 shows one example.

To test the consistency of the BATSE and triangulated positions, we have calculated S_{min}/σ_B for each burst, and plotted the distribution in Figure 2. There is no discrepancy between them.

Discussion

The error regions obtained by combining BATSE error circles with triangulation annuli have relatively small areas, but their shapes preclude deep counterpart searches with instruments whose fields of view are smaller than a few degrees. However, localizations such as these should prove useful for a number of purposes. Some examples are: searches for archival optical transients, for optical counterparts with Schmidt cameras, for EUV sources from the ROSAT WFC or EUVE all sky survey data, for soft X-ray sources using all sky survey data from ROSAT or archival Einstein or EXOSAT data, or for VHE or UHE gamma-ray counterparts. Another

Table I. Ulysses/BATSE Bursts

Date	BATSE #	$\alpha_T(2000)$ degrees	$\delta_T(2000)$ degrees	R degrees	ΔR degrees	Minimum Separation, deg.
910425	109	115.140	23.172	51.733	0.053	4.42
910429	121	116.425	22.921	59.172	0.091	0.21
910502	142	297.625	-22.679	87.120	0.011	1.39
910503	143	117.730	22.657	30.352	0.010	0.21
910507	160	299.098	-22.373	16.577	0.128	5.25
910511	179	300.194	-22.138	83.487	0.011	0.73
910523	222	124.200	21.230	26.839	0.083	2.40
910528	235	305.808	-20.843	86.119	0.032	2.33
910529	237	306.072	-20.778	72.955	0.033	7.15
910601	249	307.038	-20.538	52.936	0.011	1.46
910602	257	127.385	20.451	34.377	0.109	1.12
910619	394	312.533	-19.080	61.602	0.128	0.88
910626	444	134.553	18.504	10.698	0.149	0.76
910630	469	315.768	-18.148	52.407	0.068	0.06
910721	563	322.027	-16.197	6.720	0.439	4.13
910809	659	147.085	14.485	50.530	0.245	2.26
910809	660	327.239	-14.432	85.750	0.023	3.95
910907	764	154.319	11.852	43.469	0.131	0.95
910927	829	338.939	-10.074	68.868	0.023	0.34
910930	841	159.444	09.876	39.204	0.035	3.01
911005	869	160.397	09.501	73.282	0.023	1.75
911006	871	340.606	-09.419	50.764	0.028	0.26
911022	914	163.394	08.313	68.564	0.096	0.94
911026	938	344.080	-08.038	85.843	0.090	2.40
911031	973	344.795	-07.754	67.668	0.049	0.12
911106	1008	345.572	-07.445	29.190	0.047	0.57
911123	1114	347.387	-06.730	49.425	0.031	2.18
911127	1122	347.680	-06.618	88.465	0.021	1.76
911207	1150	348.249	-06.408	48.029	0.032	6.95
911209	1157	348.333	-06.380	82.576	0.021	0.25
911217	1190	348.478	-06.341	31.995	0.045	3.80
911221	1200	168.476	06.356	61.766	0.056	1.93
911227	1235	348.388	-06.408	61.822	0.028	4.17
920110	1288	347.669	-06.749	69.226	0.035	0.77
920130	1328	345.523	-07.676	70.312	0.273	2.78
920221	1425	161.372	08.445	42.149	0.049	1.47
920224	1432	160.728	08.515	74.987	0.061	3.12
920226	1440	340.220	-08.569	74.026	0.034	2.02
920227	1443	160.017	08.591	83.347	0.033	0.96

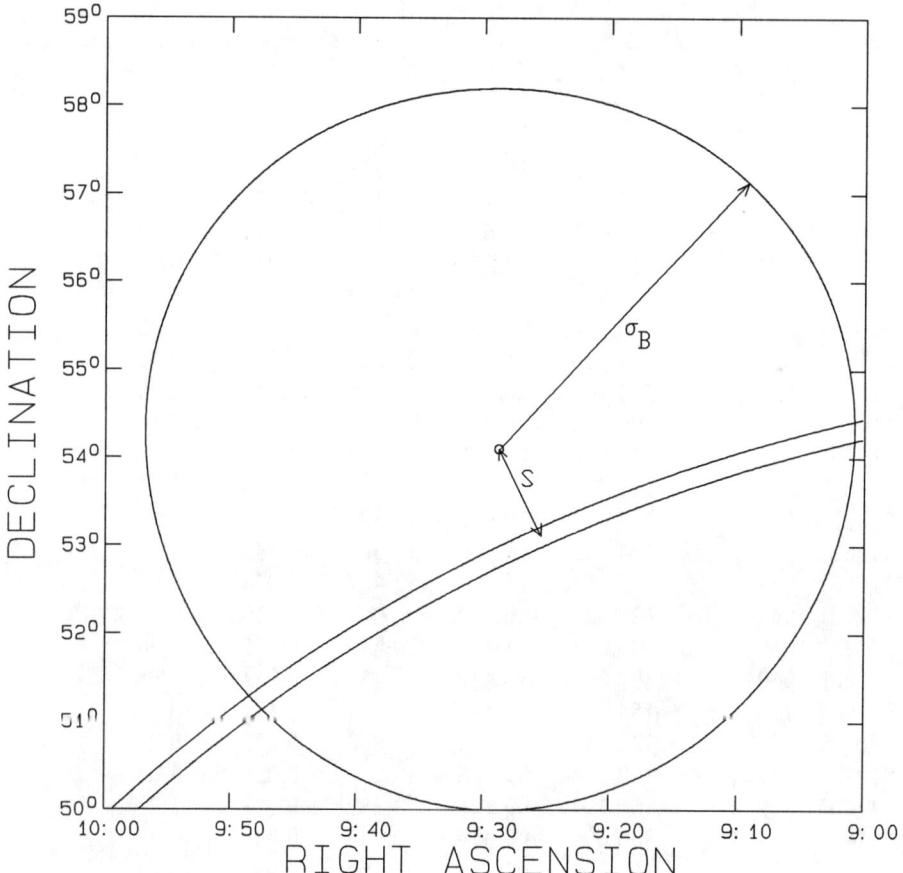

Figure 1. The BATSE error circle and triangulation annulus for event 257. The minimum separation S and the BATSE error radius σ_B are shown.

use may be to determine whether certain BATSE bursts are repeaters, as recently suggested [3,4]. Our examination of the possible repeaters reported to date has revealed at most one Ulysses detection per cluster of bursts. This is not surprising, since the Ulysses GRB experiment sensitivity is comparable to that of the small experiments which have operated over the past 15 years. If there were gamma-ray burst sources which repeated with a nearly flat luminosity function (i.e., produced numerous intense bursts over a period of years) they would almost certainly have been discovered prior to BATSE. Only recurrence involving weak bursts could have gone undetected, and these events will generally not be observable by Ulysses. Nevertheless, even the observation of a single burst in a cluster by Ulysses may be useful to check the consistency of the locations.

Over 130 BATSE/Ulysses events have been identified in the data between April 1991 and July 1993. Interested observers are invited to contact us for details of these triangulation annuli.

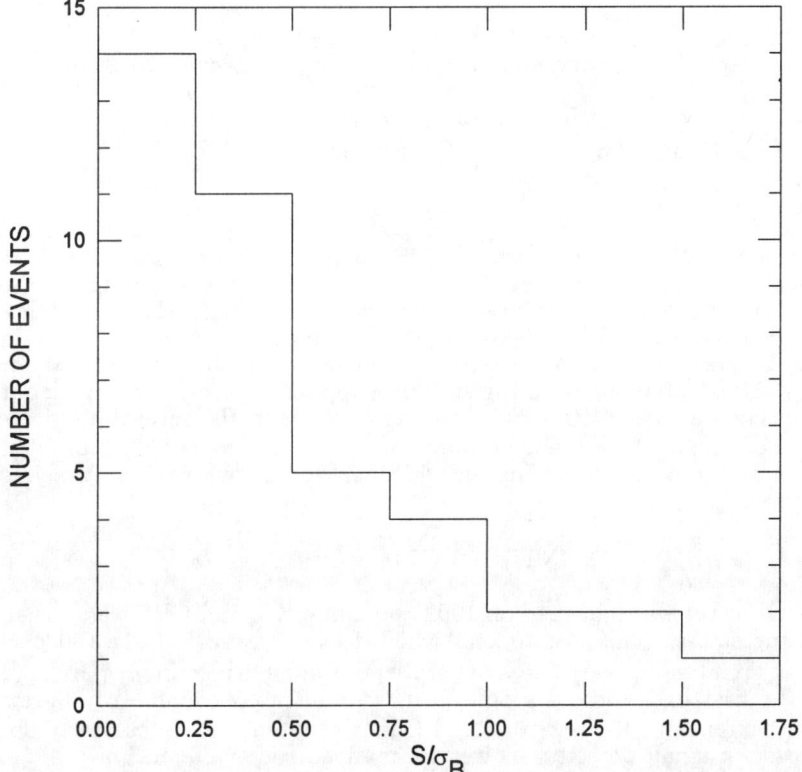

Figure 2. Distribution of S/σ_B. This distribution is consistent with the hypothesis that there are no large discrepancies between the BATSE burst positions and the triangulation annuli.

References

1. K. Hurley et al., Astron. Astrophys. Suppl. Ser. 92(2), 401 (1992)
2. K. Hurley et al., in preparation, to be submitted to the Astrophysical Journal (1993)
3. V. Wang, and R. Lingenfelter, ApJLett., 416, L13 (1993)
4. J. Quashnock and D. Lamb, MNRAS, submitted (1993)

Acknowledgments

Work on these data was supported by JPL Contract 958056 and NASA Grant NAG-1560. The Ulysses GRB experiment was built in France with support from CNES, and in Germany with support by FRG Contracts 01 ON 088 ZA/WRK 275/4-7.12 and 01 ON 88014.

STATUS OF BURST DETECTION BY THE MARS OBSERVER GRS EXPERIMENT

J.G. Laros, W. Boynton, R. McCloskey
Lunar and Planetary Laboratory, U. of Arizona, Tucson, AZ 85721

A. Metzger
Jet Propulsion Laboratory, 4800 Oak Grove Dr., Pasadena, CA 91109

R. Starr, J. Trombka
NASA Goddard Space Flight Center, Greenbelt, MD 20771

ABSTRACT

The Gamma-Ray Spectrometer (GRS) experiment on Mars Observer (MO) obtained approximately 2 months of net burst coverage during the cruise phase of the mission, before the spacecraft was lost. GRS triggered on an intense event that occurred on 1993 July 6, and the temporal data were used to direct a rapid VLA observation of the source region. A quick look at the spectral data for this event was made, but the counting statistics appear to preclude carrying out extensive spectral analysis. Sensitivity to narrow lines in the several-hundred-keV range is on the order of a few photons/cm^2. The instrument also collected data in an untriggered mode during an additional 20 known events. Analyses are in progress.

INTRODUCTION

Mars Observer was launched on 1992 September 25, and GRS was turned on shortly thereafter. This instrument, which features a fairly large (30 percent efficiency relative to NaI) HPGe crystal and a burst triggering capability[1], has made the first high-energy-resolution measurements of gamma-ray bursts since the measurements of Teegarden and Cline[2] in late 1978. Those early observations with a small detector showed hints of narrow emission lines in the several-hundred-keV range, where GRS has its maximum sensitivity. Because of its location on MO, GRS also restored our ability–interrupted when PVO entered the Venusian atmosphere in 1992 October–to derive precise burster locations using arrival-time analysis. The loss of the MO spacecraft meant that GRS never fulfilled its potential as a burst detector. Nonetheless, significant burst data were obtained during cruise, particularly in the 1993 June-July time frame. In this paper we will report on the progress to date and prospects for future analysis of the MO cruise data.

OBSERVATIONS

Because MO was designed as a low-cost, focussed-science mission with no cruise science, normal circumstances of the mission restricted the amount of available burst coverage. We were fortunate to obtain over 2 months of net data collection beginning in 1992 October, with particularly good coverage in the 1993 June-July time frame. The Germanium detector and its electronics performed well up until the time that the spacecraft was lost in 1993 August.

Burst triggering was compromised somewhat by less-than-optimum configuration of the electronic thresholds, but the continuous availability of mapping data, typically at 16 s time resolution, lessened the impact of this shortcoming.

We searched for bursts and burst coverage in the GRS data set, using the BATSE Burst Alerts and a list of Ulysses events provided by K. Hurley (private communication). Within a week of the occurrence of GB930706, we found that GRS had triggered on that event, and we had communicated the onset time to the burst community for calculation of the burster direction and a VLA target-of-opportunity observation. The VLA observation was carried out approximately 8 days after the burst (D. Palmer, this workshop). GB930706 was a rather short event, so only one significant GRS spectrum, covering 0.1–1.2 MeV with 1024 energy channels, was obtained. Because the spectrum contained only about 100 net photons, extensive spectral analysis was not warranted. Even so, the sensitivity to narrow lines in the several-hundred-keV range was a few photons per cm^2. For comparison, the possible Teegarden and Cline 740 keV line had an intensity of 20 photons/cm^2. We hope that the BATSE Spectroscopy Detector (SD) data for this event are analyzed in the near future, so we can qualitatively compare the MO and GRO spectra.

Examination of the Ulysses list showed that there was MO coverage for more than 20 events from that list. Although none of these events caused a GRS trigger, we estimate that about half of them had high enough fluences to be detectable in the GRS mapping data. However, considerable additional work will be required to put those data into a form that is amenable to burst analysis. We expect to accomplish this over the next year. Our results should include approximately 10 burst error boxes with characteristic dimensions of about one degree by a few arcmin, plus spectral analysis with a sensitivity to narrow lines of about one to 10 photons/cm^2, depending on the line energy and other factors.

SUMMARY

During its brief lifetime, the GRS Experiment on MO made significant gamma-ray burst observations. The rapid localization of one event, followed by a VLA observation, has been accomplished. During the next year we will obtain a few additional, less precise burst locations, but will concentrate on spectral analysis of lower-time-resolution burst data.

ACKNOWLEDGEMENTS

We wish to thank the GRS Team at the U of A for continuing to provide excellent analysis support under extremely difficult circumstances. This research was supported by MO Participating Scientist and GRO Guest Investigator grants.

REFERENCES

1. Metzger, A. E., et al., in *Gamma-Ray Bursts: Huntsville 1992*, ed. W.S. Paciesas & G.J. Fishman (AIP: New York, 1991), p. 353.
2. Teegarden, B.J., and Cline, T.L., Ap. J. Letts. **236**, L67 (1980).

DMSP SATELLITES AS GAMMA-RAY BURST DETECTORS

J. Terrell, P. Lee, and R. W. Klebesadel
Los Alamos National Laboratory, Los Alamos, NM 87545

J. W. Griffee
Sandia National Laboratory, Albuquerque, NM 87185

ABSTRACT

Gamma-ray burst detectors are aboard three U. S. Air Force Defense Meteorological Satellite Program (DMSP) spacecraft, in orbit at 800 km altitude, with corresponding fields of view to 117° from the zenith. A large number of bursts have deen detected by DMSP, usually confirming and supplementing data from GRO and other spacecraft. The location of a gamma-ray burst source detected by several DMSP spacecraft is considerably restricted by knowledge of the several fields of view. Often non-detection of a strong burst by one or more DMSP spacecraft is even more informative in narrowing the possible area of the burst. The DMSP data in conjunction with observations by other spacecraft can lead to reasonable positional information when more accurate positions are not available from GRO or other data.

INTRODUCTION

Three Defense Meterorological Satellites are currently in use by the U. S. Air Force, and are providing observations of a considerable number of gamma-ray bursts. They are in near-polar orbits, attaining latitudes of 81°, with an initial launch direction of 00°. These orbits were initially sun-synchronous, in near noon-midnight or dawn-dusk orbits, but do not remain precisely so after long use. Each of these spacecraft (currently DMSP 8, 10, and 11) carries two gamma-ray detectors, each with 100 cm^2 of NaI, as well as charged-particle detectors. A fuller description may be found elsewhere[1].

The relatively high (800 km) orbits allow detection of gamma-ray bursts within 117° to 118° of the spacecraft zenith, increasing the probability that at least one of the spacecraft will be in position to observe a given burst. Some of the possible observations are not usable, of course, because of non-recovery of the data or high counting rates due to passage through the north or south horns of the Van Allen radiation belts, or through the South Atlantic Anomaly. Soft gamma-ray bursts, if the photon spectrum does not reach appeciably above 50 keV, will also not be detected by the DMSP satellites.

Nevertheless, two or three of the DMSP spacecraft often manage to produce good data on a particular burst. Consideration of the overlapping fields of view of the various DMSP spacecraft then narrows the possible burst directions considerably[2]. Even better source location is possible when data from other spacecraft is available.

RESULTS

An example of the usefulness of the different fields of view of several DMSP spacecraft occurred with a gamma-ray burst on 11 February 1992. This was observed by all four of the satellites which were being tracked at that time (DMSP 8, 9, 10, and 11). GRO was not in operation during this event. The gamma-ray counting rate data from DMSP 11 are shown in Figure 1, with the 2-second resolution available with these spacecraft. The counting rates are shown as functions of transmission time for various thresholds, the highest and lowest rates corresponding to thresholds at 20 keV and 550 keV. The transmission times, after correction for an average delay of 5.50 sec in this case, give an average event time of 15775.8 sec UT for the maximum counting rate.

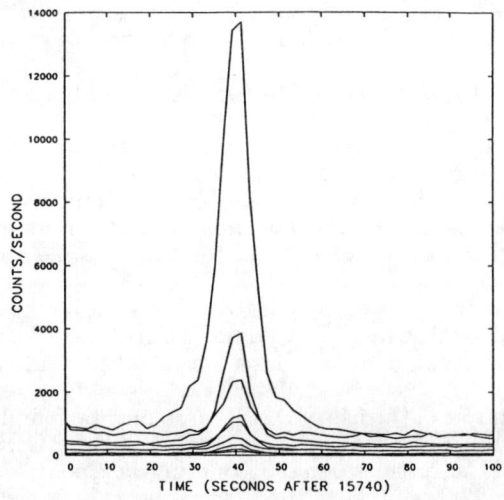

Fig. 1: DMSP 11 gamma-ray data for 2/11/92.

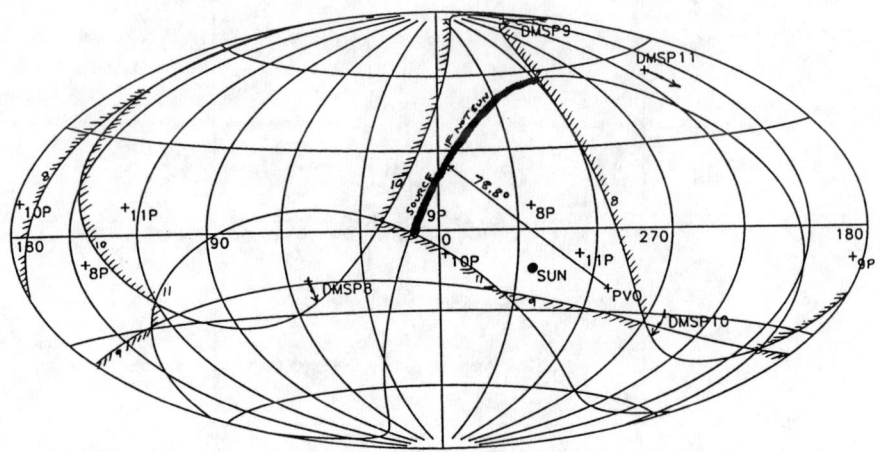

Fig. 2: DMSP fields of view (equatorial) for 2/11/92, 59406 UT.

The data for this event given by DMSP 8, 9, and 10 were very similar, although not shown here. No event triggers are currently in operation on these spacecraft, but the relative times of detection by various spacecraft may usually be determined to a fraction of a second by comparison of the counting rate data.

This event was also detected by Pioneer Venus Orbiter (PVO), at a trigger time of 15630.7 sec UT and a peak time of ~15643 sec UT. Figure 2 shows the overlapping fields of view of the DMSP spacecraft at 15775 sec UT. Orbital poles are also indicated, as one gamma-ray detector is shielded by the other for such directions. Outlined in the figure are the two portions of the celestial sphere which were viewed by all four DMSP spacecraft at that time. One of these contains the sun, which is thus a possible source of this relatively hard and short burst.

At the time of this event the sun was 492.5 light-seconds from earth and 363.3 sec from Venus, giving a difference in time of 129 seconds for a burst originating at the sun. This is slightly different from the observed time delay of ~132 seconds. If the gamma-ray burst was cosmic in origin, the relative positions of Venus and Earth at that time indicate that the source was 78.8° away from the direction of Venus (and PVO). This possibility restricts the available source locations to a narrow arc crossing one of the two possible areas in Figure 2, or to the sun.

Thus considerable information as to source location may be obtained from the combination of DMSP and PVO data in this case. There are indications from other satellites that this hard burst was a solar event (K. Hurley, private communication).

Another example of the use of DMSP data may be found in the gamma-ray burst of 19 May 1992, also not observed by BATSE on GRO. This event was observed by DMSP 10, with two major counting rate peaks as seen in Figure 3, at 59406.3 and 59422.3 sec UT.

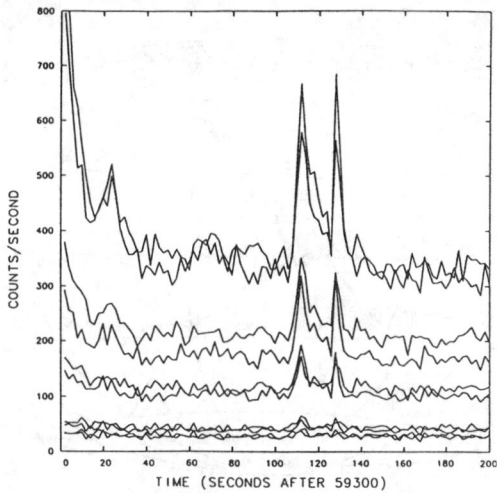

Fig. 3: DMSP 10 gamma-ray data for 5/19/92.

This event was not seen by DMSP 11, athough it was in a region of low background; DMSP 8 gave no usable data because of passage through a radiation belt. This gamma-ray burst was also observed by PVO, Phebus, and Ulysses. Triangulation using the differing signal times gave two possible directions, as seen in Figure 4. A decision between these was made possible by the directional sensitivity of Phebus, yielding (321.1°,+44.1°) as the source position (2000.0; J. Hurley, J-P. Dezalay, and C. Barat, private communication). The same result is given by the fact that the event was in view of DMSP 10, but obscured from DMSP 11, as seen in Figure 4.

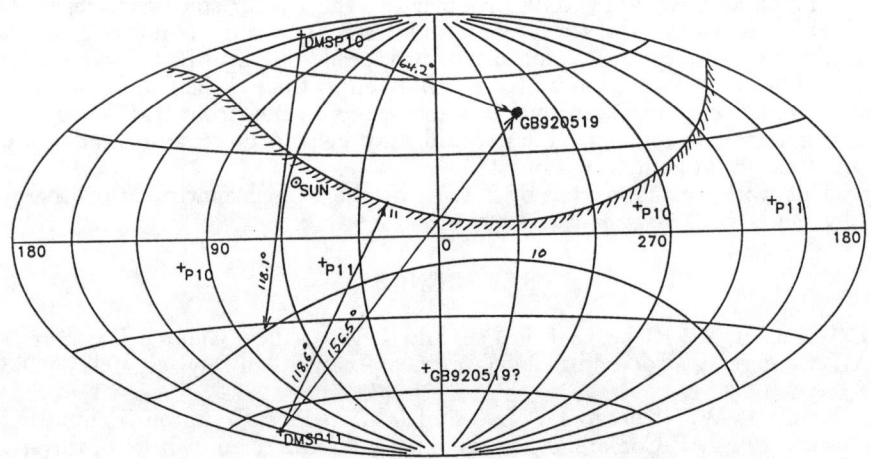

Fig. 4: DMSP fields of view (equatorial) for 5/19/92, 15775 UT.

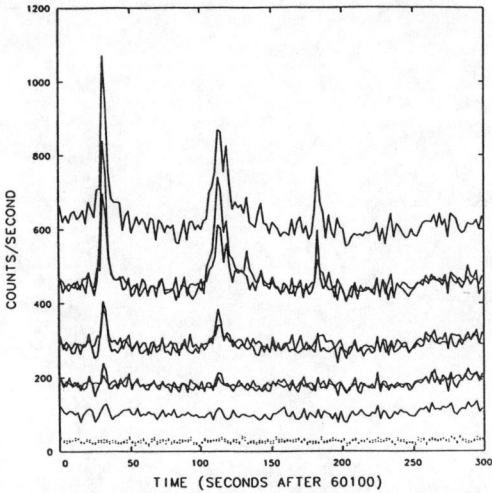

Fig. 5: DMSP 8 gamma-ray data for 2/1/93.

A final example of a more recent burst detected by DMSP 8 is of interest because of its unusual time history, displayed in Figure 5. This burst occurred on 1 February 1993, and was also detected by Batse, Ulysses, and Watch/Eureca (K. Hurley, private communication; Batse Burst 2156). There were three successive peaks as detected by DMSP 8, at 60129.0, 60211.0, and 60281.0 sec UT. These peaks are separated by ~76-sec intervals, and decrease with time. Such a time pattern raises interesting questions as to the nature of the source.

CONCLUSIONS

The gamma-ray burst data produced by the DMSP spacecraft represent a valuable resource in the study of these bursts. The value is not only in the time histories and spectral information produced, but also in the directional information due to occlusion by the earth at burst time. When combined with other observations, the directional information and times from DMSP can lead to better locations for many bursts, and may help in understanding the still puzzling nature of gamma-ray bursts.

This work was supported by NASA, by the U. S. Department of Energy, and by the U. S. Department of Defense.

REFERENCES

1. J. Terrell, R. W. Klebesadel, P. Lee, and J. W. Griffee, Gamma-Ray Bursts, AIP Conference Proceedings 265 (American Institute of Physics, 1992), p. 48.

2. J. Terrell, R. W. Klebesadel, P. Lee, and J. W. Griffee, Compton Gamma-Ray Observatory, AIP Conference Proceedings 280 (American Insitute of Physics, 1993), p. 788.

THE EFFECT OF REPEATING BURST SOURCES ON $\langle V/V_{max} \rangle$

David L. Band
CASS 0111, UC San Diego, La Jolla, CA 92093

ABSTRACT

The $\langle V/V_{max} \rangle$ test for the uniformity of gamma ray burst sources is unaffected by source repetition; the test is just as effective whether all events or only the brightest event from each source are used. However, dividing events into apparent repeaters and nonrepeaters will result in biased values of $\langle V/V_{max} \rangle$ because the difficulty of identifying repeaters increases with distance. The definition of a burst and of repeating bursts based on the temporal separation between events gives unbiased $\langle V/V_{max} \rangle$ values. Since the $\langle V/V_{max} \rangle$ test has shown conclusively that burst sources are not uniformly distributed, more appropriate tools should be used to study this nonuniformity.

INTRODUCTION

Is the diagnostic power of the $\langle V/V_{max} \rangle$ test affected if gamma ray bursts do indeed repeat? Analysis of the first BATSE catalog[1] resulted in claims of the identification of a repeating classical burst source[2,3] and of a small-separation excess in the nearest-neighbor statistic;[4] the statistical evidence for repeating sources has been disputed.[5,6,7,8] Similarly, it has been suggested that each spike within a burst which rises above the detection threshold and would have retriggered the detector should be treated as a separate burst, that is, multi-spike bursts should be considered repeating events.[9] Note that the soft gamma ray repeaters are generally considered to be a separate class from the classical bursts and are therefore not of interest here.[10] Although the possibility of repeating classical bursts is currently controversial, an obvious question is whether the true spatial distribution of burst sources could be obscured by repeating events. Specifically, given a group of sources each producing a cluster of events, what is the effect on $\langle V/V_{max} \rangle$ of including either all events or only the brightest event from each source?

In this study I assume that all events are drawn from the same luminosity function. In addition I work in the limit that a large enough sample of events and burst sources has been observed for true averages to be calculated. Therefore I do not consider the effect on $\langle V/V_{max} \rangle$ of a few repeaters with many repetitions. A more complete presentation will be published elsewhere.[11]

$\langle V/V_{max} \rangle$ AND HOMOGENEITY

The $\langle V/V_{max} \rangle$ test removes the effects of variations of the detection threshold count rate C_{min} for the purpose of testing whether the distribution of peak count rates C_{max} is consistent with homogeneity.[12,13,14,15] I define homogeneous to mean uniform in a d-dimensional flat space, usually three-dimensional Euclidean space. Thus $V/V_{max} = (C_{max}/C_{min})^{-3/2}$, which can be understood geometrically as the ratio of the volumes within which the burst originated and could have been detected. The properties of $\langle V/V_{max} \rangle$ can be studied by integrating over model source densities, luminosity functions (here the distribution of peak photon emission rate, not the energy emission rate) and detector thresholds. In the integral for the V/V_{max} distribution, V/V_{max} decouples from the luminosity function if the luminosity function is independent of the distance r of the burst source from the observer and if the source density is a power law function of this distance. Since $\langle V/V_{max}\rangle$ is the normalized first moment of the V/V_{max} distribution, under these conditions $\langle V/V_{max}\rangle$ will always be equal to the same value which is a function of the source density's power law index. If the burst sources are distributed uniformly on a d-dimensional surface (d can be fractal) then the source density averaged over spheres around the observer is a power law: $n(r) \propto r^{d-3}$. Therefore bursts from sources distributed uniformly over a d-dimensional space with a distance-independent luminosity function will always have the same $\langle V/V_{max}\rangle = d/(d+3)$. It is this characteristic which gives $\langle V/V_{max}\rangle$ its diagnostic power, usually applied for $d=3$ only. Thus for $d=1, 2$ and 3 we find $\langle V/V_{max}\rangle = 0.25, 0.4$ and 0.5. In addition, if $d=3$ then V/V_{max} will be distributed uniformly between 0 and 1, a characteristic which can be used as a secondary test of uniformity.

$\langle V/V_{max}\rangle$ is therefore invariant for homogeneous source distributions under transformations which keep the luminosity function distance-independent. If the source distribution is truly homogeneous, $\langle V/V_{max}\rangle$ will not change when bursts are selected in such a way that the luminosity function has no distance dependence.

CLUSTERS OF EVENTS

Assume I have identified a source population which produces clusters of events, each of which is drawn from the same distance-independent luminosity function. I can calculate $\langle V/V_{max}\rangle$ either by including each observed event or by selecting the brightest event from each cluster. For this study I neglect the observational difficulty of identifying all the events from a given source.

If I calculate $\langle V/V_{max}\rangle$ using all the observed events, the effective luminosity function has the same shape as the single-event luminosity function, but will be increased by the average number of events in a cluster. Therefore the diagnostic power of $\langle V/V_{max}\rangle$ will be unchanged by the presence of repeaters. Even for a source population which is not uniform in any dimension the value of $\langle V/V_{max}\rangle$ will be unchanged.

The luminosity function of the brightest event in a multi-event cluster from the same source is shifted to higher luminosities, but if the single-event luminosity function is distance-independent, so is the modified luminosity function. Therefore the power of $\langle V/V_{max}\rangle$ to test for homogeneity is unchanged. But if the source distribution is inhomogeneous, including only the brightest event emphasizes higher luminosities and $\langle V/V_{max}\rangle$ probes a larger volume. For example, for events drawn from a single-event luminosity function proportional to L^{-2} the average luminosity of the brightest event from a 10-event cluster is more than 50 times brighter than the average single-event luminosity (note that here the luminosity L is the peak photon emission rate, not the energy emission rate). If inhomogeneity increases with distance, as appears to be the case for our sources,[16] $\langle V/V_{max}\rangle$ will show greater deviations from the values expected for homogeneity.

The brightest distant events are included in $\langle V/V_{max}\rangle$ with either treatment of event clusters. However, when all events are included, the events which are faint because of their distance are diluted by nearby events which are intrinsically dim. Thus a radial decrease in the source density is less apparent. Similarly, the faint end of the intensity distribution shows a smaller deviation from the power law dependence of the bright end if all repeating events are included. Therefore calculating $\langle V/V_{max} \rangle$ or the intensity distribution by including all spikes from a multi-spike burst instead of the current method of using the peak count rate for the entire burst should not be interpreted as recovering source uniformity.

APPARENT REPEATER/NONREPEATER POPULATIONS

If and when a population of repeating burst sources is identified, a natural temptation will be to separate bursts into repeaters and nonrepeaters and then investigate the properties of each population, such as the value of $\langle V/V_{max} \rangle$. Similarly, bursts can divided into single-spike and multi-spike classes (analogous to repeaters and nonrepeaters). However, in both cases the resulting values of $\langle V/V_{max} \rangle$ will be biased. For a more distant, and thus on average fainter, repeater (or multi-spike burst) the second-brightest event is more likely to fall below the detection threshold, thereby causing the source to be misidentified as a nonrepeater (or single spike burst). In addition, even if the second event is detected, the observational difficulty of identifying it as a repetition increases as the event becomes fainter with distance; the uncertainty in source position increases as the count rate decreases. Therefore the separation into apparent repeaters/nonrepeaters is a distance-dependent operation: apparent repeaters will include predominantly nearby repeaters; and apparent nonrepeaters will also include the brightest events from distant repeaters. I therefore expect apparent repeaters (as well as multi-spike bursts) to have small values of $\langle V/V_{max} \rangle$, and apparent nonrepeaters (and single-spike bursts) to be characterized by large $\langle V/V_{max} \rangle$.

Populations based on distance-independent criteria have unbiased values of $\langle V/V_{max} \rangle$. In particular, populations defined by clustering events according to their temporal separation are valid. The only complication results from weak, undetected events bridging a long separation between observed events. However the distribution of separations between spikes within a burst drops rapidly with temporal separation, and therefore weak events effectively do not affect the population definition. Thus the current practice of considering events within a few minutes as a single burst and events separated by longer durations as separate bursts produce correct values of $\langle V/V_{max} \rangle$. Note that currently a rigid duration limit beyond which continued emission is considered a new burst is not applied to bursts.

DISCUSSION AND CONCLUSIONS

$\langle V/V_{max} \rangle$ is an effective test of the homogeneity of the source distribution because it has the same value for homogeneous sources regardless of the form of the distance-independent luminosity function or the distribution of detector thresholds. Although usually used as a test of three-dimensional uniformity, for which $\langle V/V_{max} \rangle=1/2$, this statistic can also be used to find homogeneity in any d-dimensional space. This diagnostic power is predicated on the distance-independence of the luminosity function.

If burst sources repeat with each event drawn from the same distance-independent luminosity function then $\langle V/V_{max} \rangle$ can be calculated with each event or only the brightest event from each repeating source. For the first option the luminosity function has the same functional form as the single-event luminosity function, while for the second option the luminosity function is shifted to higher luminosity, but retains the same spatial dependence. Therefore, the diagnostic power of $\langle V/V_{max}\rangle$ will be unchanged.

Because the probability of detecting the second-brightest event from a repeating source (or the second-brightest spike in a multi-spike burst) decreases as the source become more distant and therefore on average fainter, the likelihood of misclassifying repeaters (or multi-spike bursts) as nonrepeaters (or single-spike bursts) increases with distance. Therefore $\langle V/V_{max}\rangle$ is biased towards small values for apparent repeaters (or multi-spike bursts).

The current definition of a burst based on the separation between events is reasonable and produces a relatively unbiased value of $\langle V/V_{max}\rangle$. The alternatives are problematic. Attempting to remove all but the brightest burst from repeating sources is futile because faint repetitions will not be identified as originating from the repeating source. Failing to remove the weak repetitions biases $\langle V/V_{max}\rangle$ to higher values. On the other hand, the spectral evolution seen in multi-spike bursts[17,18,19] invalidates the assumption that the spikes within a burst are independent events drawn from the same luminosity function.

Bursts are clearly not homogeneous:[16] $\langle V/V_{max}\rangle < 1/2$ indicates the source distribution is not uniform in three dimensions even though the spatial distribution is isotropic; and the $\log N$-$\log P$ distribution (P is peak flux) does not have the power law dependence expected for a uniform distribution in some other dimension.[1] Therefore the $\langle V/V_{max}\rangle$ test has proved its usefulness, and should be retired from the study of gamma ray bursts. Not only does $\langle V/V_{max}\rangle$ destroy information by reducing the observations to a single number, but it also distorts the observations by convolving them with the distribution of the instrument's detection threshold.[14,15,20,21] Instead, the nature of the inhomogeneous source distribution should be investigated using techniques such as maximum-likelihood methods applied to the entire dataset,[15,22,23] survival analysis,[24] or moments of the intensity distribution.[25,26] The treatment of repeaters within these more appropriate methods will have the same effect on the luminosity function as discussed here for calculating $\langle V/V_{max}\rangle$. In addition, model intensity distributions can be compared to the observed $\log N$-$\log P$ distribution (with threshold corrections) if the treatment of repeaters and multi-spike bursts in the observed distribution is understood. Just as in calculating $\langle V/V_{max}\rangle$, the current definition of a burst is probably optimal for these studies of source inhomogeneity.

During this study I enjoyed stimulating discussions with R. Lingenfelter, and helpful comments from the rest of the BATSE instrument team. This work was supported by NASA contract NAS8-36081.

REFERENCES

1. G. J. Fishman, et al., Ap. J. Supp., in press (1994).
2. V. Wang and R. E. Lingenfelter, in the Contributed Papers of the 23rd International Cosmic Ray Conference, (1993) p. 1-93.
3. V. Wang and R. E. Lingenfelter, Ap. J. Lett. **416**, L13 (1993).
4. J. M. Quashnock and D. Q. Lamb, MNRAS, submitted (1993).
5. D. H. Hartmann, G. R. Blumenthal, E. V. Linder, and K. Hurley, Ap. J.,

submitted (1993).
6. E. Maoz, Ap. J. Lett., submitted (1993).
7. R. Narayan and T. Piran, Ap. J. Lett., submitted (1993).
8. M. A. Nowak, Ap. J. Lett., submitted (1993).
9. R. Lingenfelter, V. Wang, and J. Higdon, Ap. J. Lett., submitted (1993).
10. J. C. Higdon and R. E. Lingenfelter, Ann. Rev. Astron. Astrophys. **28**, 401 (1990).
11. D. Band, Ap. J. Lett., in press (1993).
12. M. Schmidt, Ap. J. **151**, 393 (1968).
13. M. Schmidt, J. C. Higdon, and G. Hueter, Ap. J. Lett. **329**, L85 (1988).
14. D. Band, Ap. J. Lett. **400**, L63 (1993).
15. D. Band, in Compton Gamma-Ray Observatory, AIP Conference Proceedings 280, eds. M. Friedlander, N. Gehrels and D. J. Macomb (AIP, New York, 1993), p. 734.
16. C. Meegan, et al., Nature **335**, 143 (1992).
17. J. Norris, et al., Ap. J. Lett. **301**, 213 (1986).
18. D. Band, et al., in Gamma Ray Bursts: AIP Conference Proceedings 265, eds. W. S. Paciesas and G. J. Fishman (AIP, New York, 1992), p. 169.
19. L. Ford, et al., these proceedings (1993).
20. D. H. Hartmann and L. S. The, Ap&SS **201**, 347 (1993).
21. V. Petrosian, Ap. J. Lett. **402**, L33 (1993).
22. T. Loredo and I. Wasserman, in Compton Gamma-Ray Observatory, AIP Conference Proceedings 280, eds. M. Friedlander, N. Gehrels and D. J. Macomb (AIP, New York: AIP, 1993), p. 749.
23. T. Loredo and I. Wasserman, these proceedings (1993).
24. B. Efron and V. Petrosian, Ap. J., submitted (1993).
25. J. M. Horack and A. G. Emslie, Ap. J., in press (1993).
26. J. M. Horack and A. G. Emslie, these proceedings (1993).

THE ISOTROPY OF GAMMA-RAY BURSTS: DIPOLE AND QUADRUPOLE TESTS

Michael S. Briggs, William S. Paciesas, Geoffrey N. Pendleton
Dept. of Physics, University of Alabama in Huntsville, Huntsville, AL 35899

Gerald J. Fishman, Charles A. Meegan, Robert B. Wilson,
Martin N. Brock, Chryssa Kouveliotou (USRA)
NASA Marshall Space Flight Center, Code ES-66, Huntsville, AL 35812

ABSTRACT

We search for dipole or quadrupole anisotropies in the locations of the first 743 gamma-ray bursts observed by BATSE. No convincing anisotropies have been found.

INTRODUCTION

Observations made by BATSE have shown that Gamma-Ray Bursts are both isotropic and inhomogeneous[1]. These two observations pose severe difficulties for the formerly popular galactic-disk origin theories. Herein we continue the search for large-angular scale anisotropies in the BATSE burst locations using dipole and quadrupole tests. Such tests are not sensitive to small-scale anisotropies such as clustering, but should detect any physically reasonable large-scale anisotropy.

The search is a statistical one: the values of various statistics are calculated from the locations and those values are compared to the distribution of values expected for isotropic locations. If a statistic has a value extremely improbable for isotropic locations, then a significant anisotropy has been found. The theory of dipole and quadrupole statistics is discussed by Briggs[2].

We analyse herein the first 743 Gamma-Ray Bursts observed by BATSE. The locations of the first 260 are available in the first BATSE (1B) catalog[3].

THE STATISTICS AND THE METHOD

Locations are determined by comparing the rates of BATSE's 8 Large Area Detectors[3,4]. The locations are not exactly determined because of statistical fluctuations in the rates and because of systematic errors. For the brightest bursts the location error is predominantly due to the systematic error of $\approx 4°$ while for faint bursts the statistical error of $\approx 13°$ is dominant. The systematic error in the preliminary locations of the 483 post-1B bursts is somewhat larger–this has little impact on the results because the location errors are not the limiting factor in BATSE's sensitivity to anisotropies[5], and as discussed below.

Because BATSE is on a spacecraft in low-earth orbit, the instruments's sky exposure in not uniform[3,6]. The exposure to the equatorial regions of the sky is reduced due to the time-averaged blockage by the Earth, while the instrument down-time due to the South Atlantic Anomaly reduces the exposure to Southern regions of the sky. BATSE's sky exposure is accumulated into a sky exposure map[6] so that we can determine these systematic effects, as described below.

We use six statistics because each is most sensitive to a particular kind of anisotropy. We use two galactic statistics because they are the most sensitive to galactic patterns and thus are best for testing the question of galactic versus cosmological origin. The statistic $\langle\cos\theta\rangle$, where θ is the angle between the burst and the galactic center, tests for a concentration towards the galactic center, while the statistic $\langle\sin^2 b - \frac{1}{3}\rangle$, where b is galactic latitude, tests for a concentration in the galactic plane. We use two coordinate-system independent statistics[2] to search for significant dipole and quadrupole moments in a model independent manner. The Rayleigh-Watson statistic W tests the size of the dipole moment, while the Bingham statistic B measures the deviation of the quadrupole moments from the values expected for isotropy. Finally, we use two equatorial-based statistics because they are the most sensitive statistics to the artificial anisotropies caused by the proximity of the Earth to the spacecraft. The statistic $\langle\sin^2\delta - \frac{1}{3}\rangle$, where δ is the declination of the burst, is sensitive to the quadrupole moment caused by the reduced exposure of the equatorial region and the statistic $\langle\sin\delta\rangle$ is sensitive to the dipole moment towards the North Pole caused by turning the instrument off in the South Atlantic Anomaly.

RESULTS FOR THE ENSEMBLE OF BATSE BURSTS

These six statistics and the information needed to interpret them are shown in Figure 1. The observed values of the statistics, calculated from the burst locations with no correction for the nonuniform sky exposure, are shown on the Figure with dots. The Figure shows the values for the first 40, 100, 260, 447 and 743 burst observed by BATSE. The values of each statistic for these various datasets are not independent since each dataset is a subset of the following ones. Results for the first 260[3] and 447[7,8] bursts have been reported previously. Because the locations are not exactly determined, the values of the statistics calculated from the locations are not exactly determined. The uncertainties in the statistics have been estimated by Monte Carlo propagation of the location errors–these uncertainties are depicted by the error bars on the dots.

Any particular finite sample drawn from an isotropic population will have chance anisotropies so we must consider the distribution of the statistics. For each statistic, the mean value expected for isotropic locations observed by an ideal instrument is show with a bold, dashed line and the $\pm1\sigma$ envelopes of the distribution of the statistics about their means are indicated by the nonbold, dashed lines. However, because of BATSE's nonuniform sky exposure, we should not compare the observed values with the ideal expectations. The means and distributions of the statistics expected for an isotropic population observed by BATSE have been calculated by Monte Carlo simulations using the sky exposure map and are shown with the solid lines: the bold lines indicate the expected means while the nonbold lines show the $\pm1\sigma$ envelopes of the distributions.

Several important results can be read from the Figure: 1) an anisotropy would be significant only if a statistic had a value very rare based upon its distribution corrected for BATSE's nonuniform sky exposure. This would appear on the Figure as a observed value (dot) lying far outside the corrected $\pm1\sigma$ envelope of its distribution (nonbold, solid lines). Most of the values are within 1σ of the expected mean, and all are within 2.0σ, so no significant anisotropy is found. 2) In all cases, the width of the $\pm1\sigma$ envelope of the distribution of the statistics is much larger than the uncertainties in the observed values of the statistics due to the location uncertainties. This indicates that for BATSE the

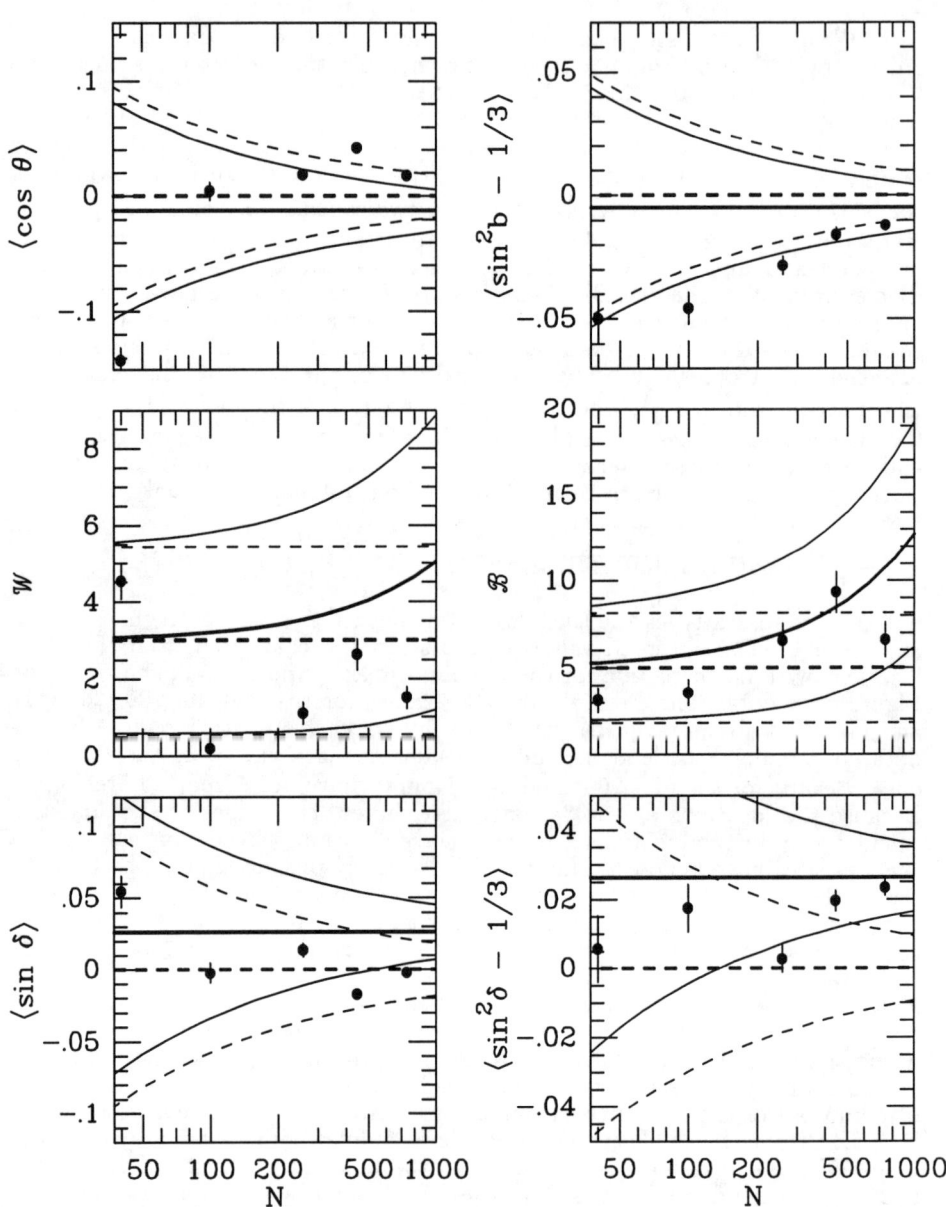

Figure 1. Location statistics for the first 40, 100, 260, 447 & 743 GRBs observed by BATSE are shown with dots and vertical lines for their error bars. The values are not independent since each set is a subset of the following. The means and $\pm 1\sigma$ envelopes of the distributions of the statistics expected for isotropic locations observed by an ideal instrument are shown with dashed lines. The expected means and $\pm 1\sigma$ distributions, taking into account BATSE's nonuniform sky exposure, are shown with solid lines.

dominant statistical limitation in detecting a dipole or quadrupole anisotropy is the finite number of bursts observed and not the location uncertainties. 3) The importance of the systematic effect of BATSE's nonuniform sky coverage is indicated by the offset between the ideal means of the statistics and the corrected means compared to the $\pm 1\sigma$ envelopes of the statistics. For the important galactic statistics and for ≤ 743 locations, the sky exposure corrections are well under 1σ and are thus not crucial. The equatorial quadrupole statistic $\langle \sin^2 \delta - \frac{1}{3} \rangle$ is expected to have a significant artificial anisotropy: the expected mean value for BASTE is, for 743 locations, 2.4σ from the ideal mean of zero. The observed value of the statistic is close to the expected mean calculated from the sky exposure map and is 2.1σ from the ideal mean, thereby indicating that the sky exposure predictions are reliable and that BATSE has detected the Earth at the 2.1σ level. It is fortunate that the equatorial and galactic planes are highly inclined, because this causes the anisotropies caused by the Earth to poorly couple to the galactic statistics.

Table 1. Location Statistics of BATSE's first 743 GRBs					
Statistic	Moment Tested	Coord. System	Value Expected for Isotropy with Uniform Sky Exposure	Value Expected for Isotropy with BATSE's Sky Exposure	Observed Value (with propagation of location errors)
$\langle \cos \theta \rangle$	Dipole	Galactic	0 ± 0.021	-0.013 ± 0.021	0.018 ± 0.003
$\langle \sin^2 b - \frac{1}{3} \rangle$	Quad.	Galactic	0 ± 0.011	-0.005 ± 0.011	-0.012 ± 0.002
W	Dipole	Indepen.	3 ± 2.4	4.5 ± 3.5	1.5 ± 0.3
B	Quad.	Indepen.	5 ± 3.2	10.7 ± 5.8	6.6 ± 1.0
$\langle \sin \delta \rangle$	Dipole	Equatorial	0 ± 0.021	0.026 ± 0.022	-0.002 ± 0.002
$\langle \sin^2 \delta - \frac{1}{3} \rangle$	Quad.	Equatorial	0 ± 0.011	0.026 ± 0.011	0.023 ± 0.002

Table 1 lists the values of the statistics for the ensemble of 743 bursts. The errors listed in the fourth and fifth columns are the 1σ finite sample fluctuations. The comparison between the next-to-last and last columns indicates the evidence for anisotropies.

RESULTS FOR SUBSETS OF THE BURSTS

We are presently examining various subsets of the bursts and have not yet found any subset with a significant dipole nor quadrupole anisotropy[9]. However, Quashnock and Lamb have found a subset of the first 260 bursts with an apparent large-scale anisotropy[10]. By considering the joint probability of a large value of $\langle \cos \theta \rangle$ and a very negative value of $\langle \sin^2 b - \frac{1}{3} \rangle$, they identified 55 "Medium" bursts ($\log_{10} V < -0.8$ & $465 < B < 1169$)[11,12] as being significantly anisotropic, which they interpret as a spiral arm signature. (Applying these selection criteria to the 1B bursts (Table 2), we remove 4 bursts from the Quashnock and Lamb dataset. These bursts were included in their dataset based upon $C_{\max}/C_{\min}$ numbers released by the BATSE team that were wrong for trigger timescales for which the instrument would not have triggered; the numbers were intended for $V/V_{\max}$ studies and were always correct for timescales on which the instrument would have triggered.)

We have applied the same B and V selection criteria to the 483 post-1B bursts of our set of 743, obtaining 51 additional bursts in the Medium category–see Table 2. (The fraction of bursts for which V may be calculated has decreased due to the CGRO tape recorder failure). The 51 post-1B bursts are an independent sample from the 51 1B bursts and are thus ideal to test a hypothesis created based upon an analysis of the 1B bursts.

The spiral arm hypothesis of Quashnock and Lamb predicts $\langle \cos \theta \rangle$ significantly above zero and $\langle \sin^2 b - \frac{1}{3} \rangle$ significantly below zero[10]. In the new set, however, $\langle \cos \theta \rangle$ is significantly below zero–it is 2.8σ different from the value the spiral arm hypothesis predicts, 0.184 ± 0.080. The hypothesis of isotropy has no difficulty with the post-1B results, differing by 1.5 and 1.2σ from isotropy, but has the the difficulty of explaining the 2.5σ and 2.6σ (not independent) deviations for the 1B dataset. Given the large number of tests applied to many subsets of the BATSE data, these deviations may be due to a statistical fluctuation. The analysis of additional burst locations will be able to conclusively settle this issue. The location errors are unimportant since the sample size is the dominant limitation in detecting any anisotropy.

Table 2. Location Statistics for "Medium" Bursts

Set	N	Isotropy $\langle \cos \theta \rangle$	Observed $\langle \cos \theta \rangle$	Isotropy $\langle \sin^2 b - \frac{1}{3} \rangle$	Observed $\langle \sin^2 b - \frac{1}{3} \rangle$
1B	51	-0.013 ± 0.080	0.184	-0.005 ± 0.041	-0.111
post-1B	51	-0.013 ± 0.080	-0.134	-0.005 ± 0.041	-0.056
all	102	-0.013 ± 0.057	0.025	-0.005 ± 0.029	-0.084

CONCLUSIONS

We have searched the first 743 gamma-ray bursts observed by BATSE for dipole and quadrupole anisotropies. Since the effect determining BATSE's sensitivity to large-angular scale anisotropies is the number of bursts in a sample, we have concentrated on analyzing the entire sample of 743 bursts. There has been a report of a subset with a significant large-scale anisotropy[10], but this anisotropy is not present in additional data. To date, no convincing dipole nor quadrupole anisotropy has been found in BASTE's data.

REFERENCES

1. C. A. Meegan et al., Nature, 355, 143, 1992.
2. M. S. Briggs, ApJ, 407, 126, 1993.
3. G. J. Fishman et al., ApJSupp, in press, 1994.
4. M. N. Brock et al., AIP Conf. Proc. 265, 1992, 383.
5. J. M. Horrack et al., ApJ, 413, 293, 1993.
6. M. N. Brock et al., AIP Conf. Proc. 265, 1992, 399.
7. C. A. Meegan et al., AIP Conf. Proc. 280, 1993, 681.
8. M. S. Briggs et al., AIP Conf. Proc. 280, 1993, 687.
9. M. S. Briggs et al., ApJ, in preparation.
10. J. M. Quashnock & D. Q. Lamb, MNRAS, in press.
11. D. Q. Lamb et al., ApJL, 413, 11, 1993.
12. C. Graziani, private communication, 1993.

MODEL INDEPENDENT CONSTRAINTS ON HELIOCENTRIC GAMMA-RAY BURST MODELS

Lawrence E. Brown, Dieter H. Hartmann, and Lih-Sin The

Dept. of Physics and Astronomy, Clemson University, Clemson, SC 29634

ABSTRACT

We present essentially model independent constraints on the distance scale for gamma-ray burst models assuming that bursts originate in circumstellar shells of known or unknown nature. The constraints described here are derived by assuming that the bursters are spherically distributed around the sun and that nearby stars must have similar distributions of bursters. To illustrate the method we consider generic Oort cloud models. Using the first public BATSE catalog (1B), we find an upper limit for the distance to the brightest bursts of $\sim 65,000 AU$.

INTRODUCTION

The gamma-ray burst brightness statistic and angular distribution on the sky essentially rules out Galactic GRB origins associated with stellar disk populations or the dark matter halo.[1,2] More exotic Galactic components, such as extended halos resulting from neutron star formation in the halo, were suggested to accommodate the geometric constraints, but these models too now appear less promising.[3] Cosmological models offer the most natural solution to the observational boundary conditions, but the opposite extreme of very nearby sources has been discussed as well. The observed isotropy requires spatial distributions to be centered on the Sun and to be at least roughly spherically symmetric. If GRBs originate in sun-centered structures such as the Oort cloud, then some simple assumptions are sufficient to constrain such models without detailed treatment of rate profiles and luminosity functions. We do not confine the possibilities to the traditional cometary cloud. We make two conservative assumptions:
1. Gamma-ray bursts are produced in a spherically symmetric region centered on the sun.
2. Other stars in the solar neighborhood will have identical burst producing regions around them.
Thus, for each burst seen from our own cloud of bursters, there will be a corresponding "cousin" burst (in a statistical sense) around nearby stars. Using this assumption, we can set statistical limits on the possible distance to the bursters. Very bright bursts must come from less than a "critical" distance (on average) or else their "cousins" would produce an excess from nearby stars, α Cen most prominently. This model applies not only to comet-related models, but, since we have not used any Oort Cloud properties, it also applies to any stellar centered model that might be proposed, but hasn't been yet (e.g. WIMPS, Alien Space Defense Network signal buoys, etc.).

METHOD

For each observed burst in the 1B catalog[2] we vary the unknown distance (D) from Earth to the burst. For each D we produce a Monte Carlo set of

10,000 "cousins" distributed isotropically around another star (located R_* away from the sun). The "cousins" are the same distance (D) from that star. Each cousin is then shifted on the sky by an error (θ_{error}) drawn from a Gaussian distribution. Our approximate function for mean error as a function of C_p/C_{lim} is shown in Figure 1.

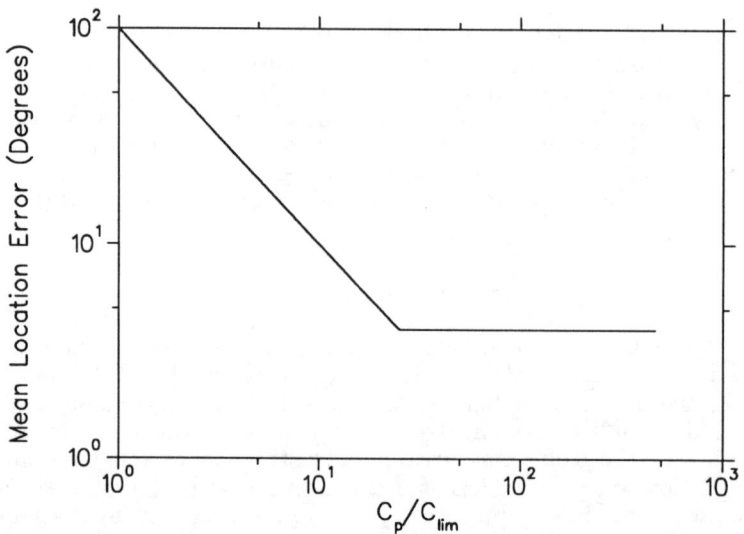

Figure 1. Approximate mean angular error of BATSE burst locations as a function of C_p/C_{lim}.

The geometry of the model is shown in Figure 2.

If a "cousin" is detectable ($C_p/C_{lim} > 1$) and if it falls within θ_{error} of the star on the sky, it is accepted. Number of accepted "cousins" over total "cousins" gives the probability of a detected cousin being produced if the real burst under consideration was actually located at distance D. The resulting probability versus D is shown in Figure 3.

The D corresponding to a 10% chance of "cousin" detection is plotted versus C_p/C_{lim} for all 241 bursts in Figure 4.

RESULTS AND CONCLUSIONS

If we assume that no direction has an excess of more than n bursts above average then the ($n \times 10$)th brightest burst in Figure 4 gives an upper distance limit. As BATSE observes more bright bursts, this limit will become stronger. The question of what n to use is left to the reader's choice. Figure 5 shows a map of the 1B catalog locations. Each half degree square pixel contains a weight which roughly indicates the probability of a burst at that location, or, more accurately, a sum of all such probabilities.

Figure 6 shows a histogram of the weights of the pixels in figure 5.

Currently, we use $n = 3$ (Horack, et al.[5]) which yields a limiting radius of $0.24 \times R_*$. For α-Cen this limit corresponds to 65,000 AU. For the Oort cloud, this limit complements the more direct, but model-dependent, limits obtained from assumptions about cloud parameters and models for how burst

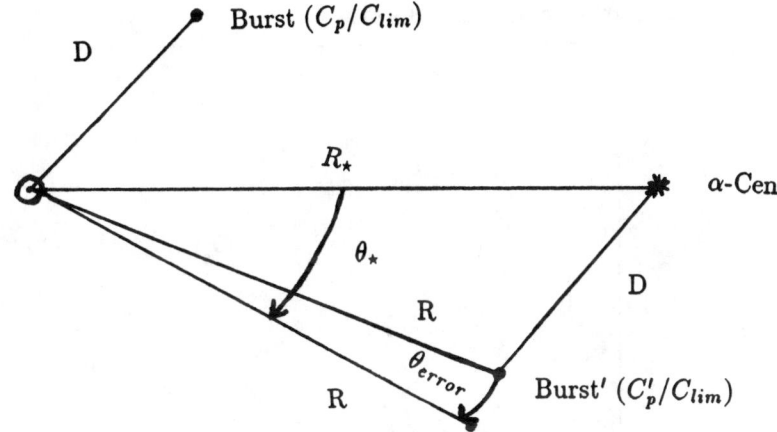

Figure 2. The geometry of our model. R_* is the distance to another star. $C'_p/C_{lim} = C_p/C_{lim} * (R^2/D^2)$. θ_{error} is the corresponding BATSE mean error. A Monte Carlo trial is "good" if $C'_p/C_{lim} > 1$ and $\theta_* < \theta_{error}$. Probability of "cousin" detection is number of good trials/number of trials.

activity might be related to comet density and other physical parameters. Recently Clarke, Blaes, & Tremaine[4] have convincingly argued against the Inner ($< 3 \times 10^3$ AU) and Intermediate ($< 3 \times 10^4$ AU) Oort Cloud regions on the grounds of event concentration towards the ecliptic the galactic poles, respectively. If, for some unknown reason, bursts could only occur beyond 3×10^4 AU, then a cometary model might appear to be consistent with the BATSE data at first sight, but the lack of event excess from nearby stars would constrain such solutions. The technique presented here provides constraints that are independent on the particular choice of radial profiles or luminosity functions. The prize we pay for gaining model independence is that the limits are not rigorous, but statistical in nature. This is because we don't know whether a source is in the local or distant shell and no burst generates a true cousin burst, but each burst only has an ensemble averaged cousin. However, with increasing burst numbers in the catalogs, such probabilistic limits become more and more reliable.

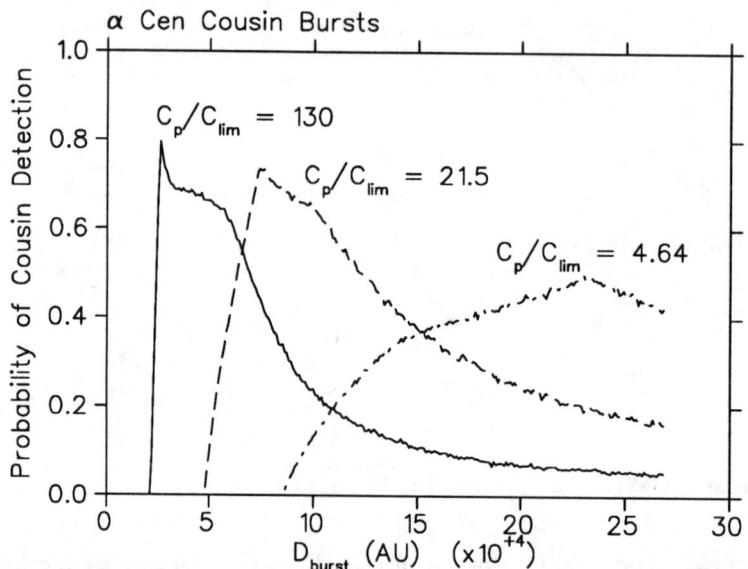

Figure 3. The probability of detecting a "cousin" burst around α Cen for a single local burst. Three values of C_p/C_{lim} are shown (corresponding to $V/V_{max} \sim 0.001, 0.01, 0.1$).

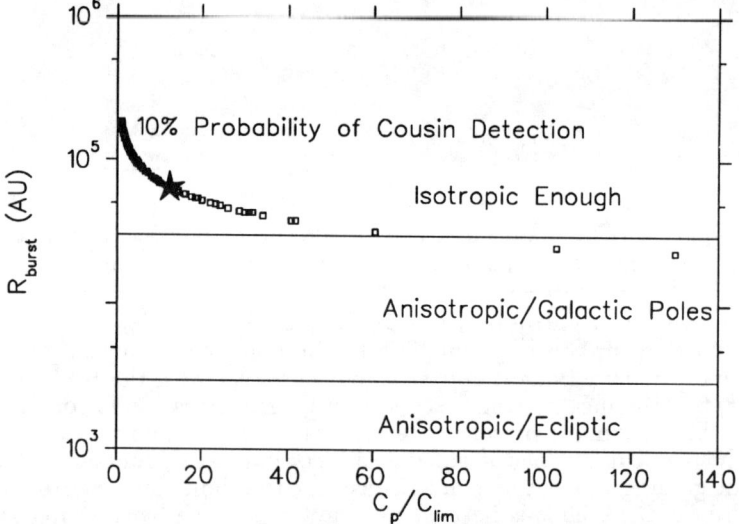

Figure 4. Distance corresponding to 10% probability of cousin detection vs. C_p/C_{lim}. The large star marks the 30th brightest burst (roughly a current upper limit to burst origin distance). The three radial Oort Cloud zones defined in Clarke, Blaes, & Tremaine[4] are shown.

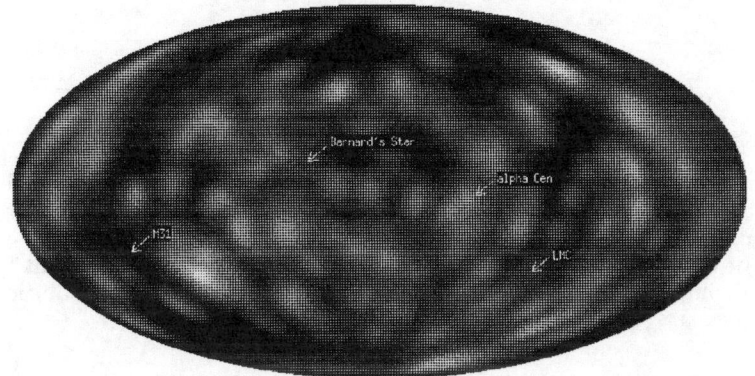

Figure 5. Each burst location point has been replaced with a Gaussian error brightness. The brightness represents total probability per steradian that a burst occurred at that point on the sky summed over the 241 BATSE bursts. Some interesting locations are marked.

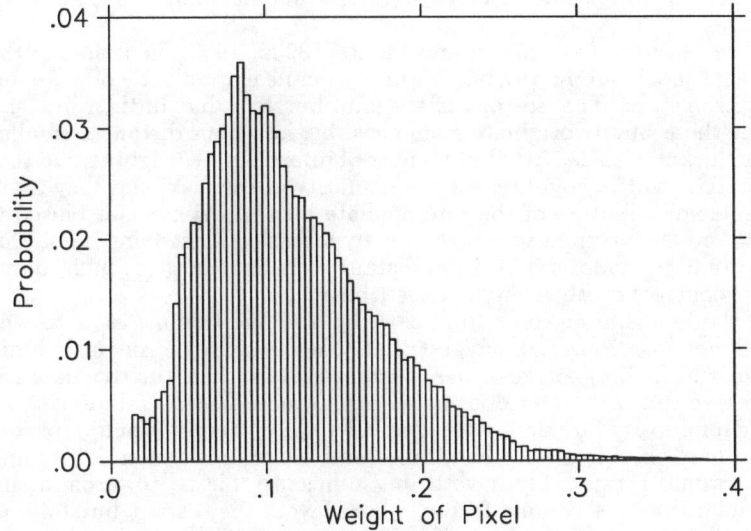

Figure 6. Distribution of burst probabilities on the sky (i.e. a histogram of the pixels in Figure 5). Mean probability=0.11, standard deviation=0.054.

REFERENCES

1. Meegan, C. A., et al., Nature **355**, 143 (1992).
2. Fishman, G. J., et al., ApJS, in press (1993).
3. Hartmann, D. H., these proceedings, (1993).
4. T.E. Clarke, O. Blaes, and S. Tremaine, preprint, (1993).
5. J.M. Horack, T.M. Koshut, R.S. Mallozzi, S.D. Storey, & A.G Emslie, ApJ, submitted (1993).

ARE GAMMA RAY BURSTS NEARBY?

David Eichler
Dept. of Physics, Ben Gurion University

ABSTRACT

The possibility is considered that the intrinsic luminosity function of gamma ray bursters has sufficient scatter that the ln N - ln S relation of observed bursts not be dominated by geometric effects favoring large volumes, but rather be dominated by nearby, intrinsically faint bursts. It is shown that the distribution of bursts on the sky would be very granular, with a significant fraction of them coming from the two or three nearest sources. Possible solutions and alternatives are briefly discussed.

INTRODUCTION

Until the BATSE results were announced, it was widely suspected that observable gamma ray bursts come from nearby ($d < 1$ Kpc) neutron stars. The discovery by BATSE that the dimmest bursts are distributed isotropically but apparently not homogeneously motivated serious consideration that they come from either the Oort cloud, a very extended Galactic halo, or from cosmological distances.

Very recently, Quashnock and Lamb (1993a, 1993b henceforth QL) have argued that most gamma ray bursts come from nearby galactic objects, presumably neutron stars. The isotropy of the faint bursts is due, in their model, to the fact that these bursts originate from close by, i.e. from distances smaller than the disk thickness. They attribute bursts of intermediate brightness to more distant bursters, which would reflect the planarity of the disk, and they argue that the angular distribution of the intermediate bursts is somewhat biased toward the disk and thus compatible with this hypothesis. Most faint bursts, in their model, would be unobservable from distances that are large enough to impose a disklike geometry on their angular distribution.

If the faint bursts come from close by, the question arises as to why they appear inhomogeneous. QL suggest that this is due to the intrinsic luminosity function, which, they suggest, has a larger scatter than the dynamic range of distances, so that the latter does not control the ln N - ln S relation. A suitable intrinsic luminosity function (hereafter ILF), they suggest, would reproduce the observed ln N - ln S relation when the volume factor is folded in. This luminosity function would presumably have to be sufficiently biased towards intrinsically faint bursts that the volume factor, which favors the distant bursts, not dominate the ln N - ln S relation. If this were not so, the distant bursts would dominate N, and the disk would show up in the burst distribution. Because nearby, intrinsically faint bursts are contributing heavily to the total, repetition from individual bursters (the closest) is implied. QL note that the data indeed suggest repetition, since the statistical test they employ suggests positive angular correlation of bursts over a scale of several degrees, the angular resolution of BATSE. The question is whether the nature of the correlation is quantitatively compatible with any ILF.

In this letter this question is considered. It is argued that, though a luminosity function can be chosen that allows significant contribution of nearby sources, this contribution would be very granular, possibly more so than current observations permit. A significant fraction of the bursts would come from the two or three closest bursters.

GRANULARITY OF THE SKY DISTRIBUTION

The distribution of bursters is assumed to be homogeneous out to a radius at which the isotropy would be spoiled by a disklike geometry. In practice, this would correspond to several Kpc or less, the maximum being attainable if the disk of neutron stars is thickened by a kick velocity at their formation.

The distribution of bursters is assumed to be Poisson. The probability that the closest burster lies within concentric spheres (centered on the observer) of volumes V and V+dV is thus $e^{-\lambda V} dV$, where λ is the space density and is set equal to unity for convenience. The probability that the ith ($i > 1$) closest burster lies between V_i and $V_i + dV$, given that the closest burster lies at V, i.e. on a sphere of volume V, is given by

$$P_i(V_i)dV_i = exp[-(V_i - V)][(V - V_i)^{(i-2)}/(i-2)!]dV_i.$$

The rate of observable bursts from a burster at distance r is the product of N(L), the burst rate per unit volume per unit ln L at luminosity (maximum luminosity, say) L, and $V_{max}(L)$ integrated over ln L from the minimum detectable luminosity at that distance to infinity. Equivalently, it is to within a constant multiplicative factor $\int_V^\infty N(V_{max})dlnV_{max}$.

We now choose the most conservative luminosity function for the purposes of this argument: we assume that the intrinsic luminosity function N(L) is simply proportional to $V_{max}^{-1}(L)$, so that neither faint, close bursts nor bright distant bursts dominate the other. If N(L) were to decrease more slowly than this with V_{max}, close bursters would contribute insignificantly leading to significant anisotropy imposed by disk geometry, and, if more rapidly than this, close bursts would dominate the total more, rendering the arguments below even stronger. For generality, we set N(L) proportional to $V_{max}^{-\alpha}(L)$, and later set α to unity.

Note that for a sufficiently soft ILF, including $\alpha = 1$, the expected burst rate from the closest bursters may diverge, which would require a low L cutoff in the ILF for any calculation of $< R_1 >$, the expected burst rate from the closest burster. We assume, however, that this cutoff luminosity is too low to be detectable even from the closest burster, so that it is irrelevant. We proceed by computing $N_j = < R_j/R_1 >$, the expected *relative* contribution of the jth closest burster ($j > 1$), normalized to the closest one (j=1). This must always be finite.

The above implies that the expected value of the relative contribution of the ith closest source (i still at least 2) normalized to that of the closest, is given by

$$N_i = \int_0^\infty dV_1 \int_0^\infty dy e^{-V_1} e^{-y} y^{i-2} V_1^\alpha /(i-2)!(y+V_1)^\alpha.$$

Reversing the order of integration, this is found to be

$$N_i = \sum_{n=0}^{i-2} (i-1)!(-1)^{i-n}/n!(i-n-2)!(i-n-1+\alpha).$$

While the above is defined only for the second closest burster (N_2) and beyond, note that $N_1 = 1$ by definition.

We now set $\alpha = 1$. The sum on the right hand side can be reduced to

$$N_i = 1/i$$

by using the identity

$$\sum_{n=0}^{i} i!(-1)^{i-n}(-1+i-n)/n!(i-n)! = 0.$$

Note that the logarithmic divergence of the series $1/i$ is expected given the logarithmic divergence of $\int N(L)V_{max}(L)dlnL$.

The statistics of the several closest bursters, as calculated above, indicates that a continuum model for the spatial distribution of bursters is inadequate. The closest burster should, on average, be 2 times as prolific as the second closest, 3 times as prolific as the third closest, etc.

Given that neutron star formation is less than one per ten years in the Galaxy, the closest is of order at least 10 pc away. If the faint bursts are assumed to lie within (liberally) 3 Kpc or so and that the bursters are neutron stars, then the most distant faint burst source is at most about e^6 or so more distant than the closest, or, in terms of V=$(4\pi/3)r^3$, $\ln[V_{furthest}/V_{closest}] \approx 20$, suggesting that the brightest burster must have contributed at least 0.05 of the total. At the time of this writing, this may be marginally consitent with the data, since the number of clusters of bursts is limited. However, the granularity of the heretofore observed bursters must be must larger than in the above estimate in the picture of QL since only $\sim 10^2$ bursts have been observed thusfar, and with, it is claimed, significant repetition among bursters at a significant number of distinct locations. The divergent series $1/i$ must be cutoff at $i < 1/m$, where m is the number of distinct clusters of bursts. If, for example, after enough observed bursts have accumulated, there are seen to be 100 distinct burst clusters of two or more bursts, straightforward summation of the first 100 terms indicates the expected contribution of the closest burster would then be of order 0.2 of the total among these 100 repeaters. At present, there are about 20 or so such clusters in the first 201 bursts (Lamb, private communication). If the correlation is real, then the contribution of the nth closest source should be of order 20/n, so that the first 20 have a significant chance of repeating. The closest one should have therefore contributed of order 20. This is uncomfortably large, but, as some of the pairs may in any case be chance, the statistical significance may not be as yet overwhelming.

The statistical significance of this prediction is, of course, limited by the fact that it relies on small number statistics, by definition. There is always the chance that the closest burster is situated $x^{1/3}$ times as far as it should have been, on average, which would lower its contribution to the observed burst set by a factor of x, and the chance of this happening is e^{-x}. But then there is always the problem of the second closest, which would also give rise to a noticeable hotspot unless also "pushed out" to a larger than expected distance. In this regard it is worth recalling that we have already fine tuned the ILF to give the smoothest possible distribution consistent with apparent inhomogeneity of faint bursts, and it is already improbable. The statistical arrangement needed to avoid excessive hotspots should not be more of a fluke than the observed angular correlation would be in a null hypothesis. We conclude by observing that while better statistics are desirable, there does not appear, in visual inspection of the burst distribution, to be any obvious candidate for the closest burster. Even allowing for considerable positional uncertainty, the most bursts among the first 200 that have been claimed to arrive from a single burster is 5 (Wang and Lingenfelter, 1993), which is at present 2 to 4 σ less than the prediction for the best case scenario. In any case the above result translates into a quantitative prediction for the model in the limit of very high number statistics.

While the present status of the observations may be debatable, given the modest angular resolution of BATSE and other factors, the existence of hot spots appears to be a natural consequence of the hypothesis that faint gamma ray bursts are occuring nearby on a Galactic scale. Constructing more complicated statistical models opens up many, perhaps unlimited, possibilities. But the rule of thumb is that postulating intrinsic differences between bursters tends to increase the variance in expected repetition number and thereby *increases* the granularity.

While the relative contributions of the very closest bursters depends in any case on imponderable assumptions about the faint end of the ILF, a more robust constraint is the requirement that N_i not decrease more slowly with i than $1/i$, for then the most distant sources would dominate, and the disk geometry would appear clearly.

Finally, note that we have not subjected the above mathematical model to a V/V_{max} test. Such a test would be an additional constraint, and would thus increase the predicted granularity relative to the model we have considered.

POSSIBLE CURES

We have considered the possibility that the soft gamma ray repeaters SGR 0526-66, SGR 1806-20, and SGR 1900+14 are simply the closest bursters, which would imply that the positional coincidences with N49 in the LMC and the Galactic center are chance coincidences. The softness of the repeating bursts has been argued to be evidence for qualitative difference between the classical bursts and the repeaters. However, it is physically plausible that intrinsically weak bursts, the ones most likely to repeat in the present model, are indeed softer. Time variability of the repetition rate, for which there is some evidence (Hurley, private communication), does not appear to alter the present arguments in any obvious way. So perhaps the soft repeaters could be interpreted as being the closest bursters. The exact ratios of N_i for small i are in any case adjustable by playing with the faint end of the ILF.

We have considered the possibility that some anthropic principle demands that burster not be too close to us. For example, the kill radius of a supernova could conceivably be of order 20 pc. But the random thermal motion of stars in the Galaxy, of order 10 km/s, would allow bursters to drift back into the kill zone - at most of order 30 pc if disk geometry is not to spoil isotropy and still permit a viable dynamic range for the isotropic component - within $3 \times 10^6 yr$. (If the burster disk is much thicker than 200 pc, then the random velocity is then also higher.) Such an anthropic principle would then have to be combined with a prescription for curtailing the lifetime of a burster. But this then lowers the space density of the active bursters, and we have been unable to construct a viable model in which the expected number of active bursters within the kill zone would be significant.

An alternative, which accomodates the reported (QL) tendency of medium brightness bursters to correlate with the disk, is that the brightest bursters are associated with the nearby portion of a disk thickened by kick velocities. The scale height of such a disk, and hence the typical distance to the close isotropic component could be of order 3 Kpc. The typical distance to the medium brightness anisotropic component could be somewhat larger, and the faint isotropic component could be in an extended halo of scale height at least tens of Kpc. An ILF that could favor this heirarchy is $N(L) \sim V_{max}(L)^\gamma$, $2/3 < \gamma < 1$, so that the source count is dominated by distant sources in a homogeneous three dimensional distribution, and by nearby sources in a two dimensional distribution.

As an alternative to a thickened disk, one could imagine a somewhat oblate halo, surrounded by a more spherical extended halo. In any case, a more quantitative construction is beyond the scope of this letter and perhaps would be more appropriate if the tendencies reported by QL standup to analysis of the larger data base that presently exists.

The possibility of repetitions is in principle just as viable if bursters are in an extended halo. However, if the repetitions are as frequent as once per year, then the total number of bursts required per burster is raised from about, say, 10^4, the number of observed bursts per Hubble time per 10^9 neutrons stars, to 10^{10}. This would severely strain the required total energy supply, since burst energies of order 10^{41} ergs are required if the bursts come from an extended halo. A possible solution to this problem would be that the bursters are formed in dense clusters of dead stars, and that a significant number remain in their cluster after formation. The repetitions could then be different bursters within a given cluster.

I acknowledge useful conversations with Drs. K. Hurley, J. Silk, D. Lamb and E. Maoz. This research was supported in part by grants from the Israeli Foundation for Basic research and the Israel - U.S. Binational Science Foundation. I acknowledge the hospitality of the Center for Particle Astrophysics at UC Berkeley, where this paper was written.

REFERENCES

Quashnock, J.M. and Lamb, D. Q. 1993a, M.N.R.A.S. (submitted)
Quashnock, J.M. and Lamb, D.Q. 1993b, M.N.R.A.S. (submitted)
Wang, V., and Lingenfelter, R..E. 1993, Ap. J. (submitted)

ADDITIONAL CONSTRAINTS ON GALACTIC CORONAL/EXTENDED HALO SOURCE DISTRIBUTIONS FROM BATSE OBSERVATIONS

J. Hakkila
Mankato State University, Mankato, MN 56002-8400

C. A. Meegan, G. J. Fishman, R. B. Wilson, M. N. Brock, J. M. Horack
NASA/Marshall Space Flight Center, ES-66, Huntsville, AL 35812

G. N. Pendleton, W. S. Paciesas, M. S. Briggs
University of Alabama in Huntsville, Huntsville, AL 35899

ABSTRACT

Galactic corona/extended halo gamma-ray burst source distributions are analyzed and compared to BATSE observations. Models have radial densities of the form $n(r) = n_c[1 + (r/r_c)^\alpha]^{-1}$, where $\alpha = 2$ for coronal dark matter, and the density is constant inside the core radius r_c. M31 is assumed to have a burst population similar to that of the Milky Way, while other Local Group galaxies are ignored.

The parameter space of all such single-population models is bounded, with $1.5 \leq \alpha < 4.0$, and with minimum and maximum sampling distances caused by the off-center location of the sun in the Milky Way, the presence of M31, and the shape of the $\log(N > F_p)$ vs. $\log(F_p)$ curve. Core radius limits are related to α and to the sampling distance. The constraints on these spatial distributions limit the allowed burst luminosity functions, with the maximum range of luminosities (representing roughly a factor of 5) occurring for $\alpha = 2$.

Two-population models are examined comprising a local Galactic disk/halo population mixed with a coronal population. The local population introduces an isotropy which allows for smaller coronal core radii, but which also forces the corona to be sampled to larger distances. The largest range of coronal parameters occurs when no disk/halo sources are present. A maximum of 30% disk/halo bursts is allowed, but these force the corona to have the narrowest range of characteristics. Smaller statistical errors resulting from more BATSE bursts will cause the parameter space of the two-population models to shrink faster than the single-population corona/extended halo models.

INTRODUCTION

Few Galactic distributions of nonrepeating gamma-ray burst sources satisfy BATSE observations, as standard Galactic distributions do not reproduce the angular isotropy and deficit of weak bursts found in this sample[1,2]. Since the visible Galaxy cannot be the main source of bursts, invisible distributions such as the dark matter corona and extended halos have been proposed[3]. There are, however, constraints on the parameters used in such models (presented[4] in Paper I), which result primarily from the off-center location of the sun in the Milky Way and from the presence of a massive neighboring galaxy (M31). The viability of extended halo/coronal models allows them to be mixed with a subpopulation of standard Galactic sources[5], thus preserving some theoretical models involving Galactic neutron stars.

The techniques used in Paper I to analyze Galactic distributions involve (1) creating Monte Carlo catalogs from burst spatial and luminosity characteristics, (2) convolving these results with the observational selection effects of sky sampling and positional errors, (3) analyzing the results in terms of angular and intensity distributions, and (4) comparing the results to BATSE observations. Many of the tests used are standard summaries of bulk characteristics, such as $<V/V_{\max}>$[6], $<\cos\theta>$, $<\sin^2 b>$[7], $\mathbf{P}$, η, ζ[8], and W, B[9]. These tests can be augmented using differential and integrated forms of angular and intensity distributions such as the two-point angular correlation function[10] and size-frequency distributions[11]. All of the aforementioned tests are valid if selection-effect corrected (see Paper I). Satisfactory models are those in which 90% of the Monte Carlo runs produce angular/intensity measurements consistent with those of BATSE.

Single and double-population burst distributions are examined here in which the principle population is a coronal/extended halo population (the second population is local disk/halo), and in which the defining parameters are allowed to have a wide range of values. The tests described above are used to identify any additional constraints.

ANALYSIS

Single-population corona/extended halo distributions and two-population models (in which one is coronal and the other is local disk or halo) are assumed to be spherical with radial densities of the form $n(r) = n_c[1+(r/r_c)^\alpha]^{-1}$, where the density is assumed to have the value n_c inside the core radius r_c, and where the density parameter α is treated as a variable (the dark matter corona is characterized by $\alpha = 2$; the halo by $\alpha = 3$). Furthermore, this distribution is finite, with an assumed "edge" defined by the tidal radius created by M31. A conservative assumption is made that M31 has a corona/extended halo identical to that of the Milky Way.

Figure 1 shows that single-population, standard candle extended halo/corona models have bounded parameter spaces. For each α, only limited core radii and sampling distances are possible, because (1) models with small core radii exhibit anisotropies from the Galactic center, (2) models with large core radii exhibit anisotropies from M31 (which is most easily observed in the two-point angular correlation function), (3) models of a particular core radius are too homogeneous if sampled to a small distance, and (4) models of a particular core radius are too heterogeneous if sampled to a large distance.

Additional constraints result from the size-frequency distributions of these models. Although an acceptable $<V/V_{\max}>$ can be produced for any value of α, the $\log(N)$ vs. $\log(F_p)$ curves of the vast majority of these are incompatible with that measured by BATSE. The curve for the limiting model with $\alpha = \infty$ has a $-3/2$ slope everywhere except near the minimum flux, where it flattens. Curves for models with $\alpha \geq 4$ are more similar to the limiting curve than they are to that measured by BATSE, and are unacceptable because they do not predict enough bright bursts. Curves for $\alpha \leq 1$ are too flat, and are unacceptable because they predict too many bright bursts. Models with $\alpha = 1.5$ are marginally acceptable for a small range of core radii and sampling distances.

The luminosity range of standard candle models can help indicate the luminosity range for a model with a broad luminosity function. Many sources

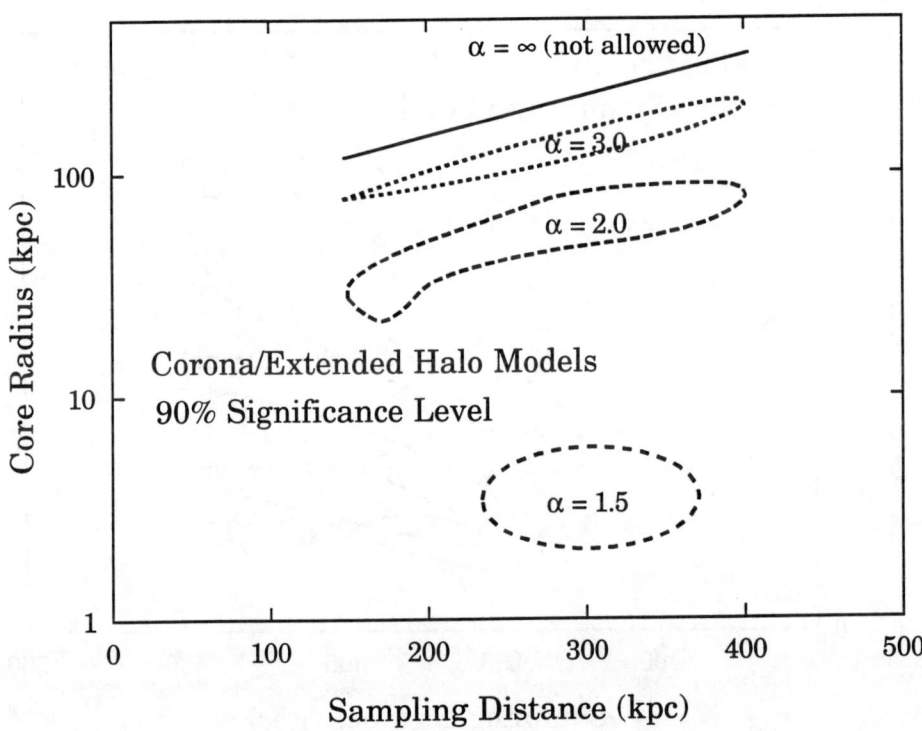

Figure 1. Standard candle corona/extended halo models satisfying BATSE observations (errors are for 452 bursts with locations and intensities).

with luminosities greater than that of the maximum sampling distance cannot be present in such a distribution, or else M31 would be observable. Many sources with luminosities smaller than that of the minimum sampling distance cannot be present, as these sources would increase $<V/V_{max}>$. Most of the bursts must lie in this small luminosity range. The maximum range of luminosities occurs for $\alpha = 2$ (Galactic corona), and corresponds to $L_{max}/L_{min} \simeq 5$.

Two-population models are chosen to have as a secondary population a local Galactic one (either disk or halo), which is inconspicuous because it is isotropic. It is also homogeneous, as heterogeneity and anisotropy are coupled for Galactic disk/halo models. The only additional degree-of-freedom needed to describe this subpopulation is the fraction of bursts drawn from it.

Figure 2 shows the parameter space for two-population models involving a standard candle coronal population combined with a local disk/halo population. The parameter spaces of these models are all bounded, as they are for the limiting case when no local sources are present (100% coronal).

The addition of a homogeneous secondary population raises the overall $<V/V_{max}>$ value. The coronal distribution's $<V/V_{max}>$ value must therefore be lowered to compensate, which can happen in three ways: by (1) limiting the

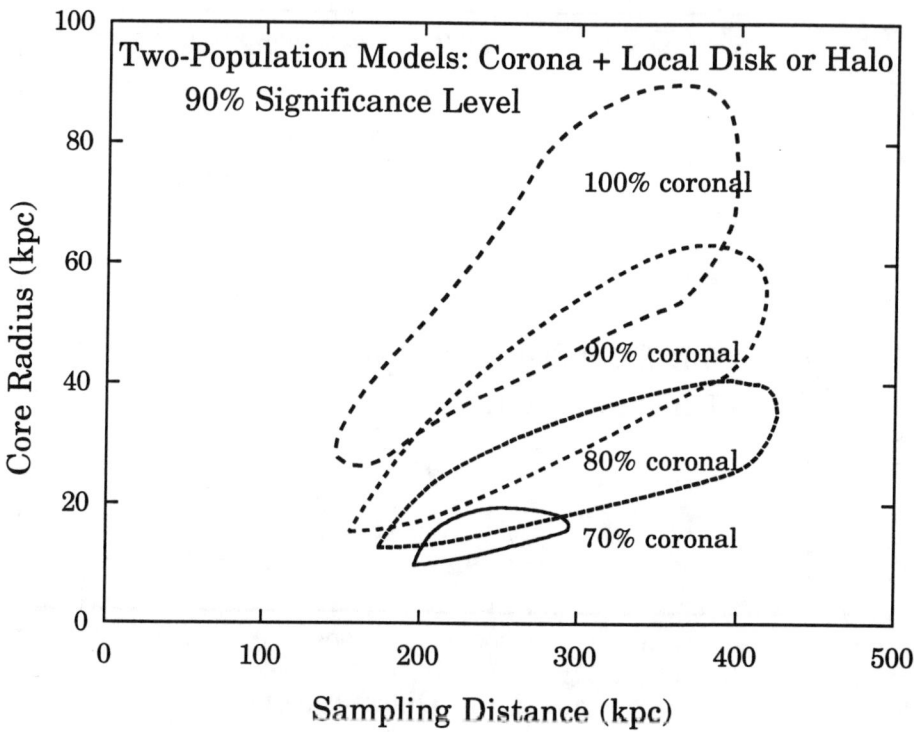

Figure 2. Constraints on standard-candle coronal populations mixed with low-luminosity disk and/or halo populations (errors are for 452 bursts with locations and intensities).

fraction of local Galactic sources, (2) increasing the sampling distance, and/or (3) decreasing the core radius. The first option provides the most important constraint, as the BATSE $<V/V_{max}>$ value can be reproduced only when the fraction of coronal sources is larger than roughly 60%. (Note that the coronal $<V/V_{max}>$ cannot get very small; M31 is a third population that contributes sources with $<V/V_{max}>$ near 1/2 when detected at around 400 kpc). The second option provides only a small amount of flexibility, as the M31 anisotropy is too large to disguise with only a small fraction of local Galactic sources. The third option is the most viable, as a Galactic core anisotropy can be disguised by adding isotropically-distributed sources.

However, many of these models with smaller core radii and larger local population fractions are excluded based upon the shapes of their $\log(N)$ vs. $\log(F_p)$ curves. This places the nearby sources at the faint end of the $\log(N)$ vs. $\log(F_p)$ curve, in agreement with the analysis of Paczyński[12]. However, this source excess is not always statistically noticeable at the faint end of a logarithmic size-frequency distribution, whereas its effect at the bright end generally is. This is because the $<V/V_{max}> = 1/2$ sources at the faint end of the curve must be balanced by low $V/V_{\max}$ sources in some other region of the curve. Since the shape of the curve for a corona/extended halo distribution with a particular α is defined, the secondary population forces the primary one to be flatter over much of its observed range. This leads to an overabundance of bright sources, which is not in agreement with the BATSE observations. For these reasons, many of the models with a moderate local population fractions and small coronal core radii are deemed unacceptable.

CONCLUSIONS

Most non-repeating Galactic bursts must belong to a coronal/extended halo population sampled to more than 100 kpc, implying that their luminosities are super-Eddington. A local secondary population with sub-Eddington luminosities does not change the super-Eddington character of the primary population. Thus a coronal/extended halo population faces some of the same theoretical difficulties as does a cosmological population.

The isotropy observed by BATSE constrains the spatial and luminosity characteristics of extended halo/coronal models. The largest range of model parameters, corresponding to only a small luminosity range, occurs when $\alpha = 2$ and when NO local Galactic disk and/or halo population is present. The narrow ranges of Galactic model parameters shrink as BATSE observes more bursts.

We thank Mankato State graduate students Van C. Vo and John G. Mayer for assistance rendered on this project. We also gratefully acknowledge valuable discussions by Dr. James N. Pierce and Dr. M. Lee Riddle of Mankato State, Dr. Dieter Hartmann of Clemson, and Dr. Richard Lingenfelter of the University of California at San Diego. This work was funded by NASA grants NAG8-192 and NRA 92-OSSA-17.

REFERENCES

1. Meegan, C. A. et al., Nature **355**, 143 (1992).
2. Fishman, G. J. et al., Ap. J. Supp. Ser. , (in press) (1994).
3. Paczyński, B., Acta. Astronomica **41**, 157 (1991).
4. Hakkila, J. et al., Ap. J. , (in press) (1994).
5. Lingenfelter, R. E., & Higdon, J. C., Nature **356**, 132 (1992).
6. Schmidt, M., Higdon, J. C., & Hueter, G., Ap. J. **329**, L85 (1988).
7. Paczyński, B., Ap. J. **348**, 485 (1990).
8. Hartmann, D., & Epstein, R. I., Ap. J. **346**, 960 (1989).
9. Briggs, M. A., Ap. J. **407**, 126 (1993).
10. Hartmann, D., Linder, E., & Blumenthal, G., Ap. J. **367**, 186 (1991).
11. Usov, V. V., & Chibisov, G. V., Soviet Astr. **19**, 115 (1975).
12. Paczyński, B., Acta. Astronomica **42**, 1 (1992).

INTEGRAL MOMENT ANALYSIS OF THE BATSE GAMMA-RAY BURST INTENSITY DISTRIBUTION

John M. Horack
NASA – Marshall Space Flight Center, ES–66, Alabama, 35812

A. Gordon Emslie
Dept. of Physics, Univ. of Alabama – Huntsville, Alabama, 35899

ABSTRACT

We have applied the technique of integral moment analysis to the intensity distribution of the first 260 gamma–ray bursts observed by BATSE. This technique provides direct measurement of properties such as the mean, variance, and skewness of the convolved luminosity–number density distribution, as well as associated uncertainties. Using this method, one obtains insight into the nature of the source distributions unavailable through computation of traditional single parameters such as $\langle V/V_{\max} \rangle$. If the luminosity function of the gamma–ray bursts is strongly peaked, giving bursts only a narrow range of luminosities, these results are then direct probes of the radial distribution of sources. As such, an integral moment analysis of the intensity distribution of the gamma–ray bursts provides for the most complete analytic description of the source distribution available from the data, and offers the most comprehensive test of the compatibility of a given hypothesized distribution with observation.

INTRODUCTION

Traditionally, two methods of analysis have been used to investigate the observed brightness distribution of the gamma–ray bursts: the $\langle V/V_{\max} \rangle$ test, and the integral number–intensity distribution. The BATSE data set have a $\langle V/V_{\max} \rangle = 0.321 \pm 0.013$.[1] From this, one deduces that the gamma–ray bursts are *not* homogeneously distributed throughout space. Because the $\langle V/V_{\max} \rangle$ test is only a diagnostic for homogeneity, one cannot discern details of the burst parent distribution from this number.

Complementing the $\langle V/V_{\max} \rangle$ test is the integral number–intensity distribution of the bursts, $\mathcal{N}(> P)$ vs. P, where P is the burst's intensity as measured by its peak photon flux. This distribution for the BATSE gamma–ray bursts is well-known.[1,2] The strong bursts are consistent with the $-3/2$ slope expected from a constant source density in Euclidean space, and the distribution shows fewer weak bursts than expected if this constant density continued out to large distances.

We submit that additional quantitative information concerning the radial distribution of bursts can be obtained through an integral moment analysis of the intensity distribution. Integral moment analyses are not new to astrophysics,[3,4] however they have yet to be applied to the gamma–ray bursts. The integral moments of the gamma–ray burst peak–flux distribution $N(P)$ can be shown to be directly proportional to moments of both the luminosity function $\phi(L)$ and the spatial density distribution function $n(r)$. Physically relevant quantities such as the mean, mean square deviation, and skewness of the convolved luminosity–density distributions, along with their associated uncertainties, can be directly computed from the moments of the intensity distribution.

An integral moment analysis also offers an excellent test for the compatibility of hypothetical burst distributions with the observed data. A consistent hypothetical distribution must reproduce *all* of the moments observed, not simply the one related to the $\langle V/V_{\max} \rangle$ value. The consistency of a certain $n(r)$ distribution can therefore be determined simply by calculating the moments of the hypothesized distribution, and comparing them to the data and their uncertainties. With the appropriate modifications for varying geometries, the same technique is applicable to all scenarios of burst distributions. This method does not require the use of large Monte Carlo simulations or statistical methods such as the Kolmogorov–Smirnov (K–S) test to examine the integral number–intensity distribution.

DERIVATION OF THE INTEGRAL MOMENTS

We define the un–normalized mth moment of the burst intensity distribution $N^{(m)}$ by

$$N^{(m)} = \int_{P_{\min}}^{\infty} P^m N(P) dP, \tag{1}$$

where P is the intensity of the burst, $P_{\min}$ is the minimum detectable intensity, and $N(P)$ is the number of bursts at a given intensity P.

By rewriting the analytic expression for $\mathcal{N}(> P)$, Horack & Emslie[5] have shown that the normalized integral moments $\{N^{(m)}\}$ of the intensity distribution are proportional to both the mth moment of the luminosity function $\phi(L)$ and the $-2m^{\text{th}}$ moment of the source density function $n(r)$,

$$\{N^{(m)}\} = (4\pi)^{-m}\{\Phi^{(m)}\}\{n^{(-2m)}\}, \tag{2}$$

provided that the luminosity is not a function of the distance from the detector.

Equation (2) was derived in the context of a Euclidean geometry, appropriate for local or galactic scenarios of gamma–ray burst distributions. In the context of cosmological bursts, however, modifications must be made to the derivaiton. The cosmological equivalent to Equation (2) assuming a Friedmann universe with $\Lambda = 0$ is[5]

$$\{N^{(m)}\} = (4\pi)^{-m}\{\Phi^{(m)}\}(c/H_o)^{-2m}$$

$$\times \int_{z_{\min}=0}^{z_{\max}} \frac{q_o^{4m-4} f(q_o,z)^{-2m} n_c(z)}{(1+z)^3 \sqrt{1+2q_o z}} \, dz \bigg/ \int_{z_{\min}=0}^{z_{\max}} \frac{f(q_o,z) n_c(z) dz}{q_o^4 (1+z)^3 \sqrt{1+2q_o z}}, \tag{3}$$

where

$$f(q_o, z) = (q_o z + (1-q_o)(1 - \sqrt{1+2q_o z}))^2. \tag{4}$$

We have also assumed bursts with no luminosity evolution, and have assigned each burst a photon spectral index of 2. The quantity q_o is the deceleration parameter, z is the redshift, and H_o is the Hubble constant.

We recognize the presence of the mth moment of the luminosity distribution $\{\Phi^{(m)}\}$ in Equation (3). However, the $-2m^{\text{th}}$ moment of the $n(r)$ distribution has become an integral over z and is no longer an actual integral moment of some physical distribution. Consequently, in this scenario we can not directly measure the moments of the radial source distribution.

Despite the inability to compute the moments of the radial distribution in the cosmological case, this approach is still an excellent means of testing hypothesized scenarios in a direct analytic fashion without the use of Monte Carlo simulations or statistical tests comparing integral distributions. The integral in Equation (3) can be evaluated given a set of model parameters, and the results can be compared directly to the BATSE normalized moments.

COMPUTATION OF THE INTEGRAL MOMENTS AND THEIR UNCERTAINTIES

We have obtained the intensities and their uncertainties for the 260 bursts in the first BATSE catalog.[2] The intensity of the burst is measured as the peak flux in photons cm^{-2} s^{-1} in an integration time of 0.256 ms.

We have calculated the moments and their associated uncertainties using all bursts with $P > 0.5$ photon cm^{-2} s^{-1}. This point lies above the intensity region where corrections need to be made for trigger inefficiencies, but also lies far below the region where the $\mathcal{N}(>P)$ curve deviates from the $-3/2$ power law, so that the interesting region of the distribution is still effectively probed. Table 1 shows the values of several normalized moments and their associated uncertainties.

m	$\{N^{(m)}\}$
-0.5	0.847 ± 0.023
-1.0	0.819 ± 0.038
-1.5	0.856 ± 0.054
-2.0	0.943 ± 0.073
-2.5	1.076 ± 0.099
-3.0	1.263 ± 0.145

Table 1 – Normalized Integral Moments $\{N^{(m)}\}$ from the observed $N(P)$ Distribution and their total associated uncertainties

The uncertainties in the moments are dominated by two contributors. The first contributor is the uncertainty due to the finite number of bursts in the sample, scaling as $1/\sqrt{N}$. The second contributor is the uncertainty in each burst's intensity. Other contributors, such as those introduced by the binning of the $N(P)$ distribution, are much smaller than these two major sources of uncertainty.

Critics of moment analyses correctly point out that a large (sometimes infinite) number of moments may be needed to specify a given distribution. In light of BATSE's ability to only measure the first few moments reliably, however, this concern should be obviated. An observed distribution cannot be specified beyond one's ability to measure it, and the integral moment analysis tells one exactly to what extent the distribution can be measured.

APPLICATION OF THE COMPUTED MOMENTS TO THE SOURCE DENSITY AND LUMINOSITY DISTRIBUTIONS

Having computed the normalized moments of the $N(P)$ distribution, we wish to use these values to explore the structure of the convolved luminosity distribution and source density distribution. We assume that the bursts are mono–luminous, although the analysis is easily generalizable to arbitrary $\phi(L)$.

Several interesting results can now be obtained concering the average, mean–square deviation, skewness, and kurtosis of the observed distribution. For example, the $m = -1/2$ moment of the $N(P)$ distribution computed from the BATSE data therefore implies that

$$\frac{\langle r \rangle}{\sqrt{L_o}} = 0.239 \pm 0.007, \tag{5}$$

where r is in units of cm, and L_o has units of photons s^{-1} in this and subsequent equations. We find for the mean–square deviation, skewness, and kurtosis, respectively;

$$\frac{\langle r^2 \rangle - \langle r \rangle^2}{L_o} = 0.008 \pm 0.004 \tag{6}$$

$$\frac{\langle (r - \langle r \rangle)^3 \rangle}{L_o^{3/2}} = -0.0003 \pm 0.0018 \tag{7}$$

$$\frac{\langle (r - \langle r \rangle)^4 \rangle}{L_o^2} = 0.0003 \pm 0.0021 \tag{8}$$

The uncertainties in the skewness and the kurtosis are both so large that their expectation values cannot be measured. The addition of more bursts to the BATSE data sample will lower the overall uncertainty in the moments, thereby reducing the uncertainties in the above quantities. However, even with a large (> 1000) sample of bursts, uncertainties will be limited by the ability to measure each burst's intensity, and will still be too large to allow a reliable mesurement of the skewness and the kurtosis of the distribution.

THE USE OF INTEGRAL MOMENTS AS A TEST FOR HYPOTHETICAL SOURCE DISTRIBUTIONS

As an example of the application of integral moments to hypothesis testing, we wish to consider the possible consistency of a simple cosmological scenario in which the number of sources per unit co–moving volume is a constant, with a standard–candle assumption and no luminosity evolution. A Friedmann universe with $\Lambda = 0$, $q_o = 1/2$, and a Hubble constant of $H_o = 75$ km s^{-1} Mpc^{-1} are also assumed.

We use Equations (3) and (4), along with the data in Table 1 to compare the moments of the hypothesized distribution to the BATSE data for various values of $z_{\max}$, the maximum visible redshift. These results are shown in the table below.

m value	Predicted	Computed	Deviation
	$-z_{max} = 0.5-$		
-1.0	(2.47 ± 0.18)	2.38	-0.51σ
-1.5	(1.42 ± 0.15)	1.29	-0.89σ
-2.0	(8.57 ± 0.169)	7.25	-1.14σ
	$-z_{max} = 1.25-$		
-1.0	(1.53 ± 0.11)	1.53	0.0σ
-1.5	(2.18 ± 0.23)	2.18	0.0σ
-2.0	(3.28 ± 0.44)	3.27	-0.034σ
	$-z_{max} = 2.0-$		
-1.0	(3.80 ± 0.27)	3.93	0.47σ
-1.5	(8.57 ± 0.89)	9.26	0.77σ
-2.0	(2.03 ± 0.27)	2.31	1.01σ

Table 2 – Comparison of computed moments from a simple cosmological scenario with those observed by BATSE

The consistency of this simple cosmological scenario with the BATSE data is apparent, yielding a best–fit z_{max} value of ~1.25. Other values of z_{max} cannot be ruled out, however they do not provide the required moments as well as those from the $z_{max} = 1.25$ scenario. These results are in agreement with other analyses[6,7,8] performed on BATSE gamma–ray burst data investigating the possible cosmological origins of the bursts. Horack & Emslie[5] have provided moment comparisons for other scenarios.

As with other model testing methods, the consistency between the moments of a given scenario with those from BATSE data does not uniquely label that scenario as the correct explanation for the gamma–ray bursts. However in contrast with other methods of hypothesis testing, the results of the integral moment analysis were obtained in an extremely straightforward, simple manner, without the use of large computer simulations, frequentist or Bayesian statistical tests comparing distributions, slope comparisons of integral number–intensity distributions, or other methods often used to examine the validity of the hypothesizd source distributions. In subsequent BATSE gamma–ray burst catalogs, we hope to provide the moments of the observed intensity distribution for use by the community in testing hypothesized models of burst distributions.

REFERENCES

1. Meegan, C. A., et al., These Proceedings , (1993).
2. Fishman, G. J., et al., Ap. J. Suppl. , Submitted (1993).
3. Brown, J. C., Ap. J. **225**, 1076 (1978).
4. Chandrasekhar, S. and Münch, G., Ap. J. **111**, 142 (1950).
5. Horack, J. M. & Emslie, A. G., Ap. J. , Submitted (1993).
6. Norris, J., et al., These Proceedings , (1993).
7. Wickramasinghe, W.A.D.T., et al., Ap. J. **411**, L55 (1993).
8. Piran, T., Ap. J. **389**, L45 (1992).

IMPLICATIONS OF THE BATSE DATA FOR A HELIOCENTRIC ORIGIN OF GAMMA-RAY BURSTS

J. M. Horack, S. D. Storey
NASA – Marshall Space Flight Center, ES–66, Alabama, 35812

T. M. Koshut, R. S. Mallozzi, A. G. Emslie
Dept. of Physics, Univ. of Alabama in Huntsville, Huntsville, AL 35899

ABSTRACT

We report here on an investigation into the implications of the BATSE data on the possible heliocentric origin of gamma-ray bursts. The study is geometric in nature, and does not concern itself with various burst production mechanisms. We have employed direct analytic calculations and Monte Carlo simulations of sources in the Oort Cloud to constrain possible heliocentric burst distributions. These can produce distributions consistent with the observed angular isotropy, $\langle V/V_{\max}\rangle$, and the observed $C/C_{\min}$ distribution of BATSE, and provide limits to burst energy of a few times $\sim 10^{27}$ ergs. The agreement of the heliocentric $C/C_{\min}$ distributions with the BATSE data is attributable, however, to the relatively limited sampling of strong, nearby bursts. These bursts are known from observation to be homogeneously distributed, yet the density of sources in the Oort Cloud is not constant in this region. Integral number-intensity distributions from the Oort Cloud for larger numbers of bursts cannot reproduce the known homogeneity of the strong bursts without modification to the computed cometary number density.

OORT CLOUD ANGULAR ISOTROPY AND $\langle V/V_{\max}\rangle$

We have previously reported the BATSE observation of no statistically significant dipole moment in the direction of the Sun, or quadrupole moment in the ecliptic plane.[1,2] Consequently, any hypothesized heliocentric distribution can effectively be generated with the detector at the center of the distribution in order to satisfy the angular isotropy required by the BATSE observations of gamma-ray bursts. Viable Oort Cloud scenarios, however, must also reproduce consistent values of $\langle V/V_{\max}\rangle$ and $C/C_{\min}$ distributions. (Even if the geometry of the distribution can be shown consistent with the BATSE observations, the challenging questions of how exactly one obtains a gamma-ray burst from a cometary body still remains unanswered.)

Maoz[3] has recently stated that a distribution of gamma-ray burst sources in the Oort Cloud cannot reproduce a value of $\langle V/V_{\max}\rangle$ consistent with the BATSE value of ~ 0.33. This conclusion can be refuted in a rather straightforward manner as we now demonstrate. Under the assumption of mono-luminosity bursts, the $\langle V/V_{\max}\rangle$ parameter is simply calculated as

$$\langle V/V_{\max}\rangle = \frac{1}{R_{\text{vis}}^3} \frac{\int_{R_{\min}}^{R_{\text{vis}}} r^3\, n(r)\, r^2\, dr}{\int_{R_{\min}}^{R_{\text{vis}}} n(r)\, r^2\, dr}, \quad (1)$$

where R_{vis} is the distance to which the BATSE instrument is sensitive to gamma-ray bursts, and $R_{\min}$ is the distance to the nearest burst. One can directly integrate Equation (1) to compute the value of $\langle V/V_{\max} \rangle$ for a given $n(r)$ when a lower and upper limit of integration, $R_{\min}$ and R_{vis}, are chosen.

We employ the cometary number density calculated in the standard model by Bailey[4]

$$n(r) = A(R_o/r - 1)^{3/2}, \qquad (2)$$

where R_o characterizes the outer extent of the cloud, approximately 10^5 AU. The variable r is the distance from the Sun, and A is a normalizing constant proportional to the total number of comets contained in the cloud.

Using this cometary number density, Equation (1) can easily be evaluated using trigonometric substitution and a table of integrals[5] to obtain

$$\langle V/V_{\max} \rangle = (\csc^6 \theta_{\text{vis}}) \frac{f(\theta_{\min}, \theta_{\text{vis}})}{g(\theta_{\min}, \theta_{\text{vis}})}, \qquad (3)$$

where $f(\theta_{\min}, \theta_{\text{vis}})$ and $g(\theta_{\min}, \theta_{\text{vis}})$ are complicated functions of $\theta_{\min}$ and θ_{vis}, with $\theta_{\min/\text{vis}}$ given by

$$\sin \theta_{\min/\text{vis}} = \sqrt{\frac{R_{\min/\text{vis}}}{R_o}}. \qquad (4)$$

We have evaluated expression (3), obtaining $\langle V/V_{\max} \rangle$ for various values of R_{vis} and $R_{\min}$. Figure 1 displays $\langle V/V_{\max} \rangle$ contours in the $R_{\min}$, (R_{vis}/R_o) plane. For this calculation, R_o, the maximum extent of the distribution described by Equation (2) is fixed at 100,000 AU, and all derived quantities will therefore scale with this number. Values of $\langle V/V_{\max} \rangle$ consistent with the BATSE measured value of ~ 0.33 are indeed found, forming a large arc from a point where $R_{\text{vis}} \lesssim 2 \times 10^4$ AU for small $R_{\min}$, and up to $R_{\text{vis}} \sim 10^5$ AU when $R_{\min} \approx 5.5 \times 10^4$ AU.

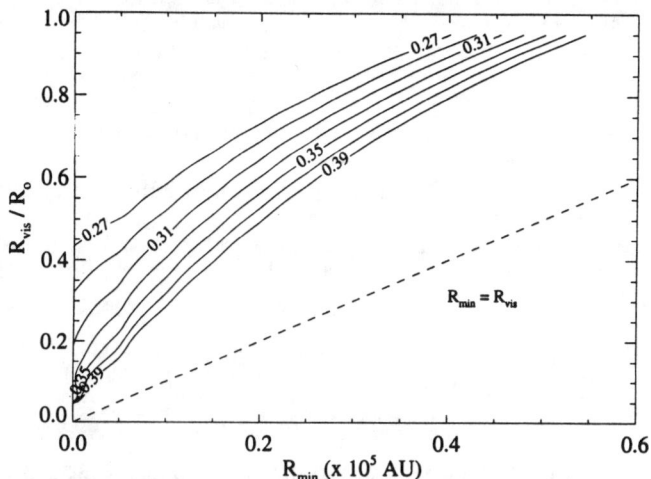

Figure 1 – $\langle V/V_{\max} \rangle$ contours for various values of $R_{\min}$ and R_{vis}/R_o

Clearly, there are a wide range of parameters which *can* produce values of $\langle V/V_{\max}\rangle$ consistent with the BATSE data when the cometary number density of Equation (2) is utilized. Therefore, these distributions cannot be eliminated as a possible source of bursts based on analysis of this parameter alone. In order to be a viable candidate, the hypothesized source population must not only reproduce angular isotropy and a consistent $\langle V/V_{\max}\rangle$, but also must reproduce the entire integral number–intensity distribution. Using $C/C_{\min}$ as a measure of burst intensity, one relates $C/C_{\min}$ to $V/V_{\max}$ by

$$C/C_{\min} = (V/V_{\max})^{-2/3}. \tag{5}$$

An investigation into the $C/C_{\min}$ distributions generated with these parameters is therefore required to determine the consistency with the BATSE data.

INVESTIGATION OF THE $C/C_{\min}$ DISTRIBUTION

Various points on the $\langle V/V_{\max}\rangle$ arc of Figure 1 consistent with the BATSE–measured $\langle V/V_{\max}\rangle$ value were chosen for further investigation into the $C/C_{\min}$ distribution. Figure 2 displays $C/C_{\min}$ distributions for two of the locations on the $\langle V/V_{\max}\rangle$ arc that were examined, along with the BATSE observed $C/C_{\min}$ distribution for 313 bursts plotted as a histogram. The inability of this distribution to fit the ~10 bursts with $C/C_{\min} > 60$ is obvious. Curiously, the computed distribution is not substantially different from a $-3/2$ slope beginning at $C/C_{\min} \sim 20$, and continuing up to $C/C_{\min} \sim 70$.

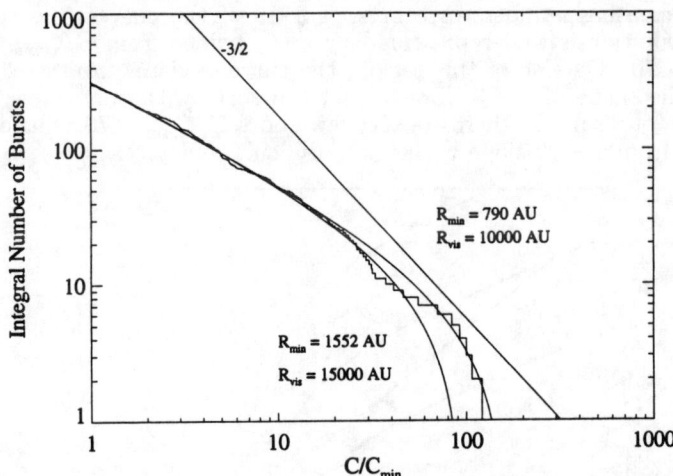

Figure 2 – Comparison of $C/C_{\min}$ distribution from the BATSE data with two heliocentric distributions having the proper $\langle V/V_{\max}\rangle$ value

The ratio $(R_{\rm vis}/R_{\min})^2$ determines the range of observed brightnesses, assuming a mono–luminous burst population. Figure 2 shows that the observed range of $C/C_{\min}$ is about a factor of 100 between the brightest and faintest bursts. Consequently, for a consistent population of bursts, one also requires

$R_{vis}/R_{min} \approx 10$ in addition to the constraints discussed previously. Only a narrow range of parameters, outlined by the intersection of the $\langle V/V_{max} \rangle$ arc of Figure 1 and a region where $R_{vis} \sim 10 R_{min}$, can therefore produce a C/C_{min} distribution that mimics the BATSE observations over a wide range of intensities.

With the relatively few number of bursts at the high–intensity end of the distribution, and the poor counting statistics that accompany these few bursts, based on the BATSE data alone, one might argue that the computed distribution and the observational data are not inconsistent. This would imply that the closest bursts are located at a distance of $\sim 1,000 - 2,000$ AU, and that BATSE is capable of detecting bursts out to a distance of $\sim 10,000 - 15,000$ AU. This would further imply an energy output of a few times 10^{27} ergs for these events. We argue, however, that the relative agreement between the BATSE data and the Oort Cloud scenario presented here is due to the fact that in its relatively short operational lifetime, BATSE has not yet sampled a large number of the strong thus infrequent bursts. This leads to a distribution at the high–intensity end which, although known to be consistent with the –3/2 slope from independent observational evidence,[6,7] cannot by itself be ruled inconsistent with some other distribution because of the poor sampling of the parent distribution in this region.

As the number of bursts observed increases, the shape of the observed C/C_{min} distribution will not change. For example, the strongest gamma–ray bursts were known to be homogeneously distributed before the activation of BATSE.[6,7] The observation of more bursts therefore should cause the integral number–intensity curve to reproduce a –3/2 slope with higher significance at the high–intensity end.

Figure 3 shows the evolution of the heliocentric model C/C_{min} distribution with increasing numbers of bursts. For all of the curves, the computed C/C_{min} distribution is well–represented by a –3/2 slope from $C/C_{min} \sim 20$ up to $C/C_{min} \sim 70$. Outside of this region, the curve deviates from a –3/2 slope because of the shape of $n(r)$. However, for the curve with only a few hundred bursts in the total sample, there are very few above $C/C_{min} \sim 70$. The deviation of the curve from a –3/2 slope is therefore difficult to detect.

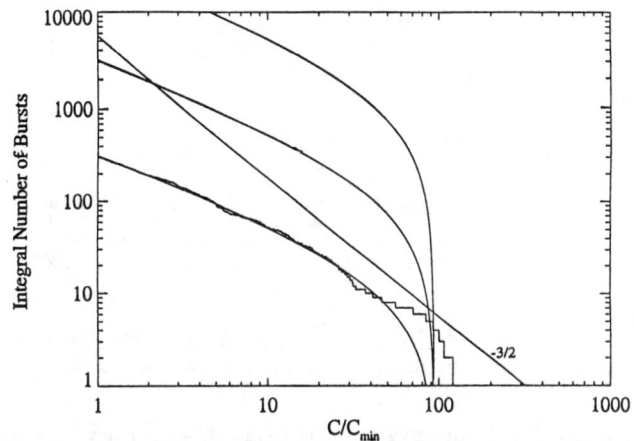

Figure 3 – Evolution of the C/C_{min} distribution with increased numbers of bursts in the model

The addition of more bursts causes the hypothesized $C/C_{\min}$ curve to move vertically upwards; the range of $C/C_{\min}$ where the slope is $-3/2$ is the same. The observational $C/C_{\min}$ curve, however, will reproduce a $-3/2$ slope with better significance as more bursts are detected, and will do so over a larger range of $C/C_{\min}$ because of the observation of more intense bursts. The hypothesized $C/C_{\min}$ distribution cannot reproduce a $-3/2$ slope over an increasingly large range of $C/C_{\min}$ with better significance as more bursts are added, and is therefore *inconsistent* with observations. To reproduce this behavior, the density of sources must be modified at small r, adding an excess of bursts over the hypothesized distribution of Equation (2), providing a more homogeneous distribution in nearby space. By itself, the cometary number density cannot generate a logarithmic slope consistent with $-3/2$ at the high-intensity end using a large number of bursts. The distribution of Equation (2) is reasonably flat, with an exponent of $3/2$. Cometary number densities with higher exponents (i.e. $5/2$, $7/2$, etc.), or scenarios[8] which require the use of $n(r)^2$ and effectively raise the exponent, will have an even more difficult time producing $C/C_{\min}$ distributions that are consistent with the BATSE observations.

CONCLUSIONS

The agreement between the model $C/C_{\min}$ distributions analyzed here and the BATSE data is due to the relatively small number of strong bursts observed by BATSE. By increasing the number of bursts in the model and comparing to an extrapolation of BATSE data based on the known homogeneity of the strong bursts, the same parameters can no longer produce a consistent result. The known $-3/2$ behavior of the strong bursts cannot be reproduced for large numbers of events using the cometary number density of Equation (2). One must hypothesize a modification to the density at small distances to generate a more homogeneous distribution that will manifest itself as a $-3/2$ slope in the integral number vs. $C/C_{\min}$ curve. Consequently, we believe that these heliocentric scenarios with density profiles similar to the expected number density of comets are unlikely explanations for the distribution of gamma-ray bursts.

REFERENCES

1. Horack, J. M., et al., Ap. J., (submitted) (1993).
2. Horack, J. M., et al., in Proc. Compton Symposium, eds. M. Friedlander, N. Gehrels, D. Macomb, (AIP : New York) , 694 (1992).
3. Maoz, E., Ap. J., **414**, 877–882 (1993).
4. Bailey, M. E., MNRAS, **204**, 603 – 633 (1983).
5. Gradshteyn, I., & Ryzhik, I., Tables of Integrals, Series, & Products , Academic Press (1965).
6. Mazets, E., et al., Astrophy. Space. Sci., **80**, 1–143 (1982).
7. Matz, S., et al., Proc. Los Alamos Wkshp. on Gamma-Ray Bursts, eds. Ho, C., Epstein, R., Fenimore, E. E. , (Cambridge Univ. Press) (1990).
8. White, S., Presentation at the 23rd ICRC , Calgary (1993).

EVIDENCE FOR THE GALACTIC ORIGIN OF GAMMA-RAY BURSTS

D. Q. Lamb, J. M. Quashnock

Dept. of Astronomy and Astrophysics, University of Chicago, Chicago, IL 60637

ABSTRACT

Adopting $(\bar{C}^{1024})_{max}$, the expected peak counts in 1024 msec, as a measure of brightness B, we find that the 54 type I bursts in the Burst and Transient Source Experiment (BATSE) 1B catalogue lying in the in the middle brightness range 490 counts $\leq B \leq$ 1250 counts (and consituting 1/3 of type I bursts) simultaneously exhibit a Galactic dipole moment $\langle\cos\theta\rangle = 0.204 \pm 0.079$ (2.6 σ) and a Galactic quadrupole moment less one third $\langle\sin^2 b\rangle - 1/3 = -0.104 \pm 0.041$ (2.5 σ). The combined values deviate significantly from those expected for isotropy (Q-value $= 6.6 \times 10^{-5}$, taking into account having divided the type I bursts into three equal samples). Performing a running average with a brightness window $\Delta\log B = 0.4$, we find that the deviations peak at $\langle\cos\theta\rangle = 0.230 \pm 0.078$ (3.0 σ) and $\langle\sin^2 b\rangle - 1/3 = -0.119 \pm 0.040$ (3.0 σ) for the 55 bursts lying in the medium brightness range 465 counts $\leq B \leq$ 1169 counts (Q-value $= 1.1 \times 10^{-4}$, taking into account having performed the running average). Considering $(\bar{F}^{1024})_{max}$, the expected peak flux in 1024 msec, as an alternative measure of brightness, we find that the 67 bursts lying in the peak flux range 0.455 counts cm^{-2} s^{-1} $< (\bar{F}^{1024})_{max} <$ 1.142 counts cm^{-2} s^{-1} (corresponding to the medium brightness range in peak counts) exhibit a dipole moment $\langle\cos\theta\rangle = 0.160 \pm 0.071$ (2.3 σ) and a quadrupole moment less one third $\langle\sin^2 b\rangle - 1/3 = -0.072 \pm 0.036$ (2.0 σ) which also deviate significantly from the values expected for isotropy (Q-value $= 4 \times 10^{-4}$). We conclude that γ-ray bursts are Galactic in origin.

INTRODUCTION

Earlier we investigated the temporal behavior of the γ-ray bursts in the BATSE 1B catalogue,[1] using the measures of burst brightness $B = (\bar{C}^{1024})_{max}$ and short time scale ($\lesssim 0.3$ s) variability $V = (\bar{C}^{64})_{max}/(\bar{C}^{1024})_{max}$. The quantities $(\bar{C}^{64})_{max}$ and $(\bar{C}^{1024})_{max}$ are the expected maximum number of counts in 64 ms and in 1024 ms, respectively. We presented evidence[2,3,4] for two distinct morphological classes of γ-ray bursts (see also Kouveliotou et al.[5]). Type I bursts (comprising $\approx 80\%$ of the bursts) are smoother ($\log V \leq -0.8$) on short time scales ($\lesssim 0.3$ s), range from faint to bright, are longer, and have softer spectra. Type II bursts (comprising $\approx 20\%$ of the bursts) are more variable ($\log V > -0.8$) on short time scales ($\lesssim 0.3$ s), faint ($B < 1900$ counts), shorter, and have harder spectra.

In a separate paper,[6] we study the clustering of γ-ray bursts. We show that all bursts and the 201 bursts for which B and V exist are significantly clustered on an angular scale $\approx 5°$. This angular scale is smaller than the typical (statistical plus systematic) error in burst locations of 6.8°, suggesting multiple recurrences from individual sources. Further, we show that the bright and faint bursts are correlated with each other, while the medium brightness bursts are not. These results imply that "classical" γ-ray burst sources repeat

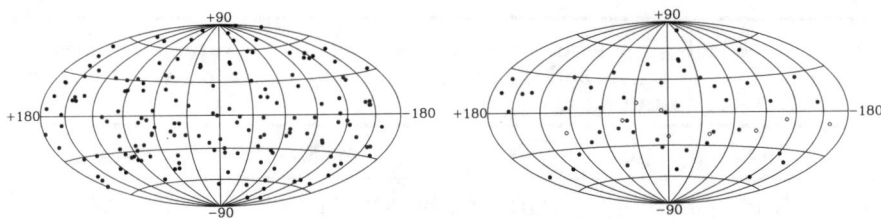

Fig. 1—(left panel) Distribution on the sky of 163 type I bursts for which the brightness B and variability V exist in the publicly available BATSE catalogue,[1] in Galactic coordinates (the Galactic Center lies at the center of the map). (right panel) Distribution on the sky of the 55 type I bursts in the medium brightness range $465 < B < 1169$, in Galactic coordinates. The bursts denoted by open circles are the nine faintest bursts in this brightness range; they are characterized by $\langle \sin^2 b \rangle - 1/3 = -0.280 \pm 0.099$.

on a time scale of months, and that many faint type I and II bursts come from the sources of bright type I bursts[6] (see also Wang and Lingenfelter[7]).

Motivated by this knowledge, we investigate here the Galactic dipole and quadrupole moments of the angular distribution of type I bursts as a function of brightness, using $B = (\bar{C}^{1024})_{\max}$ as a measure of brightness. We also consider $(\bar{F}^{1024})_{\max}$, the expected peak flux in 1024 msec, as an alternative measure of burst brightness. We report much of this work elsewhere.[8,9]

ANALYSIS AND RESULTS

Fig. 1 (left panel) shows the locations on the sky of the 163 type I bursts for which B exists in the publicly available BATSE catalogue.[1] We divide these type I bursts into three brightness samples containing equal numbers of bursts: 54 bursts which have $B < 490$ counts, 54 bursts which have 490 counts $\leq B \leq 1250$ counts, and 55 bursts which have $B > 1250$ counts. We then compute the mean values of the Galactic dipole moment $\langle \cos \theta \rangle$ and quadrupole moment $\langle \sin^2 b \rangle$ for each sample.

For an isotropic source distribution, we expect $\langle \cos \theta \rangle = 0$ and $\langle \sin^2 b \rangle - 1/3 = 0$, irrespective of brightness B; corrected for the BATSE sky exposure map,[1] these become $\langle \cos \theta \rangle = -0.0123$ and $\langle \sin^2 b \rangle - 1/3 = -0.0047$. We find that the angular distributions of the faint and bright samples are consistent with isotropy, whereas that of the middle brightness sample is not (see Table 1). For an isotropic sky distribution (the null hypothesis), the Galactic dipole and quadrupole moments are nearly independent, but not quite. We therefore evaluate the significance of the deviations of these moments from the values expected for isotropy using Monte Carlo simulations. We find that the probability by chance of an isotropic distribution of 54 bursts exhibiting values of $\langle \cos \theta \rangle$ *and* the negative of $\langle \sin^2 b \rangle - 1/3$ that equal or exceed the observed values is 2.2×10^{-5} (see Table 1). Multiplying this result by a factor of 3 in order to take into account having divided the type I bursts into three equal samples, we obtain a significance of 6.6×10^{-5}.

Having established significant evidence for a concentration of middle bright-

Table 1. Galactic dipole and quadrupole moments for various brightness samples of type I bursts and the corresponding Q-values, corrected for the BATSE sky exposure map.

Sample	Size	$\langle\cos\theta\rangle$	$\langle\sin^2 b\rangle - \frac{1}{3}$	Q-value
$490 < B < 1250$	54	$+0.204 \pm 0.079$	-0.104 ± 0.041	2.2×10^{-5} [a]
$465 < B < 1169$	55	$+0.230 \pm 0.078$	-0.119 ± 0.040	1.9×10^{-6} [b]
$0.455 < F < 1.142$	67	$+0.160 \pm 0.071$	-0.072 ± 0.036	4×10^{-4}
$465 < B < 1169$ (1B + post-1B)[c]	106	$+0.055 \pm 0.056$	-0.089 ± 0.029	1.5×10^{-4}

[a] Q-value $= 6.6 \times 10^{-5}$, taking into account having divided the type I bursts into three equal samples.
[b] Q-value $= 1.1 \times 10^{-4}$, taking into account having performed a running average.
[c] From Briggs et al.[11]

ness type I bursts toward the Galactic center and in the Galactic plane, we calculate running averages of the Galactic dipole and quadrupole moments in order to explore further their behavior as a function of B. Fig. 2 shows the running averages of $\langle\cos\theta\rangle$ and $\langle\sin^2 b\rangle - 1/3$ as a function of burst brightness B using a brightness window $\Delta\log B = 0.4$.

The running average of the dipole moment peaks at $\langle\cos\theta\rangle = 0.230 \pm 0.078$ when $B_{\rm mid} = 737$, corresponding to 55 bursts in the medium brightness range 465 counts $< B <$ 1169 counts (see Fig. 2). The running average of the deviation of the quadrupole moment from $1/3$ is $\langle\sin^2 b\rangle - 1/3 = -0.119 \pm 0.040$ for the same $B_{\rm mid}$ (see Fig. 2). These values of the Galactic dipole and quadrupole moments *each* represent 3.0 σ deviations from the values expected for isotropy (see Table 1). Fig. 1 (right panel) shows the distribution on the sky of these 55 bursts. Fifty-one of the 54 bursts in the middle brightness range 490 counts $< B <$ 1250 counts are common to the 55 in the medium brightness range 465 counts $< B <$ 1169 counts.

Considering only the 55 bursts in the medium brightness range 465 counts $< B <$ 1169 counts and again using Monte Carlo simulations which take into account the BATSE sky exposure map, we calculate that the probability by chance of an isotropic distribution of 55 bursts exhibiting values of $\langle\cos\theta\rangle$ and the negative of $\langle\sin^2 b\rangle - 1/3$ that jointly equal or exceed the above values is 1.9×10^{-6}. This significance does not take into account the large number of independent trials introduced by performing a running average over the brightness range $2 \leq \log B \leq 4.5$. We obtain a significance of 1.1×10^{-4}.

The running average of $\langle\sin^2 b\rangle - 1/3$ shows that the fainter bursts in the medium brightness range are more strongly concentrated in the Galactic plane. Indeed, the nine faintest bursts in this brightness range (denoted by open circles in Fig. 1) are characterized by $\langle\sin^2 b\rangle - 1/3 = -0.280 \pm 0.099$ (representing a 2.8 σ deviation from the value expected for isotropy for these bursts alone).

Rutledge and Lewin[10] have suggested using $(\bar{F}^{1024})_{\rm max}$, the expected peak flux in 1024 msec, as a brightness measure. Averaging $\alpha = B/(\bar{F}^{1024})_{\rm max}$ over

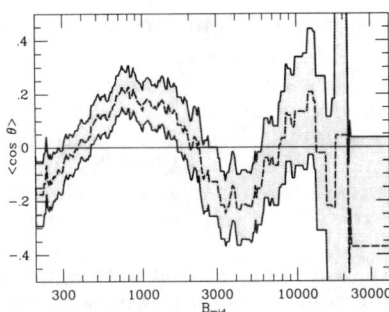 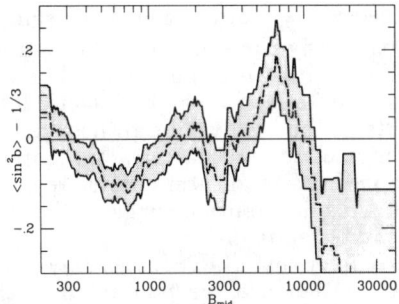

Fig. 2—(left panel) Running average of the Galactic dipole moment $\langle \cos\theta \rangle$ as a function of B. (right panel) Running average of the deviation of the Galactic quadrupole moment from 1/3, $\langle \sin^2 b \rangle - 1/3$, as a function of B.

all 163 type I bursts, we find $\bar{\alpha} = 0.9992$ cm^2. Thus the range of $(\bar{F}^{1024})_{\max}$ corresponding to the medium brightness range in B is 0.455 counts cm^{-2} s^{-1} $< (\bar{F}^{1024})_{\max} < 1.142$ counts cm^{-2} s^{-1}. Calculating the Galactic dipole and quadrupole moments for the 67 bursts in this peak flux range, we find that the combined values deviate significantly from those expected for isotropy (Q-value $= 4 \times 10^{-4}$) (see Table 1).

DISCUSSION

Adopting $(\bar{C}^{1024})_{\max}$ as a measure of brightness B, we find that the 55 type I bursts lying in the medium brightness range 465 counts $< B <$ 1169 counts (which constitute 1/3 of type I bursts) are strongly concentrated toward the Galactic center and in the Galactic plane (Q-value $= 1.1 \times 10^{-4}$, taking into account having performed a running average). Considering $(\bar{F}^{1024})_{\max}$ as an alternative measure of brightness, we find that the 67 type I bursts lying in the range 0.455 counts cm^{-2} s^{-1} $< (\bar{F}^{1024})_{\max} < 1.142$ counts cm^{-2} s^{-1} (which corresponds to the medium brightness range in peak counts) are again strongly concentrated toward the Galactic center and in the Galactic plane (Q-value $= 4 \times 10^{-4}$). We conclude that these type I bursts, and thus all type I bursts, are Galactic in origin. As mentioned earlier, in a separate paper we report evidence that many faint type I and type II bursts come from the sources of bright type I bursts.[5] We therefore conclude that all γ-ray bursts are Galactic in origin.

Briggs et al.[11] recently culled 51 medium brightness type I bursts from 483 post-1B bursts (the fraction of bursts for which V exists is smaller for the post-1B bursts due to failure of the CGRO tape recorders). These bursts exhibit no deviation of their Galactic dipole moment and only a modest deviation of their Galactic quadrupole moment from the values expected for isotropy.[11] Caution may be warranted in interpreting these values, since even a few repetitions from a single burst source can significantly distort the dipole and quadrupole moments if the repetitions fall in the medium brightness range. The sky map of the medium brightness type I bursts in the 1B catalogue suggests that the angular distribution of these bursts is not strongly affected by repetitions (see Fig. 1); however, one cannot be sure about the post-1B bursts until better burst positions are available. Nevertheless combining the 1B and post-1B bursts, the significance of the deviation of the Galactic quadrupole moment from the value expected for isotropy remains the same (3.1 σ) and the combined significance remains nearly the same (Q-value = 1.5×10^{-4}) (see Table 1).

Fig. 2 hints that the faint and bright type I bursts lie preferentially toward the anti-center and at high Galactic latitudes around us, and shows that type I bursts in the medium brightness range 465 counts $< B <$ 1169 are strongly concentrated toward the Galactic center and in the Galactic plane. The faintest of the medium brightness bursts are concentrated very strongly in the Galactic plane. This distribution suggests that most type I burst sources are associated with the Galactic disk.

We infer that BATSE sees only a fraction of the type I burst sources in the Galaxy. Otherwise, we would expect that the concentration of bursts toward the Galactic center and in the Galactic plane would be very strong, which is not the case (see Fig. 3). Further, we would expect, qualitatively, that the cumulative brightness distribution of type I bursts would have a slope $\propto -1$ throughout most of its range, whereas it is flatter at the faint end and steeper at the bright end.[4]

We therefore conjecture that BATSE is seeing type I burst sources out to $\sim 1-2$ kpc, and that the behavior of the angular distribution of these bursts as a function of burst brightness B reflects the structure of the nearby spiral arms of the Galaxy. In this picture, the bright type I bursts come primarily from the vicinity of the Orion arm, which lies around us and toward the Galactic anticenter (at distances of up to ≈ 1 kpc). The medium brightness type I bursts come primarily from the vicinity of the Sagittarius arm, which lies toward the Galactic center (at a distance ≈ 1 - 1.5 kpc). Some of the faint type I bursts may come from the vicinity of the Perseus arm, which lies toward the Galactic anticenter (at distances $\gtrsim 2$ kpc).

Figs. 2 and 3 hint that the faint type I bursts lie preferentially toward the Galactic anticenter but not in the Galactic plane. One possibility is that the faint type I bursts come from nearby sources in the vicinity of the Orion arm. In a separate paper, we present evidence that this is indeed the case, and that many of the faint type I and II bursts come from the sources of bright type I bursts.[7]

We thank Paolo Coppi for stimulating discussions about the spatial distribution of γ-ray bursts, and Carlo Graziani and Tom Loredo for informative discussions about statistical methodology. This research was supported in part by NASA grants NAGW-830, NAGW-1284, and NASW-4690.

REFERENCES

1. G. J. Fishman, et al., ApJSup, in press (1994).
2. D. Q. Lamb, C. Graziani & I. A. Smith, ApJ **413**, L11 (1993).
3. D. Q. Lamb & C. Graziani, ApJ, in press (1994).
4. D. Q. Lamb & C. Graziani, ApJ, in press (1994).
5. C. Kouveliotou et al., ApJ **413**, L101 (1993).
6. J. M. Quashnock & D. Q. Lamb, MNRAS **265**, L59 (1993).
7. V. C. Wang & R. E. Lingenfelter, Ap. J. **416**, L13 (1993).
8. J. M. Quashnock & D. Q. Lamb, MNRAS **265**, L45 (1993).
9. D. Q. Lamb & J. M. Quashnock, MNRAS, submitted (1994).
10. R. Rutledge & W. H. G. Lewin, MNRAS **265**, L51 (1993).
11. M. Briggs et al., Proceedings Huntsville Workshop, ed. J. Brainerd, G. J. Fishman, and K. Hurley, in press (1994).

COMMENTS ON GALACTIC ANISOTROPY OF GAMMA-RAY BURSTS

Hui Li, I. A. Smith

Dept. Space Physics & Astronomy, Rice University, Houston, TX 77251-1892

Robert C. Duncan

Dept. of Astronomy, University of Texas at Austin, Austin, TX 78712-1083

ABSTRACT

We analyze the spatial distribution of gamma-ray bursts and their intensities using the BATSE Burst Catalog for 260 bursts. We discuss the importance of the following effects: (1) Single versus mixed triggering timescales for the $C_{\max}/C_{\min}$ values. (2) When studying the cumulative $< V/V_{\max} >$ for a particular triggering timescale, only bursts whose durations are longer than the triggering timescale should be used. (3) The inclusion of statistical error bars. (4) The differences between using the count rate in the second brightest detector and the peak photon flux. (5) The difference between using the type I bursts of Lamb, Graziani & Smith[1] and long bursts ($T_{90} > 1$ s). Taking these effects into account, we find no compelling evidence for anisotropy in the first BATSE catalog. However, possible signals could be hidden by the small number statistics, particularly for the bright bursts, so further study with more data is needed.

INTRODUCTION

Attempts have been made recently[2,3] to look for anisotropies in the spatial distribution of gamma-ray bursts using running window averages and/or cumulative distributions of $< \cos\theta >$, $< \sin^2 b >$, and $< V/V_{\max} >$. In particular, the cumulative $< V/V_{\max} >$ distribution was used to divide the bursts into two populations with the bright bursts confined to the galactic disk[2]. On the other hand, it has been claimed[3] that the angular distribution of a sub-population containing 54 type I bursts[1] of intermediate brightness exhibit both Galactic dipole and quadrupole moments deviating from the isotropic values with a combined probability by chance of 6.6×10^{-5}.

Ideally, if there was a strong signal of anisotropy in the distribution, the results would not be sensitive to the techniques used and the quantities that are plotted; however, as we will show in the following, this is not the case, and great care is required in interpreting the results.

REANALYSIS OF CUMULATIVE $< V/V_{\max} >$ CURVES

Using the BATSE catalog for 260 bursts, Atteia & Dezalay[2] took the largest $C_{\max}/C_{\min}$ from the three BATSE triggers for each burst, and used them to calculate the cumulative $< V/V_{\max} >$ as the instrument sensitivity (in bursts yr^{-1}) is artificially decreased. Their result is shown in Figure 1. They claimed a statistically significant ($\sim 3\sigma$) "dip" at $N \approx 50$ bursts that they interpreted as an indication for two populations of gamma-ray bursts sources. Furthermore, they showed that the combined distribution containing 24 (out of

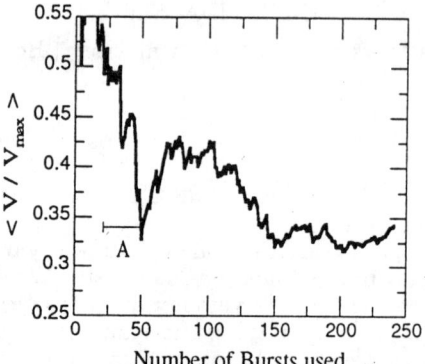

Figure 1. Cumulative $<V/V_{max}>$ for the bursts in the first BATSE catalog. For each burst, the largest C_{max}/C_{min} for the three triggers was used. A concentration towards the Galactic plane was found for the bursts in region "A".

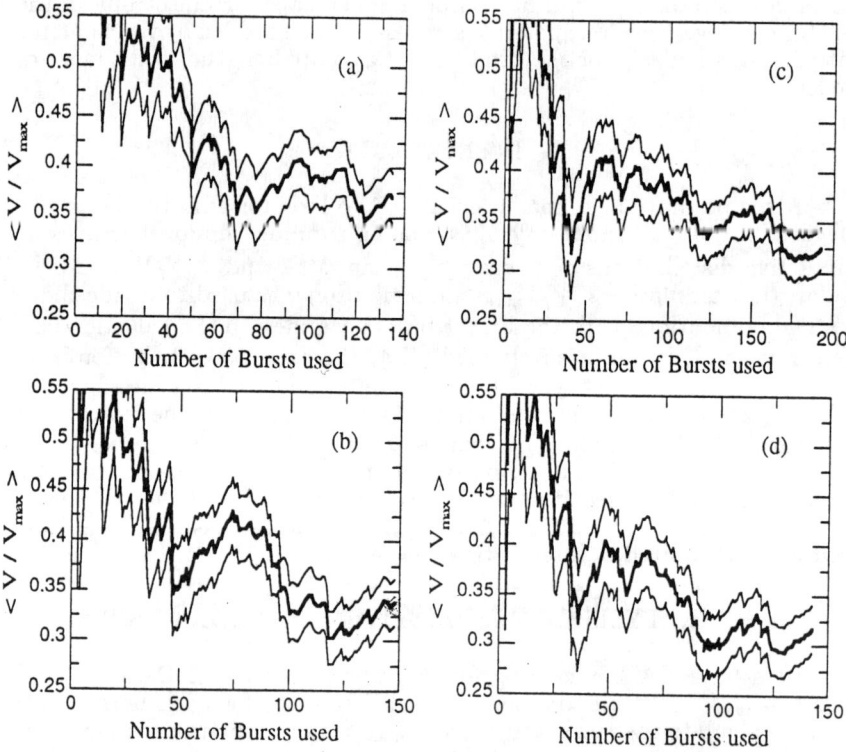

Figure 2. Cumulative $<V/V_{max}>$ for the bursts in the first BATSE catalog, using the individual triggers (thick curves): (a) 64 ms trigger, (b) 256 ms trigger, (c) 1024 ms trigger, (d) 1024 ms trigger, but only using bursts with $T_{90} > 1024$ ms. The upper and lower thin curves are standard 1 σ deviations.

51) bright BATSE bursts and 27 KONUS bursts is concentrated towards the Galactic disk with $< \sin^2 b > - 1/3 = -0.185 \pm 0.04$.

However, the three BATSE triggers have different threshold conditions and thus can have different sampling volumes for a given underlying distribution of burst time-histories. This means that one could lose information about the intrinsic burst distribution when mixing the different triggers. Here we investigate the effects of using the triggers separately.

In Figure 2, we plot the cumulative $< V/V_{max} >$ using only the C_{max}/C_{min} from the 64 ms, 256 ms, or 1024 ms trigger (thick curves). Note that only the bursts with $C_{max}/C_{min} > 1$ for the individual triggers were used (which is why the total number of bursts used in Figure 2 is always less than in Figure 1). One can immediately see that the "dip" is only present for the 1024 ms trigger. Since it only appears for this one trigger, its significance is subject to doubt.

About 25 % of the 260 BATSE bursts in the first catalog are shorter than 1024 ms. Thus $(C_{max}/C_{min})_{1024}$ systematically underestimates the brightness of those bursts. It is therefore misleading to use bursts shorter than 1024 ms when calculating $< V/V_{max} >$ for the 1024 ms trigger. In Figure 2(d), we replot Figure 2(c) removing those bursts that have $T_{90} < 1024$ ms. The "dip" is now only $\approx 1.5\sigma$, which is far less significant than in Figure 1.

If there is a real break in the intensity distribution of the bursts, the "dip" should show up in any trigger. Thus we conclude that, based on the present BATSE catalog of 260 gamma-ray bursts, we can not find any convincing evidence to support the significance of a "dip" in Figure 1; no division of the bursts based on this dip can currently be made.

REANALYSIS OF ANGULAR DISTRIBUTION VERSUS INTENSITY

Quashnock & Lamb[3] plotted the running-window averaged galactic dipole $< \cos \theta >$ and quadrupole $< \sin^2 b > - 1/3$ versus $(C_{max})_{1024}$ for the Type I bursts defined in Lamb, Graziani & Smith[1]. Their results are shown as the solid curves in Figure 3(a) and 3(b). Statistically significant anisotropy was claimed for 54 type I bursts with 490 counts $\leq (C_{max})_{1024} \leq 1250$ counts. Note that the maximum deviations from isotropy of both the dipole and quadrupole are at the same $(C_{max})_{1024} \approx 750$. Using Monte Carlo simulations, they found that the probability by chance of an isotropic distribution of 54 bursts exhibiting this anisotropy is 6.6×10^{-5}.

The peak photon flux P, which incorporates more of the burst information obtained by BATSE, is an alternative indicator of the strength of a burst. C_{max} is the count rate in only the second brightest detector whereas P incorporates the information from all the detectors with statistically significant signals, including the corrections for orientation. Both C_{max} and P contain biases, so it is important to see if the anisotropy signal is present for both of them. Figure 3 (c) and (d) are the same as Figure 3 (a) and (b) except that P_{1024} is used instead of $(C_{max})_{1024}$. Even though deviations from isotropy are still present in both figures, they occur at quite different P_{1024}; thus the joint significance is greatly reduced (see Rutledge and Lewin[4] for a detailed significance calculation.).

Instead of splitting bursts into the Type I and Type II populations of Lamb, Graziani & Smith[1], we split bursts into long ($T_{90} > 1024$ ms) and short. Figure 4 shows running-window averaged results for both $(C_{max})_{1024}$ and P_{1024}. Again, the deviations from isotropy become less significant.

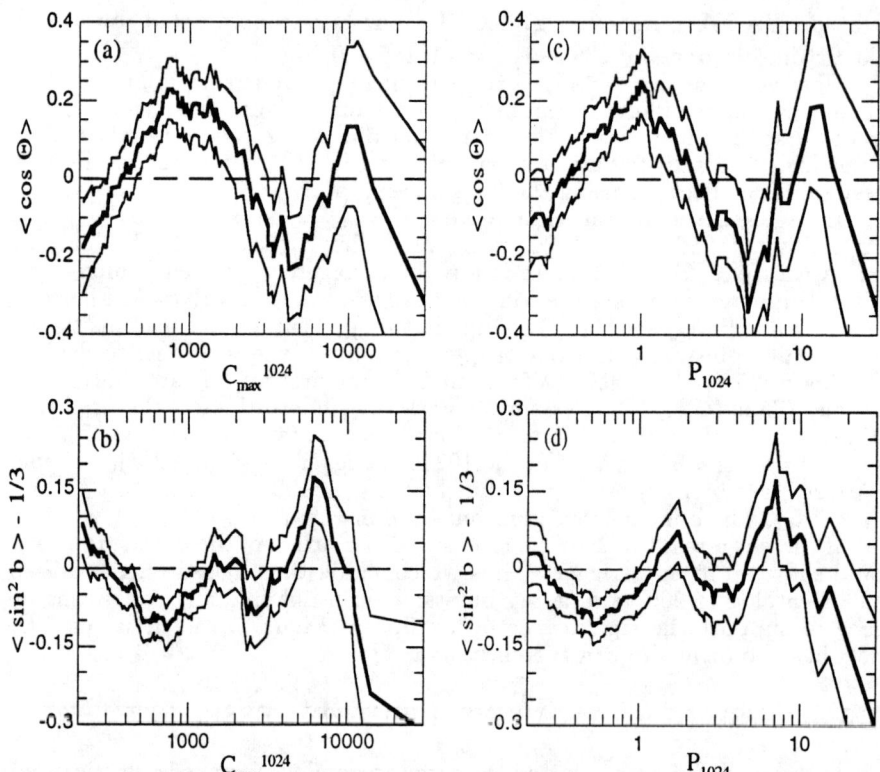

Figure 3. Running window averaged angular distribution of 163 type I bursts using a brightness window $\Delta \log(C_{\max})_{1024} = 0.4$ in (a) and (b) and $\Delta \log P_{1024} = 0.4$ in (c) and (d) (uncorrected for the BATSE sky exposure). Thin curves are standard 1σ deviations.

CONCLUSION

We conclude that there is no compelling evidence for anisotropy in the first BATSE catalog. Since these effects hinge on small number statistics, a few bursts will either enhance or diminish the signal, so it is very important to choose bursts self-consistently. We emphasize that a true, strong signal should not be very sensitive to the quantities that are plotted, which is not the case here. However, possible signals could be hidden by the small number statistics, especially for bright bursts, so further study with a larger database is needed.

This work was supported by grants NAG 5-1515 and NAG 5-2045 at Rice University, and by grant NAGW-2418 at the University of Texas.

REFERENCES

1. D. Q. Lamb, C. Graziani and I. A. Smith, Ap. J. (Letters) **413**, L11 (1993).
2. J.-L. Atteia and J.-P. Dezalay, Astr. Ap. **274**, L1 (1993).
3. J. M. Quashnock and D. Q. Lamb, M.N.R.A.S. (in press).

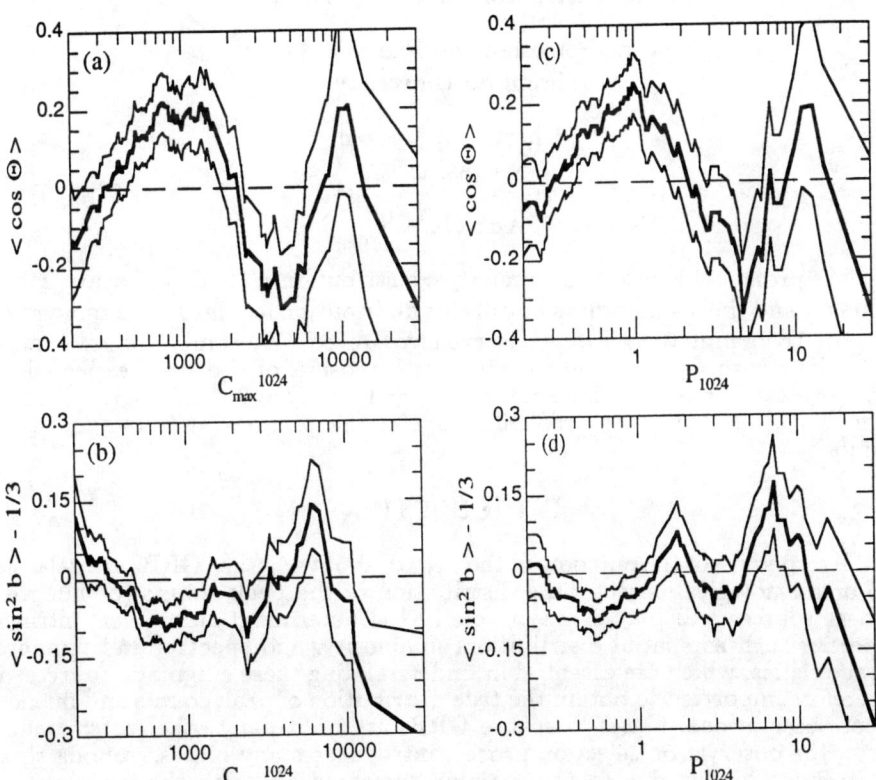

Figure 4. Same as Figure 3 except long bursts ($T_{90} > 1024$ ms) were used instead of type I bursts.

4. R. E. Rutledge and W. H. G. Lewin, M.N.R.A.S. (in press).

THE SLOPE OF THE LOGN–LOGS DISTRIBUTION FOR BATSE GAMMA-RAY BURSTS

Vahé Petrosian and Walid J. Azzam
Stanford University

Charles A. Meegan
NASA-MSFC

ABSTRACT

We present the observed cumulative distribution $N(> C_p)$ of peak photon counts C_p and the logarithmic slope of this distribution for the 428 (combined 1B and post 1B) gamma-ray bursts observed by BATSE. We compare the variation of the slope with C_p with the variation with redshift of the slope expected for standard-candle, non-evolving sources in $\Omega = 1$ and $\Omega = 0$ cosmological models. From this we determine the redshift range of the gamma-ray bursts and their luminosity.

INTRODUCTION

Irrespective of the outcome of the debate about whether GRBs are galactic-halo or cosmological objects, the distribution of the peak counts (or fluences) of these sources will play a major role in the determination of their intrinsic properties such as spatial distribution, luminosity, and spectral and temporal characteristics, which are essential in understanding these enigmatic sources. It is therefore important to obtain the true distribution of peak counts and fluences free of observational biases. Because GRBs are unlike any other astronomical source, the observation selection process introduces many biases. Among these the foremost bias is due to the variable threshold C_{lim} for the peak photon counts C_p integrated over some apriori specified finite time Δt and energy range ΔE. One way to account for this bias is to analyze the source count data in terms of the distribution $f(C_p/C_{lim})$ or, as is more customary, in terms of the integral of this distribution $F(C_p/C_{lim}) = \int_{C_p/C_{lim}}^{\infty} f(x)dx$ and one moment of this distribution, $< V/V_{max} > \equiv < (C_p/C_{lim})^{-1.5} >$. This has been a very useful tool and the deviation of $F(x)$ from $x^{-1.5}$ power law and $< V/V_{max} >$ from the value of 0.5 expected from an homogeneous, isotropic, static, Euclidean distribution (HISE, for short) in conjunction with the isotropy of angular distribution of GRBs has provided strong evidence for the cosmological distribution of these sources. Beyond this, the distribution and its moment are not of further use for comparison with theoretical models. For this task one must deal with the true distributions of the rate of occurrence of bursts with peak photon count C_p, $n(C_p)$, or its integral $N(C_p) = \int_{C_p}^{\infty} n(x)dx$.

CUMULATIVE DISTRIBUTION

As shown by Petrosian[1] we can obtain these distributions, corrected for the variation in the threshold, directly from the data. For each burst characterized by $C_{p,i}$ and $C_{lim,i}$ in a sample such as the BATSE one shown in Figure 1a, we define an associated unbiased number of bursts $\tilde{N}_i$ contained in the box with

$C_p > C_{p,i}$ and $C_{lim} < C_{p,i}$. The cumulative distribution $N(C_p)$ is obtained from this in a stepwise manner: $\delta \ln N(C_{p,i}) = \ln(1 + \tilde{N}_i^{-1})$. Figure 1b shows $N(C_P)$ for the BATSE 1B, post 1B, and 1B + post 1B samples obtained in this manner.

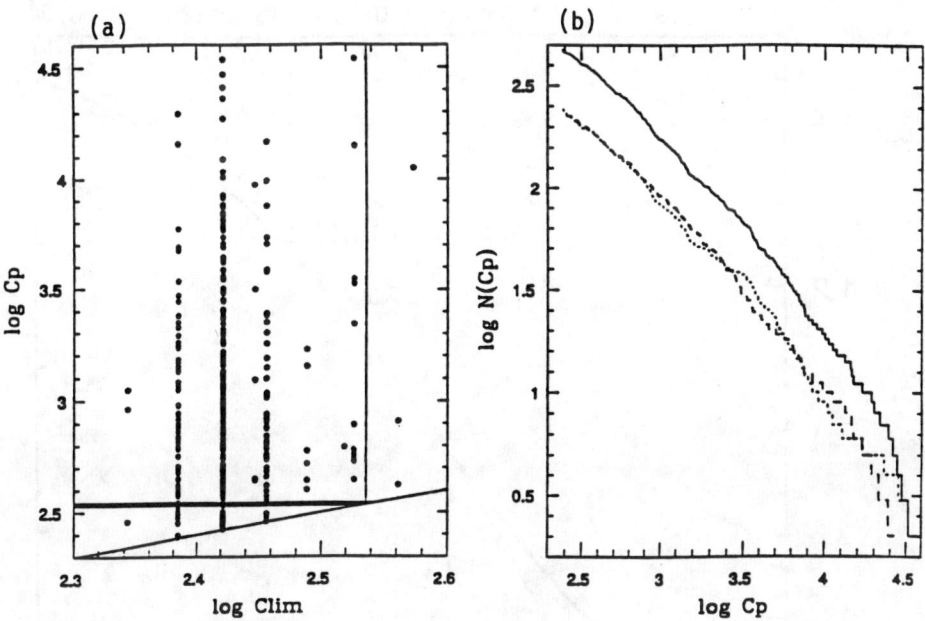

Figure 1(a). Distribution of observed peak and threshold photon count rates (C_p and C_{lim}) for a total of 428 GRBs observed by the BATSE instrument and triggered by the 1024 milisecond integration time. The horizontal lines show the ith strip containing the bursts with peak count rate $C_{p,i}$, and the left hand box contains the number $\tilde{N}_i$ of untruncated points associated with this burst. (b). The cumulative counts. The histogram with small dashes represents the cumulative distribution for the 260 publicly available data points. The histogram with long dashes is for a second set of 168 bursts with known C_p and C_{lim}. The solid histogram is for the combined 428 points whose C_p and C_{lim} are currently known.

SLOPE OF SOURCE COUNTS

The cumulative distributions are not useful for comparison with observations because the integration process can hide subtle effects and because the different points of a cumulative distribution are not independent. Therefore, it is more useful to obtain the differential distribution[2] or some other differential characteristic. It turns out that the logarithmic slope of the counts,

$s = -d\log N/d\log C_p$, referred to as the hazard rate in statistic's literature[3], is also directly related to the quantity $\ln(1+\tilde{N}_i)$ and can be evaluated at every data point $C_{p,i}$. However, for the purpose of the presentation and comparison with theoretical models it is convenient to bin the data as shown in Figure 2 for the 1B + post 1B sample.

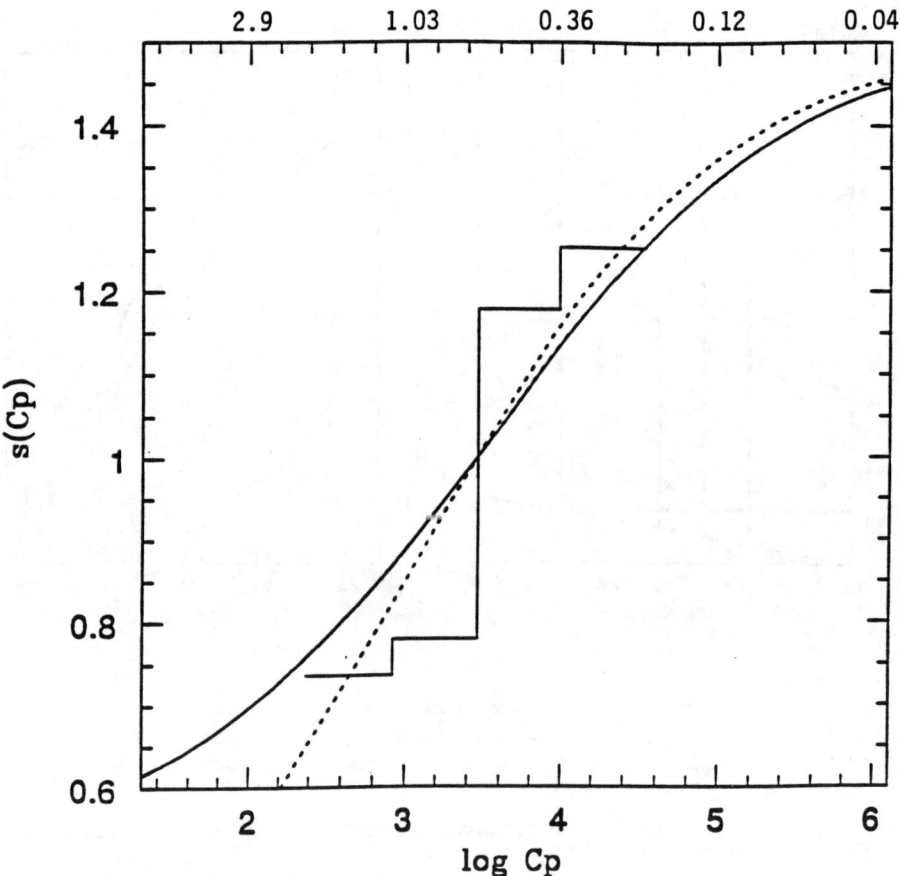

Figure 2. The logarithmic slope for the cumulative distribution of the peak count rates as a function of the peak count rate. The histogram represents the binned data for the total BATSE sample triggered by the 1024 milisecond integration time. The solid (dashed) curve is theoretically calculated for an $\Omega = 0$ ($\Omega = 1$) model for standard candles with constant comoving density. An F-test indicates that the $\Omega = 1$ model fits the data better than the $\Omega = 0$ model at the 99% confidence level. On the top axis we give the redshift, z, associated with the $\Omega = 1$ model. Note the sharp drop in slope (from 1.2 to 0.8) in the central bin. This is the primary reason the data favor the $\Omega = 1$ dashed curve.

COMPARISON WITH COSMOLOGICAL MODELS

Along with the binned data, Figure 2 also shows two curves for constant comoving density, standard candle sources in $\Omega = 1$ and $\Omega = 0$ models. The models give variation of the slope s with redshift z. The models' curves have been shifted horizontally to obtain the best fit to the data. Therefore, this comparison gives us the range of redshift of the sources and their gamma-ray luminosity for an assumed energy spectrum. (For details see Azzam, Petrosian and Linder[4].)

RESULTS

From this we can conclude that the non-evolving models are consistent with the data with the $\Omega = 1$ model providing a better fit. This is only a consistency check and does not preclude other cosmological models with different luminosity functions and/or evolution.

The range of redshifts for the $\Omega = 1$ model are indicated in the upper horizontal axis of Figure 2. The luminosity of the sources in the 50 to 300 keV range is then 1.5×10^{50} erg/sec for an assumed average photon spectral index equal to 2.

As pointed out in reference 4 and in this workshop by M. Rees, there exists a large change of the slope at BATSE count $C_p \simeq 3000$. This agrees somewhat with the variation of the slope for cosmological models for z between about 0.6 and 1. However, the observed change is more abrupt than that expected from the models. Furthermore, dispersion in spectral index and luminosity of the sources would tend to flatten the model curves making this discrepancy more serious. Whether or not this is a source of concern in the cosmological interpretation of GRBs, can be decided with further analyses of models with realistic luminosity functions and evolutions.

ACKNOWLEDGEMENT

This work is supported by NASA grant NAGW 2290.

REFERENCES

1. V. Petrosian, ApJ, 402, L33 (1993).
2. V. Petrosian, W. J. Azzam, and B. Efron, in Compton Symposium Proceedings, eds. M. Friedlander, N. Gehrels and D. J. Macomb (NY: AIP, 1993).
3. B. Efron and V. Petrosian, JASA (1993), in press.
4. W. J. Azzam, V. Petrosian, and E. V. Linder, submitted to ApJ Letters.

COMPARING AND COMBINING THE GAMMA-RAY BURSTS FROM SMM, SIGNE AND BATSE

Vahé Petrosian and Walid J. Azzam
Stanford University, Stanford, CA 94305

Kevin Hurley
UC Berkeley, Berkeley, CA 94720

Gerald Share
NRL, Washington, DC 20375

ABSTRACT

We compare the observed distribution of peak photon count rates C_p from BATSE, SIGNE and SMM. Using consideration of live times and sky coverages we determine the full sky rate of occurence of bursts observed by each instrument down to the lowest value $C_{p,min}$. Using these rates we normalize the distribution to combine the different samples. We can carry this out successfully for the BATSE and SIGNE samples. We then use the methods described in references 1 and 2 to obtain the cumulative counts $N(C_p)$ and its logarithmic slope $s(C_p)$ for the combined sample. We compare the latter with cosmological predictions as described in reference 3. We cannot find a consistent normalization for combining the SMM data which we attribute to its longer ($\simeq 16s$) triggering time.

INTRODUCTION

In the absence of direct knowledge of the distance to gamma-ray bursts (GRBs) the so-called $\log N$-$\log S$ relation is our only tool for the investigation of their spatial distribution, their distances, r, and luminosities, L, or in general their luminosity function $\psi(L, r)$. For example, the cumulative rate of occurrence of bursts with peak photon count rates greater than C_p is given by

$$N(C_p) = \int_0^\infty dL \int_0^{r_L} \psi(L,r) \frac{dV}{dr} dr, \qquad (1)$$

where $L = 4\pi r^2 C_p h(\gamma), r_L^{-2} = 4\pi C_{lim} h(\gamma)/L$, V(r) is the volume of space occupied by the sources up to distance r, C_{lim} is the detection threshold of peak photon counts and $h(\gamma)$ is some function of the spectrum of the bursts at its peak. For standard candles $\psi(L,r) = \delta(L - L_0) \times n(r)$, where $n(r)$ is the rate of occurrence of bursts per unit volume (co-moving volume in cosmological scenarios), equation (1) reduces to $N(C_p) = \int n(r)dV(r)$ so that the cumulative counts $N(C_p)$ are directly related to the spatial distribution of sources. The primary purpose for the investigation of $N(C_p)$ curves is to determine the spatial distribution by the inversion of such a relation. It is obvious that more accurate knowledge of the spatial distribution can be obtained from data with a larger number of bursts with a wider range of values of C_p. For this purpose we attempt to combine GRB data from GRS instruments of SMM, SIGNE instruments of Venera 13 and 14 and BATSE 1B data from C-GRO.

COMPARISON OF INSTRUMENTS AND DATA

In the Table below we compare the relevant parameters of the three instruments. The fifth row gives the length of observation which when multiplied by the fraction of time the instrument was capable of observing GRBs (row 3) and the fraction of sky available to the instrument (row 4) we obtain the effective full sky days of observation given in row six. Row 7 gives the number of GRBs detected during the above periods with C_p's exceeding the individual thresholds C_{lim}'s. Because the threshold count rate C_{lim} is variable, to obtain the true rate of GRBs we must determine the number of bursts with peak photon counts greater than some well defined limiting count rate.

COMPARISON OF ALL SKY RATES OF GRBs

	BATSE	SIGNE	SMM
Energy Range (keV)	50-300	50-300	350-850
Integration Time (s)	1.024	1	16
Fraction Live	0.51	0.85	0.59
Fraction Sky	0.67	1.00	0.28
Days (Total)	318	516	3396
Days (Effective)	110	440	560
$N(C_p/C_{lim} > 1)$	223	133	132
$N(C_p > C_{p,min})$	242	149	160
$C_{p,min}$ (Count/s)	240	104	16
$C_{p,max}$ (Count/s)	34000	2800	550
Rate Per Day ($C_{p,min} < C_p$)	2.2	0.34	0.28

Using the method described by Petrosian[1] we obtain the true cumulative distribution of C_p's. On Figure 1a we show these distributions for the three samples, which represent the number of bursts observed by each instrument with average peak photon count rates between $C_{p,min}$ and $C_{p,max}$ (rows 9 and 10). The total number of detected bursts with average $C_p > C_{p,min}$ (maximum value reached by each curve) is shown in row 7. Dividing these with the effective number of days we obtain the true all sky rate of bursts with average $C_p > C_{p,min}$ as shown in the last row.

COMBINING BATSE AND SIGNE SAMPLES

Having established relative rates we now must determine the relative normalization of the C_p's. This can be done by consideration of the differences between the responses of the instruments to a sample of GRBs with different fluxes, spectra and temporal variations. This task is not simple and also is affected by the angular response of the instruments. A simpler way to carry out this normalization is to use the fact that the rate of occurrence of the bursts is independent of the instrument and does not vary with time. We can then find the values of C_p's which give the same rate and shift the curves horizontally by a factor equal to the ratio of the two values. Because the integration times and spectral ranges of BATSE and SIGNE are identical, we can combine these two samples readily. On the $\log N(C_p)$-$\log C_p$ curve of the BATSE sample, we find the value of C_p^* at which $N(C_p^*)$ is equal to the SIGNE rate of 0.34 bursts/day. This BATSE $C_p^* = 3100\, s^{-1}$ then is equivalent to the $C_{p,min}$ of SIGNE. We therefore multiply SIGNE C_p's by $C_p^*/C_{p,min}^{SIGNE} = 30$ to obtain their equivalent BATSE photon count rates. The shifted curves are shown in Figure 1b where we also show the $\log N(C_p)$-$\log C_p$ relation (upper curve) for the combined data set consisting of all pairs (C_p, C_{lim}) from BATSE and pairs C_p, C_{lim} from SIGNE multiplied by the above ratio 30. In this way we obtain the variation of the rate of occurrence of bursts over 2.5 decades of peak counts.

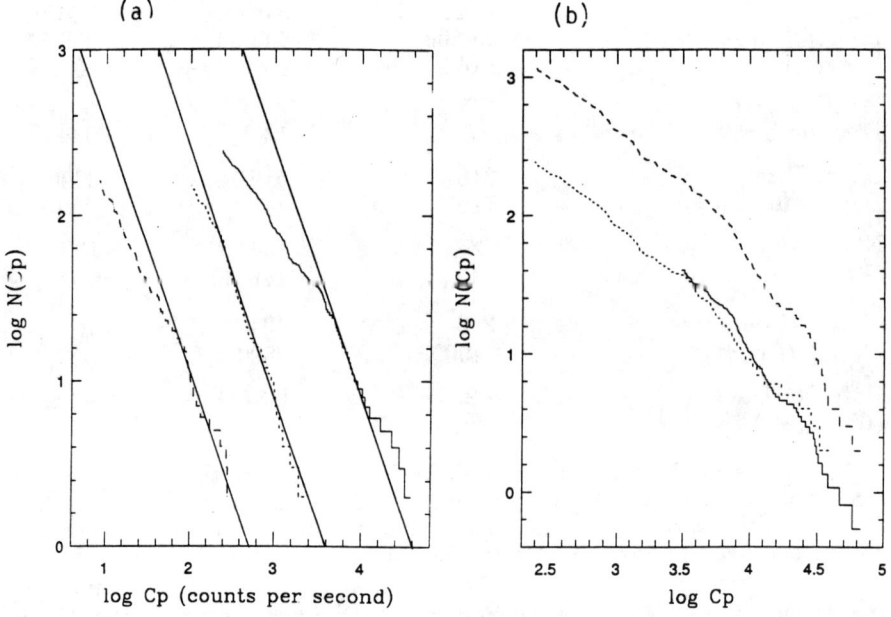

Figure 1(a). The cumulative counts for BATSE 1024 msec 1B sample (solid), SIGNE (small dashes), and SMM (long dashes). The solid lines have slope -3/2. Note flatter SMM curve compared to SIGNE. **(b).** The cumulative counts of GRBs versus the BATSE peak photon count rates for BATSE 1024 msec 1B (small dashes), shifted SIGNE (solid), and combined BATSE and SIGNE (long dashes).

The cumulative counts are useful for obtaining the correct normalization, but the differential counts $n(C_p) = dN(C_p)/dC_p$ or some other differential characteristics of the relationship are more useful for comparison with models because of the independence of the data points. As pointed out by Efron and Petrosian[2] the logarithmic slope of the counts $s(C_p) = -d\log N/dC_p$ can be obtained directly from the data. Figure 2 shows the variation of this slope with C_p for the combined data. In contrast with the same curve for the BATSE data alone[4] (see also Petrosian, Azzam and Meegan, these proceedings), the combined data clearly shows the 3/2 value expected at high values of C_p. However, as in the case of the BATSE data alone, the change of slope from 1.2 to 0.8 occurs more rapidly, over a smaller range of C_p's than that expected from simple cosmological models ($\Omega = 0$ or 1) and for non-evolving standard candles, $n(r) = n_0$. For further discussion of the comparison with the model we refer the reader to the above two papers.

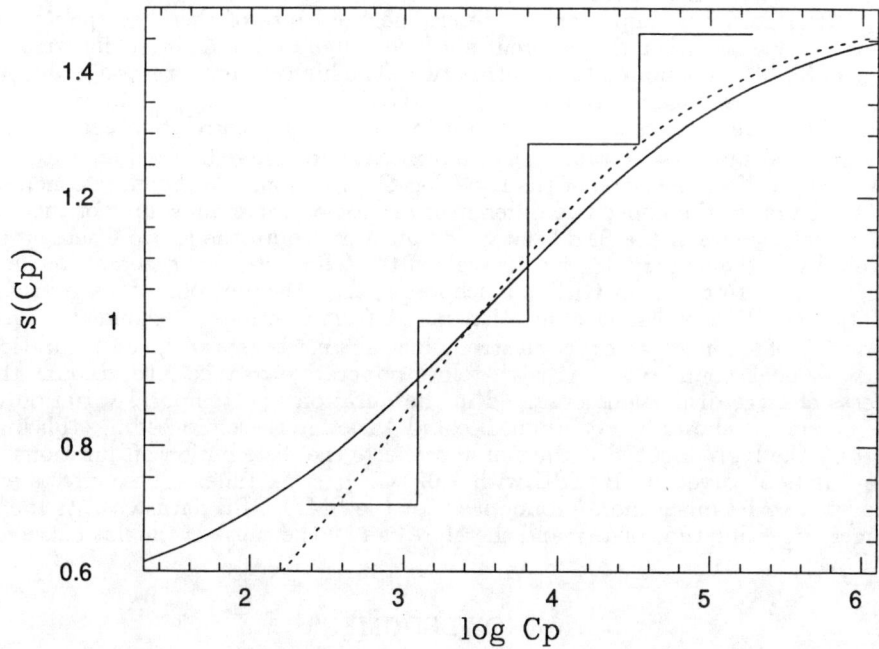

Figure 2. The histogram depicting the variation of the logarithmic slope s with the equivalent BATSE peak photon count rate C_p. The solid (dashed) curve is theoretically calculated from an $\Omega = 0 (\Omega = 1)$ model for standard candles with constant co-moving density.

SMM-GRS DATA

Combining the SMM-GRS data with BATSE or SIGNE is not as simple because of its longer ($\simeq$16s) triggering integration time and higher photon energy range. Comparison of the rate of 0.28/day of SMM with 0.34/day of SIGNE would indicate that the minimum SMM count rate $C_{p,min} = 16\text{s}^{-1}$ corresponds to the SIGNE count rate of $C_p^* = 130\text{s}^{-1}$ where the SIGNE rate is equal to 0.28/day. This means that in order to combine these two data sets we should multiply SMM peak photon count rates by a factor of 8 so that they range now from 130 to 4500s^{-1} equivalent SIGNE photon count rates.

There are about two dozen common bursts in the SMM and SIGNE samples. As expected, the ratio C_p^{SIGNE}/C_p^{SMM} for these bursts shows a large dispersion, but its mean value is about 5, which is smaller but in rough agreement with the factor of 8 obtained above. However, because of the longer observational period of SMM (∼560 days) compared to SIGNE (∼440 days) we expect SMM to sample slightly more of the brighter bursts and, therefore, more of the steeper (1.5 slope) portion of the $\log N$-$\log C_p$ curve. As can be seen in Figure 1a this clearly is not the case. The SMM curve is considerably flatter than the SIGNE curve. Because of this discrepancy we cannot combine the SMM data either with SIGNE or BATSE data.

It is, however, important to determine the cause of this discrepancy. It could be due to either the different spectral range or the different integration time of SMM as compared to the other two. The higher energy response of GRS introduces a bias against the detection of GRBs with softer spectra. However, since there does not seem to exist an obvious and strong correlation between C_p and spectral shape (spectral index or hardness ratio) of GRBs we do not expect this bias to affect the slope of the $\log N$-$\log(C_p)$ relation. We therefore conclude that the first of the above two differences cannot be the main source of the flat $\log N$-$\log C_p$ curve of the SMM data, and must attribute the above discrepancy primarily to the longer trigger time scale of the GRS detection scheme. Because the 16s triggering time of GRS is much longer than the duration of a significant fraction of GRBs which have duration ranges from fractions of seconds to many hundreds of seconds[4] we expect a strong bias against bursts with short duration and low peak count rates. It is therefore imperative to take into account the effects of large dispersions observed in the duration and temporal variation of the bursts. As shown by Petrosian, Lee and Azzam in these proceedings this bias flattens the $\log N$-$\log C_p$ distribution and affects the distribution of durations of the bursts observed by BATSE with 1.024s triggering time. These effects are expected to be much more pronounced for the SMM-GRS data with its much longer triggering time of 16s, and therefore can be the cause of the flat curve for SMM.

ACKNOWLEDGMENT

We wish to thank J. Higdon and S. Matz for their analysis of the SMM data. We also wish to thank the SIGNE group at CESR in France and at IKI in Russia for providing us with their data. The work at Stanford was supported by NASA grants NAGW 1976 and NAGW 2290. Kevin Hurley acknowledges support from NASA grant NAG5 1560. Gerald Share acknowledges support from NASA grant DPR W-17808.

REFERENCES

1. V. Petrosian, *ApJ*, **402**, L33 (1993).
2. B. Efron and V. Petrosian, *JASA*, in press. To appear in June 1994 issue.
3. W. J. Azzam, V. Petrosian and E. V. Linder, submitted to *ApJ (Letters)* (1993).
4. C. Kovelioutou, C. A. Meegan, G. J. Fishman, N. P. Bhat, M. S. Briggs, T. M. Koshut, W. S. Paciesas and G. N. Pendleton, *ApJ (Letters)*, **413**, L101 (1993).

DISTRIBUTION OF PEAK COUNTS AND DURATION OF GAMMA-RAY BURSTS

Vahé Petrosian, Theodore T. Lee and Walid J. Azzam
Stanford University, Stanford, CA 94305

ABSTRACT

The procedures generally used for detection of gamma-ray bursts were designed to obtain an unbiased estimate of the distribution of the ratio C_p/C_{lim} of peak to threshold photon count rates. This introduces a bias against detection of weak and rapidly varying or short duration bursts. We demonstrate the effect of this bias on the distributions of C_p and duration of the bursts and show how to correct for this bias and obtain unbiased estimates of these distributions. This correction steepens the $\log N$-$\log C_p$ relation and dramatically increases the number of short duration bursts.

INTRODUCTION

Because of their unusual nature, gamma-ray bursts (GRBs) have provided many new challenges in the design of instruments, in data analysis and in theoretical modeling. The primary reason for these difficulties is due to the fact that they are short duration, highly variable sources, with varied light curves. This has made it difficult to find counterparts, so that we do not know their distances and must rely on the so-called $\log N$-$\log S$ distribution for information on their spatial distribution and luminosity function. The absence of a common light curve, $C(t)$, has made interpretation of this simple test difficult.

To overcome the detection bias due to this and due to the variability of the background photon count rate, instruments such as BATSE have been designed to provide a threshold photon count rate C_{lim} and an average peak count rate $\bar{C}$ for an interval Δt; $\bar{C}(t) = \int_t^{t+\Delta t} C(t)dt/\Delta t$. For a homogeneous and isotropic distribution of sources in a static, Euclidean space (HISE for short), we expect $<V/V_{max}> = <(C_{lim}/\bar{C}_p)^{1.5}> = 0.5$ and a slope of -3/2 for the $\log N$-$\log(C_p/C_{lim})$ curve. BATSE has shown[1] clear deviations from these values which, along with the observed isotropy, has made the cosmological interpretation popular. Aside from negating HISE, the above moment or distribution is of little further use. To make any inference about the spatial distribution of the bursts we need the $\log N$-$\log C_p$ relation free of the biases due to the variable threshold, the long range of durations, and the variety of burst light curves.

In an earlier work[2] and in these proceedings[3,4], we have shown how to account for the bias due to the variability of C_{lim}. Here we discuss an additional bias due to the duration variability. Briefly, this bias arises from the fact that for bursts with duration $T < \Delta t$, the average rate $\bar{C}_p$ is not a good estimate of the true peak photon count rate C_p, but is a measure of the "fluence" $F = \int_T C(t)dt$. $\bar{C}_p = C_p$ only for bursts with $T > \Delta t$ which do not have prominent spiky features with duration $< \Delta t$. Thus, the bursts are partly selected based on the peak rate exceeding the threshold C_{lim} and partly based on their fluence exceeding the threshold fluence $C_{lim} \times \Delta t$. It is this mixing of the selection criteria which introduces the bias we consider here.

ESTIMATION OF DURATION BIAS

Because of the interconnection between duration T and C_p, the proper procedure would require determination of the joint distribution $\psi(C_p, T)$. This is beyond the scope of this paper and is complicated by the fact that the pulse shapes for bursts are variable and characterization by a single parameter T is not sufficient. In what follows we shall assume this simple characterization and make a rough estimate of the above mentioned bias.

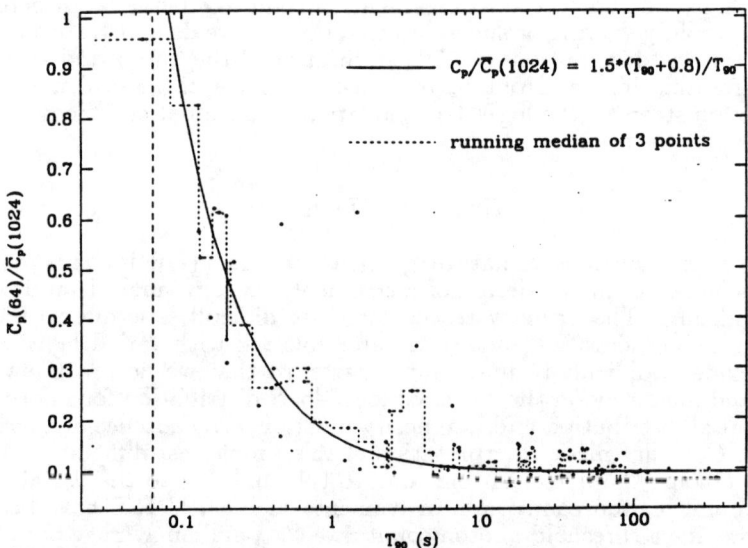

Figure 1. Ratio of average $\bar{C}_p$'s for $\Delta t = 64$ and 1024ms versus duration. The solid line gives an estimate of the ratio $C_p/\bar{C}_p(1024)$.

It is easy to show that when $T < \Delta t$ the true rate $C_p \simeq \bar{C}_p \Delta t/T$ for a variety of pulse shapes, and that $C_p \to \bar{C}_p(1 + \alpha \Delta t/T)$ for $T > \Delta t$, (e. g. $\alpha = 0, 1/2$ and 1 for square, triangular, and semi-exponential pulse shapes, respectively). Because of the finite integration time, we cannot observe C_p directly. This trend can be seen directly in the BATSE data shown in Figure 1, where we plot the ratio $\bar{C}_p(64)/\bar{C}_p(1024)$ (the so-called variability parameter[5] V_{obs}) versus the duration T_{90} defined by the BATSE team[6]. For $T_{90} > 64$ms, $C_p \simeq \bar{C}_p(64)$ so that we expect this ratio to provide a good estimate of $C_p/\bar{C}_p(1024)$ for simple pulse shapes (solid line). There is considerable dispersion around this mean relationship which is due to the fact that the pulse shapes are more complicated. For $T_{90} < 64$ms the $\bar{C}_p(64)$ also becomes underestimated and the ratio flattens out to a constant value (see dotted line). In what follows we shall use the solid line as a measure of the ratio $C_p/\bar{C}_p(1024)$.

CORRECTION PROCEDURE

On Figure 2a we show the observed $(\bar{C}_p(1024), T_{90})$ distribution for all bursts with known $\bar{C}_p, C_{lim}$ and T_{90} from the BATSE 1B data set. Because of the variability of C_{lim}, a three-dimensional representation would be more appropriate, but to simplify the discussion we cut off the data at the largest value of C_{lim} as shown by the solid horizontal line. Assuming the relation depicted by the solid line in Figure 1, we convert the rates $\bar{C}_p(1024)$ and the horizontal truncation line to C_p and obtain the truncated distribution shown in Figure 2b, where it is clear that we do not have a fair representation of the short duration, weak (low C_p) bursts. From this data we can now obtain the distribution $\psi(C_p, T_{90})$. The first task is to determine if C_p and T_{90} are statistically independent. Using a recently developed method[7] applicable to truncated data we find that there is a slight anticorrelation between C_p and T_{90}, but it is only of marginal significance so that we can write $\psi(C_p, T_{90}) = n(C_p)h(T_{90})$ and proceed to determine the two univariate distributions using the same method[2] we have used to account for the variation in C_{lim}.

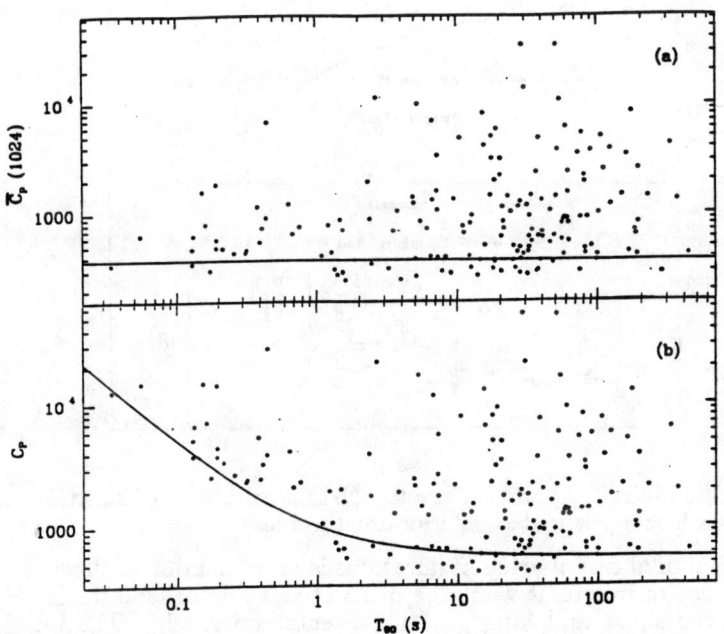

Figure 2 a) Distribution of the average peak photon count rates versus duration of all bursts in the BATSE 1B catalogue for which $\bar{C}_p(1024), T_{90}$, and C_{lim} are known. b) Same as (a) but for the true rates C_p.

Distribution of Duration. Figure 3a shows the cumulative distribution $H(T_{90}) = \int_{T_{90}}^{\infty} h(x)dx$ obtained with and without the correction for the bias. As expected, there are more bursts with short durations than is obtained if this bias were ignored. In Figure 3b we show the differential distribution $h(T_{90})$ obtained by smoothing and differentiating[8] $H(T_{90})$. Evidently the case for two populations[6] separated at $T_{90} = 2s$ is further strengthened.

Distribution of Peak Count Rates. Figure 4a shows the cumulative distributions $N(C_p) = \int_{C_p}^{\infty} n(x)dx$ of the peak rates C_p obtained without and with the correction for the bias. As expected, the correction for this bias steepens the log$N$-logC_p curve at low values of C_p. This change is not sufficiently large to allow a galactic distribution as a viable scenario. However, it makes the constrains against a halo population (see e. g., Hartmann in these proceedings) less severe. Furthermore, our estimation of this effect is very rough and uncertain because (as is evident in Fig. 3b) we know very little about very short duration ($T < 64$ ms) events from the BATSE catalog. A large population of such bursts could change all of these distributions. In addition, we have included neither the effect of this bias on complex spiky bursts nor the effect due to the so-called "Meegan bias"[5] (which probably would tend to further steepen the distribution). How these would change the current picture remains to be seen.

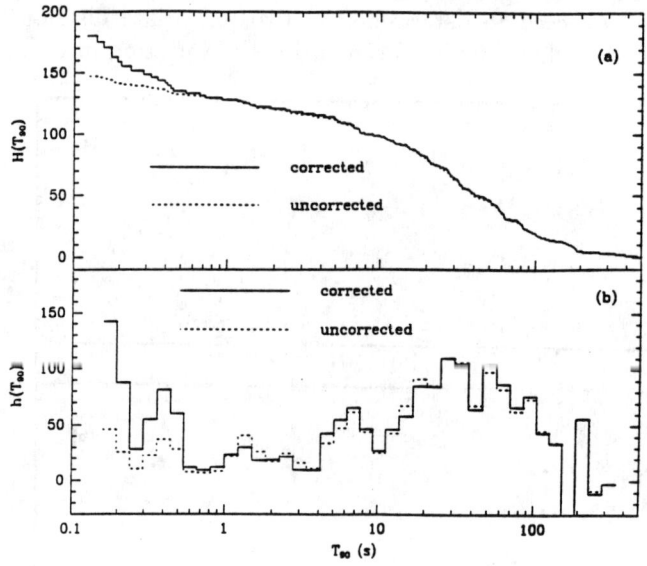

Figure 3. Cumulative (a) and differential (b) distributions of durations $H(T_{90})$ and $h(T_{90})$ with and without correction for the bias.

If the bimodal distribution of durations is an indication of the existence of two populations of bursts, it would be instructive to determine the log$N$-logC_p relation for the short and long duration events separately. The log$N$-logC_p relation for the short duration bursts ($T_{90} < 2$s, which are more strongly affected by the bias, is shown in Figure 4b. As evident this small subset of bursts is not in contradiction with the HISE hypothesis, but because there are only 23 (of 168) data points in this subset the significance of the result cannot be assessed.

SUMMARY AND CONCLUSIONS

We have discussed the effects of a selection bias when the integration time Δt is longer than the total duration of GRBs, and have shown how this affects the determination of the true distribution of peak photon count rates C_p and durations T. We have demonstrated this effect quantitatively using a subsample of bursts in the BATSE 1B catalog. The result of this analysis shows that when this bias is corrected for, the $\log N$-$\log C_p$ curve steepens and the number of short duration bursts dramatically increases. We also show the tantalizing result of $N(C_p) \propto C_p^{-1.5}$ for a small but complete sample of GRBs with $T_{90} < 2$s.

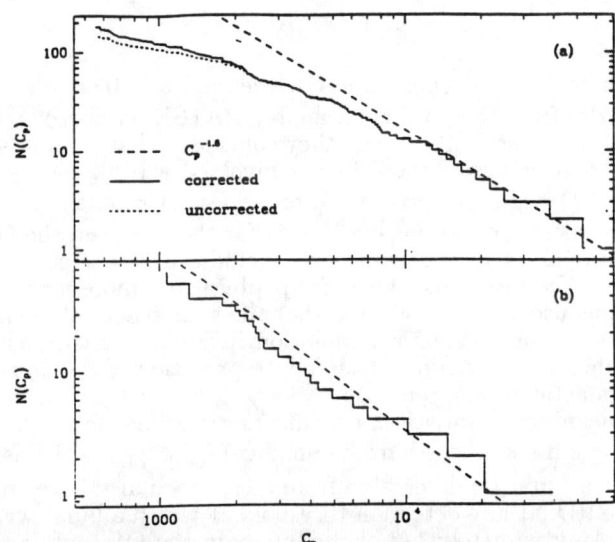

Figure 4 a) $\log N$-$\log C_p$ relation without (from Fig. 2a) and with (from Fig. 2b) the correction for the bias. **b)** Same as a but for bursts with $T_{90} < 2$ s.

More data and more detailed analysis is required before the significance of these results on the nature and distances of GRBs can be assessed. In addition to the dispersion in duration we must also consider the effect of varying pulse shapes. In particular the effects of the spiky character of long duration bursts must be included in this kind of analysis. Whether the correction discussed here or the additional ones mentioned above would make a combination halo/disk model viable remains to be seen.

ACKNOWLEDGMENT

We would like to thank G. Share for some stimulating discussions. This work was supported by NASA grants NAGW 1976 and NAGW 2290.

REFERENCES

1. Meegan, C. A. et. al, Nature **355**, 143 (1992).
2. Petrosian, V., Ap. J. Letters **402**, L33 (1993).
3. Petrosian, V., Azzam, W. J., & Meegan, C. A., these proceedings.
4. Petrosian, V., Azzam, W. J., Hurley, K. & Share, G. H., these proceedings.
5. Lamb, D. Q., Graziani, C., & Smith, I. A., Ap. J. Letters **413**, L11 (1993).
6. Koveliotou, K. et. al, Ap. J.Letters **413**, L101 (1993).
7. Efron, B. & Petrosian, V., Ap. J **399**, 345 (1992).
8. Lee, T. T., Petrosian, V., & McTiernan, J. M., Ap. J. **412**, 401 (1993).

ON THE GALACTIC DISTRIBUTION OF GAMMA-RAY BURSTS

Robert Rutledge and Walter H. G. Lewin

37-624B, Massachusetts Institute of Technology, Cambridge, MA 02139

SUMMARY

Quashnock & Lamb[1,2] (hereafter QL) defined a sub-sample of Gamma Ray Bursts (GRBs) from the publicly availably BATSE database[3] which shows clumping toward the galactic plane, and they concluded that all GRBs are galactic in origin. The selection of these bursts involved a peak count-rate (B, in counts (1024 ms)$^{-1}$) which is uncorrected for aspect. Using, as limits, the corresponding peak fluxes (in photons cm^{-2} s^{-1}) for the bursts in the QL sample, we find an additional 24 bursts, which we include in a new sample (Sample 2). We assert that the peak flux of a burst is physically more meaningful than peak count-rate, as used by QL. We find that the significance of deviation from isotropy due to a possible galactic population in Sample 2 is much less than QL's sample, which does not support QL's interpretation of the anisotropies as being due to a galactic population.

To make meaningful statistical statements regarding isotropy, burst samples must have peak fluxes above a minimum flux ($I^{1024}_{peak,\ LL}$), which is set by the requirement that a burst be detectable from any direction (above the horizon) with respect to GRO, at any detection threshold at which a burst was observed in that sample. Approximately 1/3 of the bursts in the QL sample have fluxes below $I^{1024}_{peak,\ LL}$. We split our Sample 2 into two sub-samples (Sample 3 and Sample 4) which have fluxes below and above $I^{1024}_{peak,\ LL}$, respectively. We find that Sample 4 has a marginal (2.6σ) deviation from isotropy, which we consider insufficient to justify the claim that GRBs are galactic in origin.

FLUX AS A SELECTION CRITERION

The selection criterion used by QL is B (in counts (1024 ms)$^{-1}$, uncorrected for aspect), the peak detector counts above background observed by the second most brightly illuminated BATSE Large Area Detector; this detector must have its normal directed between 35°-70° from the burst direction. Since the angular response of the BATSE LAD is $\sim \cos\theta$, (flatter for energies > 300 keV[4]) the detector counts could be different by a factor of 2.4 (= $\cos 35°/\cos 70°$) for bursts of identical flux, due exclusively to the orientation of CGRO.

The upper limits on systematic errors in the peak fluxes are 10-15 per cent[6], which is much lower than the known systematic errors, due to angular response, in peak counts as used by as QL.

In Figure 1, we show the ratio of B (in counts (1024 ms)$^{-1}$, which contains detector response) to I^{1024}_{peak} (in photons cm^{-2} s^{-1}, with detector response modeled out) to illustrate the effect of detector response on burst flux.

Since peak flux (in photons cm^{-2} s^{-1}) is more physically meaningful than B, we use the peak flux in a 1024ms bin (I^{1024}_{peak}), provided in the BATSE public

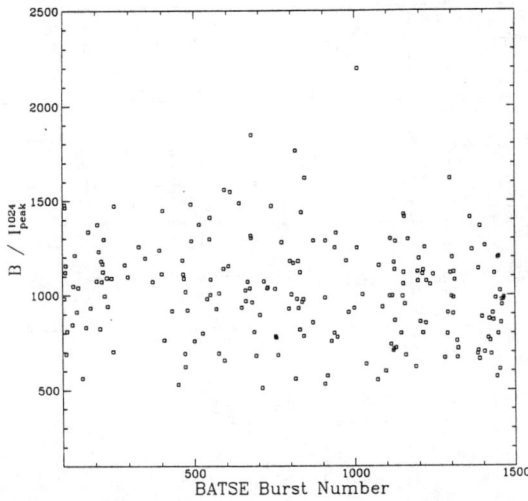

Figure 1. The ratio of B (in counts $(1024\text{ ms})^{-1}$) to peak flux vs. burst number.

database, as a selection criteria.

SAMPLES USED IN THIS STUDY

Figure 2 depicts B vs. Peak Flux for the 4 samples in this study.

Sample 1 (Fig. 2a) is identical to the burst sample used by QL. Using the B and $V(=(C^{64})_{\max}/(C^{1024})_{\max}$, the ratio of peak counts above background in a 64ms bin to peak counts above background in a 1024ms bin) values kindly provided (Carlo Graziani, private communication), bursts were selected which had $465 \leq B \leq 1169$ counts $(1024\text{ ms})^{-1}$, and $\log(V) < -0.8$. There are 55 bursts which meet these criteria.

Sample 2 (Fig. 2b) includes bursts with peak fluxes $0.396 \leq I_{\text{peak}} \leq 1.296$ photons cm^{-2} s^{-1}, with $\delta\theta_{\text{pos}} \leq 10°.77$, and with $\log(V) \leq -0.8$. $\delta\theta_{\text{pos}}$ is the total positional error box. We find 79 bursts meeting these criteria.

Sample 3 (Fig. 2c) includes bursts with peak fluxes $0.396 \leq I_{\text{peak}} \leq 0.674$ photons cm^{-2} s^{-1}, with $\delta\theta_{\text{pos}} \leq 10°.77$, and with $\log(V) \leq -0.8$. Sample 3 is a subset of Sample 2. We find 40 bursts meeting these criteria.

Sample 4 (Fig. 2d) includes bursts with peak fluxes $0.674 \leq I_{\text{peak}} \leq 1.296$ photons cm^{-2} s^{-1}, with $\delta\theta_{\text{pos}} \leq 10°.77$, and with $\log(V) \leq -0.8$. Here we have used the $I^{1024}_{\text{peak, LL}}$ as our lower flux limit. We find 39 bursts meeting this criteria. Sample 4 is a subset of Sample 2, and is complimentary to Sample 3.

ESTIMATING THE SIGNIFICANCE OF ANISOTROPY

Occultation by the earth results in unequal sky coverage at different posi-

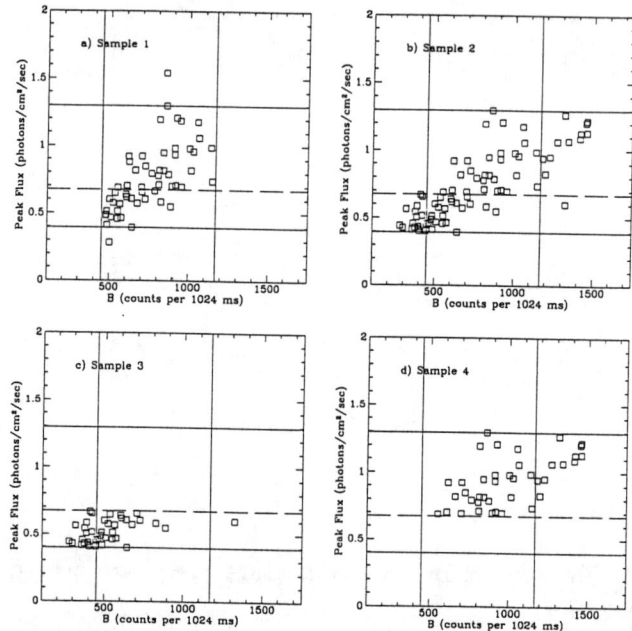

Figure 2. B vs. Peak Flux for the 4 samples used in this study.

tions on the celestial sphere, which must be taken into account when estimating the significance of observed angular anisotropy of GRB positions. The publicly available BATSE Sky Exposure Table may be applied only to bursts which can be detected by BATSE from any position on the sky which is not occulted by the earth, i.e. which have a flux above a minimum threshold – $I^{1024}_{peak, LL}$. Rutledge & Lewin[5] find this minimum flux to be roughly 0.674 photons cm^{-2} s^{-1}. Alternatively, one could find a minimum flux value from the BATSE Trigger efficiency table in the public database. A burst of flux 0.674 photons cm^{-2} s^{-1} has a trigger efficiency >99.7%. Approximately 1/3 of the bursts in the QL sample (our Sample 1) have fluxes below this level, and therefore the significance of deviation of that sample calculated by QL may not apply. Our Sample 2 and Sample 3 also have bursts which are below this flux level; we may estimate only the significance (Q) of the deviation from anisotropy in our Sample 4, which we do using a Monte Carlo method described in Rutledge & Lewin[5].

RESULTS AND CONCLUSIONS

The results of this analysis are presented in Table I. In going from a "brightness"-selected sample (Sample 1) to a flux-selected sample (Sample 2) within the flux limits of the "brightness" sample, we find that the non-Gaussian significances of $<\cos l>$ and $<\sin^2 b> -1/3$ drop considerably (from 2.9σ and 4.0σ, to 1.8σ and 2.4σ). This does not support QL's interpretation of an anisotropy in the GRB distribution as due to a galactic origin. Because the flux limits used in this sample include bursts with fluxes well below the flux completion limits of the BATSE database, it is not straightforward to estimate

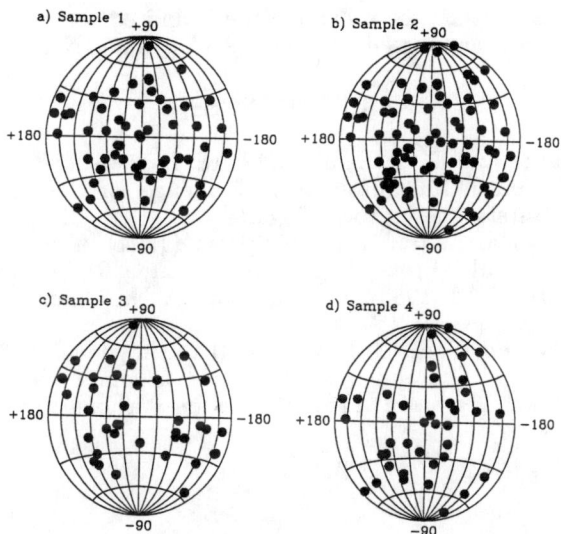

Figure 3. Maps of the 4 samples used in this study, in galactic coordinates.

Table I. Results

Sample	Flux Limits ($\frac{photons}{cm^2 \, s}$)	Number of Bursts	$<\cos l>$[a]	$<\sin^2 b> - \frac{1}{3}$[a]	Q
QL Result	-	55	$+0.230 \pm 0.078$ (2.9σ)	-0.119 ± 0.040 (3.0σ)	_[b]
1	-	55	$+0.272 \pm 0.093$ (2.9σ)	-0.119 ± 0.030 (4.0σ)	_[b]
2	$0.396 \leq I^{1024}_{peak} \leq 1.296$	79	$+0.141 \pm 0.077$ (1.8σ)	-0.070 ± 0.029 (2.4σ)	_[b]
3	$0.396 \leq I^{1024}_{peak} \leq 0.674$	40	$+0.233 \pm 0.111$ (2.1σ)	-0.035 ± 0.047 (0.8σ)	_[b]
4	$0.674 \leq I^{1024}_{peak} \leq 1.296$	39	$+0.048 \pm 0.104$ (0.5σ)	-0.106 ± 0.035 (3.0σ)	0.0090

[a] Standard deviations given parenthetically are only roughly Gaussian (see text)
[b] Probabilities not calculated (see text)

significances of this measurement in the absence of directionally specific (i.e., RA and dec.) flux-detection efficiencies.

While the probability of producing the observed anisotropies from a purely isotropic distribution of bursts on the sky is small (0.9%), we feel that it is not small enough to justify the claim that GRBs are of galactic origin.

It is notable that bursts selected in the brightness range $490 < B < 1250$ from 480 bursts observed by BATSE not included in the first BATSE GRB catalog have no significant deviation from an angularly isotropic distribution.[7]

ACKNOWLEDGEMENTS

The authors thank Chyrssa Kouveliotou and Chip Meegan for their comments, and Carlo Graziani for providing the B and corrected V values for each burst. This work was supported through NASA grant #NAGW-3234.

REFERENCES

1. J. M, Quashnock and D. Q. Lamb, MNRAS, in press, (1993).
2. D. Q. Lamb and J. M. Quashnock, "Evidence for the Galactic Origin of Gamma Ray Bursts, these proceedings.
3. G. J. Fishman et al. , Ap. J. Supp., in press, (1994).
4. G. J. Fishman et al. "Proceedings of the Gamma Ray Observatory Science Workshop April 10-12, 1989", ed., Johnson, W.N., Greenbelt, Maryland, p. 2-39
5. R. E. Rutledge and W. H. G. Lewin, MNRAS, in press (1993).
6. C. Kouveliotou and G. Pendleton, 1993, private communication.
7. C. Meegan, "Two Years of BATSE Burst Observations", these proceedings.

COMMENTS ON C_{max}/C_{min} DISTRIBUTIONS

I. A. Smith

Dept. Space Physics & Astronomy, Rice University, Houston, TX 77251-1892

ABSTRACT

Using Monte Carlo techniques, I show that there can be a large variation in the shape of the C_{max}/C_{min} distribution, and the magnitude of its values, for the brightest ~ 10 bursts; statements based on the brightnesses of the brightest bursts in a population must be made with extreme caution. I also show the effect of a time-varying C_{min} on the C_{max}/C_{min} distribution, using an extended Galactic halo model as an example. To fit the BATSE C_{max}/C_{min} distribution including a varying C_{min} requires a larger observing distance (relative to the scale-height of the halo) than for a constant C_{min}; however, the observations can still be fit using the halo models.

INTRODUCTION

The brightness distributions of gamma-ray bursts are useful for testing whether a population of bursts is homogeneous or not. While it is essential that any postulated model of the sources must be able to fit these observed distributions, it is dangerous to try to extract too much information from them. Although the burst sources may have a relatively smooth distribution in space, only a small fraction of them burst in a given time, so the distances to the brightest ones are subject to a significant amount of random fluctuation. Simulating this using Monte Carlo techniques, I show here that there can be a large variation in the slope of the C_{max}/C_{min} curve, and the magnitude of its values, for the brightest bursts.

One of the problems with doing a careful fit of BATSE's C_{max}/C_{min} distribution is that C_{min} varies with time; this complicated time-variation of the detector sensitivity has to be convolved with the hypothesized spatial distribution of the sources to see if this can fit the observed C_{max}/C_{min} distribution. The effect of BATSE's varying C_{min} on the V/V_{max} distribution was discussed by Band[1] and Hartmann and The[2]. Here I show the effect of a varying C_{min} on the C_{max}/C_{min} distribution for extended Galactic halo models.

MONTE CARLO SIMULATIONS

For illustrative purposes only, assume the gamma-ray burst sources are standard candles distributed homogeneously in space. The expected C_{max}/C_{min} distribution therefore has a $-3/2$ power law. Monte Carlo techniques are used to randomly pick bursts from this homogeneous source distribution; this mimics what happens for a "real" burst detector. In this section, C_{min} remains constant.

Figure 1 shows the C_{max}/C_{min} distributions for 50 bursts for 10 and 100 Monte Carlo runs. Figure 2 plots 200 bursts for 10 and 100 Monte Carlo runs. The "expected" $-3/2$ power law is shown as a dashed line.

It can be seen that there is a large fluctuation in the shape and magnitudes of the Monte Carlo runs. In particular, the slope of the curve for $N \lesssim 10$ is

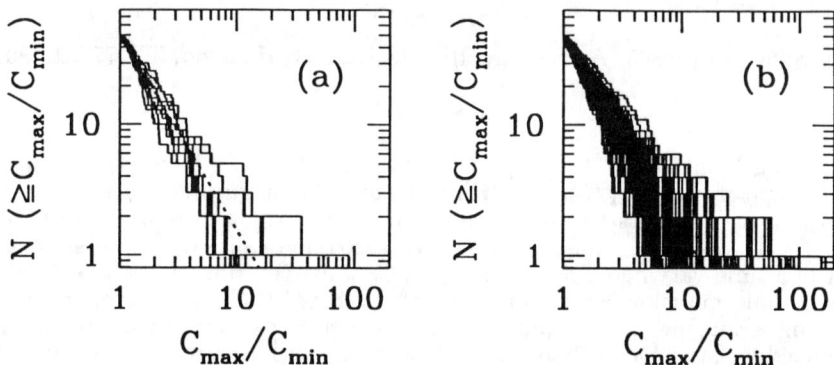

Figure 1. C_{max}/C_{min} distributions for 50 bursts for (a) 10 and (b) 100 Monte Carlo runs.

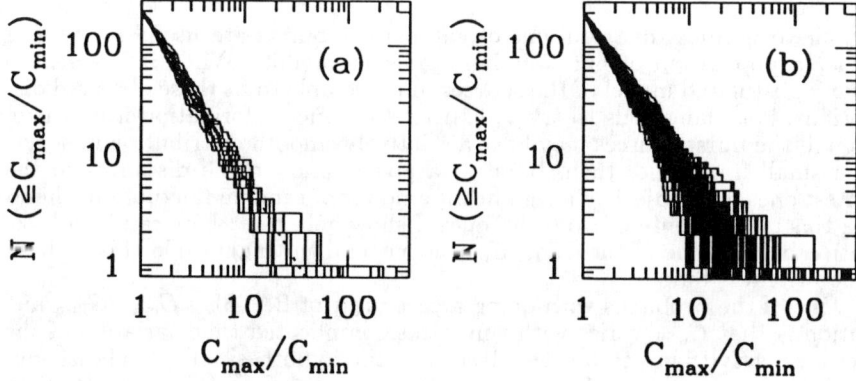

Figure 2. C_{max}/C_{min} distributions for 200 bursts for (a) 10 and (b) 100 Monte Carlo runs.

quite uncertain. In reality, we are seeing just one particular "run" when we look at an observed C_{max}/C_{min} distribution; conclusions based on the shape of this one curve should be made with extreme caution.

The brightness of the brightest burst in these distributions varies by a factor $\lesssim 100$ at the 3σ level. Therefore, it is not possible to make strong claims about the brightness of the brightest bursts in any population of gamma-ray bursts. In particular, comparisons between the brightnesses of the brightest bursts in sub-populations of the bursts must be made with extreme caution.

At $N = 10$, the range of burst brightnesses is much smaller than at $N = 1$; the range is a factor ~ 3 at the 3σ level. Thus one can make more reliable statements about the shape of the C_{max}/C_{min} curve by extrapolating the curve from $N > 10$; when fitting the observed burst brightness distribution, one must concentrate on the fainter bursts. However, one must also be careful when fitting the C_{max}/C_{min} curve near to $C_{max}/C_{min} = 1$, because threshold and sensitivity effects cause the observed distribution to be inaccurate.

For a uniform distribution of N sources in space, one expects to get[3]

$< V/V_{\max} > = 0.5 \pm (12N)^{-1/2}$; this is confirmed using the Monte Carlo simulations.

$C_{\min}$ VARIATION

To show the effect of a varying $C_{\min}$, a simple exponential halo model of the gamma-ray burst sources is used: the number density of the sources has the form $n(R) = n_h e^{-R/\bar{r}}$, where R is the distance of the source from the Galactic Center, and n_h is a constant. The sources are assumed to be standard candles distributed spherically symmetrically about the Galactic Center, and the solar system is displaced a distance $R_0 = 8.5$ kpc from the Galactic Center.

As an example, take $\bar{r} = 50$ kpc, and the distance to the faintest source that can be detected by BATSE to be $D_h = 170$ kpc. This gives $< V/V_{\max} > = 0.300$ (for $C_{\min}$ constant), $< \cos\theta > = 0.056$, and $< \sin^2 b > = 0.332$. These are consistent with the BATSE observations[4] of $< V/V_{\max} > = 0.324 \pm 0.016$ for 336 bursts, and $< \cos\theta > = 0.048 \pm 0.027$, and $< \sin^2 b > = 0.320 \pm 0.014$ for 447 bursts. In the absence of $C_{\min}$ variations, this halo gives the dotted $C_{\max}/C_{\min}$ curve shown in Figure 3. It is a reasonable fit to the BATSE $C_{\max}/C_{\min}$ curve for 193 bursts used in Smith and Lamb[5] (solid curve).

A simple model of the $C_{\min}$ variation is used here to illustrate the effect this produces. Write $C_{\max} = A/r^2$ and $C_{\min} = fA/D_h^2$, where f is a number between $f = 1$ and $f = f_{\max}$ (that varies with time), and D_h is the maximum distance at which a halo burst can be detected when $f = 1$. This gives $C_{\max}/C_{\min} = D_h^2/fr^2$. In reality, f has a complicated time dependence, but as a simple example, it is assumed that f varies uniformly between $f = 1$ and $f = f_{\max}$, so that the probability that f is between f and $f + df$ is $P(f)df = df/(f_{\max} - 1)$. The BATSE detection threshold varies by at least a factor of two[1].

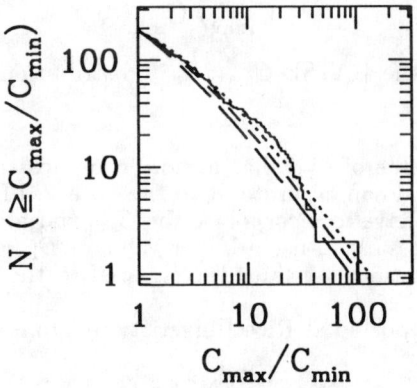

Figure 3. $C_{\max}/C_{\min}$ distributions for different $f_{\max}$ with $\bar{r} = 50$ kpc and $D_h = 170$ kpc fixed: $f_{\max} = 1$ (dotted curve, no $C_{\min}$ variation), $f_{\max} = 2$ (short dashed curve), and $f_{\max} = 5$ (long dashed curve).

In Figure 3, $\bar{r} = 50$ kpc and $D_h = 170$ kpc are kept fixed, and the $C_{\max}/C_{\min}$ distributions are plotted for different $f_{\max}$. It can be seen that as $f_{\max}$ rises (i.e. increasing $C_{\min}$ variation), the $C_{\max}/C_{\min}$ curves lie increasingly below the $f_{\max} = 1$ curve. For $f_{\max} > 1$, the brightest bursts are dimmed (relative to the case with $C_{\min}$ constant), because of the larger average $C_{\min}$.

The solid curve in Figure 4 uses $\bar{r} = 50$ kpc, $D_h = 170$ kpc, and $f_{\max} = 1$, which are the same values as the fit to the BATSE data in Figure 3 (dotted curve). For the dot-short dash curve in Figure 4, D_h was increased to 200 kpc (keeping $\bar{r} = 50$ kpc, $f_{\max} = 1$, i.e. this curve has no $C_{\min}$ variation). Applying a $C_{\min}$ variation with $f_{\max} = 2$ to this latter curve produces the short dashed curve, which is very close to the BATSE fit curve. Similarly, the dot-long dashed curve in Figure 4 ($D_h = 250$ kpc, $\bar{r} = 50$ kpc, $f_{\max} = 1$) transforms into the long dashed curve when $f_{\max} = 5$, which is also very close to the BATSE fit curve. This shows that the observed BATSE $C_{\max}/C_{\min}$ distribution can still be fit by the halo models when a $C_{\min}$ variation is included. To fit the observed $C_{\max}/C_{\min}$ distribution when $C_{\min}$ varies with time, one must sample further into the burst spatial distribution (relative to the scale-height of the halo) than if $C_{\min}$ was constant. This means that the value of $D_h/\bar{r}$ required to fit the BATSE observations must be larger for $f_{\max} > 1$ than would be inferred for the case with no $C_{\min}$ variations[1].

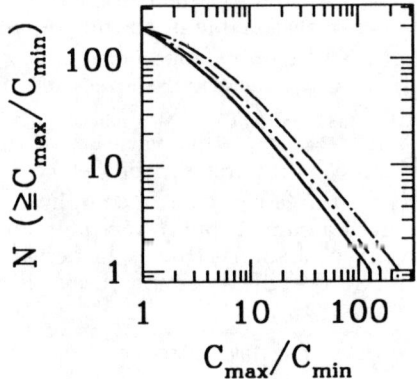

Figure 4. Fits to the BATSE $C_{\max}/C_{\min}$ distribution for different $f_{\max}$.

In practice, the value of $C_{\min}$ has a more complicated time dependence than was used here. To do an accurate fit to the observed BATSE $C_{\max}/C_{\min}$ distribution, one would have to incorporate the $C_{\min}$ variation carefully. However, it will still be necessary to use a larger value of $D_h/\bar{r}$ to fit the BATSE curve when a $C_{\min}$ variation is included, compared to the case of a constant $C_{\min}$.

This work was supported at Rice University by grant NAG 5-1515.

REFERENCES

1. D. Band, Ap. J. (Letters) **400**, L63 (1992).
2. D. H. Hartmann and L.-S. The, Ap. Space Sci. **201**, 347 (1993).
3. M. Schmidt, J. C. Higdon, and G. Hueter, Ap. J. (Letters) **329**, L85 (1988).
4. C. A. Meegan et al., in Compton Gamma-Ray Observatory (AIP, New York, 1993), p. 681.
5. I. A. Smith and D. Q. Lamb, Ap. J. (Letters) **410**, L23 (1993).

EVIDENCE THAT GAMMA-RAY BURST SOURCES REPEAT

J. M. Quashnock, D. Q. Lamb

Dept. of Astronomy and Astrophysics, University of Chicago, Chicago, IL 60637

ABSTRACT

We investigate clustering in the angular distribution of the 260 γ-ray bursts in the publicly available BATSE catalogue, using a nearest neighbour analysis. We find that while all 260 bursts are only modestly clustered (Q-value = 1.8×10^{-2}), the 202 bursts in this sample for which the statistical error in their locations is < 9° are significantly clustered on an angular scale $\approx 5°$ (Q-value = 2.5×10^{-4}). We also find a significant correlation between bright type I bursts and faint type I and type II bursts on an angular scale $\approx 5°$ (Q-value = 4.0×10^{-3}). This angular scale is smaller than the typical error in burst locations of 6.8°, suggesting multiple recurrences from individual sources. We conclude that "classical" γ-ray burst sources repeat on a time scale of months, and that many faint type I and II bursts come from the sources of bright type I bursts. Following the work of Narayan and Piran, we find a slightly positive correlation function on angular scales $\theta \lesssim 5°$ and $\theta \gtrsim 170°$, and a slight excess (Q-value = 0.14) of farthest neighbours in the aforementioned sample of 202 bursts. We show that this excess disappears when those medium type I bursts which we earlier found to be significantly concentrated towards the Galactic Center and in the Galactic Plane are removed. We conclude that the positive antipodal correlations in the BATSE catalogue are not due to some unknown systematic effect, but rather to a real physical anisotropy in the catalogue, and that they support our claim that γ-ray bursts are Galactic in origin.

INTRODUCTION

The "soft γ-ray repeaters" SGR 0526-66, SGR 1806-20, and SGR 1900+14 have been observed to produce more than 100 bursts. The bursts typically have short durations ($t_{\rm dur} \approx 250$ ms) and soft spectra (characteristic energies $E \approx 30$ keV), and are sufficiently different from "classical" γ-ray bursts that these three sources are thought to constitute a separate class.[1,2,3] In contrast, no source of a "classical" γ-ray burst has been found to repeat during nearly twenty-five years of observations. Assuming that γ-ray bursts are standard candles, Schaefer and Cline[4] estimate a recurrence time scale $t_{\rm recur} \gtrsim 10$ yrs from analysis of the locations of 89 bright bursts with relatively small error boxes, and Atteia et al.[5] derive a 3σ lower limit of $t_{\rm recur} > 8$ yrs from analysis of the locations of 84 bright bursts with error boxes determined by the Interplanetary Network. However, both limits become $t_{\rm recur} \gtrsim$ a few months if the burst luminosity function is broad.

Analysis of clustering in the angular distribution of the γ-ray bursts can indicate whether or not individual burst sources repeat. Here we investigate clustering in the publicly available BATSE catalogue[6] using a nearest neighbour analysis and the measures of burst brightness $B = (\bar{C}^{1024})_{\rm max}$ and short time scale ($\lesssim 0.3$ s) variability $V = (\bar{C}^{64})_{\rm max}/(\bar{C}^{1024})_{\rm max}$, which we introduced earlier.[7] The quantities $(\bar{C}^{64})_{\rm max}$ and $(\bar{C}^{1024})_{\rm max}$ are the expected maximum

© 1994 American Institute of Physics

Evidence that Gamma-Ray Burst Sources Repeat

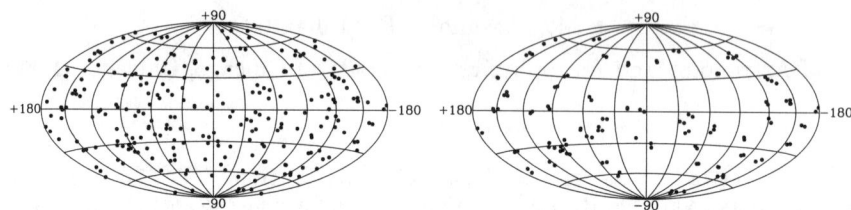

Figure 1. Locations on the sky of the 260 bursts in the publicly available BATSE catalogue (left panel), and of those bursts that lie within 5° of each other (right panel).

number of counts in 64 ms and in 1024 ms, respectively. We have reported much of this work elsewhere.[8]

ANALYSIS

In earlier studies,[7,9,10] we presented evidence for two distinct morphological classes of γ-ray bursts (see also Kouveliotou et al.[11]). Type I bursts (comprising $\approx 80\%$ of the bursts) are smoother ($\log V \leq -0.8$) on short time scales ($\lesssim 0.3$ s), both faint and bright, longer, and have softer spectra. Type II bursts (comprising $\approx 20\%$ of the bursts) are more variable ($\log V > -0.8$) on short time scales ($\lesssim 0.3$ s), faint, shorter, and have harder spectra.

Fig. 1 (left panel) shows the locations on the sky of the 260 bursts in the publicly available BATSE catalogue.[6] Because of the relatively small number of bursts, weak clustering is detectable only on angular scales $\gtrsim 10°$.[12] Yet the eye is drawn to a number of places on the sky where two or more bursts lie very close together, suggesting strong (nonlinear) clustering and hence multiple recurrences from individual sources. Fig. 1 (right panel) shows the locations on the sky of those bursts that lie within 5° (see below) of each other. Indeed, several clumps of three or more bursts are apparent.

Standard analysis techniques for investigating the clustering of objects on the sky include decomposition of the distribution of objects in terms of spherical harmonics[13] and calculation of the two-point angular correlation function $w(\theta)$.[14] However, decomposition into spherical harmonics is relatively poor at detecting sharp features (such as nonlinear clustering) because the power is spread over many high harmonics. Calculation of the two-point angular correlation function is also relatively poor at detecting nonlinear clustering because it does not include higher order correlations; in addition, it spreads the clustering over a larger angular scale because it includes all the angular separations between bursts in the cluster, averaged over all pairs of bursts.

Here we consider a related statistic, the nearest neighbour separation, which is particularly sensitive to nonlinear clustering, and thus to multiple recurrences from individual sources, because it depends sensitively on all higher order correlations. Every burst has a nearest neighbour; indeed, for an isotropic distribution, the probability that a nearest neighbour is found within an angle θ from a randomly chosen object is[15]

$$P(\theta) = 1 - [(1 + \cos\theta)/2]^N, \qquad (1)$$

where N is the total number of neighbouring bursts on the sky. Monte Carlo calculations show that for $N \approx 200$ the effect of the BATSE sky exposure map is small (of order 10^{-3}) on the angular scales ($\theta \lesssim 10°$) of interest in this work, and we therefore do not include it in $P(\theta)$.

It is important to note that for an isotropic distribution of burst sources, statistical and systematic errors in the source locations, both known and unknown, have no effect on $P(\theta)$, provided that the errors are independent of location. More intuitively, a smeared isotropic distribution is also isotropic, provided that the smearing is independent of position on the sky. Thus uncertainties in the source locations have no effect on the significance of any observed deviations from isotropy.

We can show this as follows. Let $\rho_o(\vec{x})$ be the distribution of observed positions on the sky; $\rho_s(\vec{y})$, the distribution of source positions; $E(\vec{y})$, the sky exposure; $G(\vec{x} - \vec{y}; \sigma)$, the smearing function when the location uncertainty is σ; $f(\sigma, \vec{y})$, the distribution of uncertainties on the sky. Then $\rho_o(\vec{x})$ is the convolution of all these terms:

$$\rho_o(\vec{x}) \propto \int d^2y\, d\sigma\, \rho_s(\vec{y}) E(\vec{y}) G(\vec{x} - \vec{y}; \sigma) f(\sigma, \vec{y}) \,. \qquad (2)$$

Assuming that the source distribution is isotropic, $\rho_s(\vec{y}) = c$; the sky exposure is uniform, $E(\vec{y}) = c'$; and the distribution of uncertainties is independent of position, $f(\sigma, \vec{y}) = f(\sigma)$; Equation (2) becomes

$$\rho_o(\vec{x}) \propto \int d\sigma\, f(\sigma) \int d^2y\, G(\vec{x} - \vec{y}; \sigma) = \text{constant} \,. \qquad (3)$$

Thus, for an isotropic source distribution, the observed distribution is also isotropic, and the observed distribution of nearest neighbour separations $P(\theta)$ is given by Equation (1).

We investigate the clustering of γ-ray bursts by calculating the cumulative distribution of separations of nearest neighbours within, as well as between various samples of bursts, and comparing it to that expected for an isotropic distribution with the same number of bursts, $P(\theta)$, given by Equation (1). A condition for the validity of the Kolmogorov-Smirnov test is that all the separations be independent. This condition is not satisfied for separations of nearest neighbours within a sample, since the nearest neighbour of a given burst may often have the given burst as its own nearest neighbour. In this case, we evaluate the significance of the largest deviation D of the cumulative distribution of nearest neighbour separations from $P(\theta)$ by using Monte Carlo simulations. We determine the fraction of simulations which exhibit positive or negative values of D that equal or exceed the magnitude of the observed D. We evaluate the significance of the largest deviation D of the cumulative distribution of separations of nearest neighbours between two samples from $P(\theta)$ by using the Kolmogorov-Smirnov test,[16] since in this case the nearest neighbour separations are independent.

RESULTS

Fig. 2 (left panel) shows the cumulative nearest neighbour distribution for all bursts in the publicly available BATSE catalogue. Also shown is the

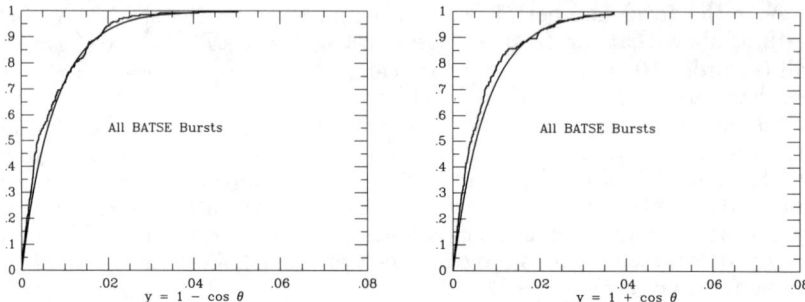

Figure 2. Nearest (left panel) and farthest (right panel) neighbour distributions for all bursts in the publicly available BATSE catalogue. Also shown is the expected distribution, given by Equation (1), for an isotropic distribution with the same total number of bursts.

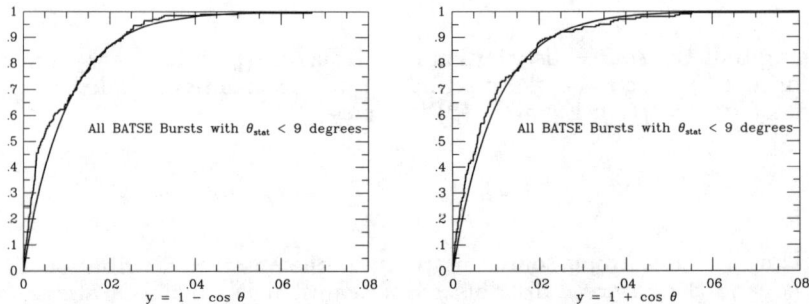

Figure 3. Nearest (left panel) and farthest (right panel) neighbour distributions for all bursts with statistical errors $< 9°$.

expected distribution, given by Equation (1), for an isotropic distribution with the same total number of bursts. Table 1 gives the significance of the maximum deviation D of each from that expected for an isotropic source distribution with the same number of bursts.

Fig. 2 (left panel) shows that the sample of all BATSE bursts has more nearest neighbours at small separations than expected for an isotropic distribution (Q-value $= 1.8 \times 10^{-2}$). If this excess is due to repeating, one expects its significance to increase as bursts with large statistical errors in their locations are removed from the sample, and then to decrease as bursts with smaller statistical errors are removed. This is the case. The maximum in the significance is 1.1×10^{-4} and occurs at a cutoff of $\approx 9°$ in the statistical error. Fig. 3 (left panel) shows the resulting cumulative nearest neighbour distribution. The corresponding total (statistical plus systematic) error of $\approx 10°$ is larger than the angular scale $\approx 5°$ of the excess of nearest neighbour separations. This suggests multiple recurrences of individual sources, since bursts with total errors larger than the angular scale of the excess then contribute importantly to the signal (see below). Using Monte Carlo simulations to take into account having chosen a cutoff in the statistical error,[8] we find a corrected significance of 2.5×10^{-4}.

Following the work of Narayan and Piran,[17] we also examine the farthest neighbour distribution, for the two aforementioned samples. Fig. 2 (right panel) and Fig. 3 (right panel) show the cumulative farthest neighbour distributions, along with those expected for isotropic source distributions with the same number of bursts, for all 260 BATSE bursts and for the 202 bursts with statistical errors $< 9°$, respectively. While both distributions show slight excesses at separations $\theta \gtrsim 170°$ ($D = 0.0949$ and 0.1078, respectively), they are not significant (Q-values=0.12 and 0.14, respectively). A corresponding positive correlation function on angular scales $\theta \lesssim 5°$ and $\theta \gtrsim 170°$ led Narayan and Piran[17] to conclude that the nearest and farthest neighbour excesses in Figs. 2 and 3 are due either to an unusual statistical fluctuation or to an unknown systematic effect. We address this question below.

Table 1 shows that faint ($B \leq 1900$ counts) type I bursts have more nearest neighbours at small separations than expected for a random distribution (Q-value = 1.3×10^{-2}), while all type I bursts and all (type I and II) bursts for which B and V exist have significantly more than expected (Q-values = 4.8×10^{-3} and 2.1×10^{-3}, respectively). Retaining only the 159 bursts in the last sample for which the statistical error in their burst locations is $< 9°$ again increases the significance of the clustering (Q-value = 7.9×10^{-4}). We do not evaluate the amount by which the last significance should be reduced in order to take into account having chosen a cutoff, because it differs only modestly from the significance for the full sample.

The increase in significance going from the sample of faint type I bursts to all type I bursts implies that faint type I bursts are correlated with bright type I bursts. Similarly, the increase in significance going from the sample of all type I bursts to all bursts implies that type II bursts are correlated with type I bursts. The latter implication is strengthened by inspection of the samples that contain type II bursts; this reveals that many type II bursts contribute to the excess in the nearest neighbour distribution at small angular separations in these samples.

Table 1. Nearest neighbour results within samples.

Sample	Number	D	Q-value	Number	D	Q-value
Faint Type I	125	0.1714	1.3×10^{-2}			
All Type I	163	0.1642	4.8×10^{-3}			
All Type I & II	201	0.1584	2.1×10^{-3}	159	0.1903	7.9×10^{-4}
All	260	0.1168	1.8×10^{-2}	202	0.1877	1.1×10^{-4}

Faint denotes bursts with $B \leq 1900$ counts. The right hand columns are for the bursts in each sample for which the statistical error in their location is less than $9°$. D is the maximum deviation of the cumulative nearest neighbour distribution from that expected for a random distribution with the same number of bursts.

The typical systematic error θ_{sys} in the locations of the bursts in the publicly available BATSE catalogue is 4°; the median statistical errors θ_{stat} for faint type I bursts, all type I bursts, and all bursts are 5.7°, 5.5°, and 5.5°.[6] The first value might seem surprising, but θ_{stat} is inversely proportional to the square root of the burst fluence, not the burst brightness (flux). Many faint Type I bursts have long durations and therefore small θ_{stat}. The typical total error $\theta_{err} = (\theta_{sys}^2 + \theta_{stat}^2)^{1/2}$ in the locations of the bursts is thus 6.8°.

One can show that N_r recurrences of an individual source produce a peak in the differential nearest neighbour distribution at $\theta_r = 1.32\, \theta_{err}/\sqrt{N_r} \approx 9°/\sqrt{N_r}$, where we have taken $\theta_{err} = 6.8°$. The burst samples we study all cluster on an angular scale $\approx 5°$. This implies $N_r \gtrsim 4$, and suggests that many of the bursts contributing to the excess in the nearest neighbour distribution at small angular separations are multiple recurrences of individual sources. This is also suggested by Fig. 1 (right panel).

In a separate paper,[18] we show that the 55 type I bursts in the brightness range $465 < B < 1169$ exhibit a Galactic dipole moment and a deviation of the Galactic quadrupole moment from 1/3 whose joint significance is high; these "medium" brightness type I bursts are strongly concentrated toward the Galactic center and in the Galactic plane. They differ in this respect from the bright ($B > 1900$ counts) type I bursts and the other faint ($B \leq 1900$ counts) type I bursts, which evidence hints lie preferentially toward the Galactic anticenter but not in the Galactic plane.

In order to investigate further the relationship between the medium type I bursts and other bursts, we have also examined the nearest neighbour distributions between these medium type I bursts and other brightness sub-samples, as well as between bright type I bursts and other brightness sub-samples.[8] Note that the statistical error is small compared to the systematic error in the locations of the bright type I bursts, and the uncertainty in their locations is therefore only about 4°. Table 2 gives the significances of each of the deviations from that expected for a random distribution, and shows that bright type I bursts have no more medium type I nearest neighbours at any separation than expected for an isotropic source distribution (Q-value = 0.71). Table 2 shows that medium type I bursts have no more nearest neighbours of all other types at any separation than expected for a random distribution (Q-value = 0.66). In contrast, bright type I bursts have significantly more faint type I and II nearest neighbours within $\approx 5°$ than expected for a random distribution (Q-value = 4.0×10^{-3}). We find that this is also the case for the sample of faint type I and II bursts, less the medium type I bursts (Q-value = 4.2×10^{-3}). Thus faint bursts cluster near themselves and near bright type I bursts, but medium type I bursts show no significant correlation with bright type I bursts or with other faint bursts.

The medium bursts are thus distinct both in their anisotropic distribution on the sky and in their not repeating. Here we examine the effect that the anisotropic medium bursts have on the farthest neighbour distributions discussed above, and on the angular correlation function near 180°. Fig. 4 (left panel) shows the cumulative nearest neighbour distribution for the 152 bursts that have errors $< 9°$ and are not medium bursts. The right panel shows the farthest neighbour distribution for the sample. Comparing Fig. 3 and Fig. 4, we find that the pronounced nearest neighbour excess near 5° (D=0.1804, Q-value=2.6×10^{-3}) still remains, but that the slight excess in the farthest neighbour distribution has disappeared (D=0.0613, Q-value=0.85).

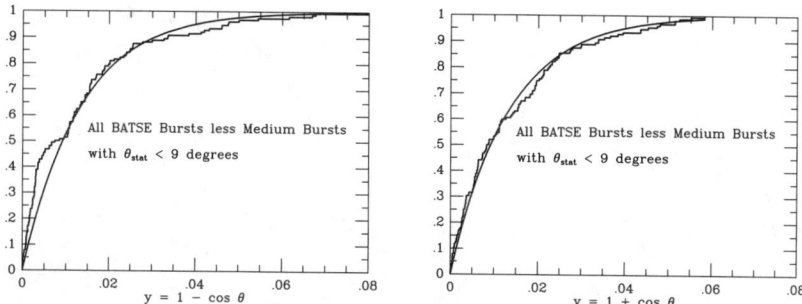

Figure 4. Nearest (left panel) and farthest (right panel) neighbour distributions for the 152 bursts that have errors $< 9°$ and are not medium bursts.

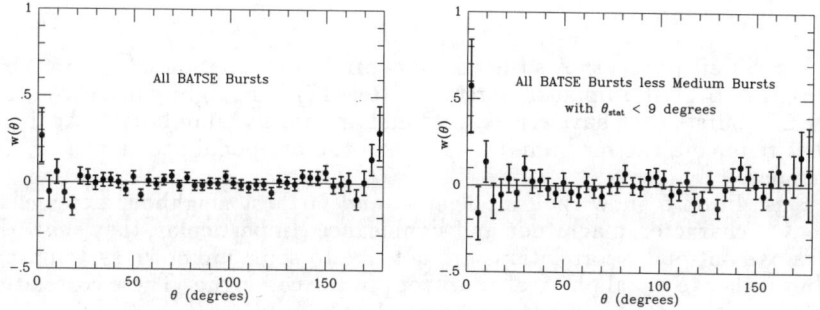

Figure 5. Angular correlation function $w(\theta)$, in $4°$ bins, for the entire BATSE sample of bursts (left panel), and for the 152 bursts that have errors $< 9°$ and are not medium bursts (right panel).

Table 2. Nearest neighbour results between samples.					
Sample 1	Number	Sample 2	Number	D	Q-value
Bright Type I	38	Medium Type I	55	0.1136	0.71
Medium Type I	55	Other Type I & II	146	0.0983	0.66
Bright Type I	38	Faint Type I & II	162	0.2862	4.0×10^{-3}
Bright Type I	38	Faint Type I & II less Medium Type I	107	0.2849	4.2×10^{-3}
Faint, bright, and medium denote bursts with $B \leq 1900$ counts, $B > 1900$ counts, and 465 counts $< B < 1169$ counts.					

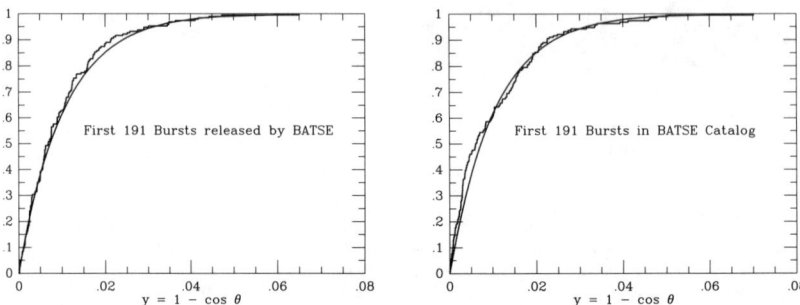

Figure 6. Nearest neighbour distribution for the first 191 bursts released by BATSE (left panel), and for the first 191 bursts in the BATSE catalogue, for which locations have been reprocessed (right panel).

Fig. 5 (left panel) shows the angular correlation function $w(\theta)$, in 4° bins, for the entire BATSE sample of bursts (cf. Ref. 17). The right panel shows $w(\theta)$ for the 152 bursts that have errors $< 9°$ and are not medium bursts. Again, we see that removing the medium bursts causes the antipodal excess at $\theta \gtrsim 170°$ to disappear.

Figs. 4 and 5 show that the nearest and farthest neighbour excesses are different in character, magnitude and significance. In particular, they show that the positive antipodal correlations are not due to some unknown systematic effect, but rather to a real physical anisotropy in the catalogue. These correlations support our claim that γ-ray bursts are Galactic in origin.[18]

Maoz[19] has suggested that a possible explanation for the peaks in $w(\theta)$ near 0° and 180° [see Fig. 5 (left panel)] is a "ring bias" in which bursts lie preferentially on great circles on the sky because of some unspecified systematic effect. While such a bias is ruled out by the lack of anisotropy in spacecraft coordinates,[20] the anisotropic distribution of the medium brightness bursts in the Galactic plane and towards the Galactic Center can produce an analogous effect. We are currently modeling the distribution of bursts on the sky and examining the effects that anisotropies and repeating have on the nearest and farthest neighbour distributions.[21]

DISCUSSION

We find that γ-ray bursts are significantly clustered on an angular scale $\approx 5°$. This angular scale is smaller than the typical error in burst locations of 6.8°, suggesting multiple recurrences from individual sources. We conclude that "classical" γ-ray burst sources repeat on a time scale of months,[8] and that many faint type I and II bursts come from the sources of bright type I bursts. The values $D = 0.1877$ from the nearest neighbour analysis within all bursts and $D = 0.2862$ from the analysis between bright type I bursts and faint type I and II bursts imply that at least $(260/202)(D/2) \approx 12\%$ (about 30) of all bursts and $D \approx 30\%$ (about 12) of bright type I bursts detected during the 10 months of BATSE observations come from sources that repeated. Wang and Lingenfelter[22] have also found evidence that five particular bursts arise from a single repeating source.

We note that previous experiments were sensitive only to bright bursts, and therefore did not detect the faint type I and type II bursts from which the evidence comes that γ-ray burst sources repeat. Confirmation of our results must come from the Interplanetary Network and/or analysis of the > 700 bursts in the new BATSE catalogue. For the latter, however, reprocessing of the burst locations to reduce systematic errors is crucial. Fig. 6 compares the cumulative nearest neighbour distribution for the first 191 bursts released by BATSE and for the first 191 bursts in the BATSE catalogue, for which locations have been reprocessed. An excess that is barely discernable at angles $\sim 10°$ shows up as a strong peak at $\approx 5°$ after reprocessing. Further reducing the systematic errors in burst locations offers the best chance of establishing unquestionably that "classical" γ-ray burst sources repeat.

Many Galactic models (such as episodic accretion, starquakes, or thermonuclear flashes involving neutron stars) predict repeated bursts, whereas most cosmological models invoke a singular, cataclysmic event (such as coalescence of neutron star-neutron star or neutron star-black hole binaries) in order to generate the tremendous amount of energy these models require. The differences between the durations, spectra, and time histories of many of the clustered bursts rules out the possibility that gravitational lensing causes the repetitions. Thus the repeating nature of γ-ray burst sources favors Galactic models.

The discovery that "classical" γ-ray bursts are Galactic in origin[18] and recur, and the recognition that many of them (i.e., the type II bursts) are of short duration,[23,11,9] suggests that the distinctions between the sources of "classical" γ-ray bursts and "soft γ-ray repeaters" are dissolving; only an apparent difference in spectral hardness remains, if that.[24]

Faint type I bursts are as much as ~ 100 times fainter than bright type I bursts, implying that the multiple bursts from individual sources have a broad luminosity function. Hence the limits on recurrence time scales derived earlier[4,5] are consistent with the repeating behavior we find.

We find that medium type I bursts show no significant correlation with bright type I bursts or with other faint bursts. They are anisotropic and are distinct in location from the sources of the other bursts, while the angular isotropy of the faint bursts reflects the fact that most come from the (nearby) sources of bright type I bursts.

Finally, we find that the excess in the farthest neighbour distribution and in $w(\theta)$ at $\theta \gtrsim 170°$, unlike the excess in the nearest neighbour distribution and in $w(\theta)$ at $\theta \lesssim 5°$, disappears when these anisotropic bursts are removed. We conclude that the positive antipodal correlations in the BATSE catalogue are not due to some unknown systematic effect, but rather to a real physical anisotropy in the catalogue. These correlations support our claim that γ-ray bursts are Galactic in origin.

ACKNOWLEDGMENTS

We thank Carlo Graziani for informative discussions about the role of locational errors in the nearest neighbour analyses and the nature of the clustering produced by repeating sources. This research was supported in part by NASA grants NAGW-830, NAGW-1284, and NAGS-4690.

REFERENCES

1. J. C. Higdon and R. E. Lingenfelter, Ann. Rev. Astron. Ap. **28**, 401 (1990).
2. A. Harding, Phys. Rep. **206**, 327 (1992).
3. K. Hurley, Gamma-Ray Bursts, eds. W. S. Paciesas and G. J. Fishman (AIP, New York, 1992), p. 3.
4. B. E. Schaefer and T. L. Cline, Ap. J. **289**, 490 (1985).
5. J.-L. Atteia, et al., Ap. J. Suppl. **64**, 305 (1987).
6. G. J. Fishman, et al., Ap. J. Suppl. (1994, in press).
7. D. Q. Lamb, C. Graziani, and I. A. Smith, Ap. J. **413**, L11 (1993).
8. J. M. Quashnock and D. Q. Lamb, MNRAS **265**, L59 (1993).
9. D. Q. Lamb and C. Graziani, Ap. J. (1994a, in press).
10. D. Q. Lamb and C. Graziani, Ap. J. (1994b, in press).
11. C. Kouveliotou, et al., Ap. J. **413**, L101 (1993).
12. D. Q. Lamb and J. M. Quashnock, Ap. J. **415**, L1 (1993).
13. D. H. Hartmann and R. I. Epstein, Ap. J. **346**, 960 (1989).
14. D. H. Hartmann and G. R. Blumenthal, Ap. J. **342**, 521 (1989).
15. D. Scott and C. A. Tout, MNRAS **241**, 109 (1989).
16. W. Press, B. P. Flannery, S. A. Teukolsky, and W. T. Vetterling, Numerical Recipes (Cambridge University Press, Cambridge, 1986), p. 472.
17. R. Narayan and T. Piran, MNRAS **265**, L65 (1993).
18. J. M. Quashnock and D. Q. Lamb, MNRAS **265**, L45 (1993).
19. E. Maoz, MNRAS (1994, submitted).
20. J. J. Brainerd, Ap. J. (1994, submitted).
21. J. M. Quashnock and D. Q. Lamb, Ap. J. (1994, in preparation).
22. V. C. Wang and R. E. Lingenfelter, Ap. J. **416**, L13 (1993).
23. R. W. Klebesadel, Gamma-Ray Bursts, C. Ho, R. I. Epstein, and E. E. Fenimore, eds. (Cambridge University Press, Cambridge, 1992), p. 161.
24. E. E. Fenimore, et al., BAAS **23**, 1322 (1992).

THE ANGULAR CORRELATION FUNCTION OF GAMMA-RAY BURSTS

G. R. Blumenthal
Board of Studies in Astronomy and Astrophysics
UCO Lick Observatory, UCSC, Santa Cruz, CA 95064

D. H. Hartmann & E. V. Linder
Department of Physics and Astronomy
Clemson University, Clemson, SC 29634

ABSTRACT

We have calculated the angular correlation function for 260 bursts from the first BATSE catalog (1B). We have also combined the data with previous catalogs to increase the sample size to N = 474. The sky exposure map of BATSE is included in the analysis. No significant correlations were found on any angular scale. If GRBs autocorrelate like galaxies the lack of observed clustering yields a minimum detector sampling depth of $\sim 200\ h_{75}^{-1}$ Mpc, corresponding to a minimum survey redshift of ~ 0.05. Cosmological models in which bursts trace the large scale distribution of galaxies to redshifts of order unity are consistent with the data. In addition, we also discuss the implications of the observed lack of clustering for Galactic distribution models. For exponential disk models the lack of large-scale clustering implies a maximum sampling depth of ~ 1 scale height. Clustering limits can also be used to constrain burst recurrence rates. This question, however, is discussed in a related paper in conjunction with the burster nearest neighbor statistic.

INTRODUCTION

The old paradigm of a Galactic neutron star origin of γ-ray bursts (GRBs) perished when BATSE observed a nearly isotropic sky distribution while clearly detecting a lack of faint sources.[1] While a few alternatives are still considered, the cosmological origin of bursts appears to be the most logical interpretation of the BATSE data.[2] If bursts indeed originate at large redshifts, the merger of compact stellar objects (neutron stars, black holes, white dwarfs) seems to be a natural GRB site. Since these events must occur in nature and since we know that they are rare, we must sample a significant fraction of the universe to be consistent with the observed event rate of $\sim 10^3$ per year. Mergers would predominantly occur inside or near galaxies, so that we expect burst events to trace the large scale structure of the universe as mapped by the distribution of luminous matter. The evolution of the large-scale structure of the universe has imprinted at least one tell-tale feature on the galaxy distribution; the two-point correlation function $\xi(\mathbf{r})$. Redshifts of γ-ray bursts are not known, so we can only analyze the angular two-point correlation function $w(\theta)$ for a flux limited sample. In principle, theoretical predictions for ξ can be projected to yield $w(\theta)$, and we can then compare angular GRB correlations with these projected models.[4] Alternatively, we can compare $w(\theta)$ of GRBs directly with the observed galaxy correlation function.

The angular correlation function is defined by the number of sources above the Poisson value in a solid angle $d\Omega$ at an angle θ around any source randomly selected from the catalog under investigation[3]

$$dP = \rho\,[1 + w(\theta)]\,d\Omega \tag{1}$$

where $\rho = N/4\pi$ is the mean surface density of sources. Hartmann&Blumenthal[4] (HB) showed that if GRBs correlate like $\xi \propto r^{-\gamma}$ with $\gamma = 2$, the product $w(\theta)\theta$ is determined by

$$w(\theta)\,\theta = \left(\frac{D_s}{D_0}\right)^{-2}, \tag{2}$$

where D_s is the sampling depth of the sample and D_0 is a normalizing length scale that is related to the correlation length, r_0, of the source population by

$$D_0 = \sqrt{4\pi}\,r_0\,h_{75}^{-1}, \tag{3}$$

where h_{75} is the Hubble constant in units of 75 km s^{-1} Mpc^{-1}. In the context of cosmological models involving mergers we can use the observed galaxy correlation length $r_c = (4/3)\,r_0\,h_{75}^{-1} \sim 7\,h_{75}^{-1}$ Mpc, i.e., we use $r_0 = 5$ Mpc. The observed galaxy correlation power index is 1.8 instead of 2 as assumed here, but this difference introduces only second order effects. If angular clustering is detected on some scale, we can use the D_s^{-2} scaling law to test whether the signal is indeed of cosmological origin. Evolution of the galaxy correlation function can be included in the analysis via

$$\xi(r,z) = \left(\frac{r_c}{r}\right)^{\gamma} (1+z)^{-(3+\epsilon-\gamma)}, \tag{4}$$

where r is expressed in comoving coordinates. If $\gamma = 1.8$, then $\epsilon = \gamma - 3 = -1.2$ corresponds to correlation properties fixed in comoving coordinates. Linear growth theory of large scale structure predicts that $\xi(r,z)$ grows as the inverse square of $(1+z)$, so that $\epsilon \sim 1$ may be a reasonable approximation, but non-linear growth could result in much larger values for ϵ. If GRBs sample the universe to redshifts of order unity, evolution of galaxy clustering should be included in the correlation analysis. Studies of this effect will be presented in a forthcoming publication.

THE DATA

We analyze the positions of 260 bursts observed by BATSE between April 21, 1991 and March 5, 1992. This 1B catalog[5] is referred to in the Figures as GRO. We enlarge the sample by adding 54 bursts with single error boxes from IPN triangulations.[6] Details of the data and our selection criteria can be found in HB. The third database added to our sample are the 160 localizations obtained from the KONUS experiment aboard the Soviet satellites *Venera* 11-14. Details and references to this database can also be found in HB. We combine these three data sets into four samples with 260 (GRO only), 314 (GRO and IPN), 420 (GRO + KONUS), and 474 (GRO + KONUS + IPN) events. No effort was made to test the various combined sets for selection biases, such as duplication of events when combining IPN and KONUS. Analysis of these raw data will therefore produce conservative upper limits.

RESULTS

The correlation function of the 1B catalog is shown in Figure 1. The relatively small effect of the nonuniform BATSE sky map is included in the analysis. The correlation function is consistent with zero, but shows some excess at small and large angles (near 180°). The small angle excess is located at $\sim 4°$, similar to the scale of systematic localization errors in the BATSE data. The excess of burst pairs with angular separations near 4° is also evident in the nearest neighbor distribution function and was interpreted as the signal of burst recurrence.[7] There are several arguments[8-11] against this interpretation and a preliminary investigation of 743 BATSE bursts suggests that the "repeater-excess" is a statistical artifact of the 1B sample.[12] The excess near 180° discovered by Narayan & Piran[10] also disappears when the larger post 1B sample is considered.[12]

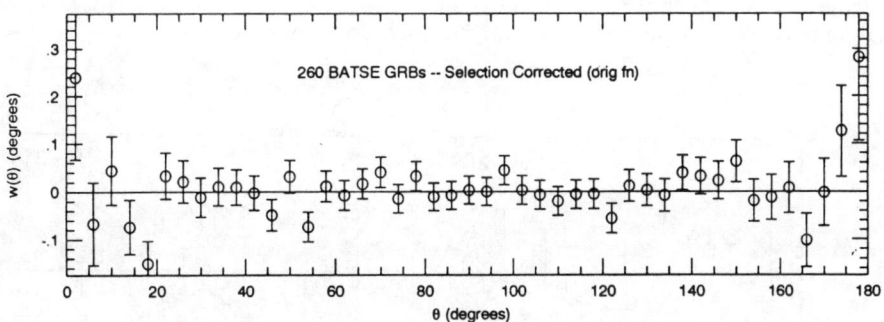

Fig. 1 — Angular correlation function of the 1B sample

Figure 2 shows the product $w(\theta)\theta$ versus the logarithm of the angular scale (in degrees). This form is used because of its convenience in constraining cosmological sampling distances (see eq. 2) and Galactic models (HB). Figures 2A to 2D show the combined catalogs of BATSE, IPN, and KONUS mentioned in the previous section. While there may be some concern about possible selection effects inherent in some of these data sets, we assume that simply combining the data does not introduce a significant bias. Based on that assumption, we obtain the following upper limits on w θ: 1.0 (1B, N = 260), 0.8 (1B + IPN, N = 314), 0.6 (1B + KONUS, N = 420), and 0.5 (1B + IPN + KONUS, N = 474). No combination of data sets shows evidence for clustering. We showed previously that IPN and KONUS data individually are not clustered.[4] Here we show that the 1B burst locations are not clustered, and that neither of the combined data sets show evidence for correlations. Combining the various data sets allows us to improve the statistical sensitivity to the point that BATSE data alone would reach after about 2 years of operation. Taking the most stringent limit derived from combining all catalogs we determine the minimum sampling redshift as

$$z_{\min} = 6.7 \; 10^{-3} \; r_0 \; (w \; \theta)_{\lim}^{-1/2} \sim 0.05 \;, \qquad (5)$$

consistent with the statistics of standard candles in a Friedman universe.[13] The large localization errors for GRBs dilute the angular correlation function below angles of ~ 5-10° and the large sampling depth derived from logN-logS statistic further reduces the expected clustering signal from galaxies.[14] Comparison of the 1B limits on w(θ) with the observed galaxy correlation function derived from the APM survey,[15] clearly shows that a significantly larger number of bursts is needed before we may detect the clustering signature expected from the merger picture of cosmological GRBs.

We can also use these constraints to place an upper limit on the sampling depth to Galactic source populations, such as exponential disks.[4] Combining Figure 2 of HB with the limits derived in this work we find that Galactic Population I scenarios for GRBs are limited to sources detected to no more than ~ 1 scale height. For such small sampling distances the corresponding uniformity of the density profile would clearly be in conflict with the low fluence turnover in the logN-logS curve. This rules out disk models, which is also obvious from multipole constraints. Here the angular correlation function does not add much new information, but supports the conclusions derived from dipole and quadrupole analyses. It is the information on cosmological distributions described above that makes correlation function analysis such an important tool.

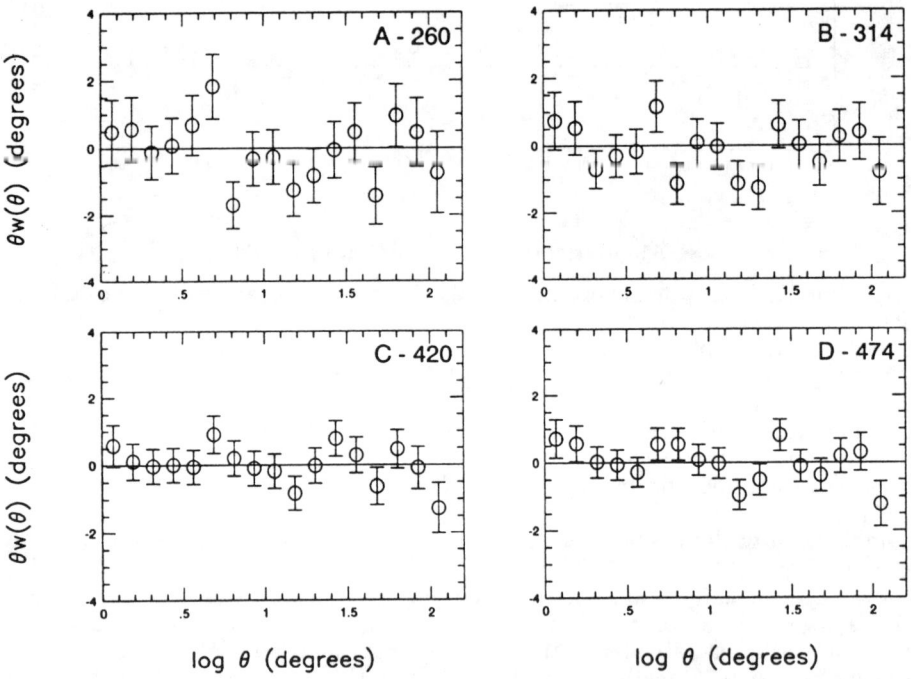

Figure 2 — Product of angle and angular correlation function versus angle for four different sample combinations: A = GRO (N=260), B = GRO + IPN (N=314), C = GRO + KONUS (N=420), and D = GRO + IPN + KONUS (N=474). The BATSE data (GRO) are corrected for sky exposure.

DISCUSSION

We have calculated the angular correlation function $w(\theta)$ of N= 260 γ-ray bursts from the first BATSE catalog[5] (1B). We have also calculated $w(\theta)$ for catalogs that combine 1B with various older burst compilations, which yields a maximum sample size of N = 474. No significant correlations on any scale were found for either of the combined samples. The excess at $\sim 4°$ in the 1B catalog is considered to be a statistical fluctuation.[12] Similarly, the excess at $\sim 180°$ is also statistical in origin. We have used the upper limits on the product $w(\theta)\theta$ to constrain cosmological and local disk models. Although the combined logN-logS and multipole statistic already rule out Galactic disk models, we use the clustering limits to be specific about the maximum sampling distance in terms of the scale height of such disk populations. The data suggest that bursts can not be sampled to more than ~ 1 scale height. For old Galactic neutron stars[16,17] this limit implies a maximum distance of about 400 pc. Within that distance, the neutron star source density is approximately uniform, which is not supported by the observed V/V_{max} statistic of γ-ray bursts.[1,5] For cosmological source distributions the minimum sampling distance is $\sim 250\ h_{75}^{-1}$ Mpc corresponding to a minimum redshift of $z \sim 0.05$. This limit is consistent with LogN-LogS fits in terms of non-evolving standard candles[13] suggesting sampling redshifts ranging from $z \sim 0.2$ for PVO to $z \sim 0.8$ for BATSE. The sample size used in this study is not large enough to actually detect the clustering signal expected if bursts trace the galaxy distribution,[18] but if BATSE continues to function well this possibility should arise in the near future. This work was supported by NASA grant NAG 5-1578.

REFERENCES

1. Meegan, C. A., *et al.*, Nature **355**, 143 (1992).
2. Paczynski, B., Acta Astr. **41**, 257 (1991).
3. Peebles, P. J. E., The Large Scale Structure of the Universe (Princeton Univ. Press, 1980).
4. Hartmann, D. H., & Blumenthal, G. R., ApJ **342**, 521 (1989).
5. Fishman, G. J., *et al.*, ApJS , in press (1994).
6. Atteia, J.-L., *et al.*, ApJS **64**, 305 (1987).
7. Quashnock, J., & Lamb, D. Q., MNRAS , in press (1993).
8. Hartmann, D. H. *et al.*, ApJ , submitted (1993).
9. Brainerd, J., in *Gamma Ray Bursts* , these proceedings (1994).
10. Narayan, R., & Piran, T., MNRAS , in press (1993).
11. Nowak, M. A., ApJL , submitted (1993).
12. Hartmann, D. H., *et al.*, in *Gamma Ray Bursts* , these proceedings (1994).
13. Fenimore, E., *et al.*, Nature **366**, 40 (1993).
14. Hartmann, D. H., Linder, E. V., & Blumenthal, G. R., ApJ **367**, 186 (1991).
15. Maddox, S. J., *et al.*, MNRAS **242**, 43P (1990).
16. Hartmann, D. H., Epstein, R. I. & Woosley, S. E., ApJ **348**, 625 (1990).
17. Paczynski, B., ApJ **348**, 485 (1990).
18. Lamb, D. Q., & Quashnock, J. M., ApJL **415**, L1 (1993).

THE NEAREST NEIGHBORS OF BATSE GAMMA-RAY BURSTS: NARROWING THE POSSIBILITIES

J.J. Brainerd,* W.S. Paciesas

Dept. of Physics, University of Alabama in Huntsville, Huntsville, AL 35899

C.A. Meegan, G.J. Fishman

NASA/Marshall Space Flight Center, ES-66, Huntsville, AL 35812

ABSTRACT

The large errors in the locations of gamma-ray bursts observed by the Bursts and Transient Source Experiment (BATSE) forces one to adopt statistical comparisons of the burst catalog to models of burst isotropy to determine if bursts are repeating or clustered. In the first BATSE catalog[1], a nearest neighbor analysis finds a deviation from isotropy.[2] In a recent article[3] It was shown that known instrumental effects cannot produce significant small scale anisotropies. It was also shown that burst repeater models can produce the observe anisotropies. In this paper we examine in more detail repeater burst models. We also show that nearest and farthest neighbor analyses of more recent bursts fail to find significant small scale clustering.

INTRODUCTION

A nearest neighbor analysis[2] of the first BATSE Gamma-Ray Burst Catalog[1] finds a maximum deviation from the isotropic nearest neighbor cumulative distribution of 2% statistical significance.[3] This contradicts an earlier analysis that examines the average value of the nearest neighbor separation and finds no significant deviation from isotropy.[4] The question arises whether the result of the first named analysis is a consequence of statistical fluctuations or of either instrumental or physical processes. The authors of one article[5] assert that the deviation must be of instrumental origin, because a similar farthest neighbor analysis of the first Catalog finds a maximum deviation from the isotropic cumulative distribution of 11% significance, which they regard as statistically significant and unphysical.

In a recent article[3] it was shown that there are no systematic effects that can produce a deviation of the magnitude seen in the data. The effect of the blockage of the sky by the earth produces a maximum deviation of $D = 0.006$ from the isotropic cumulative distribution, which is much smaller than the $D = 0.12$ deviation of the First Catalog. While one might expect a small angle anisotropy from the flux dependence of the sky map to introduce a deviation, we find from Monte Carlo simulation that the maximum deviation from the isotropic cumulative distribution is $D = 0.01$ without any spacecraft reorientations and $< 10^{-3}$ with the reorientations experienced by the Compton Gamma-Ray Observatory (CGRO). Moreover, performing the nearest and farthest neighbor analyses in the spacecraft coordinate frame produces maximum

---

* Mailing Address: Space Science Lab, ES-66, NASA/MSFC, Huntsville, AL 35812. E-mail: brainerd@ssl.msfc.nasa.gov

deviations of $\approx 50\%$ significance. If the deviation in the First Catalog is non-statistical, it must have a physical origin.

The suggestion of Lamb and Quashnock[2] that repeating burst sources are responsible for the nearest neighbor results is not invalidated by the farthest neighbor analysis.[3] In fact, if repeating burst sources are confined to the galactic plane, the resulting nearest and farthest neighbor distributions agree with the First Catalog distributions without violating the limits on the dipole and quadrupole moments.

In this paper we discuss disk confined repeater models in more detail, presenting their nearest and farthest neighbor cumulative distributions. We also present nearest and farthest neighbor analyses of a recent set of 260 gamma-ray bursts; the results are consistent with an isotropic distribution of bursts.

BURST MODELS

The nearest neighbor of a given burst is primarily determined by the local density of bursts on the sky. The magnitude of the density fluctuations therefore determines the shape of the cumulative distribution. On the other hand, both the magnitude and topology of density fluctuations strongly affect the farthest neighbor cumulative distribution. This implies that the farthest neighbor distribution is much more model dependent than the nearest neighbor distribution.

Two simple models illustrates this point. First, imagine that αN bursts are distributed on the upper celestial hemisphere and $(1 - \alpha) N$ bursts are distributed in the lower celestial hemisphere, where N, the total number of bursts, is very large. The nearest neighbor distribution is then

$$F_n = \alpha \exp\left(-\frac{1-\cos\theta}{4N\alpha}\right) + (1-\alpha)\exp\left(-\frac{1-\cos\theta}{4N(1-\alpha)}\right), \qquad (1)$$

and the farthest neighbor distribution is

$$F_f = (1-\alpha)\exp\left(-\frac{1-\cos\theta}{4N\alpha}\right) + \alpha\exp\left(-\frac{1-\cos\theta}{4N(1-\alpha)}\right). \qquad (2)$$

The isotropic nearest and farthest cumulative distributions ($\alpha = 1/2$) equal $1/2$ when $1 - \cos\theta = 2N \ln 2$. For this value, $F_n > 1/2$ and $F_f < 1/2$ for all $\alpha \neq 1/2$. Now consider a second model in which $\alpha N/2$ bursts are distributed on one half of the upper hemisphere and the same number of bursts are distributed on the geometrically opposed half of the lower hemisphere. The remaining two half-hemispheres have $(1 - \alpha) N/2$ bursts apiece. In this model the nearest and farthest cumulative distributions are both given by equation (1), proving the importance of the topology of density variations.

These simple models illustrate a second point: the nearest neighbor cumulative distribution lies above the isotropic distribution for small vales of $1 - \cos\theta$, but the farthest neighbor distribution can lie on either side of the isotropic distribution. However, the magnitude of the deviation from isotropy of the nearest neighbor distribution also depends on the density gradient, for α must be less than ≈ 0.15 to produce a deviation as large as in the left hand plot of Figure 1. This effect disappears as N increases.

The source of the anisotropy also determines whether the deviation of the cumulative distribution from the isotropy distribution changes as the number

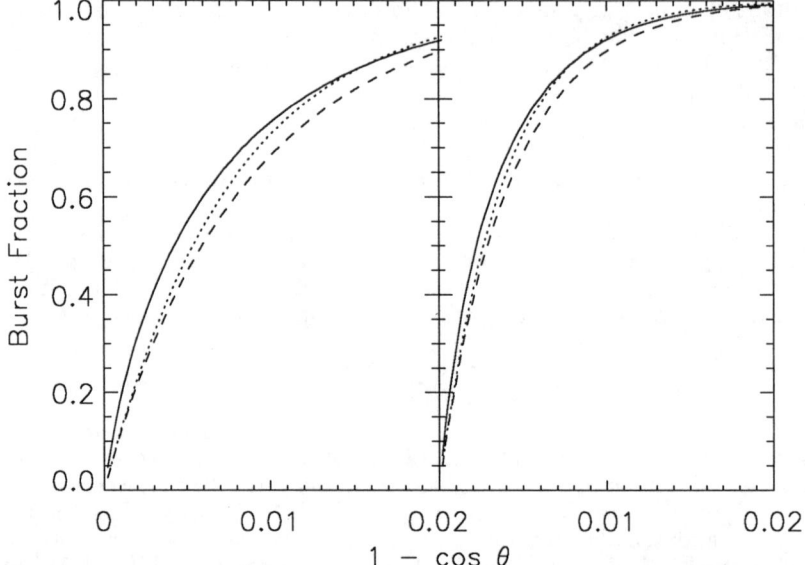

Figure 1. The nearest and farthest neighbor cumulative distributions: effect of sample size. Theoretical curves for a model in which burst sources are isotropically distributed and a fraction of the burst sources producing 4 bursts apiece. The isotropic cumulative distribution is given by the dotted line, the model nearest neighbor distribution by the solid line, and the model farthest neighbor distribution by the dashed line. Left: 200 single burst sources and 15 multiple burst sources. Right: 400 single burst sources and 30 multiple burst sources.

of bursts increases. If the anisotropy arises from the presence of a disk component, then the anisotropy persists. If repeating sources produce the anisotropy through local density enhancements over a location error angular scale, then the cumulative distributions should go to the isotropic distribution as the number of bursts increases beyond the point where the location errors about each repeating source overlaps with adjacent repeating sources. Figure 1 demonstrates the dependence of the cumulative distribution on the sample size. As a consequence, a statistical search for repeaters should use consecutive sets of bursts with a fixed number of bursts in each set.

Figure 1 shows that isotropically distributed repeating sources produce a farthest neighbor cumulative distribution that falls below the isotropic distribution. If the repeaters are confined to the galactic plane, as in Figure 2, the farthest neighbor cumulative distribution can fall above the isotropic distribution.

The model used in Figure 2 has two components, with one component an isotropically distributed set of single burst sources and the second component a set of multiply bursting sources confined to the galactic plane. Such models can reproduce the nearest neighbor distribution of the First BATSE Catalog

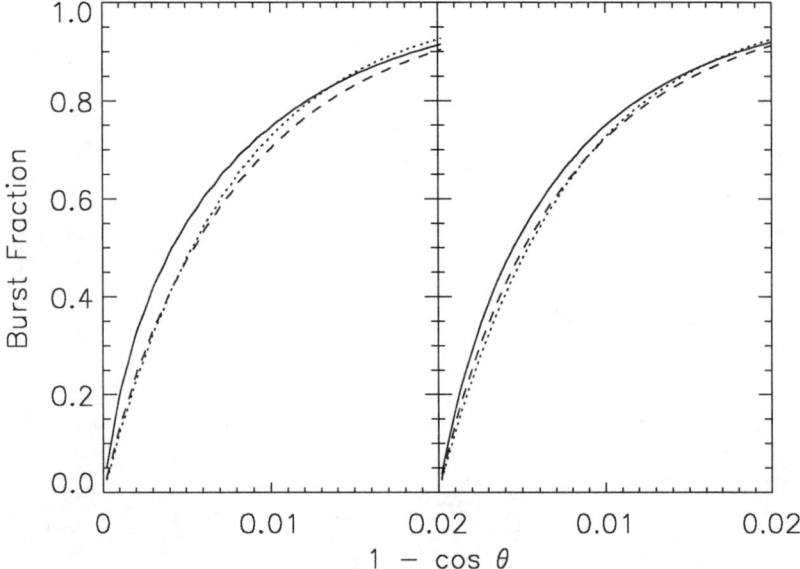

Figure 2. The nearest and farthest neighbor cumulative distributions: quadruple repeaters versus double repeaters. Two hundred single burst sources are isotropically distributed. Additional repeating sources are uniformly distributed between ±10° galactic latitude. The position of each outburst relative to the source is given by a gaussian distribution function with average offset of 5°. The lines are defined as in Fig. 1. Left: 15 repeating sources burst 4 times apiece. Right: 30 repeating sources burst 2 times apiece.

without violating constraints on the farthest neighbor distribution or the value of $\langle \sin^2 b \rangle$. The right hand plot of Figure 2 deviates from the the observed nearest and farthest neighbor distributions with significances of 68% and 31% respectively. In this model one finds $\sin^2 b = 0.26$, which is less than 3σ from the value in the First Catalog.

RECENT RESULTS AND CONCLUSIONS

The Kolmogorov-Smirnov statistic for the nearest neighbor distribution from the First BATSE Catalog[3] of 260 bursts is $N^{1/2}D = 1.86$, which has a 2% significance. A farthest neighbor analysis gives $N^{1/2}D = 1.56$, which has an 11% significance.[5] These values are consistent with both a statistical origin and a physical origin.

A more recent ensemble of 260 gamma-ray bursts disjoint from the first catalog shows no evidence for anisotropy. Their nearest and farthest neighbor cumulative distributions are given in figure 3. The maximum deviation of the nearest neighbor cumulative distribution is below the isotropic cumulative dis-

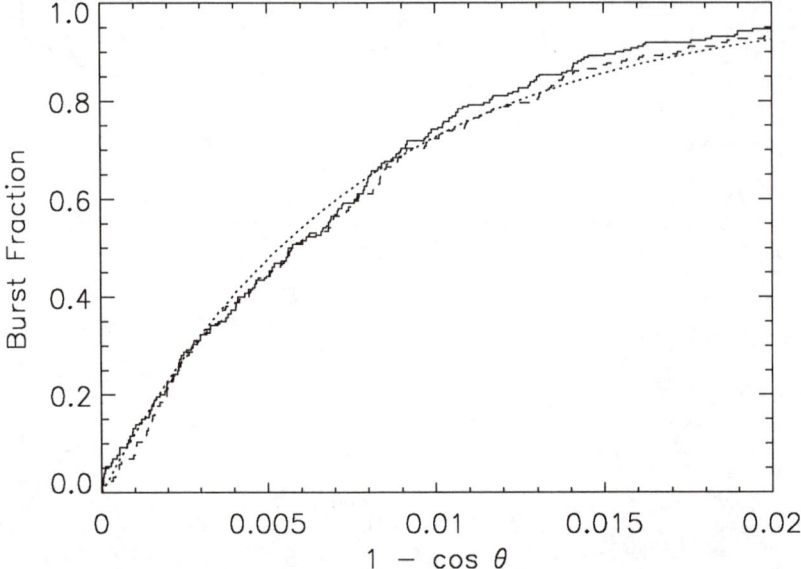

Figure 3. The nearest and farthest neighbor cumulative distributions: recent gamma-ray burst sample. A set of 260 gamma-ray bursts is selective by dropping all gamma-ray bursts with locations by the BATSE data type MAX-BC from the set of triggers between 1819 and 2494 inclusive (GRB920814b to GRB930817). These bursts will be part of the 2B and 3B gamma-ray burst catalogs. The lines are defined as in Fig. 1.

tribution, and gives $N^{1/2}D = 0.77$, with significance 83%. The farthest neighbor distribution has $N^{1/2}D = 0.85$, with significance 75%. Again the maximum deviation is below the isotropic cumulative distribution. This provides no support for the belief that the deviation found in the first catalog has a physical origin.

REFERENCES

1. G.J. Fishman, C.A. Meegan, R.B. Wilson, M.N. Brock, J.M. Horack, C. Kouveliotou, S. Howard, W.S. Paciesas, M.S. Briggs, G.N. Pendleton, T.M. Koshut, R.S. Mallozzi, M. Stollberg & P. Lestrade, Astrophys. J. Suppl. Ser., in press, (1994).
2. Quashnock & D.Q. Lamb, Mon. Not. R. Astron. Soc., in press, (1993).
3. J.J. Brainerd, Astrophys. J. Lett., submitted, (1994).
4. J.M. Horack, G.J. Fishman, C.A. Meegan, R.B. Wilson, M.N. Brock, J. Hakkila, W.S. Paciesas, G.N. Pendleton, & M.S. Briggs, in *Compton Gamma-Ray Observatory: St. Louis, MO 1992*, ed. M. Friedlander, N. Gehrels, & D.J. Macomb (AIP: New York, 1993), p. 699.
5. Narayan & T. Piran, Mon. Not. R. Astron. Soc., in press, (1993).

DO CLASSICAL GAMMA-RAY BURSTS REPEAT?

D. H. Hartmann
Department of Physics and Astronomy
Clemson University, Clemson, SC 29634

G. R. Blumenthal
University of California at Santa Cruz and
UCO/Lick Observatory, Santa Cruz, CA 95064

K. Hurley
University of California at Berkeley
Space Sciences Laboratory, Berkeley, CA94720

E. V. Linder
Department of Physics and Astronomy
Clemson University, Clemson, SC 29634

J. Hakkila
Department of Mathematics, Astronomy, and Statistics
Mankato State University, Mankato, MN 56002

G. J. Fishman, C. A. Meegan, R. B. Wilson, M. Brock, J. M. Horack
Space Science Laboratory, ES-66
NASA/Marshall Space Flight Center, Huntsville, AL 35812

C. Kouveliotou
Space Science Laboratory, ES-66
USRA/Marshall Space Flight Center, Huntsville, AL 35812

M. S. Briggs, W. S. Paciesas, G. N. Pendleton
Department of Physics, University of Alabama, Huntsville AL 35899

ABSTRACT

If classical GRBs were shown to repeat, cosmological models involving merging neutron stars inside galaxies would in effect be ruled out. We investigate the question of burst repetition by searching for the corresponding signal in the correlation function and nearest neighbor distribution. We analyze data for 742 γ-ray bursts observed with BATSE. Preliminary results derived from this study do not support the recent claim by Lamb & Quashnock that a significant fraction of bursts repeat on time scales less than ~ 1 year.

INTRODUCTION

The old paradigm of a Galactic neutron star origin of γ-ray bursts perished when BATSE observed a nearly isotropic sky distribution while clearly establishing a lack of faint sources.[1] While a few alternatives are still considered, the cosmological origin of bursts appears to be the most logical interpretation of the BATSE data.[2] If bursts indeed originate at large redshifts, the merger of compact stellar objects (neutron stars, black holes, white dwarfs) appears to be a natural GRB site. Since these events must occur in nature and since they are rare, we must sample a significant fraction of the universe to be consistent with the observed event rate of $\sim 10^3$ per year. Mergers are expected to occur at a rate of $\sim 10^{-6}$ yr^{-1} per host galaxy, so that recurrence over the lifetime of γ-ray detectors is not expected. Positional coincidences on the sky are thus expected to be random. Deviations from randomness can be caused by non-uniform sampling of the sky, spatial clustering of the underlying source population, large scale anisotropies, or burst repetition. Since the latter case would in effect rule out currently favored cosmological scenarios, the question of burst recurrence needs careful consideration.

Based on IPN localizations, Atteia et al.[3] investigated the statistics of overlapping burst error boxes. These authors derive model dependent lower limits on the recurrence time scale of order 1–10 years. More recently, however, Quashnock & Lamb[4] claimed evidence for non-linear clustering on small angular scales for the burst data in the 1B catalog[5] and suggested that multiple recurrences for a significant fraction of bursts were responsible for the observed excess. A possible correlation of bursts in both space and time was suggested by Wang & Lingenfelter[6] who isolated GBS 0855−00 as a potential location of multiple recurrences. If a significant fraction of bursts indeed repeat on short timescales, the angular distribution would be be distorted on scales comparable to the angular resolution of the instrument. Localization accuracy of BATSE is currently limited by systematic effects[5] to $\sim 4°$. Low order multipole analysis[7] does not resolve these scales, so that different methods must be employed to test for small angle clustering. Two tools suitable for this task are the angular correlation function and the nearest neighbor statistic.

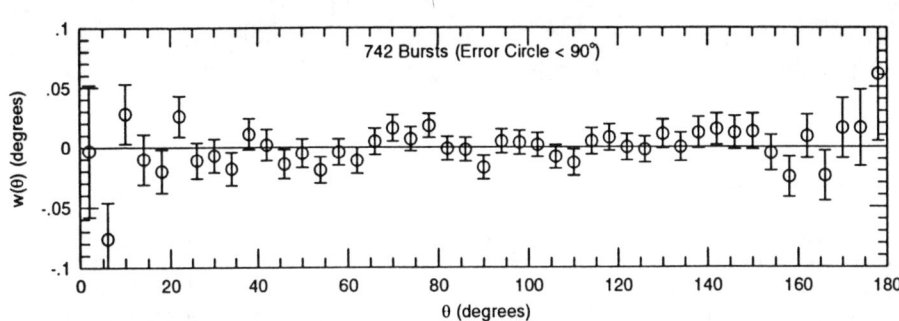

Figure 1 — Angular correlation function of 742 BATSE bursts.

SEARCHING FOR RECURRENCE

The fuzzy distribution of burst arrival directions can be analyzed by means of 2D Fourier power synthesis. In the presence of repeating sources the power spectrum would be enhanced on the angular scale corresponding to the detector resolution. However, other effects, such as global anisotropies and uneven sampling of the sky, could also yield such enhancements. To test the hypothesis of burst recurrence one must consider the possibility that deviations from randomness on the sky could resemble a repeater signal. For example, if bursts were concentrated towards the galactic plane, the mean separation between neighboring events is reduced relative to that expected for an isotropic distribution. While the effects of uneven sampling can be taken into account through the BATSE sky exposure map, anisotropic source distributions are harder to correct for. We therefore chose to be conservative by testing the data against an assumed isotropic distribution. If significantly enhanced power is found on some angular scale, one could attempt to simulate the separate contributions due to recurrence and anisotropic source distributions; the results would then be model dependent. If no power enhancements are found, the resulting limits on recurrence properties are conservative. In this study we do not include the fuzzy nature of burst localizations but instead treat bursts as point sources. In that case it is straightforward to calculate the clustering properties of GRBs, and we focus on two alternative approaches; correlation functions and neighbor separations.

The two-point angular correlation function $w(\theta)$ is defined by the number of sources above the Poisson value in a solid angle $d\Omega$ at an angle θ around any source randomly selected from the catalog under investigation[8]

$$dP = \rho \left[1 + w(\theta)\right] d\Omega , \qquad (1)$$

where $\rho = N/4\pi$ is the mean surface density of sources. To see the effect of repeaters on the correlation function consider that a fraction, f, of all observed events are drawn from a population of repeating sources. Assume that each repeating source had two appearences. In the limit of perfect angular resolution the correlation function satisfies[9]

$$(N - 1) \, w(\theta) = -f + 4\pi \, f \, \delta_2(\theta) , \qquad (2)$$

where the delta function, δ_2, operates on the unit sphere. Of course, with perfect eyesight we would not need statistics to decide whether bursts repeat or not. In the case of fuzzy eyesight, approximated by a Gaussian smearing function with resolution σ, the observed correlation function becomes

$$(N - 1) \, w(\theta) = -f + 4\pi \, f \, \sigma^{-2} \, \exp\left(-\theta^2/2\sigma^2\right) . \qquad (3)$$

To higher order we consider the possibility of more than two events per source. If j denotes the mean number of repeats one can show that the previous form of the correlation function remains unchanged, but with the substitution f → fj. In other words, if the same number of repeat events, $N_r = fN$, are arranged in a smaller number of clusters with more repetitions, the effect on the angular correlation increases (linear in j). From equation (3) it is also clear that the signal from repeaters is diluted (linearly) with increasing number of sources in the sample, and enhanced (quadratically) with angular resolution of the instrument.

The second method suitable for recurrence analysis is the statistic of angular separations between neighboring pairs of burst locations. Just as in the case of angular correlation functions we can investigate the hierarchy of nearest, second nearest, ..., second farthest, and farthest neighbor distributions. Here we consider only the lowest order statistic. The nearest neighbor (NN) statistic of a randomly selected source is described by the probability density function (pdf) of finding the closest neighboring source at an angular separation between

θ and $\theta + d\theta$. If N sources are distributed randomly on the sky the NN pdf is given by[10]

$$\text{pdf}(\theta) = (N-1)\, 2^{1-N} \sin(\theta)\, [1+\cos(\theta)]^{N-2}\ . \tag{4}$$

In the presence of repeaters the shape of the NN pdf depends on the repeater fraction f (defined as before) and the angular resolution element σ. In the small angle limit of interest here one finds[9]

$$\text{pdf}_r(\theta) = \text{pdf}(\theta)\left[1 + \pi f \sigma^{-2}(N-1)^{-1}\right]\ . \tag{5}$$

Comparing the scaling properties of this equation with that of eq. (3) shows that both the pdf and the angular correlation function give essentially the same information on small angle clustering. However, while $w(\theta)$ can be employed for arbitrary sample size (though the signal is $\propto N^{-1}$) the NN pdf is not useful for sample sizes exceeding a "critical" number $N_c = \pi\, \sigma^{-2}$ at which point it becomes exceedingly unlikely that the nearest neighbor of a repeating source is a repeat event. In other words, the mean separation of the random sources becomes smaller than the mean separation of repeaters, which is dictated by the instrumental resolution function. For BATSE parameters the critical size is by accident comparable to the size of the 1B set. To search for repeaters we analyze the positions of 260 bursts obtained by BATSE during the period April 21, 1991 to March 5, 1992. This is the 1B catalog.[5] In addition, we study the sample of 743 BATSE bursts that include events up to August 17, 1993.

RESULTS AND DISCUSSION

The correlation function of the 1B catalog has been calculated by several groups.[4,9,11] Excess correlations were noted near both ends of the defining range $(0,\pi)$. The $\sim 4°$ excess was interpreted as the signal of burst recurrence,[4] and the excess near $\theta = 180°$ was used to argue against that interpretation.[11] Figure 1 shows the angular correlation function for 742 BATSE bursts, showing no statistically significant excess on any angular scale, which implies a repeater fraction below $f \sim 10\ \%$. The NN pdf for three different samples is shown in Figure 2. The upper right panel shows the observed distribution for the 1B set in comparison to perfect isotropy. One and two sigma statistical error curves are also shown. In the upper left panel we show the expected NN pdf if a repeater fraction f is assumed; the angular resolution function is identical to that used to calculate $w(\theta)$. This particular figure shows the results for the case of two bursts per repeating source. It is clear from Figure 2 that the shape of the observed pdf does not match well the theoretically expected shape. This indicates that simple recurrence is not the cause for the observed deviations. To test the assumption that the deviations are only statistical, we calculated the NN pdf for two different samples (with N=260) drawn from the post-1B set. The lower panel of Figure 2 shows preliminary results for these sets, and it is clear that bursts observed subsequent to those in the 1B catalog support the notion that the 1B deviation was purely statistical.

In summary, we have calculated the correlation function $w(\theta)$ and the NN pdf of N = 742 γ-ray bursts observed with BATSE. No correlations above random were found. The NN pdf of two subsets drawn from post-1B data were used to test the interpretation of the small angle excess found in the NN pdf of the 1B catalog. The pdfs of both sets are consistent with random sky distributions. We thus conclude that the BATSE data are consistent with the traditional view that classical GRBs do not repeat (on time scales shorter than ~ 1 year).

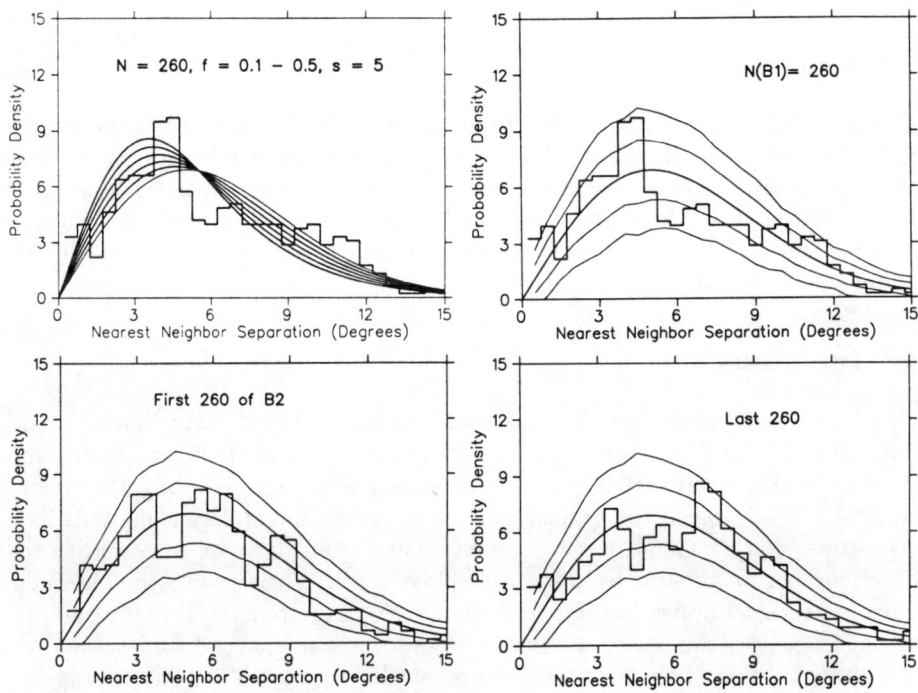

Figure 2 — Nearest neighbor probablity density. Upper left shows 1B observations and expected distributions with various repeater fractions. Upper right shows 1B data, for zero repeaters, with $\pm 1, 2\ \sigma$ statistical deviations from the expected mean. The bottom panel shows the results for two samples drawn from post-1B data.

REFERENCES

1. Meegan, C. A., et al., Nature **355**, 143 (1992).
2. Paczynski, B., Acta Astr. **41**, 257 (1991).
3. Atteia, J.-L., et al., ApJS **64**, 305 (1987).
4. Quashnock, J., & Lamb, D. Q., MNRAS , in press (1993).
5. Fishman, G. J., et al., ApJS , in press (1994).
6. Wang, V. C. & Lingenfelter, R. E., ApJL **416**, L13 (1993).
7. Hartmann, D. H., & Epstein, R. I., ApJ **346**, 960 (1989).
8. Peebles, P. J. E., The Large-Scale Structure of the Universe (Princeton Univ. Press, 1980).
9. Hartmann, D. H., et al., ApJ , submitted (1993).
10. Scott, D. & Tout, C. A., MNRAS **241**, 109 (1989).
11. Narayan, R., & Piran, T., MNRAS , in press (1993).

DO GAMMA-RAY BURST SOURCES REPEAT?

Ramesh Narayan and Tsvi Piran [1]

Harvard-Smithsonian Center for Astrophysics
60 Garden Street, Cambridge, MA 02138

Abstract

We show that the gamma-ray bursts in the BATSE 1B Catalog have an excess of antipodally related pairs of bursts, i.e. pairs separated by almost 180° on the sky. The antipodal excess is comparable to the excess of nearest neighbor pairs discovered by Quashnock and Lamb and which they interpret as evidence for repeating bursts. Since repeaters cannot produce antipodally related pairs we argue against repeaters.

1 Introduction

Quashnock and Lamb (1993, hereafter QL) have analyzed the distribution of angular positions of 260 classical gamma-ray bursts in the BATSE 1B Catalog and claim to find a significant excess of nearest neighbor pairs with angular separations $\lesssim 4° - 5°$. Since the typical position error in the cataloged bursts is $\gtrsim 4°$ they argue that a significant fraction of the bursts must have repeated during the observing period covered by the sample (one year). Their results, if confirmed, have important implications for models of gamma-ray bursts. If many gamma-ray burst sources are repeaters, then it would rule out models based on neutron star mergers, and would pose a serious constraint to most if not all cosmological models in view of the energy requirement.

We reanalyse the angular distribution of bursts using the two-point correlation function rather than the nearest neighbor statistic employed by QL. (The two approaches are closely related though not identical, e.g. Peebles 1980, pp. 162-163). We conclude from our analysis that although there may be some peculiarities in the angular distribution of bursts, suggesting possible selection effects, there is in fact no evidence for repeaters.

2 Correlation Function Analysis

We test the data against the null hypothesis that the bursts are distributed uniformly in the sky. (We ignore the fact that BATSE has non-uniform sky coverage; this can produce non-uniformities on angular scales $\gtrsim 30°$ but not on the much smaller scales considered here.) We consider all pairs of bursts in the 1B Catalog and bin them according to their angular separations. In each bin we calculate the correlation function, $w(\theta) = (N_{obs}/N_{exp}) - 1$, where N_{obs} is the number of observed

---

[1] Permanent address: Racah Institute for Physics, The Hebrew University, Jerusalem 91904, Israel

Table 1: Results of Correlation Function Analysis

Sample	N_{burst}	Bin Size	Forward Peak			Antipodal Peak		
			$w(\theta)$	No. of σ	p	$w(\theta)$	No. of σ	p
All bursts	260	4°	0.268	1.75σ	0.096	0.292	1.86σ	0.072
		5°	0.389	3.17σ	0.0024	0.327	2.71σ	0.011
$\Delta\theta_P \leq 4°$	131	4°	0.254	0.84σ	0.42	0.832	2.69σ	0.010
		5°	0.173	0.72σ	0.52	0.852	3.39σ	0.0009
$\Delta\theta_P \leq 5°$	148	4°	0.359	1.32σ	0.21	0.661	2.41σ	0.021
		5°	0.256	1.17σ	0.26	0.788	3.55σ	0.0006
$\Delta\theta_P \leq 9°$	202	4°	0.537	2.70σ	0.012	0.335	1.68σ	0.10
		5°	0.579	3.65σ	0.0003	0.424	2.65σ	0.0087

burst pairs that fall into the bin, N_{exp} is the number expected for a uniform distribution, and θ is the angular separation corresponding to the center of the bin. A positive value of $w(\theta)$ signifies an excess of bursts at the corresponding θ. We use a Monte Carlo method to determine the significance of the various positive and negative deviations in $w(\theta)$. We generate 10^4 independent sets of random burst positions and use these to estimate the probability p that any particular deviation in $w(\theta)$ can happen by chance. In addition, we also calculate the standard deviation σ of the fluctuations in each bin and divide the observed $w(\theta)$ by the corresponding values of σ to obtain the deviation as a multiple of the standard deviation. Table 1 summarizes the results.

Figure 1(a) shows $w(\theta)$ in 4° bins for the full sample of 260 bursts in the 1B Catalog. We see that $w(\theta)$ in the very first bin has a fairly large positive deviation, corresponding to 1.75 standard deviations. The probability that a deviation as large as this can occur by chance is fairly small; we find $p \sim 0.096$. These results are given in Table 1 in the columns identified as "Forward Peak." The existence of a peak in $w(\theta)$ for $\theta < 4°$ confirms QL's claim that there is an excess of nearest neighbor pairs for small angular separations. However, note that there is another peak in $w(\theta)$, in the last bin, which corresponds to angular separations $176° < \theta \leq 180°$. This peak implies that there is an excess of pairs of *antipodally* related bursts, i.e. bursts separated by nearly 180° on the sky. As Table 1 shows (look under "Antipodal Peak"), the antipodal peak is slightly more significant than the nearest neighbor peak.

Clearly, repeating bursts cannot produce an excess of antipodally related pairs of bursts. Since the two peaks in $w(\theta)$ appear to be similar in magnitude, we prefer to seek a common explanation for both peaks. Selection effects come to mind as a possibility.

One of the points that QL make in support of their repeater hypothesis is that the excess of nearest neighbors occurs on an angular scale of $\sim 4° - 5°$, which is apparently consistent with the estimated position errors of many bursts. In order to check this argument we have investigated the role that position errors play in the two peaks seen in $w(\theta)$.

The burst positions in the 1B Catalog have two kinds of errors – an rms error of $4°$ which is present in all bursts because of systematic effects, and an additional Poisson (or statistical) error $\Delta\theta_P$ which varies from $< 1°$ for the brightest bursts to $> 15°$ for faint bursts. Since the two peaks in the correlation function have widths $\sim 4-5°$, we expect that the respective signals will become even stronger if we restrict our attention only to those bursts with well-determined positions. Figure 1(b) shows an example where we eliminate all bursts with $\Delta\theta_P > 5°$, so that the remaining bursts have total position errors (systematic and statistical) under $6.4°$. This sample has 148 bursts and we have binned $w(\theta)$ in $5°$ bins. Notice that the nearest neighbor peak is significantly weaker here (1.17σ) compared to the peak in Fig. 1(a). This is surprising. It means that most of the signal in the original full-sample correlation function is produced by bursts with much poorer positional accuracy than the width of the peak, a puzzling state of affairs which seriously undermines the QL interpretation of the excess. In contrast, the antipodal peak in Fig. 1(b) is significantly stronger (3.55σ) than in Fig. 1(a). The signal in this peak at least does seem to be dominated by bursts with accurate positions.

Table 1 shows results for several other cuts of the data. Both the nearest neighbor and antipodal peaks appear with high significance in several of the data subsets, suggesting that there may indeed be some odd features in the angular correlations of the BATSE bursts. However, the nearest neighbor peak seems to be quite volatile in its strength and generally inconsistent with the position errors of the bursts.

3 Conclusions

We reject the repeater hypothesis put forward by QL for two reasons:

1. As shown in Table 1, there is roughly comparable evidence for the nearest neighbor and antipodal peaks in the angular correlation function $w(\theta)$. In our view, any explanation we advance should explain both peaks. This rules out repeaters since they can only produce the forward peak.

2. We find that the signal for the nearest neighbor peak is produced mostly by bursts with position errors much larger than the width of the peak. If this peak is due to a real spatial correlation of sources, then it is hard to see why the peak in the correlation function would be *narrower* than the position errors. We argue that this indicates a serious inconsistency in the QL interpretation of the peak in terms of repeaters. Curiously, the antipodal peak does show the right behavior as a function of position errors.

We have been unable to come up with any convincing explanation for the two peaks in the correlation function. Perhaps both peaks arise from an unusual selection effect. Alternatively, the entire effect may be caused by a statistical fluctuation, which is perhaps not so unlikely if the "signals" were obtained through a large number of "trials".

Future tests with a larger sample of BATSE bursts should clarify the situation. If the nearest neighbor peak survives with comparable amplitude to that seen in Fig. 1, and if the antipodal peak reduces significantly, then we should clearly take the repeater hypothesis more seriously. On the other hand, if the antipodal peak remains in evidence, either with or without the forward peak, then the case for selection effects will become stronger. Of course, if both peaks disappear (see the review article by C.A. Meegan in this volume), we can conclude that the signal discovered by QL and our competing antipodal peak are both the result of statistical fluctuations. In any case, on the basis of the data at hand, our conclusion is that there is no evidence for repeaters among the classical gamma-ray bursts.

Acknowledgement: This work was supported in part by grant NAGS-1904 from NASA.

Reference

Narayan, R. & Piran, T. 1993, MNRAS, in press.
Peebles, P.J.E. 1980, Large Scale Structure of the Universe, Princeton Univ. Press.
Quashnock, J.Q. & Lamb, D. 1993, MNRAS, in press.

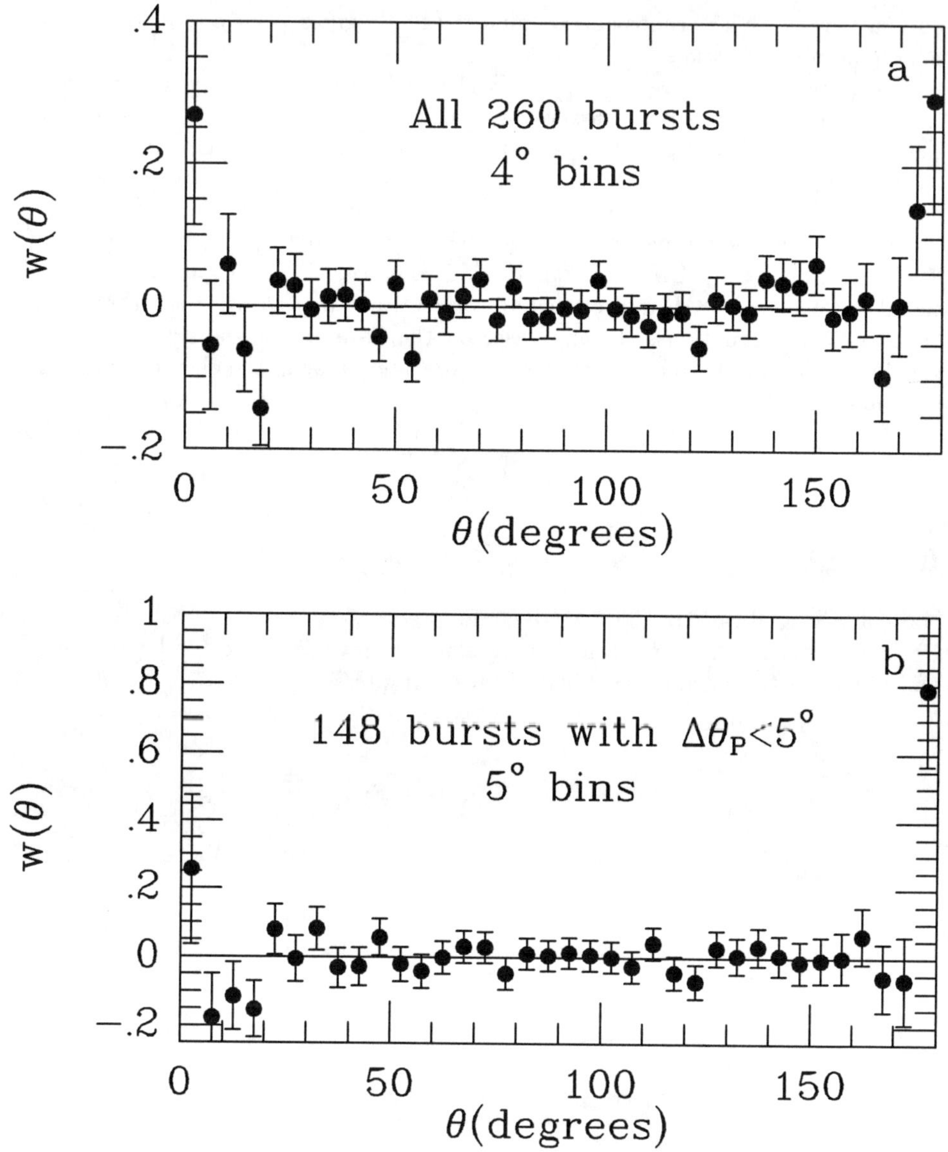

Figure 1: (a) Two-point angular correlation function $w(\theta)$ of all 260 bursts in the BATSE 1B Catalog, shown in 4° bins (taken from Narayan & Piran 1993). Note the nearest neighbor and antipodal peaks in the first and last bin respectively. (b) Correlation function for the 148 bursts with statistical position errors $\Delta\theta_P \leq 5°$, shown in 5° bins The nearest neighbor peak has reduced in amplitude but the antipodal peak appears to be highly significant.

CORRELATIONS WITH GAMMA-RAY BURSTS

Robert J. Nemiroff, Gabriela F. Marani, and Juan R. Cebral

George Mason University, CSI Institute, Fairfax, VA 22030

Jay P. Norris

NASA Goddard Space Flight Center, Greenbelt, MD 20771

ABSTRACT

We performed correlation function analyses involving the first 260 BATSE detected gamma-ray bursts (GRBs) searching for evidence of bunching or repetition. The BATSE GRB two-point angular auto-correlation function shows excesses at small and high angles as noted previously by Quashnock and Lamb (1993), and Narayan and Piran (1993). The BATSE GRB two-point temporal correlation function shows no significant excesses at any times but does show a significant dip on the time scale of hours. This dip is real and corresponds to BATSE ignoring dim bursts while processing bright bursts. Even when constrained to bursts with recorded angular separations less than 20°, no temporal excess is found. Therefore if GRBs repeat, this analysis was unable to locate any obvious repetition timescale. We have computed the general GRB - Abell cluster angular cross-correlation function and again find no significant peaks. In sum, the times and positions of the first 260 GRBs released into the public domain show no significant evidence, in our opinion, that GRBs correlate with themselves or Abell clusters.

ANALYSES AND RESULTS

We computed the two-point temporal auto-correlation function for the 260 BATSE GRBs to investigate the possible repetition of GRBs. The temporal correlation function for these GRBs without restriction on their angular separations is shown in Figure 1. We can see no evidence for significant excess, thus we find no significant indication of repetition on any timescale. We then restricted the correlation analysis to GRBs with angular separations less than 20°, the results of which are shown in Figure 2, and again see no significant excess. In both cases there is a significant dip on the time scale of hours which is due to the fact that BATSE ignores a dim burst while processing a brighter burst. We again find no significant evidence for repetition on any time scale.

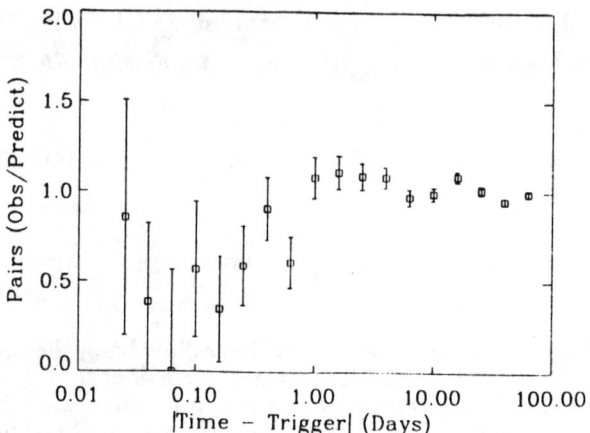

Figure 1. General GRB two-point temporal correlation function.

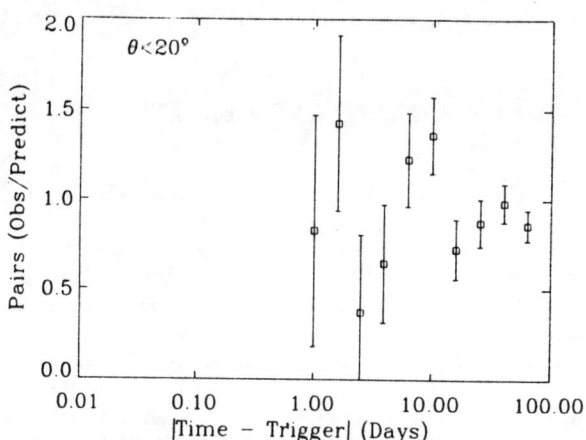

Figure 2. GRB two-point temporal correlation function for angular separations less than 20°.

Next, we computed the angular auto-correlation function for GRBs, including angular errors. For each GRB pair, a 1σ angular error circle is determined. The comparison GRB is then computationally broken up into 10 pieces which are thrown in a manner approximating a random Gaussian distribution, centered on the GRB position, each piece carrying the weight of 1/10th of a GRB.

The results, shown in Figure 3, suggest no excess of correlation for angular separations between 5° and 175°, but ~2σ peaks for angular separations less than 5° and greater than 175°. These peaks, found previously by Quashnock & Lamb[1] and by Narayan & Piran,[2] can be explained by statistical fluctuation or a possible bias in the determination of the burst position (see Maoz[3]). Thus, these results indicate that there is no clearly significant evidence for clustering or bunching of GRBs.

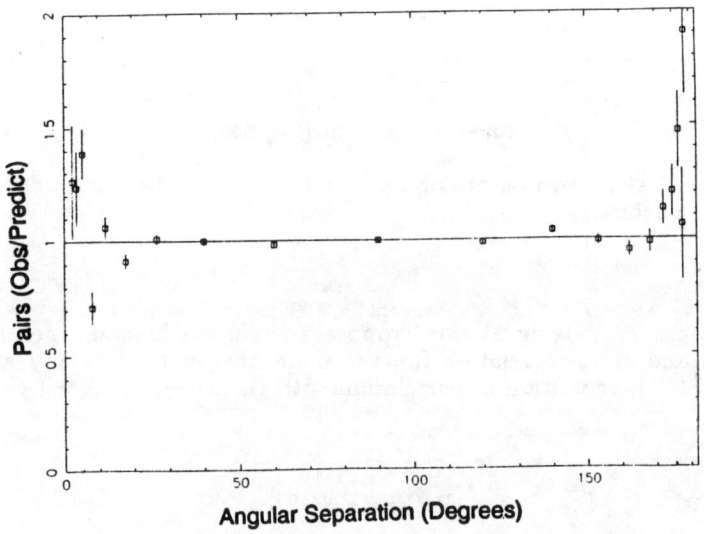

Figure 3. GRB two-point angular auto-correlation function.

Finally we searched for a correlation between GRBs and Abell clusters.[4-5] The Abell clusters lack coverage in the Galactic plane, so we didn't take into account GRBs and clusters with galactic latitude between -30° and 30°. The two-point angular cross-correlation function is shown in Figure 4. No significant peaks were found.

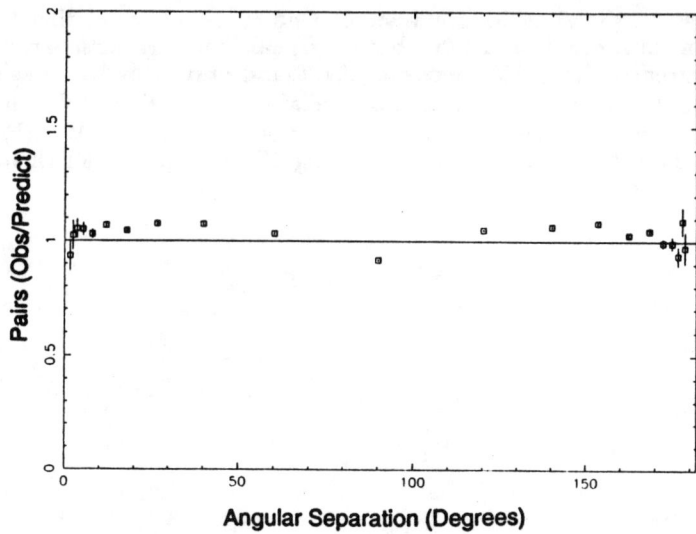

Figure 4. GRB two-point angular cross-correlation function between Abell clusters.

In summary, looking at the temporal correlation function, angular autocorrelation and cross-correlation functions, we have not found any significant evidence of GRB repetition or correlation with themselves or Abell clusters.

REFERENCES

1. J. M. Quashnock and D. Q. Lamb, In press: MNRAS , (1993).
2. R. Narayan and T. Piran, Submitted to MNRAS , (1993).
3. E. Maoz, Submitted to Ap. J. Lett , (1993).
4. G. O. Abell, Ap. J. Supp. **3**, 211 (1958).
5. G. O. Abell, H. G. Corwin Jr., and R. P. Olowin, Ap. J. Supp. **70**, 1 (1989).

ANALYSIS OF THE DENSITY FLUCTUATIONS IN THE GAMMA-RAY BURSTS DISTRIBUTION

Gely F. Zharkov and Vladimir G. Kurt

P.N.Lebedev Physical Institute, Moscow, 117924, Russia

Vladimir G. Zharkov

Institute of Biology of Gene, Moscow, 117984, Russia

ABSTRACT

Usual correlation-function analysis of the gamma-ray bursts spatial distribution shows overall no statistically significant deviations from the random distribution law. However, using the method of local statistical analysis, formulated in [1-4], we have analyzed the distribution of bursts, registered up till now by BATSE, and have found an indication, that there exists significant high-density fluctuation of events in vicinity of the Andromeda nebula. This finding may support the hypothesis of the Galactic origin of the gamma-ray bursts.

To know the sites of significant fluctuations in the observed distribution of the gamma-ray bursts may be of use in discussing the problem of their origin. The usual methods (such as correlation-function method and the like) resort to averaging the quantities over all points of the ensemble. For this reason such methods can reveal the presence of deviations from the mean values on the average only and, consequently, they are insensitive to the presence of single fluctuation (though, it may be large). We are trying to develop [1-4] another approach to this problem, use of which may reveal anomalies in the gamma-source distribution, when the usual methods are ineffective (see, for example, ref. 5). We'll present the main ideas of the new method on the simplest example, when of interest is only the information about the density distribution of events in an ensemble of observational data.

Let M be the total number of points in the ensemble, which are positioned on a two-dimensional surface of the sphere of unit radius. To every point of the ensemble belongs an area $S_1 = S_0/M$, where $S_0 = 4\pi$ ster is the total surface of the sphere. If this area is a circle, then it would be seen from the center of the sphere within a cone of angle θ_1 (the circle's radius). We shall call a window such a circle area, cut on the sphere. Thus, the mean angular distance between the points is $L = 2\theta_1$ (the circle's diameter). In a general case, the window's radius can be taken as $r_w = \varepsilon L$, where ε is an arbitrary number. The window's area will be denoted as S_w, and the area outside the window as $S_{out} = S_0 - S_w$.

If N particles (or points) are randomly thrown on the sphere, implying the equal access to every part of the sphere, then the probability of one particle to fall into the window is $p_{in} = S_w/S_0$, and the probability to to fall outside the window is $p_{out} = S_{out}/S_0$. The expected number of particles inside the window would be, on the average, $n_* = Np_{in}$.

Let us make the window's center coincide with some point j of our ensemble, and let the number of neighbors to the chosen point be n (the total number of particles inside the window is $m = n+1$). The difference between the observed value n and the expected n_* would indicate the presence of the density fluctuation in the window, located at the position j. We'll estimate below the magnitude of this fluctuation.

When $N = M - 1$ particles are randomly scattered over the sphere, the different configurations $G = \langle m, M - m \rangle$ will arise, according to the resulting window's occupation values m, the rest $M - m$ of the particles falling outside the window. The probability to obtain a configuration G is, evidently, the function of a neighbors number $n = m - 1$:

$$P(n) = \frac{N!}{n!(N-n)!} p_{in}^n p_{out}^{N-n}. \tag{1}$$

Evidently, the condition is fulfilled

$$\sum_{n=0}^{N} P(i) = 1. \tag{2}$$

The probability $P(n)$ is a local characteristic of the window's state, it depends on the window's radius r_w. For given r_w the states with smaller $P(n)$ would be more peculiar. Making the window's center coincide consecutively with all points of the ensemble and calculating $P(n)$, one can find the most peculiar configurations, as well as their positions, because every configuration is tied to the specific point j with known coordinates.

When looking through all the windows, opened around every point of the ensemble, it may happen, that some configuration $G = \langle m, M - m \rangle \equiv G(n)$ is repeated k_w times. The probability W of this configuration being realized exactly k times in M attempts of random scattering (into every of M windows) is given by the binomial distribution [6]

$$W(k) = M! \frac{P^k}{k!} \frac{(1-P)^{M-k}}{(M-k)!}, \tag{3}$$

here $P = P(n)$ is the probability of configuration (1). Evidently, the probabilities of repetitions $W(k)$ satisfy the condition

$$\sum_{k=0}^{M} W(k) = 1. \tag{4}$$

The most probable repetition number of the particular configuration $G(n)$ is $k_* = MP(n)$. A large difference between the observed repetition number k_w and the expected one k_* would also indicate, that the given configuration $G(n)$ is peculiar. The measure of it's peculiarity may be estimated as follows.

Let k_0 be integer, nearest to the expected repetition number k_*, and $\Delta = | k_0 - k_w |$. Introducing in the usual way [6] the statistical confidence

$$C(\Delta) = \sum_{k=k_0-\Delta}^{k_0+\Delta} W(k) \tag{5}$$

as the sum of probabilities, laying inside the interval $dk = \pm\Delta$, one can define the confidence interval $\Delta(\alpha)$ such, that $C(\Delta) = \alpha$, where $\alpha \leq 1$ is some number, sufficiently close to unity. If the observed repetition number k_w lays outside the confidence interval $\Delta(\alpha)$, then (with probability α) one can assert, that the observed configuration $G(n)$ can not be attributed to the random distribution law, i.e. it should be recognized as anomaly. Instead of statistical confidence $C(\Delta)$ one can consider the statistical doubt $D(\Delta) = 1 - C(\Delta)$ (compare Eqs. (3) and (4)), then the configurations with small values of $D < 1 - \alpha$ would be anomalous.

The special computer program (based on principles sketched above) was written to process the experimental data. Starting this work we had in our disposal $M = 260$ gamma-ray bursts (GRB) locations, as registered by BATSE [7]. The mean interparticle angular distance for this ensemble is $L = 14.2°$. The following procedure was used. We specified the window's radius $r_w = \varepsilon L (0.1 \leq \varepsilon \leq 4$, step $d\varepsilon = 0.1)$, and found the number of neighbors around every point and the corresponding repetition numbers k_w. The windows with maximal population were of special interest. The probability $P(n)$ to observe n neighbors was calculated according to Eq.(1) and the statistical doubt $D(\Delta) = 1 - C(\Delta)$ according to Eq.(5). The peculiar window was recognized as the one, having sufficiently small value of $D(\Delta)$.

In this way we found, that the window $\varepsilon = 0.7$ (located at the point with the Galactic coordinates $l = 92.3°, b = -31.1°$) has the smallest value $D = 7 \cdot 10^{-3}$. The observed population in this window is $m = 10$, as compared with the expected one $m_* = 1.96$. In another words, the probability to observe such a fluctuation on the basis of the random distribution law is less, than 10^{-3}. Note, that this window lays in the vicinity of the Andromeda nebula ($l_A = 110°, b_A = -20°$). (Note also, that the maximal difference $d = m - m_*$ for this window is $d = 8$, whereas the averaged over all 260 points value is $\bar{d} = 0.05$. These numbers illustrate, why the methods, using the averaging procedure, are insensitive to the presence of a single fluctuation.) Most of other windows also show the high density fluctuation in this region, though with somewhat larger doubts D.

Recently, the new portion of BATSE data ($M = 442$) was made public. We have processed the combined ensemble of $M = 260 + 442 = 702$ points ($L = 8.7°$) in the same manner and found, that the smallest value $D = 2 \cdot 10^{-3}$ for this ensemble belongs to the window with $\varepsilon = 3.7 (m = 83, m_* = 53.4)$, which is also located in the vicinity of the Andromeda nebula.

From these estimates it seems justifiable to make the conclusion, that there exists a significant high density fluctuation of GRB in the Andromeda region, which can not be explained by the random distribution law. This conclusion favors the hypothesis of the Galactic origin of gamma-ray bursts. (Note, that the analysis of angular distribution of the BATSE-data, carried out by different method in [8], has also produced some evidence of excessive bursts population in the Andromeda nebula direction.)

It is well known, that the coordinates of gamma-ray sources have been measured with experimental errors, which lay within the limits $1° \leq R_{er} \leq 25°$, where R_{er} is the positional error box radius [7]. To estimate the influence of the experimental errors on the results, obtained by using the method discussed, we adopted, for simplicity, some average radius $R_{er} = 5°$, common to all points, and randomized the positions of all M points of the ensemble within the circle, having the center at a given point and the radius $R_{er} = 5°$. The emerging new ensemble was processed by the method, presented above. It has turned out, that the high density fluctuation in the Andromeda nebula region survived, i.e. the conclusion about it's existence does not depend crucially on the precision of measurements. When the radius of random scattering R_{er} was increased gradually, this local high density fluctuation disappeared.

It should be noted, that the assessment of the experimental errors in determining the event's coordinates is liable to serious changes [7]. Besides, due to technical reasons, the sky exposition times differed significantly for different angles of observation [7]. Consequently, the number of registered events and their coordinates, probably, do not represent adequately the true picture of the GRB distribution. For these reasons, the conclusion concerning the presence of high density fluctuation in the Andromeda nebula region, which was obtained by processing the presently available data, may suffer a change, as the new information becomes available. However, in view of the importance of the problem discussed, we decided to present the method used and report the preliminary results of the investigation.

REFERENCES

1. V. G. Zharkov and G. F. Zharkov. "An approach to the study of the correlation effects in the finite ensemble of point objects", Preprint FIAN, **N82**, 1-15 (1991).
2. V. G. Zharkov and G. F. Zharkov, Short Commun. on Physics, **N7/8**, 62 (1993).
3. G. F. Zharkov and V. G. Zharkov, Short Commun. on Physics, **N7/8**, 68 (1993).
4. G. F. Zharkov and V. G. Zharkov, Astrophysical Journal , in press (1994).
5. B.V.Komberg, V.G.Kurt, Astronomical and Astrophysical Transaction , in press (1993).
6. Emlyn Lloid, ed., Handbook of applicable Mathematics, v.II (New-York: Willey, 1984).
    Probability Theory (New York: Willey).
7. BATSE bursts catalog obtained via Internet (GROSSC.GSFC.NASA.GOV), 1993.
8. A.V.Gurevich, G.F.Zharkov, K.P.Zybin, M.O.Ptitsyn. Phys. Lett., **A181**, 289 (1993).

SEARCH FOR PERIODICALLY-REPEATING GAMMA-RAY BURSTERS FROM BATSE

K. W. Chuang

Institute of Geophysics and Planetary Physics
University of California, Riverside, CA 92521

ABSTRACT

We investigate 260 triggered cosmic gamma-ray bursts, detected by the BATSE detector on the Compton Gamma Observatory from April 21, 1991 until March 5, 1992 (320 days), with the Fourier Power Spectrum method to search for possible periodically-repeating gamma-ray bursters. September 25, 1991 gamma-ray burst, GRB910925 (trigger no. 816), could be a repetitive member of SGR1806-20. Seven years after the last member of the Soft Gamma Repeater SGR1806-20 that was discovered in 1984. Seventeen candidates of burst repeaters with two-members and four of three-members were found. No significant evidence was found for periodicities of the three known SGRs and the new BATSE candidates.

1. INTRODUCTION

Two decades after the discovery of the cosmic gamma-ray bursts first announced by Klebesadel, Strong, and Olson[1], about 1500 gamma bursts have been discovered with many space probes (e.g., PVO, ISEE-3, SMM, V-11, 12, 13, and 14, GRO,..., etc); so far only three soft gamma repeaters (SGRs), SGR1806-20 (GRB790107)[2,3], SGR0526-66 (GRB790305)[4], and SGR1900+14 (GRB790324)[5] are known. One hundred and eleven members of SGR1806-20, were detected by the International Comentary Explorer (ICE) spacecraft and reported by Laros, et al. (1986)[6], Laros (1986)[7], and Hurley (1986)[8] from 1978 August 13 to 1986 June 27. The pattern of burst repetition in source SGR1806-20 appears stochastic[10]. The famous March 5, 1979 event was the initial superburst of the source of SGR0526-66 whose error box[13] lies within supernova remnant N49 in the Large Magellanic Cloud (LMC). Totally sixteen recurrent bursts in SGR0526-66, observed by KONUS experiments onboard Veneras-11, 12, 13, and 14, were discovered between 1979 and 1983. These bursts had temporal signatures essentially indistinguishable from the repeating soft bursts from the March 5 source[10]. Finally 6 events including 3 observed by V-11 and 12, GRB790324, GRB790325, and GRB790327, and 3 by BATSE[9], GRB920619a&b, GRB920708, and GRB920819 were from SGR1900+14.

Most recurrent bursts, observed between 1979 and 1984 from the three SGRs, exhibit simple light-curves with very short duration ($\sim$0.1 s) and very fast rise times ($<$ 5 ms)[9,10]. Their spectra are much softer than classical gamma-ray bursts (GRBs)[9,10] and can be well fitted with an optically-thin thermal bremsstrahlung (OTTB) function with temperatures between 30-40 ke$V^{14,15}$. The periodicity for each SGR is still unknown although Laros, et al. (1987)[3] claimed that the likely periodicity for SGR1806-20 could be 4 months.

Table 1 No. of Bursts and Repeaters in The Sky

Sections	1	2	3	4	5	6	7	8
No. of GRBs	35	34	34	25	38	29	31	34
No. of Repeaters	3	4	3	2	2	1	4	2

Two hundred and sixty gamma-ray bursts, observed by the BATSE experiments on the Gamma Ray Observatory (GRO) from April 21, 1991 to March 5, 1992 (320 days), are used to search for possible periodically-repeating gamma-ray bursters. We briefly describe the method to find the burst repeaters, and then present the results.

2. METHOD AND RESULTS OF SEARCHING FOR THE PERIODIC BURST REPEATERS

From the BATSE burst catalog covering April 21, 1991 to March 5, 1992, 260 events discovered in 320 days are collected into eight whole sky sections in galactic longitude l ($0^0 \sim 360^0$) and latitude b ($-90^0 \sim 90^0$). Sections 1, 2, 3, 4, 5, 6, 7, 8 have longitudes (deg) and latitudes (deg) of $0^0 \sim 90^0$, $0^0 \sim 90^0$; $0^0 \sim 90^0$, $90^0 \sim 180^0$; $0^0 \sim 90^0$, $180^0 \sim 270^0$; $0^0 \sim 90^0$, $270^0 \sim 360^0$; $-90^0 \sim 0^0$, $0^0 \sim 90^0$; $-90^0 \sim 0^0$, $90^0 \sim 180^0$; $-90^0 \sim 0^0$, $180^0 \sim 270^0$; and $-90^0 \sim 0^0$, $270^0 \sim 360^0$. Table 1 shows the total number of BATSE events in each section are isotropically distributed in agreement with the earlier BATSE results[17,18].

The systematic uncertainties in angular directions for BATSE's gamma-ray bursts are a few degrees[11,12,16]. Forty six events whose differences in the galactic longitude l and latitude b are $\leq 4^0$, are selected as candidates for members of burst repeaters. Table 1 also shows the total number of BATSE candidates of burst repeaters in each section,12 candidate repeaters in the northern hemisphere and 9 in the southern. With the criterion of 5.5 σ above the average background We expect 2 or 3 chance coincidence from each burst repeater in a time period of 320 days. Seventeen burst repeater candidates with two-members are shown in Table 2, eight in the northern hemisphere and 9 in the southern. Table 3 shows four candidate groups of three-members all in the northern hemisphere. The number of members (i.e., only 2 or 3 events) of each new repeater candidate are not enough to find the periodicity of each repeater candidate.

We again review the BATSE burst catalog and use the same criterion to find any other events whose directions are within the error boxes one of the three known SGRs, SGR1806-20, SGR0526-66, and SGR1900+14. One of the 260 BATSE bursts, GRB910925 (trigger no. 816) with right ascension $\alpha_{2000} \sim 268.1^0$ and declination $d_{2000} \sim -22.7^0$, has a direction near that of GRB790107 (SGR1806-20), $\alpha_{1950} \sim 271.5^0$ and $d_{1950} \sim -20.0^0$. Two kinds of durations for BATSE's gamma-ray bursts are T50 and T90. T50 is defined as the time interval during which the integrated counts go from 25% to 75% of the total integrated counts and T90 is defined from 5% to 95% instead. The T50 and

Table 2. The candidates of two-member repeaters from BATSE

910718	920203	911031	911117	910602	910607	910625	910715
910809	920209b	910430	910501	911111b	920218c	910621	911219
910918	911031b	910730	910803	910718b	911004	910502	910526
910604	910916	910629b	911101	910802	911111	910705b	910813
910905	920120						

Table 3. The candidates of triple-member repeaters from BATSE

910903	911005	920227c	910517b	920226b	920305
910523	910526b	910823	910424	910614b	911202

T90 of GRB910925 are 23.808 s and 90.944 s, respectively, much longer than the durations (FWHM) of the members of SGR1806-20 (e.g., about 0.1 s). The energy fluence of GRB910925 is a factor of 2 below the energy fluence of GRB790107. The information of the rise and fall times of GRB910925 are not available this time. We note that the event GRB910925 is a weak or soft burst with direction near the source SGR1806-20. Its time duration is much too long to be a member of SGR1806-20. Statistically we expect 2 or 3 members and see one. The GRB910925 event is likely a chance coincidence.

Table 4 gives the total number of recurrent events each year for each SGR. Many repeated events were detected from those 3 known SGRs over 10 years[6,7,8,9,10] except the time period between 1985 and 1990. The gap occured because very few spacecraft operated during that time period, PVO was functioning but only detecting strong or hard bursts for photon energy $\geq$100keV. Three candidate recurrent bursts from SGR1900+14 were detected by BATSE after 13 years, and one from SGR1806-20 after 6 years. None of the recurrent events from SGR0526-66 were found after 1983. The time period for repeating events from a source may be obtained at a high level of significance with Fourier Transform. The power spectrum is the relative power as a function of frequency. Figures 1a and 2a show the total number of recurrent events per year vs. time and the power spectrum vs. frequency of SGR1806-20. The recurrent events and the power spectrum for SGR0526-66 are shown in Fig. 1b and 2b and for SGR1900+14 in Fig. 1c and 2c.

The peaks seen on the graphs 2a, 2b, and 2c are probably due to aliasing of Fourier Transforms, because the data needs to be taken for longer time periods, assuming the same time interval. Therefore no significant evidence for periodicities in the three known SGRs are found. More years of data will have to be taken in order to see any periodicities for these sources, especially because of the irregularity now in the data. For example, there are more bursts in 1983 for SGR1806-20 than all the other events in that chart combined, that makes it hard to find a periodicity for that source. Also the SGR1900+14 only had bursts in 1979 and 1992, it is obvious that many more years of data will be

Table 4 No. of recurrent events of each year for the three known SGRs

Sources	79	80	81	82	83	84	85-90	91	92
SGR1806-20	2	3	4	0	92	10	0	0	
SGR0526-66	4	0	1	9	2	0	0	0	
SGR1900+14	3	0	0	0	0	0	0	0	3

needed to see if this could be a periodic occurrence or just a coincidence. The long term monitoring of the source by BATSE gives hopes of obtaining valuable information on its recurrence time scale.

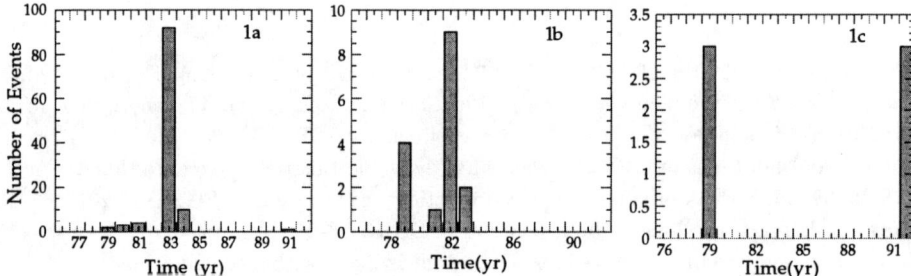

Figure 1. Total number of recurrent events per year vs. time for (1a)SGR1806-20, (1b)SGR0526-66, (1c)SGR 1900+14.

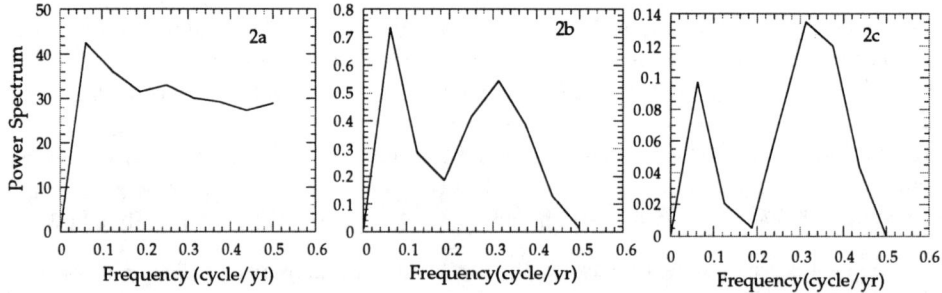

Figure 2. The power spectrum vs. frequency for (2a)SGR1806-20, (2b)SGR0526-66, (2c)SGR 1900+14.

ACKNOWLEDGEMENT

I wish to thank Dr. G. Chanan and also the physics department of UC Irvine for their hospitality. I am mostly grateful to F. Dekens for his help of developing the Fourier transform routine and helpful discussions. It is a great pleasure to acknowledge partial support for this work provided through IGPP at UCR.

REFERENCES

1. R. W. Klebesadel, I. B. Strong, and R. Olson, Ap.J.(Letters)182, L85, 1973
2. J. L. Atteia, et al., Ap.J. 320, L105, 1987
3. J. G. Laros, et al., Ap.J.(Letters) 320, L111, 1987
4. S. V. Golenetskii, et al., Nature, 307, 41, 1984
5. E. P. Mazets, et al. Soviet Astr. Lett., 5, 343, 1980
6. J. G. Laros, et al., Nature 322, 152, 1986
7. J. G. Laros, talk presented at the Taos Gamma-Ray Stars Conference Taos, New Maxico, 1986
8. K. Hurley, talk presented at the Taos Gamma-Ray Stars Conference Taos, New Maxico, 1986
9. C. Kouveliotou, et al., Proceedings of Compton Gamma-Ray Observatory Workshop, St. Louis, Mo, A.I.P. New York, 1992, 280, 882
10. J. P. Norris, et al., Ap.J.(Letters) 255, L45, 1991
11. G. J. Fishman, et al., Proceedings of Gamma-Ray Bursts Workshop, Huntsville, Alabama, A.I.P. New York, 1991, 265, 13
12. M. N. Brock, et al., Proceedings of Gamma-Ray Bursts Workshop, Huntsville, Alabama, A.I.P. New York, 1991, 265, 383
13. T. L. Cline, et al., Ap.J.(Letters) 255, L45, 1982
14. C. Kouveliotou, et al., Ap.J.(Letters) 322, L21, 1987
15. E. P. Mazets, et al., Ap. Space Sci. 80, 1, 1981
16. S. Howard, et al., Proceedings of Compton Gamma-Ray Observatory Workshop, St. Louis, Mo, A.I.P. New York, 1992, 280, 793
17. C. Meegan, et al., Proceedings of Compton Gamma-Ray Observatory Workshop, St. Louis, Mo, A.I.P. New York, 1992, 280, 681
18. C. Meegan, et al., Nature 355, 143, 1992

SEARCHING GAMMA-RAY BURSTS FOR GRAVITATIONAL LENSING ECHOES

R. J. Nemiroff[a,b,c], W. A. D. T. Wickramasinghe[d], J. P. Norris[b],
C. Kouveliotou[a,e], G. J. Fishman[d], C. A. Meegan[e], and W. S. Paciesas[e,f]

ABSTRACT

We have searched for gravitational-lens induced echoes between gamma-ray bursts (GRBs) in BATSE data. The search was conducted in two phases. In the first phase we compared all GRBs in a brightness complete sample of the first 260 GRBs with recorded angular positions having at least 5 % chance of being coincident from their combined positional error. In the second phase, we compared all GRB light curves of the first 611 GRBs with recorded angular positions having at least 55 % chance of being coincident from their combined positional error. No unambiguous gravitational lens candidate pairs were found in either phase, although a "library of close calls" was accumulated for future reference. In comparison with the QSO gravitational lens detection rate induced by the known galaxy-lens field, a limit on the distance to the brightest GRBs was derived. It is found that in a smoothly distributed $\Omega = 1$ universe where the density of galaxy-size mass clumps is approximately constant, BATSE-detected GRBs existed at an average redshift of less than 6.

INTRODUCTION

The idea of searching for a gravitational lens echo in gamma-ray burst data is not a new one. Paczynski[1-2] originally suggested that such a lens echo effect might be detectable were GRBs to lie at cosmological distances. Mao[3] estimated that the number of BATSE GRBs that need to be inspected to find a gravitational lens echo would be between 250 and 2000. Blaes and Webster[4] pointed out that GRB echoes would be expected if the universe were populated with a significant fraction of compact objects near 10^6 solar masses. Narayan and Wallington[5] discussed what information about the lens might be derived from the discovery of a gravitational lens echo.

A preliminary echo search of the first 386 GRBs detected by BATSE was conducted by Nemiroff et al.[6] without success. A study searching for lens effects within the time stream of the first 44 BATSE GRBs for massive compact dark matter[7] also did not find an echo but was able to show that either GRBs do not occur at the most likely cosmological distance implied by their brightness distribution or that a universe composed to closure density of compact objects with masses between $10^{6.5}$ and $10^{8.1}$ solar masses is marginally excluded.

[a] Universities Space Research Association
[b] George Mason University, CSI Institute, Fairfax, VA 22030
[c] NASA Goddard Space Flight Center, Greenbelt, MD 20771
[d] University of Pennsylvania, Zaccheus Daniel Fellow,
   Department of Astronomy and Astrophysics, Philadelphia, PA 19104
[e] NASA Marshall Space Flight Center, Huntsville, AL 35812
[f] University of Alabama, Huntsville, AL 35899

A gravitational lens echo is to be expected if a galaxy is near enough to the line of sight to a GRB to create more than one detectable image of the GRB. For galaxies such images would be separated by at most several arcseconds, just as lensed QSO images are. However, the time delay between images can be from hours to years, depending on the lens geometry and the relative placement of the source behind the lens. For a canonical isothermal galaxy lens, the time delay between the two brightest images is on the order of a few months, with one or two years possible when a cluster of galaxies is superposed in the field.[8] For lens geometries where more than one image is expected, the shortest time between images would be on the order of several hours, while again one or two years is roughly the maximum amount of time expected between images.[9] In general the brighter an image pair, the shorter the time delay between images, and the less probable an observer-galaxy-source alignment would be that would create this bright pair. Images themselves created by a cluster lens would typically be expected to be separated by about 100 years or more - a time period too long to search for in current GRB data.

Gravitational lens image pairs would be expected to have time profiles that are identical to within a scale factor. This assumes that any beaming inherent in the GRB progenitor scenario cannot be collimated to much better than a few arcseconds, as this is a likely angular size for the galaxy lens (at least the part that would be involved in a lens effect) on the GRB's sky.

By the equivalence principle, gravitational fields bend the light of different energies equally. Therefore, gravitational lens images would also be expected to have identical spectra. More stringently, since it is known that the spectra of most GRBs change over the duration of the burst,[10] lensed images would be expected to have identical spectra at all times during the burst.

As gamma rays traverse most material without absorption, there is not expected to be any significant absorption between images. One can not, then, rely on such a mechanism to alter one image's spectra relative to the other. It is even harder to imagine a scenario where the amount of absorption changes over the timescale of the GRB, which would be needed to alter the light curve.

Our present search differs from the lens search and subsequent dark matter limits discussed in Nemiroff et al.[7] in that here we are comparing different GRB light curves to each other in a search for galactic lenses, while Nemiroff et al.[7] searched the time stream immediately following GRBs for an echo indicative of lower mass dark matter lenses.

THE SEARCH FOR GRAVITATIONAL LENSING

Our search for lensed echoes occurred in two phases. In Phase I only the first 260 GRBs were considered, as these GRBs had measured peak fluxes. Of these 260 GRBs we used the same peak-flux complete sample of 118 GRBs as Wickramasinghe et al.[11] This sample is composed of GRBs above a limiting peak flux level to which the sample is 99% complete, and demands that the GRBs were measured during a time of normal background.

The angular positions of these 118 were compared, and those pairs with recorded angular positions having at least 5 % chance of being coincident considering their combined positional error were retained for further analysis. This left 104 pairs of GRBs for the Phase I comparison.

For the Phase II search, the first 611 GRBs were considered. This incorporated all GRBs detected by BATSE before and including 6 April 1993, with trigger numbers from 105 to 2291. The positions of all 611 were compared and those pairs that had recorded angular positions having at least 55 % chance of being coincident considering their combined positional error were kept. We excluded those GRB pairs with angular separations greater than 30°, as these involved at least one GRB that was extremely weak. This weakness created a large angular uncertainty in the GRB's position which in turn created an inordinately large number of potentially coincident GRB image pairs. In addition, many of these GRBs were so weak that they would never be very convincing in a statistical comparison. This left 1706 pairs of GRBs for the Phase II comparison.

The initial comparison procedure was visual. A hard copy of the light curve in channel 3 (100-330 keV) of each GRB was made and catalogued. Channel 3 was chosen because it usually has the highest signal and relatively low background. A single energy channel was chosen for the initial comparison so that other energy channels could potentially be used later for an independent comparison.

Light curves that bore a marked similarity were recorded for future numerical comparison. For a wide range of scale factors and time offsets, those pairs of recorded GRBs were compared via a modified χ^2 test to see if they could both have been drawn from the same parent distribution. This comparison statistic is described in more detail in Nemiroff et al.[12] A similar statistical comparison test has been suggested by Wambsganss.[13]

Further study includes comparing the GRB light curves in each of the 4 energy bands of BATSE's large area detectors. As gravitational lensing effects are independent of wavelength, each of the energy bands should show acceptable echo fits for the same scale factor f and time offset Δt, independently. In other words, the two GRBs must have indistinguishable hardness ratios as a function of time. In addition, the time streams before and after each GRB are checked for precursor or post-event emission. Such events must occur for both GRBs consistently for the pair to be a lensing candidate.

No clear gravitational lens examples were found. More specifically, no two GRB time profiles were deemed both bright enough for adequate statistical comparison and identical in a statistical comparison. A "library of close calls" (LOCC) was founded into which pairs of GRB light curves with subjectively determined coincidental similarities were placed. This library serves the function of allowing us to estimate the background against which a real gravitational lens event must be judged. Most of the LOCC members are bursts where one or both is quite dim, quite short in duration, or quite featureless in appearance.

A particularly interesting LOCC entry we show here is from the Phase II search. The GRB pair shown in Figure 1 is particularly interesting as both BATSE trigger 1733 and 1956 are highly fluent, well time resolved, and show similar although relatively smooth time structure over their whole duration. The positions are 3.7 degrees apart on the sky, corresponding to 0.6 σ of their combined positional error, and the events are separated by 55 days. This pair, however, is an example of how relatively featureless GRBs can fool our search criteria. Here we show light curves from two of the four channels of each GRB in Figures 1a-1d. In each channel the two GRBs show similarities. But upon inspection of the time profiles in different energy bands, it is clear that they have different spectra. One can see from the light curves (and the subsequent statistical analysis) that no single f factor is implied, and the widely different f factors clearly exclude a gravitational lens origin for the similarity of these GRBs.

From the null results of the gravitational lens search an upper limit to the redshift of GRBs can now be computed. If GRBs are very distant, they should undergo gravitational lensing. Lack of a convincing candidate can be translated into an upper limit on distance for BATSE-detected GRBs. More specifically, it is possible to estimate the amount of lensing of GRBs that is to be expected relative to the amount of lensing that is known to be seen for QSOs. More specifically, assuming reasonable values for the frequency of QSO lensing, the average redshift of QSOs, an average magnitude of QSO lensing, the maximum frequency of GRB lensing, and an average magnitude the GRB lens search was complete to, we can estimate the maximum average redshift GRBs can be before the number of expected lens detections is significantly greater than the null result found. A more detailed explanation of the statistics of this comparison is given in Nemiroff et al.[12]

Using the above argument, we can reasonably expect that the maximum redshift of GRBs follows $z_{GRB} < 6$. In other words, if the average redshift of the GRBs discussed above were 6 or greater, at least one gravitational lens echo from the known galaxy field would be expected, which is assumed to continue back to the epoch of the GRBs. Since we did not see an echo, their average redshift is less than 6.

Although we assumed an $\Omega = 1$ smoothly filled universe, we have found that the GRB redshift limit result may be a function of the assumed cosmology. For a smoothly distributed universe where $\Omega = 0.1$, however, the redshift limit derived above on GRBs would increase by only about 50 %. For universes with a significant fraction of closure density either in the form of compact masses or in a cosmological constant, the redshift limit derived above is not accurate. The limit also does not include magnification bias effects, which would result from a steep luminosity function of GRBs at the brightness cut-off and would increase in the probability of lensing, and so would decrease the value of the redshift limit. However, the luminosity function of GRBs is unknown, and so the minimal assumption of standard candle GRBs here is the most conservative.

An automated search procedure is being designed and should begin being implemented by the time this paper goes to press. We are hopeful this procedure will find a gravitational lens echo, although again it is still somewhat unlikely. Alternatively, if it does not, this will allow a significant decreased in the maximum redshift for GRBs.

REFERENCES

1. B. Paczynski, Ap. J. **308**, L42 (1986).
2. B. Paczynski, Ap. J. **317**, L51 (1987).
3. S. Mao, Ap. J. **389**, L41 (1992).
4. O. M. Blaes & R. L. Webster, Ap. J. **391**, L63 (1992).
5. R. Narayan, & S. Wallington, Ap. J **399**, 368 (1992).
6. R. J. Nemiroff et al., Compton Gamma-Ray Observatory, St. Louis, AIP Conference Proceedings 280, eds. M. Friedlander, N. Gehrels, and D. J. Macomb (AIP, New York, 974), p. 1993.
7. R. J. Nemiroff et al., Ap. J. **414**, 36 (1993).
8. R. Blandford and R. Narayan, Ann. Rev. Astron. Astrophys. **30**, 311 (1992).
9. D. Narasimha, private communication , (1993).
10. J. P. Norris et al., Ap. J. **301**, 213 (1986).
11. W.A.D.T. Wickramasinghe et al., Ap. J. **411**, L55 (1993).
12. R. J. Nemiroff et al., submitted to Ap. J. , (1994).
13. J. Wambsganss, Ap. J. **406**, 29 (1993).

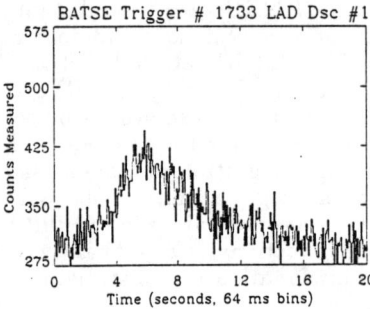
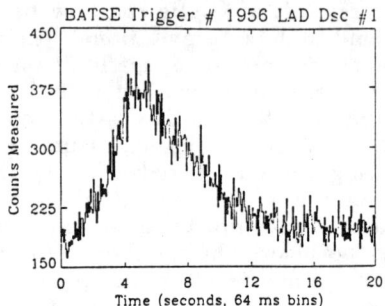
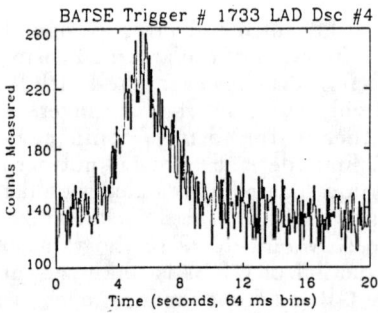
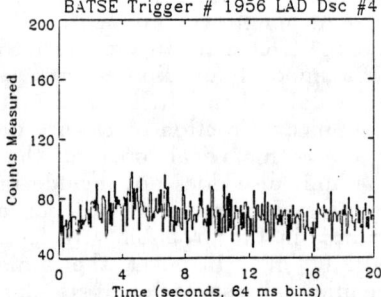

Figure 1: A comparison of the light curves for GRBs with trigger numbers 1733 and 1956. Channel 1 (25-50 keV) and channel 4 (300 keV - ~ 1000 keV) are shown. Although each GRB has similar time structure, their spectra are too different to warrant a gravitational lens interpretation.

BATSE BURST LOCATION ACCURACY AND CONSTRAINTS ON THE FRACTION OF REPEATING GRB SOURCES

Tod. E. Strohmayer, Edward E. Fenimore

Los Alamos National Laboratory, MS D436, Los Alamos, NM 87545

Juan A. Miralles

Dept. de Fisica Teorica, Univ. de Valencia, 46100 Burjassot, Spain

ABSTRACT

We use extensive simulations of GRB source repetition to investigate the ability of BATSE to detect source repetition and to place constraints on the fraction of repeating sources. From Monte Carlo simulations we find that the current uncertainty in BATSE burst locations severely limits our ability to confidently detect source repetition from distributions containing fewer than 10 -15% repeaters. A fit of our repetition model to 260 BATSE catalog bursts yields a best-fit repeating fraction of $f_r = 21\%$ with a 90% confidence region ranging from 5.5 to 32.5 %. By modifying the size of the measurement errors in our simulations we show that the location and width of the confidence region depends sensitively on the burst location errors. With BATSE's present location accuracy analysis of larger samples of bursts will not appreciably improve the constraint on the repeating fraction.

INTRODUCTION

If correct, the recent suggestion[1] that as many as 15% of the classical gamma ray bursts (GRBs) detected by BATSE arise from repeating sources has important implications for current burst source models. A significant fraction ($\geq$ 10 %) of repeating sources can be more easily accomodated in galactic neutron star models than extragalactic scenarios[1,2]. Thus the ability to constrain the fraction of repeating sources could provide crucial information distinguishing between the most currently favored source models. The excess nearest neighbor clustering could have three possible causes. 1) Burst sources are repeating. 2) Systematic errors or biases in the measured BATSE positions mimic intrinsic clustering. 3) The clustering arises from a random, source distribution purely by chance. Assuming there are no systematic biases in the BATSE positions, the probability of 3) occurring is $\approx$ 2% for the BATSE B1 catalog[3]. Recent work by several authors[4,5] warns that systematic errors may be the cause of the excess clustering. In addition, it has been shown that certain measurement biases can produce the observed excess[6]. However, at present the existence of such a bias remains unproven. Here we investigate the ability of BATSE to both detect and measure burst source repetition.

DETECTING GRB SOURCE REPETITION WITH BATSE

A test for repetition of sources is performed by computing the cumulative distribution of observed nearest neighbor separations and comparing this with the distribution expected theoretically for uniform sources[7,1]. The relevant statistic is the Kolmogorov-Smirnov (K-S) statistic D, the maximum vertical difference between the observed and expected cumulative distributions. The significance of a given measurement is the probability of obtaining from the uniform distribution a value for D greater than or equal to the measured value. Since nearest neighbor separations can be significantly correlated it is necessary to obtain this probability from Monte Carlo calculations[4,5].

We have developed a numerical model for source repetition which depends on a single parameter, f_r, the fraction of sources which repeat. Burst positions are assigned sequentially and with random sky positions unless the value of a random number between 0.0 and 1.0 sampled for each burst is less than or equal to f_r. When this occurs the next burst location is randomly selected from one of the existing burst locations, and represents a recurrent event. This procedure is repeated until N bursts have been generated. It is possible in this model to have multiple events from any given source. In table 1 we summarize some of the characteristics of this model for a sample of 260 bursts.

To investigate BATSE's ability to detect burst repetition we simulate the measurement of source distributions which contain repeating events and then compare these with the non-repeating hypothesis using the nearest neighbor method. To compute the measured source distributions we filter the model source positions with instrumental location errors[3,8,9]. Each model source position is randomly assigned one of the BATSE error circles and a measured position is then sampled from each sources error circle. We assume that each error circle corresponds to the 1σ value of a Gaussian distribution in angular separation, ensuring that 68% of the time the measured position will be within the error circle. Having simulated the measurement of a model distribution we compare it with the theoretical nearest neighbor distribution for random (non-repeating) sources. For a given model we generate many different realizations in order to compute the distribution of D which would be seen by BATSE. In figure 1 (left panel) we show the distributions of D so computed from 2000 realizations of 260 burst sources with repeating fractions of $f_r = 0$, (solid histogram), 0.1 (dotted histogram), and 0.2 (dashed histogram). Notice that the distribution with 10% repeaters looks quite random, since the clustered bursts tend to be smeared out upon measurement. Even with a 20% repeating fraction many of the measured distributions are still consistent with the $f_r = 0$ (random) histogram and upon measurement by BATSE would not be strongly identified with source repetition. For comparison, the right panel shows the same distributions but now computed under the assumption that BATSE measures positions twice as accurately as specified in the catalog. Notice now that the 20% histogram has only a small tail which extends into the random ($f_r = 0$) histogram. This graphically demonstrates the increase in sensitivity for detecting burst source repetition produced by an improvement in burst location accuracy.

CONSTRAINTS ON THE REPEATING FRACTION

We now fit our model of burst repetition to the BATSE data. We choose to investigate two samples of bursts. The full 260 burst sample from the BATSE catalog, and the sample of ≈ 200 bursts investigated in Ref. 1., and for which has been claimed the most significant clustering. We use the same Monte Carlo technique described above to compute the expected cumulative nearest neighbor distribution for a given value of f_r and a fixed number of bursts. The expected cumulative nearest neighbor distributions so computed are then compared to the observed BATSE distribution and the K-S statistic $D(f_r)$ is computed. We

Table I							
Bursts (N_b)	f_r	Sources (N_s)	N_0	N_1	N_2	N_3	N_4
260	0.0	260.0	260.0	0.0	0.0	0.0	0.0
260	0.1	234.1	212.8	17.9	2.7	0.6	0.1
260	0.2	208.2	173.2	25.1	6.3	2.1	0.8
260	0.3	182.3	139.8	26.6	8.4	3.5	1.6

Table I–Summary of the repeating source model for 260 total bursts. The column labelled N_s represents the average number of burst sources generated. The columns labelled N_i give the average number of sources which repeated i times in each model.

repeat this procedure for different f_r in order to minimize the statistic $D(f_r)$. For the 260 burst sample the minimum in D occurs at $f_r \approx 0.21$. Upon examination of the cumulative nearest neighbor distributions we find that the model with 21% repetition provides an improved fit to the observed distribution at small angular separations than the non-repeating model. We emphasize that this does not prove that sources are repeating. Rather, it establishes that burst repetition can reasonably account for the observed distribution.

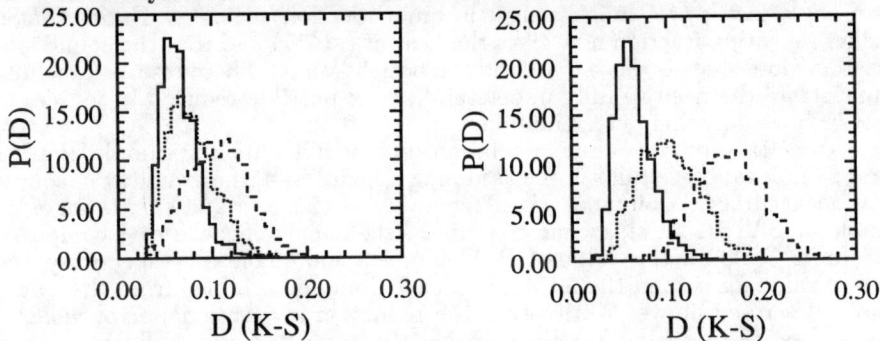

Figure 1. The distributions of the K-S statistic D obtained by comparison of simulated BATSE measurements of GRB locations with the distribution expected theoretically for random sources. The solid histograms show the random (non-repeating) model. The dotted and dashed histograms show the distributions computed from source models containing 10% and 20% repeating fractions respectively. The left panel assumed the full BATSE location uncertainties, while the right panel assumed a factor of two improvement in location measurement.

To estimate the confidence region for f_r we repeat the same fitting procedure described above but now employing simulated data sets generated from model distributions with $f_r = f_r^{best}$. Since the repeating models are most readily computed at discrete values of f_r we estimate the confidence region by tabulating the fraction of fits which yield a certain value for f_r^{best}. In figure 2 (left panel) we show the fraction of realizations which resulted in best-fit values for f_r ranging from 0.01 to 0.4 in equal steps of 0.01. This estimate of the confidence region

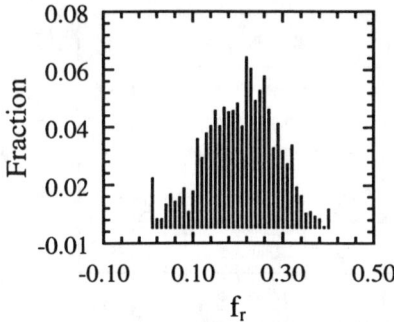 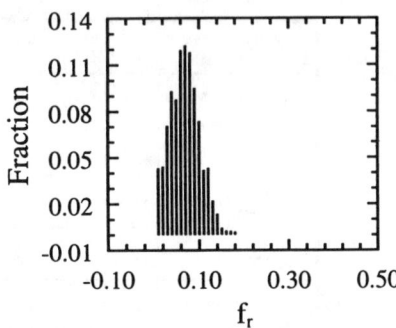

Figure 2. Estimated confidence region for f_r derived from the analysis of the 260 burst sample (left panel). Each vertical bar represents the fraction of realizations which resulted in the corresponding best-fit value for f_r. The right hand panel shows the results of a similar calculation but assuming a factor of 4 improvement in burst location accuracy.

was computed using fits to 2000 realizations. Notice that the distribution peaks near f_r^{best} as expected, but extends from a few percent to beyond 40 %. This indicates that the BATSE measurements and accompanying uncertainties do not provide a tight constraint on the fraction of repeating sources. For comparison we show in the right panel of figure 2 the results of a similar calculation, but assuming a factor of 4 improvement in burst location accuracy. Notice that a smaller repeating fraction now gives the best fit ($\approx$ 7%) and that the confidence region is more sharply peaked around the best-fit value. These results forcefully demonstrate the need to fully understand and if possible reduce the sources of error in burst location determination.

Table II summarizes our calculations for the full 260 burst sample, the 201 burst sample, and the full sample assuming a factor of 4 improvement in source location accuracy. Column 1 gives the value of the K-S statistic D for each sample of BATSE bursts, while column 2 lists the significance α_{KS} computed via Monte Carlo for this value of D. Columns 3 and 4 contain respectively, the best-fit value for f_r and the 90 % confidence limits determined from the calculations described above. Notice that the reduction in positional error yields a smaller best-fit repeating fraction and that the accompanying confidence region is more sharply peaked around f_r^{best}. This calculation emphasizes the importance of accurate burst locations for both detecting burst source repetition and constraining the fraction of repeating burst sources.

We have also investigated how burst sample size affects the constraint on f_r. We find that with BATSE's present burst location accuracy, only marginal improvement in the ability to constrain the repeating fraction is attained by analysing larger burst samples. This results from the fact that BATSE's typical error circle becomes larger than the average separation between nearest neighbors as the sample size increases. Assuming a 20% source repetition rate we estimate that analysis of a 2000 burst sample would still result in a 1σ uncertainty of ±7.36% in the measured burst repetition rate.

SUMMARY AND DISCUSSION

Although our results indicate that burst source repetition at about the

Table II				
Sample	D_{KS}	α_{KS}	f_r^{best}	$\sigma_f^{90\%}$
All bursts (260)	0.117	0.018	21%	5.5% – 32.5%
QL bursts (201)	0.159	0.0013	26%	10.5% – 34.5%
All bursts ($\sigma/4$)	0.117	0.018	7.0%	1.5% – 12.0%

Table II–Summary of the fits to our repeating source model for the full 260 burst BATSE sample, the 201 burst sample[1], and the full burst sample assuming a factor of 4 improvement in burst location accuracy ($\sigma/4$).

20% level can account for the observed excess clustering of bursts with small angular separations, they also show that the substantial burst location uncertainties severely limit our ability to place useful constraints on the fraction of burst sources which repeat. This in turn reduces the confidence with which we can exclude a given source model which might only be compatible with some moderate level of burst repetition. In addition we have shown that the derived constraint on the fraction of repeating bursts depends quite sensitively on the BATSE position measurement uncertainties. It is therefore essential that these errors be calibrated.

The results obtained from the analysis of the QL bursts would appear to provide significant evidence for burst repetition, however, several results argue for a more cautious interpretation of the data. First, excess clustering of bursts has also been identified on the largest angular scales[4,6]. This excess cannot be explained by burst repetition, and thus suggests the presence of some systematic bias in the BATSE positions[6]. Secondly, since the QL bursts are only a subset of the larger catalog, we find it difficult to assess the a priori probability that such a subset would be significant. This coupled with the marginal, 2% probability, obtained from analysis of the entire catalog suggests that the QL result be regarded as a prediction which must be tested by analysis of subsequent bursts detected with BATSE.

We thank the BATSE team for compiling the public catalog. This work at Los Alamos was supported by the United States Department of Energy, IGPP, and by the GRO guest investigator program.

REFERENCES

1. J. M. Quashnock & D. Q. Lamb, MNRAS , in press (1993).
2. B. Paczyński, Ap. J. **308**, L43 (1986).
3. G. J. Fishman et al., Ap. J. Supp. , in press (1993).
4. R. Narayan & T. Piran, Ap. J. , in press (1993).
5. M. A. Nowak, Ap. J. , in press (1993).
6. E. Maoz, Ap. J. , in press (1993).
7. D. Scott & C. A. Tout, MNRAS **241**, 109 (1989).
8. J. M. Horack et al., Ap. J. **413**, 293 (1993).
9. M. N. Brock et al., Gamma Ray Bursts 1991 (AIP Press, 1992), p. 383.

REPEATING SOURCES OF CLASSICAL GAMMA-RAY BURSTS

V. C. Wang and R. E. Lingenfelter
CASS, University of California, San Diego, La Jolla, CA 92093

ABSTRACT

From an analysis of the gamma-ray bursts in the first BATSE catalog[1], we find that there is an excess of bursts which are clustered in both space and time. These clustered bursts are separated in position by less than their positional uncertainty and separated in time by several hours to several days. This excess is nonrandom at greater than the 99.99% confidence level. We conclude that these bursts arise from repeating sources, the most prolific of which[2,3] is at R.A. 0855 and Dec. −00. Unlike most of the "soft" gamma-ray repeaters, these repeating bursts have relatively hard spectra, complex light curves, and widely varying durations, that are indistinguishable from classical gamma-ray bursts.

INTRODUCTION

We find that there is an excess of classical gamma-ray bursts in the first BATSE catalog[1] which are clustered in both space and time. These bursts are separated in position by less than their positional uncertainty and separated in time by several hours to several days. Since, there is less than a 10^{-4} probability of such an excess occuring randomly, we suggest that these bursts arise from repeating sources. The most prolific of these sources[2,3] lies at R.A. 0855 and Dec. −00. The evidence for more rapid repetition on time scales $< 10^3$ seconds is discussed separately (Ref. 4 and Lingenfelter et al. in the volume).

The search for repeating classical burst sources on time scales $> 10^3$ seconds is limited primarily by the uncertainty in determining the burst positions on the sky. The 1σ positional uncertainty of the 260 bursts in the BATSE sample, which ranges from a minimum[1] of $\sim 4°$ for the brightest bursts to $\sim 27°$ for the weakest bursts, is generally larger than the most likely separation angle of $\sim 5°$, expected between nearest neighbors in a random sample of the same size. Thus, the prospect of distinguishing repeated bursts from random associations by positional coincidence alone is not very promising. Nearest neighbor and two-point angular analyses (Ref. 5 and Quashnock & Lamb in this volume) do show excesses in the number of closely ($< 4°$), as well as antipodally ($> 176°$), clustered burst pairs in the BATSE sample, but their statistical significance is still a matter of debate (Narayan & Piran, Hartmann et al. in this volume).

We can enhance the prospects for identifying repeating sources, however, by also considering the temporal characteristics of the bursts, which can amplify the signal (repeated burst) to noise (random burst) ratio. Since, if some of the gamma-ray bursts are fast repeaters, we would also expect to see temporal clustering in burst occurrence, in addition to spatial clustering. Temporal clustering is a characteristic feature[7,8] of the soft repeaters, as well as other transient sources, which show active periods where the bursts repeat on time scales much less than the mean time between bursts averaged over the total observing time.

CLUSTERING ANALYSIS

We have searched for possible repeating classical burst sources, therefore, by restricting the sample to bursts which are clustered in time. In this search, we considered all combinations of pairs of bursts, calculating both the positional and temporal separations between each pair. With 260 bursts in the first BATSE catalog, there are $N(N-1)/2 = 33{,}670$ such combinations of burst pairs.

We studied the density of these pairs in the phase space of temporal and positional separation, where a uniform random expectation density can be defined. In particular, if we define the positional separation in terms of $\cos\theta$, where θ is the angular separation between the positions of a pair bursts, then bursts that are randomly distributed on the sky will have angular separations θ, that are uniformly distributed in $\cos\theta$. Similarly, if we define the temporal separation in terms of $\tau = 2(t/t_{max}) - (t/t_{max})^2$, where t is the separation between the occurrence times of a pair bursts, and $t_{max} = 320$ days is the total observing time, then bursts that are randomly distributed in time will have time separations t that are uniformly distributed in τ. Thus, bursts that are randomly distributed in both time and position will have temporal and positional separations that are uniformly distributed in τ-$\cos\theta$ space.

This can be seen schematically in Figure 1. While the unrelated, random bursts should be uniformly distributed, we would expect all of the bursts from repeating sources to be concentrated at small separation angles with $\cos\theta$ very close to 1, e.g. burst pairs separated by less than the minimum positional uncertainty of $\sim 4°$ will lie within $0.9976 < \cos\theta < 1$. Furthermore, all pairs of "fast" repeating bursts, which are clustered on short time scales, would be further concentrated at values of τ close to 0, as well as being concentrated at values of $\cos\theta$ very close to 1, and any "quasi-periodic" repeaters, would be clustered around their various repetition periods.

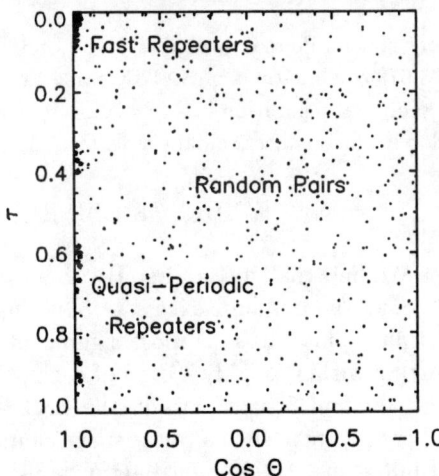

Figure 1. Temporal and positional separations of all combinations of gamma-ray burst pairs in τ-$\cos\theta$ phase space, in which random, nonrepeating bursts would be uniformly distributed and fast repeaters are concentrated close to τ of 0 and $\cos\theta$ of 1.

Thus, we see that any fast repeated bursts should all be concentrated as an excess in one corner of the τ-$\cos\theta$ phase space close to τ of 0 and $\cos\theta$ of 1. We then determine the statistical significance of any observed excess of n pairs from the Poisson probability of the occurrence of n or more pairs given an expectation of μ pairs in an element of τ-$\cos\theta$ phase space, $P(>n) = \sum_n^\infty \mu^n e^{-\mu}/n!$

We can also test the reliability of such probability estimates by comparison with the cumulative distribution of the observed pairs in all other equal phase-space elements. Since we have no specific expectation of the angular or temporal clustering scale of potential repeaters, we have optimized the binning in τ and $\cos\theta$ in order determine the maximum statistical significance of the repeater candidates.

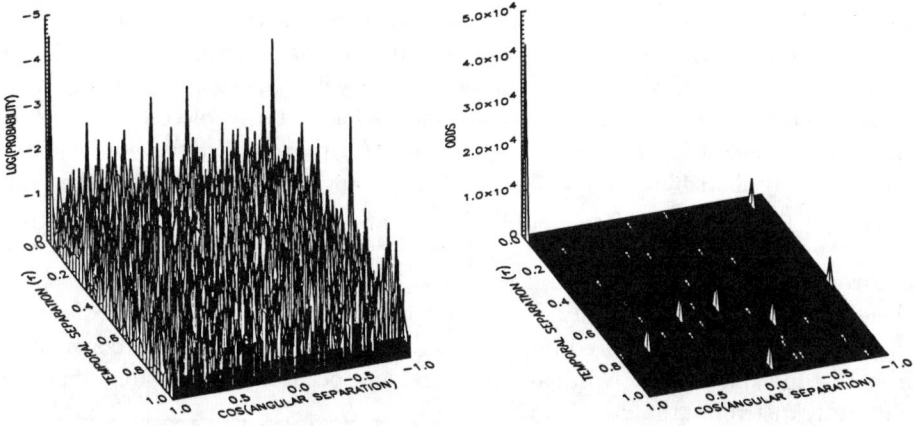

Figure 2. The Poisson probability and odds $(1/P(>n))$ for the random occurrence of all 33,670 combinations of BATSE burst pairs divided into 30,096 equal bins of angular and temporal separation in a 38×792 τ-$\cos\theta$ phase space, showing the significant excess of closely-clustered, candidate repeater pairs with τ close to 0 and $\cos\theta$ close to 1.

EVIDENCE FOR REPEATING SOURCES

We find that in the first BATSE catalog sample there is a very significant and robust nonrandom excess of burst pairs which are clustered in both space and time. This excess is most significant when we consider the pairs of bursts, occurring within roughly 4 days of each other and lying within roughly 4° of each other. We find 8 such pairs in the BATSE sample, whereas the mean expected number of pairs in such a phase space element μ is only 1.1, taking all 33,670 pairs distributed in 30,096 equal bins in a 38×792 τ-$\cos\theta$ phase space. The Poisson probability of finding 8, or more, random pairs so closely clustered in position and time is only 2×10^{-5}. The statistical significance of this excess is greater than that of the random fluctuations in any of the other equal bins, as can be seen in Figure 2, shown both in probability and in odds $(1/P(>n))$. Even though we might expect a random fluctuation at the 3×10^{-5} level in one of the 3×10^4 bins, the probability of such a random fluctuation occurring in the specific bin closest to τ of 0 and $\cos\theta$ of 1 is still only 2×10^{-5}.

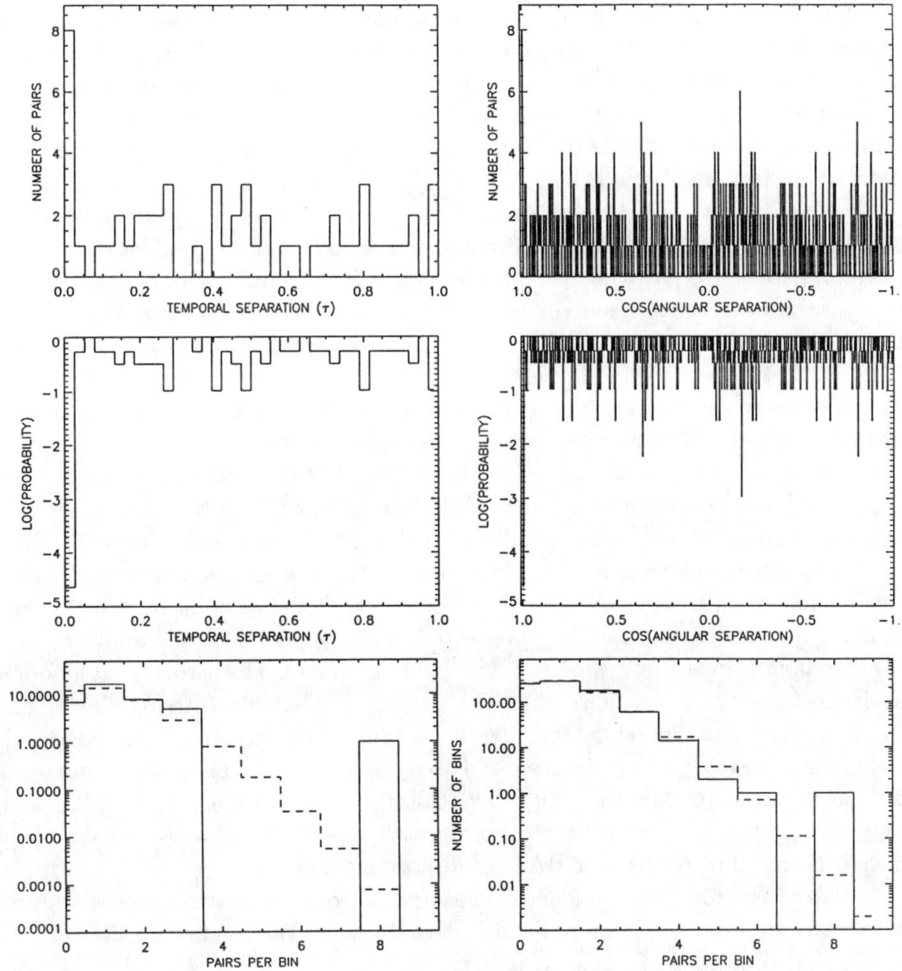

Figure 3. The excess of closely-clustered, candidate repeater pairs compared with the number per bin, the probability of random occurrence, and occurrence distributions of the other pairs in the temporal and angular phase space bins closest to τ of 0 and $\cos\theta$ of 1. The Poisson distribution (dashed line) is shown for comparison.

The high significance of this excess of closely clustered pairs can be also be seen in Figure 3, where it is compared with the number, random probability and occurrence distributions of the other pairs in the temporal and angular phase space bins closest to τ of 0 and $\cos\theta$ of 1. These statistical comparisons all very strongly suggest that the excess of closely clustered burst pairs arise from fast repeating burst sources.

This excess of closely clustered pairs is also quite robust and it does not depend strongly on the choice of bin size in τ and $\cos\theta$. We find that the observed number of pairs with small angle and time separations significantly exceeds the expected number at $> 99.9\%$ confidence level for a substantial range of angles $\sim 4°$ to $10°$ and times ~ 3 to 10 days. For a control, we also looked for the optimum binning to give the maximum statistical excess in the number of antipodal burst pairs clustered close to τ of 0 and $\cos\theta$ of -1, but found no excess of such burst pairs at greater than 90% level.

Unlike most of the "soft" gamma-ray repeaters, the 16 repeating bursts in these pairs have relatively hard spectra, complex light curves, and widely varying durations, that are indistinguishable from classical gamma-ray bursts. The locations of these 8 repeating sources on the sky are consistent with isotropy, although the statistical uncertainty is quite large. The most prolific of the sources[2,3] lies at R.A. 0855 and Dec. -00. These fast repeater candidates are also among those identified by other recent studies (Quashnock & Lamb, MacCallum et al., Chuang, in this Volume) as repeaters on longer time scales, as well.

Although we identify only 8 sources that appear to repeat on time scales of several hours to several days, corrections for detector dead time and positional uncertainty make the probable number of such sources much larger. Detector dead time, resulting from data readout, South Atlantic anomaly passages, and Earth occultation of sources reduces the BATSE coverage[9] of any point on the sky, and thus the probability of observing a repeated burst, to only 34%. In addition, with the positional uncertainties of the 260 BATSE bursts, any two repeated burst positions from the same source have only a 10% probability of lying closer than $4°$ of one another. Therefore the overall probability of detecting a repeated pair of bursts separated by less than $\sim 4°$ is only 1 out of 30. Thus, the occurrence of 8 such pairs with less than ~ 4 day separations in the BATSE sample would suggest that roughly 240 repeated bursts with separation times < 4 days could have occurred during the first BATSE observing period.

We are currently making detailed Monte Carlo simulations of repeated burst occurrence in order to set more quantitative limits on the fraction of the BATSE bursts that could arise from such sources.

We thank Duane Gruber, Chip Meegan, and Rick Rothschild for helpful discussions, and NASA for support under grant NAG-5 1597.

REFERENCES

1. Fishman, G. J., et al. BATSE Burst Catalog, on line at GROSSC (1992).
2. Wang, V. C., & Lingenfelter, R. E., Ap. J., **416**, L13, (1993).
3. Wang, V. C., & Lingenfelter, R. E., 23rd ICRC, **1**, 93, (1993).
4. Lingenfelter, R. E., Wang, V. C., & Higdon, J. C. Ap. J., submitted (1993).
5. Quashnock, J. M., & Lamb, D. Q., MNRAS, in press, (1993).
6. Kouveliotou, C., et al., Nature, in press (1993).
7. Laros, J. G., et al., Ap. J., **320**, L111, (1987).
8. Mazets, E. P., et al., Ap. Space. Sci., **80**, 3, (1981).
9. Fishman, G. J., et al. Ap. J. Supp., in press (1993).

TEMPORAL STUDIES

TWO CLASSES OF GAMMA-RAY BURSTS

C. Kouveliotou[†], C. A. Meegan, G. J. Fishman, P. N. Bhat[*]
NASA/Marshall Space Flight Center, Huntsville, AL 35812, U.S.A.

M. S. Briggs, T. M. Koshut, W. S. Paciesas, and G. N. Pendleton
Dept. of Physics, University of Alabama in Huntsville, Huntsville, AL 35899

ABSTRACT

We have studied the duration distribution of 222 Gamma Ray Bursts of the first BATSE catalog. We find a bimodality in the distribution, which separates GRBs into two classes: short events (< 2 s) and longer ones (> 2 s). Both sets are distributed isotropically and inhomogeneously in the sky. We find that their durations are anti-correlated with their spectral hardness ratios: short GRBs are predominantly harder and longer ones tend to be softer. Our results provide a first GRB classification scheme based on a combination of the GRB temporal and spectral properties.

1. INTRODUCTION

Gamma-Ray Burst (GRB) studies over the last 20 years have not succeeded in revealing tell-tale properties that would help identify the nature of their emission sites. Moreover, there exist no concrete counterpart identifications in any other wavelength within the well-defined GRB error boxes[1] that would point towards a known parent population for the phenomenon. Recent results[2] from the Burst and Transient Source Experiment (BATSE)[3] on the Compton Gamma-Ray Observatory (CGRO) have shown that the sky distribution of the GRB sources is isotropic, but not homogeneous. Any attempt to identify GRB subclasses based on similarities in their spatial, spectral or morphological properties has failed so far[4,5]. We present here a study of one of the GRB global properties, namely their duration distribution, which has led to the confirmation of their division into two subclasses. This duration bimodality is linked for the first time with a different average spectral hardness associated with each class.

2. DURATION DISTRIBUTION

The distribution of GRB durations has been studied extensively in the past[6-9,1]. Most studies agree that there is a hint of bimodality with the separation being in the 0.5 - 4 s range. There are several reasons why the previous data sets could not establish the bimodal nature of the distribution: lack of instrument trigger sensitivity to short events, low temporal resolution, difficulty of confirmation of a very short event as a burst in an often noisy data set, etc. They have all led to biases against detection of short events. The first BATSE catalog[5] presents a complete, confirmed set of 260 GRBs, detected with unprecedented sensitivity over the instrument's first year of operation (April 21, 1991 to March 5, 1992).

[†] Universities Space Research Association
[*] NRC/NASA Senior Research Associate; on leave from Tata Institute of Fundamental Research, Bombay 400 005, India

The criteria for determining a GRB duration have (widely) varied over the past, without an accepted consensus for a "duration algorithm". We have introduced[10] an unbiased and reproducible way of estimating durations. We define T_{90} as the time during which the cumulative counts increase from 5% to 95% above background, thus encompassing 90% of the total GRB counts. T_{50} is defined similarly to include 50% of the counts. The times thus defined are an intensity independent measure of duration, unlike previous definitions. In most cases the data available afforded very accurate measurements for both times. This procedure failed whenever there existed data gaps during a burst readout. Thus, out of the 260 GRBs contained in the first BATSE catalog, we have T_{90} and T_{50} values for 222.

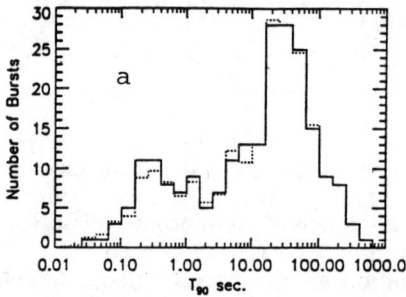 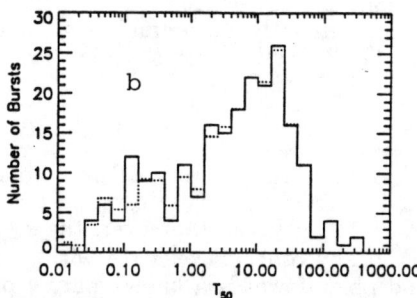

Figure 1. a) Distribution of T_{90} for the 222 GRBs of the first BATSE catalog. b) Distribution of T_{50} for the same GRB set. Solid lines are the histograms of the raw data; dotted lines are the error-convolved histograms as explained in the text.

The solid line in Figure 1a shows the uncorrected distribution of the 222 values for T_{90}. To account for the time errors δT_{90} in each histogram bin, we have assumed that each T_{90} is represented by a Gaussian of standard deviation given by δT_{90}. Each error-convolved histogram bin is then derived by adding the overlapping areas of all Gaussians that fall within its boundaries. The convolved distribution is plotted with a dotted line on Figure 1a: we notice that the inclusion of the errors has expanded the short duration range (where the uncertainties are larger), but has not affected the longer duration bursts.

Both distributions of Figure 1 show a dip around 2 seconds. The dip is not an instrumental artifact: BATSE's trigger sensitivity is maximum near 1 s, which is its longest trigger timescale. Although it is difficult to quantitatively assess the statistical significance of the dip, we estimate that convex, unimodal distributions are rejected at the 2-3 σ level. We have fit a quadratic function between the two peaks in the histogram and determined its minimum to be at $T_{90} = 1.2$ s $\pm$ 0.4 s, which rounded off to the next integer bin edge, is 2.0 s. This effectively divides the 222 GRBS into two subsets: one containing 58 short events ($T_{90} < 2.0$ seconds) with a logarithmic mean T_{90} of 0.33 s $\pm$ 0.21 s and a second of 164 longer GRBs with a mean of 26.2 s $\pm$ 1.7 s.

The fraction of short events in the data bases derived with various experiments does not seem to vary significantly: SIGNE (on Venera 11-12) reports[11] 25%, the International Sun Earth Explorer-3 (ISEE-3) shows[8] 29%, albeit with a limited sample, the Phebus instrument on Granat has[12] 27%, and for BATSE the same fraction amounts to 26% of the first catalog data. The KONUS experiments on Venera 11/12 and 13/14, however, show significantly smaller percentages, 7% and 16%, respectively. One explanation of this discrepancy could be the detection threshold for the KONUS experiments, which increased with decreasing GRB duration[7].

Comparison of the duration distributions obtained by previous observers[1] with the T_{90} distribution shows that the BATSE data have a factor of 2 higher average durations. The arithmetic mean of the T_{90} values is 37.6 s $\pm$ 2.7 s vs the mean of 18.3 s for 616 GRBs[1]. One possible explanation of this shift in mean duration could be a systematic effect of instrumental sensitivity. BATSE, with its unprecedented sensitivity, would see what previous experiments would have called an average GRB for a much longer time. If that is indeed the case, raising the instrument sensitivity would bring the average duration to a lower value. The T_{50} distribution for the same 222 GRBs (Figure 1b) effectively does this. We notice that the average value of T_{50} is 16.3 s $\pm$ 1.0 s, similar to KONUS and half of that for BATSE's T_{90}'s. Hence what we see is a convincing effect of different detection thresholds on the GRB durations, which strengthens the case of using a single experiment to derive statistics on GRB duration distributions. We also notice that the duration bimodality is not as significant in the T_{50} distribution; this again is consistent with a "tip of the iceberg" effect. We have searched for clusterings in the burst arrival times for short and long events: both samples are entirely consistent with Poisson distributions. The arrival rates are 0.2 and 0.8 per day for the short and long GRBs, resp.

In the following we will consider the set of the 58 short GRBs with $T_{90} <$ 2s as a different class and study their global properties.

3. SPATIAL DISTRIBUTIONS: Isotropy and Homogeneity tests

Figure 2 shows the sky distribution of the short events; although the sample is limited we can still see that their distribution is isotropic. The values of their various dipole and quadrupole statistics differ in most cases by about 1 standard deviation or less from the values expected for isotropy[10]. We have also examined the distribution of the angular separations of GRB pairs and found no evidence for clustering.

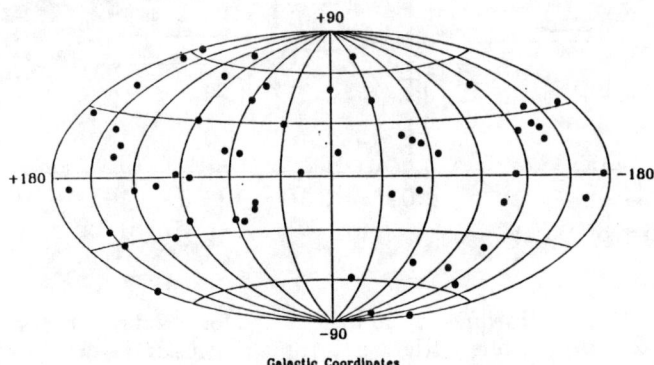

Figure 2. The sky map distribution of the 58 short (< 2 s) GRBs

Similarly, the statistics of the longer events are consistent with an isotropic distribution as expected from the overall isotropy of the 260 GRBs from the first

BATSE catalog[5].

Whenever we have data gaps during a burst accumulation we do not compute[5] the V/V_{max} for the event. We have both T_{90} and V/V_{max} values on the 64 msec trigger timescale for 48 short events and 100 long ones. The $\langle V/V_{max} \rangle$ values are 0.302 ± 0.038 and 0.367 ± 0.030, resp. There is no significant difference between the two means: they are consistent with each other and both are inconsistent with homogeneity. The same trend is evident from the log N–log P diagram for the short (58) and long (164) GRB sets[10]. We conclude that both sets are isotropic and inhomogeneous, in agreement with the overall BATSE GRB results.

4. HARDNESS RATIOS vs T_{90}

We have integrated the counts above background during T_{90} for the 222 GRBs in four discriminator channels with energy ranges of 25 – 50, 50 – 100, 100 – 300 and > 300 keV. We define as HR_{32} the ratio of total counts in the 100 – 300 keV and 50 – 100 keV energy range. Figure 4 (right pannel) shows the scatter plot of the hardness ratios HR_{32} vs T_{90}. The Spearman Rank-Order correlation coefficient[13] between HR_{32} and T_{90} is -0.375; the probability of a fluctuation causing a chance correlation at this level is $\sim 10^{-8}$. The density distributions of the hardness ratios are shown as two histograms in the left pannel of Fig 3. Short events are predominantly harder, while longer events are predominantly softer, as expected from the high correlation between hardness ratios and durations. The same trend is seen with the $HR_{3/21}$ distribution.

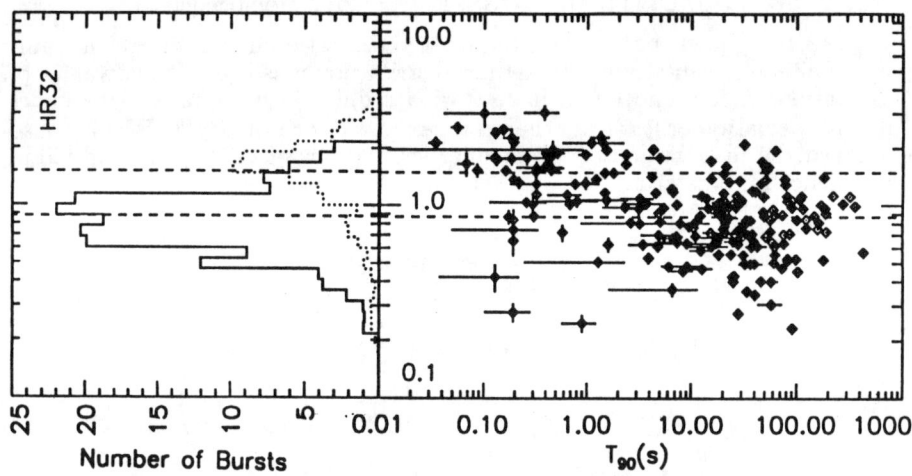

Figure 3. Left: Hardness ratio histograms for events with $T_{90} > 2$ s (solid line) and < 2 s (dotted line). Right : Hardness ratios HR_{32} vs T_{90} scatter plot. The dashed lines on both plots correspond to the mean hardness ratio of the two duration classes.

The values for the mean of the HR_{32} calculated separately for the two GRB classes previously identified are 1.49 ± 0.08 and 0.87 ± 0.03 for <2 and > 2 s, respectively. The values for the short events are clearly higher than the ones for the longer events. This result has also been seen in the Phebus data[12], albeit between higher energy ranges (0.3-7 MeV to 100-300 keV). Dezalay et al., however, were unable to detect a duration bimodality in their small sample of 66 GRBs. We believe that our data confirm these earlier results and provide the first evidence of the continuity of the hardness-duration correlation over the whole GRB observable spectrum.

5. DISCUSSION

We have linked here for the first time the duration bimodality with the hardness-duration correlation of GRBs. Previous studies have reported evidence for either the former[6-9,1] or the latter[12]. Our study shows that the two classes separated by duration are also assosiated with significantly different average hardness ratios. We find that the short events have the same peak intensity range as the longer ones; this makes the total amount of energy released by the two types significantly different. Both short and long GRBs have isotropic but inhomogeneous spatial distributions. All evidence suggests that both GRB subsets originate from the same type of objects. Different geometries of their emission sites (with respect to the observer) may be responsible for the spectral and temporal differences between the classes.

ACKNOWLEDGEMENTS

We wish to express our thanks to Bohdan Paczynski, Jan van Paradijs and Gerry Share for fruitful discussions and to J. Ostriker for suggesting to us the T_{50} algorithm as a measure for the GRB durations.

REFERENCES

1. Hurley K., in *GRBs Huntsville, AL 1991*, eds Paciesas W. S. and Fishman G. J. (AIP: New York, 1991).
2. Meegan C. A. et al., Nature **355**, 143 (1991).
3. Fishman G. J. et al., in *Proceedings of the GRO Science Workshop*, ed. W. N. Johnson (NASA/GSFC, 1989).
4. Briggs M. S., Ap. J., to be submitted, (1993).
5. Fishman G. J. et al, Ap. J. Supp. to be submitted, (1993).
6. Cline T. L., and Desai U. D., Proc. 9th ESLAB Symp., (, 1974).
7. Mazets E. P. et al., Astrophys & Space Sci., **80**, (1981).
8. Norris J. P. et al., Nature **308**, 434 (1984).
9. Klebesadel, R. W., in *Gamma Ray Bursts*, eds Ho C., Epstein R. I. and Fenimore E. E. (Cambridge Un. Press, 1990), p. 161.
10. Kouveliotou C. et al., Ap. J., **413**, L101 (1993).
11. Diyackov A. V. et al, Adv. Space Res., **3**, 211 (1980).
12. Dezalay J-P. et al., in *GRBs Huntsville, AL 1991*, eds Paciesas W. S. and Fishman G. J. (AIP: New York, 1992), p. 304.
13. Press W. H., et al., Numerical Recipes, 2nd Edition (Cambridge Univ. Press, 1992).

EXPLORATION OF BI-MODALITY IN GAMMA-RAY BURST DURATION AND HARDNESS DISTRIBUTIONS

J. P. Norris, R. J. Nemiroff,* S. P. Davis†
NASA/Goddard Space Flight Center, Greenbelt, MD 20771

C. Kouveliotou,‡ G. J. Fishman, C. A. Meegan, W. S. Paciesas§
NASA/Marshall Space Flight Center, Huntsville, AL 35812

ABSTRACT

A bimodal burst duration distribution with the minimum near 1 − 2 s, previously reported by several investigators, is now confirmed by *Compton's* Burst and Transient Source Experiment (BATSE). The burst subgroups are also distinguished by their integral spectral hardness ratios − shorter events tend to be harder.[1] After fitting 280 pulses in 35 long, bright bursts, we find that the interval distribution exhibits a single wide mode centered near 1 s. Analysis of a sample of 12 short, bright bursts indicates that they tend to have considerably fewer pulses, while their pulse width and interval distributions are peaked on the shortward end of those distributions for long bursts. Simulations patterned after these measured attributes of pulse distributions reproduce fairly well the shortward mode and minimum in the duration distribution. The average trend of spectral evolution for long, bright bursts suggests that the difference in hardness ratios might be connected with the tendency of these bursts to soften. No evidence for a relationship between pulse width and spectral hardness is found.

INTRODUCTION

Previous investigators[2,3,4,5] have found evidence for a bimodality in the gamma-ray burst (GRB) duration distribution with the minimum lying in the range 1 − 2 s. This bimodality is now confirmed in a set of 222 bursts detected with BATSE, with the additional correlation that the long bursts have significantly lower integral hardness ratios than the short ones.[1] The latter effect was also observed in a sample of bursts detected by PHEBUS on board GRANAT.[6]

Using bright BATSE bursts, we analyze fitted pulse widths and intervals between pulses, and calculate average spectral hardness trends. The bimodality in duration is examined from the point of view that the pulse generation mechanism tends to repeat on a characteristic time scale of ~ 1 s in long bursts, which average ~ 8 major pulses, while most short bursts consist of significantly fewer pulses with narrower widths and separations. We perform simulations which confirm the expectation that short bursts must have few, closely spaced pulses in order to reproduce the shortward end of the duration distribution. The difference in integral spectral hardness between the two groups might be a result of softer, later emission influencing overall burst history in long bursts. Pulse width and pulse spectral hardness are investigated and found to be uncorrelated, confirming a previous study.[7] Thus pulse width in itself does not appear to be a major contributing factor to the spectral dichotomy.

* George Mason University, Fairfax, VA 22030
† The Catholic University of America, Washington, DC 20064
‡ Universities Space Research Association
§ University of Alabama in Huntsville, AL 35899

PULSE WIDTH AND INTERVAL MEASUREMENTS

Utilizing a least-squares GRB profile fitting program, we have fitted the temporal profiles (64-ms resolution) of 35 long, bright (peak intensity > 18K counts s^{-1}) bursts in the three lowest Large Area Detector energy channels: 25–50, 50–100, 100–300 keV. We utilized bright bursts with durations longer than 1.5 s, fitting portions of profiles containing up to 16 pulses. We employ a pulse model which accommodates a wide variation in pulse shape:

$$F(t) = A \exp\{ - (|t - t_{max}| / \sigma_{l,r})^\nu \} \quad (1)$$

where t_{max} is the time coordinate of the pulse's maximum intensity, A; $\sigma_{l,r}$ are the rise ($t < t_{max}$) and decay time constants ($t > t_{max}$); and ν is the pulse "peakedness" (lower values $\Rightarrow$ more peaked). This representation is clearly not unique, but does provide some usable measure of pulse widths and intervals. Fitting was performed by two investigators and reviewed twice by a third investigator for uniformity. Addition or removal of pulses was sometimes deemed necessary in order to improve the representation of a burst. Such necessarily subjective decisions were based on the tradeoffs arising from statistical considerations – best reduced χ^2, parsimonious representation – least pulses, and indications of pulse existence in the 3 energy bands.

The fitted pulse-width and pulse-interval distributions are shown in Figures 1a & 1b, respectively. For the 35 bursts, a total of approximately 280 pulses were fitted per energy band. We recognize a fundamental limitation in the use of 64-ms data: sufficient resolution is required to distinguish (at ~ full-width half-maximum, FWHM) two closely spaced narrow pulses. As can be seen from Figure 1a, the pulse-width distribution exhibits a single mode at 200 – 600 ms, depending on energy band. In fact, narrower pulses were measured, and the falloff in number of pulses towards narrower widths is probably real since pulses are resolved with five to eight 64-ms bins. Nevertheless, there is probably some underrepresentation of narrow pulses in Figure 1a that could be addressed by use of data with higher temporal resolution.

If the mode of the pulse-width distribution (~ 450 ms, channel 2) is taken to be real, then the shortward end of the pulse-interval distribution (~ 400 ms, independent of energy) is just resolved for the most frequently occurring pulse widths. We apprehend the broad maximum in pulse interval, spanning 0.3 – 3 s, to be the so-called "characteristic time scale" between pulses, previously noted by other investigators.[8] Thus, among long bursts the mode for the duration of a two-pulse event will be

τ_{dur} ~ 1.9 s = {logarithmic mean of broad pulse-interval maximum ~ 1 s} + 2 × {mode of pulse-width distribution ~ 0.45 s} .

Since the average number of pulses in the long bursts we fitted was eight, the longward mode in the duration distribution is expected near ~ 8 × 1 s + 2 × 0.45 s ~ 9 s. We designed simulations to reproduce the measured pulse distributions; these computations yielded a mode near 9 s, far short of the observed mode of 26 s. One-fifth of the 35 fitted bursts would have required more than 16 pulses (the maximum allowed by our code) to describe the whole profile; hence the estimate of 9 s should be somewhat low. The bulk of the difference may be ascribable to the fact that the duration distribution[1] includes bursts over the entire range of peak intensities and the dimmer ones may suffer more time dilation,[9] if cosmological, whereas we fitted only bright bursts.

All bright bursts with durations shorter than 1.5 s (and where high-resolution time-tagged event data were available) were visually examined and fitted for pulses with binning resolutions ranging from 1 to 4 ms, depending on the burst intensity. This set

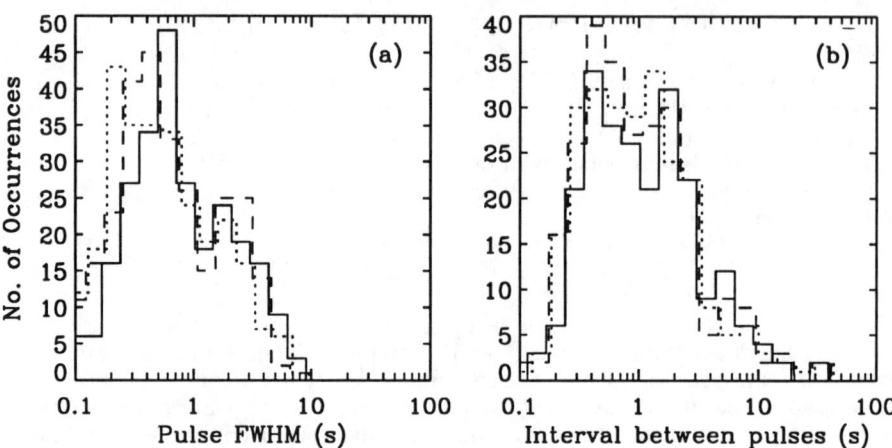

Figure 1. (a) Pulse-width distribution for 35 bright BATSE bursts with durations greater than 1.5 s. (b) Corresponding pulse-interval distribution. Energy channels: 25 – 50 keV (dotted), 50 – 100 keV (dashed), and 100 – 300 keV (solid).

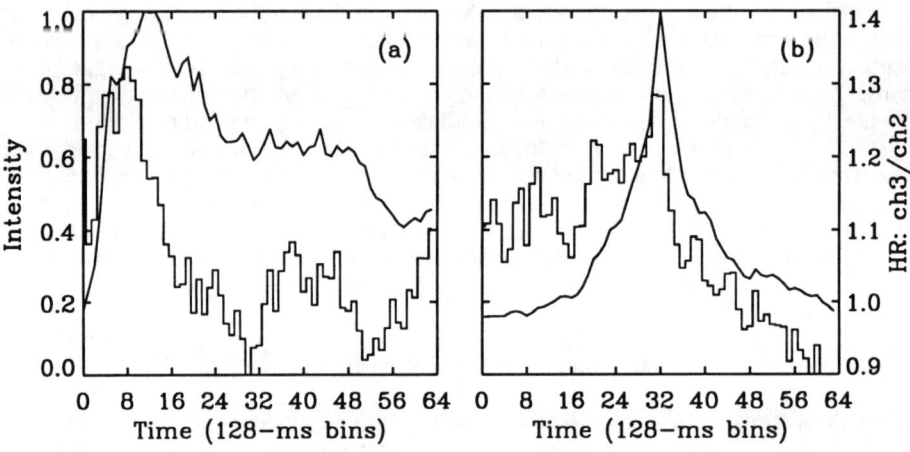

Figure 2. (a) Average trigger-aligned profile (continuous line) and hardness ratio (histogram-mode), 300 – 100 keV : 50 – 100 keV, for 41 bright BATSE bursts. Beyond 8 s, significant, but secondary, local maxima in profile and hardness ratio occur. (b) Alignment by highest peak.

consists of 12 bursts. One-third of these short bursts are well represented by a single pulse of the form described by Equation (1). One-third are representable by two pulses, and the remaining third are complex, multi-pulse events. For the 12 pulses in single- and double-pulse events, the average FWHM is 65 ms with a 1-σ dispersion of 50 ms (individual values range from 0.6 to 165 ms). Similar values were obtained for the complex group although the individual measurements are less reliable due to pulse overlap. The average pulse interval is 105 ms with a dispersion of 150 ms. A larger sample of pulses in short bursts is necessary to determine if their distributions are continuations of the distributions for long bursts (Figure 1).

The simulations we performed indicated that, to reproduce the shortward mode near 300 ms and the minimum near 1 – 2 s, short bursts are required to have on average two pulses with shorter separations than usually found long bursts. Thus, a description in terms of pulse attributes emerges which explains the appearance of the BATSE duration distribution: The shortward duration mode (~ 330 ms)[1] is a mix of short, single-pulse and few-pulse events where the latter have very brief pulse separations; the minimum near 1.5 s is populated with similar few or multi-pulse events, but the separations may be comparable to the interval mode (~ 1 s) for longer bursts (see discussion); the longward duration mode (~ 25 s) consists mainly of multi-pulse events, with pulse widths of order 500 ms, and separations drawn from the broad maximum (0.3 – 3 s) in the interval distribution.

TEMPORAL PROGRESSION OF HARDNESS RATIO

The second distinguishing characteristic of short bursts is that, statistically, they are harder than long bursts.[1,5] Kouveliotou et al.[1] found significantly different integral hardness ratios (HRs) in the BATSE data set (ch3:ch2 = 50–100 keV : 25–50 keV), 1.49 (±0.08) and 0.87 (±0.03), for bursts shorter and longer than 2 s, respectively. For a sample of bursts detected by PHEBUS, Dezalay et al.[5] report the same effect for integral HRs, and for HRs derived from the highest peak per burst.

From fitting the twelve pulses in the eight short, single- and double-pulse bursts binned to 1 – 4 ms resolution, we estimated HRs at pulse peak. For these 12 pulses we found an average peak HR (ch3:ch2) of 1.65 with a 1-σ dispersion of 0.5, within error comparable to the integral value found by Kouveliotou et al.

There is ample evidence for spectral softening trends in GRBs. To obtain a gross estimate of the magnitude of the general spectral softening trend, we computed HRs as a function of time for the longer bursts. Figure 2 illustrates the average time profile of 41 long, bright bursts – individually background-subtracted and normalized to unity – and the temporal progression of the average HR, formed by reducing the profiles in channels 2 and 3 by a factor proportional to the 4-channel peak intensity, summing over all bursts, and taking the ratio, Σch3:Σch2. The panels in Figure 2 are the trigger-aligned (a) and highest-peak-aligned (b) average profiles. Near event onset where the average trigger-aligned profile peaks, the maximum HR $\approx$ 1.3. A comparable value is evident for the rising edge and peak of the average peak-aligned profile. For both measures, the HR declines after the profile maximum.

How might these values be compared with the BATSE HRs for short bursts? It is possible that the lower integral HRs for long bursts might result from softer, later emission influencing the overall average. However, the situation may more complex: short and long bursts with very rapid rises exhibit higher than average hardness ratios.[7]

We also searched for a correlation between pulse width and HR. Figure 3 illustrates the HR (ch3:ch2) at the peak of the fitted pulse versus pulse width (channel 3). No relationship is apparent; a previous analysis also found no correlation between pulse width and HR in single-pulse events.[7]

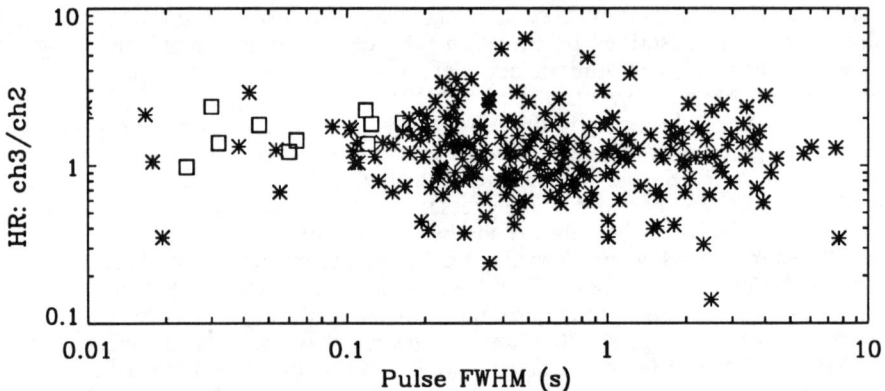

Figure 3. Hardness ratio (100 – 300 keV : 50 – 100 keV) at peak of fitted pulse versus pulse width (100 – 300 keV). Asterisks: long bursts; squares: short bursts.

DISCUSSION

By measuring pulse widths and intervals between pulses in bright bursts, we show that the bimodal duration distribution may be understood in terms of canonical time scales between major pulses. For long bursts we find a broad maximum in the pulse interval distribution spanning 0.3 – 3 s, with a logarithmic mean of ~ 1 s, and a mean pulse width of 200 – 600 ms. Short bursts tend to have fewer, narrower, more closely spaced pulses. Thus the salient question is: are the groups' combined pulse-width and -interval distributions unimodal or bimodal? The answer requires analysis of a larger sample of bursts with durations spanning the shortward mode and the minimum in the duration distribution. Simulations indicate that short bursts must have fewer, more closely spaced pulses than long bursts to reproduce the shortward mode and minimum. The average trend of spectral evolution in long bursts suggests that the difference in hardness ratios might be connected with the tendency of longer bursts to soften; however, a more complete understanding may involve the tendency of rapidly rising bursts to exhibit high HRs. As no correlation is found between pulse width and HR, we conclude that pulse width itself is not a major factor in burst spectral hardness.

REFERENCES

1. C. Kouveliotou et al., *Ap.J.* **413**, L101 (1993).
2. E.P. Mazets et al., *Ap. Space Sci.* **80**, 3 (1981).
3. J.P. Norris et al., *Nature* **434**, 308 (1984).
4. R.W. Klebesadel, in *Gamma-Ray Bursts*, eds. C. Ho, R.I. Epstein, & E.E. Fenimore (Cambridge Univ. Press: Cambridge, 1990), p. 161.
5. K. Hurley, in *Gamma-Ray Bursts*, AIP Conf. Proc. 265, eds. W.S. Paciesas & G.J. Fishman (AIP: New York, 1991), p. 3.
6. J.-P. Dezalay et al., in *Gamma-Ray Bursts*, AIP Conf. Proc. 265, eds. W.S. Paciesas & G.J. Fishman (AIP: New York, 1991), p. 304.
7. C. Kouveliotou et al., in *The Compton Observatory Science Workshop*, eds. C.R. Shrader, N. Gehrels & B. Dennis (NASA/GSFC: Greenbelt, MD, 1992), p. 61.
8. K.S. Wood et al., in *Gamma-Ray Bursts*, AIP Conf. Proc. 141, eds. E.P. Liang & V. Petrosian (AIP: New York, 1986), p. 4.
9. J.P. Norris et al., to appear in *Ap.J.* (1994).

MEASURING THE DISTRIBUTION OF TEMPORAL STRUCTURE IN GAMMA-RAY BURSTS

J. P. Norris

NASA/Goddard Space Flight Center, Greenbelt, MD 20771

ABSTRACT

Multi-resolution approaches, such as wavelet transforms, appear well suited to measuring the wide distribution of temporal structure manifest in γ-ray bursts. However, accurate interpretation of such measures requires an understanding of the intrinsic distribution of burst structure: a dilation transformation can result in structure being translated to either shorter or longer timescales, depending on the original distribution. Fits of pulses in bright BATSE bursts provide a reasonable characterization of intrinsic burst structure. Simulations are then used to illustrate how wavelet and Fourier transforms behave under dilation transformations for pulse-width distributions following either idealized burst characteristics or power-law forms.

INTRODUCTION

The isotropic celestial distribution and the inhomogeneous spatial distribution of γ-ray bursts detected and localized[1] by BATSE can be interpreted in terms of a cosmological distribution.[2,3] If the bursters are at cosmological distances, then the time profiles of bursts from more distant sources will be dilated relative to those of nearer sources.[4] Since burst durations range over more than four orders of magnitude, dilation would be detectable only in a statistical sense. Most of this range in duration arises from intrinsic variations in the rest frames of the sources; only a factor of approximately two would be expected from cosmological time dilation.

Three multi-resolution tests and two integral tests designed to search for a stretching effect yield consistent results: collectively, the dimmest bursts detected by BATSE are approximately twice as long as are the brightest bursts.[5,6] The multi-resolution approaches exploit the apparent problem of the wide distribution of burst temporal structure, revealing a signature which appears compelling in that the same stretch factor is manifest across at least seven octaves in width of temporal structure. Stretching of all time scales present in bursts by the same factor is a necessary condition for the effect to be truly cosmological time dilation.

Among orthogonal multi-resolution approaches, wavelet transforms appear well suited to measuring distributions with heterogeneous temporal structure – the situation found in γ-ray bursts: the basis functions of wavelet transforms are indexed in both position and width, allowing a parsimonious representation of broad and narrow aperiodic pulses simultaneously. This cannot be achieved using Fourier transforms since their basis functions are global (sine waves), rather than local.

If we are to utilize wavelet transforms to measure the effect of a dilation transformation on burst structure, we must have a relatively good measure of the intrinsic distribution of temporal structure in bursts, and an understanding of how dilation translates temporal structure across timescales for a given distribution. Norris et al. report the distribution of fitted pulse widths for 35 bright bursts.[7] Using simulations of synthetic bursts following pulse-width distributions with either measured burst properties or with power-law forms, it is shown how Haar wavelet transforms and Fourier transforms behave under dilation transformations.

HAAR WAVELET TRANSFORMS

The Haar wavelet is a difference operator. Among orthonormal wavelet bases, the Haar basis affords the most compact support, differencing only two adjacent intensity samples on a given scale, thereby preserving discontinuous features with the highest fidelity (however, it is not necessarily the optimum basis for characterizing the structure in γ-ray burst profiles). A complete Haar wavelet transform is a decomposition of the profile into localized intensity differences, on a range of time scales, plus a constant component – the mean value. On time scale m and at position n, the wavelet transform of the function $f(t)$ can be expressed[8]

$$T_{m,n}(f) = a_0^{-m/2} \int dt\, f(t)\, \psi(a_0^{-m}t - nb_0), \qquad (1)$$

with constants $a_0 > 1$ and $b_0 > 0$. For our analysis ψ is the Haar wavelet:

$$\begin{aligned}\psi(\tau) &= 1 & 0 \le \tau < 1/2 \\ &= -1 & 1/2 \le \tau < 1 \\ &= 0 & \text{otherwise.}\end{aligned} \qquad (2)$$

For discretely spaced data, the time series to be differenced on each successive scale can be generated by summing pairs of intensity samples ("compressing" by a factor of two). A good measure of temporal structure on a given scale is the average of absolute values of differences between pairs of intensity values. This summation over the position index of the wavelet transform may be termed a "wavelet amplitude spectrum." As in the Fourier spectrum, positional (phase) information is lost upon summation.

SIMULATIONS OF SYNTHETHIC BURSTS

Simulations of burst profiles were performed for the purpose of illustrating the effects of dilation, or stretching, transformations. The simulations were generated from idealized distributions of measured burst temporal characteristics, thus making potential comparisons with similar treatments straightforward. (For comparison with actual bursts, an approach which more nearly reproduces all measurable characteristics of bursts is desirable.[7]) At least five distributions are essential to the burst simulation procedure: burst duration, number of pulses per burst, pulse shape, pulse amplitude, and pulse width. Durations were patterned after the measured T_{90} distribution for BATSE bursts[9] truncated at 1 s and 600 s. An exponential distribution of the number of pulses per burst was assumed, using 2 and 16 pulses for the least and most pulses, respectively, per 16-s interval. Pulse were generated according to the form:

$$F(t) = A \exp\{-(|t - t_{max}|/\sigma_{r,d})^\nu\}, \qquad (3)$$

where t_{max} is the time of a pulse's maximum intensity, A; σ_r and σ_d are the rise ($t < t_{max}$) and decay time constants ($t > t_{max}$), respectively; and ν is the pulse "peakedness" (a lower value for ν results in a more peaked pulse).[5] Combined, the parameters σ_r, σ_d, and ν afford sufficient flexibility to represent virtually all pulse structure shapes observed in bursts. The peakedness parameter was modelled with a distribution the logarithm of which is Gaussian:

$$\nu = \sigma \mu^\lambda. \qquad (4)$$

Here, $\sigma = 2.0$ and $\mu = 1.75$; λ is a normally distributed random number with zero mean

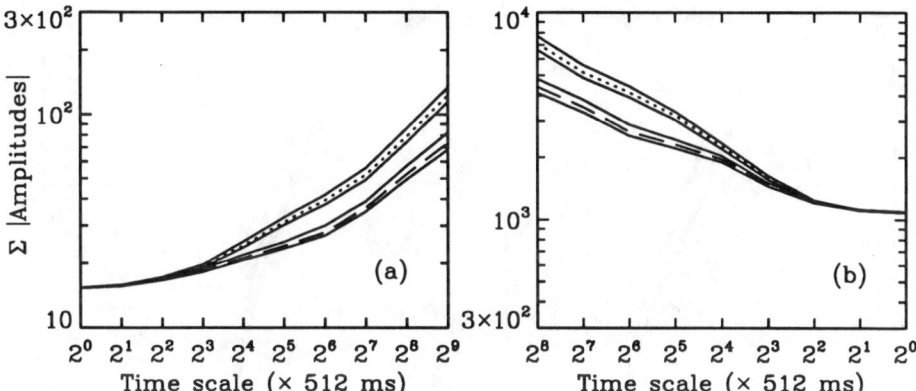

Figure 1. Average wavelet amplitude spectra (a) and Fourier spectra (b) for 200 dilated (dotted) and 200 nondilated (long dash) synthetic bursts. Longest time scale ($2^9 \times 512$ ms) is 262 s. Empirical sample errors are shown as envelopes (solid lines).

and standard deviation of unity. The ratios of rise to decay were uniformly distributed between 0.25 and 1.0. Amplitudes were uniformly distributed between 0 and 1. The full-width at half maximum (FWHM) of each pulse was chosen from the sum of two overlapping distributions of the form expressed in Equation (4) with $\{\sigma,\mu\} = \{0.5, 0.5\}$ and $\{2., 2.\}$, and probabilities of 2/3 and 1/3, respectively. This idealized FWHM distribution is comparable to the measured distribution for bright BATSE bursts (see Figure 1a, reference 7, these proceedings). Finally, pulse positions were uniformly randomly distributed within the time profile.

Two sets of 200 burst simulations apiece were performed, with and without a temporal dilation factor of two applied to the duration and the FWHM distributions; all other distributions and relevant ratios were equal for the two sets. The profiles were represented at 64-ms resolution. Each profile was normalized to a peak intensity of 980 counts s^{-1} and a constant background of level 5600 counts s^{-1} was added (these values correspond to the peak intensity — summed over DISCSC channels 1 and 2 — of the dimmest burst used in tests for time dilation, and to the average background of dim bursts[5]). Each time series was then Poissonated and truncated at 64 ms $\times 2^{12} \approx 262$ s.

Figures 1a & 1b illustrate the average wavelet and Fourier amplitude spectra, respectively, for these simulations. The outer envelopes surrounding each broader curve are the 1-σ empirical sample errors. Wavelet spectra are inherently logarithmically spaced; the Fourier frequency channels have been summed to yield a nearly exact logarithmic binning. The shortest timescale shown is 512 ms, below which both transforms converge to the noise levels, which are flat at ≈ 15.1 units (wavelet) and ≈ 525 units (Fourier). The salient aspect of both figures is that the dilation transformation results in an increase of the average amplitude over a wide range of timescales. A higher curve indicates more structure: essentially, the (fixed length) time series is on the average more filled up *per time scale* for the dilated bursts. This may seem counterintuitive; dilation might have been expected to translate a given curve to the right rather than up. In fact, either result is possible, depending on the pulse-width distribution and the dilation factor. This is demonstrated in what follows using well-understood power-law distributions for pulse width. For brevity only wavelet transforms are discussed since similar remarks apply for Fourier transforms. We note that, although the discriminatory powers of Fourier and wavelet transforms

180 Measuring the Distribution of Temporal Structure

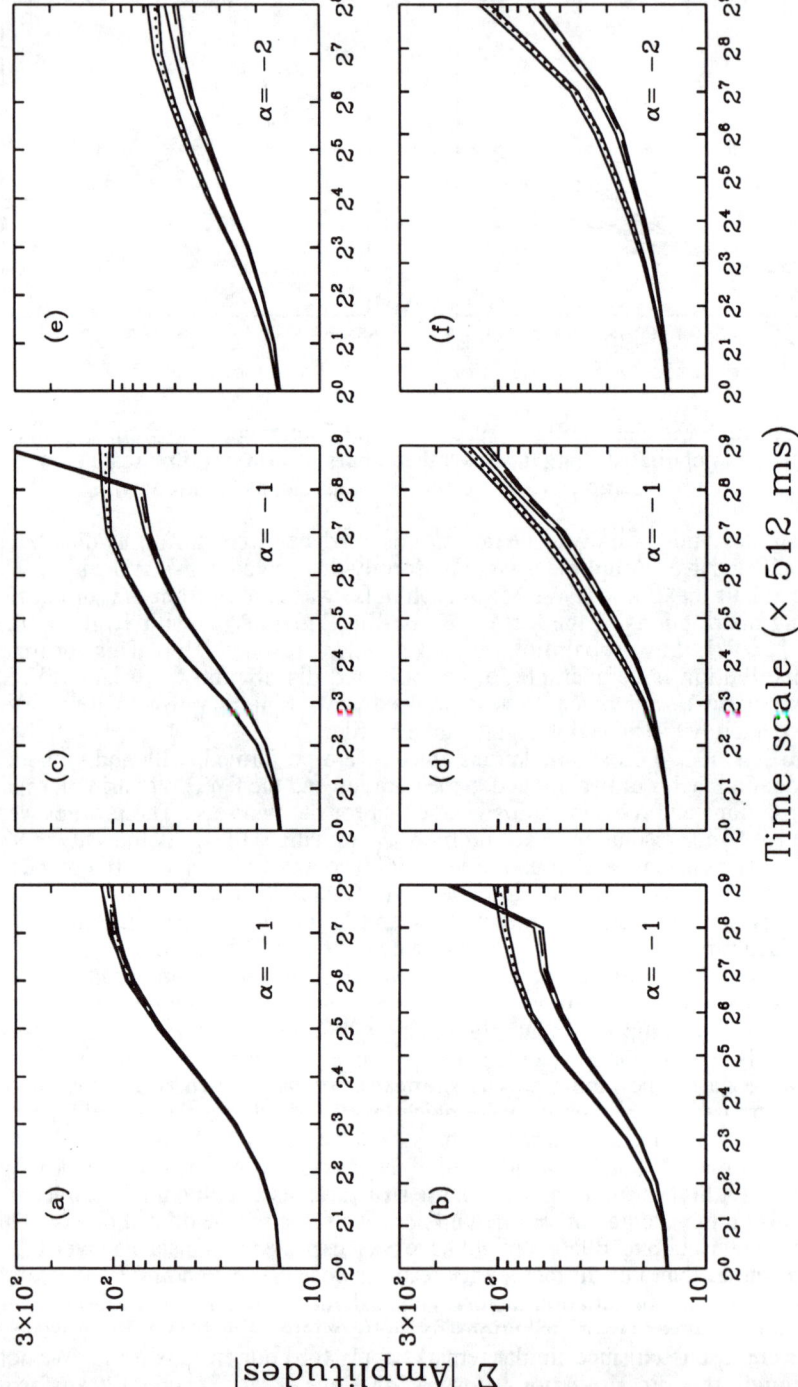

Figure 2. Similar to Figure 1a, with pulse width following power-law distribution of index α, and with various distributions of duration and number of pulses per burst. Maximum value on abscissa is $2^9 \times 512$ ms ≈ 262 s, except for panels (a) and (b) where maximum is one octave less. For each panel 400 nondilated and 400 dilated profiles were simulated.

revealed by Figures 1a & 1b appear comparable, this is often not the case for other distributions – the wavelet transform is superior for the cases discussed below.

Now consider pulse width (τ) distributions which follow power-law forms like

$$N(\tau) \propto \tau^\alpha. \tag{5}$$

For the special case $\alpha = -1$ and a dilation transformation with $\kappa = 2$, $N(\tau) \Rightarrow N'(\kappa\tau)$; in the dilated space an increased number of wavelets results, on each dyadic time scale, by a factor $\tau^\alpha/(\kappa\tau)^\alpha = \kappa^{-\alpha} = 2$. However, the total duration of the process is also dilated. For sake of illustration, assume the nondilated duration is constant (64 ms × $2^{12} \approx 131$ s) and the number of pulses per nondilated profile is constant (8); the dilated profiles are then twice as long with twice as many pulses. Figure 2a shows the average wavelet spectra for 400 profiles per set for a wavelet transform length of 131 s, containing exactly the nondilated profile interval. The spectra are indistinguishable within errors. Increasing the transform length by a factor of two to include exactly the dilated profiles then reveals the salient difference (Figure 2b): since the second half of the nondilated profiles are filled only with noise, all timescales except for the longest exhibit lower average amplitudes than for the dilated profiles. On the longest timescale the average amplitude for the nondilated set is higher because the longest wavelet represents the step between the pulse-populated first half and nonpopulated second half. For the dilated set, there is no such step – an increase in the mean value (the "mother" wavelet) represents the average pulse level above background.

Figures 2c & 2d illustrate successively more realistic simulations. In panel (c) the number of pulses is variable, and in panel (d) the durations are variable as well, both variable distributions being generated as described for the case in Figure 1. Allowing the number of pulses to vary results in a larger average amplitude difference between dilated and nondilated sets on intermediate timescales. When the durations are varied, the two curves lie more closely together, and thus the test becomes less discriminating. Of course this is the realistic case encountered in γ-ray bursts.

Figures 2e & 2f are analogous to 2a & 2d, except that the power-law index $\alpha = -2$. Now, dilation by a factor of two results in four times as many pulses on the dilated timescale, whereas the dilated durations increase only by a factor of two. Thus, wavelet transforming even the first half of each profile – the portion which exactly contains the pulses populating a nondilated profile – reveals a difference between the two distributions (panel 2e). Again, varying the durations and number of pulses per profile redistributes the average wavelet amplitudes (panel 2f), as in Figure 2d.

We conclude that – although the pulse-width distribution influences the wavelet amplitude spectrum – for $\alpha < 0$, dilation of the duration results in an upward displacement of the spectrum since fixed length intervals (long enough to encompass the longest dilated profile) will be more populated with pulses for the dilated profiles.

REFERENCES

1. C.A. Meegan, et al., Nature **355**, 143 (1992).
2. S. Mao, & B. Paczinski, Ap.J. **388**, L45 (1992).
3. W.A.D.T. Wickramasinghe, et al., Ap.J. **411**, L55 (1993).
4. B. Paczynski, Nature **355**, 521 (1992).
5. J.P. Norris, et al., accepted Ap.J. (1994).
6. S.P. Davis, et al., these proceedings.
7. J.P. Norris, et al., these proceedings.
8. I. Daubechies, I. "Ten Lectures on Wavelets" (Capital City, Philadelphia, 1992).
9. C. Kouveliotou et al., Ap.J. **413**, L101 (1993).

PULSE WIDTH DISTRIBUTIONS AND TOTAL COUNTS AS INDICATORS OF COSMOLOGICAL TIME DILATION IN GAMMA-RAY BURSTS

S.P. Davis,[*] J.P. Norris,
Goddard Space Flight Center, Greenbelt MD. 20771

C. Kouveliotou[†], G.J. Fishman, C.A. Meegan, and W.S. Paciesas[§]
Marshall Space Flight Center, Huntsville, AL.

ABSTRACT

The spatial distribution of bursts observed by BATSE is isotropic and inhomogeneous, implying that their sources may be at cosmological distances. We describe three tests that we have applied to search for the signature of time dilation. For both tests selection effects arising from intensity differences are avoided by rescaling all burst time profiles to a canonical dim peak intensity and rendering their backgrounds and noise biases uniform. These tests are (1) the distribution of pulse widths for these groups; a measure of the average rescaled integrated counts (2) within fitted pulse structures and (3) above fitted background for the burst ensembles. All tests use pulse shapes obtained from fitting major structures in temporal profiles. Results from 135 BATSE bursts indicate that the relative time dilation factor, dim to bright bursts, would be of order two, consistent with three previous tests,[1] and with cosmological interpretations of the BATSE number-intensity relation, all of which place the more distant bursts at redshifts of about unity.

INTRODUCTION

The spatial distribution of gamma-ray bursts (GRBs) observed by the Burst And Transient Source Experiment (BATSE) on the *Compton* Gamma Ray Observatory indicate that we are located in the center or close to the center of a spherical distribution of gamma-ray bursters.[2] The natural explanation is that the sources are at cosmological distances.[3,4,5,6] Recently postulated cosmological scenarios for the nature of the GRB sources involve active galactic nuclei[7] or merging massive binaries.[8] Galactic disk models in general have been ruled out by the BATSE data; they cannot explain the observed isotropy and nonuniformity in distance. Halo neutron star models are severely constrained: if the ratio of the number of halo to disk neutron stars is much less than one, then the paucity of any disk component in the BATSE data argues against the Extended Galactic Halo (EGH) model, or suggests that neutron stars in the galactic halo are much more prolific bursters.[9] Furthermore, the isotropic angular distribution and nonuniform distribution in distance[2] can be explained in terms of either a cosmological[5,10] or a heliospheric distribution of sources.[11] If the sources of GRBs are indeed at cosmological distances, then the temporal profiles of the more distant ones will be dilated, statistically, by a factor of $(1 + z) \sim 2$ relative to those that are nearby.[6,12] The implication that z is of the order of one is consistent with studies involving the BATSE number-intensity relation.[10,13]

We have performed three tests for cosmological time dilation in a sample of gamma-ray bursts observed by BATSE. The results of the tests are independent of the

---
[*]Department of Physics, The Catholic University of America, Washington, D.C. 20064
[†]Universities Space Research Association
[§]University of Alabama, Huntsville, Al 35899.

brightness of the bursts. We describe the preparation of the burst profiles, the application and results of the tests. We calibrate the results by performing the same tests on two sets of simulated burst profiles.

ANALYSIS PROCEDURE

We performed all tests using 64-ms data from BATSE's Large Area Detectors (LADs). Bursts longer than 1.5 s within the three brightness categories listed in Table 1 were analyzed. Bursts of duration less than 1.5 s were excluded from the analysis because BATSE sensitivity is lower for very short, dim bursts and because the observed bimodality in the burst duration distribution suggests that bursts shorter than about 1 s may constitute a distinct class.[14,15] All measurements were performed on temporal data from the two lowest energy channels (25 - 100 keV) of the LADs to minimize the effect of energy-dependent pulse widths which would affect redshifted dim bursts.[1] We concatenated prior to the 64-ms data four samples of a continuously available lower resolution (1.024-s) data type to avoid losing pre-trigger information in some of the dimmer bursts.

Preparation for all tests included fitting and subtraction of background, followed by scaling the signal profile to the dimmest burst observed (1400 cts s^{-1}); adding a canonical flat background (~5600 cts s^{-1}, that of the dimmest bursts); and adding noise to realize Poisson statistics.[1] These measurements virtually eliminate brightness selection effects. The time profiles were then binned to 256-ms resolution.

THE PULSE-WIDTH MEASUREMENT PROCEDURE

For our first test, pulse shapes were estimated using an interactive, multi-parameter least squares algorithm, "Pulsefit", that deconvolves temporal profiles into constituent pulses. Pulsefit calls a Fortran program (Superfit, written by W.N. Johnson), which performs the χ^2 minimization.

We utilize a pulse model which well represents virtually all pulse shapes occurring in real bursts:

$$f(t) = A \cdot \exp\{ -[|t - t_{max}|/(\sigma_{l,r})]^\nu \} \tag{1}$$

where t_{max} is the time coordinate of the pulse's maximum intensity, A; $\sigma_{l,r}$ are the rise (t < t_{max}) and decay (t > t_{max}) time constants; and ν is the pulse peakedness (lower values give more peaked pulses).[5] We fitted the real burst profiles with the form given in equation (1), compiling fitted pulse parameters for various analyses. The first test for time dilation is then a comparison of the distributions of fitted pulse widths for the dim and bright burst samples.

Because the pulse identification process involves a degree of subjectivity, we cannot always identify pulses as discrete entities – it is not always possible to distinguish random statistical fluctuations from an actual pulse structure. Furthermore, the stretching effect combined with fluctuations could often make two pulses out of one. Therefore, all measurements were calibrated by simulating a set of 100 synthetic bursts that approximate the temporal characteristics of bright bursts, and a similar set of 100 bursts with the pulses and durations dilated by a factor of two.[1] Simulations are sums of several pulses plus background. They were fitted with the model function shown in equation (1). The fitted parameters from the simulated bursts were then analysed in the same manner as were the profiles of real bursts. There are some additional caveats: the number of real pulses is not known a priori; also, the fits are nonunique in that more than one pulse model may be acceptable, possibly yielding systematically different results.

TOTAL COUNTS TESTS

The second test is a measure of the total structure within bursts obtained from fitting pulses using the "Pulsefit" algorithm and then summing the counts over all fitted structures to obtain the integrated counts for each burst. The measurement was performed over the burst profiles up to 65 s in duration, and the total counts within fitted pulses were measured for structures that exceeded the 2σ-level above background. The third test is similar, but with the summation performed for counts above the background level. These tests were also calibrated with the same two sets of synthetic bursts, again having a relative dilation factor of two. Results of the calibration procedure indicate how accurately we can recover pulse parameters from the burst profiles and how well we can measure any stretching effect.

RESULTS

The pulse-width distributions of bright and dim bursts are shown in Figure 1. The dotted line histogram is for bright bursts and the solid line is for dim bursts. The centroid of the dim sample is shifted by about a factor of 2 over a broad range of time scales, consistent with time dilation for $z \sim 1$. This shift needs to be better quantified by fitting the width distributions with a satisfactory functional form. Pulse-width distributions of dilated synthetic bursts are shown in Figure 2: Dotted lines are nondilated synthetic bursts and solid lines are the dilated synthetic bursts.

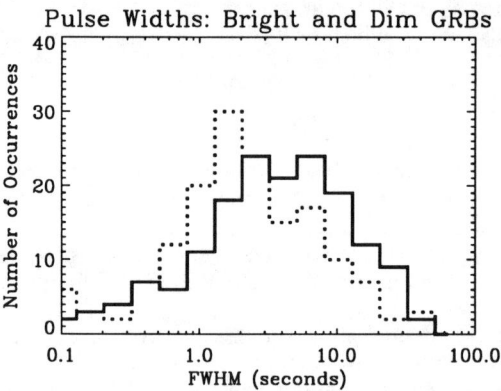

Figure 1. Fitted pulse-widths for bright (dotted) and (solid) BATSE bursts.

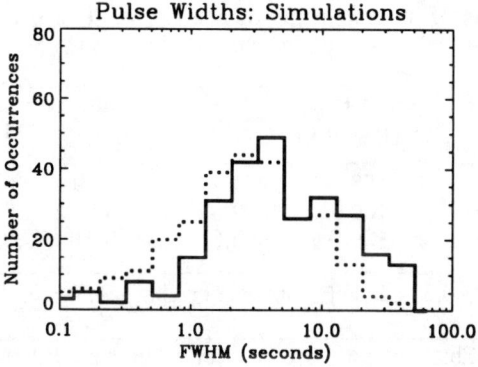

Figure 2. Fitted pulse-widths for synthetic dilated (solid) and nondilated (dotted) bursts.

Table 1 shows results of two normalized average integrated counts tests performed on real (a,b) and synthetic (c) bursts. Measurements are performed over the entire profile above background, and within fitted pulse structures. Because only pulse structures above a 2σ variation in background were selected to be fitted, the values contained in the second total counts test are some forty percent lower. The total counts ratios and their associated errors are shown in all tables. C_p is the peak count rate.

TABLE 1a. Total Integrated Counts Above Background

Brightness Group	Class	# of bursts	<Total counts>	$+1\sigma$	-1σ
Bright	$18000 < C_p < 250000$	41	4490	880	480
Dim	$2400 < C_p < 4500$	49	8520	1210	1070
Dimmest	$1400 < C_p < 2400$	45	9080	2640	1200

Ratios: dim:bright = 1.9 {+0.34, -0.44}; dimmest:bright = 2.0 {+0.62, - 0.47}

TABLE 1b. Total Integrated Counts Within Fitted Pulse Structures

Brightness Group	Class	# of bursts	<Total counts>	$+1\sigma$	-1σ
Bright	$18000 < C_p < 250000$	41	2730	360	330
Dim	$2400 < C_p < 4500$	47	5000	830	630
Dimmest	$1400 < C_p < 2400$	41	5890	1810	885

Ratios: dim:bright = 1.8 {+0.38, -0.33}; dimmest:bright = 2.2 {+0.72, -0.44}

Simulations

TABLE 1c. Total Integrated Counts Within Fitted Pulse Structures

Group	# of bursts	<Total counts>	$+1\sigma$	-1σ
Nondilated	100	2300	517	178
Dilated	98	4460	763	392

ratios: dilated:nondilated = 1.9 {+0.36, - 0.46}

Table 2 shows the results of measurements of average pulse structure widths (Full - Width Half Maximum) for real bursts (a) and simulated bursts (b).

TABLE 2a. Average Pulse-Widths Per Brightness Group					
Group	# of bursts	# of structures	<FWHM>	+1σ	-1σ
Bright	41	158	1.42	0.39	0.10
Dim	49	166	2.34	0.53	0.17
Dimmest	45	129	2.53	0.91	0.21

Ratios: dim:bright = 1.6 {+0.39, -0.47}; dimmest:brights = 1.8 {+0.65, -0.51}

TABLE 2b. Average Pulse-Width for Synthetic Bursts					
Group	# of bursts	# of structures	<FWHM>	+1σ	-1σ
Nondilated	100	281	1.37	0.19	0.08
Dilated	98	283	2.75	0.50	0.16

Ratios: dilated:nondilated = 2.0 {+0.38, -0.30}

DISCUSSION

Our results indicate a "stretch" factor between the dim bursts and the bright bursts of about 2. This signature is consistent with the dimmer bursts having durations that are twice as long as the bright bursts and with cosmological modeling of the number-intensity relation.[10] The best fit to the number-intensity relation[10] indicates a value of z ~ 1 at the BATSE 99% completeness limit,[1] which corresponds to the dimmest bursts used in this analysis.

REFERENCES

1. J.P. Norris, et. al., accepted Ap. J. (1994).
2. C.A. Meegan, et al., Nature, **355**, 143 (1992).
3. B. Paczynski, Acta Astron., **41**, 257 (1991).
4. C.D. Dermer, Phys. Rev. Lett., **68**, 1799 (1992).
5. S. Mao, & B. Paczynski, Ap. J., **388**, L45 (1992).
6. T. Piran, Ap. J., **389**, L45 (1992).
7. O.F.Prilutski, & V.V. Usov, Ap. Space Sci., **34**, 395 (1975).
8. R. Narayan, B. Paczynski, & T.Piran, Ap. J., **395**, L83 (1992).
9. D.H. Hartmann, et al., Ap. J. Suppl., accepted (1994).
10. W.A.D.T. Wickramasinghe, et al., Ap. J., **411**, L55 (1993).
11. J.M. Horack, et al., Ap. J, submitted (1993).
12. B. Paczynski, Nature, **355**, 521 (1992).
13. E.E. Fenimore, et al., 1993, Nature, accepted (1993).
14. J.P. Norris, et al., Nature, **434**, 308 (1984).
15. C. Kouveliotou, et al., Ap. J., **413**, L101 (1993).
16. J.P. Norris, et al., 23rd International Cosmic Ray Conference (1993), p.89.
17. J.P. Norris, et al., AIP Conference Proc. **280** (1993), p.959.
18. S.P. Davis, et al., Bull. AAS. **23**, 1323 (1992).
19. S.P. Davis, et al., AIP Conference Proc. **280** (1993), p.964.

THE AVERAGE TEMPORAL PROFILE OF BATSE GAMMA-RAY BURSTS: COMPARISON BETWEEN STRONG AND WEAK EVENTS

I. G. Mitrofanov, A. M. Chernenko, A. S. Pozanenko
Space Research Institute, Profsojuznaya str. 84/32, 117810 Moscow, Russia

W. S. Paciesas[1], C. Kouveliotou[2], C. A. Meegan, G. J. Fishman
NASA/Marshall Space Flight Center

R. Z. Sagdeev
University of Maryland

ABSTRACT

First results are presented from the analysis of 260 BATSE gamma-ray bursts (GRBs)[1] using a method[2,3] whereby all events are synchronized and averaged around the bins which are the brightest time intervals of each of them. For the averaged time history, a difference is found between the rise front and the back slope, and good evidence is found for the presence of hard-to-soft spectral evolution. We compare sub-sets of "strong" and "weak" events and find no evidence for time dilation in weak GRBs, as would be expected in cosmological GRB models. On the other hand, for the strong events the averaged hardness ratio is found to be larger than for the weak GRBs.

1. INTRODUCTION

As discussed in many contributions to this Workshop, our present knowledge of GRBs is full of apparent contradictions. One of these is the contradiction between the perfect isotropy of GRBs on the sky[4,5], which corresponds to a very homogeneous angular distribution, and the value of $\langle V/V_{\max}\rangle \approx 0.32$[4,5], which indicates that GRBs have a non-homogeneous radial distribution in Euclidean space. To resolve this contradiction one might suppose, for instance, that GRBs are generated in the Extended Galactic Corona[6], or that they belong to a cosmological population with redshift $z \approx 1$[7]. A definite decision between cosmological and non-cosmological models would be a crucial step in understanding the origin of GRBs.

The method of averaging time histories of cosmic GRBs around the brightest time interval was developed for analysis of events detected in the APEX experiment of the PHOBOS Mission[2]. A set of 48 events was divided into two subsets of "strong" and "weak" events, and their averaged time histories were directly compared. While differences were found in the rising parts, the back slopes were identical within statistics. The differences in the rising parts could result from a systematic difference in triggering moments for strong and weak GRBs. On the other hand, the similarity of back slopes indicated that there is no cosmological stretch of weak events with respect to strong GRBs.

---

[1] also University of Alabama in Huntsville
[2] also Universities Space Research Association

II. AVERAGE TIME HISTORY AND HARDNESS RATIO

The data from the BATSE instrument on the Compton Gamma Ray Observatory provide an excellent opportunity to study the averaged time histories of GRBs, in particular to compare subsets of strong and weak events, with much better statistics. We present herein preliminary results of this comparison.

A total of 260 GRBs from the first BATSE Catalog[1] have been studied so far. For each event we used the DISCLA count rate time histories (four broad energy channels with a time resolution of 1 s) from the large area detector which had the highest peak count rate. Those are independent of the instrument trigger, so that one has for every event a continuous time history containing both pre-burst and post-burst background. We defined each burst interval as extending from $T_0 - \Delta T$ to $T_0 + \Delta t$, where T_0 is the burst trigger time and $\Delta T = \max(T_{90}, 20 \text{ s})$, where T_{90} is a measure of burst duration[1]. Background during the burst interval was interpolated using the measurements before and after this interval, and the difference between the measured actual counts and the interpolated background curve constituted the burst time history.

The results we report are derived from two broad discriminator channels #2 (50–100 keV) and #3 (100–300 keV). For each GRB time history we determined the brightest time bin where the maximal number of counts was recorded in the sum of channels #2 and #3. The time histories (with statistical uncertainties) in the separate channels were then normalized to the count rate in the brightest bin. The result is a standardized set of GRB time histories, with time resolution of 1.024 s and maximum magnitude equal to 1.

With this data type, all events with duration shorter than 1 s are unresolved, their time histories falling within one or two bins, depending on their phase with respect to the instrument clock. At this stage, all events with $T_{90} < 1$ s were therefore excluded from the analysis. In the next analysis phase, we plan to use data with a time resolution of 64 ms, allowing use of the shorter events.

The averaged time history in the sum of discriminator channels #2 and #3 (50–300 keV) is presented in Figure 1 for 205 GRBs with $T_{90} > 1$ s from the first BATSE catalog[1]. Besides the short bursts, 8 other events from the Catalog could not be used due to data gaps. The omission of these events should not bias the final result: the set of 205 GRBs fully represent the continuous observation period of BATSE from 19 April 1991 to 5 March 1992[1]. The time scale of all figures corresponds to numbers of 1.024 s bins before and after the brightest one, which we defined as zero. Within the time interval from bin -15 to bin 20, the total number of contributing events (i.e., with significant count rate above background) is more that 90% of the total number in the averaged set. The mean duration of the averaged time profile at the level of 0.25 of its maximum height is about 7–9 s, which corresponds to 2–3 s of rise front and 4–5 s of back slope (with uncertainties of about 1 s). The difference between the averaged rise front and back slope is illustrated by the dashed curve in Fig. 1, which shows the time-reversal of the averaged light curve.

GRBs are known to have broader peaks of time histories at low energies relative to high energies. In Fig. 2 we compare the averaged time history of the full set of BATSE GRBs in discriminator channel #2·aligned around the brightest bin in channel #2 with the averaged time history in channel #3 aligned around the brightest time bin in channel #3. It is easily seen that the softer energy channel indeed has a broader peak in its averaged time history.

The averaged time history of BATSE GRBs may be compared with that for APEX GRBs, which was obtained previously by the same method[2]. The dots in Fig. 2 represent the averaged time history of 48 APEX GRBs in the broad energy channel 100–1000 keV. The two data sets agree quite well in spite of the instrumental differences between APEX and BATSE.

A comparison of "color" of GRBs along the averaged time history has been done by separate averaging of their profiles for discriminator channels #2 and #3, aligned with respect to the brightest bins for the sum of those two channels. The average time profile of the hardness ratio (HR), defined as the ratio of counts in 100–300 keV to the counts in 50–100 keV, is shown in Fig. 3. The HR has a maximum around 0.9 during the time interval from -9 to 0, during which the averaged time history rises to its maximum from the level of 0.1. Before the rise front, i.e., within time interval from -16 to -10, the averaged HR is about 0.7. After the maximum, HR decreases with the averaged flux to the value at the beginning of the rise front. Thus, we conclude that HR is a maximum during the rise front.

III. COMPARISON BETWEEN STRONG AND WEAK EVENTS

The total set of GRBs may be sub-divided into "strong" and "weak" events. We first consider a simple separation, where strong and weak events are associated with $V/V_{\max} < 0.34$ and > 0.34, respectively (for the set of BATSE GRBs $\langle V/V_{\max} \rangle = 0.34 \pm 0.02$[1]). The averaged time histories of 112 strong and 79 weak events thus defined are shown in Fig. 4 by bold and thin lines, respectively. No statistically significant difference is seen between the two profiles.

In cosmological models the time dilation stretch factor between the strong and weak event profiles should be $\sim (1 + \langle z_w \rangle)/(1 + \langle z_s \rangle)$, where $\langle z_w \rangle$ and $\langle z_s \rangle$ are the average redshifts of the weak and strong events, respectively. Strictly speaking, the above comparisons are compromised in a cosmological model by the redshift of photons from higher to lower energies. Because of the redshift, the profile of weak bursts in the observed energy channel corresponds to higher rest energies than the profile of strong bursts. Since the bursts are intrinsically narrower at higher energies (Fig. 2), this effect could wipe out any time dilation which might be present.

In order to compensate for this effect, we compared the averaged time history of 38 "very strong" events ($V/V_{\max} < 0.05$) in 100–300 keV and the averaged time history of 42 "very weak" events ($V/V_{\max} > 0.60$) in 50–100 keV. The results (Fig. 5) again show no evidence for systematic stretching of the very weak events relative to the very strong events. For comparison, we also show in Fig. 5 the stretched profile expected from cosmological models which acceptably fit the BATSE number/intensity distribution[7]. Here we assumed $\langle V/V_{\max} \rangle \approx .025$ for the very strong events and $\langle V/V_{\max} \rangle \approx .8$ for the very weak events, from which Fig. 2 of Ref. 7 predicts stretch factors of 22–67%. Fig. 5 represents a conservative test since the rest energies of the weak events are on the average lower than those of the strong events for these values of $\langle z \rangle$. Thus, the redshift in energy of photons cannot account for the absence of time dilation.

On the other hand, the averaged HR time profiles show a difference between strong and weak subsets (Fig. 6). Both profiles have rather similar shapes with a flat maximum along the rising part of the averaged time histories and a decrease just after the brightest time bin, but the averaged values of HR for strong events are larger than the corresponding averaged values for weak events. A similar correlation between intensity and hardness was found previously for

APEX GRBs[2]. This result also confirms and extends the correlation found previously for BATSE GRBs[8].

IV. CONCLUSIONS

1) The averaged time history of GRBs is non-symmetric: from the level of 0.25 of its maximum height the rise front lasts about 2–3 s and back slope about 4–5 s. The averaged HR profile has a broad maximum along the rise front of the averaged time history, and decreases along its back slope.

2) We find no evidence in this analysis for time-dilation in the averaged time histories of "weak" GRBs relative to "strong" events. If we assume a cosmological scenario, the redshift of narrower temporal structure from higher to lower energies is insufficient to account for the absence of a time dilation effect,

3) We find evidence for differences between strong and weak GRBs in the averaged hardness ratio: strong events have larger HR then weak ones not only averaged over time but also at each point in the averaged time profile. In the absence of a cosmological interpretation, we must then consider the similarities of the HR profiles to indicate their association with the same unique class of radiating objects, but with some important difference between them which leads to a hardness/intensity correlation.

Further analysis is essential to compare the sensitivity of this method with that of other analyses which have been shown to support a cosmological interpretation[9]. We will continue to investigate these issues further with various refinements to our analysis, including a larger sample of bursts, use of data with finer time resolution, and use of data in other energy ranges.

ACKNOWLEDGEMENTS

The Russian co-authors express their appreciation for access to BATSE data and the fruitful cooperation and warm hospitality of the BATSE team.

REFERENCES

1. G. J. Fishman et al., ApJS, in press (1994).
2. I. G. Mitrofanov et al., in Compton Gamma Ray Observatory: St. Louis, MO 1992, ed. M. Friedlander, N. Gehrels & D. J. Macomb (New York: AIP, 1993), p. 761.
3. I. G. Mitrofanov et al., in Gamma-Ray Bursts: Observations, Analyses and Theories, ed. C. Ho, R. I. Epstein & E. Fenimore (Cambridge: Cambridge U. Press, 1992), p. 203.
4. C. A. Meegan et al., Nature **355**, 143 (1992).
5. C. A. Meegan et al., these proceedings.
6. D. H. Hartmann, E. V. Linder & L.-S. The, in Compton Gamma Ray Observatory: St. Louis, MO 1992, ed. M. Friedlander, N. Gehrels & D. J. Macomb (New York: AIP, 1993), p. 1003.
7. W. A. D. T. Wickramasinghe et al., ApJ **411**, L55 (1993).
8. W. S. Paciesas et al., in Gamma-Ray Bursts: Huntsville, AL 1991, ed. W. S. Paciesas & G. J. Fishman (New York: AIP, 1992), p. 190.
9. J. P. Norris et al., ApJ, submitted (1993).

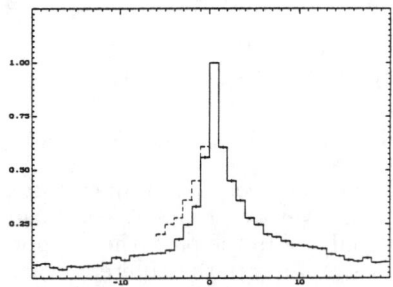

Fig. 1—Flux versus time.

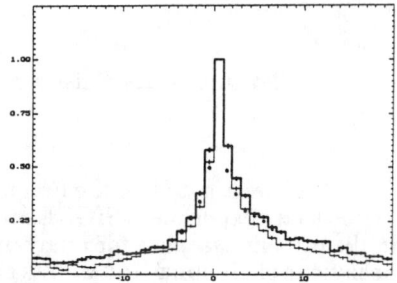

Fig. 2—Flux versus time. Bold line: 50–100 keV. Thin line: 100–300 keV. Dots represent APEX data.

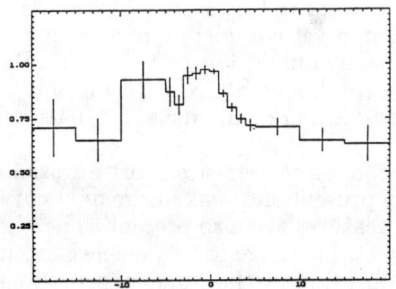

Fig. 3—Hardness ratio versus time.

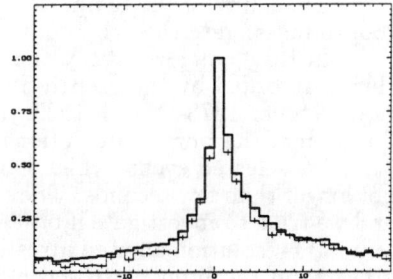

Fig. 4—Flux versus time. Bold line correspond to "strong" events. Thin line corresponds to "weak" events.

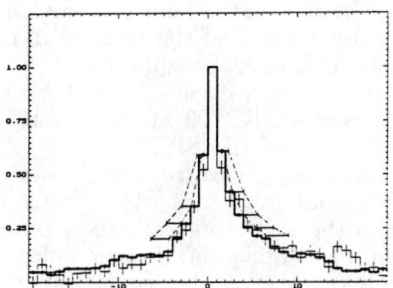

Fig. 5—Flux versus time. Bold line correspond to "very strong" events with inner ends stretched by 22% and outer ends stretched by 67%. Thin line corresponds to "very weak" events.

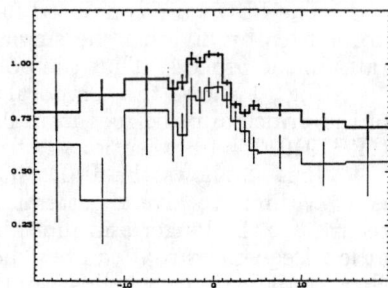

Fig. 6—Hardness ratio versus time. Bold line correspond to "strong" events. Thin line corresponds to "weak" events.

STUDY OF THE TEMPORAL BEHAVIOUR OF FOURTEEN FAST GAMMA-RAY BURSTS OF THE SIGNE EXPERIMENT

B. M. Belli

Istituto di Astrofisica Spaziale, CNR, 00044 Frascati, Italy

ABSTRACT

We investigated on the temporal behaviour of fourteen fast events recorded by the space experiment SIGNE on board the soviet Venera probes. We note that in the present case, like for many other events analysed in the past, the temporal behaviour of Gamma-Ray Bursts shows a strict similarity in the different events, the only differences being essentially due to intensity and characteristic time. This fact suggests that the physical processes at the origin of the Gamma-Ray Bursts do not allow large variations to the scenario of their emission.

INTRODUCTION

The study of the temporal structures of Gamma-Ray Bursts is a powerful tool to investigate the origin of this phenomenon yet completely mysterious.

In this paper we study the temporal behaviour of fourteen Gamma-Ray Bursts recorded by the French-Soviet space experiment SIGNE, during the period October 1978- March 1983. Temporal resolution of the data is 1/64 s and duration of the burst observations 16 s.

We selected events which have as common characteristics short durations, not longer than few seconds. They principally present one peak in the light curve and we tried to investigate if other temporal features are also present. The high temporal resolution allowed investigations on a wide range of frequencies; on the other hand the limit of 16 seconds for the duration of the burst observations, while setting a limit to the study of long burst, is largely sufficient to investigate short events and their possible tails. We perform this analysis by means the Fourier analysis .

DATA ANALYSIS

The PSD's are computed for each event by means the Fourier transform, normalized by dividing the single PSD estimates by half of the total counting rates in the process. Thus the poissonian noise level must be equal to 2.

Fig. 1 shows the temporal history of the selected events. The duration of the principal peak goes from 0.1 s, for the events GRB790406, GRB791018, GRB791013, and exceptionally to 4.8 s for the events GRB781025.

Fig. 2 shows the PSDs in the different cases. The frequency binning is logarithmic to give a better idea of the essential structures which seems to recur. For the longer and more intense events the PSDs present excess power at low frequencies that can best be fitted with a Lorentzian law having different characteristic times, which are of the same order of the event durations (Table.I).

The Lorentzian fits constitute a synthetic manner to represent Bthe behaviuor of the PSD's of the Gamma-Ray Bursts. From a particular value of frequency, that varies from case to case, the power starts to decrease down to

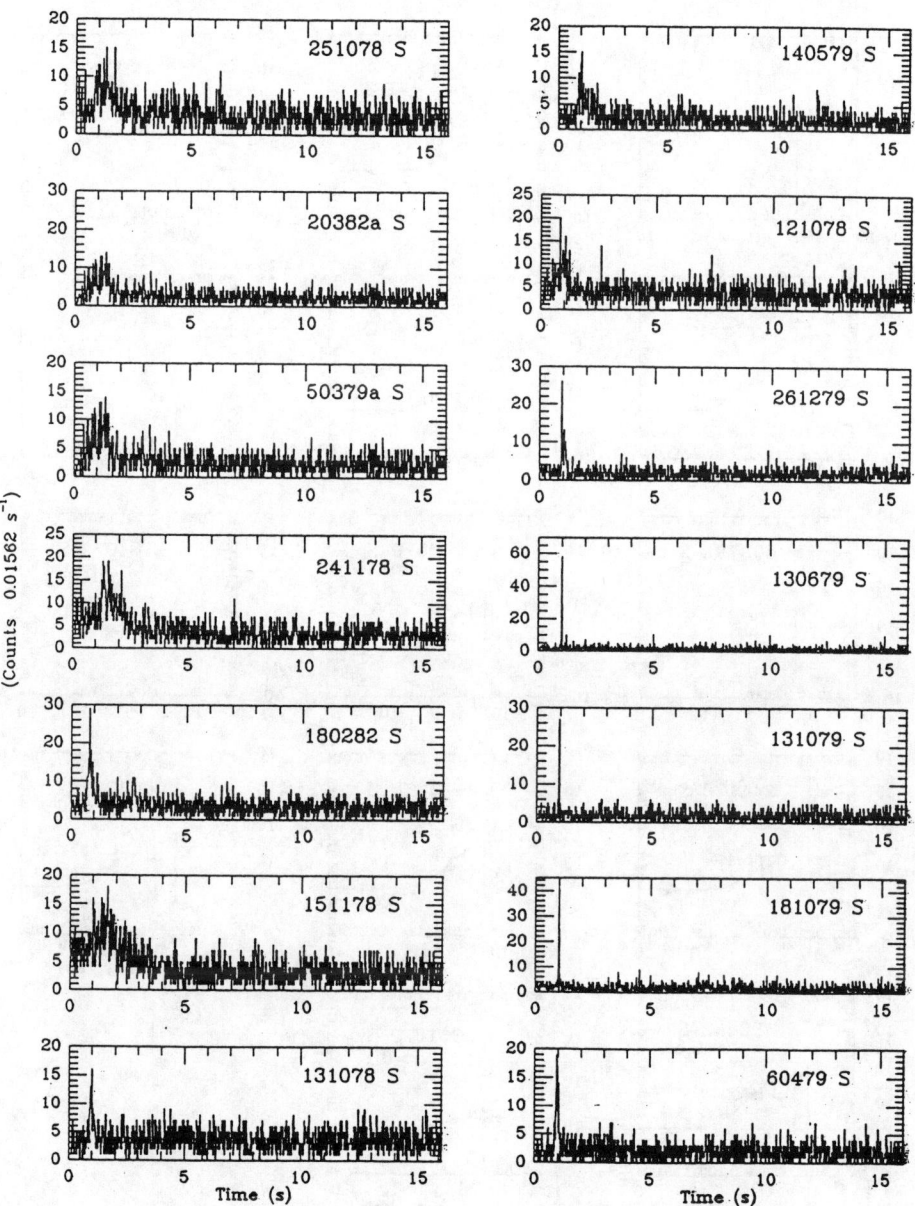

Figure 1. Light curves of 14 fast events from SIGNE experiment.

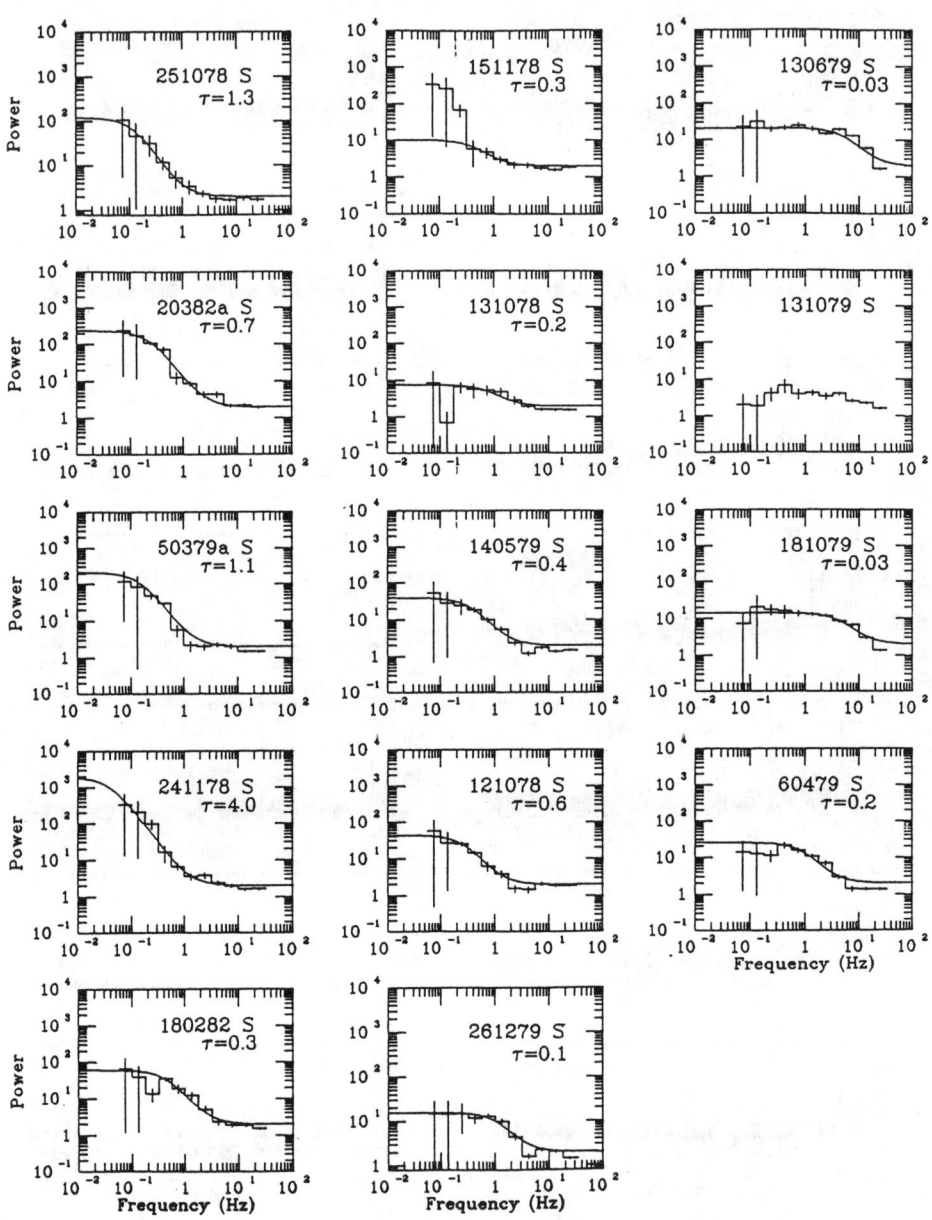

Figure 2. PSDs of selected events. Solid line is the Lorentzian best fit.

the poissonian noise level. We note that for several events the noise level appears to be lower then two, the expected poissonian level. This may be an effect due to the dead-[Btime of the experiment when the photon frequency in the burst is high. The first one or two or rarely three logarithmic bins have a very big statistical uncertainty and are also affected by the artefacts due to the box effect. We reported these bins in Fig.2, but they were neglected in the fitting. For the spike-like shorter events the PSDs are quite a constant up to a certain frequency determined by the spike duration. Also the PSD of the spike-like events may be affected by the artefacts due to the circumstance that the spike is not a real Dirac delta. It is so very difficult to distinguish the real temporal features from the artefacts of the analysis when, as in our case, the statistics is low.

Table I. Event characteristic shot time (τ) and duration.

Event Date	τ (s)	Duration (s)
GRB 251078	1.3 ±4.0	4.8
GRB 20382a	0.7 ±0.6	2.9
GRB 50379a	1.0 ±0.9	2.5
GRB 241178	4.0 ±31.0	2.4
GRB 180282	0.3 ±0.4	2.1
GRB 151178	0.3 ±0.4	1.8
GRB 131078	0.2 ±0.2	1.5
GRB 140579	0.4 ±0.3	1.0
GRB 121078	0.6 ±0.4	0.8
GRB 261279	0.13±0.08	0.3
GRB 130679	0.03±0.01	0.3
GRB 131079	-	0.1
GRB 181079	0.03±0.01	0.1
GRB 60479	0.2 ±0.1	0.1

DISCUSSION

In the present data, at the present level of the counting rate statistics, no certain substructures at shorter time scales have been detected, in the examined frequency range, if we esclude few cases such as the GRB781115 event for which the PSD should suggest a substructure that can be interpreted by means an exponential shot noise of characteristic time 0.3 s, or otherwise, for example, by a very large frequency quasi-periodic oscillation (QPO). The possibility to best fit the low frequency power (red noise) in the PSDs with a Lorentzian law is a further indication that the shape of the peak counting rate of the gamma-ray bursts is principally exponential[1] or constituited by the random superimposition of some exponential shots[2]. We observe that for the examined Gamma-Ray Bursts the red noise has always a slope of 2.

But differently happens in other phenomena which also present red noise,

e. g. in the x-ray emission from close binary systems or active galactic nuclei[3,4], where the slopes of the red noise is very variable in the range beside 2 and rarely reaches the limit value of 2.

These results suggest that the Gamma-Ray Bursts, differently from other red noise astrophysical sources[5], for which a wide spectrum of emitting situations is possible, are a phenomenon with emission mechanisms that, in most of the cases, seem to produce for the bursts exponential shapes either for single or for the multiple structures of the event.

I thank the SIGNE group in CESR for making the SIGNE time histories available.

REFERENCES

1. C. Barat, et al., Ap. J. **280**, 150 (1984).
2. B. M. Belli, Ap. J. **392**, 266 (1992).
3. B. M. Belli, A&A **97**, 63 (1993).
4. T. Belloni and G. Hasinger, A&A L33 **227**, 33 (1990).
5. L. Burderi, N. R. Robba, M. Guaninazzi, G. Cusumano, Il Nuovo Cimento **Vol16C**, in press (1993).

MORPHOLOGICAL STUDY OF SHORT GAMMA RAY BURSTS

P. N. Bhat*
Tata Institute of Fundamental Research,
Homi Bhabha Road, Bombay 400 005, India

G. J. Fishman, C. A. Meegan, R. B. Wilson
ES-66, Space Science Laboratory,
NASA/Marshall Space Flight Center, Huntsville, AL 35812

W.S. Paciesas
Dept. of Physics, University of Alabama in Huntsville, Huntsville, AL 35899

ABSTRACT

Gamma ray bursts (GRB) of durations less than about 2 sec, which presumably form a separate class, constitute nearly 26% of the bursts detected by the burst and transient source experiment(BATSE) on board the Compton Gamma Ray Observatory. A very high time resolution study of these bursts is undertaken. A few parameters to describe the complexity, its temporal and spectral evolution have been defined and computed for each burst. Results of analysis of 37 short bursts show that more complex bursts have higher average hardness ratio. It is also found that bursts with relatively higher hardness ratio exhibit sharper temporal features at higher energies. A systematic search for photon bunching during these bursts show no positive evidence. We also find that short bursts show a wide variation in spectral evolution similar to that of longer bursts.

INTRODUCTION

Gamma Ray Burst (GRB) durations vary greatly, spanning at least 5 orders of magnitude, from the shortest of < 12 ms (FWHM) for GB820405[1] to the longest of 1000 s for GB840304[2] with the range limited by possible instrumental effects on either extreme. There are reports of a break in the burst duration distribution which was first noted by Cline and Desai.[3] A distribution of burst durations from the first catalog of BATSE bursts show a clear break at ~ 2 sec supporting the hypothesis that short bursts form a separate class [4]. Based on 66 bursts detected by the PHEBUS experiment on GRANAT, Dezalay et al.[5] find that short rise times and hard spectra are common features of bursts lasting < 2 s. Kouveliotou et al.,[4] based on the analysis of 222 bursts detected by BATSE, also find that the average hardness ratio of bursts shorter than 2 sec is distinctly higher than that for longer bursts, thus making the break in duration distribution more meaningful.

In this paper we address the following questions:
(a) What fraction of short bursts contain short time (ms /sub-ms) structures in the time histories? **(b)** Is there any evidence for a coherent emission of γ-rays in a burst? **(c)** Is there spectral evolution during short bursts? **(d)** Are any of

---

* Email (Internet): pnbhat@tifrvax.tifr.res.in

the spectral characteristics related to any of the temporal features of the burst?

OBSERVATIONS

BATSE large area detectors are described elsewhere [6]. Among the various data types available, we use the time tagged event (TTE) data which have the highest time resolution (2 μs) available in any GRB experiment so far. TTE data have 4 channel spectral information; the approximate photon energies at the channel boundaries are: 25, 50, 100 and 300 keV respectively. The maximum number of events that could be recorded is 32K and hence short bursts, in general, are completely covered by TTE data while only the initial part of the longer bursts are covered.

TERM DEFINITIONS

The spectral characteristics are defined by:

Hardness Ratio (HR) which is defined as the ratio of the number of signal photons above 100 keV to that below 100 keV.

Correlation Coefficient (CC) is the simple correlation coefficient between HR and the intensity during a burst.

Energy Evolution Parameter, (EEP): In a majority of the bursts the higher energy photons lead the lower energy counter-parts. As a result, the centroid of the burst light curve at higher energy leads in time with respect to that of the lower energy light curve. We compute the slope of the straight line fitted to the centroid positions in the first 3 channels as a function of logarithm of the mean channel energies, for the 2 brightest detectors. The mean channel energy is defined as the geometric mean of the channel boundaries. The mean of the two slopes is the EEP of the burst.

The temporal characteristics are defined by:

Burst Complexity Index (CI): This is a measure of the total number of peaks in the burst time history. For this purpose the burst and the background time history of same duration & time resolution are simultaneously analysed. The number of peaks are counted by counting the number of zero crossings of the first differential of the time history. Noise due to statistical fluctuations is reduced by applying a one dimensional unweighted low pass filter of kernel size 3 to the time histories successively until the background shows a peak count of 0 or 1. The resulting number of peaks in the burst interval is CI. CI is independent of the burst intensity as the differential is expressed in units of standard deviation. The process is speeded up by using finite threshold for a peak to be counted. CI can vary depending on the burst integration time. Hence CI is computed as a function of integration time and the peak value is used. Figure 1 shows a frequency distribution of CI.

Burst Rapidity Index (RI): CI is insensitive to sharp peaks with very short rise and decay times. Hence RI is introduced to quantify the sharpness of the peaks in the time history. This is given by the ratio of the variances of the first differential of the time histories of burst and background. An advantage of RI is that one needs to search only those bursts with higher than average RI for short time structures. A frequency distribution of RI (Fig. 2) shows that a majority of the bursts have RI around 0.9. It is found that CI and RI are uncorrelated and hence independent. RI is not a sensitive function of integration time.

Channel energy at Maximum RI (CM): It is an observed feature of most of the bursts that the time histories become sharper at higher energies. The RI computed for different energy channels increases with energy and in majority of the cases shows a maximum in channels 2-4. The mean channel energy where the RI is maximum is defined as CM.

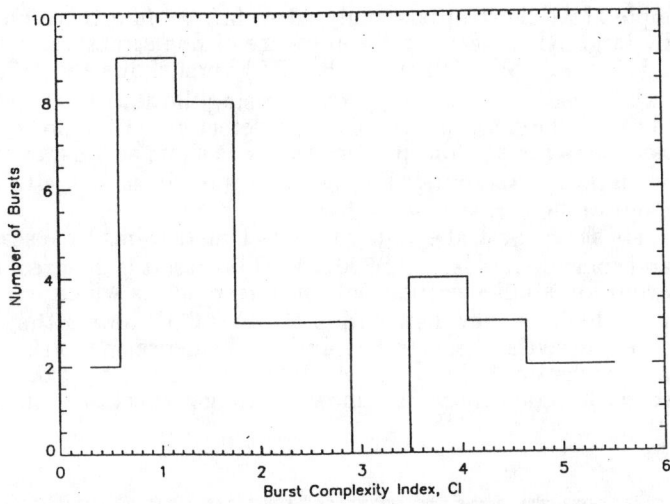

Figure 1. A frequency distribution of the Complexity Index, CI, showing that the bursts with more complex structure are less frequent.

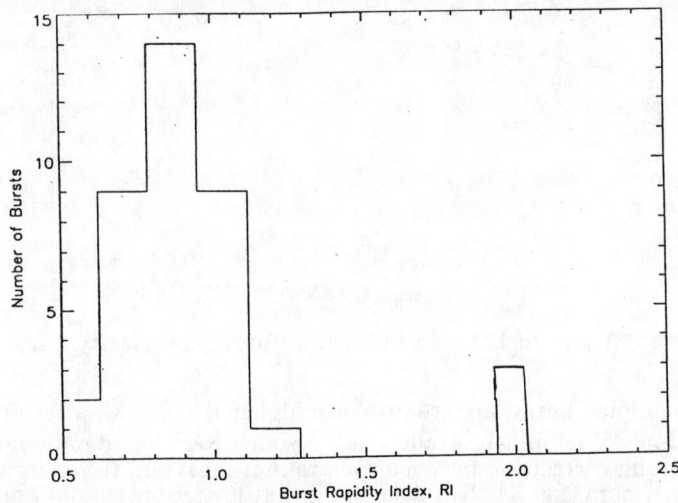

Figure 2. A frequency distribution of the Rapidity Index, RI, showing that a majority of the bursts have RI close to 1.0 while bursts which have a significantly large RI are the ones which have narrow spikes in their time history.

RESULTS

In a sample of 37 short bursts analysed so far, we find three bursts that show unusually large RI, suggesting the presence of fine structure. On further analysis we find that only one (1B910711, RI=2.0) burst shows a significant sub-ms structure, which was previously reported [7]. Using the statistical distribution of event arrival time differences in pairs of bright detectors we find no evidence for coherent emission of γ-ray photons on time scales as short as 5 μs, as suggested by Mitrofanov[8], in any of the bursts analysed, a result consistent with a similar search carried out on longer, stronger bursts[9].

Some bursts show spectral evolution even at ms/sub-ms time scales (e.g., 1B910814B) and some do not (e.g., 1B920229). This result is in agreement with the spectral evolution studies carried out on longer bursts which have similar time histories[10]. The frequency distribution of CC is flat showing that different bursts show different levels of spectral evolution. Evidence shows that spectral evolution is independent of burst duration. However there is a weak suggestion that the bursts with higher mean HR show a stronger spectral evolution than softer bursts.

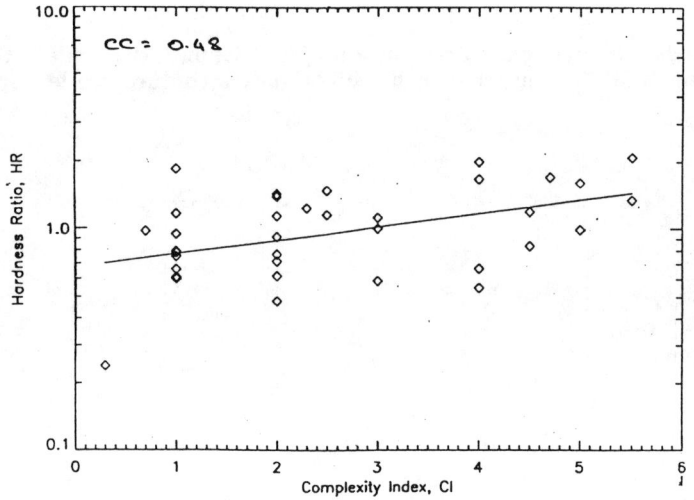

Figure 3. A plot of HR as a function of burst complexity CI.

More complex bursts are seen to have higher HR(fig. 3), a result contrary to those of Jourdain[11] and Lestrade et al.[12] which were based on longer bursts. There is a strong correlation between spectral hardness and the energy at which RI shows maximum (fig. 4). It is also found that longer bursts show larger EEP (absolute value).

CONCLUSIONS

There are certain parameters, perhaps, other than the hardness ratio and duration that distinguish short bursts. Correlation between complexity and hardness seems to be true for short bursts only. Also the correlation between hardness and CM seems to be well borne out in the case of short bursts.

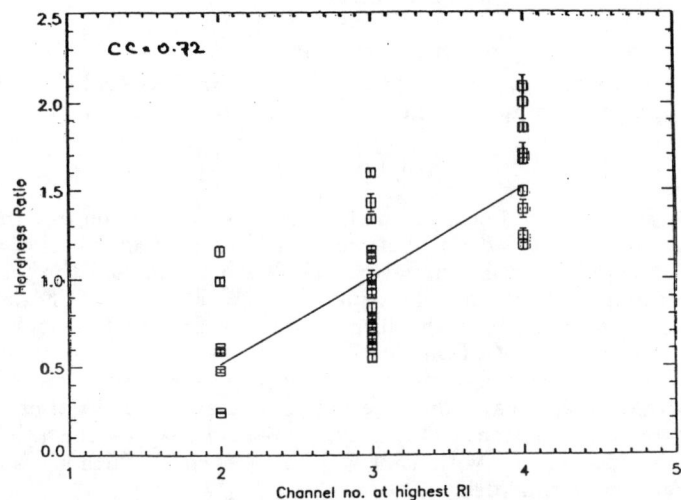

Figure 4. A plot of HR as a function of CM, again showing a strong correlation. The straight lines are least square fits to the points.

REFERENCES

1. Mazets E. P. et al., Positron Electron Pairs in Astrophysics, Ed.:M.L.Burns, A.K. Harding and R. Ramaty , 36 (1983).
2. Klebesadel, R. W., Laros, J. G. and Fenimore, E. E., Bull. Am. Astron. Soc. **16**, 1016 (1984).
3. Cline, T. and Desai U. D., Proc. 9th ESLAB symp. , 37 (1974).
4. Kouvelotou C. et al., Astrophys. J. Lett. **413**, L101 (1993).
5. Dezalay J-P. et al., Gamma Ray Bursts, Ed.: W.S. Paciesas and G.J. Fishman , 304 (1992).
6. Fishman G. J. et al., Proc. GRO Science Workshop, Ed.:W.N. Johnson , 2-39 (1989).
7. Bhat P. N., et al., Nature **359**, 217 (1992).
8. Mitrofanov I. G., Astrophys. Space Sci. **155**, 141 (1983).
9. Schaefer B.E. et al., Astrophys. J. **404**, 673 (1993).
10. Bhat P. N., et al., Astrophys. J. , May, 10 (1994).
11. Jourdain E., Ph. D. Thesis, C.E.S.R., Paul Sabatier Univ.,Tolouse, France. , (1990).
12. Lestrade J. P. et al., Compton Observatory Symposium, St. Louis Ed.: N. Gehrels, M. Friedlander and D. J. Macomb , 969 (1992).

FREQUENCY-DEPENDENT ENERGY PHASE LAGS IN GAMMA-RAY BURSTS

Eric Chipman
Laboratory for High Energy Astrophysics, NASA/GSFC, Greenbelt, MD 20771
Science Programs, Computer Sciences Corporation, Greenbelt, MD 20771

ABSTRACT

To understand the physical mechanism of pulse generation in gamma-ray bursts, we have investigated delays between hard gamma-ray emission and soft gamma-ray emission within individual bursts. We have applied the techniques of cross-correlation analysis and cross-spectral analysis to BATSE data from two bright bursts to determine the form of the hard-soft phase delays. We chose to use time-to-spill data from the GRO archives for the bursts of 910503 and 910601. We find that well-defined delays can be seen in the cross-spectral analyses, but that these delay times are not always equal to the times derived from cross-correlation analyses. One possible reason may be the existence of two or more physical regimes within a given burst, each of which gives rise to a separate characteristic time delay.

INTRODUCTION

The comparative temporal behavior of soft gamma-ray emission and hard gamma-ray emission in gamma-ray bursts can be used as a diagnostic of the physical conditions within the burst environment. If the individual pulses within bursts are due to populations of relativistic electrons which are accelerated on time scales short compared to the the pulse width and which subsequently lose energy as the gamma rays are emitted, then the hard-soft phase delays contain information on the densities, geometries or magnetic fields in the emitting regions. Alternatively, the particle acceleration may be continuous throughout each pulse, in which case the delays are due to the acceleration mechanism itself. A well known technique for measurement of the time delays between soft and hard gamma-ray emission is cross-correlation analysis, wherein the cross-correlation between the hard and soft light curves for a burst is computed as a function of time offset. In most bursts the maximum correlation shows that the soft gamma-rays lag the hard gamma-rays by a fraction of a second. This effect can be interpreted in terms of the fact that hard gamma-ray light curves typically show narrower pulses, or pulses with shorter decay times, than do the light curves for soft gamma-ray energies. Cross-correlation methods only determine a "mean" delay, averaged in intensity over temporal frequency, however. It is possible that low frequency components of the burst variability show a different hard-soft delay than do high frequency components. To investigate this possibility and to compare the results of frequency independent (cross correlation) methods with frequency dependent analyses, we have done cross-spectral analyses of the hard-soft phase lags in two selected gamma-ray bursts. This cross-spectral method has been used for gamma-ray burst analysis by Kouveliotou et al.[1], and for solar oscillation analysis by Lites and Chipman[2]. In this work we have investigated a higher range of temporal frequencies than did Kouveliotou et al., and we display our data in a mode adapted from that used by Lites and Chipman, which permits overlaying results from several different FFTs in one plot. In performing cross-spectral analyses for high temporal frequencies, it is necessary to use data for relatively bright bursts in order to maximize the signal to noise in the FFTs involved. We therefore chose to use data from the two brightest bursts seen by BATSE in its first few months of operation, bursts GRB 910503 and GRB 910601. We also wished to combine high time resolution (better than 64 msec if possible) with temporal coverage of at least 10 seconds for these bright bursts. The time-to-spill data type offers several advantages in this regard. This data divides the BATSE counts for each burst into four channels, and records the times to count each successive 64-photon set within each channel. Light curves are created by adding the delta-time intervals, with intensities given by the inverse of the spill time intervals. The resulting light curves have time resolution approaching 1 msec during times of bright emission, show nearly constant signal-to-noise of 8 for each data point, and have good time coverage relative to the time-tagged data type, for example, given the size of the BATSE data memory. A disadvantage is that the light curves have points unequally spaced in time, and the curves for the different energy channels do not have data points at the same time points. We chose to resample the data onto an evenly spaced time scale of 0.1 msec, using a simple nearest-neighbor algorithm. The four channel energy ranges are (1) 25-50 keV, (2) 50-100 keV, (3) 100-300 keV, and (4) >300 keV. In all cases, channels 2 and 3 show the highest count rates, and channel 4 appears to show a more spiky, narrow-pulsed structure than the other channels. For GRB 910601, a cross-correlation phase delay determination between channels 4 and 3 shows the soft gamma-rays (channel 3) lag the hard gamma-rays (channel 4) by 65 msec. In this case the cross-correlation curve has a central peak with FWHM of approximately 400 msec, permitting the delay to be determined to an accuracy of approximately 10 msec. Similarly, the cross-correlation delay for channels 4 and 2 was also measured, yielding a value of 165 msec (see Table 1).

CROSS-SPECTRAL ANALYSES

Figure 1 shows the cross-spectral phase delays between channels 4 and 2 for GRB 910601. We have divided the time period covered by the light curves into 25 segments of 0.4096 seconds each, and have computed the FFTs and phase lags between the channels for each FFT. A sine-squared apodization parameter was applied to each segment to reduce the discontinuities at the end points. At each frequency we plot one mark for the phase delay for each FFT, with a symbol size proportional to the amplitude of that FFT, normalized at each frequency among the 25 values computed. Larger symbols are indicative of greater power at that frequency and phase delay for that individual FFT. In this plot a constant time delay between the two light curves would appear as a set of large symbols starting at zero phase delay at zero frequency and lying along a line of constant slope, wrapping around from 180° to -180°. Note that there is apparently such a ridge of power extending from the origin to a frequency of approximately 50 Hz, which has been highlighted in the figure. The phase delay time is the period for a 360° delay, or 27 msec in this case. There are no strong candidates for additional delays, and it should be noted that the points for very high frequencies (>100 Hz) are dominated by noise, with no clear trends detectable.

Figure 2 shows a similar plot for the delays between channels 4 and 3 for this same burst. Here there seems to be a broad ridge of power as shown by the two overlaid lines, with delay times ranging from 16 to 21 msec. A comparison of the cross-correlation method delays and the cross-spectral method delays for

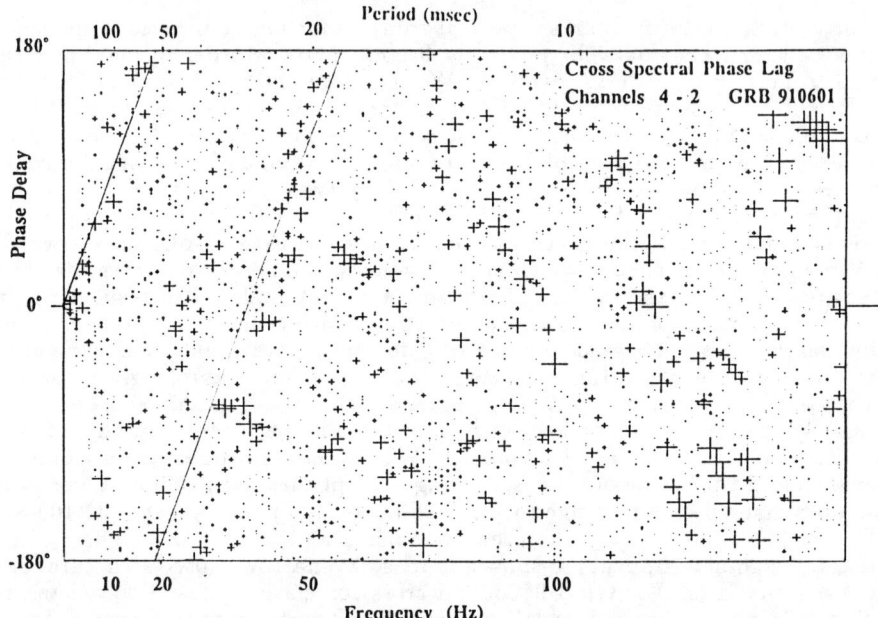

Figure 1. Cross-spectral phase delays as a function of frequency between channels 4 and 2 for GRB 910601

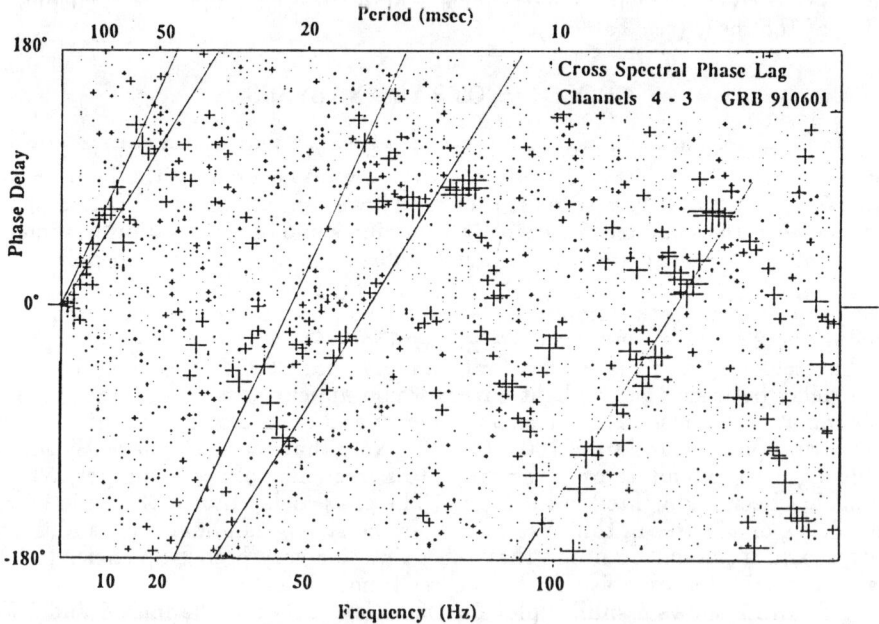

Figure 2. Cross-spectral phase delays between channels 4 and 3 for GRB 910601

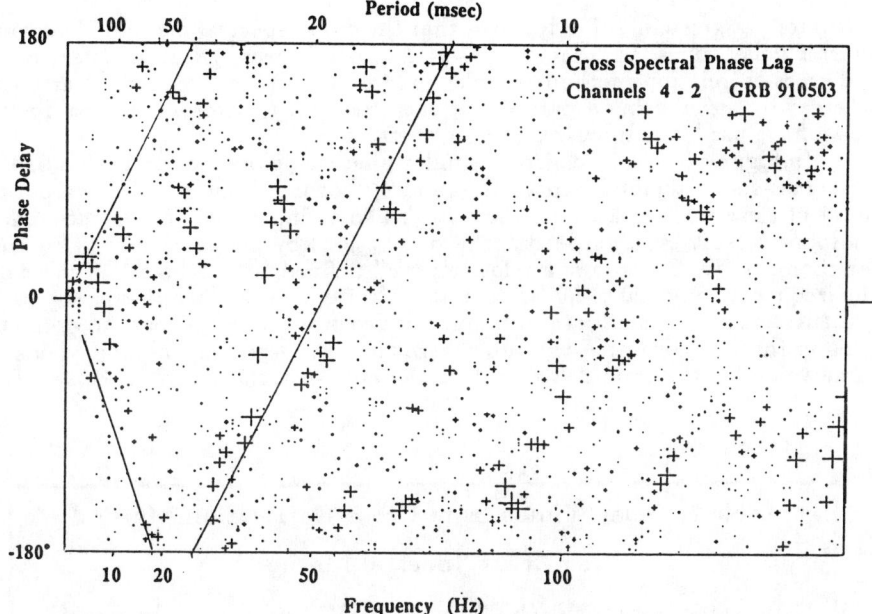

Figure 3. Cross-spectral phase delays between channels 4 and 2 for GRB 910503

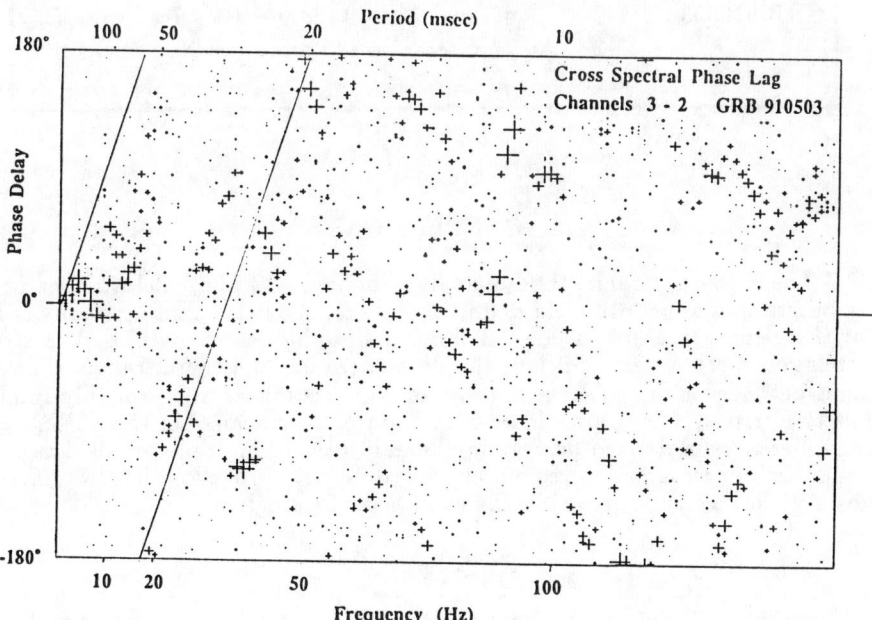

Figure 4. Cross-spectral phase delays between channels 3 and 2 for GRB 920503

this burst, as given in Table 1, shows that the cross- spectral method yields much shorter delay times, by a factor of 3-5. A possible reason for this effect is that the cross-correlation method samples predominantly low-frequency variability, whereas by normalizing separately at each frequency the cross-spectral method as we have used it emphasizes the higher frequencies.

In Figures 3 and 4 we display similar phase lag plots for burst GRB 910503. The cross-spectral plot for channels 4 and 2, shown in Figure 3, shows a positive delay of close to 19 msec, and there is also some indication of a negative delay component (soft emission leading hard emission) by about 28 msec. Figure 4, for channels 3 and 2, shows a relatively well defined time delay of 30 msec over the frequency range 20 to 50 Hz, but at lower frequencies this power ridge is not present, and much smaller (and frequency dependent) delays seem to dominate. This is the only example we find where the cross-spectral method yields the same value for the time delay as does the cross-correlation method (see Table 1).

Table I. Delay Times from Cross-Correlation Method vs. Cross-Spectral Method

Burst	Channels	CCM Delay	CSM Delay(s)	Comments
GRB 910601	4-3	65 msec	21, 16 msec	Range of values
	4-2	142 msec	27 msec	
GRB 910503	4-2	49 msec	19 -28 msec	Two values?
	3-2	27 msec	30 msec	Non-linear fit?

CONCLUSIONS

The cross-spectral method of analysis of hard-soft phase delays in gamma-ray bursts does appear to yield definite values for the delay times in many cases, but these times are not in general the same as the delays given by the cross-correlation method. In addition, the differences in the general range of delays found between our analyses and those of Kouveliotou et al.[1] probably implies that the derived delays also depend on the frequency range of the FFTs used. The differences in derived delays may be indicative of multiple physical regimes within each burst emission region, causing different pulse evolution for different pulses within a burst, or within the evolution of a single pulse.

REFERENCES

1. Kouveliotou, C., et al., in, Gamma-Ray Bursts, eds. W.S. Paciesas and G.J. Fishman (A.I.P., N. Y., 1992), p. 299.
2. Lites, B.W. and Chipman, E.G., Ap. J. **231**, 570 (1979).

DURATION VERSUS BRIGHTNESS OF GAMMA-RAY BURSTS: COMPARISONS BETWEEN SIGNE AND BATSE

V. E. Kargatis, H. Li, E. P. Liang, and I. A. Smith
Dept. of Space Physics and Astronomy, Rice Univ., Houston, TX 77251

K. Hurley
Space Sciences Laboratory, U. C. Berkeley, Berkeley, CA 94720

C. Barat and M. Niel
C. E. S. R., BP 4346, 31029 Toulouse Cedex, France

ABSTRACT

We analyze duration and brightness distributions of both the SIGNE Venera 13 & 14 and BATSE gamma-ray burst databases. We search for correlations between duration, peak brightness, and the variability measure V, proposed in Lamb, Graziani, & Smith[1]. The duration histogram for SIGNE is consistent with the BATSE histogram[2]. Estimating the instantaneous brightness by C_{64}, we find that SIGNE confirms the BATSE result that the long and short bursts have similar maximum instantaneous brightnesses. Scatterplots between duration, brightness, and V are consistent for both databases; we show that SIGNE confirms the BATSE observation that there is a lack of bursts that are both bright over 1024 ms and contain a short, bright spike.

INTRODUCTION

Recent BATSE observations have confirmed the long held suspicion[3] that gamma-ray burst (GRB) durations form a bimodal distribution (ref. 2; hereafter K93), with a short group peaking at ~ 0.3 s and a long group peaking at $10 - 20$ s. Except for a marginal difference in their spectral hardness distributions[2,4], there are no other established distinctions between the two groups separated by duration.

We are interested in checking that the observed bimodality appears in scatterplots of duration and other variables, because histogram distributions can be unreliable and strongly depend on the binning. K93 calculated both durations T_{50} and T_{90} (defined as the time over which the integrated counts increase from 25 to 75% and 5 to 95%, respectively, of the total counts above background) for each burst. In this study, we choose T_{50} as the preferable measure of duration, since it is less sensitive to the "wings" of the temporal profile and is less affected by substantial stretches of background within multi-peaked time histories.

The French-Soviet SIGNE experiments aboard the Venera 13 & 14 satellites collected gamma-ray burst data for 18 months from 1981-1983. SIGNE has two different time history memories: a 16 ms resolution, 16 s duration memory with 1 s of pre-memory (before the trigger), and a 0.5 s resolution, 64 s spectral memory. Unlike BATSE, SIGNE triggered on an 8σ fluctuation above background only in the 1024 ms window. One expects that SIGNE failed to detect many short, faint bursts on which BATSE could trigger. (See ref. 5 for more details on similar experiments aboard Venera 11 & 12.) We calculate SIGNE durations using the procedure described in K93.

DURATION HISTOGRAM

In Figure 1, we show the SIGNE and BATSE[2] T_{50} distributions. The histogram of 171 SIGNE bursts shows a single peak centered at ~ 6 s, with a long tail extending to the shortest durations. The SIGNE distribution peaks at an insignificantly lower value than the BATSE T_{50} histogram, and is entirely consistent with the combined duration histogram from several experiments[3]. The SIGNE distribution does not have a peak at small T_{50}, but this is expected because SIGNE had a much smaller detector area and triggered only on a 1024 ms integration timescale, and certainly failed to detect many of the short, faint bursts that occupy the short duration region of the BATSE histogram. The fraction of bursts seen by SIGNE with $T_{50} \lesssim 0.6$ s (the dip in the BATSE T_{50} histogram) is $\sim 15\%$, significantly smaller than the BATSE value of $\sim 25\%$.

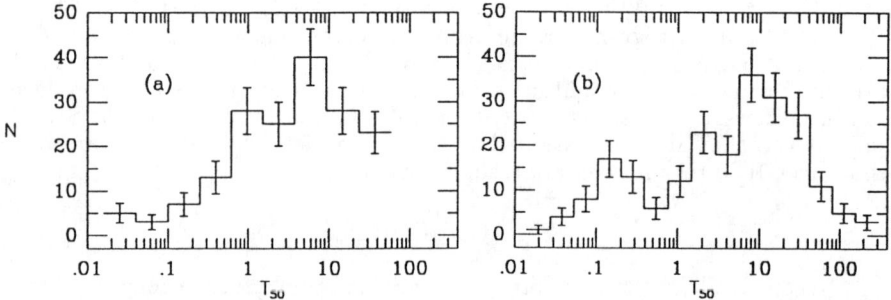

Fig. 1. T_{50} duration histogram for 171 SIGNE bursts (a) and 215 BATSE bursts (b). The SIGNE peak at large T_{50} is at a slightly smaller value of T_{50} than for BATSE, but is consistent within the bin boundaries. Also note that the SIGNE 64 s memory produces an artificial cutoff at large T_{50}, so that T_{50} is underestimated for any burst longer than 64 s.

DURATION VERSUS BRIGHTNESS

Figure 2 plots T_{50} versus C_{64} and C_{1024} for both BATSE and SIGNE, where C_{64} and C_{1024} are the peak counts in 64 ms and 1024 ms respectively. In this study, we used the 9 September 1993 revision of the BATSE catalog. Details on the normalization of SIGNE to the BATSE count rate scale can be found in ref. 6. We emphasize that the axis scaling is a coarse approximation primarily useful for visually comparing the scatterplots.

We note several features apparent in Figures 2a and 2b:

(1) Threshold edge. The left edge of the scatterplots above $T_{50} = 1$ s is roughly a vertical line for both SIGNE and BATSE, corresponding to the trigger threshold with a slightly varying background.

(2) Fluence edge. Because SIGNE triggered on a 1 s integration, bursts shorter than 1 s trigger on their fluence. As the duration decreases, C_{64} must increase to provide the necessary counts to trigger, producing an instrumental cutoff at the lower left region. The fluence edge in the BATSE plot is detailed in ref. 7.

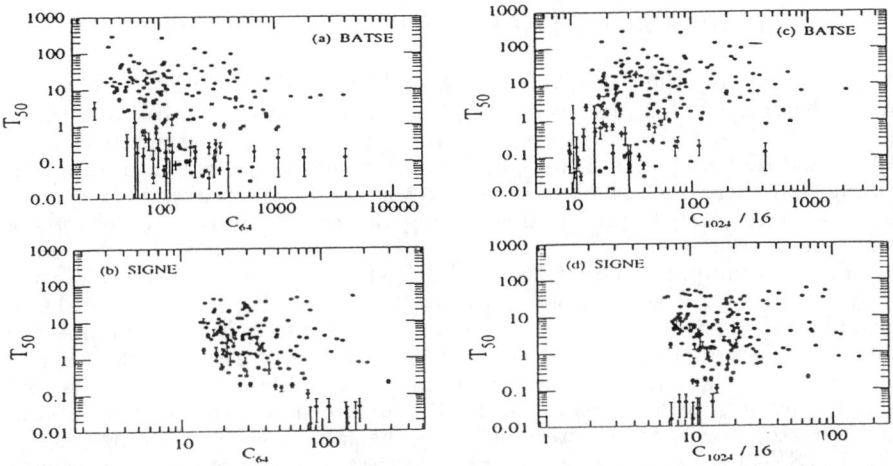

Fig. 2. Scatterplots of duration versus brightness for the BATSE (a, c) and SIGNE (b, d) databases. Figures 2a (177 BATSE bursts) and 2b (148 SIGNE bursts) show T_{50} versus estimated instantaneous brightness C_{64}. 23 of the 171 SIGNE bursts peaked after 16 s and did not have 64 ms resolution, and so could not be included in Figure 2b. 1σ error bars are shown for all bursts; most are to small to be seen on this scale. Figures 2c (181 BATSE bursts) and 2d (171 SIGNE bursts) show T_{50} versus $C_{1024}/16$. For both instruments, C_{1024} is presented scaled by a factor of 1/16 to place this plot on the same count scale as C_{64}.

(3) Duration edges. (A) The SIGNE experiment had a maximum memory of 64 s, so there is an artificial cutoff at the top of Figures 2b and 2d. The lack of long bursts in the BATSE plots is probably real since there should not be any selection effects barring their detection. (B) There are a few bursts whose durations are close to the maximum resolution for SIGNE (16 ms), so we can say little about the reality of the short duration edge. In BATSE, many more short bursts may exist than are detected, so the true percentage of short bursts should be higher than the 25% quoted in K93 (the fraction of short bursts may be ~ 40%[8]). Since so many short, faint bursts are missed, the short duration cutoff in Figure 1b may be artificial.

(4) There is a slight suggestion of a deficit of bursts in the BATSE burst distribution brighter than $C_{64} = 100$ at $T_{50} \approx 0.5$ s, where a deficit of bursts appears in the T_{50} histogram. However, the gap covers too small a dynamic range to be deemed significant. If the relative placement of the two plots is correct, SIGNE bursts would also fill in most of the apparent BATSE gap.

Figures 2c and 2d show the dilution effect that using C_{1024} as a brightness indicator has on short bursts in both databases. Bursts with duration < 1 s all move to the left relative to the longer bursts, when compared to the C_{64} plot. In such a plot, one mixes count rates for longer bursts and count fluences for shorter bursts.

COMPARISON WITH LAMB, GRAZIANI, & SMITH

To investigate the bimodality in the BATSE database, Lamb, Graziani, & Smith[1] (hereafter LGS) proposed using C_{1024} as a measure of burst brightness and the ratio $V \equiv C_{64}/C_{1024}$ as a measure of short timescale variability. They claimed a statistically significant correlation between these two parameters indicating that there was a lack of bursts that are both bright over 1024 ms and exhibit a bright, short spike (either during the bright 1024 ms or elsewhere during the burst).

Figure 3 compares the BATSE and SIGNE databases using the LGS variability measure V. Figure 3c shows the scatterplot analogous to that in LGS, which plots V versus $C_{1024}/16$ for the BATSE database; this shows an absence of bursts in the upper right quadrant of the diagram, indicating a lack of bursts that are both bright (in 1024 ms) and exhibit a bright, short ($\ll$ 1024 ms) spike. Figure 3d shows the same plot for the SIGNE data. Although the statistics are much poorer, the scatterplot shows the same basic morphology as the LGS BATSE plot. However, we emphasize that one cannot distinguish with any certainty how to separate this parameter space into two groups.

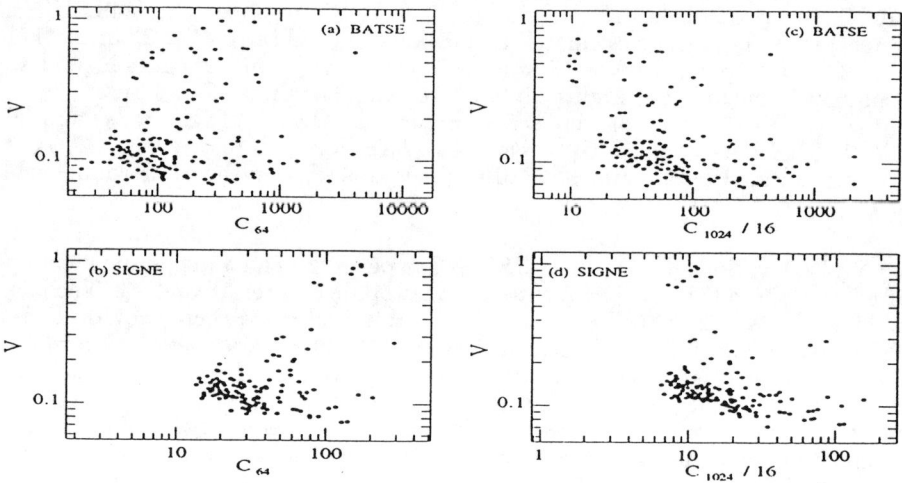

Fig. 3. Scatterplots of the LGS V parameter versus brightness for 166 BATSE bursts (a, c) and 148 SIGNE bursts (b, d). Panels a and b show V versus C_{64}. Panels c and d show V versus $C_{1024}/16$. Panel c is analagous to Figure 2 (top panel) of LGS.

Figures 3a and 3b show V plotted against C_{64}, which approximates the instantaneous brightness. As one would expect, high-V bursts are moved to the right relative to the C_{1024} plot. These plots show that in both databases the maximum brightnesses (in 64 ms) for the high- and low-V bursts are nearly equal. Both the SIGNE and BATSE plots show similar morphologies, but if the relative scaling of the count axes of the two databases is approximately correct, any apparent gaps in the BATSE plot would be filled in by the SIGNE bursts.

It can be shown[9] although the V parameter indicates some degree of true variability on the 1024 ms timescale, on shorter timescales it is largely irrelevant and primarily reflects the duration distribution. Figure 4 shows V vs. T_{50} for SIGNE and BATSE. Both plots show that almost all the bursts lie along the distribution expected from the systematic mathematical relationship between V and T_{50}; i. e. the primary factor controlling the V distribution is the duration distribution, and not a true distribution in variability. This equivalence would imply that the apparent gaps in Figures 3a and 3b are merely exaggerations of the insignificant gaps in Figures 2a and 2b, thereby removing their potential use as a means to cleanly separate the two populations of bursts.

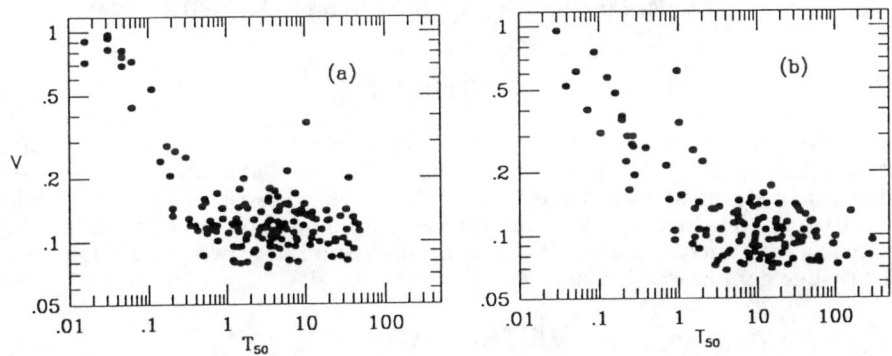

Fig. 4. V vs. T_{50} for SIGNE (a) and BATSE (b). Both databases show that V primarily follows the distribution expected from the mathematical relationship between V and T_{50} (see ref. 9 for more details), indicating that V is an insufficient measure of true variability.

We thank D. Hartmann for helpful discussions. The SIGNE experiment was developed by the scientists and specialists from the Institute for Space Research of the Russian Academy of Science and CESR in Toulouse, France. This work was supported by NASA grants NGT 50924 and NAG 5-1515. EPL acknowledges support by DOE contract W7405-ENG-48 at LLNL.

REFERENCES

1. Lamb, D. Q., Graziani, C., & Smith, I. A., ApJ **413**, L11 (1993 (LGS)).
2. Kouveliotou, C., Meegan, C. A., Fishman, G. J., Bhat, N. P., Paciesas, W. S., Pendleton, G. N., Briggs, M. S., & Koshut, T. M., ApJ **413**, L101 (1993, (K93)).
3. Hurley, K., Gamma-Ray Bursts (AIP: NY, 1992), p. 3.
4. Dezalay, J-P., Barat, C., Talon, R., Sunyaev, R., Terekhov, O., & Kuznetsov, Gamma-Ray Bursts (AIP: NY, 1992), p. 304.
5. Barat, C., et al., Space Sci. Inst. **5**, 229 (1981).
6. Kargatis, V. E., Li, H., Liang, E. P., Smith, I. A., Hurley, K., Barat, C., & Niel, M., ApJ Lett (1994, in press).
7. Li, H., Liang, E. P., Smith, I. A., & Kargatis, V., 1993 Huntsville Gamma-Ray Burst Workshop, AIP, NY (1994).
8. Mao, S., Narayan, R., & Piran, T., ApJ (1994, submitted).
9. Meegan, C. & Kouveliotou, K., 1993 Huntsville Gamma-Ray Burst Workshop (AIP: NY, 1994).

GAMMA-RAY BURST VARIABILITY:
A SEARCH FOR CORRELATIONS WITH OTHER PARAMETERS

John Patrick Lestrade
Dept. of Physics and Astronomy, Miss. State Univ., MS 39762

M. Briggs and W. B. Paciesas
Univ. of Alabama at Huntsville, Huntsville, Al

G. J. Fishman, C. A. Meegan, R. B. Wilson
ES-64, Marshall Space Flight Center, Al, 35812

ABSTRACT

We present an analysis of BATSE GRB time profiles using an algorithm that is based on the statistics of runs up and down. This algorithm has been shown to be a good measure of burst "structure" (Lestrade, 1994) and is less sensitive to burst distance than algorithms that depend directly on the intensity of peaks in a burst. The results show no indication of a two-population source for cosmic gamma-ray bursts.

INTRODUCTION

One important question that is hotly debated in gamma-ray burst circles is the number of different phenomenological sources. The pure isotropy of the angular distribution (Meegan[1]) has led many to abandon galactic models. On the other hand, the observation of possible cyclotron absorption lines as well as a red-shifted 511 keV emission line in a small number of GRB spectra beg for a galactic neutron source for at least a fraction of gamma-ray bursts (Katz[2]).

Recently Lamb et al.,[3] studying GRB time profiles, used the ratio of the maximum count rates in the 64-msec bins to that in the 1024-msec bins as a type of burst "variability". They used this parameter to conclude that there are two different source distributions. We wondered if a different variability parameter, one presented by Lestrade[4] which measures structure over the entire burst, would lead us to the same conclusion of a dual-population source model.

In the latter paper, it is shown that a reasonable parameter to measure the variability of a profile is the number of times the profile shows long runs of monotonically increasing (or decreasing) bin-to-bin differences. Highly-structured profiles more often show long runs of monotonically increasing bin-to-bin differences, while the runs in smooth profiles are broken up by noise. This idea, although new to the area of GRB's, was heavily studied in the 1940's by researchers in agronomy and meteorology (cf. Besson[5] and Wallis and Moore[6] – hereafter WM[6]). We applied this "runs" statistic to 173 BATSE GRB profiles (using the 64-msec. DISCSC data type) looking for any unexpected correlations between structure and burst angular position, hardness, duration, and V/V_{max}. Such correlations have thus far eluded researchers but, if found, could be an important step in GRB source identification. We also looked for groupings which could imply distinct morphological classes.

RUNS UP AND DOWN

Consider a gamma-ray burst time profile containing n bins of data. Let the number of counts in the ith bin be b_i. Consider the sequence of signs whose ith element is the sign of $b_{i+1} - b_i$, $(i = 1, 2, \ldots, n-1)$. A subsequence of p consecutive signs (either $+$ or $-$) is called a run up ($+$) or down ($-$). The notation that I use differs slightly from that used by Knuth[7] in that, in this paper, a run of length p involves $p+1$ original bins of data. For example, consider the series of gamma ray counts registered in the following 8 64-msec. bins: 501, 502, 504, 510, 503, 502, 503, 501. This series contains four runs (up, down, up, down) of lengths 3, 2, 1, and 1, respectively.

Let r_p be the number of runs of length p, and r'_p be the number of runs of length p or greater, and r be the total number of runs in a series. Obviously r'_1 equals r. Bienaymé[8] was the first to show that the average number of runs (i.e., $\langle r \rangle$) in a random series of length n is $(2n-1)/3$. The expected values for the number of runs of other lengths are given by WM[6] as:

$$\langle r_p \rangle = 2n \frac{p^2 + 3p + 1}{(p+3)!} - 2\frac{p^3 + 3p^2 - p - 4}{(p+3)!}, \quad (1)$$

$$\langle r'_p \rangle = 2n \frac{p+1}{(p+2)!} - 2\frac{p^2 + p - 1}{(p+2)!}. \quad (2)$$

A series of data may be tested for randomness by counting the number and lengths of runs. WM[6] construct a statistic through the usual procedure of squaring the difference between observed and expected, dividing by the variance and summing these ratios. However, as they point out, since these runs are *not* independent, the resultant sums are not distributed according to the usual χ^2 statistics. In order to calculate this statistic, we need the variances, $\sigma^2(r_p)$ and $\sigma^2(r'_q)$. The formulae for these and other co-variances have been provided by Levene and Wolfowitz.[4,9]

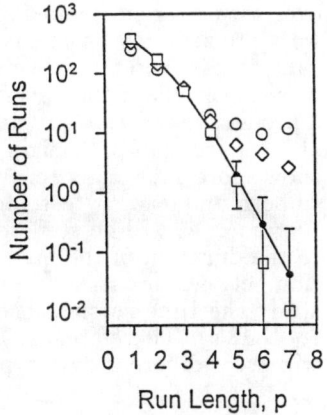

Fig. 1. The Number of runs per minute in time profiles.

It might be thought that the number of runs in a random series is dependent on the underlying distribution, e.g., uniform or Poisson. Wallis and Moore show this not to be the case. While their arguments were based on non-repeating counts, repetition occurs infrequently enough in our profiles so as to not change the conclusion. Certainly the size of the bin-to-bin differences is affected by the statistical distribution, but the frequency and lengths of runs are not. Our own calculations with Monte Carlo simulations bear this out.

THE PROFILE VARIABILITY PARAMETER, S_7

Figure 1 presents the number of runs per minute (64-msec. data) observed for (a) two typical BATSE burst profiles (open circle and diamond), (b) post-burst background data (open square), and (c) the expected numbers from (1) and (2) (line/dot/error bars). The error bars are the standard deviations from

Lestrade.[4] As shown, the numbers seen in the background agree with the statistical predictions. In a burst profile however, we see increased numbers of runs for longer lengths ($p > 3$) and usually a smaller number of short runs. It is this difference that we exploit to measure a profile's structure. Following the lead given by WM,[6] we have used the expected values and variances to construct a statistic that measures the variability of GRB profiles. The statistic is given by

$$S_7 = \sum_{p=4}^{6} \frac{(r_p - \langle r_p \rangle)^2}{\sigma^2(r_p)} + \frac{(r'_7 - \langle r'_7 \rangle)^2}{\sigma^2(r'_7)}. \quad (3)$$

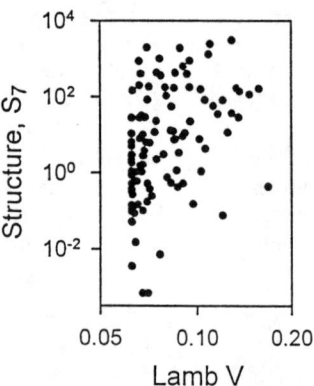

Fig. 2. S_7 vs. Lamb Variability, V.

Note that the r'_7 term accounts for all runs of length 7 or greater. Since we are not strictly trying to measure the deviation from statistical chance, but rather trying to measure the structure of a profile by its frequency of long runs, we sum only over terms larger than $p = 3$ and, in addition, we set to zero any terms in (3) where the observed value is less than the expected value. Not doing so would overinflate the variability parameter when a profile showed *less* structure than random noise at some value of p. A highly-structured burst will show more long runs (e.g., r_6, r_7, etc.) than a smooth one leading to a higher value of S_7. Furthermore, as will be seen below, S_7 is insensitive to the duration of the burst. Figure 2 shows that there is little, if any, correlation between our structure parameter and the Lamb variability, V. Note, in this study the time resolution of the data limited us to long-duration bursts (i.e., $t_{90} > 13$ s). Therefore, the Lamb variability is limited to $0 - 0.2$. A future study will use a different data type to include shorter bursts.

THE PEAK HEIGHT PARAMETER, H_7

In addition to the number of runs in a profile, another measure of structure is the sum of heights of a profile's spikes. Our code for the analysis of profiles keeps a record of the positions of runs as well as their heights (in counts). We have the option of rejecting runs that don't meet a height filter threshold; although in this study the threshold was set at zero.

One of the appealing attributes of S_7 is that, statistically, it is not directly dependent on the heights of spikes, i.e., the distance to the burster. As a comparison, we have constructed a second parameter that is directly dependent on peak heights. Following the lead given by (3), we formulate a height parameter given by

$$H_7 = \sum_{p=1}^{6}(h_p - \langle h_p \rangle) + (h'_7 - \langle h'_7 \rangle) \quad (4)$$

where as above, the sum includes values only when the observed height exceeds the expected height. Note that h_p in (4) is the sum of heights for *all* runs of length p in a profile. In short, the value of H_7 is the sum of all run heights minus that which you would find in a Poissonian profile of the same length.

A single-spike burst (Fast Rise, Exponential-Decay or FRED) would ideally be composed of two runs (one up and one down). The value of H_7 would then be twice the height of the FRED. Another burst with two equal FRED's would have twice the value of a single FRED. Because of this definition of H_7, longer bursts will have higher values as will be seen in the correlation below. Since this height parameter is per force highly dependent on the burst distance, it is not as appealing as the structure parameter, S_7. Still, it gives us another aspect in the structure parameter space to search for correlations.

CORRELATIONS

While H_7 does depend on burst duration, S_7 does not. This is seen in Figure 3. (As a measure of duration we take the values of T_{90}.[10]) On the other hand, both S_7 and H_7 depend, to different degrees, on the burst distance or $V/V_{\max}$. The structure, being dependent on the number of runs and not their heights, shows less dependence on distance than does H_7 (Figure 4).

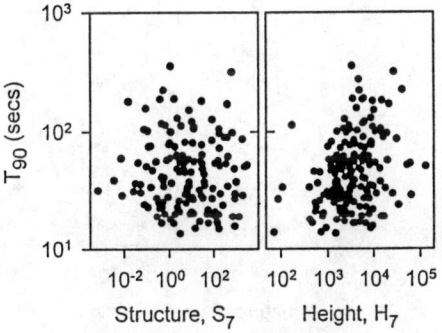

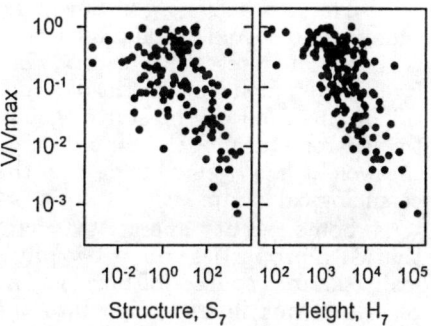

Fig. 3. Burst Duration versus Structure and Spike Heights

Fig. 4. Burst V/Vmax versus Structure and Spike Heights

In order to remove the dependence on distance in the correlations, we normalized the burst profiles by reducing all of them to the same value of $C_{\max} = 138$ counts. The algorithm to correctly handle the reduction and addition of Poissonian noise was obtained from Norris and Nemiroff.[11]

Fig. 5 presents the graph of duration vs. S_7 and H_7 for the normalized profiles. With the distance dependence removed, we see much more clearly the dependence of H_7 on duration. There is no overall dependence of S_7 on duration.

Figures 6, 7, and 8 present correlations with galactic longitude, latitude, and burst hardness, respectively.

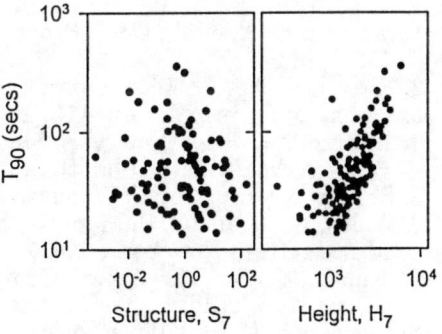

Fig. 5. Burst Duration versus S_7 and H_7 for normalized profiles.

None shows evidence of grouping nor correlation. The hardness ratio is approximately equal to the flux above 100 keV divided by the flux below 100 keV (down to the threshold of ≈ 25 keV).

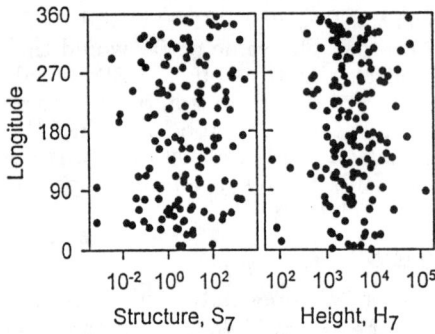

Fig. 6. Longitude vs. S_7 and H_7

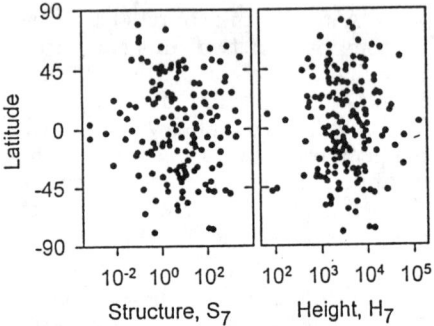

Fig. 7. Latitude vs. S_7 and H_7

CONCLUSION:

The debate raging between the galactic and cosmological modelers of GRB has an interesting history. Before GRO launch, the weight of scientific opinion fell on the side of galactic neutron star models. Since launch the weight has been shifting into the cosmological camp.

Some recent papers[2,3,12] claim that GRB properties hint at two physically distinct sources of bursts. The results of our investigation into the behavior of GRB variability, however, show no evidence of a two-population source.

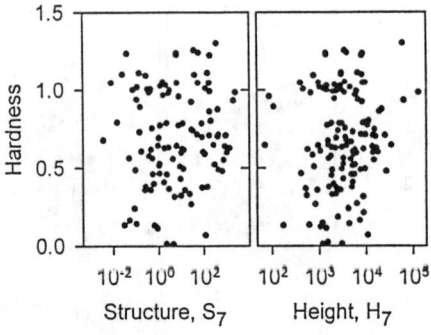

Fig. 8. Hardness versus S_7, H_7.

REFERENCES

1. Meegan, C. A. et al., Conf. Procs. 265: Gamma-Ray Bursts (New York: AIP, 1991), p. 61.
2. Katz J. I., Ap. J., **422**, in press, (1994).
3. Lamb, D. Q., Ap. J. Let. **413**, L11-L14 (1993).
4. Lestrade, J. P., A New Variability Parameter for Gamma-Ray Burst Time Profiles, Ap. J. Let., submitted, (1994).
5. Besson, L. (transl. by E. Woodward), Monthly Weather Rev. **48**, 89-94 (1920).
6. Wallis, W. A. and G. H. Moore, A Significance test for time series (Nat. Bureau of Econ. Res., New York, 1941).
7. Knuth, D. E., The Art of Computer Programming, Seminumerical Algorithms, 2nd (Addison Wesley, Reading, Mass., 1981).
8. Bienaymé, J., Bull. de la Société Mathématique de France **2**, 153-154 (1874).
9. Levene, H. and J. Wolfowitz, Annals of Math. Stat. **15**, 58-89 (1944).
10. Fishman, G. J., et al., The First BATSE Gamma-Ray Burst Catalog, Astrophysical Jour. Supp. Ser., accepted, (1993).
11. Norris, J. and R. Nemiroff, Private Communication, (1993).
12. Atteia, J-L. and J-P. Dezalay, Astron. and Astrop. **274**, L1 (1993).

EFFECTS OF PULSE SHAPE ON THE DETAILS OF THE "FLUENCE EDGE" FOR SHORT BURSTS

H. Li, E. Liang, I. A. Smith, V. Kargatis

Dept. Space Physics & Astronomy, Rice University, Houston, TX 77251-1892

ABSTRACT

We analyze the shape of the fluence edge (Kargatis et al.[1]) taking into account the various time-history morphologies of gamma-ray bursts; we study a range of time profiles from a simple rectangular pulse to the "worst" case in which the burst barely triggers on 64 ms without triggering on 256 ms or 1024 ms. We also discuss the effects of Poisson noise, and the temporal variation of C_{min}. The number of short bursts that exist below the fluence edge (but are not detected by BATSE) is estimated to be as high as 10 − 20% of the total number of observed bursts; thus the slope at the faint end of the $\log N - \log P_{64}$ curve could be modified significantly.

INTRODUCTION

BATSE triggers on the number of integrated counts over 3 time scales, 1024, 256 and 64 ms. For very short bursts to trigger, their brightness must be larger than a certain level to compensate for their short durations. This creates a cutoff edge for short faint bursts with a negative slope in the duration versus peak counts plot (Kargatis et al.[1]), which we call the fluence edge. Since we do not know the intrinsic duration distribution of gamma-ray bursts, the detection of short bursts is subject to a strong selection bias because of both the time resolution and triggering criterion. We are interested in finding the exact shape of this fluence edge, because it affects the estimate of the number of "missing" short bursts and the $\log N - \log P_{64}$ distribution of the short bursts[1].

EFFECT OF PULSE PROFILES ON FLUENCE EDGE

We have considered several different pulse profiles, since the bursts have a variety of time histories; as we will show, the shape of the fluence edge depends on the pulse profile. We have calculated the fluence edge for the following pulse shapes, as sketched in Figure 1: a) Simple rectangle; b) Symmetric triangle; c) Spline; d) "Worst" case. The observed bursts have much more complicated time profiles than those shown here, but these shapes mimic some of the basic features of the bursts.

The fluence edges were obtained analytically by integrating the pulse profile to find the durations when a weak burst of that shape could trigger on 64 ms, 256 ms or 1024 ms. The fluence edges for the above profiles are plotted in Figure 2. The "kink" in the curves corresponds to the transition between triggering on 256 ms and 1024 ms. The vertical line in Figure 2 is for the "worst" case profile in which the burst has a spike with C_{64} equal to $(C_{min})_{64}$, but it never triggers on 256 ms or 1024 ms. If a short faint gamma-ray burst can be described by one of these time profiles, it must lie above the appropriate edge to be detected by BATSE. Figure 2 shows that for very short bursts ($T_{90} < 128$ ms), the fluence edges are vertical as expected; the value of C_{64} for

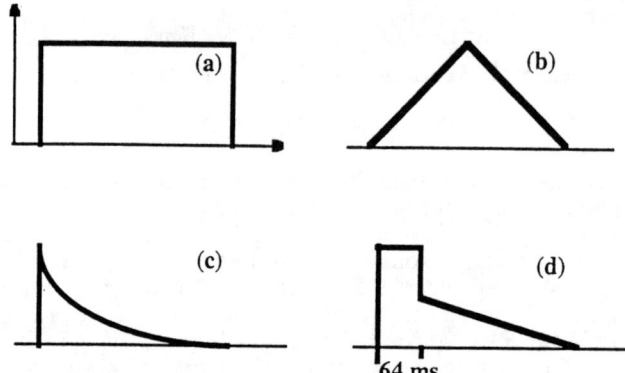

Figure 1. Schematic plots of the time profiles (count rate vs. duration) used to calculate the fluence edges: a) rectangle; b) symmetric triangle; c) "spline"; d) "worst" case.

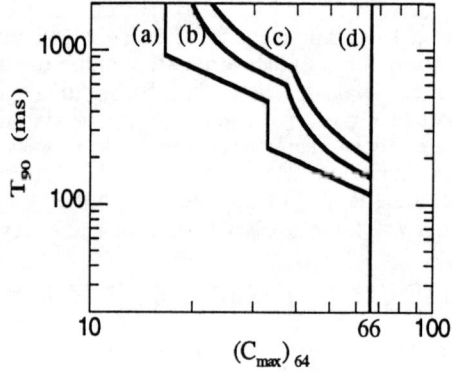

Figure 2. Fluence edges in the duration (T_{90}) versus $(C_{max})_{64}$ diagram for the four profiles sketched in Figure 1. $(C_{min})_{64} = 66$ was used.

this edge is given by $(C_{min})_{64}$.

We have considered other interesting pulse profiles, such as one with multiple peaks. However, these will give fluence edges that are below the "worst" case curve. For a real burst, the fluence edge will typically be between the "spline" case and the "worst" case.

EFFECT OF DIFFERENT C_{min}

Figure 3 illustrates the shift of the fluence edge as $(C_{min})_{64}$ changes (the background count-rate varies with time). We only plot curves for the rectangular and the "worst" case time profiles to show how much phase space has been changed by varying C_{min}. We use the largest and smallest values of $(C_{min})_{64}$ that have been seen by BATSE to generate the upper and lower curves shown:

in the first BATSE catalog of 260 gamma-ray bursts, the most common value of $(C_{min})_{64}$ is 66 (for 127 bursts) and the variation in $(C_{min})_{64}$ is from 60 to 77 (except for the overwrite bursts).

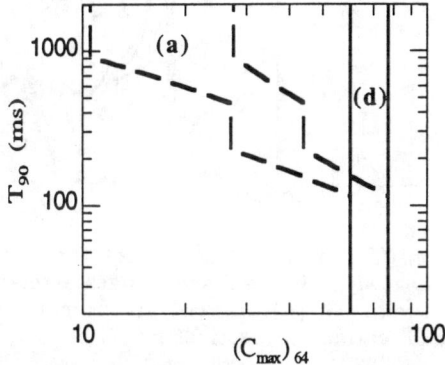

Figure 3. Variation in fluence edge due to the variation in $(C_{min})_{64}$. Only the rectangular (a) and "worst" case (d) time profile are plotted.

COMPARISON OF FLUENCE EDGES WITH BATSE BURSTS

Figure 4 is the same as Figure 2 except that bursts from the first BATSE catalog are added. Figure 4(a) uses the bursts in the catalog prior to the revision on September 9, 1993, while Figure 4(b) uses the revised catalog, which removed many of the faint bursts. $(C_{min})_{64} = 66$ was chosen for the fluence edge curves in Figure 4. 1σ error bars are added for each burst. We have extended the duration to cover all the BATSE bursts, but in principle the fluence edge effect stops around 1024 ms.

One might expect that the short weak bursts would scatter around these fluence edges, but above curve (a), because some bursts might be too weak to trigger on 64 ms and instead trigger on 256 ms or 1024 ms; also, short bursts could have a variety of morphologies, with some smooth bursts close to curve (a). Surprisingly, most of the short (duration less than 1024 ms) bursts are above the fluence edges considered here, even the "worst" case (the burst below the "worst" case edge has $(C_{min})_{64} = 60$). The possible reasons for the lack of bursts below the "worst" case fluence edge are: a) Not enough bursts have been detected yet. So far, all the detected bursts in the first catalog shorter than $T_{90} \approx 922$ ms have $(C_{max}/C_{min})_{64} > 1$, thus these burst must be above $(C_{min})_{64}$. As BATSE detects more bursts, some of the short faint bursts might only trigger on 256 ms or 1024 ms and have $(C_{max}/C_{min})_{64} < 1$. b) Possion fluctuations (Lamb, Graziani & Smith[2]); for bursts longer than 64 ms, $(C_{max})_{64}$ will select the largest C_{64}, which might be higher than the intrinsic brightness. This will shift the bursts to the right in Figure 4. c) Real burst time profiles for the short faint bursts are close to the "worst" case considered above. It will be interesting to study some of the short and faint BATSE bursts to see if this is the case. A brief search of the catalog[3] supports this possibility, but high

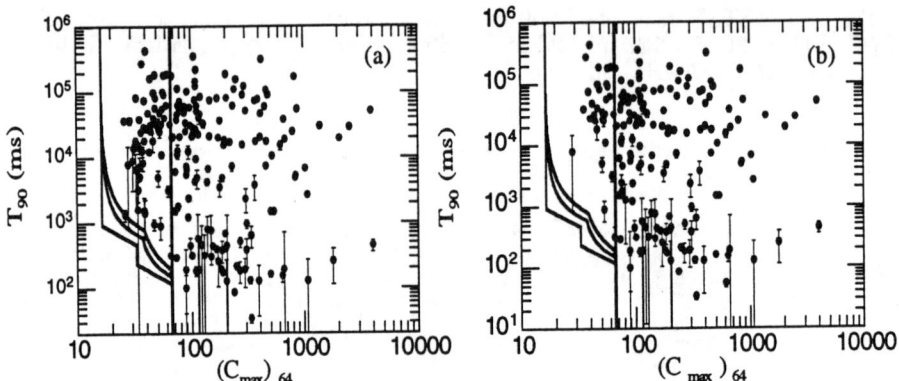

Figure 4. Duration (T_{90}) versus $(C_{max})_{64}$ for the bursts from the first BATSE catalog with all the fluence edges extended to longer durations. Data points in (a) and (b) are from the old and the revised first BATSE catalog respectively.

resolution data is needed.

We note that there are a couple of bursts with duration longer than 1024 ms that have very large errors in their durations, and some of these might actually be below the fluence edge curves.

IMPLICATIONS FOR $\log N - \log P_{64}$ DISTRIBUTION OF SHORT BURSTS

Although the short ($T_{90} < 922$ ms) bursts in the first BATSE catalog have $(C_{max}/C_{min})_{64} > 1$, we believe that future observations will detect short bursts with $(C_{max}/C_{min})_{64} < 1$ that are below the "worst case" curve in Figures 2 and 4. It is impossible to know exactly how many short bursts are under the fluence edges, since we do not know the real duration distribution of the short bursts. However, a simple estimate can be made by assuming the "number density" remains constant across the edge. The number of "missing" short faint bursts could be as high as $10 - 20\%$ of the total number of bursts. This percentage will be refined as more bursts are added to the catalog.

Most of these "missing" bursts will have P_{64} (peak photon flux in 64 ms) less than 0.8 photons cm^{-2}s^{-1}. Currently, all the bursts that have $P_{64} < 0.8$ photons cm^{-2}s^{-1} are deleted from the $\log N - \log P_{64}$ distribution[3], because their fluxes are unreliable. In the future, if reliable values for the fluxes of the bursts with $P_{64} < 0.8$ photons cm^{-2}s^{-1} can be obtained, it will be important to include an estimate of the number of bursts that are "missing" below the fluence edge, because this will affect the slope of the $\log N - \log P_{64}$ distribution.

CONCLUSION

As we have shown in Figures 2, 3, and 4, there is an obvious selection effect against detecting short faint bursts. This effect has been quantitatively described assuming different pulse shapes for the gamma-ray bursts. The distribution of bursts from the first BATSE catalog is found to be consistent with the fluence edges calculated here. We emphasize that the fluence edge can lead to an incompleteness in the number of short bursts observed. A detailed high time resolution study of the profiles of the short faint bursts might shed some light on the intrinsic distribution of gamma-ray bursts.

This work was supported at Rice University by grant NAG 5-1515.

REFERENCES

1. V. Kargatis et al., Ap. J. (Letters) (in press).
2. D. Q. Lamb, C. Graziani and I. A. Smith, Ap. J. (Letters) **413**, L11 (1993).
3. G. Fishman et al., Ap. J. Suppl. (in press).

GAMMA-RAY BURST REPETITION AND THE DEFINITION OF BURSTS

R. E. Lingenfelter, V. C. Wang and J. C. Higdon*
CASS, University of California, San Diego, La Jolla, CA 92093

ABSTRACT

We point out a basic inconsistency in the traditional definition of "classical" gamma-ray burst durations, and we suggest an observational definition to correct it. All classical bursts occurring within an arbitrary data accumulation time are traditionally, defined as a single "multipeaked" burst, even when individual "peaks" are separated by periods of subthreshold emission and would retrigger an unbiased detector. Such peaks would also be identified as separate bursts, if either the assumed accumulation time was shorter, or the burst time scale was longer. This is a fundamental flaw in the traditional definition that seriously biases the identification of classical bursts and the statistical analysis of their properties. We suggest instead an observationally self-consistent definition for classical bursts, simply defining burst duration as the time during which the emission, averaged on the triggering time scale, is continuously observable above the trigger threshold flux. This observational definition provides a consistent measure of both the number of bursts and their duration, that directly reflects the detector trigger threshold and gives an unbiased sample above that threshold. We adopt this definition of burst duration for classical bursts, and we find that a large fraction ($\sim$ 1/5 to 1/3) of the classical bursts repeat on rapid ($\sim$ 1 to 10^2 s) time scales. These repeated bursts clearly show that classical bursts are not standard candles. They also remind us that the observed intensity and spatial distributions are of burst occurrence, not of burst sources, because we can not identify individual burst sources in large samples (e.g. KONUS and BATSE) where the uncertainty in position is greater than the random separation.

INTRODUCTION

There is a fundamental inconsistency in the traditional definition of classical burst durations that seriously biases both the identification of classical bursts and the statistical analysis of their properties, and also precludes the recognition of rapidly repeating bursts. Traditionally, all bursts occurring within an arbitrary data accumulation time are defined[1-4] as a single, "multipeaked" burst. This is done even when individual "peaks" are separated by periods of subthreshold emission and would retrigger an unbiased detector, and such peaks would also be identified as separate bursts, if either the assumed accumulation time was shorter, or the burst time scale was longer. This is a fundamental flaw in the traditional definition. Studies of soft gamma-ray repeaters, however, make no such assumptions, and adopt a simple observational definition of burst duration.[5-7]

* Visiting from Joint Science Dept., Claremont McKenna College, Claremont, CA 91711

We adopt such an observational definition of gamma-ray burst duration for the study of classical gamma-ray bursts and explore its implications. Recognizing the importance of the detector trigger threshold in setting a level for complete sampling that is needed for statistical studies, we define the duration of a burst observationally as that time during which the emission, averaged on the triggering time scale, is continuously observable above the trigger threshold flux.

The traditional definition, of course, simply recognizes that closely associated bursts almost always come from a common source. Thus, whether we refer to multiple peaks or separate bursts may seem like a trivial semantic distinction. But, as we show, this traditional definition seriously biases the statistical analysis of gamma-ray bursts.

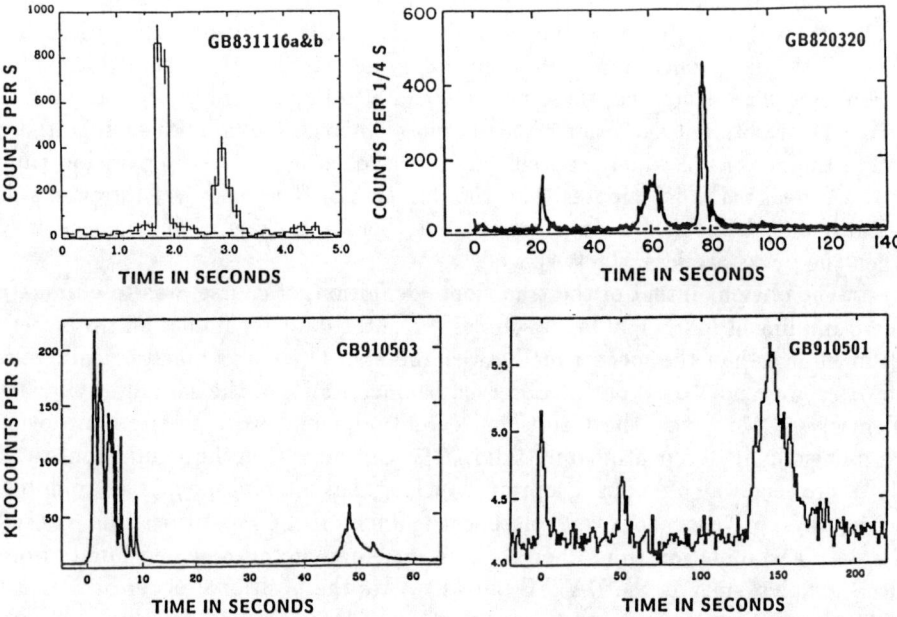

Figure 1. Time profiles of gamma-ray bursts illustrating the difference between the traditional and observational burst definitions. Two close bursts[8] (GB831116a&b) from the soft repeater source GBS 1806−20 are recognized as separate bursts, because they were identified using an observational definition of burst duration. In contrast, widely spaced "peaks" in three classical bursts,[9,10] GB820320, 910501 & 910503, are not recognized as separate bursts, because as classical bursts they were traditionally defined only as the "multiple peaks" of a single burst if they occur within an arbitrary data accumulation time. These bursts clearly illustrate the inconsistency of the traditional definition, because if the occurrence time scale were simply increased or the assumed accumulation time were decreased, these "peaks" would then be counted as separate bursts using the same traditional definition. This inconsistency is removed, however, if we adopt the observational definition of burst duration.

The difference between the traditional and observational definitions can be seen in Figure 1. Here, we see (GB831116a&b) two very close bursts[8] ~ 1 s apart from the soft repeater source GBS 1806−20 that are recognized[6] as separate bursts, as are many others separated by tens of seconds or more, because they are identified using an observational definition of burst duration. In contrast, we see (GB820320, 910503 & 910501) several classical bursts[9,10] ~ 10 to 100 s apart, separated by periods with no emission above threshold, that are not recognized as separate bursts, because as classical bursts they were defined only as the multiple peaks of a single burst, even when each of these peaks would independently trigger an unbiased detector, since their flux rises above the trigger threshold well after that of the previous peak fell below.

Although these examples have longer than average separations between peaks, they clearly illustrate the inconsistency of the traditional definition, because, if the occurrence time scale were simply increased or the assumed accumulation time were decreased, these peaks would also be counted as separate bursts using the traditional definition. This is most obvious in examples such as these where the peaks are widely-spaced bursts in the sense that the separation time between peaks is much greater than the duration of the peaks, yet they are still traditionally defined as a single event; but there is no less of an inconsistency when the peaks are less widely spaced.

The obvious intent of the traditional definition, of course, was to eliminate over-sampling of individual burst sources, since the data accumulation times were all much less than the mean time between bursts. These accumulation intervals, however, were not based on the observed characteristics of the gamma-ray bursts themselves. Moreover, the traditional definition would work to eliminate over-sampling only if all repeating bursts did, in fact, occur within the assumed interval. But there is no way of knowing a priori how long that interval should be – minutes, hours, days, or years – or even whether it is finite. If the repetition times exceed the data accumulation time, then we can not eliminate all repeated bursts from large samples, such as the BATSE bursts, where the positional uncertainties, 1σ of at least 4° for the brightest bursts and $\sim 27°$ for the weakest bursts[10], are comparable to the mean random separation angle for the sample on the sky, because we can not determine which bursts come from separate sources. Thus, we need to recognize that the observed spatial distributions are those of burst occurrence, not of burst sources, and adopt a burst definition that gives a more complete sampling of their occurrence.

A DEFINITION

As Mazets & Golenetskii[11] and others have previously emphasized, not only the detection and the duration, but many other burst properties, are detector threshold dependent. For meaningful statistical analyses, therefore, we need to obtain as complete and unbiased a sampling as possible above the threshold level. We suggest an observationally self-consistent definition that is directly tied to the detector threshold.

The start of a burst has generally been defined observationally as the time at which the observed flux, or more specifically the detector count rate within some trigger sampling time and some energy band, rises above a statistically significant threshold value. If we make no a priori assumptions about the duration of bursts, it then follows that the end of a burst should also be defined observationally as the time at which the observed flux, or detector count rate, within the same trigger sampling time and the same energy band, falls below the same trigger threshold value. Thus, a burst is defined to be that period of flux emission during which the flux, or count rate, averaged over the trigger sampling time, continuously exceeds the trigger threshold, and its duration is the length of that period.

This observational definition provides a consistent measure of both the number of bursts and their duration, that directly reflects the detector trigger threshold and gives a complete and unbiased sample above that threshold. This definition also corrects another flaw in the traditional definition of the duration[1,4] which has been to include all (or up to 90% for BATSE) of the flux that is observable above the 1-σ noise level, even though it is well below the detection threshold. Such a definition, which can be dominated by long periods without detectable emission, is clearly not consistent with either the observed emission above the threshold, or even that observed above the noise level. The observational definition, on the other hand, does not have these inconsistencies. It provides a complete sample of bursts above the detector threshold, and it directly measures the duration of those bursts above that threshold.

IMPLICATIONS

When we apply the observational definition of burst duration, given above, to the classical gamma-ray bursts, that have been traditionally defined as single, multipeaked bursts, we recognize each peak that is separated by periods of subthreshold emission and would retrigger the detectors as a separate burst independent of the time scale, as long as it is greater than the trigger time. We find that a large fraction of the classical bursts are clustered on time scales of ~ 10 to 10^3 seconds. Since the probability of such clustering occurring randomly is quite negligible, these bursts come from rapidly repeating sources that can produce repeated bursts within a few seconds to few hundred seconds. There is also much recent evidence (Refs. 12, 13, and Wang & Lingenfelter, Quashnock & Lamb, MacCallum et al., Chuang, and Castro-Tirado et al., all in this Volume) that they also repeat on much longer time scales, as well. The range of repetition times observed for the soft repeaters, in fact, spans many orders of magnitude[6] from 1 to 10^8 s.

We are presently making a detailed analysis of both the KONUS and BATSE data samples using the observational definition and the results will be presented in a separate paper. From a preliminary analysis, however, we find that there are multiple peaks in a large fraction ($\sim 1/5$ to $1/3$) of the observed bursts that would have retriggered an unbiased detector. In the first BATSE catalog sample of 260 bursts we find roughly 80 traditionally defined single multipeaked bursts which included one or more peaks that would be recognized as separate bursts under the observational definition, and in the KONUS sample of 142 burst we find roughly 30 such multipeaked bursts. Inclusion of these repeated burst will substantially increase the number of bursts and provide a complete and unbiased sample.

But most important, the rapid repetition of many classical bursts reminds us that what we measure are gamma-ray bursts, not burst sources. Thus, the spatial distributions implied by luminosity and position are the distributions of burst occurrence, not burst sources. This distinction has generally been ignored in the past, but considering the uncertainties in both the range of repetition times and the positional determinations, we realize that each burst position does not necessarily represent a separate source. We can not know the distribution of burst sources in the large samples (e.g. KONUS and BATSE), because we can not identify bursts from individual sources where the uncertainty in position is greater than the random separation.

Further considering the intrinsic range of luminosities shown by the rapid repeaters from individual sources, which approaches the range $\sim 10^2$ of the full burst sample, we also see that the classical bursts are not standard candles and their intensities do not give us a direct measure of distance. Thus, the distributions of burst occurrence, which necessarily include all repeating bursts, are the only ones with which burst models can meaningfully be compared. Such distributions obviously give added weight to rapidly repeating sources, but that, in fact, properly reflects the nature of burst occurrence in both luminosity and position.

We thank Chip Meegan, David Band, and Rick Rothschild for helpful discussions, and NASA for support under grant NAG-5 1597(REL,VCW) and NAG-5 2010 (JCH).

REFERENCES

1. Mazets, E. P., et al., Ap. Space. Sci., **80,** 3, (1981).
2. Matz, S. M. PhD thesis, Univ. New Hampshire, 132 pp, (1986).
3. Murakami, T., et al., Pub. Ast. Soc. Japan, **41,** 405, (1989).
4. Fishman, G. J., et al. BATSE Burst Catalog, on line at GROSSC (1992).
5. Atteia, J.-L., et al., Ap. J., **320,** L105, (1987).
6. Laros, J. G., et al., Ap. J., **320,** L111, (1987).
7. Norris, J. P., et al., Ap. J., **366,** 240, (1991).
8. Kouveliotou, C., et al., Ap. J., **322,** L21, (1987).
9. Golenetskii, S. V., et al., Fiz. Tekh. Inst. Ioffe Report 1819, (1983).
10. Fishman, G. J., et al., Ap. J. Supp., in press, (1993).
11. Mazets, E. P., & Golenetskii, S. V. Astronomia, **32,** 16, (1987).
12. Wang, V. C., & Lingenfelter, R. E., Ap. J., **416,** L13, (1993).
13. Quashnock, J. M., & Lamb, D. Q., MNRAS, in press, (1993).

EVIDENCE FOR TWO DISTINCT MORPHOLOGICAL CLASSES OF GAMMA-RAY BURSTS

C. Graziani, D. Q. Lamb

Dept. of Astronomy and Astrophysics, Univ. of Chicago, Chicago, IL 60637

ABSTRACT

We describe evidence for two distinct morphological classes of γ-ray bursts based on their short time scale variability.

INTRODUCTION

We have analyzed the 260 bursts in the Burst and Transient Source Experiment (BATSE) 1B catalogue,[1] using $B = (\bar{C}^{1024})_{\max}$ as a measure of burst brightness and $V = (\bar{C}^{64})_{\max}/(\bar{C}^{1024})_{\max}$ as a measure of short time scale ($\lesssim 0.3$ s) variability. The quantities $(\bar{C}^{64})_{\max}$ and $(\bar{C}^{1024})_{\max}$ are the expected maximum number of counts in 64 ms and 1024 ms, respectively. We have identified two distinct morphological classes of γ-ray bursts on the basis of their short time scale variability.[2-4] Kouveliotou et al. have independently identified two classes of γ-ray bursts on the basis of their durations.[5]

ANALYSIS AND RESULTS

The quantity V measures an extreme property of the burst (short time scale variability) at a particular moment during the burst (the peak of the burst). Thus it is complementary to global measures of burst variability, such as power spectra[6] and wavelet analyses.[7] The behavior of V can be illustrated by considering two different burst time histories. If the burst time history is relatively smooth, $(\bar{C}^{64})_{\max}$ is roughly one-sixteenth of $(\bar{C}^{1024})_{\max}$, and $V \approx 1/16$. On the other hand, if the burst time history has a bright spike with duration $t^{\text{spike}}_{\text{dur}} \lesssim 64$ ms, $(\bar{C}^{64})_{\max}$ is almost as large as $(\bar{C}^{1024})_{\max}$ and $V \approx 1$.

Figure 1 shows the distribution in the (B, V)-diagram of the 201 bursts for which V is available in the BATSE 1B catalog.[1] The distribution has been corrected for a bias that arises from the fact that we define V in terms of peak counts, rather than average counts.[2] The striking feature of this diagram is the correlation between burst brightness B and variability V. In particular, the upper right-hand quadrant of the diagram is empty, implying that there is a lack of bursts which are bright *and* have a bright, short spike.

Gamma-ray bursts exhibit a bewildering variety of time histories.[1] However, if the brightest spike in each time history were flat (or exponential) in shape *and* there were equal numbers of bursts per logarithmic interval in variability, diagonal lines running from the upper left to the lower right in the (B, V)-diagram would be uniformly populated.[8] The fact that this is not the case, due to the lack of bursts in the upper left hand quadrant, again implies the existence of two classes of γ-ray bursts.

From Figure 1 alone, it is not possible to determine whether the classes are defined by dividing the bursts according to their brightness B or according to their variability V. However, by considering other burst properties, it becomes

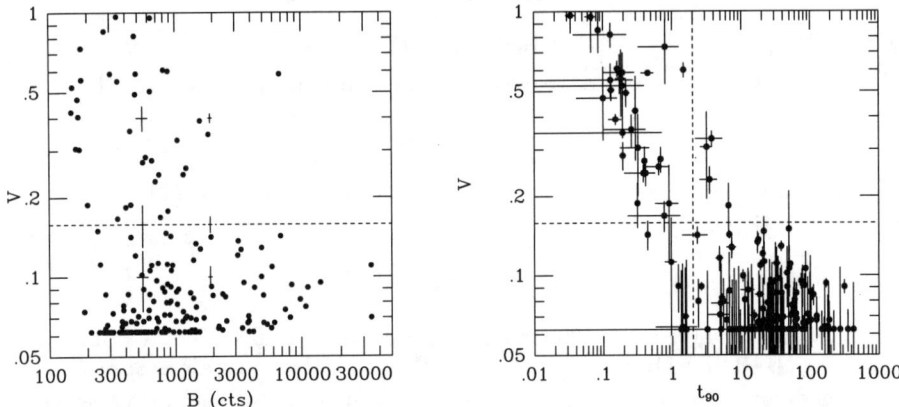

Fig. 1. Distribution of 201 bursts in the (B, V)-diagram. The dashed line corresponds to $\log V = -0.8$, the cut which separates the bursts into two classes on the basis of variability.[2]

Fig. 2. Distribution of 173 bursts in the (t_{90}, V)-diagram. The horizontal dashed line corresponds to $\log V = -0.8$, the cut which separates the bursts into two classes on the basis of variability.[2] The vertical dashed line corresponds to $t_{90} = 2$ s, the cut which separates the bursts into two distinct classes on the basis of duration.[5]

clear that the variability cut is the appropriate one.

We make the cut, as shown by the dashed line in Figure 1, at $V = 0.16$. Bursts below the cut are relatively smooth on short time scales ($\lesssim 0.3$ s) and we denote them as type I bursts, while bursts above the cut are relatively variable on short time scales and we denote them as type II bursts. In order to qualify as a type II burst, a burst must satisfy

$$t_{\text{dur}}^{\text{spike}} \lesssim 64(1.2V)^{-1} \text{ ms} = 0.3(V/0.16)^{-1} \text{ s} ; \qquad (1)$$

that is, it must have a bright spike of duration $t_{\text{dur}}^{\text{spike}} \lesssim 0.3$ s.

Figure 2 shows the distribution in the (t_{90}, V)-diagram of the 173 bursts for which V and t_{90} are available in the BATSE 1B catalogue[1]. The empty lower left hand corner is a consequence of the definitions of V and t_{90}. Koveliotou et al[5] have identified two classes of bursts by making a cut in duration at $t_{90} = 2$ s, corresponding to the vertical dashed line. We have independently identified two classes by making the cut in variability at $\log V = -0.8$, which is orthogonal to the cut in duration in this diagram. Seven smooth (type I) bursts are short ($t_{90} < 2$ s) and four variable (type II) bursts are long ($t_{90} > 2$ s). Nevertheless, the two classes identified by Koveliotou et al. are basically the same as the two classes that we have identified.

There has been some discussion of whether or not variability and duration are distinct properties of γ-ray bursts. In order to shed some light on the matter, consider the time histories of the six variable (type II) bursts shown in Figure 3. All six of the bursts exhibit one or more bright, short spikes with $t_{\text{dur}}^{\text{spike}} \lesssim 0.3$ s, while the durations t_{90} of the bursts themselves range from ≈ 0.25 s to ≈ 50 s. Now consider bursts 936, 1204, and 1997: if these bursts were

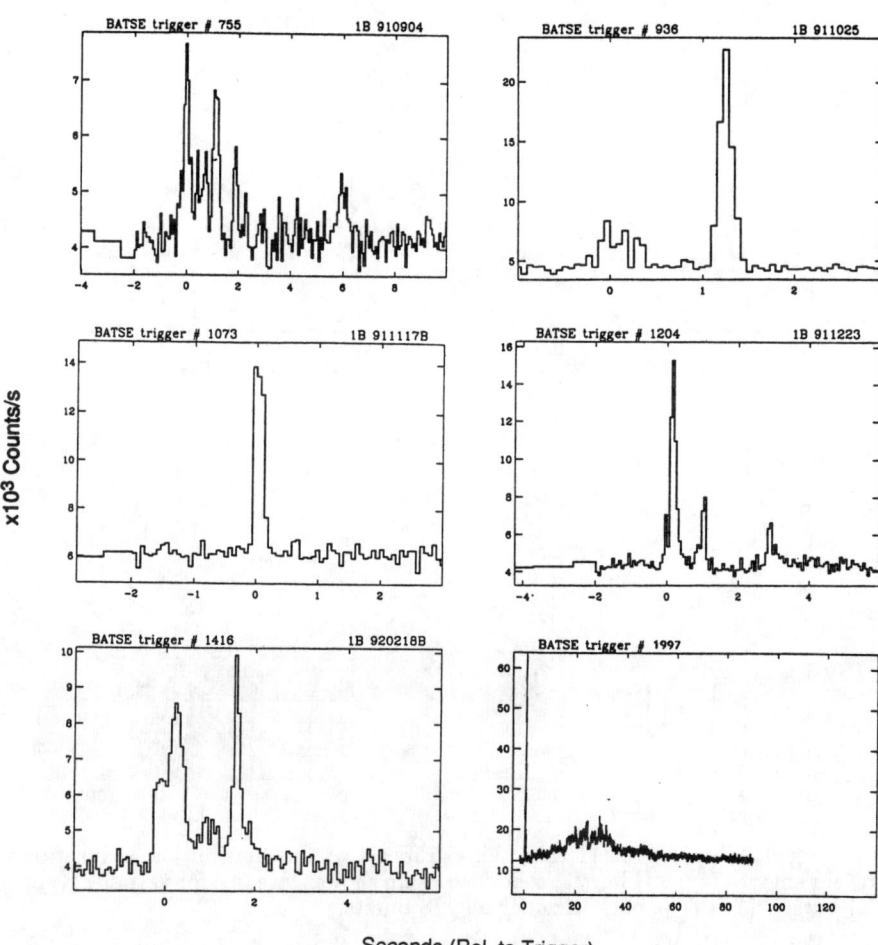

Fig. 3. Time histories of six variable (type II) bursts; all except burst 1997 are from the BATSE 1B catalogue.[1]

intrinsically fainter, or originated farther away, their extended emission would become lost in the background, and their time histories would become a single, bright spike, resembling burst 1073. We know of no evidence to contradict the supposition that *all* short bursts have extended emission, but that it is often too faint to be distinguished from the background.

Note also that if we modified our criterion for type II bursts so as to accept only bursts whose *durations* are less than 0.3 s, rather than bursts with spikes of duration less than 0.3 s, five of these six bursts would not make the cut. Indeed, of the bursts we classify as variable, 19 consist only of a short, bright ($\lesssim 0.3$ s) spike while 15 consist of a short ($\lesssim 0.3$ s), bright spike and fainter emission lasting $t_{90} > 0.3$ s.

Having made the cut in V, we can look at other properties of the bursts. First consider burst duration. Figure 4 shows histograms of the durations for

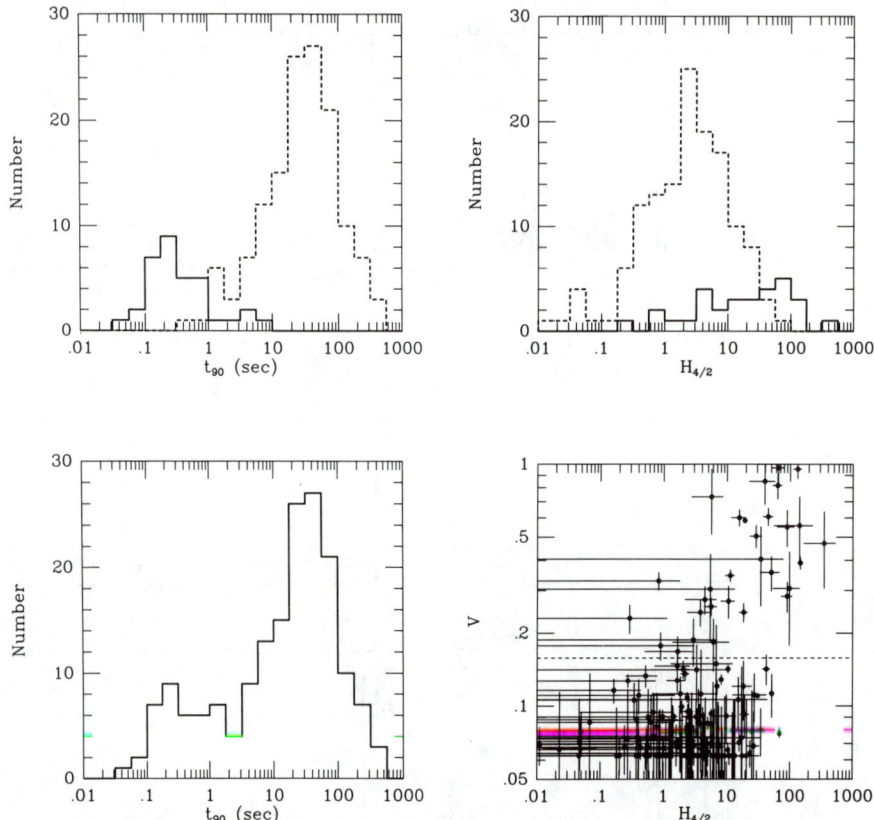

Fig. 4. (Upper panel) Dashed histogram shows the distribution in duration t_{90} for smooth (type I) bursts; solid histogram shows the same for variable (type II) bursts. (Lower panel) Same for all 173 bursts.

Fig. 5. (Upper panel) Dashed histogram shows the distribution in the hardness ratio $H_{4/2}$ for smooth (type I) bursts; solid histogram shows the same for variable (type II) bursts. (lower panel) Distribution of 201 bursts in the $(H_{4/2}, V)$-plane.

those bursts for which both V and t_{90} are available in the BATSE 1B catalog.[1] The dashed line in the upper panel is a histogram of the durations of smooth (type I) bursts, while the solid line in the upper panel is a histogram of the durations of variable (type II) bursts. The lower panel is a histogram of the durations of all bursts. Figure 4 shows that the cut in variability V cleanly separates the bimodal distribution of burst durations into two unimodal components.

Second, consider burst spectral hardness. Figure 5 shows the distribution of hardnesses $H_{4/2} \equiv S(> 300 \text{ keV})/S(50 - 100 \text{ keV})$, where the S are the burst fluences in the specified photon energy ranges. The dashed line in the upper panel is a histogram of the hardness ratios of smooth (type I) bursts,

TABLE I
MORPHOLOGICAL CLASSES OF GAMMA-RAY BURSTS

TYPE I (80% of bursts)	TYPE II (20% of bursts)
Smooth on timescales $\lesssim 0.3$ s	Variable on timescales $\lesssim 0.3$ s
Long (0.3 s $\lesssim$ Durations $\lesssim 10^3$ s)	Short (10^{-2} s $\lesssim$ Durations $\lesssim 10$ s)
Softer spectra (esp. above 300 keV)	Harder spectra (esp. above 300 keV)
Most are not uniformly distributed in space	Most are uniformly distributed in space

while the solid line in the upper panel is a histogram of the hardness ratios of variable (type II) bursts. Figure 5 shows that the smooth (type I) bursts are, on average, much softer than the variable (type II) bursts. The difference is highly significant (Q-value = 4.8×10^{-7}).

CONCLUSIONS

We find significant evidence for two distinct morphological classes of γ-ray bursts. Table I summarizes the properties of the two classes. Type I bursts (comprising $\approx 80\%$ of the bursts) are smoother ($\log V \leq -0.8$) on short time scales ($\lesssim 0.3$ s), range from faint to bright, are longer, and have softer spectra. Their brightness distribution implies that the sources of most of the type I bursts detected by BATSE are not uniformly distributed in space. Type II bursts (comprising $\approx 20\%$ of the bursts) are more variable ($\log V > -0.8$) on short time scales ($\lesssim 0.3$ s), faint ($B < 1900$ counts), shorter, and have harder spectra. Their brightness distribution is consistent with most of the type I bursts detected by BATSE being uniformly distributed in space. The brightness distributions for type I and type II bursts suggest that $(\bar{C}^{64})_{max}$ is similar for both, and that the difference between the two distributions is a result of the greater distance to which BATSE samples type I bursts compared to type II bursts.

We thank Chryssa Kouveliotou, Chip Meegan and the BATSE team for providing to us the time history of burst 1997. This research was supported in part by NASA grants NAGW-830, NAGW-1284, NASW-4690, and NGT-50617.

REFERENCES

1. G. J. Fishman, et al., ApJSup, in press (1994).
2. D. Q. Lamb, C. Graziani & I. A. Smith, ApJ **413**, L11 (1993).
3. D. Q. Lamb & C. Graziani, ApJ, in press (1994).
4. D. Q. Lamb & C. Graziani, ApJ, in press (1994).
5. C. Kouveliotou et al., ApJ **413**, L101 (1993).
6. C. Kouveliotou, et al., in Gamma-Ray Bursts, eds. W. S. Paciesas & G. J. Fishman (New York: AIP, 299), p. 1992.
7. J. P. Norris, et al., in Gamma-Ray Bursts, eds. W. S. Paciesas & G. J. Fishman (New York: AIP, 294), p. 1992.
8. R. Rutledge & W. H. C. Lewin, these proceedings (1994).

ON BURST CLASSIFICATION USING PEAK RATE RATIOS

Charles Meegan
NASA/Marshall Space Flight Center

Chryssa Kouveliotou
USRA, NASA/Marshall Space Flight Center

ABSTRACT

We examine the variability parameter V introduced by Lamb, Graziani, & Smith[1] to classify gamma-ray bursts. The correlation between V and burst duration is found to be an artifact of the mathematical relation between these parameters, and not a indication of burst morphological classes.

INTRODUCTION

Lamb et al.[1] (LGS) have advocated a burst classification scheme based on a parameter they call variability (V), which is a ratio of peak count rates on two different time scales. It was claimed that this parameter efficiently separated bursts into two classes - Type 1, which are short, variable, and have generally harder spectra, and Type 2, which are longer, smooth, and have generally softer spectra. This definition of variability is linked to the duration of the burst. We examine the nature of this link to determine whether the correlation between V and duration is a direct consequence of the mathematical relation between these parameters.

DEFINITIONS

We define C_t as the peak counts above background, integrated over energy, in a time interval t during a burst. These can be determined for BATSE bursts for time intervals of 64 ms, 256 ms, and 1024 ms, using the public catalog at the GRO Science Support Center. The variability V has been defined in LGS as C_{64}/C_{1024}. The time intervals of the two peak rates need not be coincident. The range of V is 1/16 to 1. LGS compute a correction factor for V to account for the bias introduced by statistical fluctuations. While this correction is important when considering V as a function of burst intensity, we neglect it here.

As a measure of burst duration, we use T_{50}, which is the time interval during which the counts increase from 25% to 75% of the integrated number of counts in the burst. T_{50} is computed for events in the 50 to 300 keV energy range; V includes all counts. This difference in energy ranges does not significantly affect the analysis.

LOWER BOUND ON V

There is a lower bound to V that follows trivially from the definitions. Consider a burst with total counts C. There are $C/2$ counts within T_{50}. If T_{50} is greater than 64 ms and less than 512 ms, then these $C/2$ counts are spread among ($T_{50}/64$) bins of width 64 ms. The 64 ms peak C_{64} must be at least as large as the average, which is $32C/T_{50}$. The maximum value for C_{1024} is obtained by having all C counts fall within 1024 ms. Thus, $V > 32/T_{50}$. If T_{50} is less than 64 ms, then the counts outside of the T_{50} interval can be allowed to fall outside the 64 ms peak as well, and $V > 1/2$. If T_{50} is greater than 512 ms, then $V > 1/16$. These lower bounds do not depend on any assumed temporal structure of the burst. Note also that the lower bound is easily generalized to other time scales than 64 and 1024 ms, and that it depends quite strongly on the selected time intervals.

V VERSUS T_{50} FOR VARIOUS PULSE SHAPES

An understanding of the relation between V and duration can be gained by considering the maximum value of V for bursts with specific temporal structure. A maximum V is obtained for any particular pulse shape by assuming that the 64 ms window is centered on the peak, and that the 1024 ms interval is phased to give the lowest possible maximum for C_{1024}. Figure 1 shows V as a function of T_{50} for pulses with the following shapes: square, triangular, gaussian, and instantaneous rise followed by exponential decay. The lower bound curve is also shown. All the curves show a similar behavior. The variability parameter is high for short events and low for long events. If any time history is stretched uniformly, V decreases. Note that a cut at $V = 0.2$ will automatically divide the bursts into two classes by duration at about 1 second.

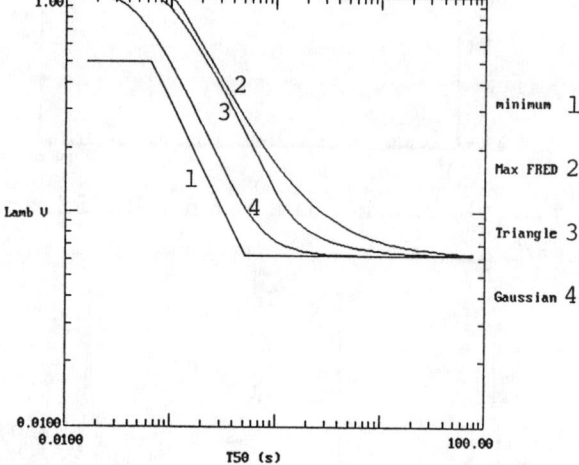

Figure 1. Computed V vs. T_{50} for square, triangular, gaussian, and exponential pulses.

COMPARISON TO DATA

Figure 2 duplicates Figure 1, with the addition of data points from the first BATSE catalog. The computed curves are seen to predict the observations fairly well. The computed lower bound explains the absence of events in the lower left area of the plot. There are several events with high V at about 1 s duration that appear to be outliers. The time histories of these events are shown in Figure 3. In this, and all subsequent figures, count rate, in units of 10^3 cps, is plotted against time in seconds relative to trigger time. In each case, the burst consists of two narrow pulses (thereby giving high V) separated in time by about a second (thereby defining T_{50}). LGS emphasized the lack of high V events with long duration, which is quite evident in Figure 2. We see, however, that this is to be expected, given the mathematical relation between duration and V. To have a high V, a burst must have at least one high 64 ms rate, but a low 1024 ms rate *throughout the event*. With increasing burst duration, the latter requirement is increasingly harder to satisfy, given the wide range of temporal behavior of bursts. Figure 4 shows an example of such a rare event, discovered from among bursts after the first BATSE catalog. It has a dominant short spike (giving high V) and smoother, lower intensity emission lasting tens of seconds (giving large T_{50}). The point is that the rarity of such events should not be surprising.

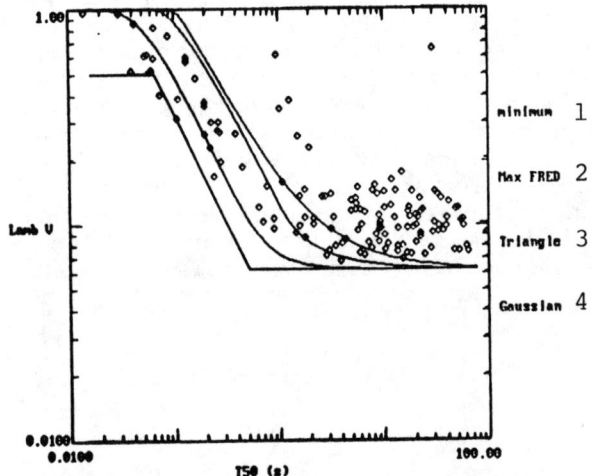

Figure 2. V vs. T_{50}, including data from first BATSE catalog.

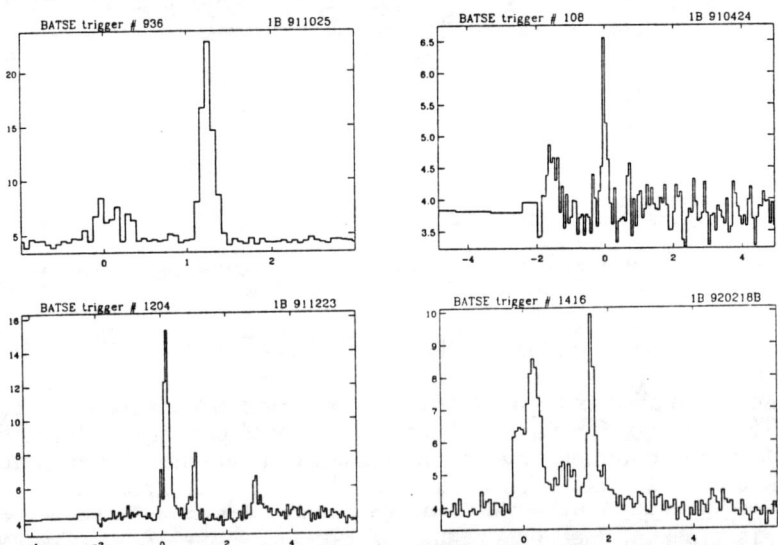

Figure 3. Time histories of high V events with T_{50} of ≈ 1 s.

The high V and low V events are referred to in LGS as "variable" or "smooth", respectively, since they interpret V as a measure of variability. Figures 5 and 6 show the 4 highest V and the 4 lowest V bursts from the first BATSE catalog[2]. Clearly, V does not distinguish events by morphology, but by duration.

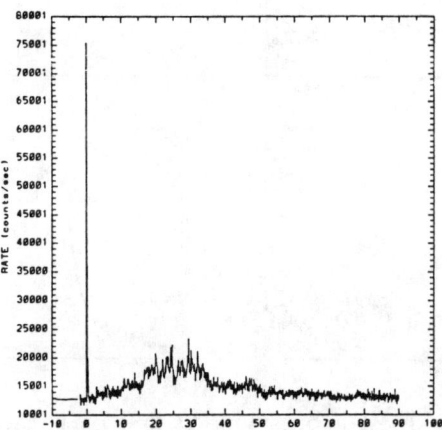

Figure 4. Time history of a burst that has high V and high T_{50}.

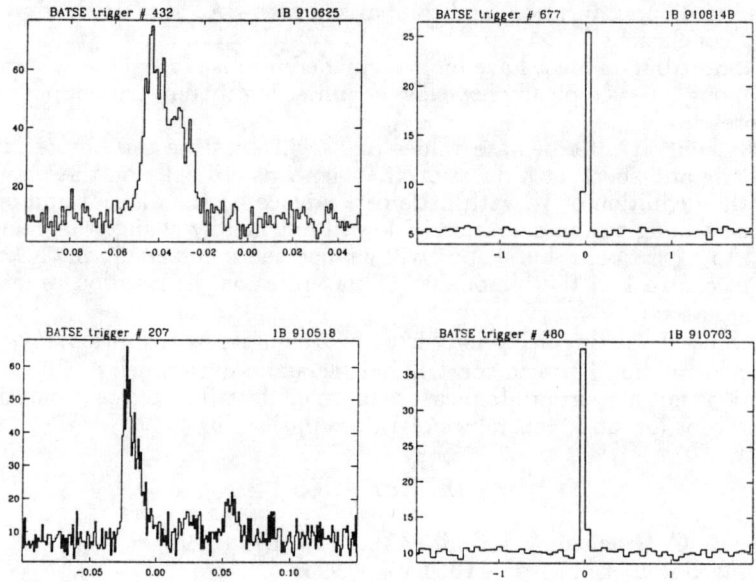

Figure 5. The 4 highest V bursts in the first BATSE catalog.

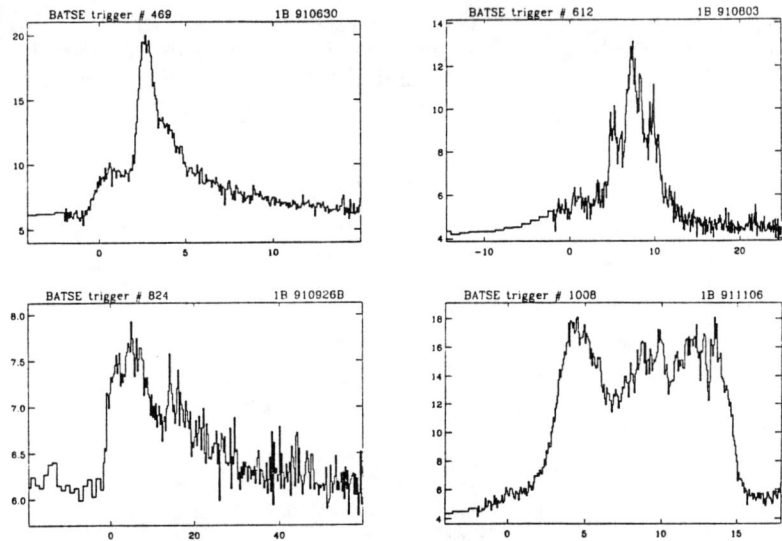

Figure 6. The 4 lowest V bursts in the first BATSE catalog.

CONCLUSIONS

Analysis of the mathematical relation between the variability parameter V of LGS and burst duration, and comparison with BATSE data, lead us to the following conclusions.

1. Short bursts must have high V, by definition.
2. Long bursts typically have low V, unless highly unusual temporal structure is present.
3. A cut at intermediate values of V will separate the bursts into two classes, long and short, at a duration that depends critically on the time scales used in the definition of V, with little dependence on the distribution of burst durations. Changing the definition of V will not merely change the value of V assigned to a particular burst, but will change the membership in the classes. This is in contrast to the bimodality of the duration distribution reported by Kouveliotou et al.[2,3]
4. Since V distinguishes long from short bursts by definition, it will also show the correlation between spectral hardness and duration.[2].
5. V is not a appropriate measure of variability. The terms "smooth" and "variable", for low and high values of V, are misleading.

REFERENCES

1. D. Lamb, C. Graziani, & I. Smith, ApJ **413**, L11 (1993).
2. C. Kouveliotou et al., ApJ **413**, L101 (1993).
3. C. Kouveliotou et al.,"Bimodality of Burst Duration Distribution", these proceedings.

TIME ASYMMETRY IN GAMMA-RAY BURST LIGHT CURVES

Robert J. Nemiroff [a,b,c], Jay P. Norris [b],
Chryssa Kouveliotou [a,d], Gerald J. Fishman [d],
Charles A. Meegan [d], and William S. Paciesas [d,e]

ABSTRACT

A simple test for time-asymmetry is devised and carried out on the brightest gamma-ray bursts detected by the Burst and Transient Source Experiment (BATSE) on board the *Compton* Gamma Ray Observatory. We show evidence that individual bursts are time-asymmetric on all time scales tested, from a time scale shorter than that of pulses which compose GRBs to a time scale similar to a greater envelope that contains these pulses. We also find bursts which manifest significant asymmetry only on time scales comparable to the duration of burst, and bursts for which no clear asymmetry on any time scale is present. The sense of the asymmetry is that bursts and/or component structures rise in a shorter time than they decay. We also find that our whole sample of bursts taken together is time-asymmetric, in that there are significantly more bursts and pulses where the rise is more rapid than the decay, on all time scales tested and for all energy bands tested. When our whole GRB sample is binned at 64-ms and integrated over all BATSE energies, the statistical significance is at the 6 σ level. Models that predict time-symmetry are therefore excluded.

INTRODUCTION

There is no present consensus as to the cause or location of gamma-ray bursts (GRBs). GRBs were first discovered in data from the Vela satellites as reported by Klebesadel, Strong, and Olson[1], and since have been detected by numerous spacecraft. Recent results by the Burst and Transient Source Experiment (BATSE) on board the *Compton* Gamma Ray Observatory have confirmed their isotropic angular distribution and "confined" peak brightness distribution.[2] These results generate an interesting puzzle. Currently, there are more than 100 models for the origins of GRBs.[3] To help eliminate inapplicable models, discriminating statistical tests are needed.

Some GRB models depend on processes or geometries that are time-symmetric. A certain amount of relativistic motion is essential to many GRB models to avoid being dominated by collisions between gamma-rays. Were these models to rely on a relativistic beam crossing our line of sight in order to generate the time structure inherent in GRBs, this reliance would make the resulting light curves, when taken together, time-symmetric.[4] This is because there is an equal chance that the beam sweeps past us going in one direction as there is in the other direction.

[a] Universities Space Research Association
[b] CSI Institute, George Mason University, Fairfax, VA 22030
[c] NASA Goddard Space Flight Center, Greenbelt, MD 20771
[d] NASA Marshall Space Flight Center, Huntsville, AL 35812
[e] University of Alabama, Huntsville, AL 35899

Time-asymmetry in GRB light curves has been discussed previously by several authors.[4-9] Mitrofanov et al.[7] computed an asymmetry in the durations and hardness ratios between the rising parts of bright GRB light curves and the falling parts. Link, Epstein and Priedhorsky[4] used a *skewness function* which measured temporal asymmetry for a majority of 20 bright GRBs taken from the first 48 detected by BATSE. This skewness function is similar to that used by Weisskopf et al.[10] on Cygnus X-1 data.

Our asymmetry test differs from that in Link, Epstein and Priedhorsky[4] in that its statistical significance can be more easily computed. Also, our test involves a cumulative sum which is taken over a complete sample of GRBs, allowing us to see if an asymmetry accumulates over many GRBs. Our asymmetry test differs from Mitrofanov et al.[7] in that our test can be easily conducted over a wide range of time scales. This proceedings contribution is based on a paper now accepted to the Astrophysical Journal.

STATISTICS, DATA, AND RESULTS

The time-asymmetry parameter we use is the ratio of the number of times where the counts in the previous time bin are higher to the number where the counts are lower - a statistic we call Υ. The number counts that Υ is based on begin with the first recorded bin of the GRB time series that is 9 σ above a fit background level, and continue to a time where the GRB has subjectively been deemed to have concluded. At this time a background fit begins and typically runs for 50 seconds.

The Υ statistic is particularly easy to compute and test for significant deviation from symmetry, for which Υ is unity. A time-symmetric profile would have an equal number of rising and falling bins. The statistics of Υ is a direct extension of the statistics of flipping a coin, which is well understood in terms of binomial and normal distributions.

We use data with intrinsic 64-ms time resolution (BATSE PREB + DISC-SC data types[11]) to compute Υ for a wide range of time scales: from 64 ms to 4096 ms. 64 ms is the shortest accumulation trigger time scale employed by the BATSE instrument. Above the upper limit of 4096 ms, the statistical significance of the test begins to drop.

We have applied this test to a sample of the brightest GRBs detected by BATSE up to 1993 March 10. Specifically, we utilize bursts with peak count rates greater than 18000 counts sec^{-1}, and durations greater than 1 second. These brighter GRBs are easy to distinguish and give good statistics.

After computing Υ for the GRB data, it is clear that some GRBs are seen to have a statistically significant excess of falling bins over rising bins, while others do not. A GRB that is time-symmetric would have a ratio of unity, and a sample of time-symmetric GRBs would be evenly distributed about unity. None of the GRBs yielded Υ significantly greater than unity. It is straightforward to compute a cumulative Υ statistic for all the GRBs in the sample, to see if the asymmetry averages out over many GRBs, or accumulates to become more significant. To compute a cumulative Υ, we counted the total number of rising and falling bins for all the GRBs in the whole sample, and then computed the ratio. The resultant value for Υ on the 64-ms time scale and integrated over the complete BATSE energy range (above 25 keV) is 0.898 $^{+0.017}_{-0.016}$. Thus, for this group of GRBs, the time-asymmetry result is significant at the 6 σ level, which corresponds to better than one part in 10^{-8}.

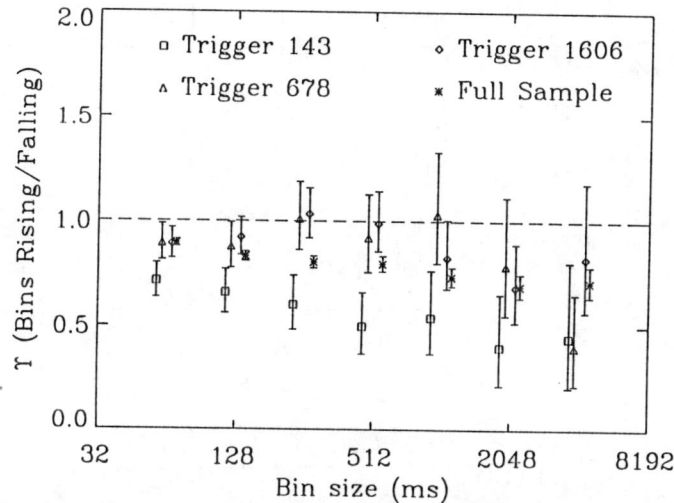

Figure 1 shows Υ for three individual GRBs and for the complete sample for the energy band 50 - 300 keV. The light curves for these GRBs can be found are shown in Figures 2a, 2b, and 2c. The size of the bins was varied from 64 ms to 4096 ms. The points plotted in Figure 1 were artificially staggered in time (slightly) so that the error bars would be clearly visible for each.

As seen in Figure 1, BATSE trigger 143 shows significantly more falling bins than rising bins on all time scales. The asymmetry of this GRB is also clearly evident in Figure 2a, from the asymmetric shape of the first envelope of gamma-ray pulses. A general trend is evident in this burst: pulse intensity decreases as the burst progresses. This behavior drives the asymmetry on the 1.024, 2.048 and 4.096 second time scales. On shorter time scales, the asymmetry is still evident because the individual pulses themselves are time-asymmetric, a result previously reported by Norris et al.[12] In contrast, Link, Epstein and Priedhorsky[4] reported asymmetry only on longer time scales.

Next we inspect trigger 1606. From inspection of Υ and this GRB's light curve in Fig. 2b, we find no significant evidence for time-asymmetry in this GRB, either on time scales characteristic of pulses or in the general shape of the burst profiles or envelopes.

Finally, we inspect trigger 678. This GRB is especially interesting since the general envelope of pulses is significantly asymmetric (evident from a visual inspection of Fig. 2c), while there is no strong evidence from the asymmetry parameter that the individual pulses themselves are asymmetric. This is shown in Fig. 1 by the lack of a significant deviation of Υ from unity on shorter time scales, but the onset of significant asymmetry at the longest time scale, 4096 ms. A separate 16-ms time scale was run on MER data for this burst, showing only a slight asymmetry with an Υ not significantly different from the 64-ms result shown.

The GRB population tested as a whole is significantly time-asymmetric on all time scales from 64 ms to 4096 ms, as Fig. 1 also shows. The error bars denote the one standard deviation errors for the cumulative number counts for rising and falling bins.

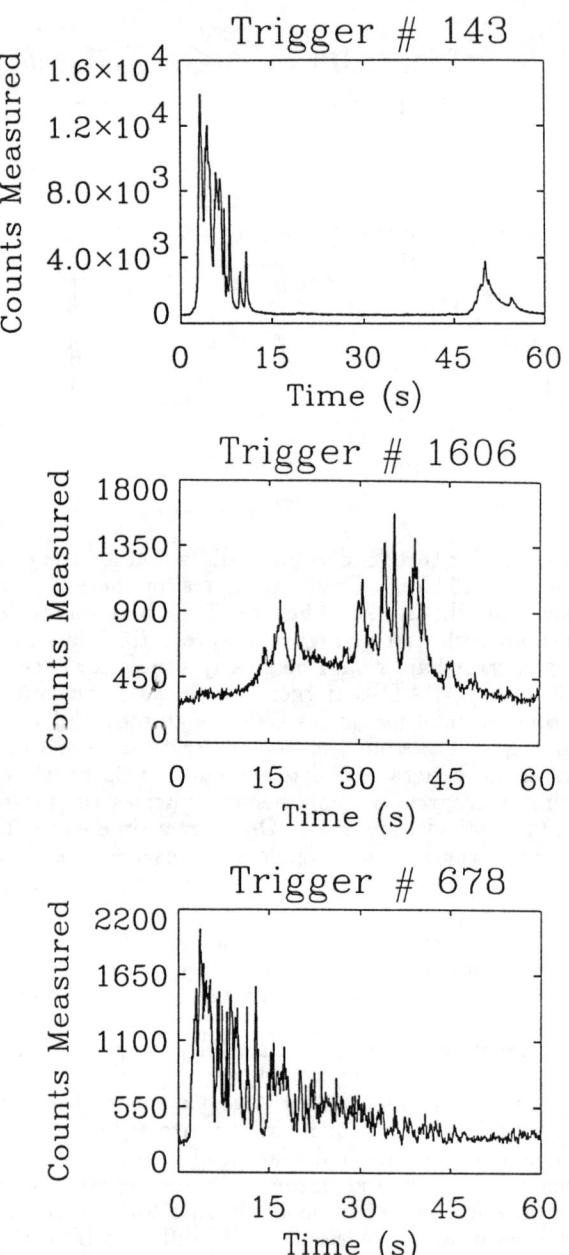

Figure 2: Light curves for three specific gamma-ray bursts.

REFERENCES

1. R. Klebesadel, I. B. Strong, R. A. Olson, Ap. J. **182**, L85 (1973).
2. C. A. Meegan, G. J. Fishman, R. B. Wilson, W. S. Paciesas, G. N. Pendleton, J. M. Horack, M. N. Brock, and C. Kouveliotou, Nature **355**, 143 (1992).
3. R. J. Nemiroff, in press: Comm. Astrophys. , (1993).
4. B. Link, R. I. Epstein, and W. C. Priedhorsky, Ap. J. **408**, L81 (1993).
5. C. Barat, R. I. Hayles, K. Hurley, M. Niel, G. Vedrenne, I. V. Estulin, and V. M. Zenchenko, Ap. J. **285**, 791 (1984).
6. J. P. Norris et al., Adv. Space. Res. **6**, 19 (1986).
7. I. G. Mitrofanov, et al., Gamma-Ray Bursts, AIP Conference Proceedings 265, eds: W. S. Paciesas and G. J. Fishman (New York, AIP, 1992), p. 163.
8. C. Kouveliotou, W. S. Paciesas, G. J. Fishman, C. A. Meegan, and R. B. Wilson, The Compton Observatory Science Workshop, eds: C. R. Shrader, N. Gehrels, and B. Dennis (NASA Conference Publication 3137, 1992), p. 61.
9. G. J. Fishman, Presentation at the Compton Observatory Symp. (Washington University, Oct. 15, 1993).
10. M. C. Weisskopf, P. G. Sutherland, J. I. Katz, and C. R. Canizares, Ap. J. **223**, L17 (1978).
11. G. J. Fishman et al., Gamma Ray Observatory Science Workshop, ed: W. Neil Johnson (10-12 April 1989, Greenbelt, MD, 1989), p. 2-39.
12. J. P. Norris, S. P. Davis, C. Kouveliotou, G. J. Fishman, C. A. Meegan, R. B. Wilson, and W. S. Paciesas, Compton Gamma Ray Observatory, AIP Conference Proceedings 280, eds: M. Friedlander, N. Gehrels, and D. J. Macomb (New York, AIP, 959), p. 1993.

ON THE USEFULNESS OF "VARIABILITY" IN GAMMA-RAY BURSTS

Robert Rutledge and Walter H. G. Lewin

37-624B, Massachusetts Institute of Technology, Cambridge, MA 02139

ABSTRACT

We investigate the usefulness of $V = C_{max}^{64}/C_{max}^{1024}$ which has been introduced by Lamb, Graziani and Smith[6] (hereafter L93) as a criterion to separate gamma-ray bursts into two classes, called "variable" and "smooth". Bursts with durations ≤ 0.4 sec must have values of V which make them "variable" according to the criterion used by L93. Therefore, the majority ($\sim$ 22 of 34) of the bursts designated as "variable" do not deserve to be so called any more than their smooth bursts do; their selection according to V is based on their durations. It is therefore no surprise that there exists a correlation between burst duration and "variability"; neither is it a surprise that there exists a correlation between burst hardness and "variability", since this reflects the known correlation between spectral hardness and burst duration.[1,3]

SUMMARY

The separation of Gamma Ray Bursts (GRBs) into distinctive classes using physically meaningful burst characteristics (for instance, duration, flux, and spectrum) is a useful analytical tool which may aid the development of models. Dezalay et al. [1] and Kouveliotou et al. [3] demonstrate a bimodal distribution in the duration of GRBs and a tendency of short bursts ($\leq$ 2 sec) to have harder spectra.

Lamb, Graziani and Smith[6] (hereafter L93) attempt to classify bursts using a parameter to indicate "variability" in bursts, defined, using data from the publicly available BATSE database[2]:

$$V = \frac{C_{max}^{64}}{C_{max}^{1024}}$$

where V is the "variability", and C_{max}^{t} is the maximum number of counts in t ms measured by the second most brightly illuminated BATSE detector.

It is important to note that:

- A burst of constant count-rate C with a duration longer than 1024 ms will have a V= C × 64ms/(C × 1024ms) = 0.063 (ignoring Poisson effects in the burst as well as in the background).
- A burst of constant count-rate C with a duration t ms which is shorter than 1024ms and longer than 64ms will have V=64 ms/t ms; a burst of constant count-rate with a duration shorter than 64ms will have V= 1 (again, ignoring Poisson effects).
- Of bursts of all possible different profiles, all lasting t ms, those with constant count-rate have the lowest possible value of V (ignoring Poisson effects).

© 1994 American Institute of Physics

L93 arbitrarily chooses V=0.16 as a dividing line (after correcting for Poisson effects), calling bursts with higher values "variable" and bursts of lower values "smooth".

All bursts of duration t< 400ms, regardless of their profiles, must have V> 0.16, and are therefore considered "variable" by the criterion of L93. It is not possible to distinguish between the "variable" property with the duration property in bursts of duration t< 400ms. Due to Poisson noise, bursts of slightly longer duration, but less than 1024 ms, are more probable to be "variable" (according to L93) than are bursts longer than 1024 ms.

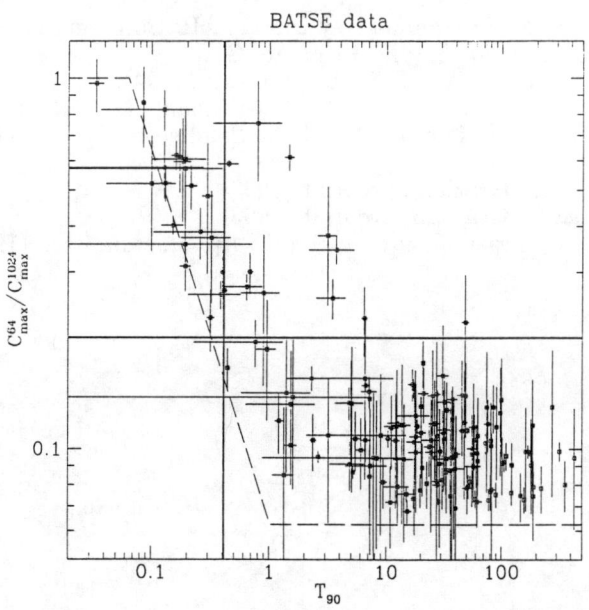

Figure 1. $C_{max}^{64}/C_{max}^{1024}(=V)$ vs T_{90} of GRBs from the first BATSE catalog (as available in the public BATSE database).

In Fig. 1, we show $C_{max}^{64}/C_{max}^{1024}(=V)$ vs T_{90}, from data from the publicly available BATSE data base. The horizontal line corresponds to the "variability" criterion used by L93; that is, all bursts above this line on this diagram are considered "variable" by L93. The vertical line is at 400 ms. The broken line is the minimum "variability" (described above) as a function of burst duration. The error bars on $C_{max}^{64}/C_{max}^{1024}$ assume the background is known to infinite precision (although subject to poisson uncertainties), and is not variable on the timescale of the burst (in practice, it is known to $\approx 0.6\%$).

There are ~ 6 bursts which are $>1\sigma$ within the upper-right hand quadrant defined by the L93 "variability" criterion and the 400ms line.

Since 2/3 of the "variable" bursts found by L93 (22 of 34) have durations t< 400ms, and 6 of the remaining 12 bursts have durations less than 1 second, we believe that this sample is highly contaminated with bursts which are merely short.

We conclude that that the selection made by L93 is largely, though perhaps not exclusively, a seletion based on burst duration. It is therefore no surprise that Lamb & Graziani[4] find a correlation between "variability" and duration. It is also no surprise that Lamb & Graziani[5] find a correlation between "variability" and hardness, since this reflects the known correlation between burst duration and spectral hardness.[1,3] Our conclusions are not affected by changing the name of "variable" bursts to "type II" and "smooth" bursts to "type I".

REFERENCES

1. J-P. Dezalay et al., in Gamma Ray Bursts, AIP Conference Proceedings 265, 1991, eds. Paciesas W. S., and Fishman, G. J.
2. G. J. Fishman et al., Ap. J. Supp., in press, (1994).
3. C. Kouveliotou, C. A. Meegan, G. J. Fishman, N. P. Bhat, M. S. Briggs, T. M. Koshut, W. S. Paciesas, & G. N. Pendleton, Ap. J. Lett. **413**, L101 (1993)
4. D. Q. Lamb & C. Graziani, preprint (1993a).
5. D. Q. Lamb & C. Graziani, preprint (1993b).
6. D. Q. Lamb, C. Graziani, & I. A. Smith, Ap. J., submitted (1993)

SPECTRAL STUDIES

THE SEARCH FOR GAMMA-RAY BURST SPECTRAL FEATURES IN THE COMPTON GRO BATSE DATA

D. M. Palmer, B. J. Teegarden, B. E. Schaefer, T. L. Cline
NASA/Goddard Space Flight Center, Code 661, Greenbelt, MD 20771

D. L. Band, L. A. Ford, J. L. Matteson
CASS, UC San Diego, La Jolla, CA 92039

R. D. Preece
NAS/NRC, MSFC, ES-62, Huntsville, AL 35812

W. S. Paciesas, G. N. Pendleton, M. S. Briggs
University of Alabama at Huntsville, Huntsville, AL 35899

G. J. Fishman, C. A. Meegan, R. B. Wilson
NASA/Marshall Space Flight Center, ES-62, Huntsville, AL 35812

J. P. Lestrade
Mississippi State University, MS, 39762

ABSTRACT

The Spectroscopy Detectors (SDs) of the Burst And Transient Source Experiment (BATSE) on the Compton Gamma Ray Observatory (GRO) have seen no convincing line features in Gamma-Ray Burst (GRB) data, in contrast to expectations based on results from other instruments. In this paper we discuss the search for lines in the SD data. The search has examined 148 bursts, of which ~17 were intense enough that lines like those seen by the Ginga satellite would have been visible between ~20 and ~100 keV. Crude calculations show that the Ginga and BATSE results are consistent at the ~3-13% level, depending on the definition of consistency.

INTRODUCTION

One of the primary scientific motivations for the BATSE Spectroscopy Detectors (SDs) was to search for absorption lines in Gamma-Ray Burst (GRB) spectra. These lines have been seen by instruments such as Konus,[1,2] HEAO,[3,4] and, most convincingly, by Ginga.[5,6,7,8] The Ginga observations include pairs of absorption lines spaced a factor of two apart in energy, consistent with the first and second harmonics of cyclotron absorption by electrons in a teragauss magnetic field. Because the only place where such fields are known to exist is in neutron star magnetospheres, these features are taken as evidence that GRB events involve neutron stars. However, the BATSE Spectroscopy Detectors (SDs) have not seen any convincing evidence for line features.

The SDs provide a medium-resolution spectral capability to complement the low-resolution BATSE Large Area Detectors (LADs). Each of the 8 SDs is a NaI(Tl) scintillator crystal, 12.7 cm in diameter by 7.62 cm thick, optically coupled to a photomultiplier tube of the same diameter, and housed in a 2 mm thick aluminum can with an 8.3 cm diameter beryllium window.[9] The window provides low energy response (down to ~15 keV) with a roughly cosine-law dependence on the angle of the burst from the detector axis (axis angle), while transmission through the aluminum housing combined with the thickness of the scintillator gives sensitivity at all axis angles up to 90° at higher energies. Each SD is associated with one of the 8 BATSE LADs, with an 18.5° offset between their axes for mechanical reasons.

This paper describes our search for these lines. We will discuss what would constitute convincing evidence for spectral features, taking into account the large number of spectra which have been searched and the characteristics of the SD detectors. We will also describe our current search technique and why we believe that we would have detected strong lines if they were in our brightest bursts. We have found candidate line features in some bursts, but these have failed various consistency tests. These tests will be discussed, with more details given elsewhere[10,11] in these proceedings. A brief calculation of the consistency between the Ginga and BATSE results is included in this paper, but a more detailed analysis is given by Band et al.,[12] also in this volume. An automated line search is being developed,[13] and this will allow us to validate our search, quantify our results, and place upper limits on the fraction of bursts containing lines of a given strength.

LINE DETECTION CRITERIA

The criteria by which we will test the reality of any discovered candidate feature should be specified in advance. Because of the large number of spectra examined in this search, the criteria must be strict enough to eliminate false positives due to statistical fluctuations. To determine the significance of line features we use an F-test, which compares the χ^2's of model fits of spectra with and without a line.[14] Our current criterion is that, for some time range and combination of detectors, the F-test probability of the resulting χ^2 reduction must be $P_F \leq 10^{-4}$. This significance threshold may result in a few false positive detections of line features. We have examined $> 10^3$ independent spectra, but not all of these spectra were strong enough to support a line feature—or a statistical fluctuation which mimics a line feature—with high significance. A better estimate for the number of strong spectra would be in the neighborhood of 100-250. The energy range over which we expect lines to be visible is roughly 10-20 times as wide as the detector's energy resolution, and so our search has looked at roughly 1000-5000 independent samples (the automatic search described below will quantify this further). Therefore, we expect a 10-50% chance of a false positive detection of a line above our F-Test threshold.

The detection of a line feature must be consistent in all detectors which would be expected to see it. This provides additional protection against false positives in most bursts. BATSE's eight LADs and eight SD's are arranged so that any point in the sky is observed by at least three detectors of each type, although not all of these will be useful for examining a line feature in any specific burst. For each burst, data is available from four detectors of each type, selected by the count rates in the LADs at trigger time,[15] but the triggering system does not necessarily select the optimal detectors for viewing a burst because it cannot compensate for excess flux produced by scattering from the Earth or the spacecraft. The LADs are typically less sensitive to narrow line features than the SDs.[11] The SDs have adjustable gains which are set at different levels to achieve various scientific and technical objectives, and so an SD may not cover the energy of the candidate feature. However, if an SD detects a feature at $P_F \leq 10^{-4}$, any other SD with the appropriate energy range and orientation should also show the feature, although not necessarily with as much significance.

The final criterion for a line detection is that it 'looks physical'. The feature in the count spectrum should be well-formed, at least as wide as the energy resolution of the detector, and consistent with or above the background level at the center. The observation should be clean, and there should be no unexplained bumps or wiggles in the background used for subtraction. Detectors which are pointed more directly at the burst will tend to show the line with greater significance, although counting statistics may produce a higher significance in a less-well situated detector. Systematic effects such as gain shifts, detector nonlinearities, and nuclear activation could cause spurious spectral features, and these must guarded against.

No candidate features have met all of these criteria.

OUR LINE SEARCH TECHNIQUE

Currently we use a visual search technique, examining spectra by eye to look for features. Our search has been applied to 148 bright bursts as of October, 1993. Only the brightest 15-20% of the ~800 bursts detected by BATSE LADs are strong enough that a search for spectral features using the SDs is useful, and in most of those, lines would be significant only if their equivalent widths were much greater than those seen by Ginga.

Each bright burst is divided into successive non-overlapping intervals, usually based on the accumulation of a fixed number of counts. For most bursts, the interval used is the on-board accumulation time for the third-brightest detector (using the 'SHERB' data type[15]). This typically results in an accumulation time of 0.5-2 seconds during the brightest part of the burst. The full duration of the burst and of each emission episode (for bursts with separated emission episodes) are also used as time intervals.

Count spectra are obtained over these intervals, and then a background spectrum is subtracted. The background for each energy bin is based on a fit to a Taylor expansion (up to third order) of the data in that bin for a period up to a thousand seconds before and after the burst time. The background interval chosen depends on the behavior of the background, the availability and quality of the data, and the strength and duration of the burst.

The spectra are typically oversampled in energy, so they are rebinned with a resolution of roughly half the FWHM detector resolution. This produces data points with enough energy resolution to detect narrow lines while maintaining adequate statistical significance for each point.

The rebeinned count spectra are then examined by eye for features. We have found that statistical fluctuations with $P_F \approx 10^{-2}$ are easily visible. Also apparent are features which exist in multiple detectors (at the same energy) and those which persist over consecutive intervals. These multiple and consecutive detections are combined to maximize the significance of the feature in further analysis. A few of the strongest bursts have also been searched comprehensively, examining all possible time intervals rather than the pre-selected, non-overlapping intervals of the standard search. An automatic search, described below, will extend this comprehensive search to all of the bright bursts.

When a candidate feature is found, the spectrum is fitted, at full telemetry resolution, to a continuum model with and without an absorption line to minimize χ^2_{c+l} and χ^2_c, respectively. The absorption line model used (*e.g.*, a Gaussian distribution of optical depth) prevents the model flux from becoming negative. The energy range of the fit is wide enough that the continuum level is well-determined on both sides of the line. The difference in χ^2 values for the fits determines the feature's significance, using the probability distribution $P_F(F > F_{\text{obs}}, N_{\text{line}}, \nu_{c+l})$ for

$$F_{\text{obs}} = \frac{(\chi^2_c - \chi^2_{c+l})/N_{\text{line}}}{\chi^2_{c+l}/\nu_{c+l}}$$

where P_F is the F-test probability distribution.[14] N_{line} is the number of parameters in the line model (*e.g.*, $N_{\text{line}} = 2$ if the model has an energy and an equivalent width, 3 if the model includes a true width as well, or a correspondingly larger number if multiple lines were fitted). ν_{c+l} is the number of degrees of freedom in the fit with both line and continuum. For uniformity, all P_F's in this paper are calculated in this way.

The time interval and the selection of detectors used in the fit are adjusted to maximize the significance of the line. For a physical line feature, this adjustment should improve the significance of the line by increasing the exposure time and the effective area of the detector. For a spurious statistical fluctuation, the increase in significance will usually be small.

VERIFICATION OF THE SD LINE SENSITIVITY

This visual search has not revealed any convincing line features in the 148 BATSE bursts that have been searched, in contrast to measurements by other instruments, discussed below. This raises the question of whether the BATSE SDs are sensitive to line features.

Simulations have propagated the line spectra measured by Ginga through the detector response function of the BATSE SDs.[16] These simulations predict that BATSE would see lines if they had the strength and energy found by Ginga. Lines near 40 keV were easily visible, although BATSE was less sensitive, and more dependent on the GRB's axis angle, to 20 keV lines, due to the combination of the SD's aluminum housing and beryllium window.

Simulations can only model an instrument as it is understood—they do not test the actual function of the instrument. Only actual measurements can verify that the instrument is working properly and has the calculated line sensitivity. Measurements of the background spectrum by BATSE show a smooth continuum with well-formed emission lines (from cosmic ray scattering and activation, etc.) found at their known energies,

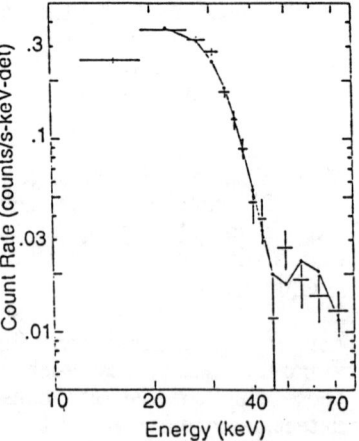

Figure 1. Preliminary results of the count spectrum of Her X-1, measured by the BATSE SDs. (From Briggs et al.[17])

which gives us confidence that the detectors are working properly, but the detection of a cyclotron absorption line would be a clearer proof of our understanding of the system.

Hercules X-1 is an accreting pulsar with a cyclotron line. Figure 1 shows a preliminary count spectrum obtained by using the BATSE SDs in 'pulsar' mode and subtracting the off-pulse data from the on-pulse data. The line is detected with $P_F = 1.3 \times 10^{-3}$. (Our $P_F \leq 10^{-4}$ threshold is not relevant in the case where we are looking at a single spectrum of an object already known to have a line feature.) A complete analysis of this and further observations of Her X-1 using the SDs will be covered in Briggs et al.[17]

Since both simulations and observations show that the BATSE SDs could detect absorption lines, we are confident that our non-detection of lines is due to the absence of strong lines in the bright bursts we have observed.

A STRONG BUT PROBABLY SPURIOUS CANDIDATE

Figure 2 shows the count spectra for our strongest candidate line so far. The upper plot shows a 3.8 second accumulation from burst 930506 (trigger number 2329) using spectroscopy detector SD2. The line feature has $P_F = 6 \times 10^{-5}$ and is well-shaped in the count spectrum. Its centroid energy is 56 keV and its equivalent width is 4 keV.

However, this GRB is at an axis angle of 74° from SD2. Another detector, SD7, has a 56° axis angle, and it also covers the appropriate energy range. (SD3 is even closer, but its low energy threshold of 200 keV excludes the line energy.) SD7 also sees less scattering from the Earth than SD2 (with axis angles to the center of the Earth of 132° and 84°, respectively). Therefore, SD7 should show the line with greater significance than SD2.

The lower plot of figure 2 shows the count spectrum for SD7 over the same time interval and energy range as the upper plot. This spectrum shows no line feature near 56 keV, and a model fit verifies this ($P_F = 0.73$). These two spectra are inconsistent at the 99% level, and so this feature does not meet our detection criteria. The LADs also show no line near 56 keV, although the significance of this discrepancy is probably limited by systematics which cannot be quantified

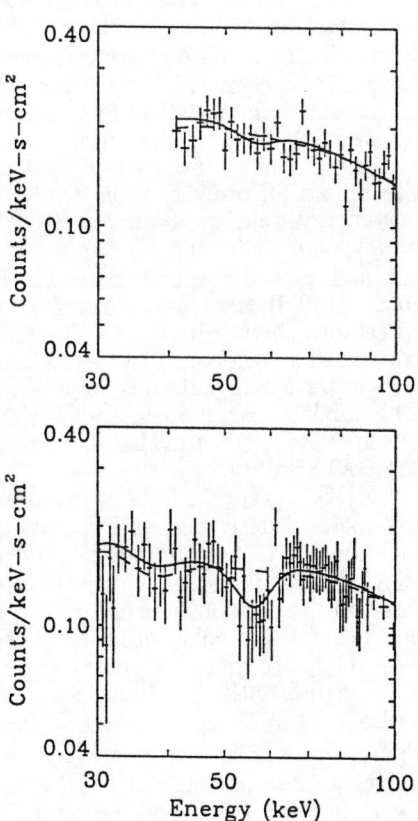

Figure 2. Count spectra for GRB 930506 for SD2 (top) and SD7 (bottom). Dashed line: continuum fit; Solid line: absorption feature fit. (From Ford et al.[10])

at this time. More details are given elsewhere in this volume for the SD[10] and LAD[11] analyses.

If this discrepancy were due to instrumental problems, then either the line is there, but SD7 and the LADs don't see it, or the line is not there, and SD2 has the problem. There is no indication in the background spectra nor during other periods of this burst that any of the detectors are malfunctioning. Specifically, the background line shapes and continuum look good, and the GRB continuum during other time periods of the burst shows no evidence of lines.

Therefore, this feature is probably a statistical fluke. With the large number of independent samples measured, a feature with $P_F = 6 \times 10^{-5}$ has a 6-30% *a priori* probability of occurring by chance. The rejection of this line shows the value of using multiple detectors to eliminate false positives.

CONSISTENCY WITH OBSERVATIONS BY OTHER INSTRUMENTS

The non-detection of absorption lines by BATSE must be reconciled with the detection of lines by other instruments. Lines have been reported for the Konus instruments on various spacecraft,[1,2] the HEAO A-4 instrument,[3,4] and Ginga.[5,6,7,8] Some of these reported features have low significances which are consistent with statistical fluctuations, and others would not have been detected by BATSE, which makes inter-instrument comparisons difficult.

For Konus, low energy "absorption features" are reported for 15-20% of the bursts. However, some (but not all) of these features are merely broad low-energy deficits of the spectrum relative to the spectral forms used in the fit (typically $E^{-\gamma} \exp(E/E_0)$), and would probably be compatible with other commonly-used spectral forms which were not included in the Konus analysis.[2] In the BATSE analysis we use more general continuum forms, which are unlikely to require the addition of a line feature at our P_F threshold unless there is a count deficit compared to the continuum both above and below the line energy, and so we would not consider these spectra to be evidence of absorption lines. The number, parameters, and statistical significances of the Konus features that are line-like are not available to us, which prevents a detailed comparison with the BATSE data.

HEAO reports lines in two bursts out of the 21 it observed.[3,4] For 780325, the method described earlier gives $P_F = 6 \times 10^{-4}$ for the decrease in χ^2 for including a 49 keV line over a 12 second interval. The other burst, 780608, shows a 3σ equivalent width ($P = 3 \times 10^{-3}$) line at 67 keV over a 20 second interval. Neither of these exceed our P_F threshold, but a relaxed standard (based on the smaller number of HEAO bursts) might deem the 780325 line significant.

Ginga presents the most convincing evidence for cyclotron absorption lines. Most significantly, in Ginga Burst GB880205, a pair of lines was found at 1× and 2× 19 keV over a 5 second interval. For the reported change in χ^2, the significance of the detection is $P_F = 2.5 \times 10^{-5}$. GB870303 shows a 21 keV line during a 4 second interval '(S1)' with $P_F = 1.5 \times 10^{-7}$. A pair of lines at 1× and 2× 21 keV for a 9 second period '(S2)' later in the same burst is significant only in the context of the S1 line, with $P_F = 1.1 \times 10^{-3}$. A third reported pair of line features, at 1× and 2× 24 keV over a 2 second time interval in GB890929, has $P_F = 2.7 \times 10^{-3}$, which we do not consider significant.

Thus, we do not have enough information to determine our consistency with Konus, and the HEAO data has no lines at our required level of significance.

However, the Ginga line detections in GB880205 and GB870303(S1) can be compared with the BATSE data. Band et al.[12] gives a more detailed consistency calculation for the Ginga and BATSE data elsewhere in these proceedings, but here is a crude estimate.

A simulation[16] of the BATSE response to the spectrum found by Ginga in GB880205 shows that a detector with a 30° axis angle would have seen the 38 keV line with a typical significance of $P_F = 2 \times 10^{-7}$. The detection sensitivity depends on the intensity of the burst and the persistence of the line. If all bursts had a 38 keV line with a 7 keV equivalent width during the brightest 5 seconds of the burst, BATSE would have seen the feature with $P_F < 10^{-4}$ in 17 bursts out of those observed through October 1993. If the line persisted through the brightest 10 seconds, the number of bursts goes to 24, and if the line is present throughout the burst, the number is 39. A similar calculation by Fenimore et al.[18] shows that Ginga would have seen such a line with 5 second persistence in 5.4 of the GRBs that it observed.

When the BATSE response to the GB870303(S1) spectrum was simulated, we found that a detector with an axis angle of 30° would see the 21 keV line with $P_F < 10^{-4}$ less than half of the time, and that the detection probability drops significantly as the axis angle increases. This is partially due to the angle-dependence and decreased sensitivity of the SDs at low energies, and partially due to the fact that the continuum flux during GB870303(S1) was much lower than during the GB880205 line period. In contrast, Ginga saw the GB870303(S1) line with higher significance than it saw the GB880205 lines, due in part to Ginga's low-energy proportional counter which provides sensitivity down to 2 keV. We will assume for this calculation that BATSE could have seen GB870303(S1)-type lines half as often as GB880205-type lines, but Ginga could have seen GB870303(S1)-type lines twice as often as GB880205-type lines.

Thus, BATSE and Ginga respectively saw GB880205-type lines in 0/17 and 1/5.4 GRBs, and they saw GB870303(S1)-type lines in 0/8.5 and 1/10.8 GRBs. The question arises: Are these results consistent? As is often the case, the answer depends on how the question is asked. One possible way to ask the question is: If one line of each type is observed, what is probability that both lines are observed in the GINGA data? The probability that the GB880205-type line is in one of the 5.4 Ginga bursts, rather than among the 17 BATSE bursts, is $5.4/(5.4+17) = 0.24$, and likewise $10.8/(8.5+10.8) = 0.56$ for GB870303(S1), so the joint probability is $P_{\text{consistency}} = 0.24 \times 0.56 = 13.5\%$. However, this calculation depends on the fact that exactly two bursts were seen to have lines, which may in itself be unlikely. Another way to ask the consistency question is: Given some distribution of line parameters, $p(L)$, what is the probability of obtaining at least one Ginga line detection for each of the two types, but no BATSE detections? This question is dependent on $p(L)$, but one possible choice is to adjust $p(L)$ to maximize consistency with the data, which gives a consistency probability of $P = 3.0\%$.

A Bayesian analysis of the consistency calculation is given in this volume[12] with more details given in Band et al.[19] This analysis shows that the BATSE and Ginga results are consistent. Specifically, the data are consistent with the null hypothesis: 1) BATSE and Ginga are viewing burst populations with the same line probabilities 2) BATSE could detect lines of the type seen by Ginga and 3) the Ginga lines are not spurious.

THE AUTOMATIC LINE SEARCH

We are currently developing an automatic system to search GRB spectra for lines. The automatic search could find persistent lines which last much longer than the integration time used in the visual search, but not long enough to be seen when the spectrum over the entire burst is examined. Details are given elsewhere in this volume.[13]

The automatic search is comprehensive (searching overlapping time intervals of all lengths), objective, consistent, and quantitative. This allows us to compare the distribution of the features found with that expected for statistical fluctuations. A deviation from the expected distribution could indicate systematic effects in our detectors or analyses. An excess of features of intermediate significance could also indicate that some fraction of bursts have lines which, although not individually exceeding our detection criteria, can be detected statistically.

The automatic search will calculate best-fit line intensities as a function of energy by convolving model photon spectra through the detector response functions to get model count spectra and comparing them to the data. Therefore, it will effectively search photon rather than count space. This will give it greater sensitivity to lines near well-understood detector nonlinearities. It will also make it susceptible to false positives due to poorly-quantified effects, but each candidate detection will be carefully and individually examined, so false positives are tolerable if they can be eliminated through further analysis. Although a feature which appears only in the photon spectrum would not be considered believable, any candidate feature which exceeds our detection criteria would be visible in the count spectrum.

The automatic search can only find what it is programmed to detect. For example, if a line varies in energy over the course of a burst it would not be detected automatically, although it would be detected by the visual search. Spectrum peculiarities, which could indicate either problems with the detector or physically interesting phenomena, might also be missed by the automatic search.

Therefore, we will continue to visually search each burst even after the automatic search program is operational.

CONCLUSIONS

BATSE has the ability to see lines with the characteristics reported by Ginga.

No convincing evidence of GRB lines have been seen in the BATSE data, indicating that none of the bright GRBs seen by BATSE have lines like those found by Ginga. We have seen what are probably statistical fluctuations that mimic lines, but not significantly more often than would be expected by chance. There is no statistically significant discrepancy between Ginga and BATSE.

We are continuing to receive GRB data, and we will continue to search the bursts visually and, soon, automatically. These future bursts may contain lines.

REFERENCES

1. E. P. Mazets *et al.*, Nature **290**, 378-382 (1981).

2. E. P. Mazets et al., AIP Conference Proceedings #101 (Positron-electron Pairs in Astrophysics, Goddard Space Flight Center, 1983) (AIP Press, N.Y., 1983), p. 36-53.
3. G. J. Hueter, AIP Conference Proceedings #115 (High Energy Transients in Astrophysics, Santa Cruz, CA, 1983) (AIP Press, N.Y., 1984), p. 373-377.
4. G. J. Hueter, Ph.D. Thesis (UCSD, 1987).
5. T. Murikami et al., Nature **290**, 378-382 (1990).
6. A. Yoshida et al., PASJ **43**, L69 (1991).
7. J. C. L. Wang et al., PRL **63**(15), 1550 (1989).
8. C. Graziano et al., Gamma-Ray Bursts, ed. Ho, Epstein & Fenimore (Cambridge University Press, 1992), p. 407.
9. J. M. Horack, Development of the Burst and Transient Source Experiment (BATSE)—NASA Reference Publication 1268 (NASA, 1991).
10. L. Ford et al., These proceedings (AIP Press, N.Y., 1993).
11. R. Preece et al., These proceedings (AIP Press, N.Y., 1993).
12. D. Band et al., These proceedings (AIP Press, N.Y., 1993).
13. B. Schaefer et al., These proceedings (AIP Press, N.Y., 1993).
14. B. R. Martin, Statistics For Physicists (Academic Press, London, 1971).
15. G. Fishman et al., Proceedings of the Gamma Ray Observatory Science Workshop, April 10-12, 1989 (Greenbelt MD, 1989), p. 2-39.
16. D. Palmer et al., AIP Conference Proceedings #280 (Compton Gamma Ray Observatory, St. Louis, MO 1992) (AIP Press, N.Y., 1992), p. 892.
17. M. Briggs et al., In preparation , (1994).
18. E Fenimore et al., AIP Conference Proceedings #280 (Compton Gamma Ray Observatory, St. Louis, MO 1992) (AIP Press, N.Y., 1992), p. 917.
19. D. L. Band et al., In preparation , (1994).

CYCLOTRON LINE SEARCH: BATSE-GINGA CONSISTENCY

D. Band, L. Ford, J. Matteson
CASS, UC San Diego, La Jolla, CA 92093

D. Palmer, B. Teegarden, B. Schaefer
NASA/GSFC, Code 661, Greenbelt, MD 20770

M. Briggs, W. Paciesas, G. Pendleton
University of Alabama at Huntsville, Huntsville, AL 35899

R. Preece
NRC-NASA/MSFC, ES-62, Huntsville, AL 35812

ABSTRACT

We have adopted a new Bayesian comparison of hypotheses regarding the consistency between the Ginga detections and BATSE nondetections of absorption features. The apparent discrepancy between the two instruments is insufficient to lead us to conclude there are serious deficiencies in either instrument. We therefore conclude there is no inconsistency.

INTRODUCTION

No absorption lines have been discovered in the spectra of ~150 gamma ray bursts observed by BATSE's Spectroscopy Detectors[1,2,3] while two sets of absorption lines were detected in the 23 bursts observed by Ginga.[4,5] Earlier missions also claimed to detect absorption lines,[6,7] although these reports did not provide full details as to the significance of the lines and the ensemble of searched bursts. These absorption features between ~15 and ~75 keV are thought to be cyclotron lines created in teragauss magnetic fields; since neutron stars are the only astrophysical objects known to possess such fields, the presence of strong magnetic fields is a powerful constraint on models of the source of gamma ray bursts.[8,9] Here we will consider whether the absence of BATSE detections is seriously discrepant with the Ginga detections, requiring a search for a deficiency in detector performance, or is merely a statistical fluke.

Four sets of lines in the Ginga bursts have been reported, but only two sets exceed our detection threshold (using the F-test, a probability less than 10^{-4} that the feature is a fluctuation of a featureless continuum). The two sets of lines in the Ginga data are the harmonically spaced lines at 19.3 and 38.6 keV in GB880205,[4] and the single line at 21.1 keV in the S1 segment of GB870303.[5] Although the line in GB870303 is formally very significant, the low signal-to-background of the continuum and the small final χ^2 make this feature suspect. Because the two line sets are very different, we treat them separately; we cannot assume they have the same frequency of occurrence.

In previous studies of the consistency between the Ginga and BATSE line observations[10] we used as the consistency statistic the probability of two or more detections in any of the Ginga bursts and none in the BATSE bursts, that is, the probability of results as discrepant. When we maximized this consistency

© 1994 American Institute of Physics

statistic with respect to the unknown line frequencies (i.e., the probability that a line is actually present in any given burst, independent of whether it is detectable), we found values of order ~5%. However, we have since realized that this consistency statistic measures the likelihood of the current pattern of detections and nondetections, but not whether there is an inconsistency between the two instruments. The current results may be equally unlikely under various hypotheses concerning instrumental deficiencies.

We have therefore reformulated our analysis within a Bayesian framework. The astronomical use of Bayesian statistics is controversial as a consequence of the community's unfamiliarity with Bayesian concepts. While there are fundamental conceptual differences between Bayesian and "frequentist" (i.e., standard) statistics, in our case the Bayesian formulae make intuitive sense. The Bayesian approach has a number of virtues for our analysis. First, the Bayesian methodology permits the comparison of two hypotheses in light of both quantitative prior information (and more qualitative expectations) and new data. This is possible since the Bayesian framework expands the definition of "probability" to a measure of our certainty in the truth of a proposition.[11] Thus the observations may be unlikely for both hypotheses, yet may still favor one over the other. Second, the observed data are used without assumptions about the population of results from which they may have been drawn. Third, an uninteresting variable is eliminated by integrating over all possible values ("marginalizing" a "nuisance" variable), weighted by the variable's probable values. In our case the line frequencies are uninteresting in answering the consistency question.

METHODOLOGY

The Bayesian approach provides a prescription for updating our assessment of the validity of hypothesis H after obtaining new data (represented by the proposition D). We explicitly represent the dependence on additional information (e.g., our understanding of the Ginga and BATSE detectors, the laws of physics, etc.) by the proposition I. The *a posteriori* probability of the truth of H, $p(H \mid DI)$ (the conditional probability of H given D and I), is based on our assessment of H's truth prior to the new data (the "prior"), $p(H \mid I)$, and the probability of obtaining the data given H, $p(D \mid HI)$ (also called the likelihood of H). In general, probabilities expressing expectations developed independent of the new data are considered priors. Using Bayes' Theorem (a basic relationship between probabilities) the odds ratio compares hypotheses H_0 and H_x:

$$O_H = \frac{p(H_0 \mid DI)}{p(H_x \mid DI)} = \frac{p(H_0 \mid I)}{p(H_x \mid I)} \frac{p(D \mid H_0 I)}{p(D \mid H_x I)} . \qquad (1)$$

The second term, the Bayes factor $B = p(D \mid H_0 I)/p(D \mid H_x I)$, is the ratio of the probabilities of obtaining the observed data under the two hypotheses. It does not matter whether these probabilities are large or small, just which hypothesis makes the observations more likely. In our case the Bayes factor will tend to indicate that there is an inconsistency between the Ginga and BATSE observations. The first term, $p(H_0 \mid I)/p(H_x \mid I)$, the ratio of the hypothesis priors, expresses our expectation as to the relative validity of the two hypotheses, and as such is somewhat subjective. Since we believe that both instrument teams understand their instruments, this term will favor consistency between Ginga and BATSE, and therefore establishes a threshold which the Bayes factor must exceed before we conclude there is an inconsistency. Thus the odds ratio incorporates information obtained before the new observations.

The basic probability $p(D \mid fHI)$ is the product of the burst-by-burst probability of the line detections or nondetections represented by the proposition D. The probability of detecting a line is $p_l = \alpha f + \beta(1 - f)$ where: f is the line frequency; α is the probability a line is detectable if present; and β is the probability of a spurious detection. The detector-dependent probabilities α and β differ from burst to burst, while the line frequency f is instrument-independent. These quantities must be calculated for each line type considered.

Since the unknown line frequency f is not of interest in determining whether Ginga and BATSE are discrepant, we marginalize f,

$$p(D \mid HI) = \int_0^1 df\, p(f \mid HI) p(D \mid fHI) \quad , \qquad (2)$$

using the prior for f, $p(f \mid HI)$, the distribution of possible values of the line frequency based on information obtained before the new data. Since we are comparing the Ginga and BATSE results, and Ginga data are represented in D, we should not use the Ginga results to calculate the prior for f. We therefore use a uniform prior between $f = 0$ and 1, or $p(f \mid HI) = 1$.

In addition to the detailed calculation of the detection probability α for each burst, we set $\alpha = 1$ for a simplified heuristic calculation since the integrals can then be done analytically. Note that each line type has a different number of bursts in which the line is detectable. This simplified calculation allows us to study the dependencies on the number of bursts, and develop a deeper understanding of the results.

RESULTS

We perform three sets of hypothesis comparisons. In each case the first hypothesis H_0 states that Ginga and BATSE are consistent: we understand the detection capabilities of both instruments; lines exist; and the detection threshold has been set high enough to eliminate false positives. In the first comparison, we use as a generalized inconsistency hypothesis H_1 the supposition that the Ginga and BATSE bursts are characterized by different line frequencies. If H_1 is favored we would not believe there actually are different line frequencies, but instead would conclude the instruments are not well understood. Differences between an instrument's true and calculated line detection capabilities can indeed be modeled by changes in the line frequency. For the second comparison we use the hypothesis H_2 that BATSE is unable to detect lines, even if present. Thus we set BATSE's line detection probability α to zero. Finally, the third comparison is between H_0 and the hypothesis H_3 that there are no lines, and the reported Ginga lines are all false positives.

We have not yet performed a definitive calculation for the Ginga and BATSE data, in part because not all the detection probabilities have been calculated consistently. In addition, there are many possible assumptions about the line distribution within a burst. However for illustrative purposes we present in the following table the Bayes factor, the ratio of the probabilities of obtaining the data, for each set of hypotheses using the lines in GB880205 alone, the line in GB870303, and both line sets together (the column labeled "Joint"). The odds ratio is the product of the Bayes factor and the ratio of the hypothesis priors (our expectation as to the validity of each hypothesis). For each line set we use the listed model values of N_G and N_B, the number of Ginga and BATSE bursts, respectively, in which lines are detectable. Given the ~ 150 bursts searched, the number of BATSE bursts in which Ginga-like lines would have been found is surprisingly small. Most bursts are not sufficiently strong for line detection. In addition, the BATSE spectra frequently do not extend to low enough energies for lines at ~ 20 keV to be detected, hence the particularly small number of bursts in which GB870303-like lines could be discovered. The hypothesis H_3 that there are no absorption lines and the Ginga detections were merely false positives depends on the currently unknown rate of finding such false positives; we minimized the Bayes factors with respect to this unknown rate.

	Table: Illustrative Example			
	GB880205	GB870303	Joint	
	Effective Burst Number			
N_G	10	15		
N_B	35	10		
	Bayes Factors			Prior Ratio
H_0/H_1	1.9	4.1	7.8	~ 10
H_0/H_2	0.053	0.37	0.020	~ 100
H_0/H_3	0.058	0.10	0.0065	~ 100

We see from the table that the Bayes factors favor the hypotheses regarding instrumental deficiencies (H_2—BATSE is unable to detect lines—and H_3—the Ginga detections are spurious because lines do not exist), and therefore our conclusions are based on the ratios of hypothesis priors which favor our assertion that the instruments are understood. We are quite confident that BATSE could detect lines if present,[1,2,3] and therefore we assign a high value (~ 100) to the prior ratio of H_0 relative to H_2. Similarly, an unrealistically large value of the rate of spurious detections β minimizes the Bayes factor comparing H_0 to H_3; the pre-Ginga reports of line detections[6,7] lead to a prior favoring the existence of lines, although the Ginga observations (which cannot be used to calculate priors since Ginga data is part of D) were much more convincing that these earlier claims. We assign a value of ~ 100 to the prior ratio of H_0 relative to H_3.

The Bayes factor is surprisingly greater than 1 for the comparison of H_0 to the hypothesis H_1 that the Ginga and BATSE bursts are characterized by different line frequencies. Even if the number of BATSE bursts in which lines would be detectable is increased, the Bayes factor remains fairly close to unity; for a single detection from among $N_G = 10$ Ginga bursts, $N_B > 90$ strong BATSE bursts are required for the Bayes factor $B < 1$. Of course, we assume we understand our instruments and therefore assign prior values favoring H_0 over H_1 (a ratio of ~ 10), although not by as large a factor of H_0 relative to H_2 or H_3 since H_1 is not as extreme as these other two hypotheses. H_1 is not as extreme since errors in the instruments' detection rates can be modeled by different line frequencies. The surprisingly large value of the Bayes factor results from the structure of the space of the data probability, $p(D \mid f_G f_B H_1 I)$ (H_1 implies $f_G \neq f_B$), which is sampled as the line frequencies are marginalized. The maximum of the data probability occurs for $f_B = 0$ and $f_G \neq 0$, but there is a range $f_G = f_B = f \sim 0.0035 - 0.07$ where the data probability is non-negligible. It is this range of line frequencies which keeps the consistency hypothesis H_0 viable. Our conclusions do not change when we repeat this analysis with line frequency priors derived from the Ginga and BATSE data (formally improper since the new data are used to determine our prior expectations) instead of a uniform prior. Further details will be provided elsewhere.[12]

SUMMARY

We adopted a new Bayesian methodology to determine whether the two Ginga detections of absorption lines and the absence of any BATSE detections are inconsistent. This methodology permits us to compare specific hypotheses through an odds ratio which is the product of a quantitative Bayes factor, the ratio of the probabilities of obtaining the data given the hypotheses, and a more subjective factor quantifying our expectations. We find that the Bayes factors favor the hypotheses that BATSE is unable to detect lines or that there are no lines and the Ginga detections are spurious, but not by large enough factors to exceed our confidence in our understanding of our instruments. On the other hand, the Bayes factor alone favors consistency compared to the hypothesis that the Ginga and BATSE bursts are characterized by different line frequencies. Given the tests to which the Ginga and BATSE instruments have been subjected, we conclude there is not an inconsistency between the observed pattern of detections and nondetections.

We thank Tom Loredo for his assistance with the Bayesian methodology. This work was supported in part by NASA contract NAS8-36081 (UCSD group).

REFERENCES

1. B. Teegarden, et al., in Proceedings of the Compton Symposium, eds. M. Friedlander, N. Gehrels and D. J. Macomb (AIP, New York, 1993), p. 860.
2. D. Palmer, et al., these proceedings (1993).
3. D. Band, et al., in the Contributed Papers of the 23rd International Cosmic Ray Conference (1993), p. 1-105.
4. T. Murakami, et al., Nature **335**, 234 (1988).
5. C. Graziani, et al., in Gamma-Ray Bursts, eds. C. Ho, E. Fenimore and R. Epstein (Cambridge University Press, Cambridge, 1992), p. 407.
6. G. J. Hueter, PhD Thesis (UC San Diego, 1987).
7. E. P. Mazets, et al., Nature **290**, 378 (1981).
8. J. C. L. Wang, Phys. Rev. Lett. **63**, 1550 (1989).
9. A. K. Harding, Physics Reports **206**, 327 (1991).
10. D. Band, et al., in the Contributed Papers of the 23rd International Cosmic Ray Conference (1993), p. 1-120.
11. T. J. Loredo, in Maximum Entropy and Bayesian Methods, ed. P. Fougère (Kluwer Academic Publishers, Dordrecht, 1990), p. 81.
12. D. Band, et al., in preparation (1993).

A CANDIDATE ABSORPTION FEATURE FROM A GAMMA RAY BURST OBSERVED BY BATSE

L. Ford, D. Band, J. Matteson
CASS, UC San Diego, La Jolla, CA 92093

D. Palmer, B. Schaefer, B. Teegarden
Goddard Space Flight Center, Greenbelt, MD 20771

R. Preece
NRC-NASA/MSFC, Huntsville, AL 35812

M. Briggs, W. Paciesas, G. Pendleton
University of Alabama at Huntsville, Huntsville, AL 35899

ABSTRACT

A highly significant but probably spurious absorption feature was discovered in spectra from a single BATSE spectroscopy detector viewing GRB930506. The probability that the feature was a statistical fluctuation in one detector is 6.1×10^{-5} but confirming this observation with other BATSE detectors is complicated by uncertainties in their channel-to-energy conversion. When reasonable but $ad\ hoc$ energy calibration corrections are applied, features are not seen in the other detectors. Therefore, although the feature appears significant in one detector, inconsistency between detectors seems to indicate that the feature was a rare but expected statistical fluctuation. This example shows the advantage of using multiple detectors to determine the reality of spectral features.

INTRODUCTION

Absorption lines in gamma ray bursts have been widely reported[1,2,3] but in its first two years of operation, BATSE has yet to see a definitive feature.[4] Because such lines could contain a wealth of information about burst sources and environments, line searches have been a high priority for spectral research. To ensure that any feature is real and not a statistical fluctuation, we require the F-test probability that a feature is a chance fluctuation be less than 10^{-4} and that other detectors viewing the burst be consistent with the observation.

A comprehensive visual search of spectra from GRB930506 revealed an absorption feature at 55 keV in Spectroscopy Detector 2 (SD2) which met our F-test criterion. Five BATSE detectors (three SDs and two Large Area Detectors (LADs)) had favorable viewing angles (Table 1) but only SD2 saw a clear absorption feature. Of the others, SD3 was in a gain state with a low energy cutoff of around 200 keV and the rest had channel-to-energy conversion uncertainties. We report here on the investigation of the feature as seen by the SDs (for the LADs, see Preece et al.[5]).

SPECTROSCOPY DETECTOR 2

The feature seen by SD2 is the most significant absorption feature found in the BATSE data to date. A time history of the burst as seen by SD2 is

Table 1. Veiwing Angles	
Detector	Viewing Angle
LAD3	28.8°
LAD7	42.8°
SD3	43.7°
SD7	56.2°
SD2	73.7°

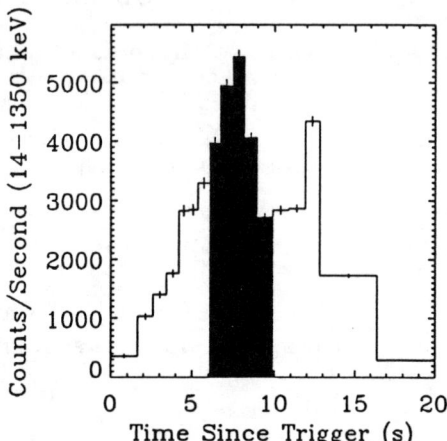

Table 1. Detector viewing angles for GRB930506. The angle is between the source direction and detector normal.

Figure 1. Time history from SD2 for GRB930506. The interval during which the feature was present is shaded.

given in Figure 1 and the count spectrum in which the feature is strongest is in Figure 2, showing a clear feature at about 55 keV. The significance of the feature was determined by evaluating the F-test probability that the improvement in χ^2 from a continuum fit to a continuum-plus-line fit was due to a statistical fluctuation in the data[6]. The energy range of the fit was from 30 to 100 keV and the continuum model used was a power law,

$$N_E(E)\left(\frac{\text{photons}}{\text{keV-s-cm}^2}\right) = A\left(\frac{E}{100 \text{ keV}}\right)^\alpha, \qquad (1)$$

where A and α were fitted parameters. A model which allowed continuum curvature was tried but no curvature was required. The continuum-plus-line model included a multiplicative absorption feature,

$$N_E(E) = A\left(\frac{E}{100 \text{ keV}}\right)^\alpha \exp\left[-\tau \exp\left(-\frac{(E-E_0)^2}{\Delta E^2}\right)\right], \qquad (2)$$

where A, α, τ, E_0, and ΔE were fit. The best fit line parameters were $\tau = 60$ (τ was constrained to be ≤ 60), $E_0 = 55.8 \pm 1.4$ keV, and $\Delta E = 0.94^{+2.45}_{-0.29}$ keV. The equivalent width of the feature was 4 keV, much less than the detector resolution at this energy (about 11 keV). A Voigt profile was also tried but the fit went to the Gaussian limit.

The fitted count and photon spectra appear in Figure 2. Since the best fit line is black and the width is much less than the detector resolution, the true profile of the line cannot be determined and fitted profile degenerates into a nearly rectangular form. Therefore, a fit of centroid and equivalent width is all

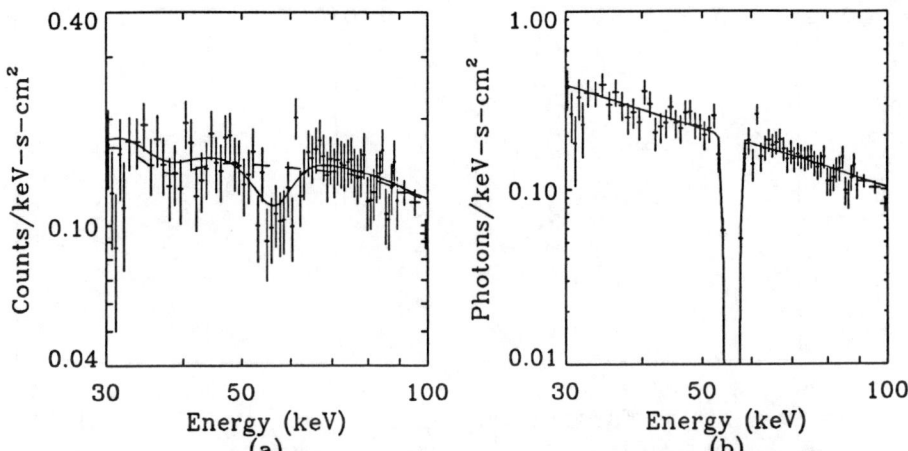

Figure 2. (a) The count spectrum from SD2 showing the candidate feature. The accumulation interval is from 6.08 to 9.92 seconds after the trigger. The dashed line indicates the best fit continuum while the solid curve shows the effect of adding a line to the model. The small dip around 35 keV is due to the iodine K-edge. (b) The photon spectrum for the best fit line. Since the energy resolution is $\sim$ 11 keV, the true line profile cannot be determined.

that is required and only two additional parameters are added by the line. For two additional parameters, the probability that the line is a statistical fluctuation is 6.1×10^{-5}. The F-test for three additional parameters gives 1.5×10^{-4}.

SPECTROSCOPY DETECTOR 7

While the line looks quite convincing in SD2, consistency with other detectors is required. A check for consistency between SD2 and SD7 is difficult because the Spectroscopy detector Low Energy Distortion (SLED)[7] in SD7 was centered around 50 keV and had a width of about 10 keV. Briefly, the SLED is caused by the addition of a noise pulse from the Lower Level Discriminator (LLD) firing. For small pulses, the pulse height analyzer gate closes before the noise gets through and for large signals the entire pulse is added. In between the noise pulse is partially included, introducing a nonlinearity into the channel-to-energy conversion which depends on the LLD setting. SLED corrections have been determined from a set of in-orbit background measurements at different detector gains. For SD7 however, the LLD setting was changed after the corrections were calculated. Therefore, an *ad hoc* correction was made in order to study SD7 (definitive corrections will be made after a new set of in-orbit measurements). SD2 did not have a SLED problem since it was in a gain state which moved the SLED below 20 keV.

The *ad hoc* correction assumed that the background between 40 and 60 keV could be represented by a power law extrapolated from the background between 60 and 65 keV. This is expected to produce reasonable results since our experience with SD7 is that the background in this energy range is a featureless

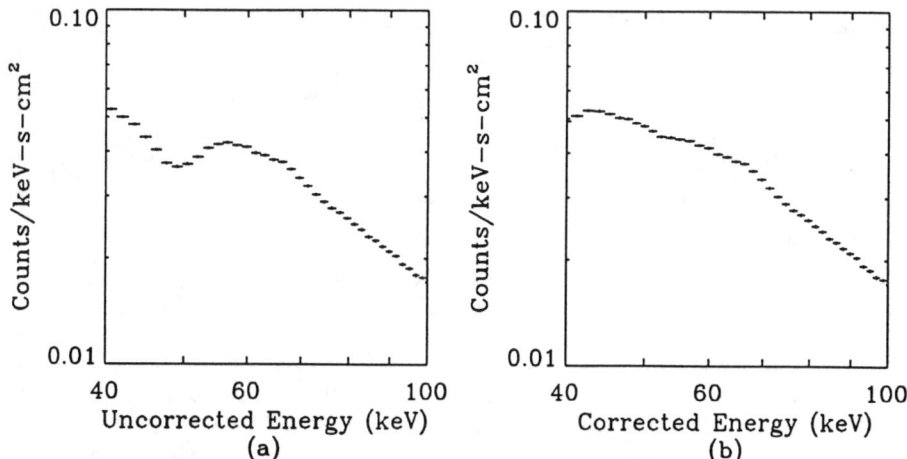

Figure 3. (a) A background accumulation showing the effect of the SLED. (b) The same spectrum with an *ad hoc* correction applied. The turnover at 40 keV is caused by the LLD cutoff.

power law when the detector's gain state moves the SLED below 20 keV. SLED correction parameters were varied until the corrected background resembled the assumed background. Figure 3 shows how the correction straightened out the background, although a slight ripple remains. Considering the strength of the feature in SD2 and the fact that SD7 had a better viewing angle, the line should show up clearly in corrected SD7 spectra.

No feature stands out in SD7's corrected count spectrum (Figure 4). Using the same models as for SD2 (but fitting from 45 to 100 keV), a very weak line is found at 55.9 keV with has a probability of being a statistical fluctuation of 0.73. This is less significant than an average statistical fluctuation. To determine the likelihood that a line would be seen in SD2 but not SD7, a Monte Carlo simulation of 200 spectra containing a line with SD2's fitted parameters was run for SD7. These simulations show that if the line in SD2 was real, then the probability of seeing the line in SD7 with an F-test probability of 0.73 or larger is less than one percent (Figure 5). Since SD7 had a better viewing angle to the burst than SD2 and should therefore have a clearer signal, there is a serious inconsistency between SD7 and SD2. Since the LADs do not see a feature either[5], we cannot claim that the feature seen in SD2 is real.

A true test of the feature's significance would be to fit all detectors at once rather than treating individual detectors independently. However, we have not yet implemented joint fits into our software. Given that the line in SD2 barely meets our significance threshold and that the other detectors see no evidence for a line, the other detectors would likely dilute the significance to a value well below our detection criterion. Therefore, a joint fit would not provide any new information regarding the reality of the feature.

DISCUSSION

The observation of a highly significant linelike feature in only one BATSE

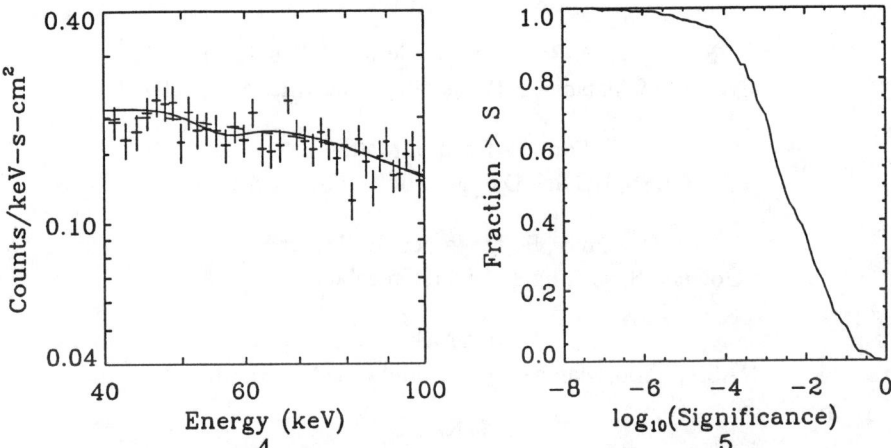

Figure 4. The count spectrum from SD7 over the same interval as SD2. The dashed curve corresponds to the continuum fit and the solid curve shows the best fit line.

Figure 5. The result of the Monte Carlo simulation showing the probability of obtaining a given F-test result in SD7 for the line parameters fit to spectra from SD2. Significance refers to the probability that an observed feature is a statistical fluctuation.

detector may seem troubling. Considering the number of burst spectra which have been searched for features though, a statistical fluctuation of the size seen in SD2 is expected.[4] Despite this, if it had not been for the other BATSE detectors, we might have considered the feature to be real since lines have been detected by other experiments. Given the importance of spectral features in deriving information about burst sources, an erroneous conclusion could have resulted in mistaken inferences in attempting to ascribe physical meaning to a statistical fluke. Therefore, despite the negative result of this analysis, it clearly demonstrates the important advantage of having multiple burst detectors for verifying the reality of linelike features.

BATSE work at UCSD is supported by NASA contract NAS8-36081.

REFERENCES

1. G.J. Hueter, Ph.D. Thesis, UCSD (1988).
2. E.P. Mazets et al., Nature **290**, 378-382 (1981).
3. T. Murikami et al., Nature **335**, 335-336 (1988).
4. D. Palmer et al., these proceedings (1993).
5. R. Preece et al., these proceedings (1993).
6. B.R. Martin, Statistics for Physicists (Academic Press, London, 1971).
7. D. Band et al., Experimental Astronomy **2**, 307-330 (1992).

CONSISTENCY ANALYSIS OF A CANDIDATE LINE FEATURE IN GRB930506

R. Preece, M. Briggs, W. Paciesas, G. Pendleton
University of Alabama at Huntsville, Huntsville, AL 35899

L. Ford, D. Band, J. Matteson
CASS, UC San Diego, La Jolla, CA 92093

D. Palmer, B. Teegarden, B. Schaefer
Goddard Space Flight Center, Greenbelt, MD 20771

M. Brock
NASA, ES66, Marshall Space Flight Center, AL 35812

ABSTRACT

A highly significant candidate absorption-type line feature has been observed in several spectra obtained by one BATSE spectroscopy detector (SD) for GRB930506. The ability of the BATSE large area detectors (LADs) to detect *Ginga*-like absorption features has been demonstrated previously to be quite good at the fit energy (55 keV) of the candidate line feature. We present an analysis of the consistency between a model derived from a fit to the SD data and spectra from the two brightest LADs. The low-end of the current channel-to-energy conversion routine for the LADs is known to be inadequate. A new determination of the LAD calibration was made by relaxing the energy edges of 16 channel occultation data for the Crab Nebula spectrum until a good fit was obtained. We show that the re-calibration of the LAD channel-to-energy conversion function at the low end produces spectral fits inconsistent with the presence of an absorption feature.

INTRODUCTION

With the controversy surrounding the question of the distances to the source population of gamma ray bursts (GRBs) provided by the BATSE GRB source location and log N / log (C/C_{min}) distributions[1], it is natural to look to spectroscopic analysis for additional clues. One of the most important of these is the use of spectral line features as a diagnostic of the source object. Unambiguous detection and confirmation of cyclotron-like absorption features is a goal of BATSE spectral analysis, in view of such observations by detectors on *Ginga*[2]. Two harmonically-spaced features are clear evidence for the presence of strong ($\approx 10^{13}$ G) magnetic fields, which in turn implicates neutron stars, and thus a galactic source population. Indeed, a single absorption feature, with a high significance, was detected in data obtained by BATSE SD #2 during some portions of GRB930506[3], and is the most significant such feature reported by BATSE to date[4]. There are some problems with the interpretation of the reality this feature, however. Among the eight BATSE SDs, the feature is present only in data from the detector which received the third brightest illumination from the source. Difficulties in the analysis of the data from detectors which received more illumination are fully discussed in Ref. 3. Due to their larger

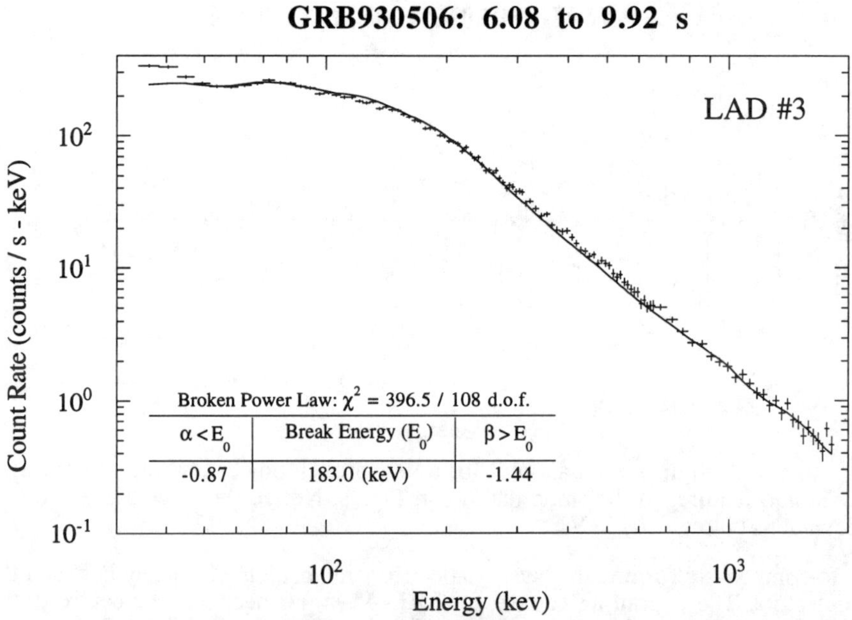

GRB930506: 6.08 to 9.92 s

LAD #3

Broken Power Law: χ^2 = 396.5 / 108 d.o.f.

$\alpha < E_0$	Break Energy (E_0)	$\beta > E_0$
-0.87	183.0 (keV)	-1.44

Figure 1: Best fit to a broken power law continuum plus a Gaussian absorption line from SD #2 applied to data from LAD #3 for the same time interval. Only the amplitudes of the two components were allowed to vary.

collecting area, the LADs can be quite sensitive to narrow spectral features of sufficiently high energy, in spite of their relatively poor energy resolution[5]. However, systematic effects are known to produce line-like features in LAD count spectra at low energies. In this paper, we would like to cover the results from analysis of data taken from the BATSE LADs for the same burst. First, we take up the issue of consistency between fits to the reported line feature in the SD data and LAD data. Then we discuss the effort to recalibrate the low end of the LAD channel-to-energy conversion algorithm, which is the major source of the systematic features. We apply the new algorithm to a scientific question, that of the confirmation or not of the 55 keV candidate line feature by analysis of the LAD data.

INITIAL CONSISTENCY ANALYSIS

In order to investigate consistency between the data obtained by the BATSE SDs and LADs, we first fit the data from SD #2 in the best interval for observing the candidate feature, which was from 6.08 to 9.92 s after the trigger. The model used was a broken power law plus a Gaussian line, with all parameters allowed to vary except for the width of the feature (the feature is intrinsically narrower than the detector energy resolution). We obtained a global fit between the energies of 12.3 and 1290 keV, with a χ^2 of 230.6 for 230 d.o.f. The line centroid energy was 55.0 ± 1.4 keV. This model was then applied to data obtained by the two brightest source-facing LADs over the same time interval and approximately the same energy interval. The result for LAD #3, including the fit information, is shown in Figure 1. In this case, we allowed only the overall amplitudes of the model and the Gaussian component to

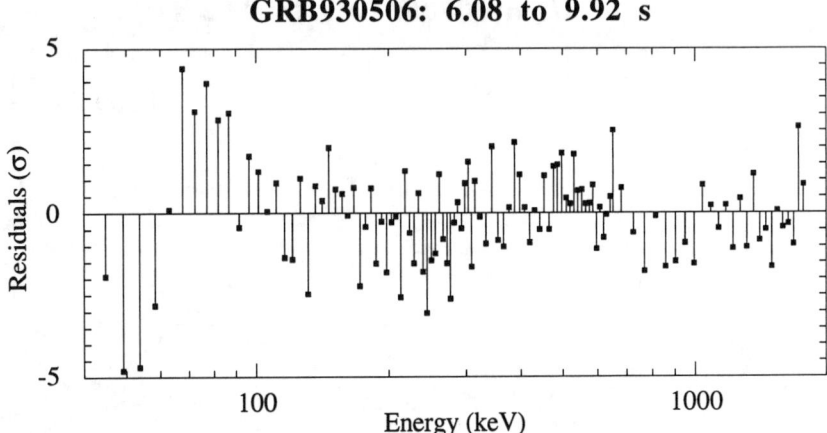

Figure 2: Sigma residuals plot for a best broken power law fit, without a Gaussian feature, to the same data as in Fig. 1. Notice the large systematic deviations below ~100 keV.

vary, to compensate for possible systematic errors in absolute efficiency between the two detectors. The overall agreement for LAD #3 is fair near the line centroid; the Gaussian component does improve the fit. Although large contributions to χ^2 are due to systematic differences between the count spectrum and the model above 100 keV, the model also departs from the data below 50 keV. A 'pull' plot (counts − model residuals divided by the model errors) for the continuum model without a line component is shown in Figure 2. The case of LAD #7 (not shown) is even worse: the goodness of fit is poor, and although the count spectrum displays a pronounced depression near the line centroid, the modeled line does not lie on top of the data points. Thus, the presence of the systematic features make the LAD data inconclusive in verifying the SD #2 line feature.

RECALIBRATION

The amount of spectral variation routinely observed in the low end of LAD spectra indicates a problem with the routine used to convert the data bin number to a set of energy boundaries for the bin. Any number of effects could result in the observed systematic deviations, which have a unique pattern in each detector. The current channel-to-energy conversion scheme[6], assumes that the energy edges could be fit with constant, linear and one-half power terms in the linear channel number, with the overall gain determined by the position of the 511 keV background line centroid. It is clear from inspection of data from solar flares, the Crab nebula, and various other sources that this scheme produces systematic deviations of the data from a smooth spectrum at low energies, although it works quite well above ~150 keV.

Pendleton[7] describes a new method for determining a channel's energy edges which is not dependent upon a particular functional form. Rather, it is an empirical method, arriving at a set of energy edges by iteratively fitting a smooth input (source) model to data obtained in orbit. The standard broken power law model[8] of the Crab Nebula spectrum is fit to LAD occultation data using detector response matrices with

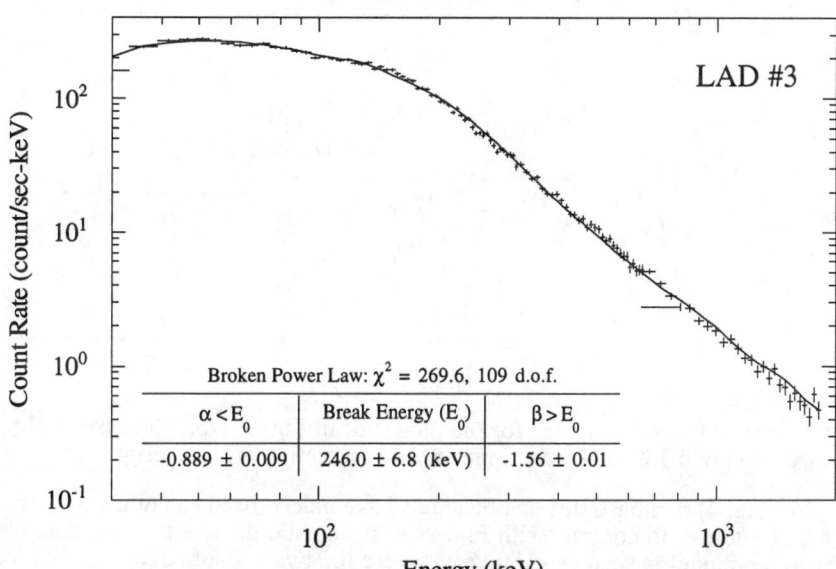

Figure 3. Best broken power law fit to the same data as in Fig. 1 after recalibration of the energy edges. Parameters of the fit are detailed in the plot.

bin edges produced by the original channel-to-energy conversion scheme described above. The edges are allowed to float to obtain the best spectral fit to the power-law model. A new matrix is calculated for the new edges, and the fit is performed again, adjusting the edges again. After several iterations, the edges settle into an equilibrium position that best fits the data to the assumed model. Several viewings of the Crab were obtained for each detector, to determine how the calibration routine is affected by the angle to the source. Only statistically insignificant differences were found, so the sets of edges were averaged over the viewings to produce a standard set of edges. The spectra derived from Earth occultation data have 16 energy channels; for the purposes of burst spectroscopy, these must be translated into edges appropriate for the 128 channel High Energy Resolution Burst (HERB) data type. We fit the 17 edges to a cubic spline and interpolated to obtain 129 edges. We can compare spectral fits between the LADs and the SDs for consistency, as described above, so we tested the splined edges on a well-studied burst, GRB930131. We obtained the best results when we use the new edges below ~150 keV and the old edges above, since statistical errors in the Crab data at high energies limit the effectiveness of the new technique.

IMPROVED CONSISTENCY ANALYSIS

Once we had obtained some experience in applying the new routine to burst data, we returned to the question of verifying the GRB930506 candidate line. A count spectrum with the new edges is shown in Figure 3, which is to be compared to Fig. 1. The best fit to a broken power law model is also displayed, without the Gaussian line component which is present in Fig. 1. Although the fit is still poor, adding a Gaussian line to the model does not improve the fit, as it had before recalibration. Here, the

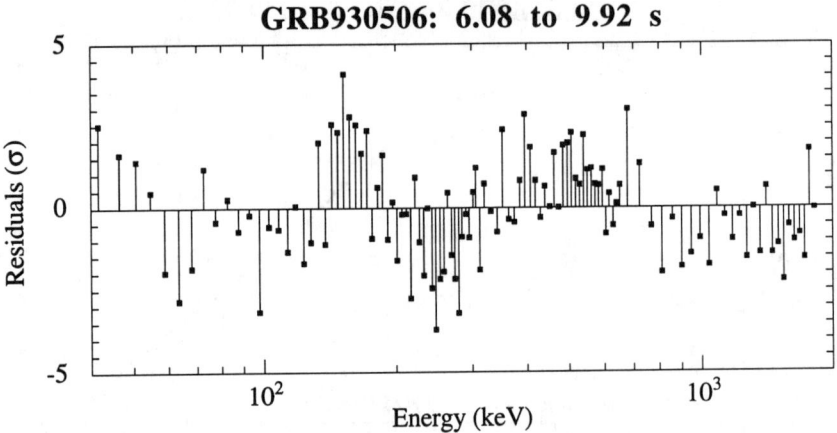

Figure 4. Residual sigmas for the model fit in Fig. 3. Note the systematic 'pulls' at around 150 keV, due to incomplete correction of the energy edges.

'pull' plot (Fig. 4) demonstrates that residuals have indeed been smoothed out for the energies of interest, in contrast with Fig. 2. A systematic deviation in the data with respect to the model is seen at ~150 keV, where the two sets of edges were spliced together. This may be responsible for much of the large residual χ^2 in the fit; however, the broken power-law model may simply be inappropriate. We hope to improve upon the 128 channel energy edges in the near future by constraining the edges above 150 keV with additional Crab data and by testing with Solar Flare data.

SUMMARY

Overall, the evidence displayed in Figs. 3 & 4 points away from an interpretation of the candidate feature as an absorption line. Ford *et al.* have looked at the next brightest SD data (#7), which similarly required some improvement in the channel-to-energy calculation, and have not detected the candidate feature[3]. We conclude that the candidate feature is spurious.

REFERENCES

1. C. Meegan *et al.*, Nature **355**, 143-145 (1992).
2. T. Murakami *et al.*, Nature **335**, 335-336 (1988).
3. L. Ford *et al.*, these proceedings (1993).
4. D. Palmer *et al.*, these proceedings (1993).
5. R. Preece *et al.*, in Compton Gamma-Ray Observatory: St. Louis, MO 1992, ed. M Friedlander, N. Gehrels & D. J. Macomb, 867-871 (New York: AIP, 1993).
6. J. Lestrade, NASA internal memo (1991).
7. G. Pendleton *et al.*, in The Second Compton Symposium, ed. C. E. Fichtel, N. Gehrels & J. P. Norris, to be published (New York: AIP, 1994).
8. G. Jung *et al.*, Ap. J. **338**, 972 (1989).

BATSE CYCLOTRON LINE SEARCH PROTOCOL

Bradley E. Schaefer, Bonnard J. Teegarden, David M. Palmer,
Thomas L. Cline, Sandhia Mitruka
NASA/Goddard Space Flight Center, Greenbelt, MD 20771

Lyle A. Ford, David L. Band, James L. Matteson
CASS, UC San Diego, La Jolla, CA 92093

Michael S. Briggs, William S. Paciesas, Geoffrey N. Pendleton
University of Alabama, Huntsville, AL 35899

Robert D. Preece
NAS/NRC, NASA/MSFC, Huntsville, AL 35812

ABSTRACT

One of the primary goals of BATSE Spectroscopy Detector data analysis is the search for cyclotron lines. We have implemented an automated line search which can provide an exhaustive and objective measure of the presence of significant cyclotron lines during a burst. This procedure will be run for all bright bursts for line energies between typically 20 keV and 100 keV.

INTRODUCTION

Do gamma-ray bursts have cyclotron line absorption in their spectra? This is one of the primary questions for this conference. If cyclotron lines are present, then we learn that bursts may have high magnetic fields and we can learn many details about the physical conditions in the emission region. In addition, it is difficult to form cyclotron lines with cosmological models, so a line detection would have implications for the burster distance scale.

In 1988, the Ginga satellite observed low energy absorption lines during two bursts[1], one of which displayed harmonically spaced lines. These confidently detected lines provided a primary support for the (then current) galactic neutron star paradigm. The common expectation was that BATSE would find more lines.

But so far, BATSE has not confidently detected any spectral line features in bursts (see papers by Ford et al., Preece et al., Band et al., and Palmer et al. in these proceedings). Since this is contrary to our expectations based on the Ginga rate for line features, questions can be raised whether bursters have cyclotron lines.

It is therefore important to examine the BATSE data in many ways, first to detect lines if present, and second to provide confidence in the validity of the search. To these ends, the BATSE team has developed an automated line search routine which will supplement the current visual examination routine.

This computerized algorithm is similar in some ways to that of Messina and Share[2]. They searched for lines with $E_{line} > 300$ keV from 177 bursts with the GRS on SMM. The SMM line search was much simpler than the planned BATSE search because SMM had many fewer spectra than BATSE (16 second time accumulations versus ≥ 128 msec accumulations) and because SMM performed a single continuum fit to each spectrum and searched the residuals for lines, while we do a continuum and line fit at each trial line centroid.

SEARCH PROTOCOL

The BATSE automated line search will examine all BATSE bursts with adequate data that are sufficiently bright that a line feature such as reported by Ginga can be seen. Currently, there are roughly 40 bursts which are bright enough for Ginga-like lines to be visible.

The search will utilize the BATSE spectroscopy detectors for which the burst direction is within 80° of the detector axis and for which the LLD (lower level discriminator, roughly the lowest energy with valid data) is below 50 keV.

Besides searching bright spectra where we could detect a Ginga-like line, we will also search spectra which are not expected to have lines. These spectra will serve as controls. Such spectra include time intervals of bright bursts in which there is apparently no burst emission, faint bursts, solar flares, and background data.

The automated search will utilize the SHERB data type, which consists of full resolution (256 channels) spectra with good time resolution (ranging from 128 msec to roughly one second) and extends from the BATSE trigger for typically 100 seconds. The available spectra for an individual detector will consist of many consecutive spectra, which can be combined together to create spectra for smaller time intervals. All combinations will be made by adding the N consecutive spectra with every possible start spectra for N equaling 1, 2, 3, 5, 7, 10, 14, 20, 28, 40, 56, and 80. (So for example, if the time interval covers five spectra, labelled A, B, C, D, and E, then acceptable combinations would include A, B, C, D, E, AB, BC, CD, DE, ABC, BCD, CDE, and ABCDE.) However, if there is too much overlap in time with the previous spectrum, then a combination will be ignored. Here, we consider there to be too much overlap if $D/T_1 > 0.8$ and $D/T_2 > 0.8$, where D is the time duration of the overlap, T_1 is the time duration of the previous group, and T_2 is the time duration of the prospective group.

Count spectra are extracted, background spectra are interpolated, and background subtracted by the usual procedures[3] or slight modifications thereon. The detector response matrices are created with an Earth scattering correction. The continuum model of choice is the 'Comptonized' model, which is simply a power law times an exponential cutoff:

$$dN/dE = Ce^{-E/E_o}(E/100\text{keV})^\alpha,$$

where α is the power law index and E_o is the cutoff energy. The Comptonized model has sufficient flexibility to fit most gamma-ray burst spectra from 20 to 150 keV. In any cases where it is inadequate, we will use a more flexible model, such as the 'Gamma-Ray Burst' model (Ford et al., these proceedings). The continuum-plus-line model is a Gaussian line added to the continuum-only model:

$$dN/dE = (dN/dE)_{\text{continuum}} + A\exp(-[E - E_{\text{line}}]^2/2W^2).$$

Here, W is the line width which is taken to be half the detector resolution. This choice implies that the automated search is primarily for narrow line features, although there will be sensitivity to moderately wide (relative to the instrumental resolution) features with a lower efficiency. Broad lines are more difficult to evaluate because of uncertainties in the true continuum. E_{line} is the energy of the line center, and is taken to step through the entire range under 100 keV at intervals of one-third the detector resolution. A is the line amplitude and may be positive or negative. For any particular model fit, all continuum parameters are allowed to vary while the only line model parameter allowed to vary is A.

The significance of the line will be calculated for every continuum-plus-line fit by means of an F-Test. The prescription by Martin[4] will be used:

$$F(\nu_1, \nu_2) = ([\chi_c^2 - \chi_L^2]/\nu_1)/(\chi_L^2/\nu_2),$$

where χ_c^2 is the chi-square of the continuum-only fit, χ_L^2 is the chi-square of the continuum-plus-line fit, ν_1 is the number of additional parameters in the line model, and ν_2 is the number of degrees-of-freedom of the continuum-plus-line fit. This F-Test parameter can then be converted to a probability, which is the probability that of obtaining a statistical fluctuation at least as significant as the observed line if the line is not real. If this probability is very small, then the addition of the line to the continuum model significantly improves the fit and we have good evidence for the reality of the line.

While we do each line fit by varying only one parameter (the line amplitude), the effective number of varying line parameters depends upon the question being asked. If we ask whether a line exists at 20 keV (or any other specific energy), then we are in fact using only one additional parameter. If we ask whether a line exists at any energy in a given spectrum, then we have added two parameters, line amplitude and centroid. Although we do not vary the line centroid in a particular fit, we have optimized the result versus line centroid by doing many fits with closely spaced centroids.

A convenient quantity for evaluating the significance of a line feature is A/σ_A where A is the best fit line amplitude and σ_A is the one sigma uncertainty of A. For example, $A/\sigma_A = +6$ would indicate a significant emission line, while $A/\sigma_A = -1.2$ would imply an insignificant absorption line. A histogram of A/σ_A should have a Gaussian distribution for random fluctuations (cf. Messina and Share[2]) and this is a test of the fit procedures.

IMPLEMENTATION

The BATSE automated line search has been implemented independently at Goddard Space Flight Center (led by B. Schaefer), University of California at San Diego (led by L. Ford), and at Marshall Space Flight Center (led by M. Briggs). The three programs are named LSRCH, SOAR, and WINGSPAN respectively. One reason for this redundant effort was to provide cross-checking on each of the codes. And indeed, a variety of small problems were identified in this manner in each of the three programs. Another reason for the three programs is so that the line search can be run at each institution within each programming environment.

One of the inevitable problems of an exhaustive search is that it takes a long time to examine all bursts. This is due to the large number of spectra that each must have many model fits. Roughly, we will examine 40 bright bursts, each with two detectors, each with 100 spectra combinations, and must perform 24 model fits for each spectra. In practice, the total computing time required is 22 hours per burst on average. So to examine the bright bursts to date will take 37 CPU days. By spreading the computation among the three institutions, the total time will be manageable.

Currently, four bursts (BATSE trigger numbers 143, 1541, 2083, and 2329) have been examined. The line candidate in trigger 2329, discovered by the visual search method and not considered to be real (see Ford et al. and Preece et al., these proceedings), was found by the automatic search method at a probability of roughly 10^{-4}. Full evaluation of the exhaustive search algorithm is still in progress. The results will be reported in future publications.

STRENGTHS AND WEAKNESSES

The automated line search routine is intended to work in parallel with our visual line search, so that both are running simultaneously. The reason for having two search routines is that an automated algorithm has several advantages (and disadvantages) when compared to our visual search:

There are four advantages of the automated search over the visual search: (1) An automated search can examine many more spectra than a visual search, so that more time interval combinations can be tested. In particular, the visual search has a default time scale and so might be less sensitive for line features lasting significantly longer than the default time bins. (2) An automated search will examine photon spectra and thus prevent the masking of line features by well understood detector non-linearities. (3) An automated search will use an objective criterion that does not depend on human judgement and will never tire after looking at many spectra. (4) An automated search will allow for the systematic collection of statistics, such as the distribution of line significances. The automated search will be run not only on burst data but also on solar flare and background data where we do not expect to find narrow spectral features. The distribution of line significances for background and solar flares will test our understanding of the instrument and our line search procedures. If the distribution is not Gaussian, then the systematic effects responsible for this can be studied quantitatively.

Nevertheless, the automated search technique has several disadvantages. (1) The search of photon spectra (instead of count spectra) can create false line detections associated with poorly modelled detector non-linearities. Fortunately, these effects are not likely to be time variable, and thus should be easily identifiable since they will appear in many bursts as well as the background. (2) False line detections can be created if the continuum model does not closely match the actual burst continuum. That is, if a burst has a more curved spectral shape than can be obtained from the fit model, then the line model will try to compensate for the imperfect continuum by increasing the line strength. However, further analysis would identify this case due to the broad fitted line width. Furthermore, line centroid energies used in the automatic search are inset from the energy range used for the continuum fit by 2×FWHM, to avoid fitting a spurious line where the continuum fit is most likely to fail. (3) A computer program will detect only what it is programmed to detect, whereas a human might notice a constantly shifting line or some other spectral peculiarities.

REFERENCES

1. T. Murakami et al., Nature **335**, 234 (1988).
2. D. C. Messina and G. H. Share, Gamma-Ray Bursts (AIP, NY, 1992), p. 206.
3. B. E. Schaefer et al., Ap. J. Supp. , in press (1994).
4. B. R. Martin, Statistics for Physicists (Academic Press, 1971).

SPECTRAL PROPERTIES OF GAMMA-RAY BURSTS OBSERVED BY COMPTEL

L.O. Hanlon, K. Bennett, O.R. Williams, C. Winkler
Astrophysics Division, ESTEC, NL-2200 AG Noordwijk, The Netherlands

W. Collmar, R. Diehl, J. Greiner, V. Schönfelder,
H. Steinle, A. Strong, M. Varendorff
Max-Planck-Institut für Extraterrestrische Physik, 85740 Garching, Germany

R. van Dijk,* J.W. den Herder, W. Hermsen, L. Kuiper
SRON-Leiden, P.O. Box 9504, 2300 RA Leiden, The Netherlands

A. Connors, R.M. Kippen, M. McConnell, J. Ryan
University of New Hampshire, Durham NH 03824, USA

ABSTRACT

During the first year of operation, the COMPTEL instrument on board the Compton Gamma Ray Observatory detected 22 γ-ray bursts within its field of view. Spectra and time histories for the strongest 7 of these bursts have been obtained from both the main instrument (0.75–30 MeV) and the burst modules (0.1–10 MeV). The deconvolved photon spectra for the majority of bursts are fit by a single power law model with spectral index between -1.6 and -2.8. One strong burst, GRB 910814, exhibited significant curvature and could not be fit by a single power law model. A broken power law model with a break in slope at $\sim$ 2 MeV is a good fit to the time averaged spectrum of this burst. There is evidence, at the 2.8σ level, for a change in the break energy of GRB 910814, from above 2 MeV to below 1 MeV during the first 9 s of the burst.

INTRODUCTION

COMPTEL is an imaging γ-ray telescope on board the Compton Gamma Ray Observatory (CGRO) operating in the energy range 0.75–30 MeV. The instrument consists of an upper layer of seven detectors (total area 4188 cm^2), made from the low-Z liquid scintillator material NE213A along with a lower layer of fourteen NaI cells (total area 8620 cm^2). Two independent modes of operation are employed by COMPTEL for the detection of γ-ray bursts.

The "double scatter" (or "telescope") mode, which is the normal imaging mode of the telescope[1], is used to produce images, spectra and time histories of bursts which occur within the field of view (FOV).

The "single detector" (or "burst") mode[2] is triggered upon receipt of a signal from CGRO-BATSE that a burst has commenced. In this instance, two of the lower NaI detectors (called D2-7 and D2-14) accumulate 6 high time resolution ('burst') spectra for an integration time of 0.5 s, followed by up to 255 'tail' spectra, each of 6 s integration time. Thereafter the burst modules return to 'background mode', integrating spectra for 100 s and awaiting the next BATSE trigger.

* Also: Astronomical Institute, University of Amsterdam, Kruislaan 403, NL-1098 SJ, Amsterdam, The Netherlands

The burst modules (each with a detecting area of 615 cm^2) operate in overlapping energy regions. The low range detector (D2-14) is sensitive between 0.1 − 1.3 MeV while the high range module (D2-7) covers the energy interval 0.4 − 10 MeV (122 and 128 energy channels per module respectively). The spectral resolution is 9.6% at 0.5 MeV and 7.0% at 1.5 MeV.

This paper presents a summary of spectroscopic results from the burst module data of γ–ray bursts which were detected in the FOV of the COMPTEL instrument in the first year of operation.

OBSERVATIONS

Table I Bright COMPTEL FOV Bursts				
BATSE Trigger	Date	Seconds (UT)	Zenith (Degrees)	Fluence (> 0.6 MeV) (ergs cm^{-2})
109	910425	2265.75	44.8	5.5×10^{-5}
143	910503	25452.7	23.0	1.5×10^{-4}
249	910601	69734.55	8.0	3.5×10^{-5}
451	910627	16157.78	10.0	2.6×10^{-6}
503	910709	41602.13	36.7	2.8×10^{-6}
678	910814	69273.04	29.2	1.3×10^{-4}
1085	911118	68258.06	35.7	6.6×10^{-6}

The BATSE all–sky monitors detected $\sim$ 300 cosmic γ–ray bursts between April 17 1991 and April 16 1992. The COMPTEL burst detectors (modules D2-7 and D2-14) are, in principle, 4π sensitive. In practice, however, their field of view is obstructed by a considerable amount of mass from both COMPTEL itself and the rest of the CGRO spacecraft. Only bursts with a zenith angle less than 45° are considered in this paper because the sensitivity of the instrument in the forward direction is high and relatively uniform so the same detector response matrix may be used in the spectral deconvolution of these events. COMPTEL detected 22 events within its FOV during the first year of the CGRO mission, of which 7 were bright enough to be subjected to detailed spectral analysis[3] (see Table I).

RESULTS

Single power-law, broken power-law (BPL) and optically thin thermal bremsstrahlung (OTTB) functions were used as trial photon models. The latter two were only applied when a single power law resulted in an unacceptable fit. The single power law model is of the form **a** × E(MeV)$^{-\alpha}$, where **a** is the normalisation, or flux at 1 MeV (in photons/(cm^2 s MeV)). The low range module (D2-14) was switched off between May 25 1991 and May 13 1992, so low range data are only available for GRB 910425 and 910503. The results of the spectral fitting using a power law model are shown in Table II.

Table II: Results of spectral fitting					
Date	Duration	0.3–1.3 MeV		0.6–10 MeV	
	(sec)	α	χ^2/ν	α	χ^2/ν
910425	39	2.45 ± 0.23	28/38	2.1 ± 0.23	13/15
910503	57	2.03 ± 0.09	99/81	2.12 ± 0.09	39/36
910601	33	–	–	2.82 ± 0.27	9.5/10
910627	2	–	–	$1.6^{+1.1}_{-0.8}$	–
910709	0.5	–	–	2.0 ± 0.42	7/5
910814	33	–	–	2.06 ± 0.07	134/38
911118	9	–	–	$2.64^{+0.56}_{-0.38}$	7/6

DISCUSSION

The γ-ray bursts shown in Table II are adequately described by a single power law model over their entire duration, with the exception of GRB 910814. The best fit to the single detector data in this case was obtained using a BPL or an OTTB model. Neither model is rejectable on the basis of χ^2 statistics. The spectral break is confirmed by BATSE[4] and SIGMA[5]. The best fit BPL photon spectrum, averaged over the entire 33 s duration of the event, is shown in Figure 1.

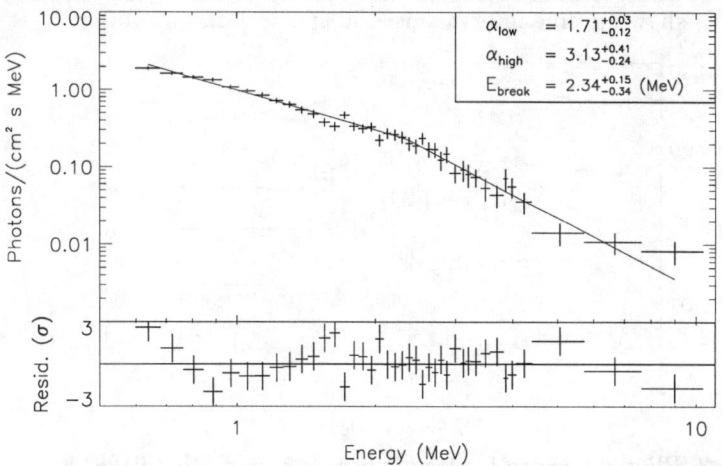

Figure 1. Photon spectrum of GRB 910814 derived from the best fit BPL model. 1σ errors for one interesting parameter are given for the fit results.

GRB 910814 was intense enough to be analysed in individual time intervals as shown in Figure 2. The intervals marked 1, 2 and 3 represent 1 s time bins, while 4, 5, 6 and 7 are each 6 s in duration. These intervals have been used in the time-resolved spectral fitting of the burst mode spectra to investigate the spectral evolution in this burst.

The spectral curvature, requiring a BPL or OTTB model, is only significant during the first 9 s of the burst (intervals 1 to 4) after which time a single power law model is an acceptable fit. The deconvolved spectra for intervals 1 through 5 are shown in Figure 3. The variation in break energy, from above 2 MeV to below 1 MeV during the first 9 s of the burst is significant at the 2.8σ level.

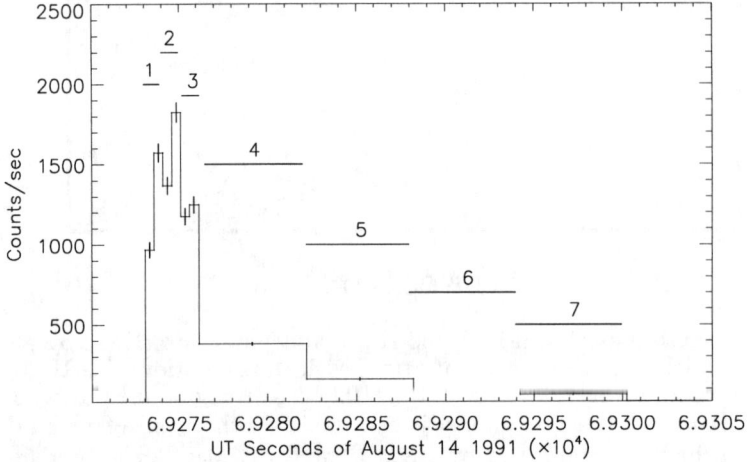

Figure 2. Background subtracted time history of GRB 910814 (0.6–10 MeV) showing time intervals used in spectral deconvolution

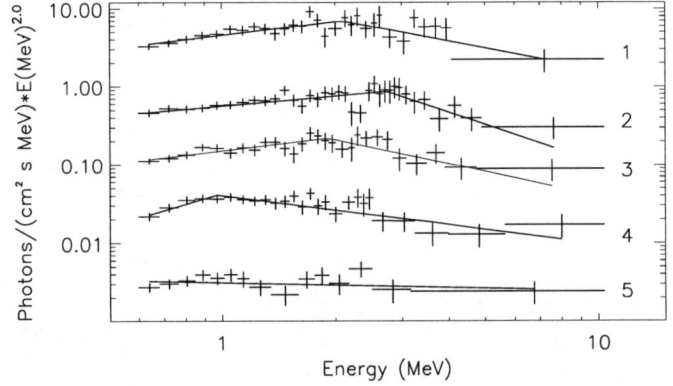

Figure 3. Photon spectra for time intervals 1–5 of GRB 910814

The time resolved spectral analysis of GRB 910814 leads to the possibility that the spectral break energy decreases in the course of the burst, such that after the first 9 s, it has shifted to below the energy range of the instrument.

The EGRET data[6] from this burst support this conclusion, since for the first 7 s of the burst, a single power law fits their data only if the first few points (∼2 MeV) are not included in the fit. In subsequent time intervals, a single power law fits the data well over the whole energy range. Thus, the combined COMPTEL and EGRET results provide clear evidence for a time dependent spectral break at MeV energies in this γ–ray burst.

CONCLUSIONS

The majority of bright GRBs observed by COMPTEL had single power law spectra, with $1.6 \leq \alpha \leq 2.82$. Only one of the seven showed definite curvature, unlike the BATSE sample[4] in which roughly a quarter of bursts exhibited curvature. These are not incompatible statistics, since some of the breaks observed by BATSE are below the COMPTEL energy range (e.g. GRB 910601 and GRB 910709). One bright burst, GRB 910814, exhibited a spectral turnover at ∼2 MeV which is confirmed by other experiments. There is evidence in the COMPTEL data, supported by EGRET observations, that the spectral break energy of this burst is time dependent.

ACKNOWLEDGEMENTS

L.H. gratefully acknowledges an ESA fellowship. This research was supported in part by the Deutsche Agentur für Raumfahrtangelegenheiten (DARA) under the grant 50 QV 90968.

REFERENCES

1. Schönfelder, V., et al, Ap. J. Suppl. **86**, 629 (1993).
2. Winkler, C., et al, Adv. Sp. Res. **6**, 113 (1986).
3. Hanlon, L.O., et al, Accepted for publication in A&A , (1993).
4. Schaefer, B.E., et al, Ap. J. **393**, L51 (1992).
5. Pelaez, F., et al, Proceedings of INTEGRAL Workshop , (1993).
6. Kwok, P.W., et al, Compton Symposium (AIP 280, 1993), p. 855.

CROSS CALIBRATION OF BURST SPECTRA WITH BATSE, EGRET, AND COMPTEL FOR GRB910503

Bradley E. Schaefer*, Bonnard Teegarden, Thomas Cline, Brenda L. Dingus*
NASA/Goddard Space Flight Center, Greenbelt, MD 20771

Gerald J. Fishman, Charles A. Meegan, Robert B. Wilson
NASA/Marshall Space Flight Center, Huntsville AL 35812

William S. Paciesas, Geoffrey N. Pendleton
University of Alabama, Huntsville, AL 35899

David L. Band
CASS, UC San Diego, La Jolla, CA 92093

E. J. Schneid
Grumman Aerospace Corp., MS A01-26, Bethpage NY 11714-3580

Ping W. Kwok
NAS/NRC, NASA/GSFC, Greenbelt MD 20771

V. Schonfelder
Max-Planck Institut, D/W-8046 Garching Germany

C. Winkler
Space Science Dept., ESA/ESTEC, NL-2200 AG Noordwijk, Netherlands

W. Hermsen
Lab. for Space Research, Leiden, PB 9504, NL-2300 RA Leiden, Netherlands

R. M. Kippen
University of New Hampshire, Durham NH 03824

ABSTRACT

Nine photon spectra for GRB910503 from BATSE, EGRET, and COMPTEL are extracted and compared. We find that all spectra agree well with each other. In addition, we present a composite spectrum from 20 keV to 200 MeV.

RESULTS

The Compton GRO instruments all have different detection modes for Gamma Ray Bursts than are used for steady sources. For steady sources, in-flight calibration can be made by observations of the Crab or other standard sources. For burst sources, the in-flight calibration must also be made by observations of the same burst. The intercomparison of data from the GRO in-

---

* Also Universities Space Research Association

struments is complicated by the wide variation in energy ranges and integration times.

The bright GRB910503 is a good case for cross calibrating BATSE, EGRET and COMPTEL since its source was inside the field of view for all three instruments. Spectra from this burst have been previously reported for all three instruments[1,2,3], but a direct comparison can only be made only for the same time intervals. A total of nine GRO detectors (BATSE LAD4, LAD2, SD6, SD4, SD2, COMPTEL telescope, D2-low, D2-high, and EGRET TASC) have been used to extract directly comparable spectra for the first pulse (10 sec duration) of GRB910503.

There is remarkably good agreement between all nine spectra, in that the photon spectra overlap within their uncertainties with no adjustments. For example, all nine spectra provide a spectral flux measure at 1 MeV, and these all agree with an rms scatter of 7%. This is strong and gratifying evidence that the GRO burst spectra are well calibrated.

A similar test was performed for five bright bursts detected with just the BATSE detectors. Once again the intercomparison of photon spectra from different detectors showed remarkable agreement. Over seven energies from 30 to 2000 keV, the rms scatter between the BATSE LADs and SDs was 9% with a maximum deviation of 22%.

With this assurance that the spectra for GRB910503 are well calibrated, we have constructed a composite spectrum for the first pulse from 20 keV to 200 MeV. This was created from a weighted average of all nine spectra and is displayed in the figure on this page. This burst also was seen brightly in the EGRET spark chamber, and a ~10 GeV photon was detected (Dingus et al., these proceedings). This very high energy spectral flux can be estimated by taking the effective area to be ~10 cm^2 and the energy band to be ~10 GeV, so the spectral fluence is of order 10^{-8} photons cm^{-2} keV^{-1}. This value can be compared to the composite spectrum (multiply by 10 s to convert from flux to fluence) and it lies a factor of ~10^3 above a simple extrapolation of the composite spectrum, but uncertainties are large.

This spectrum covers four orders of magnitude in energy, and is a great improvement over spectra available in the past. A burst spectrum over a restricted energy range will appear as a power-law-with-curvature, and virtually all spectral models could fit such. But now our composite spectrum has a sufficiently broad energy range that complex structure is visible. It is our hope that theorists will address their spectral models to this composite.

1. B. E. Schaefer et al., Ap. J. **393**, L51 (1992).
2. E. Schneid et al., Astron. Ap. **255**, L13 (1992).
3. C. Winkler et al., Astron. Ap. **255**, L9 (1992).

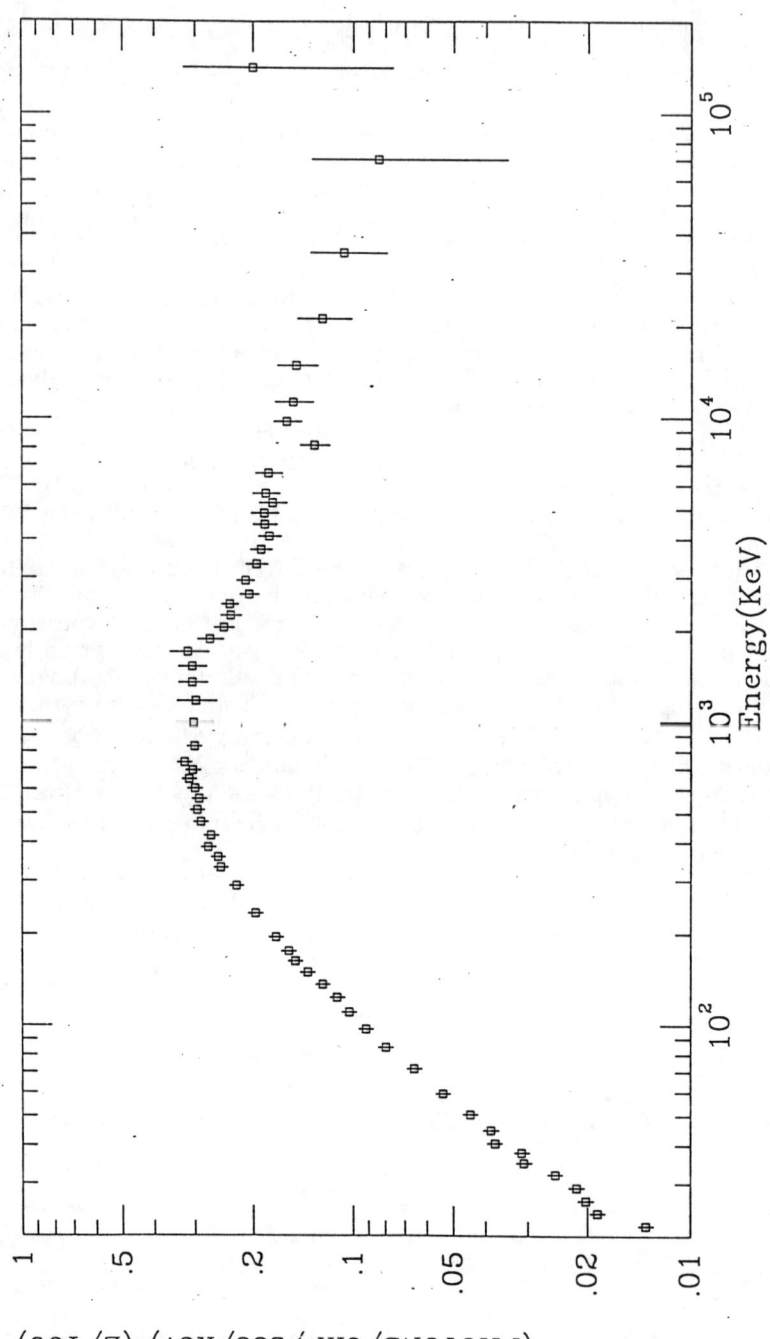

COMPARISON OF BATSE, COMPTEL, EGRET, AND OSSE SPECTRA OF GRB 910601

G.H. Share, W.N. Johnson, J.D. Kurfess, R.J. Murphy
E.O. Hulburt Center for Space Research, Naval
Research Laboratory, Washington, DC 20375, USA

A. Connors
Univ. of New Hampshire, Durham, NH 03824, USA

B.L. Dingus, B.E. Schaefer
Univ. Space Research Assoc., NASA/GSFC, Greenbelt, MD 20771, USA

D. Band, J. Matteson
UC San Diego, La Jolla, CA 92093, USA

W. Collmar, V. Schönfelder
Max Planck Inst. for Extraterrestrial Physics, 8046 Garching, Germany

C.E. Fichtel, P.W. Kwok,* B.J. Teegarden
NASA/Goddard Space Flight Center, Greenbelt, MD 20771, USA

G. Fishman
NASA/Marshall Space Flight Center, Huntsville, AL 35812, USA

L. Kuiper
SRON-Leiden, 2300 RA Leiden, The Netherlands

G.V. Jung
Univ. Space Research Assoc., NRL, Washington, D.C. 20375, USA

S.M. Matz
Northwestern University, Evanston, IL 60208, USA

P.L. Nolan
Stanford University, Stanford, CA 94305, USA

E.J. Schneid
Grumman Aerospace Corp., Bethpage, NY 11714, USA

C. Winkler
Astrophys. Div. ESA, ESTEC, 2200 AG Noordwijk, The Netherlands

* NRC Research Assoc.

ABSTRACT

GRB 910601 was well observed by all three broad-field experiments on the COMPTON Observatory; it was also at a known position within the narrow aperture of the OSSE instrument. This has permitted us to compare spectra observed from all four of the COMPTON Observatory instruments. The burst lasted for about 40 s and was observed into the MeV region by all of the detectors; it was not detected at energies above 10 MeV. Time-integrated spectra from COMPTEL, EGRET and OSSE are in good agreement. Spectra from both the BATSE large area and spectroscopy detectors during the early part of the burst are also in good agreement with the OSSE spectrum.

INTRODUCTION

The burst of 1991 June 1 provided the first opportunity to compare the spectral responses of the four COMPTON Observatory experiments over a broad dynamic range. Schaefer, et al.[1], Winkler, et al.[2], Kwok, et al.[3], and Share, et al.[4] describe measurements individually made by the BATSE, COMPTEL, EGRET, and OSSE instruments, respectively. This burst occurred in the field of view of two of the OSSE detectors. Its location is depicted in Figure 1 which displays the individual error boxes derived from the OSSE and COMPTEL instruments; also shown is the arc derived from the interplanetary network (see Hurley et al. these Proceedings). The best source location is derived from the intersection of the COMPTEL error circle with the IPN arc. This position was used to derive the spectrum of the burst from OSSE; we estimate that there is $\sim 10\%$ error in the flux due to the uncertainty in position.

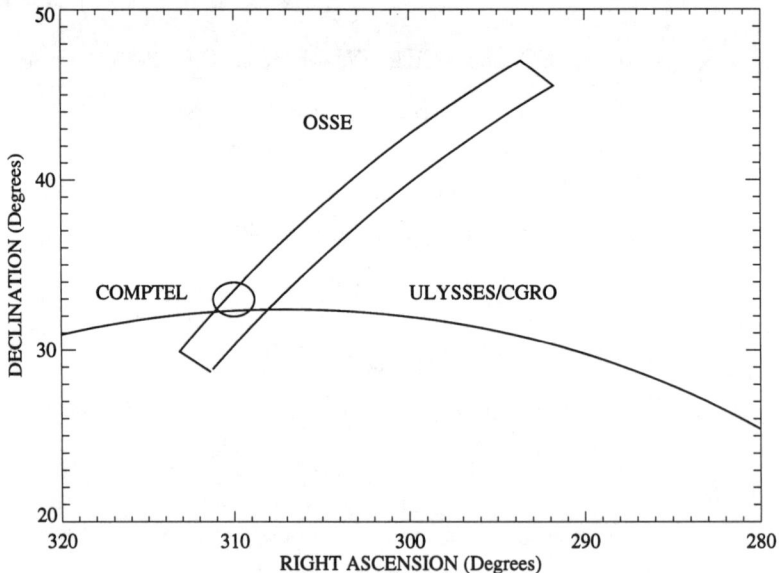

Figure 1. Positions for the source of the June 1 burst derived from COMPTEL (oval), OSSE (error box) and CGRO/Ulysses (arc).

The burst history recorded >100 keV by the summed shield elements of OSSE is displayed in Figure 2. BATSE triggered on the small peak at 4 s. OSSE accumulated spectral data in three 16.38 s intervals during the burst. Spectral comparisons with BATSE were made during interval #1 and are described in the next section. OSSE spectral comparisons with COMPTEL and EGRET were from data summed over all three intervals; the time interval over which the EGRET spectrum was accumulated is not identical with OSSE but does encompass a bulk of the emission shown in Figure 2.

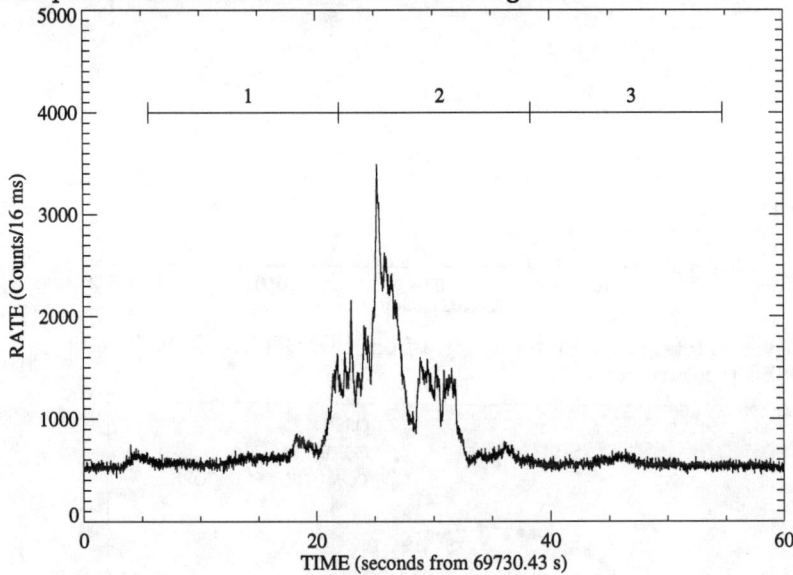

Figure 2. Count rates >100 keV recorded at 16 ms resolution by OSSE annular shields. Intervals in which spectra were accumulated are shown.

SPECTRAL COMPARISONS

Figure 3 shows the spectra measured by two of the OSSE detectors (summed together), the large EGRET NaI crystal (TASC), COMPTEL's D2 NaI detector, and the COMPTEL telescope; upper limits are shown for the EGRET spark chamber over this time interval. The agreement between all the measurements is good, although the COMPTEL spectra appear to be harder than those of either OSSE or EGRET. A clearer comparison is revealed in Figure 4 where the differential spectral points are multiplied by E^2. The solid line is a fit to the OSSE data with a model consisting of the sum of an exponentiated power law and a high-energy power law.

Spectral data from BATSE are only available for the early part of this burst. BATSE detector module #2 had minimum exposure to scattered radiation from the earth; this was thus used for comparison. Spectra from both the large area detector (LAD) and spectroscopy detector (SD) are compared with the OSSE spectrum in Figure 5. The SD data are in excellent agreement with OSSE over the full range in energy. The LAD and OSSE data agree quite well

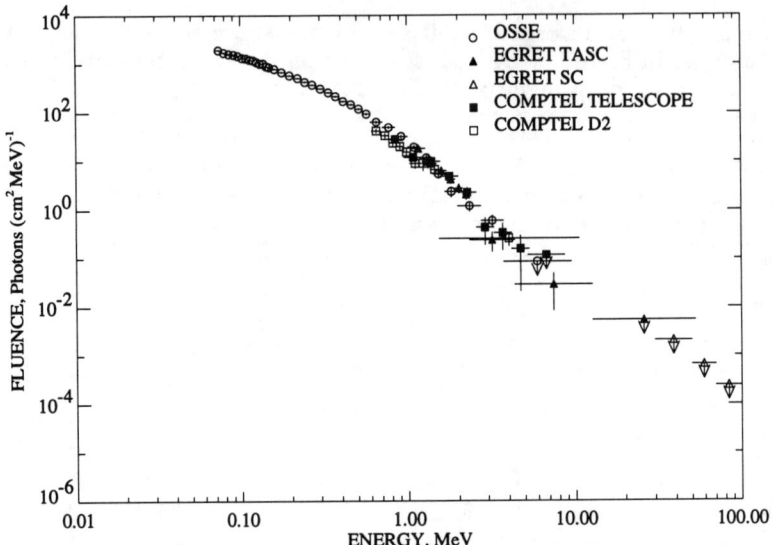

Figure 3. Integrated photon spectra from COMPTEL, EGRET, and OSSE measurements.

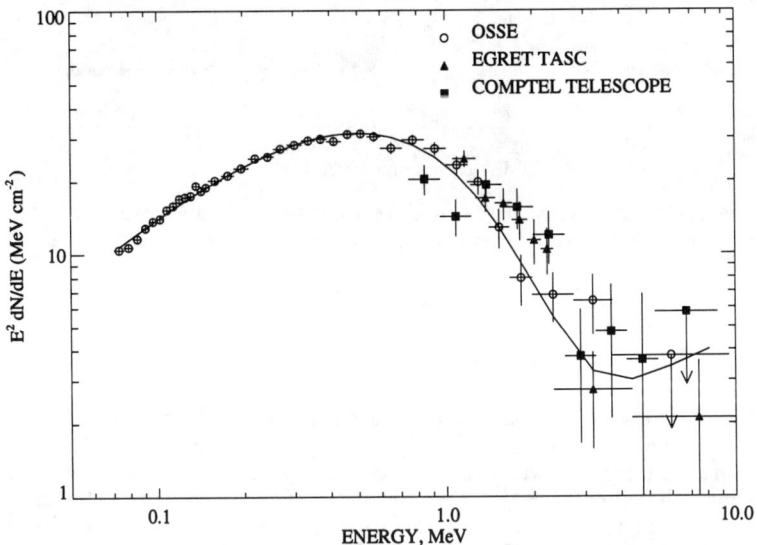

Figure 4. Integrated photon spectrum from COMPTEL, EGRET, and OSSE measurements.

<0.6 MeV; the LAD data points fall systematically below those of OSSE at higher energies. This could be due to uncertainties in the instrument response function for these relatively thin detectors near 1 MeV. The solid curves show a fit to the OSSE data.

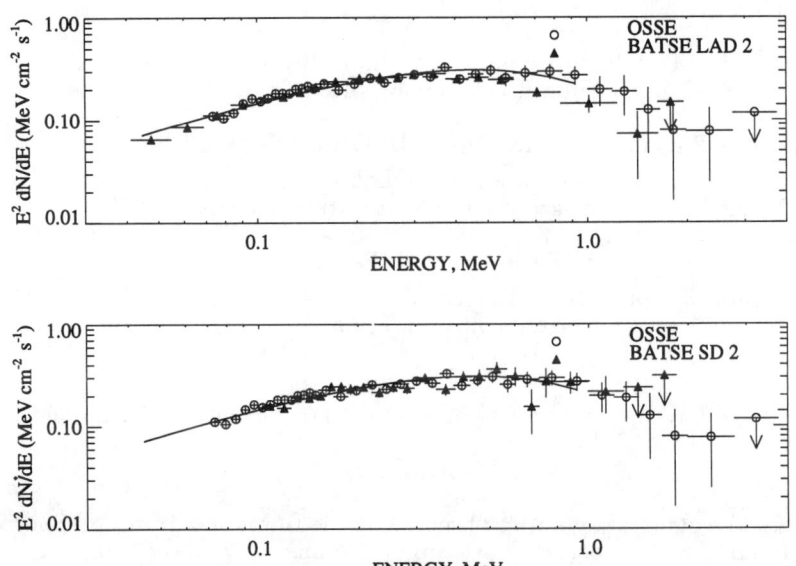

Figure 5. Comparison of BATSE LAD and SD spectra with OSSE spectra early in the burst.

These comparisons indicate that the full complement of detectors on the COMPTON Observatory can be used to construct spectra of bursts over a broad range in energies, from ~ 0.05 to ~ 1000 MeV. Further evidence for the good agreement between BATSE, COMPTEL, and EGRET spectra of bursts is given by Schaefer, et al. in these Proceedings.

It is also clear from Figures 3-5 that there is no evidence for any spectral features superimposed on the broad continuum in this burst. The measured fluence of the burst is $\sim 1.2 \times 10^{-4}$ erg cm^{-2} >70 keV.

We wish to thank Rick Mackinnon for assistance in the analysis of the OSSE data.

REFERENCES

1. Schaefer, B.E., et al., Ap. J. Lett **393**, L51 (1992).
2. Winkler, C., et al., Compton Gamma-Ray Observatory (Amer. Inst. of Physics, N.Y., 1993), p. 845.
3. Kwok, P.W., et al., Compton Gamma-Ray Observatory (Amer. Inst. of Physics, N.Y., 1993), p. 855.
4. Share, G.H., et al., Gamma-Ray Bursts (Amer. Inst. of Physics, N.Y., 1992), p. 32.

SPECTRAL EVOLUTION OF A SUB-CLASS OF GAMMA RAY BURSTS

P. N. Bhat*
Tata Institute of Fundamental Research,
Homi Bhabha Road, Bombay 400 005, India

G. J. Fishman, C. A. Meegan, R. B. Wilson, C. Kouveliotou†
ES-66, Space Science Laboratory,
NASA/Marshall Space Flight Center, Huntsville, AL 35812

W.S. Paciesas, G. N. Pendleton
Department of Physics, University of Alabama in Huntsville,
Huntsville, AL 35899

B. E. Schaefer
NASA/Goddard Space Flight Center, Code 661, Greenbelt MD 20771

ABSTRACT

Among the gamma ray bursts observed by the Burst and Transient Source Experiment (BATSE) on board the Compton Gamma Ray Observatory we define a sub-class of bursts based on similar morphology: a sharp rise followed by a longer decay time. About 7% of all the Gamma Ray Bursts observed by BATSE fall in this sub-class. We study the spectral evolution of these bursts by fitting models to time segmented burst spectra and find no clear distinction between the spectral evolutionary properties of this sub-class and those of other bursts. Further we study the high time resolution spectral evolution of this sub-class of GRB's using their spectral hardness ratios. A majority of the bursts show hardness ratio leading the counting rate and also display a continuous hard to soft evolution. The time lag between the counting rate and the hardness ratio is found to be directly correlated with the rise time of the counting rate profile. We also find for the first time, evidence for spectral variation in a time scale of 64 ms.

INTRODUCTION

One of the characteristics of a "classical" gamma ray burst (GRB) is its hard energy spectrum, often with a very high energy tail extending up to tens of MeV[1]. These spectra are remarkable in that nearly all of the emission is at γ-ray energies. The power per logarithmic energy interval rises steeply in all burst spectra at low energies and often peaks around 100 keV. A study of the evolution of the spectral characteristics is expected to lead to an understanding of the possible emission mechanisms. A hard-to-soft spectral evolution in the 50 keV- 2 MeV energy range has been observed in many GRB's with durations greater than 1 s.[2] The hardness ratio decreases monotonically from the burst rise through the decay phase. However there are several exceptions to this general

---

* Email (Internet): pnbhat@tifrvax.tifr.res.in
† University Space Research Association.

behaviour. Continuum spectra of GRBs seem to be variable on time scales as short as the detector time resolution and there is some indication of a correlation between temporal variability of luminosity (derived from the count spectra) and a parameter measuring the hardness, like the temperature[3,4]. Also, some bursts begin with hard spectra while in others hardness seem to be correlated with intensity[5] and do not show an unambiguous correlation of luminosity with temperature[6].

In 1990 Jourdain[7] analyzed the count spectra of several bursts observed by the APEX experiment and concluded that spectral evolution has no correlation with the time history. Thus spectral evolution varies significantly over the distribution of observed GRB's. We chose a sample of bursts which have short rise time (≤ 0.5 s) and a nearly exponential decay time. Some of the GRBs chosen have a smooth profile while others are highly structured.

OBSERVATIONS

Each of the eight BATSE Large Area Detectors (LAD's) consist of 1.27 cm thick NaI(Tℓ) crystal of size 50.8 cm in diameter. The details of the trigger criterian and various data types available are discussed elsewhere[8].

ANALYSIS

A background spectrum is generated by a polynomial fit to the spectra before and after each burst and then interpolating to the burst interval. Each burst belonging to this sample is subdivided into segments such that a background subtracted spectrum for this segment has statistically significant signal in the energy range of 100 - 2000 keV. Each spectrum is then fitted to either one or a combination of the following functions: (a) optically thin thermal Bremsstrahlung spectrum (OTTB) (b) power-law with an exponential cut-off function (COMP) (c) black-body spectrum and (d) power-law function. The choice of spectral function is merely to parameterize the spectrum rather than to study the physical process responsible for the photon emission at the burst source. In some cases like the burst 1B910814, the spectrum is not well represented by the OTTB function while either a COMP function or a combination of a black-body spectrum and a power-law function fit the spectra best. In cases where more than one function fits the spectrum, that function with lesser number of parameters is chosen. We used the Batse Spectral Analysis Software (BSAS) for deriving the spectra and the functional fits. The detailed simulation studies[9] of the detector response at various energies and source angles have been used in deconvolving the count spectra before fitting to standard functions.

To study the spectral evolution over short time scale ($\sim$ 64 ms) we used the hardness ratio (defined as the ratio of the number of counts above 100 keV to those below 100 keV) computed using 4-channel spectral data. Thus we could study the spectral evolution over different time-scales during each burst.

RESULTS

A majority of the bursts with varying pulse profiles show a good correlation of the spectral hardness with count rate (correlation coefficient ranging from 0.64 to 0.98). In cases where more than one model function was fitted, different hardness parameters of different models (like the 'temperature' and the power law index show similar or identical evolution during the burst. Hence the correlation does not depend on the spectral-model chosen to fit the data. This result is also borne out by a similar analysis carried out on the data from the second brightest detector for the same burst, which also rules out possible systematic effects due to differences in the detector response functions for different burst incident angles.

However there are a few (2 among 19 in the present sample) bursts which do not show any spectral evolution at all. There are a few bursts (2 among 19 in the present sample) which show a negative correlation during the rising part of the burst and a positive correlation with the burst intensity during the decay phase (Fig. 1).

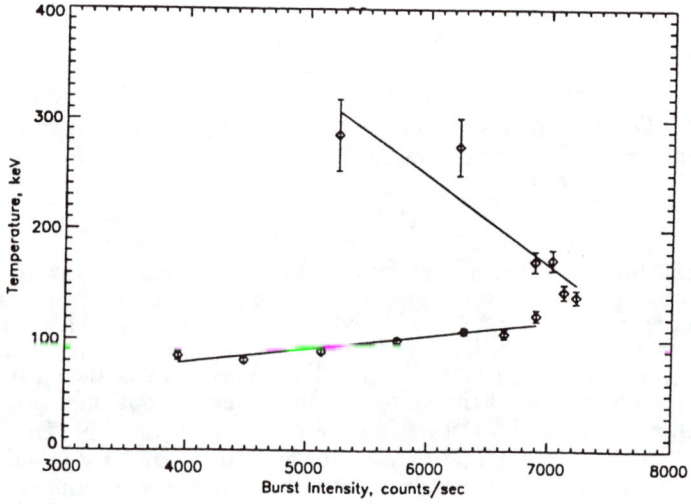

Figure 1. A plot of the OTTB temperature in keV as a function of the burst intensity for 1B920216. The rising part of the burst displays an anti-correlation while the decay phase displays strong correlation.

Figure 2 shows the time histories of the spectral hardness ratio and the intensity (background corrected count rate; dotted line), suitably scaled, on the same plot for 1B911031. It may be noted that for bursts which show a correlation between fitted hardness parameter and burst intensity for most part of the burst, the two time histories are similar in shape but shifted in time. In a majority of the cases the hardness leads the counting rate. In 2 cases (among 19 in this sub-set) the hardness lags. Except for this difference, these bursts behave similarly to most other 15 bursts showing a similarity in the time histories. In 2 other cases (1B910421 and 1B910602) the hardness ratio does not show any significant variation during the entire burst.

The hardness ratio time profile has its own rise and decay time. The time lag between the hardness ratio profile and the count rate profile has been estimated for those bursts which show a positive lag. This time lag, τ_l is proportional to the rise time, τ_r of the burst profile and the correlation coefficient

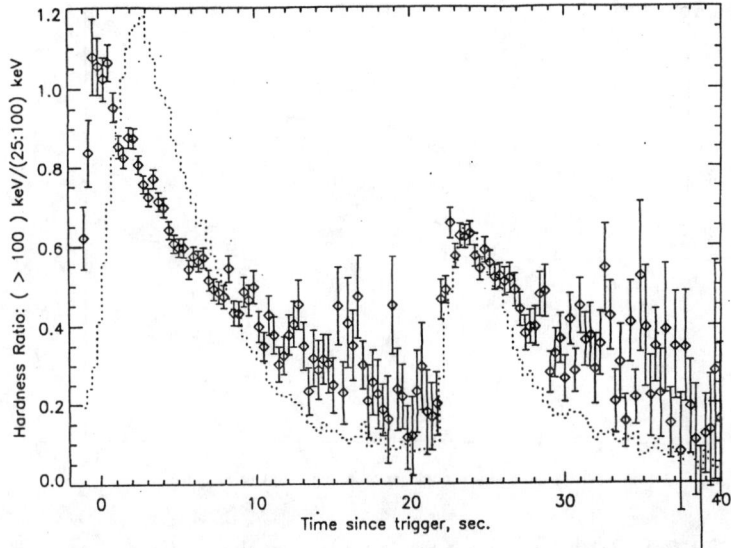

Figure 2. A relative comparison of the evolutions of the hardness ratio and burst intensity for 1B911031. Hardness leads burst intensity significantly.

is 0.8. This relationship can be written as : $\tau_l = K \times \tau_r^{(0.9\pm 0.2)}$, where K is a constant. (Fig. 3)

DISCUSSION

We studied a sample of 19 bursts using their hardness ratios as their spectral hardness parameters. For 7 randomly chosen bursts, the analysis was also carried out using the conventional spectral fitting techniques in order to confirm the consistency between the two procedures. It is clear from this analysis that the GRB's which belong to this sub-class do not share common spectral evolution properties. The results from the two methods establish that the spectral evolution of gamma ray bursts does not depend significantly on the parameters chosen to represent the spectral hardness.

A correlation between the spectral hardness and burst intensity has been suggested to imply that the instantaneous value of the source luminosity is determined by the temperature in the emitting region[10] However this is a model dependent conclusion. The soft lag observed in a majority of the GRB's in this sub-set is not specific to these bursts. A similar behavior was reported for short gamma ray bursts by Bhat et al.[11,12] who showed that the time lag varies as log(E) where E is the mean photon energy.

The fast spectral variability seen in 1B910814 could place a limit on the size of the production region. The observation of spectral variability on the 64 *ms* time scale would imply that the size of the effective emission region is $\leq 2 \times 10^4$ km, assuming no reltivistic effects.

REFERENCES

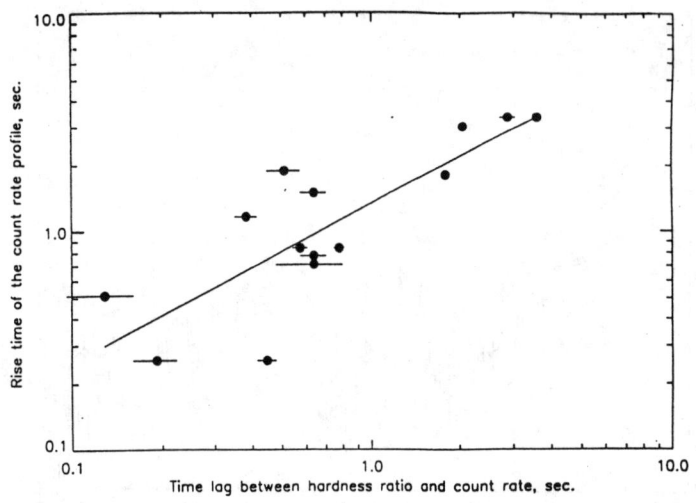

Figure 3. A plot of the burst profile rise-time as a function of soft lag. The correlation coeft. is 0.8.

1. S.M. Matz *et al.*, Ap. J. Letters **288**, L37 (1985).
2. J.P. Norris *et al.*, Ap. J. **301**, 213 (1986).
3. E.P. Mazets *et al.*, Astrophys. and Space Sci. **82**, 261 (1982).
4. S.V. Golenetskii *et al.*, Nature **306**, 451 (1983).
5. K. Hurley *et al.*, Gamma Ray Bursts, Huntsville (Ed.: W.S. Paciesas and G.J. Fishman, 1991), p. 195.
6. V.E. Kargatis *et al.*, Gamma Ray Bursts, Huntsville (Ed.: W.S. Paciesas and G.J. Fishman, 1991), p. 201.
7. Jourdain, E., Ph. D. Thesis (C.E.S.R., Paul Sabatier University, Tolouse, France, 1990).
8. G.J. Fishman *et al.*, Gamma Ray Observatory Science Workshop; Greenbelt (Ed.:W.N. Johnson, 1989), p. 2-39.
9. G.N. Pendleton *et al.*, Gamma Ray Observatory Science Workshop; Greenbelt (Ed.:W.N. Johnson, 1989), p. 4-547.
10. E. P. Mazets *et al.*, Positron Electron Pairs in Astrophysics, (Ed: M. L. Burns, A. K. Harding and R. Ramaty, 1983), p. 36.
11. P. N. Bhat *et al.*, Compton Observatory Symposium, St. Louis, (Ed: N. Gehrels, M. Friedlander and D. J. Macomb, 1992), p. 953.
12. P. N. Bhat *et al.*,, Astrophys. J **1994**, (May, 10).

DECOMPOSITION OF A COSMIC GAMMA-RAY BURST INTO TWO NON-CORRELATED RADIATION COMPONENTS

Anton M. Chernenko and Igor G. Mitrofanov
IKI, Moscow

ABSTRACT

Emission of gamma-ray burst GB 881024 is decomposed into two physically distinct noncorrelated components by means of a technique called Rate-Rate diagrams. One of the components dominates at the energies below 500 keV and the other one — above 500 keV. Time histories and rate/spectrum evolution of the components are considered. The softer component manifests clear hardness/rate correlation while the harder one does not. Possible astrophysical consequences are discussed.

INTRODUCTION

It has become clear that statistics of GRBs can not alone resolve the problem of their origin[1]. In this situation any new phenomenology that could lead to clarification of physical mechanism of the emission is very important. In this respect we consider possible identification of distinct spectral components in the emission of GRBs as a significant step towards understanding of the emission physical mechanism.

For this purpose we analyzed the continuum variability of the most intense GRB recorded by Soviet-French APEX experiment (see Figure 1). This analysis involved a technique called Rate-Rate diagrams.

IDENTIFICATION OF COMPONENTS ON RATE-RATE DIAGRAMS

Recently a simple method of study of spectral variability of GRBs called Rate-Rate diagrams (RRDs) was proposed[2]. On the RRD each time interval is represented by a point, which coordinates X, Y correspond to photon rates in two different energy channels. In Figure 2 an RRD is presented for GB 881024. X-axis corresponds to a soft photon rate $R[75 - 255]$ measured in the energy range 75–255 keV, Y-axis corresponds to the hard photon rate $R[925 - 7750]$. All time-to-spill intervals of spectral measurements, that contain statistically significant number of photons above the background are displayed.

There are two clusters of points on the RRD: a cloud (A) of points that occupy the upper part of the RRD, and group (B) of points concentrated along a certain curve in the lower part of the RRD. From Figure 1 one can conclude that time intervals of cloud A constitute the most intense peak of the burst.

Let us first consider time intervals that group along the curve (B). It is clear that hard and soft photon rates over these intervals are highly correlated. To study this correlation in more details, we determined photon rates R_j in 8 broad energy channels: 75–140 keV, 140–255 keV, 255–512 keV, 512–925 keV, 925–1880 keV, 1880–3040 keV, 3040–4980 keV, and 4980–7750 keV. Then we considered 8 RRDs in logarithmic representation (e.g. see Figure 3). A broad channel 75–255 serves as a 'reference' channel for all of them. It is clear that the rate in this reference channel $R[75 - 255] = R_1 + R_2$, so it is labeled $R_{[1-2]}$.

294 Decomposition of a Cosmic Gamma-Ray Burst

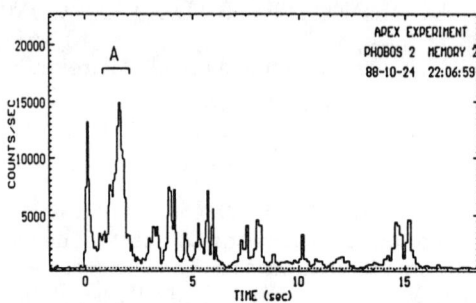

Figure 1. Time history of the GB881024.

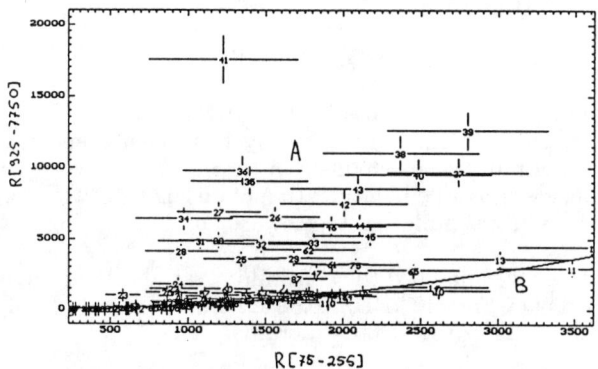

Figure 2. Rate-Rate diagram for the entire GB881024. Each time interval is labeled by its number. $R[75-255]$ and $R[925-7750]$ are expressed in phot/s. Time intervals that constitute the cloud (A) are indicated in the time history (Figure 1).

It is apparent that the correlations are linear between all $\lg(R_j)$ and $\lg(R_{[1-2]})$. Thus, we came to the following models:

$$R_j = a_j \cdot \left(R_{[1-2]}\right)^{b_j}, \qquad (1)$$

that contain pairs of free parameters (a_j, b_j) associated with each channel j. These parameters were found by minimizing corresponding χ^2 functionals[3].

For any time interval models (1) permit to determine the photon rate in any of 8 energy channels from the single parameter — soft photon rate $R_{[1-2]}$. In this sense one may state that there exists a single soft emission component over all intervals beyond the most intense peak of the burst.

Hard and soft photon rates over the most intense peak of the burst do not show any evident correlation. To study this correlation pattern we considered correlations between rates in all 8 channels. In Figure 4 we graphically present a matrix of linear correlation coefficients (LCC) corrected for Poissonian fluctuations[3]. Upper/left part of the matrix presents the LCC themselves, dark color here corresponds to the $LCC > 0.8$. The lower/right part of the

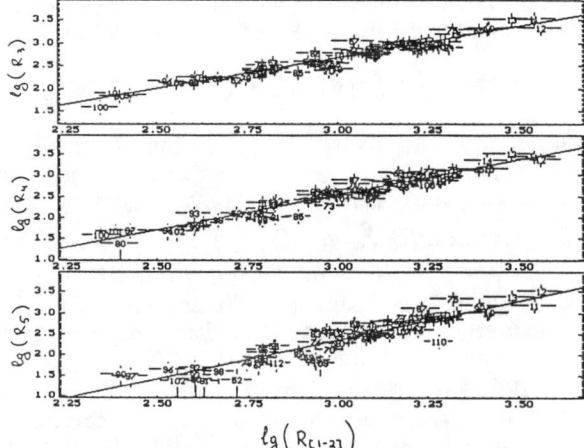

Figure 3. RRDs for time intervals beyund the most intense peak in logarithmic representation.

matrix presents the level of statistical significance of found correlation. Dark color corresponds to the significance of the correlation higher than 3σ level.

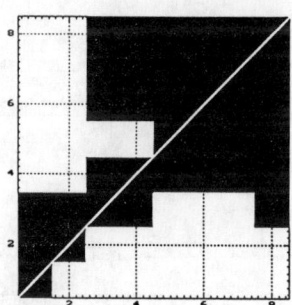

Figure 4. Graphical representation of correlation matrix for photon rates within the most intense peak (see text).

The following conclusions could be drawn from the consideration of this matrix:

1. It is evident that photon rates in channels 4–8 manifest a high degree of mutual linear correlation. This resembles the behavior of soft component beyond the most intense peak. Therefore it is quite natural to suppose that during the most intense peak of the burst there exists a hard emission component, which dominates above 500 keV.

2. There is definitely no correlation between rates in hard 4-8 and soft 1-3 energy channels. It confirms that hard and soft photon rates are probably independent. This implies that the hard component does not contribute significantly to channels 1-3. So a hypothesis may be proposed that there exists a softer component dominating in channels 1-3.

3. It could be assumed that there is a high degree of linear correlation between rates in channels 1-3. So, it is very natural to suggest that the soft

component identified outside the most intense peak (intervals of group (B), see Figures 1, 2) persists also during the peak and dominates in channels 1–3.

ANALYSIS OF INDIVIDUAL COMPONENTS

Let us assume, in accordance with the conclusion (2) above, that the hard component does not contribute at all to channels 1, 2 during the most intense peak. In this case, following the conclusion (3) above, we can by means of models (1) determine the contribution $R_{S,j} = a_j \left(R_{[1-2]}\right)^{b_j}$ of the soft component to the channels with numbers $j > 2$. If this contribution is removed the residuals $\tilde{R}_j = R_j - R_{S,j}$ in energy channels $j > 2$ represent the emission of hard component.

In Figure 5 we present photon rates of both components in the energy channel [3–5] that corresponds to the energy range 255–925 keV. Both components contribute significantly to this channel. The calculated rate of the soft component is denoted by $R_{S,[3-5]}$, and the rate of the hard component — $\tilde{R}_{[3-5]}$. One may see that the soft component persists over all burst while the hard component dominates in the second peak on the time history only. On the other hand there are a few other time intervals where it is also present.

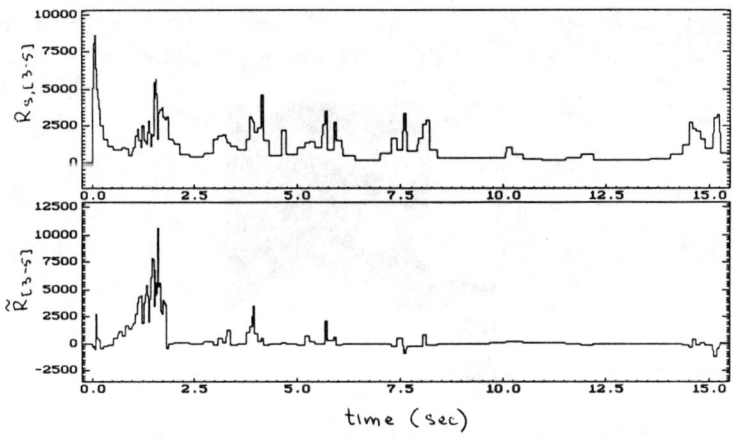

Figure 5. Time histories of the components in the energy range 255–925 keV. The upper panel corresponds to the soft component.

For both components new sets of parameters (a_j, b_j) that associate rates $R_{S,j}$ and $R_{H,j}$ with corresponding reference rates in channel [3–5] were estimated. The knowledge of 8 photon rates as a function of reference photon rate is equivalent to the knowledge of the spectrum. In Figure 6 we show how spectra of the components change as rates of the components in channel [3–5] increase from 500 to 10000 ph/s. Uncertainties of the spectral density were calculated by error propagation from the uncertainties of coefficients (a_j, b_j).

All spectra are well approximated by a common spectral shape function $f(E) = E^{\gamma} \exp(-E/E_0)$. Here is the summary of spectrum–rate evolution of the components in terms of parameters γ and E_0:

Soft component with increasing photon rate in the reference channel [3–5] the main trend is the increase of γ from -1.25 up to -0.35 which corresponds to

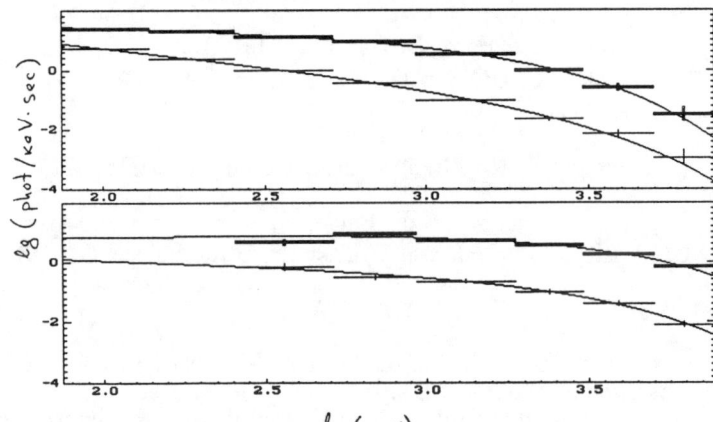

Figure 6. Evolution of the spectra of the components as the rate in the range 255-925 keV varies from 500 to 10000 ph/s. The upper panel corresponds to the soft component.

the positive hardness/rate correlation. The energy of cutoff E_0 slightly decreases from 1500 keV down to 1100 keV.

For the hard component both spectral index γ and the high energy cutoff E_0 do not vary significantly with the rate in the reference channel [3-5]. Actual value of γ is about 0.0 and E_0 is about 1.8 MeV.

DISCUSSION

Identification of two physically different components indicate that at least some of the burst sources should comprise several distinct emitters of gamma-rays. At the same time a closer inspection of time histories shows that peaks of the components generally coincide. This may imply that there existed a single source of energy for both emitters. In this case NS scenario for GRBs could involve two different active regions in the magnetosphere. Cosmological scenario could involve two different interacting objects.

Analysis of other APEX events[3] and of BATSE data[4] has shown that multicomponent structure of GRB emission is rather common phenomenon. Therefore further investigation is possible to find physical mechanisms that could produce the sort of spectrum/flux evolution described above.

REFERENCES

1. C. Meegan, these Proceedings, (1993).
2. A. Kozlenkov et. al., Proceedings of the Los Alamos Workshop on Gamma-Ray Bursts (Cambr. Univ. Press, 1990), p. 255.
3. A. Chernenko et. al., Preprint IKI No. 1856 (Moscow, 1993).
4. G. Pendleton, these Proceedings, (1993).

CONTINUUM EVOLUTION OF BRIGHT GAMMA RAY BURSTS OBSERVED BY BATSE

L. Ford, D. Band, J. Matteson
CASS, UC San Diego, La Jolla, CA 92093

B. Teegarden
Goddard Space Flight Center, Greenbelt, MD, 20771

W. Paciesas
University of Alabama at Huntsville, Huntsville, AL, 35812

ABSTRACT

We investigate the evolution of the peak energy of νF_ν for bright gamma ray bursts observed by the BATSE spectroscopy detectors. We find that the peak energy is correlated with the intensity of the burst, that this energy softens over the burst as a whole and within individual intensity spikes, and that intensity spikes which come late in a burst tend to be softer than earlier spikes within the same burst. We also find evidence that bursts in which the main emission comes well after the triggering event tend to be softer and evolve less than bursts in which the main emission comes promptly after the trigger.

INTRODUCTION

Gamma ray burst continuum spectra are a reflection of primary burst processes. Therefore, an understanding of the structure and evolution of continua should provide vital clues for solving the mystery of gamma ray bursts. The BATSE spectroscopy detectors were designed with this goal. This set of eight 5" diameter by 3" thick NaI detectors provide a series of spectra covering two decades in energy which are analyzed into 256 quasi-logarithmic energy channels. A time-to-spill accumulation mode gives integration intervals which vary from ~ 0.1 seconds (high intensity) to > 1 second (background).

Previously, the most common measure used for spectral evolution was the hardness ratio,[1,2] which is the flux ratio in two adjacent energy bands. Although this ratio traces the hardness of a spectrum, its value does not have a quantitative physical meaning. With this problem in mind, Kargatis et al.[3] used physical models to fit burst spectra and described evolution in terms of the fitted parameters. However, the true physics of continuum generating processes is not known rendering their discussion of continuum evolution somewhat model-dependent and the physical significance of their parameters ambiguous.

To describe burst continuum spectra we use the empirical model[4]

$$N_E(E) \left(\frac{\text{photons}}{\text{keV-s-cm}^2}\right) = \begin{cases} A\left(\frac{E}{100 \text{ keV}}\right)^\alpha e^{-E/E_0}, & E \leq (\alpha - \beta)E_0 \\ A'\left(\frac{E}{100 \text{ keV}}\right)^\beta, & E > (\alpha - \beta)E_0 \end{cases} \quad (1)$$

where $A, \alpha, \beta,$ and E_0 are fit to observed spectra and A' is chosen so that the function is continuously differentiable everywhere. A convenient feature of this

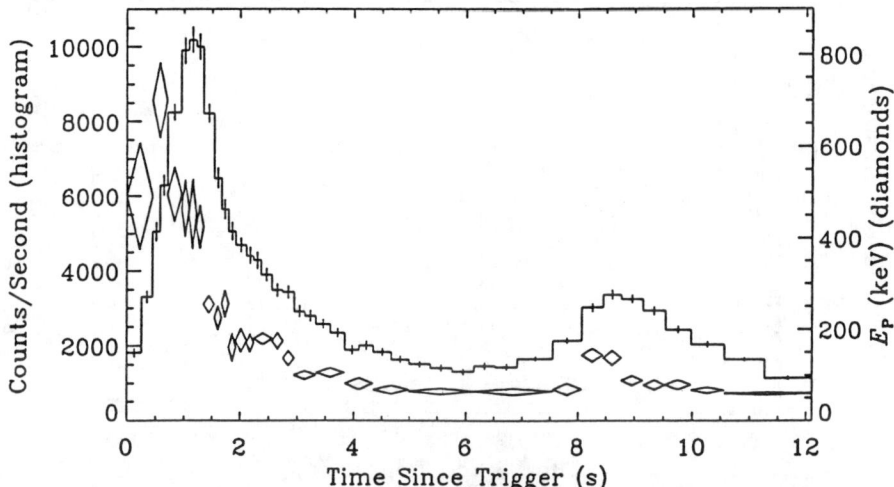

Figure 1. The evolution of E_P for GRB921207 plotted over a time history from spectroscopy detector 0 (20.2-1283 keV). The height of the diamonds represents the 1σ error in E_P.

model is that it has the flexibility to mimic most simple physical models such as thermal bremsstrahlung, power law, photon exponential, *etc.*, and therefore accommodates our ignorance of the continuum physics. Although this model is phenomenological, we can make statements about spectral evolution based on physically meaningful parameters derived from it. Here we consider the evolution of peak energy flux per logarithmic energy band (peak in $\nu F_\nu \equiv E_P$) during a burst. Since $E^2 N_E(E) \propto \nu F_\nu$, E_P is just $(2+\alpha)E_0$ (assuming $\beta < -2$).

ANALYSIS

We analyzed bright, long gamma ray bursts observed by the BATSE spectroscopy detectors in high gain states (energy range ~15-4000 keV). In order to ensure E_P was well determined, we averaged consecutive spectra until a signal-to-noise ratio (S/N) of at least 15 was achieved in the 60–200 keV band. The entire energy range above the SLED[5] (an electronic artifact) was fit. One difficulty with using high gain detectors was that occasionally the upper spectral index β was larger than -2 which means either E_P occurs at an energy outside the fit range or that the high energy signal is insufficient to fix β. For this work, β was constrained to be less than -2 for all fits. For those few bursts where $\beta > -2$, this condition holds for most of the burst. If β is poorly determined, it should be ignored, but if β is indeed greater than -2, the constraint imposes a uniform bias.

RESULTS

Of the ~ 750 BATSE bursts detected as of September 1993, only 33 had at least eight spectra which met our S/N criterion. For these bursts, E_P was plotted over the count rate and various properties were noted. Figures 1 and 2 show two bursts which typify our results. In Figure 1, it is apparent that

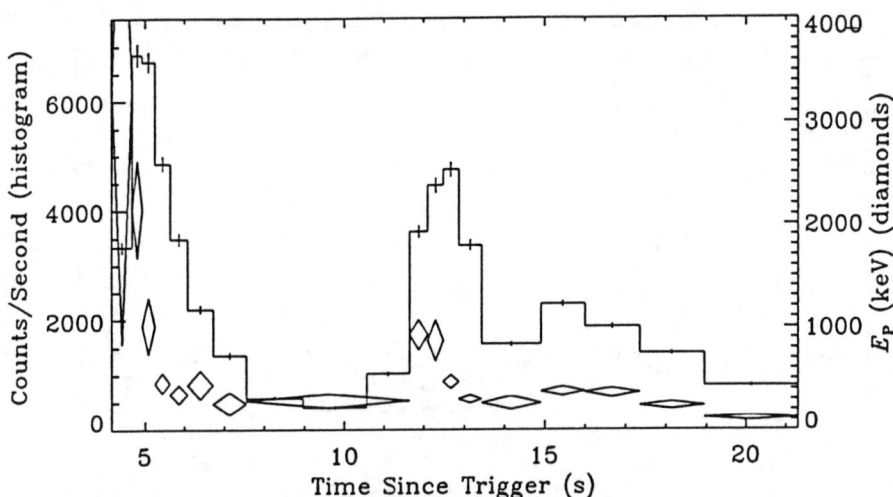

Figure 2. The evolution of E_P for GRB920525 plotted over a time history from spectroscopy detector 5 (15.4-1327 keV). The height of the diamonds represents the 1σ error in E_P.

E_P softens over the whole burst and that it increases along with the intensity 8 seconds after the trigger. Figure 2 shows a burst in which we were able to resolve two intensity spikes. E_P softens within both spikes and the later spike is softer than the first.

Table 1. Results for 33 Bursts		
# Observed	# Possible	Property
20	21	E_P– intensity correlation
16	26	E_P softens over whole burst
2	26	E_P hardens over whole burst
15	18	E_P softens within intensity spikes
1	18	E_P hardens within intensity spikes
6	12	Later spikes softer than earlier ones
1	12	Later spikes harder than earlier ones

Characteristics for the sample as a whole appear in Table 1. In this table, the first column refers to the number of bursts in which the trait in question was seen and the second gives the number of bursts in which we could have seen the trait. Not all bursts could be classified due to large errors in E_P, unresolved structure, poor time resolution, *etc.* Table 1 shows clear evidence for an E_P-intensity correlation but since we were unable to resolve fine time structure,

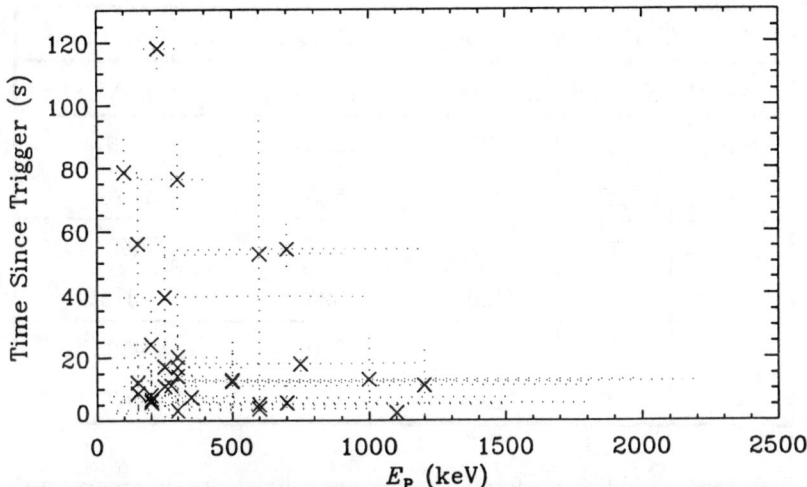

Figure 3. The range in E_P for a given emission time. The horizontal bars give the range in E_P (max-min) and the vertical bars show the period during which significant emission occurred relative to trigger. The crosses mark median values.

a quantitative study (*e.g.* lags or leads) of this correlation was not possible. The table also shows that most bursts soften both overall and within individual intensity spikes and that there is a trend toward softness in successive spikes.

The amount of spectral variability in bursts as a function of when the main burst emission occurred was also studied, where main emission is defined as the time during which spectra met our S/N criterion. In Figure 3 we show the time of main emission relative to trigger plotted against the range in E_P for each burst. This shows that our sample was devoid of bursts which were hard, evolved considerably, and whose emission came well after the BATSE trigger. Since this interpretation depends strongly on the bursts in which the emission was delayed substantially, we divided the bursts into two groups based on the ratio of the time at which significant emission began relative to trigger and the duration of main emission. This is preferable to the absolute timescale used in Figure 3 since it is a relative measure of time delay. Table 2 shows the number of bursts in each group for several time ratios as well as values for the range in E_P ($\Delta E_P = \max(E_P) - \min(E_P)$ within a burst). It can be seen from this that the group in which the emission comes late tends to show less evolution in E_P.

DISCUSSION

We showed in this work, that E_P is correlated with a burst's intensity and that bursts whose main emission occurs at relatively long times after the trigger tend to be softer and have less evolution in E_P. These results have interesting implications for gamma ray bursts. It appears that spectral models must account for the length of time between the triggering event and any significant emission. A possible interpretation might be that more time between an initial output of energy and significant observed emission allows the energy to be distributed among more particles or be degraded by long timescale processes.

Table 2. Bursts Divided According to Emission Time

$\frac{\text{start time}}{\text{duration}}$	Number	min(ΔE_P)	median(ΔE_P)	max(ΔE_P)
> 0.5	8	80	300	950
≤ 0.5	25	120	525	2075
> 0.9	7	80	300	950
≤ 0.9	26	120	525	2075
> 1	5	80	300	950
≤ 1	28	120	440	2075

This work might also explain the apparent hardness difference in the two duration classes of gamma ray bursts. It has been shown that there is a bimodality in the duration distribution of gamma ray bursts and that shorter bursts tend to be harder than longer ones.[6,7] The hardness measures used for the bimodality studies were averaged over the whole burst. Our study indicates that time averaged long bursts tend to be softer than shorter ones since spectra from longer bursts have time to evolve. As Norris et al.[8] also suggested, the first two seconds of long duration bursts may be no softer than short bursts.

The data in Figure 3 include several bursts in which the emission is prompt but soft and slowly evolving. An interesting possibility is that BATSE missed initial event of these bursts and triggered when the (late) main emission occurred. If this is indeed the case, then the amount of evolution in E_P which occurs in a burst may be even more strongly related to the time elapsed since the initial burst event than our analysis indicates.

We wish to thank M. Briggs, G. Pendleton, and R. Preece for invaluable assistance with software. BATSE work at UCSD is supported by NASA contract NAS8-36081.

REFERENCES

1. J.P. Norris et al., ApJ **301**, 213-219 (1986).
2. N. Bhat et al., these proceedings (1993).
3. V. Kargatis et al., ApJ, in press (1994).
4. D. Band et al., ApJ **413**, 281-292 (1993).
5. D. Band et al., Experimental Astronomy **2**, 307-330 (1992).
6. C. Kouveliotou et al., ApJL **413**, L101-L104 (1993).
7. D.Q. Lamb et al., ApJL, in press (1994).
8. J.P. Norris et al., these proceedings (1993).

A STUDY OF CONTINUUM SPECTRA OF SHORT-DURATION GAMMA-RAY BURSTS OBSERVED BY BATSE

T. M. Koshut, G. N. Pendleton, R. S. Mallozzi, W. S. Paciesas, M. S. Briggs

Dept. of Physics, University of Alabama in Huntsville, Huntsville, AL 35899

ABSTRACT

We use 4 and 16 channel data from the BATSE Large Area Detectors to compare the continuum spectral characteristics of the short-duration bursts with those of short spikes within long-duration bursts. We compare frequency distributions of the resulting best-fit model parameters for the short bursts and for the short spikes within the long bursts. These distributions are different for the two types of events. The best-fit model parameters are used to calculate the peak energy E_{max} of the νF_ν spectrum for each type of event. The resulting distributions of E_{max} are consistent within a cosmological scenario if the redshift of the short spikes within long-duration bursts is larger than the redshift of the short bursts.

INTRODUCTION

The existence of two subclasses of classical gamma-ray bursts, defined by durations ≤ 2.5 seconds and durations > 2.5 seconds (see reference 1, and references therein), has been confirmed.[2] The time histories of the long-duration events often contain substructures with pulse widths that are very similar to the durations of the short-duration bursts. These time histories can be very diverse, ranging from those that are highly variable over the duration of the burst to those having an impulsive spike riding on top of gradually-varying emission. By treating each such substructure as a separate event, we hope to investigate the possibility that each substructure is a manifestation of the same emission mechanism producing the short-duration bursts.

METHODOLOGY

BATSE consists of eight uncollimated Large Area Detectors (LADs), spanning an energy range of $\sim$ 20–2000 keV. Details of BATSE instrumentation can be found elsewhere.[3] Five BATSE data-types are used in this study: Medium Energy Resolution (MER), Discriminator Science (DISCSC), Preburst (PREB), Continuous (CONT), and LAD Discriminator (DISCLA). MER data consist of count rates in 16 energy channels, with a time resolution of 16 ms for the first $\sim$ 32 seconds after the burst trigger time t_o, changing to a time resolution of 64 ms for the next $\sim$ 128 seconds. MER data begins to accumulate $\sim$ 30 ms after t_o, usually making it unsuitable to study the short-duration bursts (which are often over by t_o). The PREB and DISCSC data consist of count rates in 4 energy channels, with a time resolution of 64 ms, starting $\sim$ 2 seconds prior to t_o and lasting until $\sim$ 240 seconds after t_o. CONT data consist of count rates in 16 energy channels, with a time resolution of 2.048 seconds. DISCLA data consist of count rates in 4 energy channels with a time resolution of 1.024 seconds. CONT and DISCLA data are only used for the purpose of background subtraction in this study. Only the 16-channel data were corrected for dead-time effects; the

4-channel data would not have suffered any significant dead-time during the bursts used in this study.

A total of 29 short-duration bursts and 13 long-duration bursts were examined in this study. A count spectrum was integrated over the duration of each short-duration burst. Additionally, within each long-duration burst, a spectrum was integrated over short spikes with a pulse width less than 2.5 seconds. This criterion resulted in 55 short spikes among the long-duration events.

The count spectra were deconvolved to produce photon spectra using a forward-folding model-fitting technique. A detector response matrix was used for each burst, correcting for the detector's energy response function, angular response function, and for flux scattered by the Earth's atmosphere and the CGRO spacecraft.[4] The photon spectrum model employed in the fit was a Gaussian as a function of $\log_{10}(E)$, given by

$$\frac{dN}{dE} = \frac{A}{\sqrt{2\pi}\sigma_g} \exp\left\{-\frac{1}{2}\left[\frac{\log_{10}(E) - \log_{10}(E_{\text{cent}})}{\sigma_g}\right]^2\right\} \quad (1)$$

where σ_g is defined as

$$\sigma_g = \frac{\log_{10}(FWHM)}{2.35482} \quad (2)$$

Three parameters were allowed to vary in this model: A, E_{cent}, and σ_g. A represents the amplitude of the Gaussian, E_{cent} is the center of the Gaussian, and σ_g measures its full-width at half-maximum. Though this model is unphysical, it served as a good functional representation for most of the spectra analyzed, making it useful for comparison purposes. The reduced χ^2 for the model-fits to the 4-channel data was, on average, higher than that for the 16-channel data; this is attributed to systematic errors. The uncertainties in the two parameters of interest, E_{cent} and σ_g, were estimated *jointly* by assuming a 68% confidence region in the shape of an ellipsoid, centered on the measured values.[5]

To address the issue of consistency of results using different datatypes we analyzed both 4-channel and 16-channel spectra integrated over the same time intervals for a small sample of bursts. In general, the best-fit parameters agreed within their uncertainties. The amplitude of the 16-channel data was usually slightly larger than that of the 4-channel data, which we attribute to slightly over-correcting the 16-channel data for dead-time effects.

RESULTS

Figure 1a shows a plot of σ_g vs E_{cent} for the set of 29 short bursts. Figure 1b shows the same information for the 55 short spikes within the 13 long-duration bursts. Though it is difficult to draw conclusions from such crowded plots, it does seem clear that there is a large concentration of short spikes within long-duration bursts at low values of E_{cent}. The frequency distributions of E_{cent} and σ_g are given in Figures 2a and 2b, respectively. The distributions for the short-duration bursts are shown as solid lines, while the long-duration bursts are represented by dotted lines. Assuming normally distributed errors, the uncertainties for each bin are simply the square-root of the number of events in that bin.

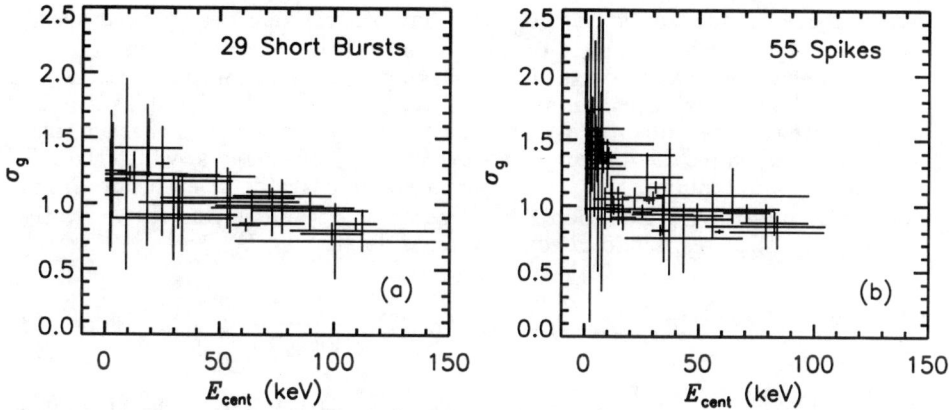

Fig. 1 (a and b). 1a shows a plot of the best-fit model parameters, along with their uncertainties, for the short-duration bursts. 1b shows the same information for the short spikes within the long-duration bursts.

Figure 2a shows that the short spikes within the long-duration bursts and the short-duration bursts sample the same range of parameter space of E_{cent}. However, their distributions are clearly different; i.e., the short spikes in long-duration bursts tend to concentrate at low values of E_{cent} while the values for the short bursts seem to be distributed uniformly over the entire range. This is in contrast to the result of comparing the distributions of σ_g for each type of event, shown in Figure 2b. It indicates that the short bursts do not sample the same range of available parameter space as the short spikes, and moreover, that the range of values of σ_g for the short bursts is a subset of those sampled by the short spikes. Further, there is a hint of a bimodality in the distribution of σ_g in Figure 2b for the short spikes, thus suggesting the interesting possibility of two classes of short spikes within long-duration bursts, based on spectral parameters alone. At this point, no work has been done to identify which long-duration bursts would makeup the two populations shown in Figure 2b, nor has any

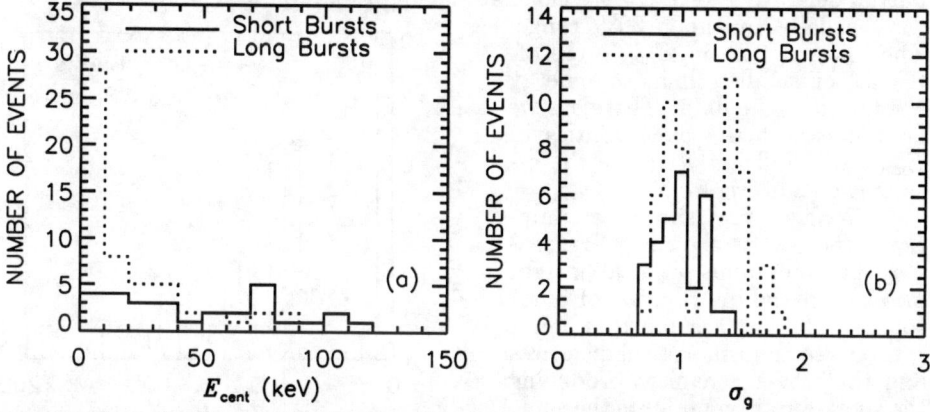

Fig. 2 (a and b). 2a and 2b show the frequency distributions of E_{max} and σ_g, respectively.

attempt been made to correlate σ_g with any other properties (peak intensity, duration, sky distribution, etc.) of the spikes or their "parent" bursts.

It was found that the three freely varying parameters of the function used to model the photon spectrum are significantly correlated. Therefore it is not appropriate to independently use any one of these parameters to characterize a spectrum. To reduce the impact of this correlation, the energy at which the νF_ν spectrum is a maximum, $E_{\max}$, was calculated analytically for each burst. This is done by multiplying Equation 1 by a factor of E^2 and setting the derivative of the result with respect to E equal to zero. Solving for E_{max} gives

$$\log_{10}(E_{\max}) = \log_{10}(E_{\text{cent}}) + \frac{2\sigma_g^2}{\log_{10}(e)} \qquad (3)$$

$E_{\max}$ is then used to characterize each spectrum, with the hope that the effect of the correlation between the parameters has been minimized.

The resulting distributions of $E_{\max}$ for the two types of bursts are similar and are shown in Figure 3. The mean value of $E_{\max}$ for the short bursts is 394 keV while for the short spikes within the long bursts it is 292 keV. This is consistent with the previously-known result that the short bursts tend to be harder than the long bursts.[2] The uncertainties for each bin in Figure 3 may be approximated as the square-root of the number of events in that bin, assuming that the errors in $E_{\max}$ are distributed normally. The uncertainty in $E_{\max}$ was calculated in a standard fashion.[6] Assuming that the errors in E_{cent} and σ_g are small, the uncertainty in $E_{\max}$ was found using

$$(\delta E_{\max})^2 = \begin{pmatrix} \dfrac{\partial E_{\max}}{\partial E_{\text{cent}}} & \dfrac{\partial E_{\max}}{\partial \sigma_g} \end{pmatrix} \begin{pmatrix} (\delta E_{\text{cent}})^2 & \text{cov}(\delta E_{\text{cent}}, \delta\sigma_g) \\ \text{cov}(\delta\sigma_g, \delta E_{\text{cent}}) & (\delta\sigma_g)^2 \end{pmatrix} \begin{pmatrix} \dfrac{\partial E_{\max}}{\partial E_{\text{cent}}} \\ \dfrac{\partial E_{\max}}{\partial \sigma_g} \end{pmatrix} \qquad (4)$$

where δE_{cent} and $\delta\sigma_g$ are the uncertainties in E_{cent} and σ_g, respectively, and the partial derivatives of $E_{\max}$ are calculated using Equation 3. The relative errors were usually in the 5–30% range. The uncertainties in $E_{\max}$ were often at, or smaller than, the energy resolution of the data. Therefore we do not expect the uncertainties in $E_{\max}$ to significantly change the results shown in Figure 3.

With a few simple assumptions, the results shown in Figure 3 may provide some physical insight into the spatial distribution of burst sources. Let us assume that burst sources are at cosmological distances and that the mechanism producing the short-duration bursts is the same mechanism that produces the short spikes within the long-duration

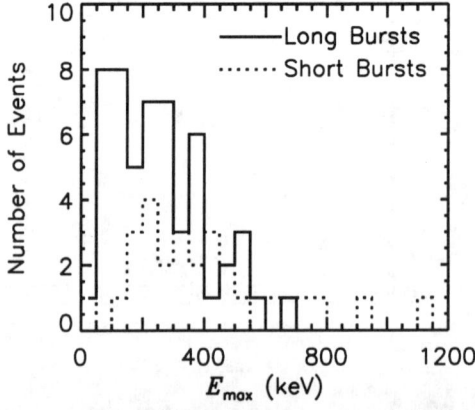

Fig 3. Frequency distribution of $E_{\max}$

bursts. By using the mean peak energy of the νF_ν spectrum for each distribution as a characteristic energy, we have calculated the cosmological redshift of the long-duration bursts z_L relative to the redshift of the short-duration bursts z_S. We find that under these assumptions $z_L = 1.35 z_S$. However, a cosmological interpretation is not *required* to explain Figure 3; the fact that short bursts tend to be harder than long bursts, convolved with any tendency for hard-to-soft spectral evolution, would give the same results.

A cautious approach must be taken in interpreting Figures 2 and 3 due to the low statistical significance of the data. The uncertainties in the best-fit parameters may not be normally distributed; thus, the errors on the frequency distributions should be calculated directly from the best-fit parameter uncertainties. It is clear that the uncertainties by themselves will not be able to wash out the strongly peaked distribution of E_{cent} for the short spikes within long-duration bursts shown in Figure 2a. However, the bimodal distribution of σ_g must not be taken too seriously until the uncertainties are more closely examined. Also, it should be pointed out that Equation 4, used to calculate δE_{max}, breaks down when the uncertainties δE_{cent} or $\delta \sigma_g$ are large because second-order terms in the Taylor expansion have been neglected.

FUTURE WORK

There are a number of ways to improve the procedure used in this study. An attempt to choose a larger sample of bursts will be made, allowing more confidence in the interpretation of any interesting results. A function must be chosen for the photon spectrum model in which the freely varying parameters are less correlated. The limitation here is that the only BATSE data types available to study the short-bursts often consist of 4-channel energy spectra; thus the model must contain no more than 3 free parameters to ensure at least one degree-of-freedom in the fitting procedure. It may be possible to use a more complicated model with one parameter fixed to a value that is fairly representative of most of the bursts being studied. Additionally, the assumption of a symmetrical ellipse for the shape of the 68% confidence region when determining the uncertainties in the best-fit parameters may not be accurate. Software to examine these regions for each fit is in place and will be utilized in the future. Finally, the uncertainties in the best-fit parameters should be propagated through the frequency distributions in a more rigorous way. This will indicate how serious one can take features such as the bimodality in Figure 2b.

REFERENCES

1. K. Hurley, in Gamma-Ray Bursts, p. 3, (Huntsville, AIP), (1991)
2. C. Kouveliotou et al., Ap. JLett **413**, L101 (1993)
3. G. Fishman et al., in Proceedings of the GRO Science Workshop, 2-39 (1989)
4. G. Pendleton et al., in Proceedings of the GRO Science Workshop, 4-547 (1989)
5. W. Press, B. Flannery, S. Teukolsky, W. Vetterling, Numerical Recipes, p. 687–693, (Cambridge University Press), (1986)
6. L. Lyons, Statistics for Nuclear and Particle Physicists, p. 62–63, (Cambridge University Press), (1986)

THE ENERGY EMISSION OF GAMMA-RAY BURSTS AND SOLAR FLARES

R.S. Mallozzi, G.N. Pendleton, T.M. Koshut, W.S. Paciesas, M.S. Briggs

Dept. of Physics, University of Alabama, Huntsville, AL 35899

ABSTRACT

A study of the continuum spectra of gamma-ray bursts (GRBs) and solar flares (SFs) is performed to investigate the photon energy flux of these events. BATSE 16 channel Large Area Detector (LAD) data are fit using a forward-folding model-fitting technique to produce photon spectra. The integral of the fit of each spectrum is used to compute the energy flux over consecutive intervals of the event, resulting in a time history of the power output (ergs/sec). These are compared to search for similarities among the emission mechanisms in GRBs and SFs. Comparison with the solar flare energy emission may yield information about the gamma-ray burst production site.

DATA AND ANALYSIS

The Burst and Transient Source Experiment (BATSE) consists of eight independent detector modules mounted on the corners of the Compton Gamma-Ray Observatory (CGRO), and is capable of nearly full-sky observations. This detector system, explained in detail elsewhere[1,2], provides continuous monitoring of the x-ray and gamma-ray sky. Each of the detector modules consists of two NaI(Tℓ) scintillation detectors: a Large Area Detector (LAD), optimized for temporal resolution, and a Spectroscopy Detector (SD), optimized for energy resolution.

We use the BATSE Continuous (CONT) and Medium Energy Resolution (MER) datatypes from the LADs. The CONT datatype, received in every telemetry packet, provides 2.048 second resolution in 16 energy channels spanning the range ~20–1800 keV; the MER datatype, accumulated only in burst mode and beginning ~30 ms after a trigger occurs, consists of 16 energy channels providing 16 ms resolution for ~32.8 seconds followed by 64 ms resolution for ~130 additional seconds. Thus MER data does not provide any pre-trigger information. Events which had significant emission before the trigger were, in general, analyzed with CONT data; events with fine time structure were analyzed with MER data to enable resolution of shorter spikes within longer events.

The data were deconvolved using a standard forward-folding model-fitting technique to produce photon spectra. Each event was divided into time intervals of equal duration and a photon spectrum was produced for each interval. The spectra were integrated over the fitted energy range (~20–1800 keV for GRBs and ~30–225 keV for SFs) to obtain the total photon and energy output during each interval, resulting in a time history of the photon and energy flux for that event. The flux histories of gamma-ray bursts and solar flares were compared to search for similarities among the two types of events. We find that the solar flare photon and energy flux histories are of almost identical shape, indicating little or no spectral evolution for these events. The gamma-ray bursts sometimes show significant differences between the photon flux and and the energy flux. This gives some information regarding the emission mechanism of the burst sources, suggesting that the burst emission environment is not the same as that of solar flares.

The gamma-ray burst data were fit with a normal distribution in $\log_{10}$ energy (E) with variable (linear function of energy) width over the ~20–1800 keV range:

$$\frac{dN}{dE} = \frac{A}{\sigma_g \sqrt{2\pi}} \exp\left\{-\frac{1}{2}\left[\frac{\log_{10}(E) - \log_{10}(E_{\text{cent}})}{\sigma_g}\right]^2\right\} \quad (1)$$

where

$$\sigma_g = \frac{\log_{10}(FWHM(E))}{2.35482},$$

and

$$\log_{10}(FWHM(E)) = \sigma(E_{\text{cent}}) + \Delta\sigma(E_{\text{cent}}) \times (\log_{10}(E) - \log_{10}(E_{\text{cent}})).$$

There are four varying parameters in this model: A denotes the amplitude, E_{cent} the energy of the centroid, $\sigma(E_{\text{cent}})$ the full width at half maximum ($FWHM$) of the Gaussian, and $\Delta\sigma(E_{\text{cent}})$ the change in the full width with energy. This function fits the wide range of observed gamma-ray burst spectra well, keeping in mind that we are not strictly concerned with the model that is used, but only that the function represents the data reasonably well. Therefore this functional form is adequate to ensure that the integral is descriptive of the actual energy output. The solar flare data were fit with a simple power law over the ~30–225 keV range:

$$\frac{dN}{dE} = A\left(\frac{E}{E_0}\right)^{-\alpha} \quad (2)$$

with E_0 fixed at 100 keV. A is the amplitude of the function, and α is the power-law index (generally ~3–6 for the solar data examined here). Limiting the energy range over which the solar flare data were fit enabled a straight power law to sufficiently describe the spectra of these events. The power law fits to the solar data sometimes exhibited large χ^2 during the most intense portions of the events due to the excellent counting statistics at these times.

RESULTS

The analysis was performed for 15 solar flares and 11 gamma-ray bursts observed by BATSE. Although the sample sizes were limited, we attempted to select events which differed significantly in their time histories. Eight of the 15 solar flares were visible above ~100 keV, the remainder showing emission only below ~100 keV. Seven of the 11 gamma-ray bursts showed significant emission above ~300 keV.

For emission produced by a single mechanism, the absence of extreme spectral evolution implies that the energy flux history closely mimics the photon flux history. The solar flares examined exhibited this property, supporting observations made with the Gamma-Ray Spectrometer (GRS)[3] on the Solar Maximum Mission (SMM), which showed that the hard x-ray (~10–140 keV) and γ-ray (~10–100 MeV) fluxes peak within ~one second of each other. It has

310 Energy Emission of Gamma-Ray Bursts

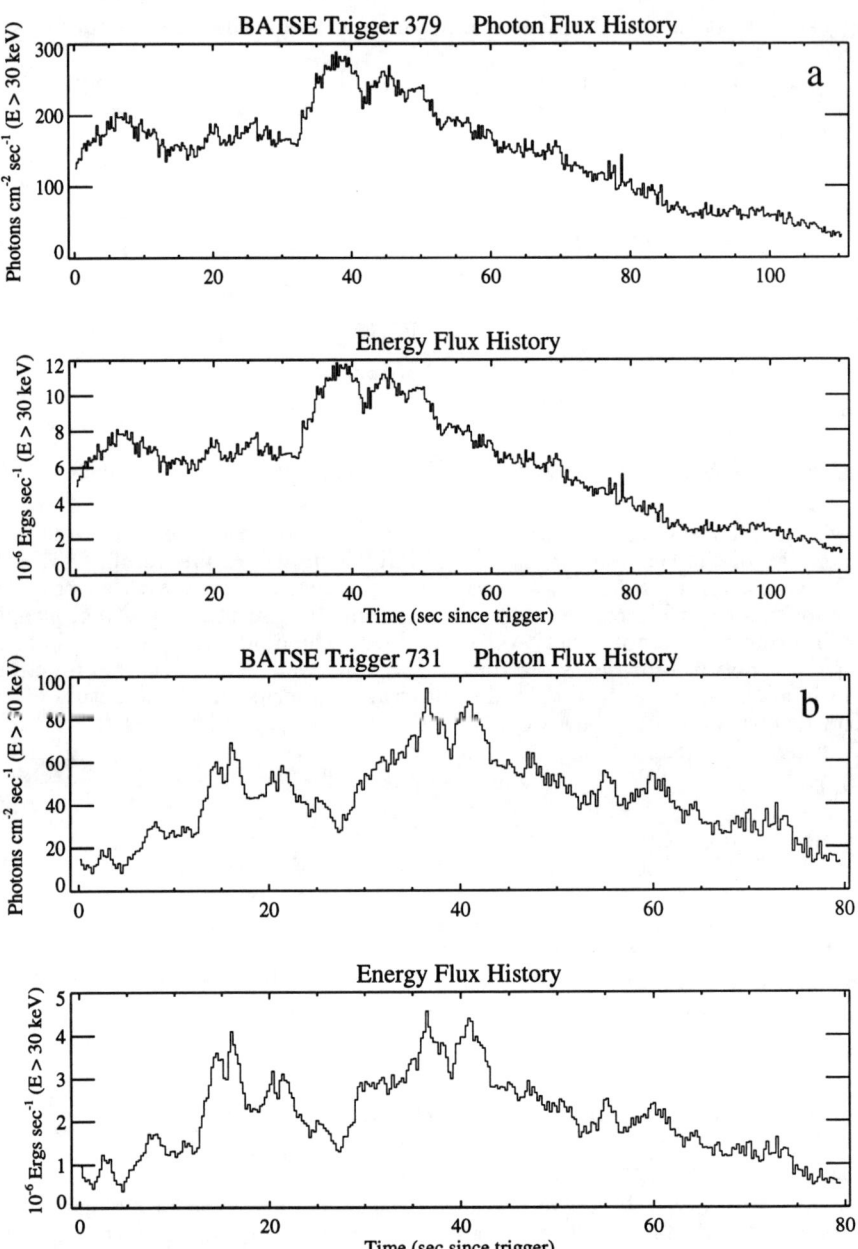

Figure 1. Solar Flares

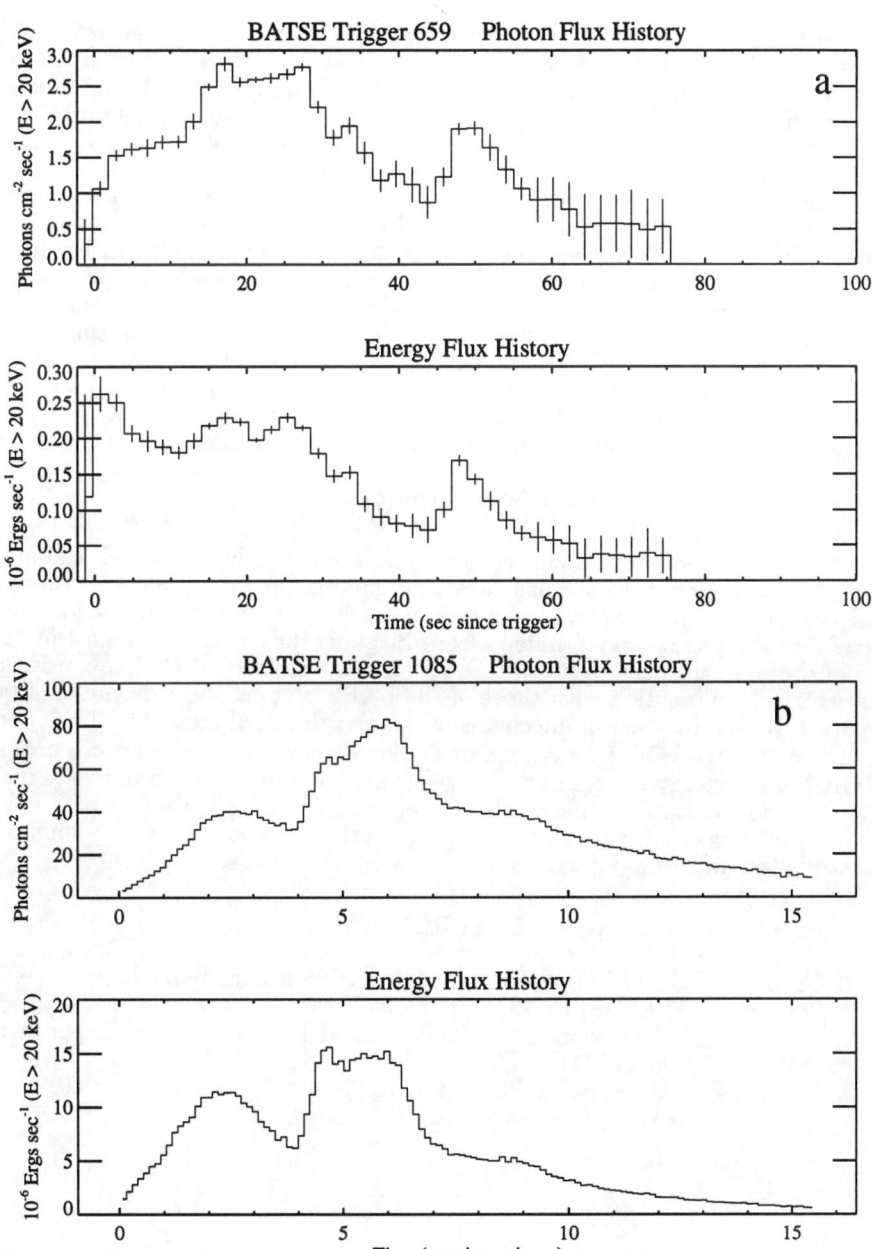

Figure 2. Gamma-Ray Bursts

also been shown that the radio flux peaks near this time as well (see reference 4 and references therein). Figure 1 shows two typical examples of the flares which were analyzed, demonstrating the excellent correlation between photon flux and energy flux. It should be noted that it is possible for the two varying parameters of the power law model used to fit the solar data to change in such a manner that the energy evolution follows the photon evolution.

Several of the gamma-ray burst profiles appear to exhibit a more pronounced rise in the energy flux followed by a steady decay even though the photon flux may be increasing (see Figure 2a). The burst shown in Figure 2a reaches its maximum in energy at the start of the event, while the photon flux peaks at approximately 20 seconds later. The smooth gamma-ray burst shown in Figure 2b reveals particularly intriguing features in its energy output evolution. The energy reaches maximum quickly, then remains relatively constant for several seconds, implying sustained energy injection or lack of energy dissipation in the burst environment. There is also a striking exponential decay in energy at the end of the event, although the count rate history also shows feature.

CONCLUSIONS

Although the photon and energy fluxes are strongly correlated, examination of the energy output of solar flares and gamma-ray bursts suggests that the emission environments of the two types of events are, in general, not identical. Several of the gamma-ray bursts reveal intriguing differences between the photon output and the energy output. Observations with the BATSE Spectroscopy Detectors in conjunction with those of the LADs support the hypothesis of two separate photon production mechanisms in solar flares: thermal ($\lesssim 10$ keV) and thick-target ($\gtrsim 10$ keV) Bremsstrahlung. The diversity in the temporal profiles of GRBs with energy suggests that these events are also the result of two or more photon production mechanisms, albeit mechanisms which appear to be distinct from those responsible for solar flares. Comparison of the solar and gamma-ray burst photon and energy flux histories support this suggestion.

REFERENCES

1. Fishman, G.J. et al., Proceedings of the GRO Science Workshop, ed. W.N. Johnson (NASA/GSFC, 1989), p. 2-39.
2. Horack, J.M., Development of the Burst and Transient Source Experiment, NASA-RP 1268 (1991).
3. Forrest, D.J. et al., Solar Phys., **65**, 15-23 (1980).
4. Aschwanden, M.J., et al., ApJ, **417**, 790-804 (1993).

Continuum Spectral Characteristics of Bursts Measured with the BATSE Large Area Detectors

G. N. Pendleton, W. S. Paciesas, M. S. Briggs
R. S. Mallozzi, T. M. Koshut

University of Alabama in Huntsville

G. J. Fishman, R. B. Wilson, C. A. Meegan

NASA/Marshall Space Flight Center

C. Kouveliotou

Universites Space Research Association

ABSTRACT

The continuum spectra of bursts are studied with particular emphasis placed on the search for spectral subcomponents. The spectra observed in individual bursts show significant evolution throughout their durations. This study addresses the degree to which hard and soft spectra are combined in bursts and the effect this has on the complexity of the observed spectra. In particular it is shown that the presence of broad cusps in spectra in the 40-100 keV range can be explained by the superposition of hard and soft spectra in the bursts. This interpretation is not meant to explain narrow spectral absorption lines in bursts. It is simply an alternate explanation for some of the broader features observed in the continuum spectra of GRBs.

INTRODUCTION

Gamma-ray bursts are known to exhibit a wide range on continuum spectra[1]. Spectral variation is not only seen between bursts but within bursts as well[2,3]. Bursts have been observed with narrow absorption and emission lines in them as well as with broader features of this type[1,2,4,5,6]. This work shows that broad absorption features in the 40-100 keV range can be explained by the superposition of hard and soft burst continuum spectra. These results are important for estimating the frequency with which absorption line features should be observed in gamma-ray burst continuum spectra. It also has implications for the temporal behavior of the energy release mechanisms operating in bursts.

PROCEDURE AND DATA ANALYSIS

The medium energy resolution (MER) data from the BATSE[7] instrument, consisting of 16 channel Large Area Detector (LAD) data in the 20-2000 keV energy range with up to 16 ms time resolution, were used for the bulk of this analysis. The LAD spectra were compared for consistency to the spectroscopy (SD) detector data and were found to be in agreement.

The main spectral continuum model used here is a generalized form of the lognormal distribution. It is a gaussian of the log of the energy where the full width half maximum (FWHM) is a linear function of the log of the energy. This is a very flexible functional form that will fit spectra that do not have a change in the sign of their curvature in the interval being fit. Its mathematical form is presented in detail elsewhere[8].

The channel to energy conversion for the LAD 16 channel data has been optimized[9] using in-flight calibration data to remove systematic effects from the spectral analysis.

Figure 1 shows two very different spectra observed in the same burst fit with the model described above. Figure 1A shows the continuum spectrum for the first 1.2 s of BATSE trigger 907 (1B 911007) while figure 1B shows a 1.2 s spectrum 117.5 s into the burst. The superposition of these two spectra would produce a cusp or broad absorption feature in the resulting spectrum.

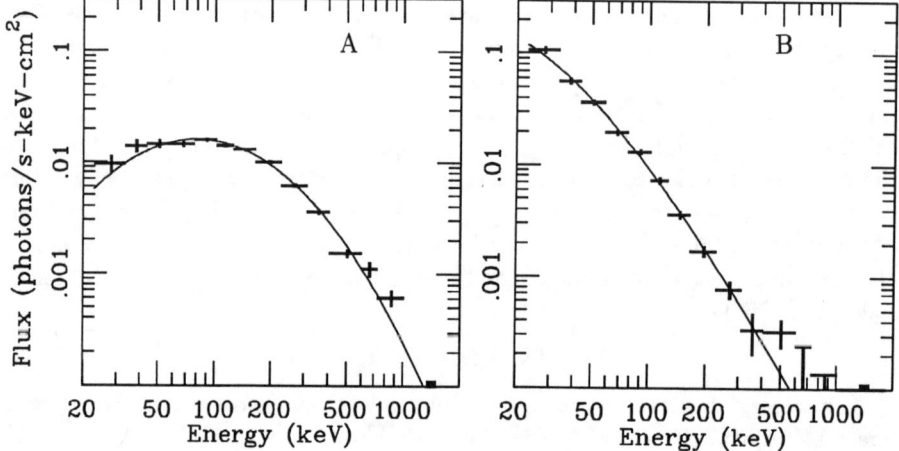

Figure 1: Spectral fits for two separate intervals of trigger 907 (1B 911007)

The obvious next step is to inspect the time profiles of the bursts to find intervals where the onset of hard Figure 1A type emission is preceded by softer emission and see whether significant cusps are visible in the spectra. Figure 2 shows the time profiles of the counts spectra of trigger 1025 (1B 911109) in four broad energy channels. These are the LAD discriminator data. In this burst there is an episode of hard emission 2s after the burst trigger lasting 0.3 s, as is evident in the $E > 300$ keV time profile. The 20-50 keV time profile shows a fairly gradual rise in counts from the trigger time until the hard outburst. Furthermore no feature corresponding to the hard episode is visible in the 20-50 keV range.

In order to test for the significance of two spectral component superposition in any of the spectra comprising this burst the MER data were summed into 0.2s intervals over the duration of the event. These spectra were then fit with the generalized lognormal described above for each interval. Then the Sunyaev-Titarchuk Comptonization model was added to the generalized lognormal and

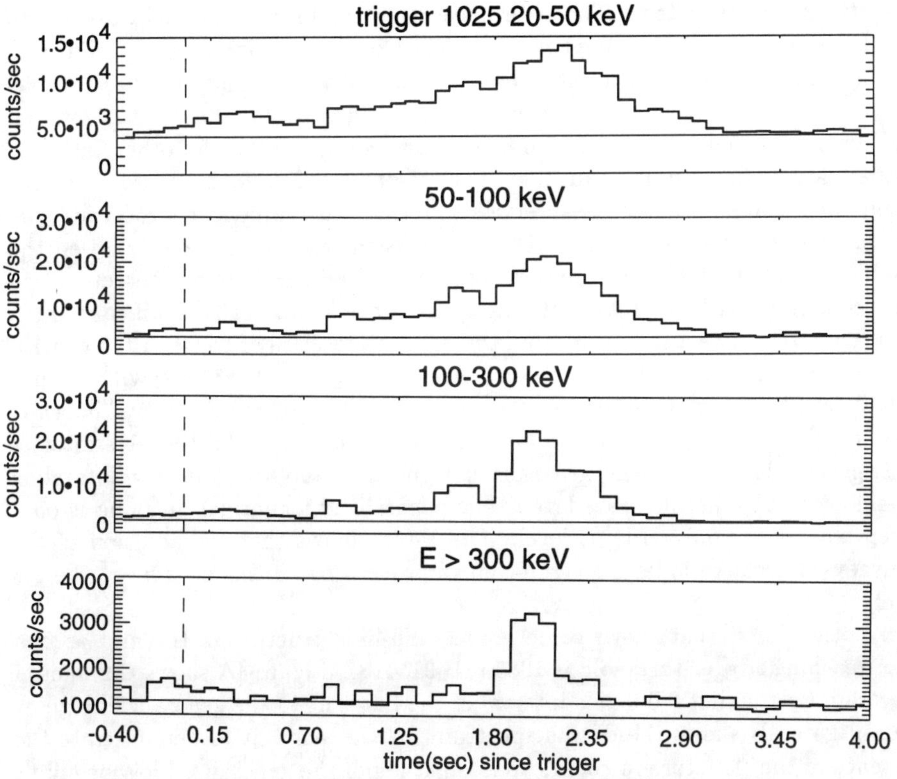

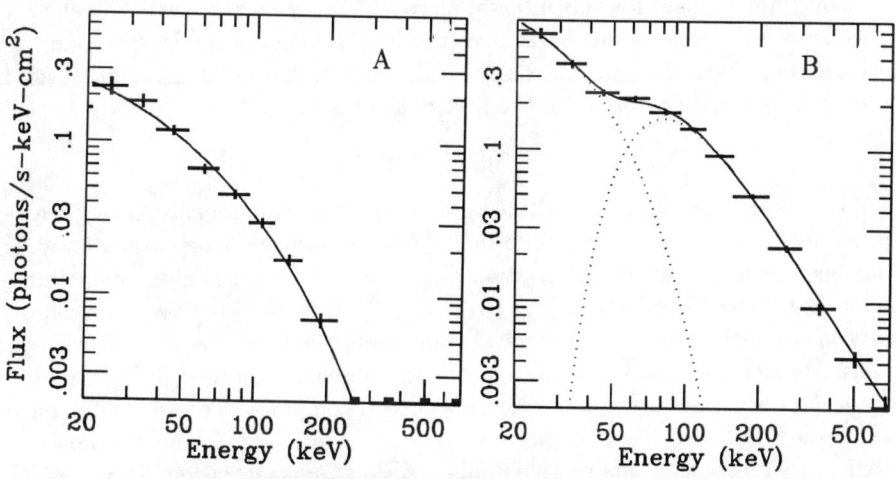

Figure 2: Time profiles of trigger 1025 (1B 911109)

Figure 3: Spectral fits for two separate intervals of trigger 1025 (1B 911109)

the fitting was repeated. The F-test was then performed on each interval to see if the addition of the Sunyaev-Titarchuk Comptonization model produced a significant improvement in the fit. The improvement was significant at the 1% level (i.e. a 1% chance that the improvement was incidental) for the 2.0-2.2s interval and at the 5% level for the two intervals preceding it. All the other intervals showed no significant improvement even though many of these intervals had source count rates comparable to the intervals showing two component structure. The 16 channel, 2.048 s resolution LAD data for each separate detector showed the significance of the added continuum component as well. The SD detector data were consistent with the LAD data although not as statistically significant.

Figure 3A shows a spectrum from the 1.2-1.4 s interval of trigger 1025 fit with the generalized lognormal acceptable at the 10% level (i.e. the model will produce a data set like this at least 10% of the time). Addition of the Sunyaev-Titarchuk model yields no significant improvement. Figure 3B shows the 1.9-2.1 s interval of trigger 1025 fit with the generalized lognormal and the Sunyaev-Titarchuk model acceptable at the 10% level. The generalized lognormal fit alone is only acceptable at the 5.0×10^{-5}% level. The F-test shows that the addition of the Sunyaev-Titarchuk model is a significant improvement for this interval at the 1% level.

Another burst that shows some broad cusp-like structure in the first second after reaching trigger threshold is 1085 (1B 911118). Figure 4A shows the photon spectrum for the first 0.0-0.25 s interval of the burst fit to the generalized lognormal. Figure 4B shows the counts spectrum for the same fit to demonstrate the presence of the structure in counts space. Although the generalized lognormal fit in the 20 to 300 keV range cannot be ruled out since it is acceptable at the 8% level, the F-test shows that the addition of the Sunyaev-Titarchuk model is a significant improvement for this interval at considerably greater than 1% level. The pattern of evolution of this burst over the first second since trigger is one where a hard component grows in intensity until it buries the softer emission obscuring the cusp-like feature present in the first quarter second.

DISCUSSION

This is not the first time that multiple spectral components have been proposed for bursts[10] or the first time that the superposition of spectral components has been proposed as an explanation for some of the broad, absorption line-like features observed in bursts[11]. The BATSE LAD data provide us with an opportunity to study the evolution of spectral components and absorption lines with good statistics and see which explanation fits the observations most convincingly. It seems unlikely that the superposition of spectral components could explain narrow absorption features that have been observed with other instruments[5], however it could explain some of the more complicated looking spectra found in the Konus Catalog[1]. Determining how many of the previous observations of lines are clearly narrow absorption lines vs. broad cusps is important for determining how likely it is that BATSE should see narrow absorption lines in the SD continuum spectra. Studying the comparative evolution of hard and soft components in bursts may yield valuable insights into the physical mechanisms that produce the observed spectra. For instance, the time profile of trigger 1025 clearly shows the onset of hard emission after the burst process has been initiated. This shows that the bursts do not evolve uniformly from hard to soft and indicates that hard emission processes can trigger repeatedly during a burst's output. Studying the development of the bursts' spectra should help map out the characteristics of this as yet unexplained physical phenomenon.

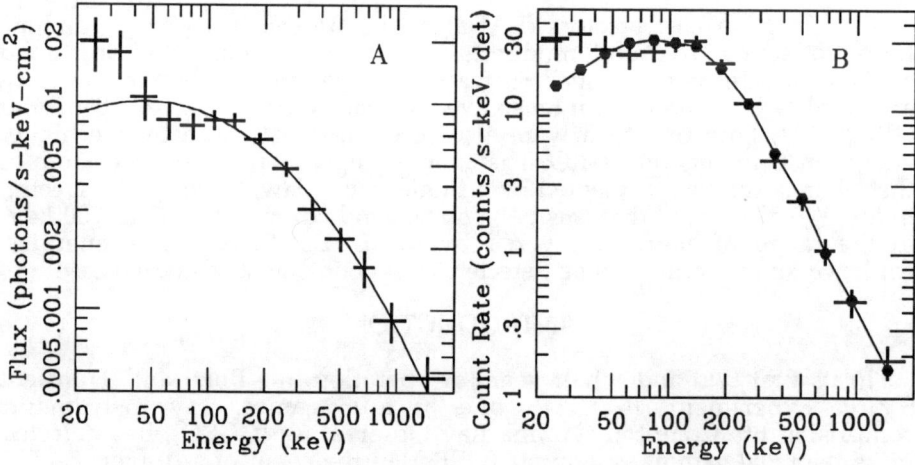

Figure 4: Spectral fits for the same interval of trigger 1085 (1B 911118) in units of photons and counts.

REFERENCES

1. Mazets et al, 1981, Astophys.Space Sci. 80:3-143
2. Mazets et al, 1982, Astophys.Space Sci. 82:261-282
3. Mitrofanov et al, 1984, Sov. Astron. AJ 28:547-49
4. Teegarden et al, 1980, Ap J. Lett. 236:L67-70
5. Murakami et al., 1988, Nature 335:234-35
6. Golenetskii et al 1986,Astrophys. Space Sci. 124:243-78
7. Fishman et al, 1989 in proceedings of the GRO Science Workshop, ed. N. Johnson, pp. (2)39-2(50)
8. Mallozzi et al, these proceedings.
9. Pendleton et al, proceedings of the second compton symposium. Sept 20-22, 1993, U. Maryland, in press.
10. Mitrofanov et al, 1992 in Gamma-Ray Bursts: Observations, Analyses, and Theories, Ed. C. Ho, R. Epstein, and E. Fenimore, pp. 209-216
11. Fenimore et al, 1982 in Gamma Ray Transients and Related Astrophysical Phenomena, Ed. R. Lingenfelter, H. Hudson, and D. Worrall

SPECTRAL CURVATURE IN HIGH-ENERGY GAMMA RAY BURSTS OBSERVED BY THE BATSE LARGE AREA DETECTORS

R. D. Preece, M. S. Briggs, W. S. Paciesas, G. N. Pendleton
University of Alabama at Huntsville, Huntsville, AL 35899

C. Kouveliotou[†], M. N. Brock
NASA, ES66, Marshall Space Flight Center, AL 35812

ABSTRACT

The Large Area Detectors (LADs) of the BATSE instrument combine large collecting area and moderate energy resolution, providing good statistics for detection of spectral features in gamma ray bursts. We have analyzed two of the most intense events seen by BATSE, GRB910503 and GRB930131, both of which were also serendipitously seen by the higher energy instruments on CGRO. The average high-energy spectra of both of these bursts can be represented by a single power law, fit with a −2 spectral index. We also show that this behavior extends down to ≈200 to 700 keV, in the range of energies covered by the LADs. Below these energies, evidence for curvature can be detected with good statistics in each case.

INTRODUCTION

In its first two and a half years of operation, the Burst and Transient Source Experiment (BATSE)[1], one of four gamma ray instruments comprising the Compton Gamma Ray Observatory (CGRO), has detected more than 800 gamma ray bursts (GRBs). In the event of a trigger, BATSE sends a trigger signal to the other experiments onboard CGRO enabling simultaneous observations of GRBs. In two unique cases, where BATSE triggered on extremely intense events, EGRET and COMPTEL were able to observe the high-energy photons from each burst. The first event occurred at 7h 4m UT on 3 May, 1991 (GRB910503) and the second at 18h 57m UT on 31 January 1993 (GRB930131). Detailed spectral analysis was carried out by the EGRET and COMPTEL instruments for each burst[2-4], with the result that a power-law fit was acceptable in most cases, with a spectral index consistent throughout much of each burst with a value of −2. These results imply a smooth continuum spectrum for intense GRBs in the energy range from ≈0.9 MeV to ≈1000 MeV. In particular, our results compare well with those presented in Ref. 5.

PROCEDURE

The BATSE instrument consists of 8 LADs, arranged at the sides of a regular octahedron, which triggers when the count rates in 2 or more detectors exceed a preset threshold of 5.5 s at the end of either a 64, 256 or 1024 ms time interval. Table 1 lists the characteristics of the detectors

---

[†] Universities Space Research Association

receiving the highest illumination from the source for the two most intense events observed by BATSE, including their trigger times relative to the beginning of the day. Hereafter, all times cited will be relative to the trigger times presented in Table 1. In the present work, we make use of the High Energy Resolution Burst (HERB) data, which is accumulated in a time-to-spill mode, beginning just slightly after the burst trigger (1 - 3 × .064 s). Up to 128 spectra total are accumulated over times that are multiples of 64 ms (but no less than 128 ms). Detectors which were ranked by the on-board computer as being brighter receive shorter accumulations. Because of extreme deadtime and pulse pileup effects in the initial pulse of GRB930131, the on-board computer ranked the source-facing detector (#4) as having the least flux of the four selected detectors. As a consequence, the accumulation times for spectra from this detector were quite long. Although the time resolution of some of the other LAD data types for this burst was at least as good, or better, the HERB data has the best energy resolution. For details of the temporal structure of this burst, see Ref. 6.

Table 1. Detector Description				
Burst	Trigger Time (s)	Detector #	Burst Angle	Earth Angle
GRB910503	25452.678	6	28.3	136.0
GRB930131	68231.682	4	27.5	151.4

Note: All angles are measured in degrees from the vector which is normal to the detector's face. The LAD energy range spans ≈28 to 1800 keV.

For each of the two bursts, a background model was constructed by fitting selected background spectra channel-by-channel to a quartic polynomial. Since these were unusually intense events, the source-to-background ratio was quite high in each spectrum studied, thus we have good confidence that the background-subtracted spectra we examined are free of systematic errors arising from an incorrectly determined background.

In both GRB910503 and GRB930131, the high count rates may have affected the lower-energy bins, producing spurious "ripples" in the spectra. Such features were also seen in other intense events, thus pointing to their systematic, rather than intrinsic, nature. In order to avoid contamination by these problems, we chose to fit the data from ≈120 to 1800 keV. This has some consequences for the determination of parameters in some of the continuum models, as we shall discuss below. In order to make a comparison with the EGRET results, the HERB data for GRB910503 was binned in time to match the three Total Absorption Shower Counter (TASC) spectral accumulation periods of 0 to 1, 1 to 3 and 3 to 7 s, relative to the BATSE trigger signal. A background-subtracted HERB spectrum for the first of these periods is shown in Fig. 1. The data are fit to a broken power law model, which is one of the simplest which results in an acceptable fit. The technique of forward-folding a model through the BATSE LAD response matrices was used, minimizing the value of χ^2 in the model variances. The detector response matrices were calculated using the best determined source position angles from our burst location algorithm[7] and include the effects of atmospheric scattering[8]. The high-end behavior of this fit indicates that the energy of the peak in emission power has not been reached by the last usable data bin (power law index is significantly greater than −2). Data from the TASC for this

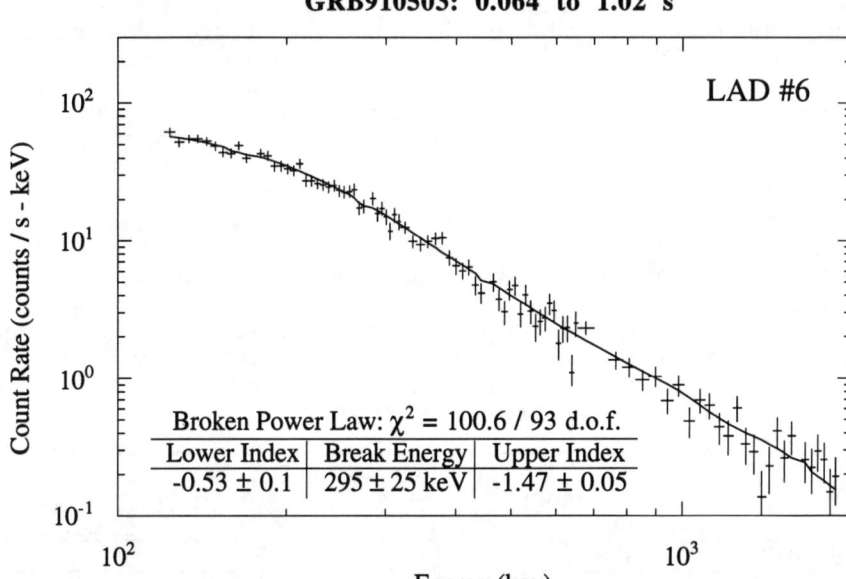

Figure 1. Count rate spectrum for GRB910503; the solid line is a plot of the fit to a two-segment power law.

time interval can be fit to a single power law < -2.0 above ~ 1 MeV, indicating that a turn-over in the power-per-decade spectrum exists near the high end of the BATSE data.

The brightest portion of the burst occurred during times corresponding to the second accumulation in the TASC. With more counts, spectral fits to the BATSE data are more constrained. We fit a number of different models to the data, to try and determine which, if any, might be a reasonable description of the continuum spectrum. Several models yielded nearly equally acceptable values of reduced $\chi^2 \lesssim 1$. A common characteristic of these models was that each required a considerable amount of curvature in the photon spectrum; i.e. a single, or even a two segment, power law results in a poor fit to the data. More interesting is the fact that nearly all of the acceptable models are consistent with a high-energy power law (above ≈ 700 keV) with index ≈ -2. Figure 2 shows a fit to the 'GRB' model, consisting of two power law segments which are smoothly joined at the break energy, with a continuous derivative[5]. The low-energy power-law index for this model and the break energy are highly correlated, especially since, in our data, the dynamic range in energy is so small below the break energy (120 – 300 keV). A physically-motivated argument in favor of this model is that spectral evolution during the time of accumulation of the data will tend to smooth out a source spectrum which is sharply broken at any instant. Indeed, detailed analysis of this burst on shorter time scales clearly indicates that spectral evolution is present.

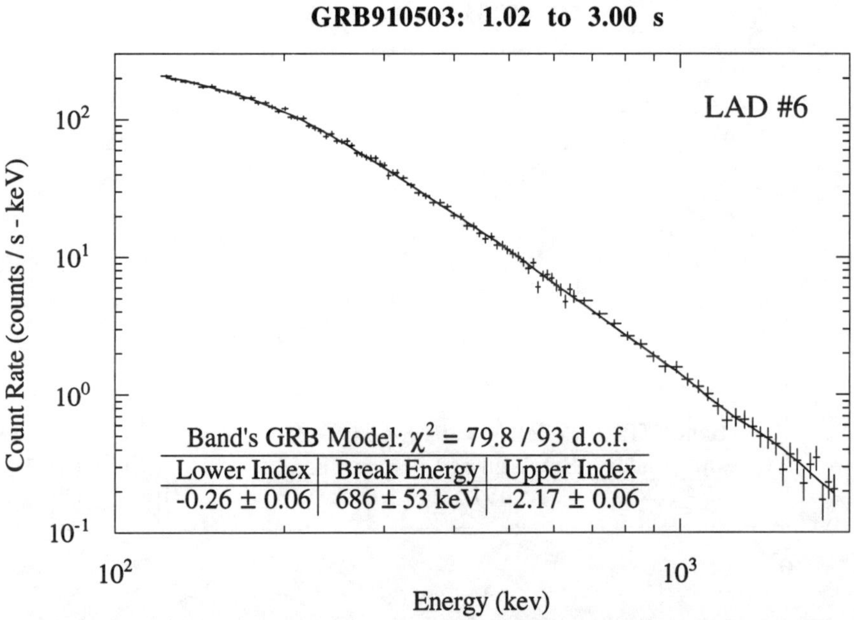

GRB910503: 1.02 to 3.00 s

Figure 2. Count rate spectrum for GRB910503; the solid line is a plot of the fit to the 'GRB' model described in the text.

Band's GRB Model: χ^2 = 79.8 / 93 d.o.f.

Lower Index	Break Energy	Upper Index
-0.26 ± 0.06	686 ± 53 keV	-2.17 ± 0.06

The lowest energy point from the EGRET TASC spectrum for the time interval corresponding to the spectrum presented here falls only ≈30% above the corresponding point in the photon spectra derived from the fits to our data. The reported power law index from the EGRET data was −2.24 ± 0.03, which is consistent with both models.

In Figure 3 we show a spectrum from a ~2 s accumulation obtained during GRB930131. Due to the difficulties in obtaining high time resolution data from the source-facing detector in this burst, as mentioned above, we cannot very well match the TASC accumulation times for this event. Therefore, we have chosen an interval which covers all the peak output from this event. Several spectral models were fit to the resulting background-subtracted count spectrum, with the simplest, a single power law, being rejected as having essentially zero probability for the data exceeding the obtained chi-squared, assuming the model was valid. Similar results were obtained for blackbody and OTTB spectral shapes. Models with intermediate curvature and a power-law tail fared quite well, including the Band 'GRB' model, discussed above, as shown in Fig. 3, as well as a broken power law and Log normal. All of these had roughly 50% probabilities and fitted high-energy power-law indices were consistent with −2.

SUMMARY

We find good evidence, for these intense bursts, at least, that a single power law fit describes the data adequately from 100s of keV up to 100s of

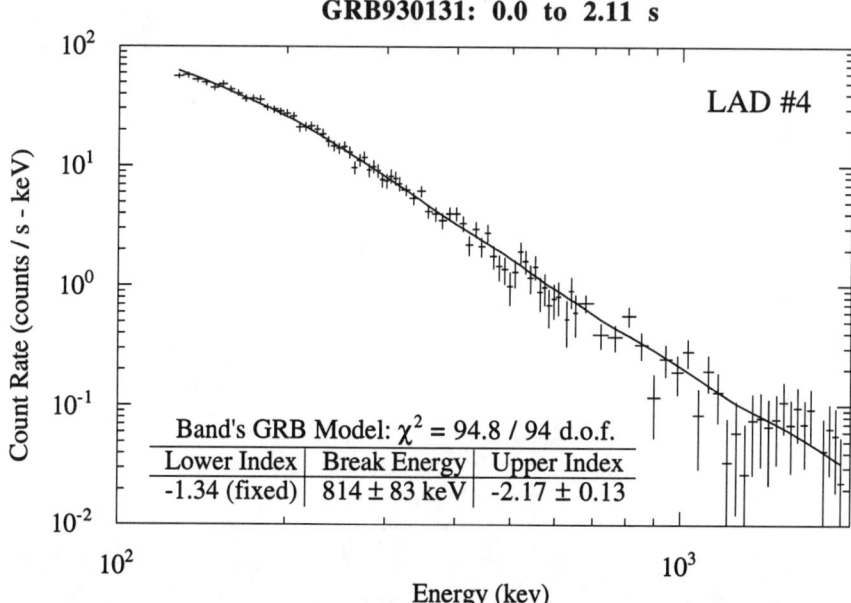

Figure 3. Count rate spectrum for GRB930131. One of the better-fit spectral models is presented, along with the count spectrum.

MeV, roughly 3 orders of magnitude in energy. We have also shown good evidence of spectral curvature in the 0.1 to 1 MeV energy band. These two results from observations of two bursts by several instruments will surely add to the growing debate concerning the theoretical explanation underlying GRB spectra.

REFERENCES

1. G. Fishman et al., in *Proc. of the GRO Science Workshop*, ed. W. N. Johnson (GSFC: NASA, 1989).
2. C. Winkler et al., A. A. **255**, L9 (1992).
3. E. Schneid et al., A. A. **255**, L13 (1992).
4. M. Sommer et al., Ap. J. (Letters), in press (1994).
5. D. Band et al., Ap. J. **413**, 281 (1993).
6. C. Kouveliotou et al., Ap. J. **422**, L59 (1994).
7. M. Brock et al., in "Gamma-Ray Bursts", ed. W. S. Paciesas & G. J. Fishman (New York: AIP, 1992).
8. G. Pendleton et al., in *Proc. of the GRO Science Workshop*, ed. W. N. Johnson (GSFC: NASA, 1989).

THE COLOR OF GAMMA-RAY BURSTS

Francesco De Paolis
Dipartimento di Fisica, Università di Lecce, I-73100 Lecce, Italy
I.N.F.N., Sezione di Lecce, I-73100 Lecce, Italy
I.C.R.A., P.le A. Moro 2, I-00185 Roma, Italy

Stefano Pezzuto
Dipartimento di Fisica, Università di Lecce, I-73100 Lecce, Italy

Marco Tavani
Physics Department, Princeton University, Princeton, NJ 08544

ABSTRACT

We present preliminary results of a study of BATSE spectra of gamma-ray bursts (GRBs). Our ultimate aim is to investigate short-duration as well as long-duration GRBs and to focus the analysis on possible similarities and differences between the two populations. We present here spectral properties obtained from both *time-integrated counts* and from *time-variable counts* for a subset of long-duration GRBs. A study of trends in the 'color-color' diagrams and/or in 'color-intensity' diagrams is of crucial importance for modelling the emission mechanisms of GRBs.

INTRODUCTION

We investigate spectral properties of gamma-ray bursts (GRBs) using LAD detector data available in the phase 1 GRO/BATSE public archive. Our study is motivated by the possible existence of two populations of GRBs characterized by short ($\tau_{90} \leq 2$ s) and long ($\tau_{90} \geq 2$ s) durations, respectively. Previous analysis of GRBs considered time-integrated spectral properties of a sample of GRBs showing different characteristics for the two populations [1,2]. It is currently not clear whether the two populations of bursts belong to different classes of physical objects. The difference seen in the time-integrated spectral characteristics of the two populations (short events appear to be predominantly harder than long events) might indicate different emission properties for the two populations. However, previous studies showed a non-universal but definite *hard to soft* trend of GRB spectra [3]. The apparent difference in the time-integrated spectral quantities for the short and long events [1,2] might be due to the hardness-duration correlation * noted by Norris et al. [3] and recently reconsidered for the BATSE LAD data [4] and for the BATSE SD data [5]. The use of information on time-integrated quantities makes ultimately impossible to decide whether the short events are systematically harder than the long events.

In order to uncover different spectral behaviors between the populations of short and long events, we started a systematic analysis of the time behavior of GRB spectra. We restrict the results reported here on the spectral time evolution of a few bursts of long duration with the use of the 64 msec time resolution. We

* We acknowledge an enlightening conversation with J. Norris.

plan to complete our exploratory study of GRB spectral properties with the use of high time resolution information for the study of short bursts. The study of short bursts is in progress.

In the following, we adopt the standard LAD energy channels: *channel 1:* 25-50 keV; *channel 2:* 50-100 keV; *channel 3:* 100-300 keV; *channel 4:* $E \geq 300$ keV. We correct the count rates by taking into account the LAD spectral response per channel [6].

TIME-INTEGRATED SPECTRAL PROPERTIES

We consider first the time-integrated properties of short and long bursts. The corresponding 'fluence-colors' give the gross spectral properties of the events. We define the fluence-colors $H_{i/j}$ as ratios of fluences, or time-integrated counts, for the channels i and j. $H_{i/jk}$ indicates that the denominator is the sum of time-integrated counts for the channels j and k. We obtain 'fluence-color-color' diagrams for the colors $H_{3/12}$ and $H_{4/123}$ vs. $H_{2/1}$. Short bursts appear more scattered than long bursts in the color $H_{2/1}$. The opposite is true for the color $H_{4/123}$, indicating that short events appear on the average harder than long events. No obvious effect is obtained for the color $H_{3/12}$. Note, however, that this hardness-duration correlation can be due to a systematic lack of emission detected in channel 4 for short events. We therefore confirm the hardness-duration correlation noted in previous investigations [1,2], even though this result should be taken *with a grain of salt* because it could be due to instrumental effects.

For a subclass of archival GRBs with a well defined 'fluence - color' it is possible to deduce average properties of the spectrum, assuming a power-law of the form $N(E) = E^{-\alpha}$. The index α gives the gross properties of the spectra, and in several cases is consistent with what found at higher energies with the BATSE SD detectors [7,8]. From the ratio of fluences per channel (fluence-colors) we can compute values of α with the assumption that the photon spectrum is a power law. It is convenient to define different values of α for each different color, i.e., α_1, α_2 and α_3 for the colors $H_{2/1}$, $H_{3/2}$, $H_{4/3}$, respectively. We obtain the average value of the α_i's defined to be $\bar{\alpha} = (\alpha_1 + \alpha_2 + \alpha_3)/3$. The mean value of $\bar{\alpha}$ for the GRBs shorter than 2 sec is $\bar{\alpha}_{short} \sim 1.12 \pm 0.48$, while for the long GRBs is $\bar{\alpha}_{long} \sim 1.74 \pm 0.57$.

TIME EVOLUTION OF COLOR-COLOR DIAGRAMS

A previous study showed the possible existence of three types of BATSE LAD's color-color diagrams: crescent, island-like and flat [8]. This study was based on 30 GRBs selected for a relatively large flux in channel 4.

We selected a sample of GRBs by adopting a set of selection criteria to simplify the analysis aimed at obtaining time information for 'count-color-color' diagrams. We use the 64 msec time resolution files and we sum at least 5 consecutive bins of the light curve in order to improve S/N ratio. The exact number of summed bins was fixed by the constraint of binning every light curve in 30 bins indipendently from burst duration. Therefore, it turns out that the minimum duration of a burst to satisfy our constraint is 9.6 seconds with a time resolution of $\delta t \sim 320$ msec. In our analysis, longer bursts have typically a time resolution larger than $\delta t \sim 320$ msec.

For a good estimation of the background we selected only bursts lasting no more than ~ 200 sec after trigger. We take into account instrumental uncertainties that influence the color determination by correcting the background subtracted counts for the different spectral response of the LAD channels.

As an example of the results we obtained, we show in Figs. 1 and 2 our results for two selected bursts: the corrected light curve, the 'color - color' diagram $C_{3/2}$ vs. $C_{2/1}$ and two 'color - time' plots. We note that in the GRB #109 (910425) it is evident an *anticorrelation* between the burst intensity and its colors, i.e. the colors increases while the intensity is decreasing. However, for the burst #130 (910430), the colors appear to be *correlated* with burst intensity. Both bursts #109 and 130 have a shape of the color-color diagram of the 'crescent'-type.

REFERENCES

1. C. Kouveliotou, et al., Ap. J. **413**, L101 (1993).
2. D. Q. Lamb and C. Graziani, Ap. J. inpress, (1993).
3. J. P. Norris, et al., Ap. J. **301**, 213 (1985).
4. P. N. Bhat, et al., In AIP Conference Proceedings no. 280, *Compton Gamma-Ray Observatory*, eds. M. Friedlander, N. Gehrels, D. Macomb (New York, AIP, 1993), p. 912.
5. Kargatis, V.E., et al., In AIP Conference Proceedings no. 280, *Compton Gamma-Ray Observatory*, eds. M. Friedlander, N. Gehrels, D. Macomb (New York, AIP, 1993), p. 907.
6. B. Schaefer, BATSE Spectral Analysis Software (User's Guide, 1991).
7. D. Band, et al., Ap. J. **413**, 281 (1993).
8. B. Schaefer, et al., Ap. J. **393**, L51 (1992).
9. C. Kouveliotou, et al., A&A Suppl. Series **97**, 55 (1993).

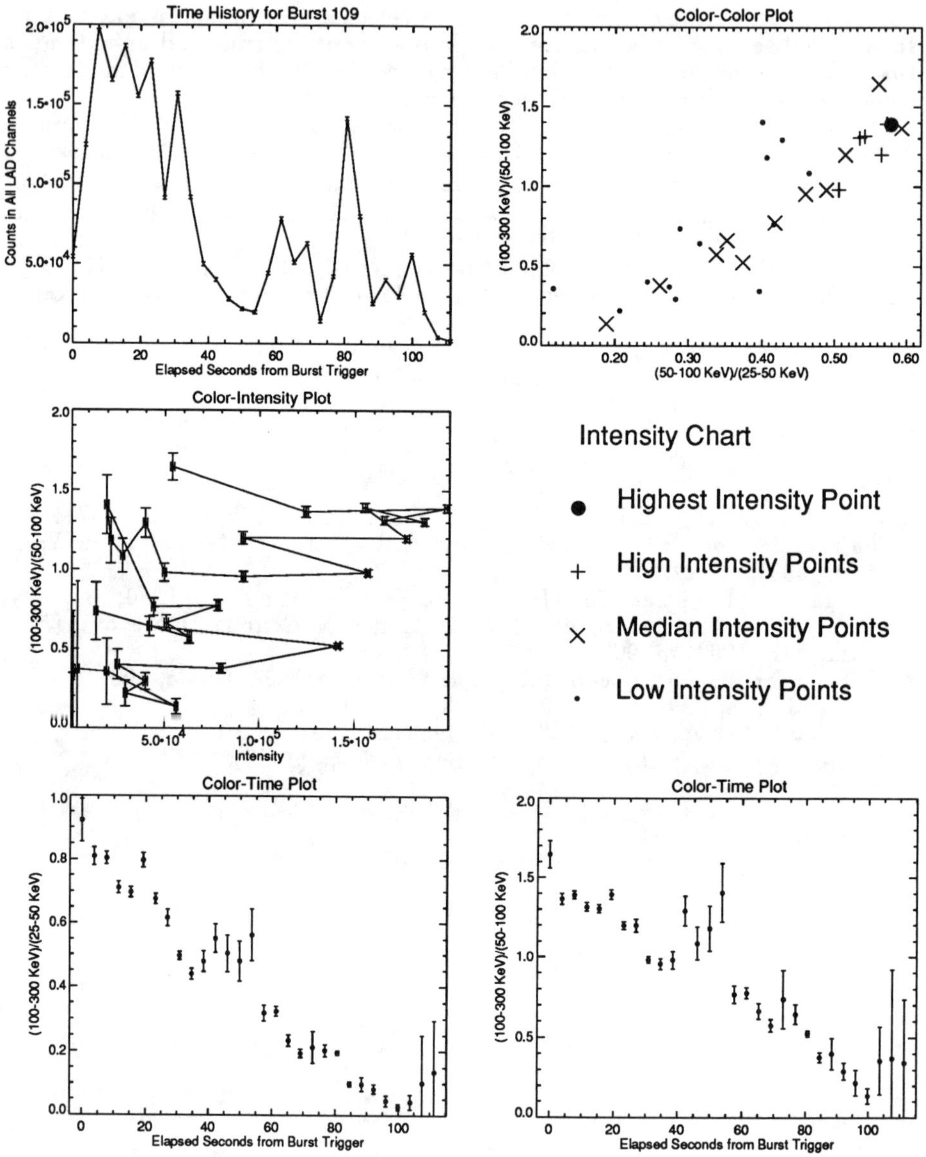

326 The Color of Gamma-Ray Bursts

Figure 1. For the burst no. 109 (910425) we show: a) the corrected light curve (for the sum of all LAD channels); b) 'color-color' diagram $C_{3/2}$ vs. $C_{2/1}$; c) Time evolution of the color $C_{3/2}$ as a function of burst intensity; d) 'color-time' diagram for $C_{3/1}$; e) 'color-time' diagram for $C_{3/2}$. In figures d) and e) it is evident the *hard to soft* evolution of the colors; the softening trend is not smooth and it appears to be *anticorrelated* with burst intensity. The color-color diagram appears to be of the 'crescent' type.

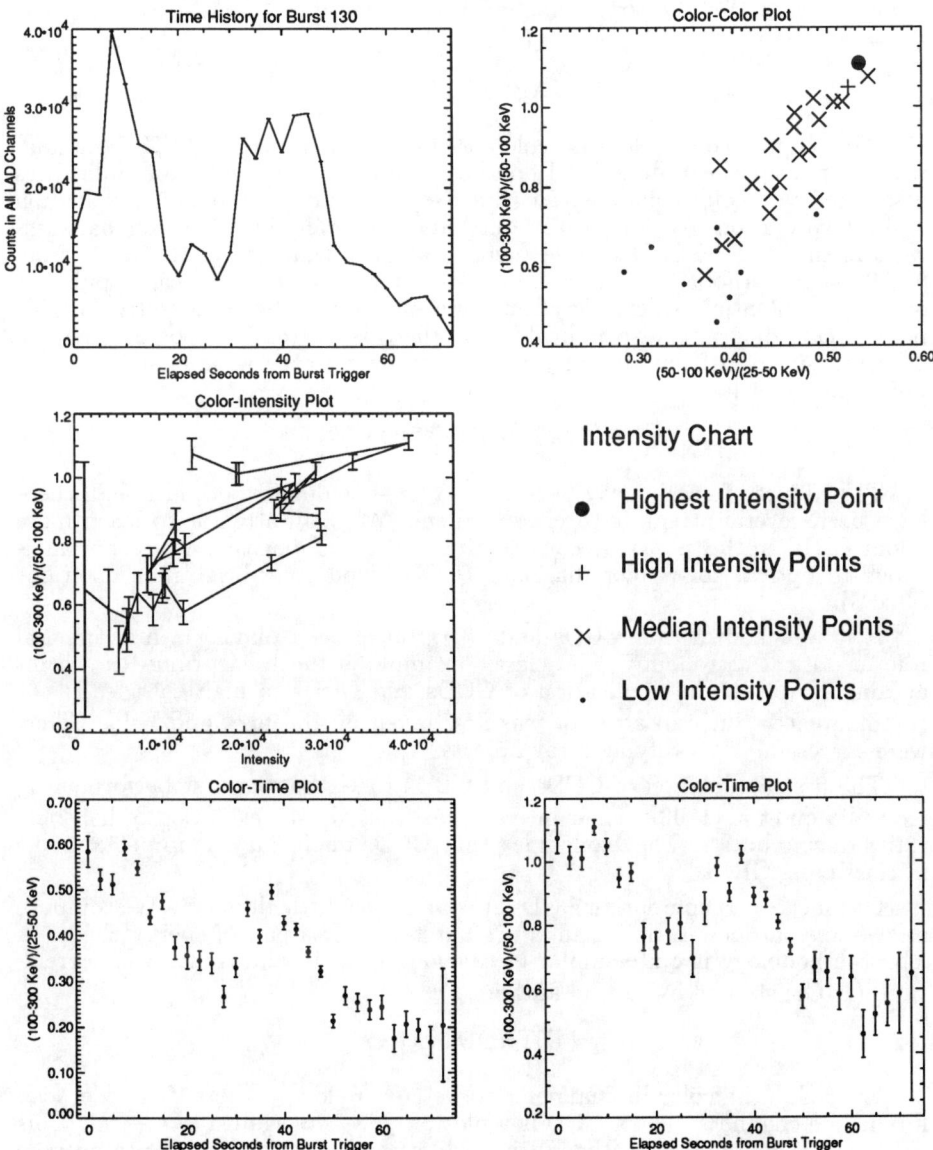

Figure 2. Same as in Fig. 1 for the burst no. 130 (910430).

COLOR DIAGRAMS OF GAMMA-RAY BURSTS

W.A.D.T. Wickramasinghe,[1] R.J. Nemiroff,[2,3,4] J.P. Norris[3]
C. Kouveliotou,[4,5]

ABSTRACT

We compute color-color and color-brightness diagrams for BATSE gamma-ray bursts using peak fluxes and fluences. We describe some of the difficulties associated with color-color diagrams due to systematic effects of data acquisition and interpretation. We find that faint bursts generally have a softer hardness ratio in the BATSE LAD middle energy channels, which cover the range from 50-300 keV. Although this may be consistent with a cosmological population of bursts, potential systematic effects prevent us from being definitive at this point. Our diagrams also indicate that there is a great deal of diversity of observed spectra of bursts and confirm that bursts cannot be explained by a universal power law.

INTRODUCTION

The nature of gamma-ray bursts is, so far, not understood, although there have been several attempts to classify them. We shall attempt to learn more about GRBs in this paper by constructing a series of diagrams which are analogous to optical color-color diagrams (CCDs) and color-brightness diagrams (CBDs).

It is well known that CCDs and CBDs have been playing a fundamental role in optical astronomy. The classic example is the Hertzprung-Russell diagram. However, the application of CCDs and CBDs to high energies is not commonplace. Such an attempt was first taken by Hasinger and Kilis.[1] They were successful in classifying x-ray objects.

The first application of CCDs and CBDs to GRBs was first performed by Kouveliotou et al. 1993.[2] Their interest was to study the evolution of hardness ratios during bursts. They concluded that CCDs might play an important role in classifying GRBs.

Our method of computing CCDs and CBDs is a little different. We compute an average hardness-ratio for an entire burst and for a pair of energy channels. This methodology is quite similar to calculations of hardness-intensity correlations (HICs) studied by several authors.[3,4,5]

METHODOLOGY

We define the color in channel x to be $chx := \log f_x$ where f_x is the peak flux in the channel. Thus a CCD is a plot of $chi - chj$ against $chk - chl$. This definition is quite similar (though it is different because we compute average

---

[1] Zaccheus Daniel Fellow, Dept. of Astronomy, University of Pennsylvania, Philadelphia, PA 19104
[2] George Mason University, CSI Institute, Fairfax, VA, 22030
[3] NASA Goddard Space Flight Center, Greenbelt, MD 20771
[4] Universities Space Research Association
[5] Marshall Space Flight Center, Huntsville, AL 35899

colors) to the hardness ratios defined in standard literature.[3] There are serious difficulties associated with any definition of colors, HRs or HICs.[6] We shall discuss these methodological problems later.

We have chosen to work primarily with the public domain BATSE flux table because it has been corrected for relative spacecraft orientations to GRBs and reflections of γ-rays from Earth.

To be sure we have not created a implicit mathematical correlation in the HR plots, we constrained ourselves to HRs with standard deviation much smaller than the mean.[3] We have therefore chosen our sample such that SN $\geq$ 5 and discarded all GRBs whose fluence is "zero" in either of the energy channels being considered.

There is a major disadvantage in using fluences for computing colors. Since GRBs durations span several orders of magnitude and the calculation of fluence involves GRB durations, one cannot say for sure that fluence is directly proportional to brightness: for example, less fluent GRBs are not necessarily faint. The choice of fluence or peak flux in colors can affect the appearance of CCDs greatly but it is unclear if either is better correlated with distance.[6]

DISCUSSION

We present our CCDs and CBDs in Figures (1) through (7). We shall discuss the problems associated with our methodology first. In computing CCDs, we used chx or $chx - chy$ as defined before. We use the fluence as well as the peak rates in computing CCDs. Though all of GRBs in our sample satisfy SN $\geq$ 5, the observational threshold might depend on color itself. Therefore, in principle, weak bursts can preferentially be either hard or soft (see Ref. 6 and references therein for a full discussion of this effect). Thus it is difficult to judge the dependence of "color" (or HR or "softness") from CCDs.

Luminosity function of GRBs can also play a very important role in CCDs.[6] A luminosity function can smear out a CCD and effectively mask other systematic effects. However, the degree of smear depends upon the width of the luminosity function, which is unknown.

We use peak fluxes in Fig. (2). There is a problem associated with the use of peak fluxes in CCDs.[6] Weak GRBs can systematically appear brighter while this effect is less important for bright bursts. If this effect correlates with the hardness, then a systematic effect should appear in Fig.(2).

In principle, redshift is also present in CCDs or CBDs. Due to the curving shape of the GRB spectrum (Figure 1), bright bursts should be systematically harder than faint ones in the 25 - 100 keV range, if GRBs are cosmological in origin. It is very easy to show that cosmological GRBs (or non-cosmological) should accumulate to a point (spot) in CCDs if GRBs can be represented by a single universal power law. However, if they can be represented by spectra with slowly varying slopes (see Eq. (1) of Ref. 7), then the cosmological bursts should define a straight line (or a locus depending on the intrinsic spectral law) with some dispersion around it on CCDs.

Since we are calculating an average color for a given burst, any variation (from burst to burst) of the pattern in which color changes within a burst itself can smear CCDs. This happens because after all this pattern determines the average color between two channels of a given burst. It has been claimed that these patterns indeed vary from burst to burst.[8]

In Fig. (1) we present a large area (BATSE) detector (LAD) 4-channel spectrum. The ordinate represents the fluences of individual GRBs in our sample, and cumulative fluence in each LAD energy channel. In Fig. (2), we plot the color $ch2 - ch3$ against the linear addition of the fluences in channels two and three (labeled as $ch(2+3)$), which can be viewed as a magnitude. These energy ranges were chosen because channels 2 and 3 determine the trigger criterion of the BATSE experiment, and thus any inclusion bias created by these energy bands is more easily evident.

In producing Fig.(2), GRBs were divided into four categories:bright, medium, dim, weak.[9] The division on the x-axis of these groups is artificial and just represents the difference in the brightness between the brightest group and the three dimmer groups. We see that weak bursts are *marginally* softer, on the average than bright bursts. This might be consistent with a cosmological origin hypothesis but the present weakness of the correlation coupled with the potential presence of systematic effects prevents us from being definitive at this time.[6]

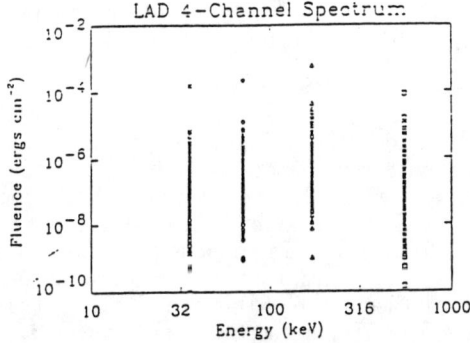

Figure 1: All the fluences for each channel. The top point represents the total fluence of all the bursts in a particular channel.

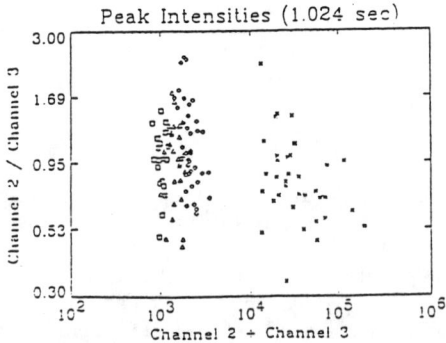

Figure 2: Color between channels 2 and 3 against total peak intensity in those channels for faint and bright bursts. Intermediate bursts are not visible because of our sampling criteria. For all bursts with signal-to-noise ratio ≥ 5.

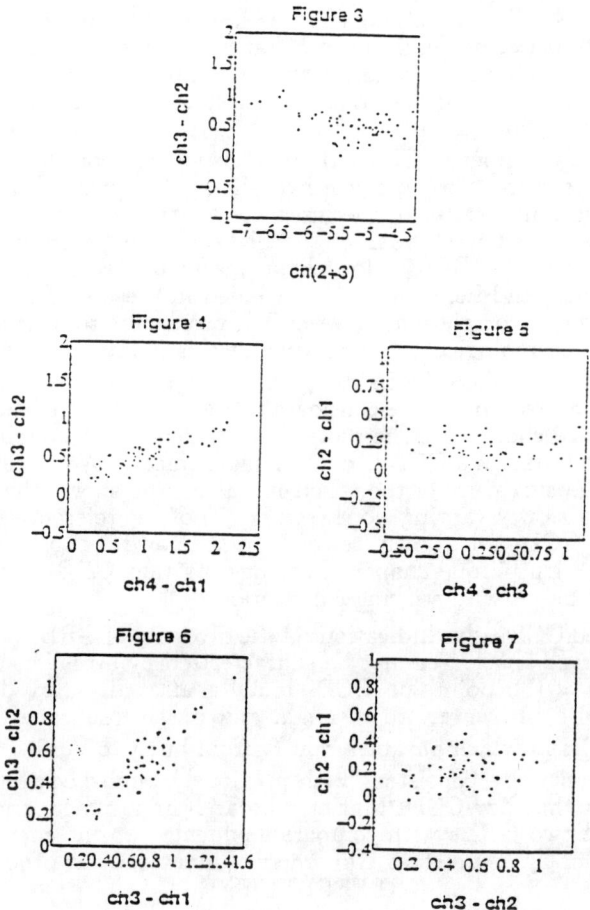

Figure 3: Color between channels 2 and 3 against total fluence in those channels for all the bursts with signal-to-noise ratio ≥ 5.

Figures 4 through 7: Color-color diagrams for our sample: signal-to-noise ratio ≥ 5. Notice that these colors are similar to hardness ratios between channels.

One clearly sees that a similar effect is not evident in Fig. (3) in which fluences have been used in place of peak fluxes. Strangely, Fig. (3) has more scatter, and seems to imply that bursts with high fluence have *softer* spectra. It is difficult to attribute this fact to a known effect other than statistical fluctuation, and we are continuing to study it.

Figure (4) shows a linear CCD (with some dispersion) correlation, which also may be consistent with a cosmological hypothesis. However, we are suspicious about the correlation because of the integrated nature of channel 4. Channel 4 typically shows the lowest fluence, the highest errors, and even a marked inclusion bias, in that GRBs with low channel 4 fluence were typically listed as having zero fluence, and hence were not included in these CCDs. Fig. (5) also shows a similar small correlation, however it is premature to conclude that this correlation has a cosmological interpretation at this point.

Figures (6) and (7) show CCDs using only BATSE channels 1-3. The CCDs made of these channels should be considerably more accurate because these are the highest fluence channels. The plots show evidence of a rough correlation. The position of a burst on a CCD is determined primarily by two factors: power law index and the curvature of the spectrum. It can be shown that for a given power law with a slowly varying slope (see Eq.(1) of the reference 7), there is a unique locus on CCDs for a given set of indices. However, these CCDs imply that there is no such unique shape. This suggests that GRBs indeed do have diverse spectral shapes with varying curvature.

The displayed CCDs give indications of the diversity of GRB spectra[10] They imply that observed spectra cannot be characterized by a universal power law. The scatter and of the points on CCDs imply a markedly curved spectra, on average, for GRBs. However, there might still be a closed analytic form for many of the GRBs in which parameters vary from burst to burst.[10] Our results on spectral diversity are consistent with previously published results.[3,11,12,13] It is also evident that our CCDs hint at a broad luminosity function of GRBs, in the sense that two GRBs with comparable fluences at one energy may have fluences that differ by as much as two orders of magnitude at other energies.

REFERENCES

1. G. Hasinger & M. van der Klis, A&A, **225**, 79 (1989).
2. C. Kouveliotou et al., A&A (Sup.), **97**, 55 (1993).
3. G. J. Laros et al., ApJ, **286**, 681 (1984).
4. J. P. Norris et al., ApJ, **301**, 213 (1986).
5. W. S. Paciesas et al., Gamma-Ray Bursts, ed. W. S. Paciesas & G. J. Fishman (New York: AIP), 190 (1992).
6. B. E. Schaefer, ApJ, **404**, L87 (1993).
7. B. Paczyński, Nature, **355**, 521 (1992).
8. K. Hurley et al., Gamma-Ray Bursts, ed. W. S. Paciesas & G. J. Fishman (New York: AIP), 195 (1992).
9. J. P. Norris et al., ApJ, in press (1993).
10. D. Band et al., ApJ **413**, 284 (1993).
11. M. Katoh et al., High Energy Transients in Astrophysics, ed. S. E. Woosley (New York: AIP), 390 (1985).
12. P. L. Nolan et al., High Energy Transients in Astrophysics, ed. S. E. Woosley (New York: AIP), 399 (1984).
13. B. E. Schaefer et al., ApJ, **393**, L51 (1992).

A SPECTRAL STUDY OF AN "X-RAY RICH" GAMMA-RAY BURST

Atsumasa Yoshida
The Institute of Physical and Chemical Research (RIKEN)

Toshio Murakami
Institute of Space and Astronautical Science

ABSTRACT

The *Gamma-ray Burst Detector* on the *Ginga* satellite observed a peculiar soft burst on January 26, 1990. This burst showed an X-ray precursor with a black-body-like shape spectrum as previously reported. A further spectral analysis on this burst unveiled an unusually softer spectral nature than those of typical classical gamma-ray bursts; (1-10 keV band)/(1-400 keV band) ratio~ 0.41, in striking contrast to a typical value of $\lesssim 0.1$.

INTRODUCTION

Classical Gamma-ray Bursts (GRBs) are known as a phenomena that emit high energy radiation mostly in the gamma-ray band[1] in contrast with other known high energy phenomena as X-ray bursts or X-ray novae. Spectra of GRBs are roughly represented by Optically Thin Thermal Bremsstrahlung spectra (OTTB) with $kT \gtrsim 100$ keV, and are very often time variable. They show various time profiles; some have complex multi-peaks, others show a single peak structure. GRBs have not been identified with any known astronomical object so far.

Another class of gamma-ray transients known are the Soft Gamma-ray Repeaters (SGRs) which typically show time-independent softer spectra represented by OTTB with $kT \sim 30$ keV and are known to show repeating bursts. One of three known SGRs, SGR1806-20, has been recently identified with a newly discovered X-ray source AX1805.7-2025 with ASCA satellite[2,3] at a position consistent with the radio observation of SNR G10.0-0.3[4].

We will present here a peculiar soft burst observed with *Ginga* satellite. It emitted the radiation mostly in the X-ray band below 30 keV.

GINGA OBSERVATION OF AN "X-RAY RICH" BURST GB900126

Gamma-ray Bursts Detector carried by *Ginga* satellite was the first instrument which have steadily monitored GRBs in the X-ray band as well as in the gamma-ray band. The GBD consisted of two instruments and could measure spectra from ~ 1.5 keV to ~ 30 keV by the gas proportional counter (PC) and from ~ 14 to ~ 400 keV by the NaI scintillation counter (SC) in standard observation mode: there are altogether 48 energy channels between ~ 2 and ~ 400 keV. The effective areas of the PC and SC counters are ~ 60 cm^2 and ~ 63 cm^2, respectively [5].

During nearly five years of the *Ginga* mission life, GBD recorded 119 bursts; however about a half of them were occurred out of the field-of-view of PC, hence no X-ray data were available for them. Among the *Ginga* GRB data, GB900126 is one of events showing the most striking features; it was found to show an X-ray emissions during an interval of ~10 seconds before the γ-ray onset[6], while the entire burst activity lasted for ~ 20 seconds. Murakami et al. reported that a spectrum during the X-ray precursor can be well represented by a black-body model with a temperature of kT=1.6 keV. This burst was also observed with WATCH on the GRANAT satellite by which the source was localized to ($\alpha = 8^h 45^m, \delta = -38°$) [7].

SPECTRAL ANALYSIS AND RESULTS

To see spectral softness, we introduce *Softness* defined as *(Total PC counts) / (Total SC counts)*. The average value of *Softness* for 56 samples of bursts for which PC data were available is 1.32, while *Softness* of GB900126 is 2.83. Although highly time-variable spectra were seen during GB900126[5], we could see a spectral property for the entire burst with an averaged spectrum of the burst. We made a spectrum from the data for 17.5 seconds (real time) including the peak.

Several models were applied to fits of this averaged spectrum; Power Law (PL), Optically Thin Thermal Bremsstrahlung (OTTB), and a two component model, a combination of PL and OTTB (PL-OTTB).

Since GBD/*Ginga* did not have any collimator either for PC or SC, we could not know the incidence angle of photons to the detectors. For GB900126, fortunately, WATCH/GRANAT localized the burst position, hence, we could employ response matrices with a right incidence angle of 50.8° for fitting.

A low energy absorption (ABS) with an absorption column of several $\times 10^{22}$ is required, in the fitting, for any models mentioned above to represent an observed averaged spectra. The fitting results are summarized in Table 1; where dof, α and kT denote degree-of-freedom, the power-law index, and the temperature of OTTB model respectively.

Model	χ^2	dof	α	kT	$\log N_H$	X/γ Fluence Ratio (1-10/1-400 keV)
PL*ABS	140.1	35	2.12		22.88	
OTTB*ABS	1020	35		18.8	20.94	
PL-OTTB*ABS	105.9	33	2.02	11.5	22.76	0.41

Table 1. Fitting results of averaged spectrum of gb900126.

GB900126 was one of the most intense bursts that *Ginga* recorded, and exhibited highly time-variable spectra during the burst activity. Therefore an average spectrum must be a mixture of spectra of different shapes. This may be a reason that an acceptable fit could not achieved. Although the fit is poor, we can roughly represent a spectral shape and could estimate an X/γ fluence ratio from the fit with PL-OTTB*ABS model; we found from the fit that (1-10 keV band)/(1-400 keV band) ~ 0.41 and an Energy versus νF_ν plot peaks at around 15 keV.

DISCUSSIONS

GB900126 exhibited the time variability commonly seen in GRBs, and showed spectra harder than those of X-ray bursts but much softer spectra than those of typical GRBs ever seen with conventional GRB instruments sensitive to $\gtrsim 30$ keV band. Our common understanding on GRBs is that X/γ ratio is small, typically $\lesssim 0.1$ [8,9]. GB900126, contrary to this, emitted the radiation mostly in X-ray band and showed four times larger X/γ ratio (in *Ginga* band). With respect to the spectral hardness, GB900126 resembles SGRs rather than GRBs.

Castro-Tirado mentioned that about 10% of GRBs that WATCH saw were soft events, and that WATCH detected another burst inside the 3σ error box of GB900126[10]. This might suggest the possibility that the source generated bursts recurrently.

Using the BATSE catalogued data, Quashnock and Lamb suggested, based on the nearest neighbor analysis,[11] that some kind of GRBs do repeat, and Wang and Lingenfelter reported the possibility that five bursts came from a single source[12] ("classical gamma-ray repeater"). Quashnock and Lamb's conclusion, however, has come from the angular correlation between bright and faint bursts. On the contrary, GB900126 and the WATCH event were both bright[10]; the peak flux was $\sim 10^{-5}$ erg sec^{-1} cm^{-2} for GB900126. The five bursts reported by Wang and Lingenfelter showed harder spectra; The νF_ν plots have maxima > 100 keV. The spectral nature is much different from GB900126, therefore, GB900126 may not be in a family of a "classical gamma-ray repeater".

The present analysis could not conclude whether this event belongs to a separate class of high energy transients or just a soft extreme of GRBs having a divergent nature in spectral hardness. Further studies are necessary to solve this problem. Clearly, it is crucially important that future observations are performed with instruments capable of localizing burst sources as well as being sensitive in both the X-ray and the gamma-ray bands. All the above requirements can be realized by HETE mission[13] which is now under development preparing for launch in 1995.

REFERENCES

1. J. N. Imamura and R. I. Epstein, Astrophys. J. **313**, 711 (1987).
2. Y. Tanaka and the ASCA team, I.A.U. Circ. , 5875 (1993).
3. T. Murakami et al., Nature , submitted (1994).
4. S. A. Kulkarni and D. A. Frail, Nature **365**, 33 (1993).
5. T. Murakami et al., Publ. Astron. Soc. Japan **41**, 405 (1989).
6. T. Murakami et al., Nature **350**, 592 (1991).
7. N. Lund, Proceedings of the Los Alamos Workshop on Gamma-Ray Bursts, ed. C. Ho, R. I. Epstein and E. E. Fenimore (Cambridge, 1992), p. 188.
8. J. G. Laros et al., Astrophys. J. **286**, 681 (1984).
9. A. Yoshida et al., Publ. Astron. Soc. Japan **41**, 509 (1989).
10. A. J. Castro-Tirado et al., these proceedings (, 1993).
11. J. M. Quashnock and D. Q. Lamb, MNRAS , submitted (1993).
12. V. C. Wang and R. E. Lingenfelter, Astrophys. J. **416**, L13 (1993).
13. Ricker, G. R., et al., Proceedings of the Los Alamos Workshop on Gamma-Ray Bursts, ed. C. Ho, R. I. Epstein and E. E. Fenimore (Cambridge, 1992), p. 288.

THE EFFECT OF AN INTRINSIC COLUMN ON GRB SPECTRA

Alan Owens
Department of Physics, Leicester University, Leicester LE1 7RH, U.K.

ABSTRACT

We present results of a study to ascertain if material surrounding a GRB site would prevent a measurement of the neutral hydrogen column density to the site and thus an estimate of the distance scale. Using a 1-D photo-ionization structure model in conjunction with simple input spectra, it is found that un-ionized matter densities near the burst site cannot exceed ~ $2 \times 10^{22} cm^{-2}$ otherwise the resulting roll-over in the spectrum would have been observed. Furthermore, by varying the matter density and its ionization state, we find that the spectrum in the ~ 0.5 to few keV region experiences considerable distortion long before the cut-off due to neutral hydrogen is affected, - no matter how low the column to the source is. When viewed in the light of existing measurements, the data indicate that the column to the source can be derived for any scenario.

INTRODUCTION

Recent results from the Burst and Transient Source Experiment on-board the Gamma Ray Observatory[1] indicate that the long standing question of burster origins may not be answered by gamma-ray measurements alone. When coupled with the lack of an identifiable counterpart at any wavelength (or even a convincing correlation with any astronomical population) the data have led to a confused proliferation of models, covering everything from magnetospheric effects in nearby neutron stars to superconducting cosmic strings. Whereas models are vague on much of the burst details, they can at least be grouped according to a distance scale. Schaefer[2] suggested that this scale may be best established by measuring the column densities of a representative sample of burst spectra at XUV wavelengths and pointed out that, even if the result does not lead to a unique identification, it will eliminate whole classes of models by providing the first quantitative description of the 3-dimensional spatial distribution of bursters. Owens and Schaefer[3] explored this idea and demonstrated that the expected inter-model column variations should be easily resolved with current silicon detector technology.

Obviously, when measuring columns, the question of a column intrinsic to the source is crucial, since it may not be possible to isolate the column to the burst site. Although no GRB model requires or suggests any material around the burst site, several related studies have considered the effect of matter near a burst. For example, Fencl et al.[4] considered a thick uniform shell of matter around a burst source and showed that a gamma-ray line afterglow may be visible for some time after the burst. Band and Hartmann[5] considered the flash photo-ionization of material to predict observable Hα emission long after the burst. However, both these studies were exclusively concerned with establishing a diagnostic with which to locate the burst *via* its interaction with its environment, rather than the effect of the matter on the burst itself. Given the paucity of X-rays or fluorescence features in GRB spectra, it is reasonable to expect that if the burst passes through any matter, it must be extremely tenuous and metal deficient. For a neutron star progenitor, one might envision that such matter exists in the form of a thin supernova shell. Alternately, the burst itself may spew out matter. Previous work on the flash ionization of nebulae and the ISM suggest that such matter will have little effect, since the burst itself will ionize all matter to great distances around the burst site, thus preventing photoelectric absorption (*e.g.*, see Jennings[6]).

In this paper, we have extended our original study[3] to include matter surrounding the burst source itself. We refer to this as the intrinsic column and the intervening matter between the burst site and the observer, as the extrinsic column. By propagating a 'gamma-blast' through matter of various densities and comparing the emergent spectrum with an "average" GRB spectrum (see the next section), we have determined a self-consistent set of solutions that not only reproduce our test spectra, but can explain some of its subtler nuances. Indeed, within the allowable range of parameter space we predict there may be observable effects, depending on the ionization parameter. Lastly, for the matter densities explored, it is found that the extrinsic column density can be deduced for all distance scales and therefore the technique proposed by Schaefer[2] and Owens & Schaefer[3] is still valid.

GRB SPECTRA

In order to test various attributes of the model, it is necessary to compare the emergent spectra with an "archetypal" spectrum measured at the Earth. Whilst considerable variation exists between individual bursts, a generic form can be ascribed based on observations in different energy regions. We consider each decade separately. Above 1000 keV, the spectral slopes measured by EGRET, BATSE and COMPTEL are all close to -2 for those bursts bright enough to give a statistically significant signal. In the 100 keV to 1 MeV range, the range of slopes varies from -1.3 to around -4. From the BATSE spectroscopy detector catalog[7], the average is ~ -2 although a value of -1.5 is often quoted in the literature. The spectral slope from 10 keV to 100 keV ranges from -0.25 to -2.0. The dispersion is fairly broad, but a slope of -1.0 is a good average. In the 2 keV to 10 keV range, there is little information[8-10]. The slopes for photon spectra, range from -0.5 to -0.9 for three bursts only, and hence a value of -0.7 would seem prudent. Below 2 keV, there are no measurements, but assuming no interstellar absorption, it is reasonable to expect a hardening of the spectra since all GRB show an acceleration of curvature with decreasing energy. Therefore, an $E^{-0.7}$ extrapolation is probably bordering on optimistic, whereas an E^2 Rayleigh-Jeans rollover (which must occur at some energy due to self absorption), is probably unduly pessimistic. Therefore, from 0.1 keV to 1 keV, we consider $dN(E)/dE \propto E^0$ to be conservative. Summing up, our standard 'unabsorbed' spectrum has the form;

$$\frac{dN}{dE} = \begin{cases} \beta & 0.1\,\text{keV} \leq E \leq 1\,\text{keV} \\ \beta(E/E_o)^{-0.7} & 1\,\text{keV} < E < 10\,\text{keV} \\ 2\beta(E/E_o)^{-1.0} & 10\,\text{keV} \leq E < 100\,\text{keV} \\ 200\beta(E/E_o)^{-2.0} & E \geq 100\,\text{keV} \end{cases} \quad (1)$$

where $E_o = 1$ keV and β (photons cm^{-2}s^{-1}keV^{-1}) is related to the burst fluence by,

$$\beta = S/2.5 \times 10^{-6} \Delta T \quad (2)$$

where S is the burst fluence above 30 keV at the Earth in units of ergs cm^{-2}s^{-1} and ΔT is its duration. From the review of Hurley[11] we assume a median $<\Delta T>$ of 10s. Lastly, we note there is a weak hardness/duration correlation, with the softer bursts having duration's > 1 sec[12]. Within a single burst, the evolution of hardness is always complex and highly variable[13]. For the purposes of calculation, we assume it is constant.

THE MODEL

Obviously, if matter around the burst site is totally ionized, it will have no effect on the emergent spectra and so we must investigate the effects of partially ionized matter. We have used the one dimensional ionization model of Done et al.[14] which calculates the equilibrium photo-ionization structure of illuminated matter. It considers 10 elements; H, He, C, N, O, Ne, Mg, Si, S and Fe, and determines a simple ionization balance under the assumptions that neither collisional ionization nor radiative transfer are important. For GRB studies, the first assumption should have no direct relevance whereas the second will be true, since from gamma-ray opacity arguments, the encountered optical depth must be « 1. For the present study, we have assumed an infinite slab of absorber whose thickness d is much less that the distance to the observer; i.e., d « R. In this case, the gamma-ray beam may be considered plane parallel. The primary GRB beam is allowed to shine through this absorber and not onto it, otherwise reflection features should be present in measured GRB spectra, - which are not observed. For the initial set of calculations, the absorber was assumed to be composed of uniformly mixed matter of solar system abundances[15]. Actually, the calculations should be insensitive to composition, since a large burst will more than likely destroy heavy nuclei. The pertinent model parameters are; the column density between the absorber and observer N_{ex}(cm^{-2}), temperature T (K) of the absorbing material and the ionization parameter $\xi = L/(N_e r^2)$, where L is the luminosity between 5 eV and 300 keV, N_e is the number density of electrons and r is the distance from the source to the ionized material. This is essentially the ratio of ionizing photons to particles in the absorbing material. Note, the working energy range was chosen as follows - the lower bound is less than the ionization potential of Fe and Mg, while the contribution from photons above the upper bound is negligible because of the v^{-3} dependence of photoelectric cross-sections. Lastly, from the work of Schaefer[16] we assume that cosmological redshift effects on the column determination are negligible.

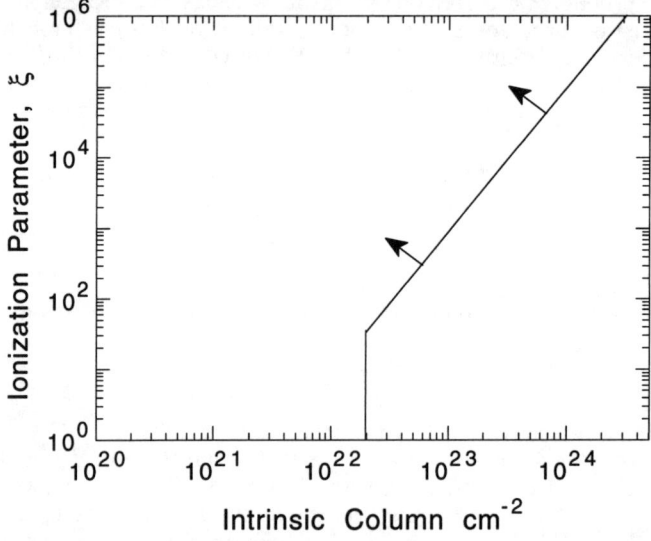

Figure 1. The allowable ranges of intrinsic column and ionization parameter.

RESULTS

For ease of calculation, the input spectrum was assumed to vary as E^{-1} below 100 keV. The gamma-rays were then propagated through the material and the output spectra compared to our "generic" spectrum. The column and ionization parameter were changed under the condition that the emergent spectrum plus interstellar absorption to the Earth did not differ significantly from our trial spectra above 2 keV. The benchmark on which this is based, were the published spectra of a few Hakucho and GINGA spectra which extend down to ~ 1.5 keV. It was required that the absorbed spectrum be smooth and continuous (at the ~10 % level) down to 2 keV. However, because of the limited statistical precision around 2 keV, the spectra were allowed to roll-over by as much as 20% and still be considered to be consistent with the measurements. We next explored the boundaries of parameter space by varying each of the input parameters whilst satisfying the above criteria. The results are shown in fig. 1 in which we plot the intrinsic column, N_{in}, versus ξ. From purely absorption arguments, the N_{ex} cannot exceed ~ 2×10^{22}cm^{-2}, since the spectra at the Earth becomes noticeably different from the comparison spectra. Likewise, in the absence of ionization, N_{in} can't exceed the same value. The solid line in the figure delineates the limiting intrinsic column for a worst case extrinsic column of 10^{18} cm^{-2} H - the region to the left of this line representing permissible parameter space. Assuming, $N_{in} \sim N_e$, the maximum intrinsic column can be expressed in terms of burst luminosity,

$$N_{in} < \sqrt[3]{3 \times 10^{43} (L/r^2)} \qquad (3)$$

Interestingly, a hardening of the spectrum below 10 keV to a slope of ~ - 0.7 could be induced for intermediate values of ξ and N_{in}. In reality, the low energy portion of the spectrum is even less sensitive to ξ and N_{in} than implied by fig. 1. For example, fig. 2 shows a family of curves in ξ, for which the extrinsic and intrinsic columns were set at $N_{ex} = 10^{20}$cm^{-2} and $N_{in} = 10^{23}$cm^{-2} respectively. These values aren't arbitrary. The value of N_{ex} was chosen because it is low enough to be very sensitive to the matter conditions at the source and yet realistic for local models. The value of N_{in} was chosen because in the absence of ionization it should have already been detected. The important point is that the low energy cut-off is virtually independent of ionization parameter for $\xi > 20$, even though considerable structure is introduced in the ~ 0.5 - 4 keV region of the spectrum for low ξ values. In fact, measurable absorption and line features would be observed long before the column begins to affect the low energy cut-off and thus the extrinsic column determination. Even then, the effect on the cut-off is catastrophic, causing a complete collapse of the low energy component (e.g., see $\xi = 20$ in fig. 2). In a GINGA-type measurement, these effects would manifest themselves as a roll-over near 4 keV indicating a column of 10^{23}cm^{-2}. Since this is not observed experimentally, we may conclude that for this case, the low energy cut-off must be dominated by the extrinsic column. This illustrates that existing instrumentation with high energy thresholds can still make important contributions in constraining the burster distance scale. In fact, the limited existing data would appear to exclude extragalactic plus host galaxy absorption models. The only possible scenario for a dominant intrinsic column would be an un-ionized column of 10^{21}-10^{22}cm^{-2} and an extrinsic column of 10^{20}cm^{-2}. However, it is easy to show that even a marginal state of ionization results in the return of low energy emission - in which case the extrinsic column can again be determined by modeling. In order to be unresolvable, the matter shell would have to be located $> \sim 10^8$cm from the burst source (e.g., the Vela SNR; $N_{in} \sim 2 \times 10^{21}$cm^{-2}, $r \sim 10^{19}$cm)[17]. Even then, N_{in} can be deduced by the apparent constancy in the measured columns of a range of bursts across the sky and N_{ex} isolated by the temporal evolution of the derived column during the burst, or from a statistical analysis of many burst positions.

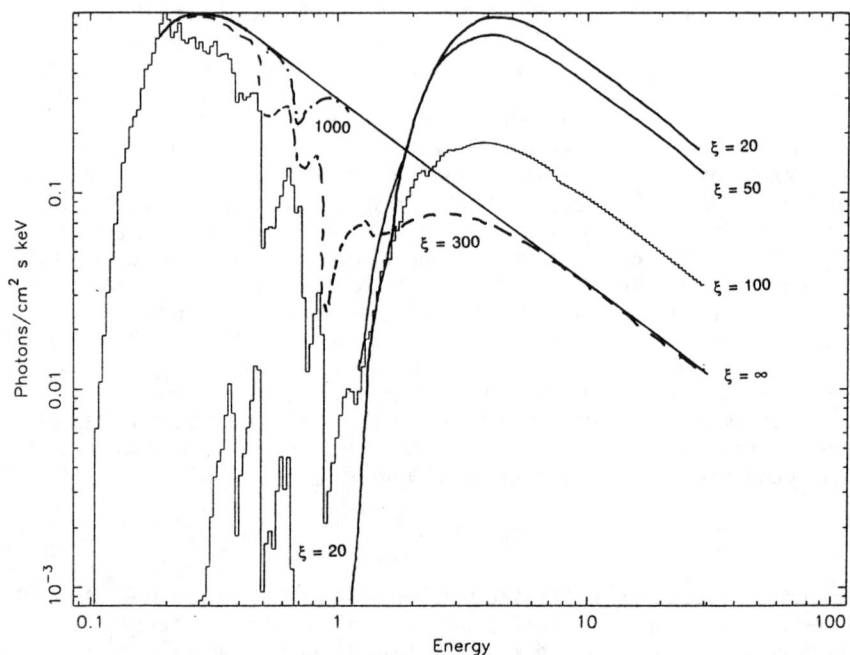

Figure 2. The effect of varying ionization parameter on a burst spectrum of extrinsic and intrinsic columns, 10^{20}cm^{-2} and 10^{23}cm^{-2}, respectively.

Lastly, we point out that the detection of any structure around 0.5 - 2 keV would be an important diagnostic of the state of matter around the source and could provide an independent estimate of burster distances if significant redshifts are involved.

REFERENCES

1. C.A. Meegan et al., Nature, **355**, 143 (1992).
2. B.E. Schaefer, Compton Gamma-Ray Observatory, AIP, New York, 803 (1993).
3. A. Owens & B.E. Schaefer, Comments on Astrophys., in press.
4. H.S. Fencl, et al., Gamma-Ray Bursts, AIP New York, 267 (1992).
5. D. Band, & D.H. Hartmann, Astrophys. J., **386**, 299 (1992).
6. M.C. Jennings, Astrophys. J., **273**, 309 (1983).
7. B.E. Schaefer et al., Astrophys. J. (Lett.), **393**, L51 (1992).
8. M. Katoh et al., 'High Energy Transients in Astrophysics', AIP, 390 (1984).
9. J.G. Laros et al., 'High Energy Transients in Astrophysics', AIP, 378 (1984).
10. T.M. Murakami et al., Nature, **335**, 234 (1988).
11. K. Hurley, Proc. 14th Texas Symposium on Relativistic Astrophysics,, Annals of the New York Academy of Sciences, **571**, 442 (1989).
12. D. Band et al., in Gamma-Ray Bursts, AIP New York, 169 (1992).
13. K. Hurley et al., in Gamma-Ray Bursts, AIP New York, 195 (1992).
14. C. Done, et al., Astrophys. J., **395**, 275 (1992).
15. K.R. Lang, Astrophysical Formulae (New York: Springer, 1974).
16. B.E. Schaefer, this workshop.
17. B. Aschenbach, Proc. Satellite Sym. 3, ESA ISY-3, 41 (1992).

DISTANCE TO GAMMA RAY BURSTS FROM THEIR SOFT X-RAY SPECTRA

Bradley E. Schaefer *
NASA/Goddard Space Flight Center, Greenbelt, MD 20771

ABSTRACT

The distance scale to Gamma Ray Bursts can be derived from the measure of absorption in their soft x-ray spectra due to neutral gas in the intervening interstellar medium. This idea holds strong promise of definitively solving the distance scale problem and so has attracted much recent experimental development. This paper addresses several questions relating to the requirements and limitations of the method. First, the observation of absorption edges can provide a distinct diagnostic for the cause of any low energy spectral turnover. Second, these edges can be used to uniquely identify absorption from a host galaxy as well as to measure its red shift. Third, it may be possible to distinguish far halo models from cosmological models where bursters reside outside galaxies. Fourth, unionized gas close to the burster will be flash ionized by the burst itself and cannot contribute significant absorption.

INTRODUCTION

The single most important parameter for gamma ray burst (GRB) models is the burst energy. But since the distance scale to GRBs may be from 30 AU (Oort cloud models[1]) to 10 billion light years (the edge of the Universe[2]), the uncertainty in burst energies is 26 orders of magnitude! So the determination of the GRB distance scale is the single most important question for the field.

The quickest answer is to find a quiescent counterpart whose distance we can measure by other means, but searches have been notable for their lack of success[3]. Another measure that could establish the distance scale is a gravitational lensing event, but none have been seen and they are predicted to be extremely rare[4]. So how will we solve the distance scale problem?

At the GRB workshop in Stanford in 1984, I proposed that observation of interstellar photoelectric absorption in the extreme ultraviolet could yield a column density of intervening neutral gas and hence a distance scale. At the First Compton Symposium in St. Louis in 1992, I extended this proposal to the soft x-rays and pointed out that this experiment was feasible with a small dedicated satellite experiment[5]. Such an instrument could also discover cyclotron lines with $B \leq 10^{12}$ G, detect iron or oxygen lines, test the GINGA soft precursors and tails for a blackbody spectral shape, obtain good spectra for Soft Gamma Repeaters, and search for the ultra-soft transients possibly discovered by the Einstein satellite.

This proposal has prompted several groups around the world to propose, design, or build instruments to measure the soft x-ray spectrum of GRBs. Two such proposals (called BALLERINA [N. Lund P.I.] and RULER [A. Owens P.I.]) were submitted to ESA as part of the M3 mission and both were rated in the top three. Unfortunately, neither was ultimately selected for programmatic reasons,

---

* Also Universities Space Research Association

although it is clear that these proposals will keep at the fore in contention for launching. In the meantime, a group lead by V. Kurt (Lebedev Physical Institute, Moscow) is starting to build an instrument to be on the Russian RADIOASTRON satellite scheduled for a 1996 launch. A fourth group, lead by E. Mazets (Ioffe Physical-Technical Institute, St. Petersburg) will start on a soft x-ray spectrometer for GRBs to be launched in 1995 on a Russian satellite.

MEASURING THE DISTANCE SCALE

The idea is to measure the soft x-ray spectrum of at least a dozen GRBs, to derive the characteristic column density from photoelectric absorption by intervening interstellar matter, then to deduce the distance scale from the column densities and their pattern with position on the sky. Each of the proposed distance scales for GRBs has a distinct prediction how the measured column densities should vary. Thus, Oort Cloud and heliosphere models will be sharply distinguished by the utter lack of any absorption. Galactic disk models predict that the measured column density will have a characteristic dependence on brightness and galactic latitude. Galactic halo models predict that the absorption will be consistent with the path length through our Milky Way, except that the bright and/or low latitude bursts will have smaller absorption. If bursts are at cosmological distances, then the observed absorption will be for the entire path length through our galaxy plus any additional absorption for the path through the host galaxy.

This idea is so powerful that even the measurement of one soft x-ray spectrum may be sufficient to solve the GRB distance scale problem. For example, if a single GRB spectrum is measured to 0.1 keV with no absorption, then all galactic and cosmological models would be rejected. Or if one burst has a column density significantly less than that through the Milky Way, then we could eliminate cosmological models. Or if one high galactic latitude GRB where seen with a column density of 10^{23} cm^{-2}, then this would be proof that bursters were in distant galaxies. While in practice more than one burst is needed for confidence in a deduced distance scale, nevertheless the above scenarios show that a small number of soft x-ray spectra can yield an unambiguous answer.

SOFT X-RAY SPECTRA OF GAMMA RAY BURSTS

There is little spectral information for GRBs at energies below roughly 15 keV. The satellite P78-1 had its lowest channel covering 3-6 keV[6], the Hakucho detectors yielded a lowest energy channel plotted from 2-10 keV[7], while both the Apollo x-ray spectrometer and the GINGA detectors record 7 energy channels from 2-10 keV[8,9]. For purposes of calculating fluxes and detector sensitivities, I have constructed a typical GRB spectrum as:

$$dN/dE = \begin{cases} \eta \, (E/1\text{keV})^{-0.7}, & \text{for } E < 10 keV; \\ 2\eta \, (E/1\text{keV})^{-1.0}, & \text{for } 10 keV \leq E \leq 100 keV; \\ 200\eta \, (E/1\text{keV})^{-2.0}, & \text{for } 100 keV < E. \end{cases}$$

The 30-1000 keV fluence of this 'standard' burst is $9.6 \times 10^{-7} \, \eta \, D$ in units of erg $\times$ keV, where D is the burst duration. The number of photons from 0.1-0.3 keV for an unabsorbed spectrum and detector area A is 0.65 η A D keV.

There is no guarantee of significant flux below 2 keV in the intrinsic GRB spectrum. However, the available data show GRB spectra to have a power law with a slope near -0.7 from 2-10 keV. Since there is no reason to expect a spectral cutoff around 1 keV from the source, it is reasonable to presume that the intrinsic GRB spectra is a power law into the soft x-ray regime.

Photoelectric absorption by the interstellar medium produces a relatively sharp cutoff with a distinctive spectral shape. The spectrum will be lowered by a factor of $e^{-\sigma N_H}$, where σ is the photoelectric cross section per hydrogen atom and is roughly proportional to $E^{-2.5}$ for normal abundances and N_H is the column density. The typical N_H through our galaxy ranges from 10^{20} cm^{-2} at the galactic poles to 10^{23} cm^{-2} near the galactic center.

To distinguish between near-Sun models and galactic models, there must be significant absorption at the lowest measured energy for the lowest expected column in the galactic models. For a column density of 10^{20} cm^{-2}, the photoelectric absorption will be a factor of 0.0015 at 0.1 keV, so that the existence of a cutoff can be clearly distinguished from any plausible intrinsic spectral curvature. Thus, any instrument to measure the distance scale from the soft x-ray spectrum should reach to approximately 0.1 keV in energy range.

The photoelectric cross section is not a smooth function of energy, but has discontinuities caused by the K edges of various elements in the interstellar medium. The most prominent edge is that of oxygen at 0.53 keV (with a characteristic width of 0.2 keV) across which the cross section changes by a factor of 2.0. The presence of this and other edges will provide proof that the spectral cutoff is due to photoelectric absorption.

COSMOLOGICAL REDSHIFTS

In the rest frame of the absorbing gas, the cutoff energy for photoelectric absorption will depend only on the column density. But if the absorber is moving away at high velocities, then the cutoff energy will be red shifted to lower energies, and may be interpreted as a lower column density.

The cutoff energy will scale as $(1+z)^{-1}$ for the absorber, where z is the red shift. Since the photoelectric cross section varies as roughly $E^{-2.5}$, the deduced column density will be in error by a factor of $(1+z)^{2.5}$. For all absorption in our Milky Way, this red shift will be negligible. For the most popular cosmological models, the bright bursts (for which the soft x-ray spectrum will be measured) will have a $z \sim 0.1$ so that the red shift effect will also be negligible. The bright GRBs would have to be farther than $z \sim 3$ before the error in the deduced column would have an effect in the detectability of extinction from a host galaxy.

Any cosmological effects should also be identifiable by a red shift of the absorption edges. For example, if a burst near the galactic pole has large absorption and an edge at 0.40 keV, then we would deduce a distance of $z=0.3$ to the burster.

GRBs OUTSIDE GALAXIES

A problem arises because there are not large amounts of neutral gas between the edge of our galaxy's disk and some random distant point outside of any galaxy. Along the path from halo to cosmological distances, the absorption does not increase appreciably, and hence the measured absorption will poorly distinguish the two populations.

Nevertheless, there is neutral gas in our galaxy's halo, and this may be used to distinguish the two populations. York[10] reviews the gas in our galactic halo, and finds there to be a 'disk' extending roughly 10 kpc away from the galactic plane which is filled with clouds with hydrogen density of order 1 cm^{-3}, with perhaps 5% in a neutral state. This corresponds to a column density of $\sim 10^{21}$ cm^{-2} out to 10 kpc from the Earth. While this is probably an over-estimate, many clouds of column density $\sim 3\times 10^{18}$ cm^2 have been observed. In a halo model with the faintest BATSE bursts 200 kpc distant, the edge of the halo gas distribution corresponds to a fluence of roughly 4×10^{-5} erg cm^{-2}. Thus, for halo GRB models the nearest (brightest) bursts will systematically have a less absorption than the farthest (faintest) bursts, whereas for cosmological models this trend will not exist. This trend will not be easy to measure, since only at high galactic latitude will the galactic disk absorption be small enough for the added halo component to be apparent.

The intergalactic medium does contain neutral gases, as revealed by the absorption lines in quasar spectra. However, these clouds have low column densities of neutral hydrogen (typically $\sim 10^{14}$ cm^{-2} for the Lyman-α clouds and $\sim 10^{18}$ cm^{-2} for the metal-line systems), so that even at cosmological distances there will be little absorption from the few intervening clouds.

In general, cosmological models of GRBs presume them to be distributed inside galaxies, as this is the logical place where interesting objects can be formed. Indeed, there are few cosmological models that place GRBs outside galaxies. B. Paczynski has raised the possibility that GRB systems might be created inside galaxies only to be ejected before they burst. Nevertheless, there is little theoretical reason to expect that bursters are outside galaxies.

NEUTRAL GAS NEAR THE BURSTER

A cutoff in a GRB soft x-ray spectrum could be due to intervening interstellar material which would lead directly to the distance scale or equally to material local to the burster which would not. Thus, the existence of local material could make the apparent absorption higher than would be indicated by the distance scale, and could fool us into believing that GRBs are farther away than in reality. This will not be a problem for any of several reasons:

First, no GRB model predicts that there will be substantial unionized material near any burster, so the problem of local matter is unlikely to arise. Many burst models require bursters to be in isolated collapsed stars of great age, so that no clumps of ISM will be associated. Even the models that use accretion onto neutron stars from the ISM require the number density to be small. Any old supernova shell would be too dispersed to be noticed. The only exceptions for nearby unionized matter is if the material is in a companion star or an accretion disk, but then the column would then either be near zero (when the line of sight is away from the object's edge) or extremely large (such that no GRB event would be seen). Intermediate cases would occur only over a small solid angle for the viewing direction and hence will be of rare occurrence.

Second, if any neutral gas is near the burster, it will rapidly be ionized by the burst radiation so that the bulk of the event will appear unabsorbed. To be quantitative, I have calculated the column density which will be ionized for a given fluence. Schwarz[11] presents a set of calculations at a variety of energies for the evolution of the ionization structure in a slab of gas exposed to a flash of x-rays. A characteristic length scale for this problem is the distance from the inner edge of the slab that would be formed if all photons were absorbed within a sharp boundary between completely ionized and completely neutral material. This length is easily calculated by balancing the input photons with ionized atoms. Schwarz shows that the ionization front depth in realistic calculations is within a factor of two of this characteristic length. With the 'standard' burst spectrum, the column density ionized will be 5.0 η D keV. Here D is the duration of the pulse and η is the brightness parameter. The unabsorbed burst brightness at Earth will be $(R_{shell}/D_{burst})^2$ times smaller than at the gas shell, where R_{shell} is the burster-to-gas distance and D_{burst} is the burster-to-Earth distance. If the gas shell is sufficiently far from the burster, then it can remain unionized. This critical shell radius will depend on the GRB distance scale, the column density of the shell, and the observed 30-1000 keV fluence S. The critical shell radius is $D_{burst}(5.2\times10^6 \, S \, N_H^{-1} \, erg^{-1})^{0.5}$. The maximum value for the column density before the burst is 10^{23} cm^{-2}, since if the value were any larger then the absorption would already have been seen. If we set the fluence equal to 10^{-6} erg cm^{-2}, then any burst for which a column density can be measured will have the gas shell ionized for the bulk of the burst duration. The critical shell radii for distance scales of 10^{15} cm (Oort Cloud), 10^{23} cm (galactic halo), and 10^{27} cm (cosmological) are 0.1 km, 0.05 AU, and 0.2 pc respectively.

Third, the case of matter local to the GRB can easily be distinguished by temporal and positional patterns. That is, if the measured column density decreases systematically as the spectrum evolves over time as a linear function of the fluence up to that time, then the existence of local material will be proven. A model fit to this time variation will quickly yield the size of the local and interstellar column densities. Also, if the measured column density is consistent with a galactic component plus a variable component, then the additive component would be identified with local absorption. So for example, if the measured column density is evenly distributed from 10^{18} cm^{-2} to 10^{22} cm^{-2} for all galactic latitudes, then GRBs must be near our Solar System with some local absorption up to 10^{22} cm^{-2}. As another example, if the column density for high latitude bursts vary between $10^{20.5}$ cm^{-2} to 10^{24} cm^{-2}, then GRBs must be in distant galaxies with the host galaxy providing the local absorption.

REFERENCES

1. K. F. Bickert and J. Greiner, Compton Gamma-Ray Observatory (AIP, N. Y., 1993), p. 1059.
2. B. Paczyński, Ap. J. **335**, 525 (1988).
3. B. E. Schaefer, this volume , (1994).
4. S. Mao, Ap. J. **389**, L41 (1992).
5. B. E. Schaefer, Compton Gamma-Ray Observatory (AIP, N. Y., 1993), p. 803.
6. J. G. Laros et al., High Energy Transients in Astrophysics (AIP, N. Y., 1984), p. 378.
7. M. Katoh et al., High Energy Transients in Astrophysics (AIP, N. Y., 1984), p. 390.
8. A. E. Metzger et al., Ap. J. **194**, L19 (1974).
9. A. Yoshida et al., PASJ **41**, 509 (1989).
10. D. York, ARAA **20**, 221 (1982).
11. J. Schwarz, Ap. J. **182**, 449 (1973).

THE COMPTON ATTENUATION MODEL OF COSMOLOGICAL GAMMA-RAY BURSTS

J.J. Brainerd*

Dept. of Physics, University of Alabama in Huntsville

ABSTRACT

A gamma-ray burst spectrum can be modeled by passing a power law spectrum through a medium that is optically thick to Compton scattering. The roll-over of the Klein-Nishina cross section at high energies produces a break in the spectrum at several hundred keV if the gamma-ray source is at $z \approx 1$. Photon-photon pair production limits the scattering region's size to greater than ≈ 0.1 parsec. Because the optical depth is greater than unity for a length scale of several parsecs, the density must be of order 10^5cm^{-3}, which occurs only in molecular clouds at the centers of galaxies. This model therefore precludes source models employing objects that are common to the galactic plane—for example, merging neutron stars. The model spectrum is described by five free parameters: a cosmological red shift, a Thomson optical depth, a power law index, a metallicity, and an amplitude. The attenuation model can be tested by comparing the consistency of model parameters derived at different times over the duration of a burst. Small angle scattering of x-rays by dust within the molecular cloud produces an x-ray afterglow. A consequence of this model is that any optical or ultraviolet radiation is heavily absorbed, making detection of gamma-ray bursts at these energies unlikely.

INTRODUCTION

A long standing problem in constructing models of gamma-ray bursts is modeling the non-thermal burst spectrum. These spectra are generally a power law above 1MeV, and emit their peak power at several hundred keV.[1,2] Before BATSE, when the predominate models were emission from neutron stars, the small fraction of energy in the x-ray band caused considerable theoretical difficulty, and became known as the x-ray paucity problem. Now we have cosmological models which have no predefined characteristic energy, implying that the characteristic spectral energy peak is still a problem.

The root of the difficulty with cosmological models is the large Lorentz factors of $\gamma > 100$ required to overcome photon-photon pair creation. This makes the characteristic energy in the rest frame of the emitter ill-defined, because the value of the Lorentz factor is not physically constrained. The likely radiative processes—photoelectric absorption, cyclotron emission, Compton scattering—are further unconstrained in optical depth, electron distribution, magnetic field strength, and seed photon density and energy.

These considerations suggest that the characteristic energy of gamma-ray bursts is set by processes in the rest frame of the host galaxy and is associated with the electron rest mass energy. This article examines Compton attenuation as the mechanism responsible for producing the observed spectra. More detailed

* Mailing Address: Space Science Lab, ES-66, NASA/MSFC, Huntsville, AL 35812. E-mail: brainerd@ssl.msfc.nasa.gov

analyses of this model are presented elsewhere.[3,4]

THE COMPTON ATTENUATION MODEL

Because the purpose of this model is to explain the characteristic peak values of the νF_ν curve independent of the Lorentz factor of the source, I must assume that the source spectrum is a power law. In the model the source is embedded in a cool, static medium with Thomson optical depth much greater than unity. Compton scattered radiation can annihilate with other burst photons and produce electron-positron pairs. If significant scattering occurs closer than $\approx d_s = 0.04\,(\tau/10)\exp{(\tau/2-5)}$ pc to the source, where τ is the Thomson optical depth of the scattering region, the number of pairs created raises the optical depth by at least 1, which produces a rapid and permanent extinction of the gamma-ray burst. As this contradicts the observations, the scattering region must be larger than d_s. The travel time between scatterings is therefore $> 10^6$s, so the scattered radiation does not contribute to the burst.

A second process affects the x-ray spectrum: photoelectric absorption. Because the scattering region is primarily composed of hydrogen, this process affects the spectrum at the lower energy limit of the BATSE instrument (approximately 20keV). The energy of unity optical depth is pushed far below the instrument threshold for a cosmological red shift of $z \approx 1$. As a consequence, photoelectric absorption is generally unimportant for fitting the spectra of bursts observed by BATSE.

Figure 1 shows the Compton attenuated spectrum for several values of the Thomson optical depth. At high energies the scattering optical depth falls to zero because of the ν^{-1} dependence of the Klein-Nishina cross section above 1MeV. The transition from the Thomson cross section to the Klein-Nishina cross section produces a "broken" spectrum below 1MeV. Below 10keV the effects of photoelectric absorption appear. In the absence of photoelectric absorption the spectrum follows a power law with the index of the source power law spectrum.

The high Thomson optical depth in this model limits it to sources at the centers of galaxies. Because the scattering region is large, it must be associated with the interstellar medium. But the optical depth through the plane of nearby spiral galaxies is much less than unity. This leaves molecular clouds with densities of order 10^5cm^{-3}, which exist only at the centers of galaxies. This model is therefore inconsistent with sources common to the planes of galaxies, such as merging binary stars. The source must in fact be unique to the cores of galaxies, which suggests that the energy source in this model is a massive black hole.

X-RAY AFTERGLOW

A small number of bursts have x-ray emission lasting up to several hundred seconds after the gamma-ray emission falls below background.[5,6,7] The afterglow spectrum falls rapidly and can be fit by a black body continuum with a temperature of approximately 1keV to 2keV. To this point the only interpretation offered for this radiation is thermal emission from the burst source. For extragalactic sources this implies a source size of $\approx 10^{10}$cm for $\gamma = 1$, which is larger than the size derived from the rise time.

Because the Compton attenuation model requires a dense molecular cloud for the scattering region, it provides a natural explanation for the x-ray afterglow—coherent scattering with dust.[4,8,9,10] X-rays produced in the gamma-ray

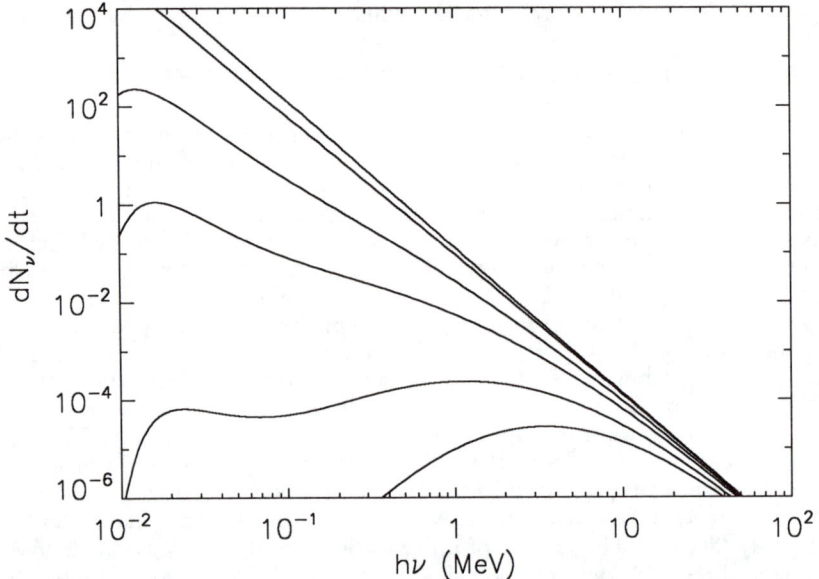

Figure 1. Spectra produced by Compton attenuation of a power law with index $\delta = -3$. The scattering region has 0.5 of the solar abundance of metals. The Thomson optical depths from top to bottom are $\tau = 0, 1, 5, 10, 20,$ and 30. No cosmological red shift is present.

burst are scattered by a small angle into the observer's line of sight. The longer propagation path introduces a delay in the x-ray emission relative to the gamma-ray emission. At several keV the scattering angle is of order 10^{-4} radian for a dust particle diameter of 0.01μm, which produces a time delay of 100 seconds if the scattering occurs 10pc from the x-ray source.

I assume the distribution function for dust grain size is

$$f(a) = \frac{3(4-m)\rho_m}{4\pi a_0^4 \rho_g} \left(\frac{a_0}{a}\right)^m, \qquad (1)$$

where $a < a_0$ is the grain radius, ρ_g is the mass density within each grain, and ρ_m is the mass density of metals in the molecular cloud. The distribution function is zero when $a \geq a_0$. The total cross section for coherent scattering of x-rays by dust is then[4]

$$\alpha_d = 6.7 \times 10^4 \left(\frac{Z}{A}\right)^2 \frac{4-m}{5-m} a_{0\mu m} \rho_m \rho_g \left[\frac{F(E)}{Z}\right]^2 E_{keV}^{-2} \text{ cm}^{-1}. \qquad (2)$$

In this equation the function in brackets is of order unity. For the molecular clouds under consideration in this model, one expects the energy of unity optical depth to be near 10keV. For these equations and the differential scattering cross section, I ran a Monte Carlo simulation of 10^5 photons to produce Figure 2.

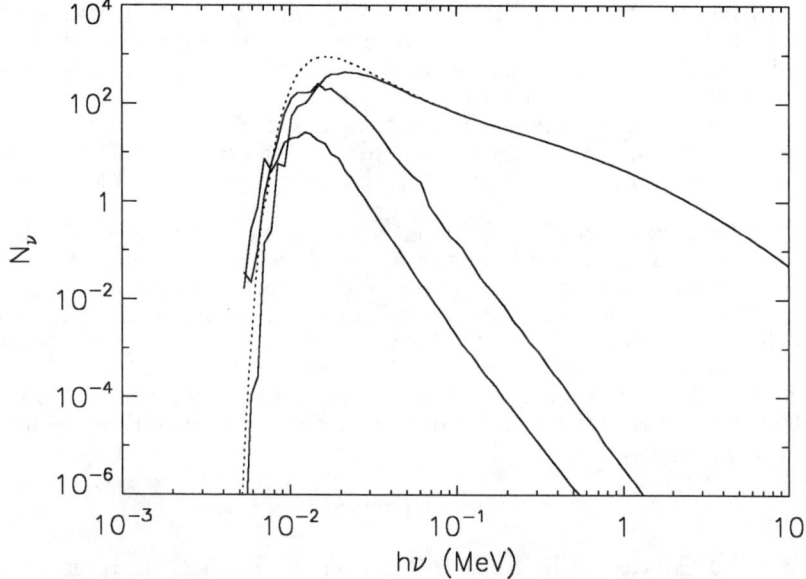

Figure 2. X-Ray afterglow. The dotted curve gives the Compton attenuated burst spectrum in the absence of dust for $\tau = 10$. The remaining parameters are those for Fig. 1. The uppermost solid curve gives the direct spectrum in the presence of dust, where all metals are assumed to be in dust. The dust distribution is given in the text. The remaining two solid curves give the delayed x-ray spectrum. The upper curve is the delayed emission integrated over time 100 s after the burst. The lower curve is the emission integrated between 100 s and 10^3 s.

The dust decreases the direct x-ray flux by a significant factor below 30keV. The afterglow spectrum is quite steep above unity optical depth, having a spectrum given by the burst spectrum times E_{keV}^{7-m}. At low energies the optical depth to scattering exceeds unity, causing the delayed spectrum to be flat. While the spectra in Figure 2 resemble thermal spectra with temperatures of $\approx$ 5keV, which is higher than observed, dust scattered spectra at a cosmological red shift of ≈ 2 will have the correct shape to fit the observations. One characteristic of this model is that the peak of the x-ray afterglow should shift down in energy with time.

OBSERVATIONAL TESTS

The Compton attenuation theory must satisfy a stringent test: model fits to a burst's spectrum at different times over the duration of the burst must yield the same cosmological red shift z. One does not generally expect other models to produce a constant value of z. For instance, if the energy of the spectral break is dependent on the Lorentz factor γ of the emitter, and if γ decreases with time, the spectral break energy should fall with time, leading to a z that rises with time. The constancy of z with time can also be tested by fitting the color-color diagrams of individual bursts[11] with model color-color curves found by allowing the power law index to vary.

A second test of this model is the comparison of the values of z derived from spectral fits with the allowed values of z implied by each burst's position on the $\log N$–$\log P_{max}$ curve. High values of z must correspond to low values of P_{max}.

The effects of dust offer additional opportunities for testing this theory. The spectrum and time profile of the x-ray afterglow is dependent to some extent on the spectrum and time profile of the direct gamma-ray burst. One should be able to model this correlation and test for its presence. A second test is the presence of an optical flash at the time of the gamma-ray burst. Optical emission during the gamma-ray burst will be heavily absorbed by the dust, making the observation of gamma-ray bursts at these wavelengths unlikely. The observation of an optical flash would pose great physical difficulties for the Compton attenuation model.

REFERENCES

1. D. Band, J. Matteson, L. Ford, B. Schaefer, D. Palmer, B. Teegarden, T. Cline, M. Briggs, W. Paciesas, G. Pendleton, G. Fishman, C. Kouveliotou, C. Meegan, R. Wilson, & P. Lestrade, Astrophys. J. **413**, 281 (1993).
2. C. Winkler, K. Bennett, L. Hanlon, O.R. Williams, W. Collmar, R. Diehl, V. Schönfelder, H. Steinle, M. Varendorff, J.W. den Herden, W. Hermsen, L. Kuiper, D.N. Swanenburg, C. de Vries, A. Connors, D. Forrest, M. Kippen, M. McConnell, & J. Ryan, in *Compton Gamma-Ray Observatory: St. Louis, MO 1992*, ed. M. Friedlander, N. Gehrels, & D.J. Macomb (AIP: New York, 1993), p. 845.
3. J.J. Brainerd, Astrophys. J., in press, (1994).
4. J.J. Brainerd, Astrophys. J. Lett., submitted, (1994).
5. J.G. Laros, et al., Astron. Astrophys. **286**, 681 (1982).
6. M. Katoh, T. Murakami, J. Nishimura, T. Yamagami, M., Fujii, & M. Itoh, in *High Energy Transients in Astrophysics*, ed. S.E. Woosley (AIP: New York, 1984), p. 390.
7. T. Murakami, H. Inoue, J. van Paradijs, E.E., Fenimore, & A. Yoshida, in *Gamma-Ray Bursts: Observations, Analyses and Theories*, ed. C. Ho, R.I. Epstein, & E.E. Fenimore (Cambridge University Press: Cambridge, 1992), p. 239.
8. J.W. Overbeck, Astrophys. J. **141**, 864 (1965).
9. C.W. Mauche & P. Gorenstein, Astrophys. J. **302**, 371 (1986).
10. C.S.R. Day & A.F. Tennant, Mon. Not. R. Astron. Soc. **251**, 76 (1991).
11. C. Kouveliotou, W.S. Paciesas, G.J. Fishman, C.A. Meegan, & R.B. Wilson, Astron. Astrophys. **97**, 55 (1993).

A COLD ABSORPTION MODEL OF GAMMA RAY BURST SPECTRA

Edison P. Liang

Department of Space Physics & Astronomy
Rice University, Houston, TX 77251-1892

ABSTRACT

Motivated by the lack of x-rays below ~ 200 keV in most GRB spectra, we consider a spectral model in which an underlying source of power-law of gamma rays is shielded by an optically thick layer of absorbing circumburster material rich in iron-group elements whose photoelectric opacity exceeds Thomson opacity below ~ 200 keV. For reasonable distributions of line-of-sight optical depths we find that the absorbed spectrum can indeed mimic the typical observed GRB spectra.

INTRODUCTION & SUMMARY

SMM results show that the typical GRB spectrum above ~ 300 keV fits a simple power law of photon index 2 - 2.5. Yet all lower energy measurements, including BATSE results, show flattening below ~ 200 keV with asymptotic x-ray indices typically < 0.5-1. Some spectra are even consistent with complete turnover (<0) in the x-rays. This lack of x-rays has been a persistent mystery in the origin of gamma ray burst spectrum.

Here we explore a possible *cold absorption* origin of the spectral flattening or turnover of the GRB x-ray spectrum. Noting that the Fe-group elements have photoelectric (bound-free) absorption opacity exceeding Compton opacity below ~ 200 keV, we study the emergent spectrum of an underlying simple power law continuum passing through a cold layer of Fe-rich absorbing material. We find that *the typical GRB spectral shape can indeed be produced by the absorption of a simple power law continuum by a distribution of cold Fe absorption depths*. Here we show sample results corresponding to two simple models of absorption depth distribution functions: power law with upper cutoff and Gaussian.

A consequence of the cold absorption model is that it *predicts a simple hard to soft spectral evolution*: as the absorbing layer is heated or blown away during the gamma ray burst, its absorption depths are reduced, resulting in increasing exposure of the soft photons below ~ 200 keV in the underlying continuum, so that the overall spectrum continues to soften. This may be relevant to the almost universal hard-to-soft spectral evolution of the FRED (Fast Rise Exponential Decay, cf. contributions in this volume by the BATSE team) bursts which constitute over a quarter of the observed bursts.

An important observational prediction of the cold Fe absorption model is the likely presence of *an absorption edge at ~7 keV and the K-alpha emission feature at ~6.7 keV*, similar to the computed supernova spectra (e.g. Pinto and Woosley 1988). Even though thermal, dynamical and inhomogeneity effects will likely broaden and smear such features in real situations we hope that they may still be recognizable in some subset of bursts due to special conditions or view angles. High resolution x-ray spectrometers will be needed to search for such features in future missions.

THE MODEL

We use cold Fe opacity derived from the LLNL ENDL Tables (Cullen et al 1989) : The absorption depth is

$$\tau_{ab} = 4\times 10^{-20}\, nh\, (\upsilon/7\text{keV})^{-2.825} \qquad \upsilon > 7\text{ keV}$$

where nh is the Fe column density in cm^{-2} and we concentrate on x-rays above 7 keV. At the hard x-ray energies Compton scattering results in minor distortion of the power law shape (cf. Fig.1, see also Arons 1971, Pozdnyakov et al 1983). As a first approximation we will ignor it and use a simple power law input spectrum. If I_0 is the input power law the emergent absorbed spectrum is given by:

$$I = I_0 \exp(-\tau_{ab}) = I_0 \exp(-\tau\, (\upsilon/7\text{keV})^{-2.825})$$

where $\tau = 4\times 10^{-20}\, nh$ is the absorption column depth at 7 keV. For pure Fe^{26} this equals to 2313 times Thomson depth. The effects of absorption for different τ are illustrated in the Figures.

The observed GRB spectrum is likely the composite of spectra from different lines of sight (LOS) through media of varying absorption depths. We can model this distribution of τ by a normalized probability distribution $P(\tau)$ ($\int d\tau P(\tau) = 1$). Assuming that the input power law continuum is the same for all LOS, the composite absorbed spectrum becomes a Laplace transform of P:

$$I = I_0 \int d\tau P(\tau) \exp(-\tau(\upsilon/7\text{keV})^{-2.825})$$

We consider two simplest forms of P:
1. Power Law Distribution:

$$P \sim \tau^{-\beta} \quad (\beta<1) \qquad \tau \leq \tau_{max}$$
$$= 0 \qquad \tau > \tau_{max}$$

2. Gaussian Distribution:

$$P \sim \exp(-\beta\tau^2) \quad \beta > 0.$$

Sample absorbed spectra for both cases are presented in Figures 3-5.

CONCLUSIONS

1. Fe absorption of simple power law continuum spectrum can reproduce the typical observed GRB spectral shape.
2. Gaussian distribution of absorption column depths leads to cutoff of hard x-rays steeper than seen in most bursts.
3. Power law distribution of absorption column depths with upper cutoffs leads to broken power laws resembling the typical GRB spectrum. The break energy is related to the maximum column depth. The spectral index γ below the break is related to the absorption depth distribution index $\beta(<1)$ and spectral index α above the break via:

$$\gamma = \alpha + 2.825\,(\beta - 1).$$

When Compton scattering is included we expect the break to be somewhat smoother than that given here.

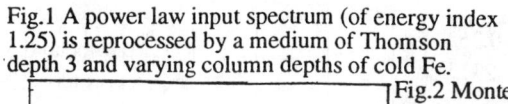

Fig.1 A power law input spectrum (of energy index 1.25) is reprocessed by a medium of Thomson depth 3 and varying column depths of cold Fe.

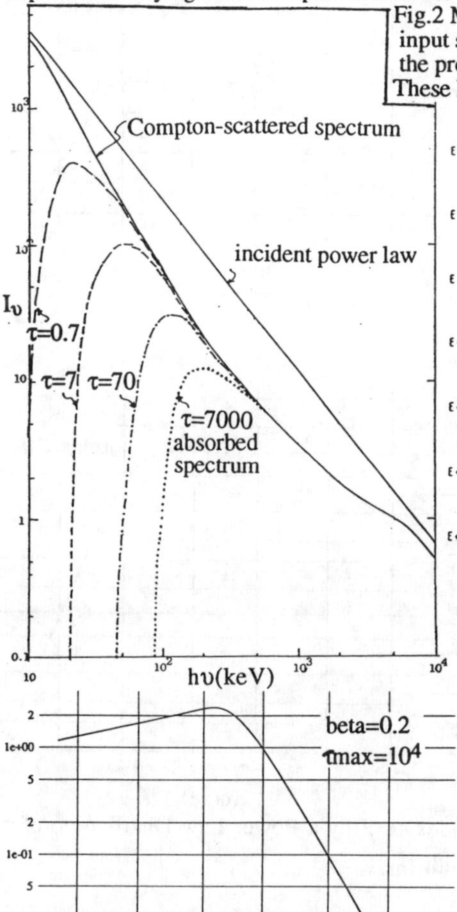

Fig.2 Monte Carlo simulation of an inverse Compton input spectrum absorbed by cold Fe. Note the prominent K-edge and the K and L emission lines. These lines are broadened by Thomson downscattering.

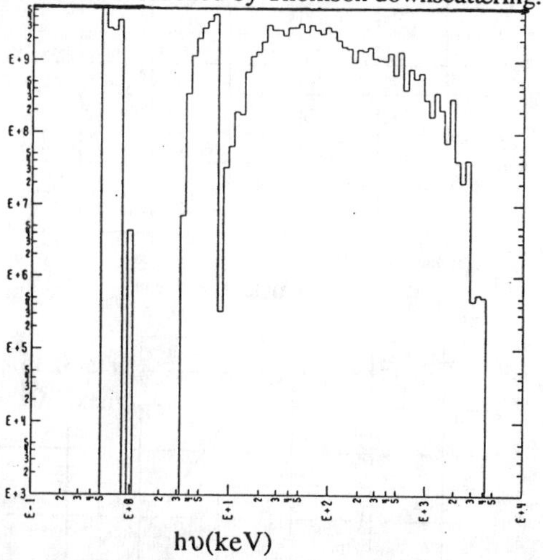

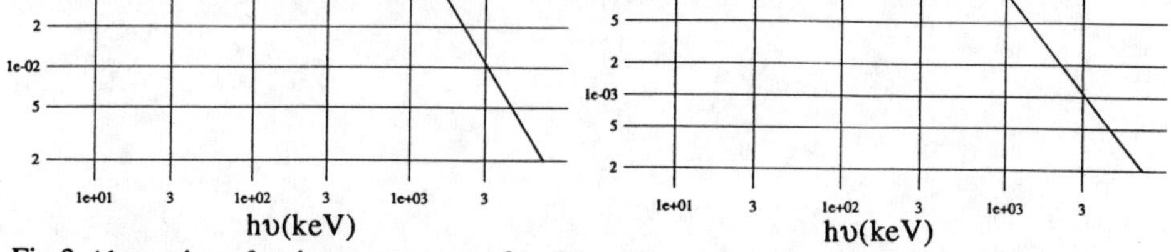

Fig.3 Absorption of an input spectrum of $\alpha=2$ by a Power Law Distribution of τ with varying index β (here $\tau_{max}=10000$).

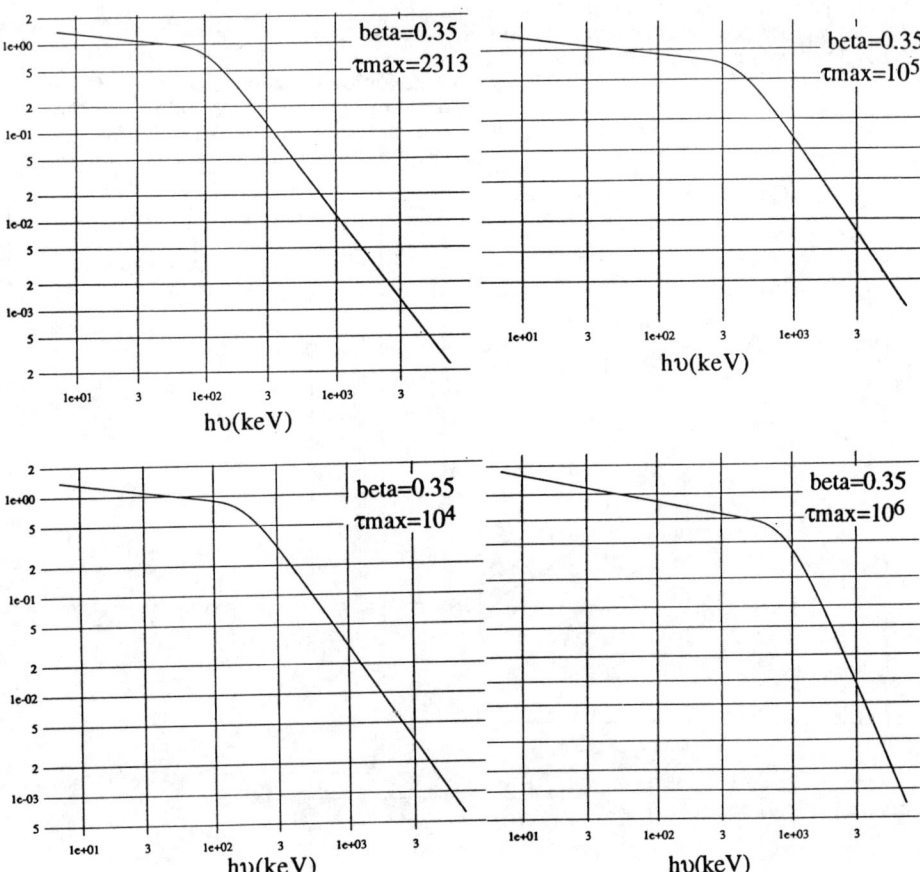

Fig.4 Absorption of an input spectrum of index $\alpha=2$ by a Power Law Distribution of τ ($\beta=0.35$). Note how the break energy varies with τ_{max}.

4. The cold absorption model is consistent with a hard-to-soft spectral evolution found commonly in FRED bursts.

5. The Fe absorption model predicts that the K-edge at 7 keV and Kα emission at ~ 6.7 keV may be detectable for some bursts.

This work was partially supported by NASA Grant NAG 5-1515. EPL also acknowledges partial support from LLNL as summer faculty fellow where part of this work was done. LLNL is operated by the University of California for the US DOE under contract No. W-7405-ENG-48

REFERENCES

Arons, J. 1971, Ap.J. 164, 437.
Cullen, D.E. et al. 1989, UCRL-50400, 6, 170 (LLNL, Livermore, CA).
Pinto, P. and Woosley, S.E. 1988, Ap.J. 329, 820.
Pozdnyakov, L.A. et al. 1983, Sov. Ast. & Sp. Phys. Rev. 2, 189.

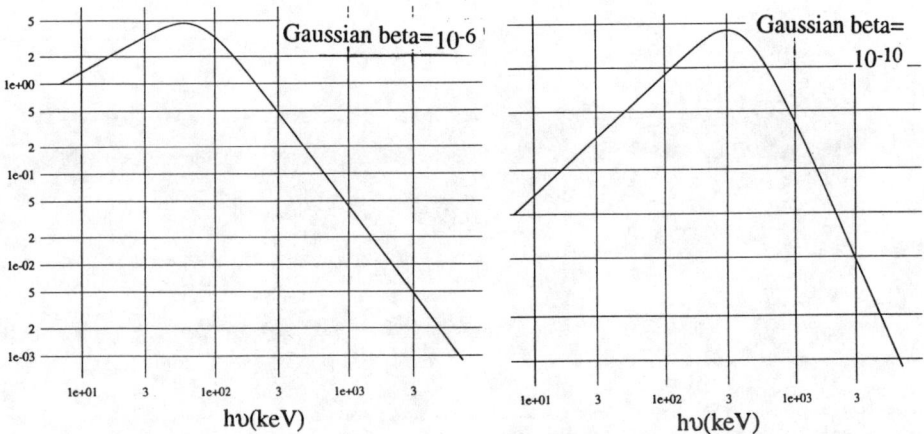

Fig. 5 Absorption of an input spectrum of α=2 by a Gaussian Distribution of τ with varying Gaussian width β. Note that all spectra turnover sharply due to the lack of power at low column depths in the distribution.

BURST LOCALIZATIONS AND SEARCHES FOR COUNTERPARTS

Burst Localization
Counterparts—General
Optical Counterparts
X-Ray Counterparts
Very High Energy Counterpart Searches
Soft Gamma-Ray Repeaters—Counterparts

PRECISE LOCALIZATIONS AND COUNTERPART SEARCHES OF GRBS FROM THE 2ND INTERPLANETARY NETWORK

F. Hack, K. Hurley
University of California Space Sciences Laboratory, Berkeley, CA 94720

J.-L. Atteia, C. Barat, M. Niel
Centre d'Etude Spatiale des Rayonnements, 31029 Toulouse Cedex, France

T. Cline, B. Dennis
NASA GSFC, Greenbelt, MD 20771

C. Kouveliotou
NASA MSFC, Huntsville, AL 35812

R. Klebesadel, J. Laros
Los Alamos National Laboratory, Los Alamos, NM 87545

V. Kurt, A. Kuznetsov, V. Zenchenko
IKI, 11780 Moscow, Russia,

Abstract

The second interplanetary network of GRB detectors operated between 1981 and 1984. It consisted of the Venera 13 and V14 spacecraft (SIGNE detectors), PVO, ICE, and SMM (HXRBS detector). Many of the approximately 90 cosmic GRBs it detected can be localized to a high accuracy. Such localizations will make a better estimate of the burster recurrence time possible and contribute to the search for quiescent counterparts. We will present a number of localizations to illustrate the potential of this network, discuss the techniques we use to perform multi-wavelength counterpart searches, and compare present and promising future localization techniques.

Introduction

Twenty years after their discovery, source identifications and the distance scale to bursters remain contentious. Precise localizations of burst sources can help identify counterparts and constrain source models. Correlations between error boxes and unusual astrophysical objects would present a breakthrough, but even if no such correlations are found, it may be possible to place faintness limits on possible counterpart galaxies to help constrain cosmologically distant source models. While about a thousand bursts are at present localized to an accuracy of several degrees, only a few dozen source locations have an accuracy of several arcminutes and better. Here we describe work in progress using the triangulation method on GRBs detected by up to five spacecraft located in the inner solar system between November 1981 and April 1983. The spacecraft are Venera 13 and 14 (SIGNE detectors), PVO, ICE, and SMM (HXRBS). We have data on 269 bursts, with 39 bursts detected by four or more spacecraft, and 49 by three.

Discussion

The triangulation method requires, at minimum, precise timing (~10 ms) and ephemeris (~1000 km) data from three spacecraft. These alone result in two mirror-image positions, but for the 2nd IPN the KONUS detectors on Venera 13 and 14 offer many localizations at an accuracy of degrees which we may use to resolve the ambiguity in some cases. Also, the two SIGNE detectors on Venera 13 and 14 faced opposite directions, so in some cases a comparison of their responses may yield the unique location. The absolute clock accuracy on the spacecraft is ~1 ms, and ephemeris data are good to ~1000 km or better, equivalent to ~3 ms timing error. The baselines are 0.5 -1.5 AU. The fundamental limitations on localization accuracy are timing uncertainties in cross-correlating the arrival times of the GRBs at different detectors. Even with ~15 ms bin widths, low count rates and/or lack of sharply varying time history curves can degrade timing information to a conservative ~300 ms. This yields location accuracies of order 1 degree, but it will be possible to improve this considerably by refining cross-correlation techniques.

At present we compute the lag between bursts at different spacecraft by extremizing a 'goodness of fit' function between the two time histories. After regrouping and vernier fitting the smaller time base into the larger one, we compute a reduced chi-square and the statistical correlation function. For the former an estimate of the sensitivity ratio between the detectors as well as the measured backgrounds are needed. The latter is background-subtracted but inherently insensitive to the sensitivity ratio. The time histories are then stepped by 1/256 s and the process is repeated. Difficulties in the procedure include bursts with different spectral time histories being compared in instruments with different spectral response characteristics, violating the assumption that a single, true time history underlies all detector responses. Other difficulties include bursts for which different instruments trigger at different times, leaving little or no overlap at mutual high time resolution. We use the timing and ephemeris data to construct the timing annuli and, by overlapping them, source error boxes. As it is difficult to statistically estimate the confidence of lag timing errors, we have chosen a conservative value of 300 ms.

Figure 1 shows an example of two time histories aligned to produce a minimum chi-squared. The experimental counts are background subtracted and have been regrouped into common 47 ms bins. The higher peak is Venera 13. Figures 2 through 4 show three error boxes (1950 coordinates). The error boxes are the enclosed, deformed polygons. Localizations where only one baseline is long result in long, narrow error boxes.

We will carry out multi-wavelength counterpart searches in astronomical catalogs, and make our error boxes available to the wide community for radio, optical, etc. observations. For this paper we used the Astrophysics Data System and the SIMBAD data base in non-comprehensive searches of the three error boxes. Although catalog searches have been carried out in the past, the availability of the ADS is a promising new aspect. In the ADS we searched:

Einstein IPC Sources
ADS Master X-Ray Catalog Database
Hewitt & Burbidge Cat. of QSO's 1989
General Cat. of Variable Stars
HEAO 1 A-3 MC LHSS Cat. of X-Ray Sources
Cat. of Abell Clusters of Galaxies
Cat. of Galaxies & Clusters of Galaxies
Cataloged Galaxies & Quasars Observed by IRAS version 2

For May 11 we also included:

Cat. of Zwicky Clusters of Galaxies
Seyfert Galaxies
Revised New General Cat. of Nonstellar Astron. Objects
CfA Redshift Catalogue
2nd Ref. Cat. of Bright Galaxies
Dixon Master List of Radio Sources
Cat. of Extragalactic Radio Source ID's
Unresolved X-Ray Sources Within or Nearby SNR
The New Revised IUE Merged Observing Log
Einstein IPC Source list from EOSCAT
HEAO A-1 All Sky Catalog

It is not surprising that many objects have turned up in these relatively large, preliminary error boxes. In the March 13 1982 error box we found the 14th magnitude galaxy NGC 1118. The March 20 1982 box contains 6 IRAS sources, V392 Ori, a variable A5 close binary with period 6.5 days, AG+19 548, a magnitude 10.1 K0 star, BD+18 1117, a type B magnitude 11.3 Wolf-Rayet star, and BD+19 1249, a magnitude 9.5 star. An IPC X-ray source with flux 2×10^{-12} erg cm^{-2} s^{-1} is near the edge of the box and may fall outside once the timing is refined. The small May 11 1982 box contains LS I+63 13, a type B star of magnitude 13.1.

Conclusion

We will produce a 2nd IPN catalog in which the precision of the relative lag times is improved, Monte Carlo techniques are used to establish confidence limits, and the interdependence of lags for different spacecraft pairs is taken into account. Confidence limits on error ellipses will also be given. The number of precise locations available for counterpart searches will perhaps double as a result of this effort.

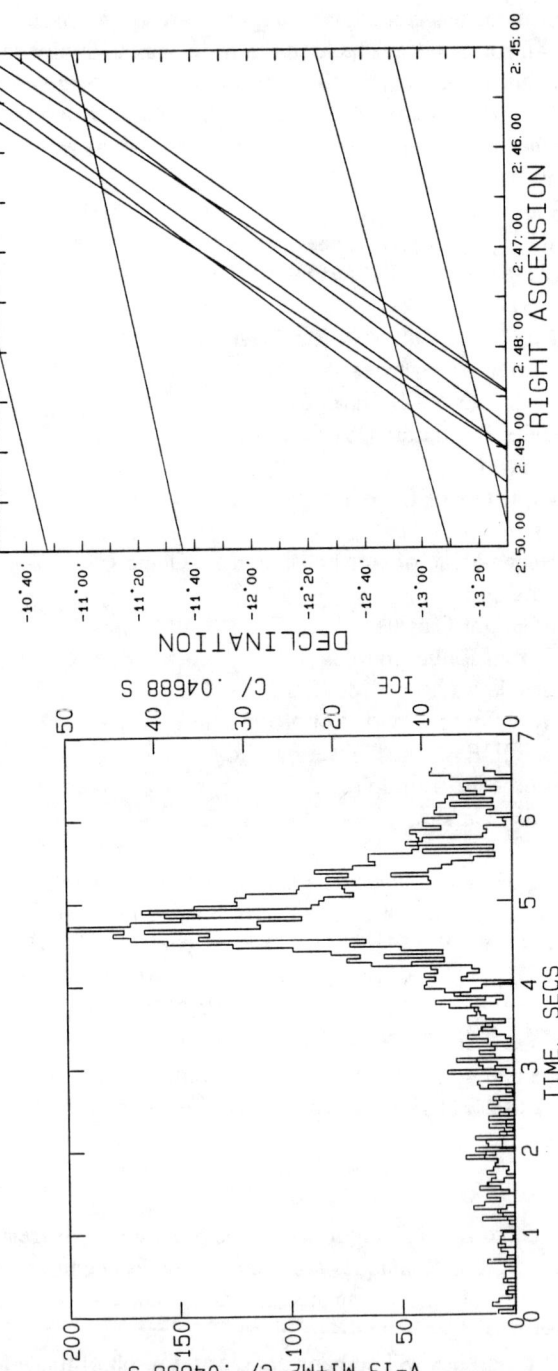

Figure 1. Venera 13 (higher peak) and ICE time histories for GRB 820406B. They are aligned for minimum chi-squared and regrouped into common 47 ms bins.

Figure 2. GRB 820313A error box (the stretched, enclosed polygon). The longest baseline is 0.5 AU.

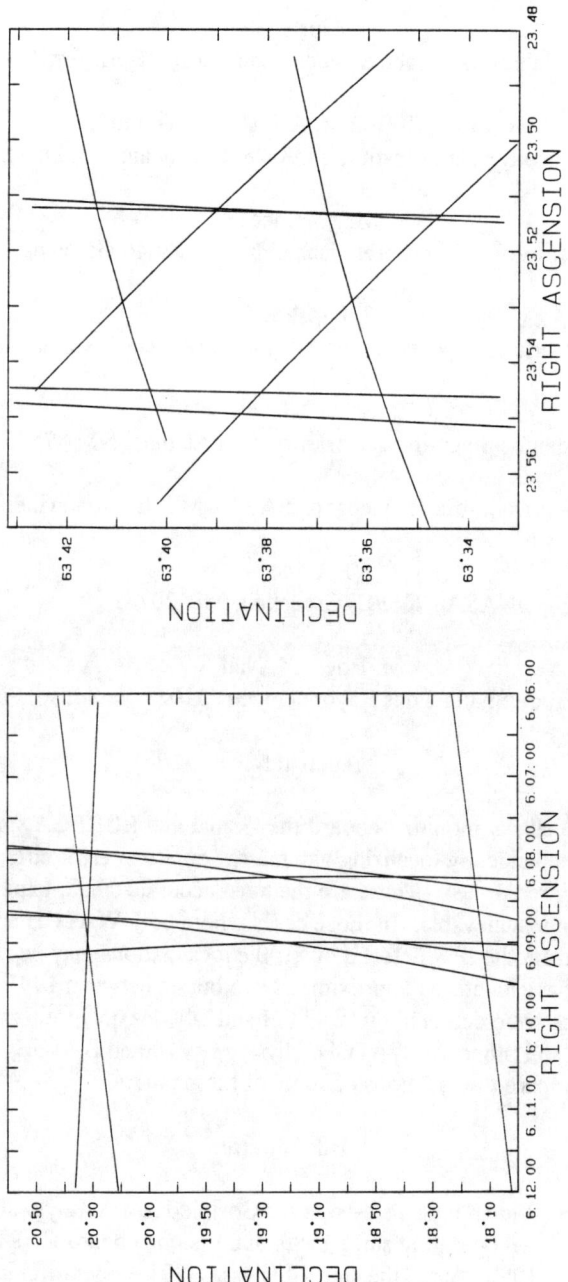

Figure 3. GRB 820320 error box. The longest baseline is 0.6 AU.

Figure 4. GRB 820511B error box. The longest baseline is 1 AU. The localization precision significantly exceeds that of Figures 2 and 3.

COMPARISON OF WATCH AND IPN LOCATIONS OF GAMMA-RAY BURSTS

K.Hurley
University of California, Space Sciences Laboratory, Berkeley, CA 94720

N. Lund, S. Brandt, A.J. Castro-Tirado
Danish Space Research Institute, DK-2800 Copenhagen, Denmark

M. Sommer
Max-Planck Institut für Extraterrestrische Physik, D8046 Garching, Germany

I. Lapshov
Institute for Space Research, 117810 Moscow, Russia

J. Laros, R. Klebesadel
Los Alamos Scientific Laboratory, Los Alamos, NM 87545

G. Fishman, C. Kouveliotou, C. Meegan, NASA - MSFC, Huntsville, AL 35812

T. Cline
NASA - GSFC, Greenbelt, MD 20771

M. Boer, M. Niel
Centre d'Etude Spatiale des Rayonnements, 31029 Toulouse, France

Abstract

The WATCH all sky monitors aboard the Granat and EURECA spacecraft have the capability of independently localizing gamma-ray bursts to error circles whose 3 sigma radii are 1 degree or less. These are the most accurate single-experiment localizations currently achievable. In those cases where both WATCH and one or more experiments from the IPN detect a burst, the localizations may be refined considerably. We have identified approximately 35 bursts between 1991 and 1993 in this category. Some were detected by WATCH and Ulysses only, others by WATCH, Ulysses, and PVO, still others by WATCH, Ulysses, PVO, and BATSE, and so on. We present and compare the locations of some of these bursts.

Introduction

The WATCH (Wide Angle Telescope for Cosmic Hard X-Rays) all-sky monitor aboard Granat[1] has been operating since 1989, and the one aboard EURECA from August 1992 to June 1993. Since the start of Ulysses GRB[2] operation in November 1990, it has detected about 35 gamma-ray bursts which were also observed by Ulysses

(Table I). WATCH is unique among current burst detectors in that it can provide the smallest single-instrument positions, typically error circles of radius 1°. The burst positions determined by WATCH/GRANAT are still beset with significant systematic errors, mostly due to the unknown deviations between the actual spacecraft attitude at the time of the bursts, and the mean attitude (averaged over several hours around the bursts). For most events it will be possible to eliminate this source of error eventually. In those cases where Ulysses and one or more near-earth instruments (including WATCH) observe a burst, it is possible, as we show below, to obtain error boxes whose dimensions are of the order of 2°x20" and above. Typically the procedure is to use Ulysses and BATSE to obtain the best possible triangulation annulus, taking advantage of BATSE's excellent statistics and time resolution, and superpose it on the WATCH error circle. When this is not possible, Ulysses/WATCH triangulation often yields a narrow annulus. In four cases, the SIGMA telescope has observed bursts within its partially coded field of view[3], giving an independent error box whose characteristic dimension is ~1°. Finally, in ten cases, Pioneer Venus Orbiter detected the burst, resulting in a three-spacecraft location. Table 1 lists all the Ulysses/WATCH events found to date; for the majority of them, an independent WATCH location can be derived, although this has not yet been done in all cases. Where the position determination can be improved using the observations of other spacecraft (i.e., BATSE, COMPTEL, PVO, or SIGMA), this is also indicated (additional observations by non-imaging near-earth instruments are not included in the table). N/O indicates that the burst was not observable, a blank means status unknown, and RI means rate increase (as opposed to a trigger).

Discussion

Four of the results are shown in figures 1-4 (J2000 coordinates have been used). At best, it is possible to obtain ~30 square arcminute error boxes with Ulysses, WATCH, and BATSE. With the recent disappearance of Mars Observer, this demonstrates that WATCH-GRANAT will remain a particularly valuable resource at least until the full interplanetary network can be re-established with Mars '94.

References

1. S. Brandt, N. Lund, and A. Rao, Adv. Space Res. 10(2), 239 (1990)
2. K. Hurley et al., Astron. Astrophys. Suppl. Ser. 92, 401 (1992)
3. A. Claret, on behalf of the SIGMA team, private communication (1993)
4. Courtesy of the SIGMA groups in Saclay, Toulouse, and Moscow (1993)
5. F. Vrba et al., Ap. J., accepted (1993)

Acknowledgments

Work on these data was supported by JPL Contract 958056 and NASA Grant NAG-1560. The Ulysses GRB experiment was built in France with support from

CNES, and in Germany with support by FRG Contracts 01 ON 088 ZA/WRK 275/4-7.12 and 01 ON 88014.

Table I. WATCH/Ulysses burst observations by other instruments

DATE	T,SEC	BAT	COM	PVO	SIG
22 JAN 91	54829	N/O	N/O	YES	YES
19 FEB 91	42324	N/O	N/O	YES	
10 MAR 91	46925	N/O	N/O	YES	
8 APR 91	81835	N/O	N/O	N/O	
14 APR 91	28523	N/O	N/O	YES	
27 JUN 91	16159	YES	YES	YES	
17 JUL 91	16386	YES		YES	
16 OCT 91	39696	YES		RI	
2 DEC 91	73731	YES		YES	
9 DEC 91	66959	YES		N/O	
11 MAR 92	08426	YES		YES	YES
14 JUL 92	47069	YES		YES	
18 JUL 92	52860	YES		N/O	
18 JUL 92	77563	YES		N/O	
23 JUL 92	72188	NO		YES	YES
24 AUG 92	39183	YES		N/O	
2 SEP 92	01736	YES		N/O	
3 SEP 92	05837	NO		N/O	
3 SEP 92	84540	NO		N/O	
25 SEP 92	73841	NO		N/O	
25 SEP 92	82154	NO		N/O	
4 OCT 92	49107	NO		N/O	
22 OCT 92	55259	YES		N/O	
18 NOV 92	79904	YES		N/O	
7 DEC 92	57647	YES		N/O	
18 DEC 92	09002	YES		N/O	
6 JAN 93	56259	YES		N/O	
1 FEB 93	60115	YES		N/O	
17 FEB 93	53078	YES		N/O	
10 MAR 93	26360	YES		N/O	
18 MAR 93	44993	YES		N/O	
10 APR 93	51547	NO		N/O	
26 APR 93	45632	YES		N/O	
2 MAY 93	49972	YES		N/O	N/O
14 JUN 93	13230	YES		N/O	

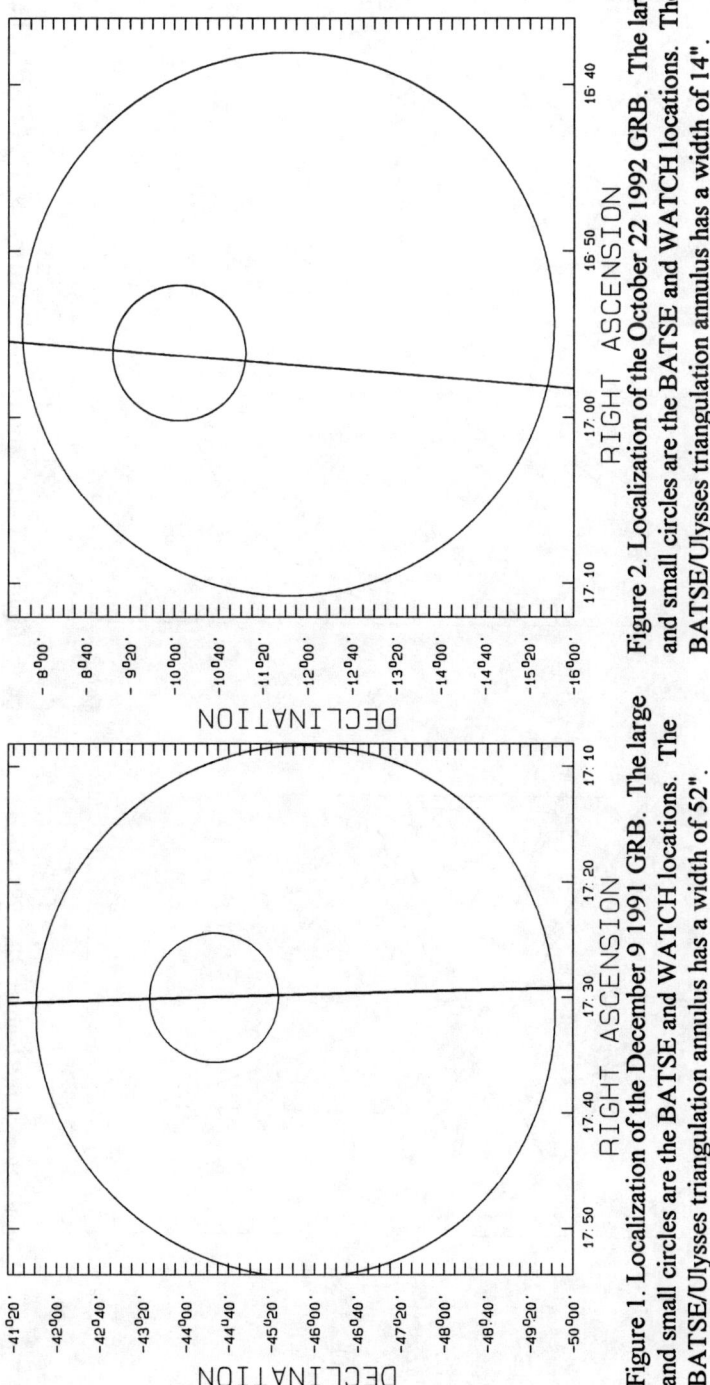

Figure 1. Localization of the December 9 1991 GRB. The large and small circles are the BATSE and WATCH locations. The BATSE/Ulysses triangulation annulus has a width of 52".

Figure 2. Localization of the October 22 1992 GRB. The large and small circles are the BATSE and WATCH locations. The BATSE/Ulysses triangulation annulus has a width of 14".

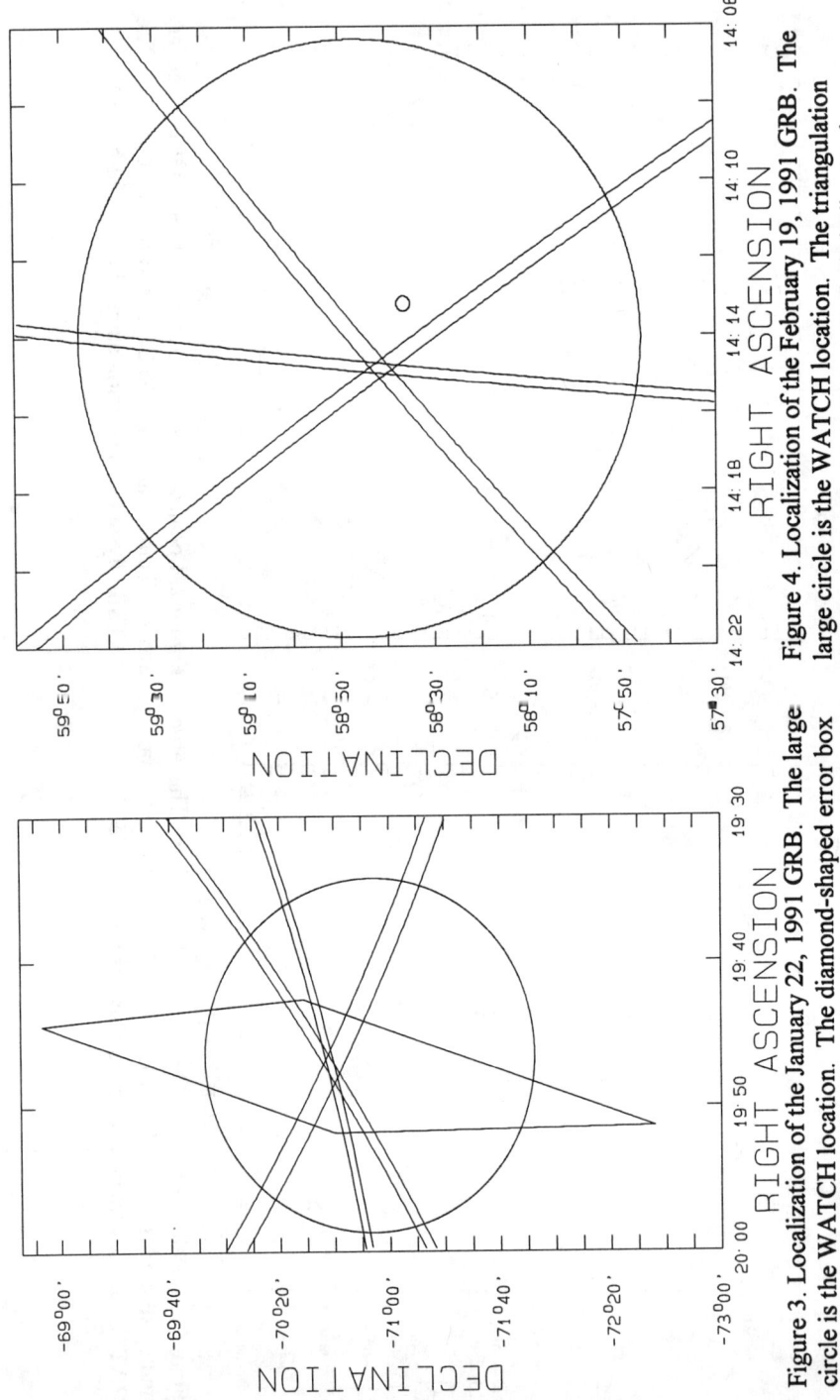

Figure 3. Localization of the January 22, 1991 GRB. The large circle is the WATCH location. The diamond-shaped error box gives the approximate boundary of the SIGMA error box[4]. Triangulation annuli are 2' - 6' wide.

Figure 4. Localization of the February 19, 1991 GRB. The large circle is the WATCH location. The triangulation annuli are 2' - 3' wide. The small circle indicates the position of an optical transient[5].

PRECISE TRIANGULATION OF THE JAN 31 1993 ("SUPERBOWL") BURST

K.Hurley
University of California, Space Sciences Laboratory, Berkeley, CA 94720

M. Sommer
Max-Planck Institut für Extraterrestrische Physik, D8046 Garching, Germany

G. Fishman, C. Kouveliotou, C. Meegan
NASA - MSFC, Huntsville, AL 35812

T. Cline
NASA - GSFC, Greenbelt, MD 20771

M. Boer, M. Niel
Centre d'Etude Spatiale des Rayonnements, 31029 Toulouse, France

Abstract

The January 31 1993 gamma-ray burst was not only one of the most intense events in recent years, but also had an exceptionally fast rise time. This, along with the fact that Ulysses was over 2100 light-seconds from Earth, makes it possible to determine an annulus of arrival directions to an accuracy of 20", comparable to that of the March 5, 1979 burst. We present the BATSE-Ulysses triangulation annulus and compare it to the positions independently determined by other GRO instruments. We discuss how the location accuracy would scale if a similar burst were detected by a 4th Interplanetary Network consisting of missions at distances ≳40 AU.

Introduction

The gamma-ray burst of January 31 1993 reached the highest peak flux of any burst since the launch of the Compton Gamma Ray Observatory. It was observed and independently localized by three instruments on that spacecraft: BATSE[1], COMPTEL[2], and EGRET[3] (localization accuracies ≈4°, 3°, and 1°, respectively). In addition, this event was detected by the Ulysses GRB experiment; the burst detector aboard Mars Observer was unfortunately not operating in burst mode at this time[4]. Based on the intial CGRO positions and the IPN triangulation arc, numerous multi-wavelength counterpart searches were initiated[5].

The initial triangulation arc was based on the quick-look scientific data from the Ulysses spacecraft, the predict spacecraft ephemeris, and spacecraft clock data which were uncalibrated for long term drifts. The overall uncertainties were conservatively estimated to be in the 300 ms range, leading to an annulus width of ~3'. Final data

© 1994 American Institute of Physics

were received about one month after the event, and have been used to generate a more accurate annulus which we present here.

Time Histories

The Ulysses and BATSE time histories for the burst are shown in Figure 1, aligned for the lowest χ^2 [6]. These are the raw count rates before background subtraction or dead time correction. The time resolution of the BATSE data is 64 ms, and that of the Ulysses data, 31.25 ms. The application of the χ^2 procedure includes rebinning the data to a common time resolution. The results of this procedure are not shown in the figure; however, the rebinned time histories are essentially identical for the first pulse, but differ in their peak intensities for the second pulse by an amount

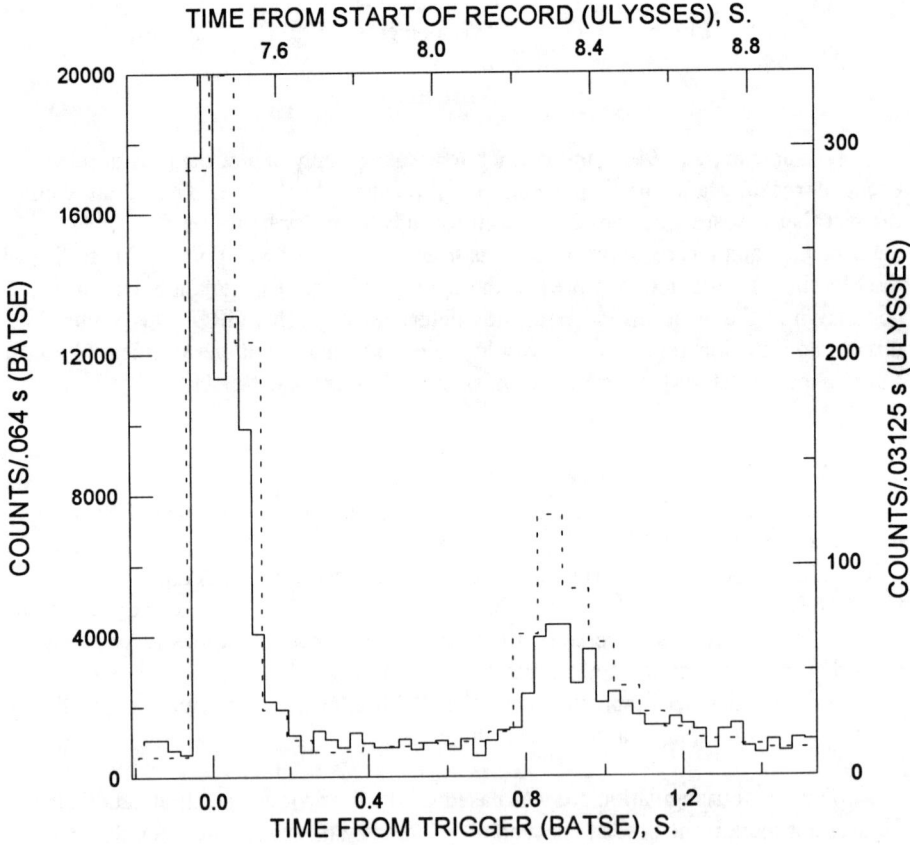

Figure 1. Ulysses (solid line) and BATSE (dashed line) count rates for the January 31 1993 gamma-ray burst. The rates are uncorrected for background and dead time.

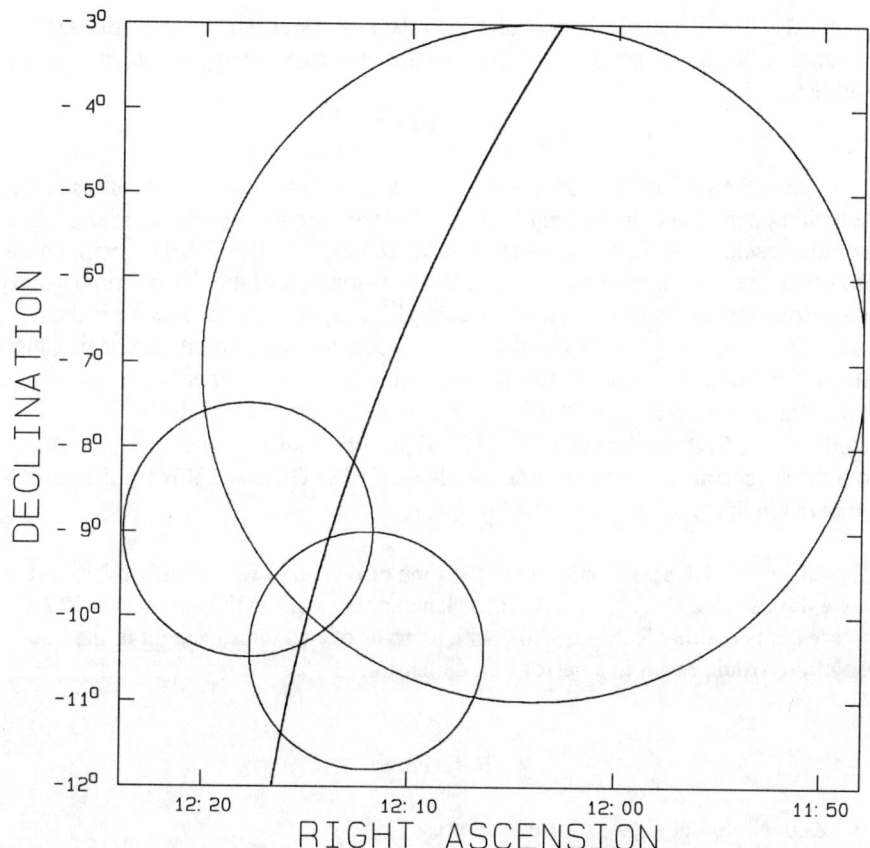

Figure 2. Localizations of the January 31 1993 event. From top to bottom, the error circles are from BATSE, COMPTEL, and EGRET. The 20" IPN triangulation arc is too narrow to be resolved.

which exceeds the statistical uncertainties in the raw count rate values. A possible explanation for this is saturation of the BATSE detectors, leading to a relatively large dead time.

Localization

When the statistical errors in the count rates become small, as is the case here, the limiting uncertainty in aligning two time histories approaches the smaller of the two time resolutions. Because we have used the BATSE DISCLA data, with 64 ms resolution, the limit in this case is the 31.25 ms resolution of the Ulysses time history data. However, an improvement of a factor of 2 or more may be possible if the BATSE time-tagged event (TTE) data can be used; we are currently examining these data, and hope to be able to reduce the location uncertainty further in the near future. At present, the annulus is ~20" wide, centered at $\alpha(2000)=161.4107°$, $\delta(2000)=-14.5632°$, with radius $22.2558°$. The contribution to the annulus width from the spacecraft location uncertainties is ~3". The GRO and IPN localizations are compared in figure 2.

The GRO-Ulysses separation at the time of this burst was slightly over 4 AU. As we discuss elsewhere[7], a fourth interplanetary network with baselines of 40 AU and greater is entirely feasible. This, with a factor of two improvement in the time resolution, would result in an error box dimension of 1".

References

1. C. Kouveliotou et al., Ap. J. Lett., in press (1993)
2. J. Ryan et al., Ap. J. Lett., in press (1993)
3. M. Sommer et al., Ap. J. Lett., in press (1993)
4. J. Laros, private communication (1993)
5. B. Schaefer et al., Ap. J. Lett., in press (1993)
6. K. Hurley, these proceedings (1993)
7. K. Hurley and T. Cline, these proceedings (1993)

Acknowledgments

Work on these data was supported by JPL Contract 958056 and NASA Grant NAG-1560. The Ulysses GRB experiment was built in France with support from CNES, and in Germany with support by FRG Contracts 01 ON 088 ZA/WRK 275/4-7.12 and 01 ON 88014.

STUDY OF THE PRECISION OF THE GAMMA-RAY BURST SOURCE LOCATIONS OBTAINED WITH THE ULYSSES/PVO/CGRO NETWORK

T. L. Cline
NASA / Goddard Space Flight Center, Code 661, Greenbelt, MD 20771

K. C. Hurley
University of California, Space Sciences Laboratory, Berkeley, CA 94720

M. Sommer
Max-Planck Institut fur Extraterrestrische Physik, Garching, Germany

M. Boer and M. Niel
Centre d'Etude Spatiale des Rayonnements, Toulouse, France

G. J. Fishman, C. Kouveliotou, C. A. Meegan, W. S. Paciesas and R. B. Wilson
NASA/Marshall Space Flight Center, ES-62, Huntsville, AL 35812

J. G. Laros and R. W. Klebesadel
Los Alamos National Laboratory, MS D436, Los Alamos, NM 87545

ABSTRACT

The interplanetary gamma-ray burst network of the *Ulysses*, *Compton-GRO* and *Pioneer-Venus Orbiter* missions has made source localizations with fractional-arc-minute precision for a number of events, and, with auxiliary data, will provide useful annular-segment loci for many more. These studies have, thus far, yielded one possible counterpart, a *Rosat* x-ray association with the 92 May 1 burst. Similar to the historic '78 Nov. 19 burst / *Einstein* association, this possibility gives hope that network studies will provide a fundamental source clue for 'classical' bursts, just as a second supernova remnant in a network-defined source field has done for sgr events.

INTRODUCTION

The third interplanetary burst network, based on the *Ulysses*, *CGRO*, *PVO* and *Mars Observer* missions, functioned for over 1.5 years. Although the *PVO* and *MO* missions are no longer operational, the network's utility should not be considered to be significantly crippled: sources defined with two widely separated spacecraft are of great value, given auxiliary data. This is demonstrated with the *ASCA* observation of the flaring soft gamma repeater[1] that confirmed a distant, galactic snr to be the source[2] of a known sgr. Prompted by a *CGRO* snr activity alert,[3] that discovery required the source accuracy provided by an earlier network study[4,5] with only one distant vertex.

The present network, given auxiliary data from the experiments on *C-GRO*[6] and other spacecraft[7], and soon to be from the *GGS-Wind* and *Spectrum-X-Gamma* experiments, will also help continue to supply annular-segment source fields until the *Mars 96*, a redefined *MO*, or some other planetary or deep space mission may be launched to upgrade the network to full capability. In fact, advanced instruments are now being developed for the possibility that next-generation space probes, presently under consideration for spectacularly distant applications, may ultimately be flown.[8]

We report here on the network capabilities, highlighting with one source study.

PROCESS

Preliminary source studies from this network e.g., [9,10,11] and the calibration and analysis techniques and other details are outlined in the literature e.g., [12,13,14].

Throughout the development of the interplanetary networks, one concern has centered on the impossibility of independent verification of a precise burst source location. The network measurement technique is based on observations from too few (widely separated) vertices to provide the redundancy that could give the potential for precise self-verification. Although the spacecraft clock timing can be calibrated[12], the intensity profiles can be demonstrated to be free enough of spectral distortion for their reliable comparison[13], and the clerical errors can be minimized with redundant labor, there was no way to be sure of exact and correct calibration: it remained to be proved that the source studies of the networks were as accurate as in our claims.

The 1979 March 5 event source localization, fitting within the extent of the LMC supernova remnant N49, was the most precise source determination of the previous gamma-ray burst networks[15]. That source association was never fully accepted as a source identification due to the added hypothesis of the distinct nature of this event, as based on its character[16] and on the appearance of the subsequent sgr series[17] from approximately the same direction, that was necessary to divorce it from the problem of the isotropy of common or 'classical' bursts. The *ASCA* identification[1] of a second distant snr, for which the sgr source position had been also determined by an earlier network, gives the indirect benefit of new confidence in the precision of our measurement technique. The direct benefits, of course, also include the vindication of the N49 source identity and the renewed opportunity for theoretical study and understanding that this provides.

Analyses of the three-spacecraft events of the Ulysses-based network have been carried out with great attention to the elimination of all possible 'bugs' in the data and in the technique. Duplications of efforts with independence in approach and style have given complete confidence in the new source locations. Varying the profile comparison methods, which always seem to be the most subjective and thus the weakest link in the process, confirms that the outcome does not depend on a 'philosophical' choice; in fact, one technique appears to be genuinely objective[13].

RESULTS

The reanalysis of the 1972 May 1 gamma ray burst yields an order-of-magnitude smaller source region than that described in earlier reports[18,19], which outline the possibility of its association with a *Rosat* x-ray source. The source locus is also curiously located between two radio sources; these do not appear to be mutually related, and so their proximity to the burst location may be accidental, as well. The reduced burst source locus, in fact, remains consistent with the x-ray source location, as illustrated in Figure 1. Tempering this relation is the galactic latitude of this region, close to zero, where the likelihood of source confusion is greatest. However, the density of other *Rosat* x-ray sources in this general region is quite low. This situation may be similar to the '78 Nov. 19 gamma-ray burst - Einstein x-ray source overlap[20]. These two juxtapositions may not be merely coincidences.

Investigating for possible x-ray associations with all precise burst source fields is clearly desired. This need is amply illustrated by the recent clarification of the sgr puzzle, fitting our original speculation that the mid-1980's sgr series may have its source near the galactic center, to be compatible with the '79-Mar-5 sgr sequel series coming from N49 (with the typical intensities of the two series varying as the inverse

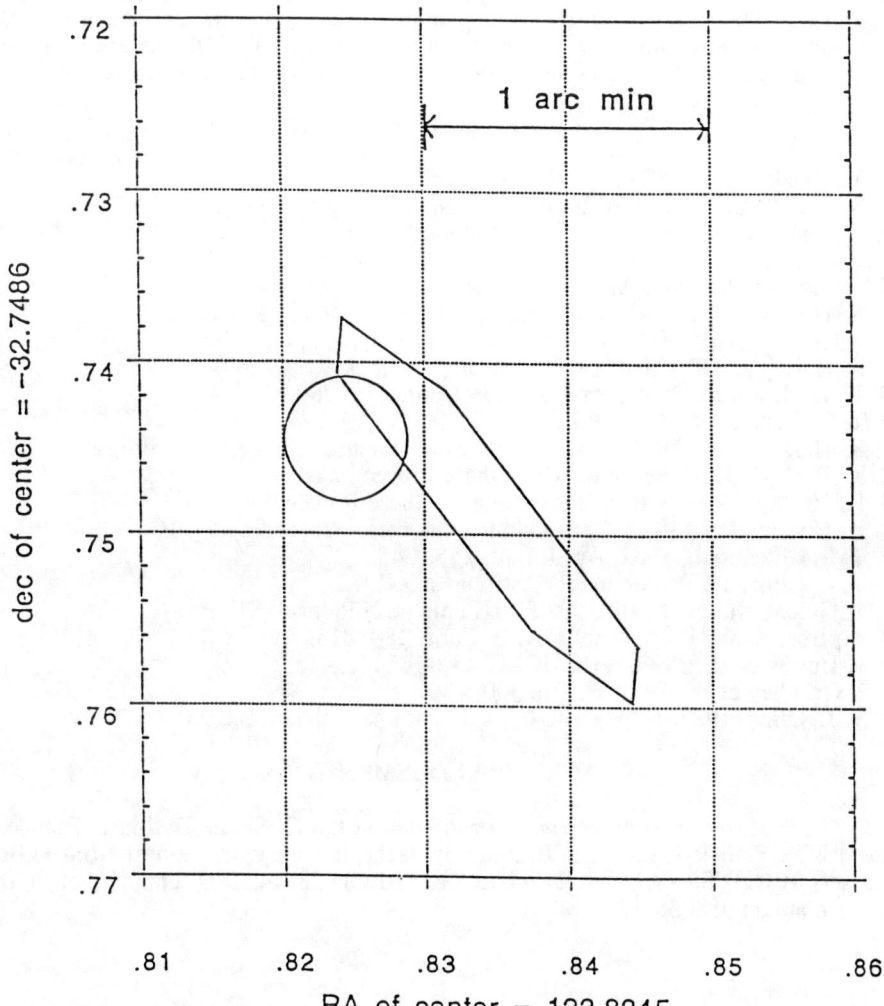

Figure 1. The source field of the 1992 May 1 gamma ray burst (hexagon) is shown, reduced in area by about an order of magnitude from its earlier, preliminary value. The coordinates of the center of this source field hexagon are also indicated. A weak, target-of-opportunity, *Rosat* x-ray source field is plotted as the circle. The questions of whether any association of 'classical' bursts with x-ray sources may be accidental, or real, and of the implications of that relationship, must be further investigated.

squares of the distances to the GC and to the LMC)[21]. This pattern has no parallel for 'classical' bursts. In fact, the absence of an anisotropy, however weak, forms much of the basis for the current support for cosmological models. If gamma ray bursts do originate at cosmological distances, hopes for success of the network studies may be groundless. The unexploited range of potentialities, however, continues to fuel optimism that the networks will make a central contribution to the solution of the puzzle of gamma ray bursts, as they have done for the soft gamma repeaters.

REFERENCES

1. Y. Tanaka et al., 1993 I. A. U. Circular 5880.
2. S. R. Kulkarni and D. A. Frail, 1993, Nature 365, 33.
3. C. Kouveliotou et al., 1993, I. A. U. Circular 5567.
4. J. -L. Atteia et al., 1987, Ap. J. Lett. 320, L105
5. J. Laros et al., 1987, Ap. J. Lett. 320, L111.
6. K. Hurley et al., 1993, 'Precise Triangulation ...', these Proceedings.
7. K. Hurley et al., 1993, 'Comparison of WATCH ...', these Proceedings.
8. K. Hurley and T. Cline, 1993, these Proceedings.
9. K. Hurley et al., 1992, Proc. I. A. P. Conf. 280, 769.
10. T. L. Cline et al., 1992, Proc. I. A. P. Conf. 280, 774.
11. K. Hurley et al., 1993, 'The Ulysses Supplement ...', these Proceedings.
12. K. Hurley and M. Sommer, 1993, these Proceedings.
13. K. Hurley, 1993, 'Cross-Correlating ...', these Proceedings.
14. K. Hurley et al., 1993, 'Two Oddities ...', these Proceedings.
15. T. L. Cline et al., 1982, Ap. J. Lett. 255, L45.
16. T. L. Cline, 1980, Comments Astrophys. 9, 13.
17. S. Golenetskii et al., 1983, P. T. I. (Leningrad) Preprint 813 .
18. M. Boer et al., 1992, Proc. I. A. P. Conf. 280, 813.
19. K Hurley et al., 1993, Proc. ICRC, OG-1, 116.
20. J. Grindlay et al., 1981, Nature 300, 730.
21. T. L. Cline, 1987, NASA 2464, 'Essays in Space Science', 295.

ACKNOWLEDGEMENTS

The *Ulysses* instrument was constructed at the CESR in Toulouse, France, and at Max Planck Institut fur Extraterrestrische Physik with support from FGR contracts 01 ON 088 ZA/WRK 275/4-7.12 and 01 ON 88014. It is supported at UCB by JPL contract 958056.

A SEARCH FOR HIGH ENERGY GAMMA-RAY BURSTS IN THE EGRET DATA UTILIZING SPACE-TIME CORRELATION

R. Buccheri, M.C. Maccarone

CNR/Istituto di Fisica Cosmica e Applicazioni dell'Informatica, Palermo, ITALY

J.R. Mattox, D.J. Thompson

NASA/Goddard Space Flight Center, Greenbelt, USA

G. Kanbach

Max Planck Institut für Extraterrestrische Physik, Garching, GERMANY

U. Camerini, W.F. Fry

Dept. of Physics, University of Wisconsin, Madison, USA

ABSTRACT

A Maximum Likelihood Method is described which was used to make a search for gamma-ray bursts in 3 dimensions: 2 space coordinates and time, without relying on any previous knowledge about the event. Extensive Montecarlo simulations have been performed to evaluate the statistical significance of potential bursts utilizing the EGRET experimental conditions. Applications to the EGRET observational data show that a) the 3May91 burst is detectable by our method at a 6.5σ significance level and b) a burst of the intensity of the "superbowl" burst is detectable at a 14σ significance level.

INTRODUCTION

The investigation of the nature of gamma-ray bursts is one of the most exciting topics in modern astrophysics. A catalogue of 731 of these mysterious objects has been recently published by the group working on BATSE aboard the Compton Gamma Ray Observatory, all detected in the hard X-rays, the BATSE working range[1]. At higher energies, EGRET, making use of the information obtained by BATSE, has detected between 30 and 300 MeV, three of the most intense and hard bursts detected by BATSE[2]. Apart from showing an almost perfect spacial isotropy and a variable range of durations, the time histories and the spectral characteristics of the whole population show a large variety of forms which does not allow a definitive identification of the sources of the bursts with known physical objects in the sky.

We consider it of utmost importance to obtain information about bursts in the energy range well above BATSE, on their number, time duration, energy distribution etc.., which could lead to a better physics understanding of the nature of the processes which produce these events. Using the favourable occasion of the presence of EGRET aboard the CGRO, a method has been devised to search for bursts in the EGRET data, independently of BATSE. This would allow the detection of new events which occur predominantly in the high energy gamma-ray region.

THE METHOD

Details on the mathematical formulation of the method are given in Buccheri et al.[3]. We will summarize here the main points. Let $f(\alpha,\theta,t,E/n=0)$ be the probability density function (pdf) per unit time, energy and solid angle to detect by chance a photon with parameters α, θ, t and E in the hypothesis of absence of a burst and $f(\alpha,\theta,t,E/n,\alpha_o,\theta_o,t_1,t_2,\phi_s)$ the pdf in the hypothesis of presence of a burst with n counts from an angular direction (α_o,θ_o) within the time interval $(w=t_2-t_1)$ and energy spectrum according to a distribution $\phi_s(E)$. The likelihood ratio is:

$$F = \prod_i^N f(\alpha_i,\theta_i,t_i,E_i / n,\alpha_o,\theta_o,t_1,t_2,\phi_s) / \prod_i^N f(\alpha_i,\theta_i,t_i,E_i /n=0) \tag{1}$$

In the absence of bursts or other structures of the background, $f(\alpha,\theta,t,E/n=0)$ is proportional to the product $\phi(E)\cdot S(\alpha,\theta,E)$ between the energy distribution of the detected photons and the known variation of the experiment sensitive area $S(\alpha,\theta,E)$. In the presence of a bursting source, $f(\alpha,\theta,t,E/n,\alpha_o,\theta_o,t_1,t_2,\phi_s)$ is given by:

$g\cdot\phi(E)\cdot S(\alpha,\theta,E) + r\cdot\phi_s(E)\cdot PSF(\beta,E)$ for $t_1 < t_i < t_2$

$\gamma\cdot f(E)\cdot S(\alpha,\theta,E)$ for $t_i < t_1$ or $t_i > t_2$

where g, r and γ are scaling factors and $PSF(\beta,E)$ is the Point Spread Function at (α_o,θ_o) for the energy E. The method consists in the maximization of the log-likelihood ratio $\lambda=2\cdot\ln(F)$ with respect to the 5 parameters t_1, w, n, α_o and θ_o over the whole list of selected photons: in correspondence with the maximum value of λ, we have the group of photons which accumulate in time and space (compatible with the source Point Spread Function of the experiment) with minimum probability for chance occurrence.

THE MONTECARLO SIMULATIONS

In order to evaluate the statistical significance of the values of λ_{max} obtained from the analysis of the EGRET data, we generated pseudo-random data sets having statistical characteristics the same as those of the observational data under simulation. Fig.1 shows the distribution of λ_{max} derived applying our method to 2500 sets of times simulated in the experimental conditions of the CGRO viewing period (VP) 4.0. In particular, the characteristics included in the simulation were:

- the angular distribution according to $S(\alpha,\theta,E)$;
- the energy distribution of the photons according to that measured for VP 4.0;
- a poissonian distribution of arrival times. Periodic data gaps according to the satellite orbit were included;
- the angular dispersion of point sources according to the experimental Point Spread Function.

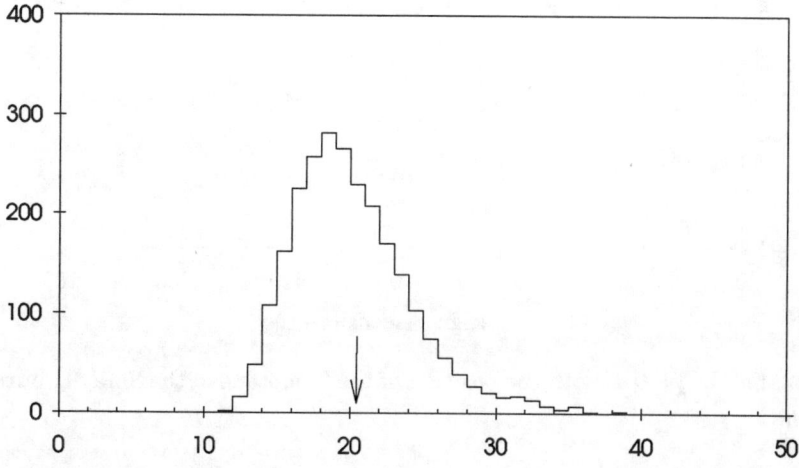

Fig.1. Distribution of λ_{max} derived after the simulation of 2500 set of times in the experimental conditions of the VP 4.0. The arrow indicates the value of λ_{max} (20.54) found after analysis of the photon arrival times of VP 4.0.

APPLICATIONS TO TEST THE SENSITIVITY OF THE METHOD

The following applications demonstrate the sensitivity of our technique:

The viewing period 0.4 (May 1 to May 4, 1991) containing the 3May91 burst (discovered by BATSE in the energy range 20 to 300 KeV and confirmed by EGRET above 30 MeV[4]) was analyzed with our method which does not take into account the information given by BATSE.
As a result, all the 7 photons of the burst were correctly identified out of the 5757 photons selected in the 40° FoV of the experiment, the statistical significance of the detection reaching the 6.5σ confidence level (λ_{max}= 46.79). Fig. 2 shows the direction in the sky of the 7 photons of the burst within the EGRET FoV.
A second application of the method was envisaged in order to quantify the statistical significance at which the "superbowl" burst of Jan. 1993 would be

detected by our method. Because the viewing period containing the "superbowl" burst is not yet publicly available, the known burst events[5] were merged with the 8305 photons of the VP 3.0 (June 15 to June 28, 1991) which viewed the same region of the sky. The photons of the burst were then correctly identified by our method at a 14σ confidence level (λ_{max}= 77.97).

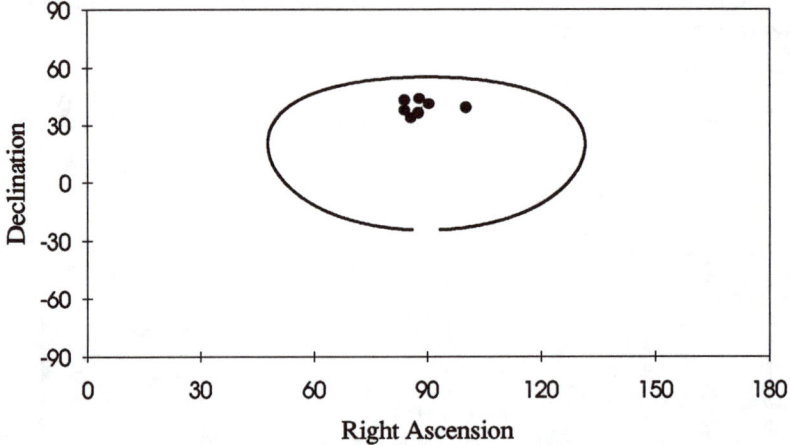

Fig.2. FoV of VP 0.4 with the position of the 7 photons of the 3May91 burst.

Finally, an upper limit (at 0.001 level) to the detection of bursts *on axis* was derived in the experimental conditions of the VP 4.0. To do this, a number of synthetic photons, with arrival directions compatible with a point source located at the direction of the EGRET experiment axis, was added to the data of VP 4.0 such to reach the required statistical significance of 0.001 for various burst widths. Fig.3 shows the result which is compared with the positive detections of the 3May91 and the "superbowl" bursts.

AKNOWLEDGEMENTS

This research has made use of data obtained through the Compton Observatory Science Support Center GOF account provided by the NASA-Goddard Space Flight Center.

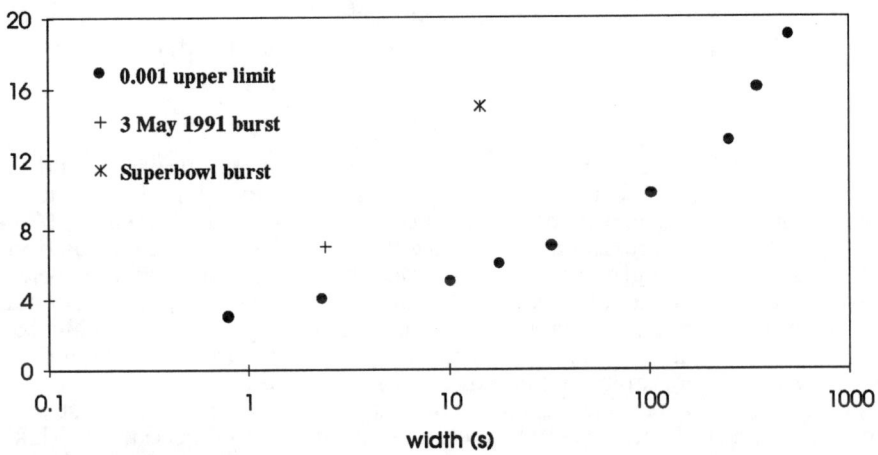

Fig.3. Minimum number of photons required to detect a burst of a given width at a 0.001 significance level. The "superbowl" and the 3May91 bursts are superimposed for comparison.

REFERENCES

1. G.J. Fishman, C.A. Meegan, R.B. Wilson, M.N. Brock, J. Horack, C. Kouveliotou, S.Howard, W.S. Paciesas, M.S. Briggs, G.N. Pendleton, T.M. Koshut, R.S. Mallozzi, M. Stollberg, J.P. Lestrade; Ap J., (1994), in press

2. E.J. Schneid, B.L. Dingus, P. Sreekumar, D.L. Bertsch, C.E. Fichtel, R.C. Hartman, S.D. Hunter, D.J. Thompson, G. Kanbach, H.A. Mayer-Hasselwander, C. von Montigny, M. Sommer, Y.C. Lin, P.L. Nolan, P.F. Michelson, J. Chiang, J. Fierro, D.A. Kniffen, J.R. Mattox, these proceedings

3. R. Buccheri, W.F. Fry, M.C. Maccarone, A&A, **277**, 353-359 (1993)

4. E.J. Schneid, D.L. Bertsch, C.E. Fichtel, R.C. Hartman, S.D. Hunter, G. Kanbach, D.A. Kniffen, P.W. Kwolk, Y.C. Lin, J.R. Mattox, H.A. Mayer-Hasselwander, P.F. Michelson, C. von Montigny, P.L. Nolan, K. Pinkau, H. Rothermel, M. Sommer, P. Sreekumar, D.J. Thompson, A&A, **255**, L13 (1992)

5 M. Sommer et al., Ap.J. Lett., (1994) in press

SEARCH FOR GAMMA RAY BURST COUNTERPARTS

Bradley E. Schaefer *
NASA/Goddard Space Flight Center, Greenbelt, MD 20771

ABSTRACT

The confident detection of a Gamma Ray Burst counterpart would likely provide the much needed breakthrough in our understanding of the cause and site of bursts. As such, a lot of work has been expended since the early 1970's in attempts to find bursts at energies below 2 keV. These searches can be divided based on the wavelength and the time (since the burst) of observation. Searches have been made in the soft x-ray, ultraviolet, optical, infrared, and radio bands. The counterpart might be detected while flaring (during the burst), while fading (soon after the burst), or in quiescence (long after the burst). This review gives an extensive bibliography and summary of these results.

This review also highlights five recent results: First, in the last several years the speed of burst position measures has greatly increased, and this has allowed for deep, fading counterpart searches within roughly a day of the burst, although no counterparts have been identified. Second, the lack of galaxies and active galactic nuclei in small error boxes puts severe constraints on any extragalactic model. Third, a consensus has developed that the optical transient images inside small burst error regions on archival photographs are of real astrophysical flares, however the relationship of these events to the burst phenomenon is unclear. Fourth, a study of the ROSAT all-sky survey data has been unable to reproduce the detection of 'ultrasoft transients'. Fifth, the recent discovery of counterparts for Soft Gamma Repeaters has identified the type of system (a young and magnetized neutron star in a supernova remnant) out of the many possibilities, and this advance should provide strong encouragement in the search for classical burst counterparts.

Several ideas for future instrumentation offers good hope that the burst distance scale will be resolved with counterparts. One such idea is that a cosmological distance could be confirmed by observing a delay of the radio burst caused by dispersion. Another idea is to detect counterparts by hooking sensitive telescopes up to the BATSE and HETE data steams so that they can look in the right direction during the burst. A third idea is that if the soft x-ray burst spectra can be measured, then the derived column density of the intervening neutral gas will uniquely identify the burst distance scale.

INTRODUCTION

The quickest solution to the Gamma Ray Burst problem is to find a counterpart. Our community has long realized this[1-6], and at this meeting, 24% of all papers are on counterparts. As a result, a large effort has been made in counterpart searches in years past and the activity appears to be increasing exponentially. This review will cover the past work, highlight important recent results, and give the ideas of the future.

I will take a counterpart to be any source detected below 2 keV. This review

---

* Also Universities Space Research Association

will only briefly discuss Soft Gamma Repeaters (SGRs) and Optical Transients (OTs) which occur outside Gamma Ray Burst (GRB) error regions.

Counterparts can be divided on the basis of the time scale after the burst as 'flaring', 'fading', and 'quiescent' sources. A flaring counterpart emits low energy radiation given off during the gamma ray event. A fading counterpart is a transient low energy source that decays away on a time scale of perhaps minutes to years after the burst. A quiescent counterpart is a steady source which is left behind long after the burst is over. To draw an analogy with a camper observing a camp fire; the flaring counterpart would produce infrared light to heat hands and marshmallows, the fading counterpart would be visible as a glow from cooling embers, and the quiescent counterpart would be the cold ash and ring of stones.

Counterparts can also be divided on the basis of the wavelength of observation as 'x-ray', 'ultraviolet', 'optical', 'infrared', and 'radio'. With the divisions by time scale and wavelength, we have 15 types of GRB counterparts. These divisions have utility for description, as each type has different theoretical expectations, observational techniques, search limits, and counterpart candidates.

PAST WORK

The theoretical expectations for counterparts depend critically on the time scale and wavelength of observation. Table 1 presents a summary of processes that could result in a GRB counterpart, and is divided into 15 boxes based on the time scale and wavelength divisions. The references are indicated by numerical superscripts in the table and are presented immediately below the table.

At this time, we do not know the nature of classical GRBs, so we cannot say which of the theoretical predictions are valid. Most GRB models require that a counterpart should be detectable at some wavelength on some time scale. For a given model, the predictions would tell us which box will yield the greatest chance of finding a counterpart. Unfortunately, we do not know which model is correct, so searchers should explore all 15 counterpart types. Nevertheless, a moral to be drawn from Table 1 is that all the boxes are full, and thus there is plenty of reason to expect that counterparts can be discovered. I take this as a strong encouragement that counterpart searches must be pursued with vigor as the best hope of solving the GRB puzzle.

Observing techniques also strongly depend on the wavelength and time scale (see Table 2). Many of the boxes are empty due to technological limitations, for example because it is difficult to built an ultraviolet all-sky camera that can catch flaring transients.

A vital part of any counterpart search is knowing the background. (Otherwise it is like looking for a needle in a haystack without knowing what the hay looks like.) The background for quiescent and fading sources consists of the normal steady and variable objects for which many studies have been made as part of classical astronomy. For flaring counterpart candidates, a case can be made to associate the event with the GRB if all backgrounds can be eliminated. But background flaring sources (say a flare star or a glint off a satellite) are relatively rare so it is difficult to get good rate estimates or property distributions. I know of no flare background rate studies for radio flares, but x-ray[7] and optical[8-14] studies have appeared. In general, it is impossible to conclusively prove that no background event could have caused the flare, even though the evidence can become quite strong. The only final proof to connect an observed flare with a GRB is temporal and directional coincidence with a gamma ray event.

Table 1. Theoretical Expectations

	FLARING	FADING	QUIESCENT
X-RAY	•Reprocessing off neutron star[6] •Compton scattering off companion star[3] •Plasma radiation[7]	•Cooling of heated neutron star surface[6]	•Accretion onto neutron star •Host AGN
UV	•Reprocessing off accretion disk •Cyclotron reprocessing in magnetosphere[5] •Plasma radiation[7]	•Transient accretion disk	•Accretion disk
OPTICAL	•Cyclotron reprocessing in magnetosphere[5] •Reprocessing off accretion disk[4,10] •Reprocessing off companion star[8,9,15] •Comets on white dwarf companion[16] •Proton cyclotron[1]	•Transient accretion disk	•Accretion disk[4,10] •Hα from ISM photoionization[2] •Thermal emission from neutron star[11] •Host galaxy •Companion star
IR	•Cyclotron reprocessing in magnetosphere[5]	•Transient accretion disk	•Accretion disk[10] •Low mass companion star
RADIO	•Synchrotron radiation in outer magnetosphere[14]	•Ejecta hitting ISM[13]	•Pulsar[12] •Host AGN

[1]Apparao & Chitre 1984, preprint
[2]Band & Hartmann 1992, ApJ, **386**, 299
[3]Dermer, Hurley, & Hartmann, 1991, ApJ, **370**, 341
[4]Epstein 1985, ApJ, **291**, 822
[5]Hartmann, Woosley, & Arons 1988, ApJ, **332**, 777
[6]Imamura & Epstein 1987, ApJ, **313**, 711
[7]Liang 1985, Nature, **313**, 202
[8]London & Cominsky 1983, ApJLett, **275**, L59
[9]Melia, Rappaport, & Joss 1986, ApJLett, **305**, L51
[10]Melia 1988, ApJLett, **324**, L21
[11]Melia 1989, Nature, **338**, 322
[12]Melia & Fatuzzo 1993, ApJLett, **398**, L85
[13]Meszaros & Rees 1993, preprint
[14]Paczynski & Rhoads 1993, ApJLett, **418**, L5
[15]Rappaport & Joss 1985, Nature, **314**, 242
[16]Tremaine & Zytkow 1986, ApJ, **301**, 155

Just as the observing techniques vary greatly from box-to-box, so must the search thresholds. Table 3 presents the best available limits, although these are usually available for just one or a few burst positions. The limits are such that modern technology has been pushed as far as currently possible. This means that to find a counterpart we must either (1) continue the work and hope for some bright or unusual event or (2) get new technology with deeper limits.

So what do we have to show for all the work that is summarized in Tables 1-3? The answer is remarkably little. This is one of the primary reasons why we have no definite knowledge about the nature of GRBs. Table 4 presents a list of the few counterpart candidates identified to date. Unfortunately, each entry has problems in relating the counterpart to the GRB phenomenon. In particular, the archival optical transients are of events many decades before the gamma ray

Table 2. Past and Current Search Techniques

	FLARING	FADING	QUIESCENT
X-RAY	•EINSTEIN[12,13] •ROSAT[5,19]	•EINSTEIN[11] •ROSAT[5]	•EINSTEIN[21] •EXOSAT[4] •ROSAT[5,18]
UV			•HST
OPTICAL	•Archival plates[1,7,14,16,17,27] •Patrol plates[8,9,10] •Explosive Transient Camera[30] •Rapidly Moving Telescope[2]	•Schmidt cameras[3,6,29]	•Deep CCD images[22,23,26,28,31]
IR			•IR photometry[24] •IRAS[24]
RADIO	•Continuous radio pulse search with wide FOV antenna[15,20]	•VLA[20]	•VLA[20,25]

[1] Atteia et al. 1985, A&A, 152, 174
[2] Barthelmy et al. 1993, AIP#280 (1st Compton Symp) p 1137
[3] Barthelmy, Schaefer & Palmer 1993, these proceedings
[4] Boer et al. 1988, A&A, 202, 117
[5] Boer et al. 1993, A&ASupp, 97, 69 and later unpublished observations
[6] Castro-Tirado, Brandt & Lund 1993, these proceedings
[7] Flohrer et al. 1986, Adv Space Res, 6, 55
[8] Greiner, Wenzel, & Mohlmann 1990, Adv Space Res, 10, 195
[9] Greiner et al. 1992, AIP#265 (Huntsville Conf.), p 327
[10] Grindlay, Wright, & McCrosky 1974, ApJLett, 192, L113
[11] Helfand & Long 1979, Nature, 282, 589
[12] Helfand & Vrtilek 1983, Nature, 304, 41
[13] Helfand, Gotthelf, & Hamilton 1992, Compton Observatory Science Workshop, 317
[14] Hudec et al. 1987, A&A, 175, 71
[15] Inzani et al. 1982, AIP#77, 79
[16] Karnashov et al. 1991, Sov Astron, 35, 256
[17] Moskalenko et al. 1992, Gamma-Ray Bursts, eds Ho, Epstein, & Fenimore, p. 127
[18] Owens et al. 1993, AIP#280 (1st Compton Symp) p798
[19] Owens et al. 1993, MNRAS, 260, L25
[20] Palmer 1993, in preparation
[21] Pizzichini et al. 1986, ApJ, 301, 641
[22] Ricker, Vanderspek, & Ajhar 1986, Adv Space Res, 6, 75
[23] Schaefer 1986, Adv Space Res, 6, 47
[24] Schaefer et al. 1987, ApJ, 313, 226
[25] Schaefer et al. 1989, ApJ, 340, 455
[26] Schaefer 1990, ApJ, 353, L25
[27] Schaefer 1990, ApJ, 364, 590
[28] Schaefer 1992, Gamma-Ray Bursts, eds Ho, Epstein, & Fenimore, p. 107
[29] Schaefer et al. 1994, ApJLett, in press
[30] Vanderspek, Doty, & Ricker 1992, AIP#265 (Huntsville Conf.), p 404
[31] Vrba, Hartmann, & Jennings 1993, this conference

event, the 1905OT may not lie inside the timing annulus, the fading x-ray source in the small GRB781119 box was of marginal significance and never confirmed, the association with a bright star raises many problems, and several variables which appear in the 1928OT region cannot all be the counterpart.

Table 3. Counterpart Search Limits

	FLARING	FADING	QUIESCENT
X-RAY	• $3 \cdot 10^{-10}$ erg cm^{-2} (ref. 5)	• $2 \cdot 10^{-13}$ erg cm^{-2} s^{-1} (38 days)[3]	• 10^{-13} erg cm^{-2} s^{-1} (1,7)
UV			• U~25[11]
OPTICAL	• B~4 (patrol plates)[2] • B~10 (archival)[8]	• B=23 (40 hours)[12]	• B=25[11]
IR			• K=19.03[9] • Flux(25μ)~0.5 Jy[9]
RADIO	• Flux(0.4GHz) ~10^4 Jy[4] • Flux(151MHz) ~200 Jy[6]	• Flux(20 cm) ~1 mJy (9 days)[6]	• Flux(6 cm) ~100μJy[10]

[1]Boer et al. 1988, A&A, **202**, 117
[2]Greiner et al. 1993, AIP#280 (1st Compton Symp) p828
[3]Helfand & Long 1979, Nature, **282**, 589
[4]Inzani et al. 1982, AIP#77, 79
[5]Owens et al. 1993, MNRAS, **260**, L25
[6]Palmer 1993, in preparation
[7]Pizzichini et al. 1986, ApJ, **301**, 641
[8]Schaefer 1981, Nature, **294**, 722
[9]Schaefer et al. 1987, ApJ, **313**, 226 and later observations
[10]Schaefer et al. 1989, ApJ, **340**, 455
[11]Schaefer et al. 1992, unpublished HST and ground-based observations
[12]Schaefer et al. 1993, ApJ, in press

Table 4. Counterpart Candidates

	FLARING	FADING	QUIESCENT
X-RAY		• Einstein source & GRB781119[4]	
UV			
OPTICAL	• Archival OTs[2,3,5,7,11] 1928OT (Harvard)[9] 1905OT (Harvard)[6]	• HDE249119 and GRB790929[1]	• Variables in 1928OT area of GRB781119[8,10]
IR			
RADIO			

[1]Borovicka, Hudec, & Dedoch 1992, A&A, **258**, 379
[2]Flohrer et al. 1986, Adv Space Res, **6**, 55
[3]Greiner et al. 1987, Ap Space Sci, **138**, 155
[4]Grindlay et al. 1983, Nature, **300**, 730
[5]Hudec et al. 1987, A&A, **175**, 71
[6]Hudec et al. 1993, A&A, in press
[7]Moskalenko et al. 1992, Gamma-Ray Bursts, eds Ho, Epstein, & Fenimore, p. 127
[8]Pedersen et al. 1983, ApJLett, **270**, L43
[9]Schaefer 1981, Nature, **294**, 722
[10]Schaefer, Seitzer, & Bradt 1983, ApJ, **270**, L49
[11]Schaefer et al. 1984, ApJLett, **286**, L1

RECENT RESULTS

Rapid Response. Until recently, virtually no searches have been made for fading counterparts on a time scale as short as a day after the burst. This is because GRB positions were derived long after the event, so any images just after the burst had to be serendipitous. The ability to respond rapidly to new bursts has been greatly improved now that BATSE and COMPTEL positions are available within roughly 4 hours after the burst[15]. The WATCH cameras also provide rapid and accurate burst positions[16]. In addition, the Interplanetary Network has greatly improved its speed with its fastest response being about one day[17]. Unfortunately, the loss of Mars Observer means that the triangulation positions will only be narrow annuli for the near future.

The error boxes available soon after the burst have typical size scales of several degrees, so that any rapid response instrument should have a wide field-of-view. Large Schmidt telescopes admirably fit this requirement and can record stars to 21 mag on a single photograph. To date, there have been 14 bursts with Schmidt plates taken within 100 hours, including 8 WATCH burst positions recorded on the ESO and Calar Alto Schmidts[18], three deep images of the very bright GRB930131 burst on the CTIO, UK, and KPNO Schmidts[19], and three other BATSE bursts obtained by the Goddard counterpart group[20]. The three fastest reaction times are 6 hours for GRB930807[20], ten hours for GRB930720[20], and 12 hours for GRB911016[18]. The bottom line is that no fading counterpart candidates have been identified.

Palmer[21] has used the VLA to look look at the small error region of GRB930706 with a nine day delay and saw no source to 1.0 mJy. Observations such as these will test the general predictions of strong synchrotron radio emission from an ultra-relativistic fireball[22].

No Host Galaxies. Most cosmological models require GRBs to be in host galaxies. If so, then deep counterpart searches should reveal the host galaxy. But in general, the small GRB error regions do not contain any galaxies to fairly deep limits[23]. This can be used to constrain either the total energy of a burst (for some assumed host galaxy brightness) or the luminosity of the host galaxy (for some assumed distance scale). The limits vary greatly from GRB to GRB, but I have collected x-ray, optical, infrared, and radio limits on the brightest source in each of 26 GRB boxes. These can then be compared to the predictions for various host types and distance scales.

The first conclusion from these studies[23] is that bursts cannot come from a normal galaxy population at the usual cosmological distance scale. (This scale is based on a burst luminosity of roughly 6×10^{50} erg sec^{-1} derived from the LogN-LogP relation[24] and on the available energy in a neutron star.) Thus, the most popular cosmological model (one or two compact stars catastrophically releasing most of their binding energy within a normal galaxy) cannot be correct. Three dodges have been proposed, but each has its problems: First, the burster could reside outside any host galaxy, but then there will be difficulty in constructing plausible mechanisms to get all bursters to be ejected into or born in the intergalactic space. Second, the host galaxy may be systematically subluminous, but then there must be some plausible reason to explain why bursters preferentially reside in these galaxies. Third, the host galaxies could be much farther than indicated by the usual cosmological distance scale, but then there is a problem getting sufficient energy and in explaining the -3/2 slope region of the LogN-LogP curve. In any case, theorists who advance any cosmological model must identify the means by which they dodge the no-host-galaxy limits.

The second conclusion is that GRBs cannot come from active galactic nuclei. The reason is that such galaxies are very bright in many bands and can be seen to great distances. For example, the GRB790325B box is empty to around 60 μJy at 6 cm so that a radio-QUIET quasar host (with a power of 10^{25} W/Hz) must have a luminosity distance of greater than 135 Gpc (i.e., long past the edge of the Universe).

Archival Transients. The optical transient images found on archival photographic plates are the best evidence to date for a GRB counterpart. Around one dozen such images have been been reported in the literature[25], but these have varying degrees of confidence for association with the GRB phenomenon. Recently, Hudec et al.[26] have identified a particularly strong flash candidate. The image appears on a plate exposed in 1905 (this rules out airplane strobes and satellite glints), is untrailed on a trailed plate (this rules out supernovae, novae, and dwarf novae), does not appear on a plates 1 day before and 10 days after (this rules out many classes of variable stars as well as comets and asteroids), appears 5.5 mag above the plate limit (this is a strong indication that it is not a plate defect), is at an empty position on the Palomar Sky Survey (which rules out all known classes of variable stars except novae and supernovae), shows coma image distortion (this can arise only for light passing through the telescope optics and rules out plate defects), and shows a halo due to fog (this proves that the light came through the lower atmosphere and also rules out plate defects of all types). Thus, they have eliminated all known background sources, so we are left with the conclusion that a bright, short duration, optical transient of a previously unkown type occurred in the middle of a GRB error box.

Another good OT/GRB candidate is the 1928OT[27]. This image also was exposed before strobes and satellites were invented, does not appear on plates taken 45 minutes before and 45 minutes after, is untrailed on a trailed plate, is 5.8 mag above the plate limit, is at an empty position on the Palomar Sky Survey, and shows coma. This image has been extensively studied[27-32], with some concluding that the image is a plate defect. The first basis for this claim is the statement that the 1928OT image does not show coma. But this statement is easily seen to be false since all contour plots of the OT and nearby stars show the characteristic elongations towards the west and the south[27-30,32]. Indeed, a systematic study of three dozen nearby stars shows that the orientations of the 1928OT elongations to be exactly as expected for coma[32]. The second basis is that a detailed microscopic examination of the grain distribution with depth in the emulsion appeared to show the OT image to be different from that of surrounding normal stars[30]. However, recently Hudec, Pravec, and Borovicka[32] reconcile all earlier workers by pointing out that the structure with depth is not different if compared to stars of similar local density. Thus, there are no longer any grounds for considering the 1928OT to be anything other than a bright, short duration, optical transient which flashed inside a small GRB region.

The weakness of the OTs is not whether they are astrophysical in nature, but in how they are related to GRBs. The connection can be made in two ways; first by ruling out all known background flash sources (see above), and second by observing that a large control region outside of GRB error regions does not contain OT images[29]. Although strong, neither of these arguments is proof positive. The one case that would overcome this problem is to detect an OT at the same time and direction as a GRB event detected in the gamma rays.

Ultrasoft Transients. Helfand and Vrtilek[33] reported on ultrasoft transients (T~10^5 °K, duration~1 sec) detected in the Einstein IPC data. Owens et al.[7] looked through the ROSAT all-sky survey to find either these ultrasoft transients or normal GRBs. None were found even though their coverage and sensitivity were comparable with Einstein (which saw 42 ultrasoft transients), so the astrophysical nature of the ultrasoft transients must now be questioned.

Soft Gamma Repeaters. Until recently, little was known about the nature of SGRs beyond the suggestion that they were possibly related to neutron stars and of Population I. The counterparts of SGRs have also been extensively searched for. In fact, these searches have been too successful, since there are eleven proposed counterparts for the three known SGRs. In October 1993, a confident counterpart was discovered for SGR1806-20 by the ASCA x-ray satellite[34], and their accurate position coincided with the center of a plerionic supernova remnant[35]. When combined with the old problematic identification of the SGR0526-66 with the N49 supernova remnant[36] (especially with a possibly plerionic x-ray hot spot in N49[37]) and the fact that remnants appear in the error box for SGR1900+14[38], it now appears certain that SGRs are caused by young magnetized neutron stars in supernova remnants. Thus, the discovery of a SGR counterpart has collapsed the many possible models down to one tightly constrained setting. This is strong encouragement for further classical GRB counterpart searches.

THE FUTURE

Very Rapid Response. To date, the only low energy observations during a burst have come from sky patrols or all-sky cameras of relatively low sensitivity. The sensitivity of searches for flaring counterparts can be greatly increased if a small field-of-view instrument can be made to look at the right place at the right time. But GRBs occur randomly in both time and direction.

A way to solve this dilemma is to communicate a burst position to a fast slewing telescope. This is the idea behind HETE and its worldwide distribution of accurate burst positions[39]. HETE has a scheduled launch date in April 1995 and will carry an omnidirectional gamma ray spectrometer (6 keV to >1 MeV), a wide-field x-ray monitor (2-25 keV), and an ultraviolet transient camera array (5-7 eV). The ultraviolet camera may directly detect a flaring counterpart. In addition, the HETE burst position can be transmitted to ground based telescopes in real time allowing very rapid response counterpart searches.

Since the failure of the tape recorders on GRO in early 1991, BATSE data has been transmitted to ground with virtually no delay, and this opens up the possibility of using this data stream to guide ground based counterpart searches. The BACODINE system[40] does just this, and will report BATSE burst positions an average of 4.5 seconds after the BATSE trigger. This will allow fast slewing telescopes to start recording the right region of sky while most bursts are still bright. Not only can the ground based telescopes look during the burst, but they can also look soon after with long integrations in attempts to find faint fading counterparts. BACODINE will feed its positions to at least three telescope systems (GTOTE[40], GROSCE[41], and TOASTS[42]).

Radio Dispersion. It is quite plausible that GRBs give off bright radio emission during a burst. If the radio luminosity is just 10^{-8} of the gamma ray luminosity, bright GRBs would appear with flux larger than 1 Jy and be easily detectable by a radio telescope that happens to be pointing correctly. If in addition, GRBs are at cosmological distances, then the radio signal should be delayed (relative to the gamma rays) due to dispersion in the intervening ionized gas. For a distant GRB, Palmer[43] derives a typical dispersion measure of 500 pc cm^{-3} and notes that the resultant delay is 20 seconds for the VLA at 90 cm wavelength, or 90 seconds for the Cambridge Low Frequency Synthesis Telescope at 151 MHz, or one hour for the Arecibo telescope at 25 MHz. Such long delays might provide the critical time to slew a radio telescope to a prompt GRB position. In addition, the characteristic sliding of a radio flare to low frequencies will provide a unique signature for the GRB radio flare as well as proof of the cosmological distance to the burster.

X − ray Absorption. GRBs have only been seen as low as 2 keV[44-47], but in all cases the spectra appear to be a power law to the lowest energy. Thus it appears reasonable that the spectrum intrinsic to GRBs continues well into the soft x-ray regime. As this light travels from the burster to Earth, it must pass through the interstellar medium and possibly the intergalactic medium, which will strongly absorb the soft x-rays. Photoelectric absorption produces a characteristic spectral turnover where the cutoff energy is a simple function of the column density of neutral gases along the path. So if we could obtain GRB soft x-ray spectra, then we could directly measure the amount of gas from the burster to the Earth and consequently get a distance estimate to the burster. While any such measure will have poor accuracy, the distance scale to GRBs is unknown by roughly a factor of 10^{13}, so a poor but reliable distance will still solve the second biggest question in our field. For example, if the column density is shown to be less than 10^{18} cm^{-2}, then all galactic and cosmological models must be rejected in favor of Oort Cloud or heliosphere models. Or if high latitude bursts are seen to have column densities as high as 10^{23} cm^{-2}, then the bursts must lie in distant host galaxies which provide the extra absorption.

I have presented details of this proposal and a feasibility study in other papers[48,49]. The measurement of GRB soft x-ray spectra requires a relatively simple and inexpensive satellite experiment. As such, four groups around the world have started designing or building the necessary instruments, including RULER[50] and BALLERINA[51].

PROSPECTS

The state of our knowledge of GRBs is poor, with nothing definite known. Much of the blame for our ignorance comes from not having discovered any counterparts that can be reliably tied to GRBs. So the most likely solution to the GRB enigma is to find a counterpart. Even though this has proved to be a hard task, the motivation to continue and improve counterpart searches is strong, as it may be the only way to break the logjam of theories. The new ideas (very rapid response, radio dispersion measures, and soft x-ray absorption) promise good opportunities to score the breakthrough that is so desperately needed.

REFERENCES

1. Santa Cruz Conference Panel, Gamma Ray Transients and Related Astrophysical Phenomena (AIP, 1982), p. 497.
2. K. Hurley, Adv. Space Res. **3**, 203 (1983).
3. J. Katz, Electron-Positron Pairs in Astrophysics (AIP, 1983), p. 65.
4. S. E. Woosley, High Energy Transients in Astrophysics (AIP, 1984), p. 709.
5. NASA, Gamma Ray Astrophysics to the Year 2000, Report of the Gamma

Ray Program Working Group (NASA, 1988).
6. N. Lund, Gamma Ray Bursts (Cambridge Univ. Press, 1992), p. 489.
7. A. Owens, C. Page, S. Sembay, and B. Schaefer, MNRAS **260**, L25 (1993).
8. B. E. Schaefer, R. Vanderspek, H. Bradt, and G. Ricker, Ap. J. **283**, 887 (1984).
9. B. E. Schaefer, A. J. **90**, 1363 (1985).
10. B. E. Schaefer, H. Pedersen, C. Gouiffes, J. M. Poulsen, and G. Pizzichini, Astron. Ap. **174**, 338 (1987).
11. B. E. Schaefer et al., Ap. J. **320**, 398 (1987).
12. B. E. Schaefer, Ap. J. **337**, 927 (1989).
13. B. E. Schaefer, Ap. J. **353**, L25 (1990).
14. B. E. Schaefer, Ap. J. **366**, L39 (1991).
15. R. M. Kippen et al., these proceedings (AIP, 1994).
16. N. Lund, S. Brandt, and A. J. Castro-Tirado, Gamma-Ray Bursts (AIP, 1992), p. 53.
17. K. Hurley et al., Compton Gamma-Ray Observatory (AIP, 1993), p. 769.
18. A. J. Castro-Tirado, S. Brandt, and N. Lund, this volume (AIP, 1994).
19. B. E. Schaefer et al., Ap. J. , in press (1993).
20. S. D. Barthelmy, B. E. Schaefer, and D. M. Palmer, this volume (AIP, 1994).
21. D. M. Palmer et al., Ap. J. , in preparation (1993).
22. B. Paczynski and J. E. Rhoads, Ap. J. **418**, L5 (1993).
23. B. E. Schaefer, Gamma Ray Bursts (Cambridge Univ. Press, 1992), p. 107.
24. E. E. Fenimore et al., Nature , in press (1993).
25. R. Hudec, Astron. Lett. Comm. **28**, 359 (1993).
26. R. Hudec, A. Dedoch, P. Pravec, and J. Borovicka, Astron. Ap. , in press (1993).
27. B. E. Schaefer, Nature **294**, 722 (1981).
28. A. Zytkow, Ap. J. **359**, 138 (1990).
29. B. E. Schaefer, Ap. J. **364**, 590 (1990).
30. J. Greiner, Astron. Ap. **264**, 121 (1992).
31. B. E. Schaefer, Gamma Ray Bursts (Cambridge Univ. Press, 1992), p. 133.
32. R. Hudec, P. Pravec, and J. Borovicka, Astron. Ap. , in press (1993).
33. D. J. Helfand and S. D. Vrtilek, Nature **304**, 41 (1983).
34. Y. Tanaka et al., IAU Circ , 5880 (1993).
35. S. R. Kulkarni and D. A. Frail, Nature **365**, 33 (1993).
36. T. L. Cline et al., Ap. J. **255**, L45 (1982).
37. R. Rothschild et al., Compton Gamma-Ray Observatory (AIP, 1993), p. 808.
38. D. A. Frail et al., this volume (AIP, 1994).
39. G. Ricker et al., Gamma Ray Bursts (Cambridge Univ. Press, 1992), p. 288.
40. S. D. Barthelmy et al., this volume (AIP, 1994).
41. C. Akerlof et al., this volume (AIP, 1994).
42. J. Bonnell et al., this volume (AIP, 1994).
43. D. M. Palmer, Ap. J. **417**, L25 (1993).
44. J. G. Laros et al., High Energy Transients in Astrophysics (AIP, 1984), p. 378.
45. M. Katoh et al., High Energy Transients in Astrophysics (AIP, 1984), p. 390.
46. A. E. Metzger et al., Ap. J. **194**, L19 (1974).
47. A. Yoshida et al., PASJ 41, 509 (1989).
48. B. E. Schaefer, Compton Gamma-Ray Observatory (AIP, 1993), p. 803.
49. B. E. Schaefer, this volume (AIP, 1994).
50. A. Owens et al., this volume (AIP, 1994).
51. N. Lund et al., this volume (AIP, 1994).

Rapid Optical Follow-up Observations of Three Recent Gamma Ray Bursts

S. D. Barthelmy[1]
D. M. Palmer[2]
B. E. Schaefer[1]

NASA - Goddard Space Flight Center
Code 661
Greenbelt, MD 20771

1) Universities Space Research Association
2) NRC Fellow

ABSTRACT

Optical follow-up observations using wide field of view Schmidt telescopes have been made on 3 recent gamma ray bursts (GRB). We present results of searches in the optical band made on the GRB 930131, 930614, and 930706 position error boxes. Recent developments in the rapid dissemination of burst location information by the GRO BATSE (G. Fishman, PI) and Comptel (V. Shonfelder, PI) teams, the WATCH EURECA (N. Lund, PI) instrument, and the Third Interplanetary Network team (K. Hurley, PI) plus good fortune have allowed these observations to be made in a timely fashion. The response times range from 35 hours to 8 days. The sensitivity of the searches extends down to 20 mag.

INTRODUCTION and MOTIVATION

During the 24 years since their discovery[1], over 1000 Gamma Ray Bursts (GRBs) have been detected by numerous spacecraft. Some of these detections have produced rather small positional error boxes (most notably the various forms of the Interplanetary Network (IPN) have yielded several boxes on the arcminute-square scale), but to date no source object for classical GRBs has been identified.

Because of the difficulty to determine the precise direction of gamma rays, especially for instruments with the extremely large fields-of-view (FOV) necessary for the detection of GRBs, we have been working on making observations of the GRBs in the optical and radio bands. Since some theories[2-6] predict post-GRB optical emission (with time scales of minutes to days), we feel that there is a reasonable chance of observing this emission if the response time can be kept small. And since positional accuracy in the optical band is about 1 arcsec, any optical emission associated with the GRB will almost assuredly allow an identification of the quiescent source object of the GRB. The alternative strategy of making truly simultaneous optical band observations has the problem of poor sensitivity due to the very large FOV necessary to yield a reasonable rate of observations[7].

OUR PROGRAM

We currently have three Targets-of-Opportunity (ToO) programs to make rapid response follow-up optical and radio observations on localized GRBs. Whether we invoke these ToO's depends on the size of the preliminary GRB position error box and the type of detector (CCD or Plates) mounted on the two Schmidts. Table 1 lists the characteristics of the various instruments and detectors. There are no hard and fast rules concerning the decision to make follow-up observations, but example guidelines can be given. If a Schmidt is configured for plates, then we would only invoke the ToO if it were a bright well localized GRB detected by BATSE only; i.e. the positional accuracy of a bright BATSE-only detection is comparable to the 5° FOVs of plates. However, if the Schmidt is configured with CCDs, then we require additional information such as COMPTEL, EGRET or WATCH positioning or an IPN arc. The IPN arcs, although not as desirable as IPN boxes, are still usually quite narrow, ~1arcmin, and as such, the area of the arc inside the BATSE error circle is small enough to justify invoking with a CCD by tiling the arc segment with a few exposures. Of course since this is a ToO situation, we must use whatever instrument is mounted even if it is overkill in coverage, i.e. the use of plates for the GRB930614. The range of pixel sizes in Table 1 for the VLA at 20cm is due to the different array configurations. In the past 9 months we have invoked one or more of the 3 ToO's for a total of 4 times.

TABLE 1: INSTRUMENT and DETECTOR CHARACTERISTICS

Instrument: Detector	FOV	Pixel Size	Sensitivity
KPNO Schmidt:			
Plates	5.2 x 5.2°	~2"	20 mag in 30 min
2Kx2K CCD	1.0 x 1.0°	1.8"	20 mag in 20 min
4Kx4K CCD	0.85x0.85°	0.75"	19 mag in 15 min
CTIO Schmidt:			
Plates	5.2 x 5.2°	~2"	20 mag in 30 min
2Kx2K CCD	0.4 x 0.4°	0.7"	20 mag in 20 min
VLA at λ =20cm	30' FWHM Dia	1.4-44"	1 mJ in 20 min

OBSERVATIONS

Table 2 lists a synopsis of the optical observations made on the three GRBs. The error box width for GRB930131 is determined by width of IPN arc and its length by the length of the arc segment in the EGRET/COMPTEL/BATSE localization[8]. The error box width for GRB930607 is determined by width of IPN arc and its length by the diameter of the WATCH EUREKA error circle (N. Lund, private communication). The error box width for GRB930706 is determined by the IPN (Ulysses/BATSE/Mars Observer) localization (K. Hurley, T. Cline & J. Laros private communication). The term limiting magnitude means the magnitude to which the scanner was conservatively sensitive to changes (see below) in the field stars in the various exposures and the Palomar prints. Typically, the actual exposure limit was about 1 mag fainter.

Figures 1, 2 & 3 are diagrams of the relative positions and areas covered by the various CCD and plate exposures and IPN arcs or boxes for the three GRBs.

TABLE 2: THE GRBs OBSERVED

GRB Date Time UT	Time Delay [days]	Data Type	RA Dec (J2000)	Error Box	Limiting Magnitude
930131 18:57:11	1.46	1 CTIO Plate IIa-O, unfiltered	183.9 -10.2	43" x 0.7°	20.5
930614 03:40:33	4.10	2 KPNO Plates IIa-O, unfiltered	170.2° +23.1°	10' x 1.0°	19.5
930706 05:13:30	8.07	KPNO $4K^2$ CCD B & R Frames	281.2° -19.9°	0.95' x 2.08'	18.5

ANALYSIS and RESULTS

In short, no counterpart was identified for any of the 3 GRBs. In all cases, the comparison exposures for the GRB plate and CCD exposures were the Palomar Sky Survey prints. The scanning was done manually with 7x magnification by two of us (SDB & BES). The plates and CCD frames were scanned far beyond the stated error boxes so as to (1) cover out to a few sigma beyond the stated 1-sigma localizations, and (2) to get a measure of the background rate of changes. Changes are here defined as (1) stars that drop below the limiting magnitude of the Palomars (disappearances), (2) star that appear on the current exposure but not on the Palomars (appearances), and (3) stars present in both the Palomar and the current exposures that change by more than 1 magnitude.

ACKNOWLEDGMENTS

For 930131: We would like to thank Dr. M. Phillips of NOAO for granting discretionary time on the CTIO Schmidt.
For 930614: We would like to thank Dr. J. Laird of Bowling Green State U. for forgoing 2 hours of his regularly scheduled observing time during our ToO observation on the Kitt Peak Schmidt.
For 930706: We would like to thank Dr. T. Boroson of NOAO, for taking the observations on the Schmidt with the "new" f4ka1 CCD, and Kitt Peak National Observatory.

REFERENCES:

1. R.W. Klebesadel, I.B. Strong, R.A. Olsen; Ap.J.Let.; **182**, L85; (1973)

2. Jennings, M.C.; ApJ; **273**, 309; (1983)

3. S.A. Rappaport & P.C. Joss; Nature; **314**, 242; (1985)

4. D. Eichler & A.F. Cheng; ApJ; **336**, 360; (1989)

5. H.S. Fencl, R.N.Boyd, & D.H. Hartmann in "Gamma-Ray Bursts", eds: W.S.

Paciesas & G.J. Fishman (New York: AIP), 267; (1992)

6. D.L. Band & D.H. Hartmann in "Gamma-Ray Bursts", eds: W.S. Paciesas & G.J. Fishman (New York: AIP), 342; (1992)

7. J. Greiner, et al.; Proceedings of the Compton Symposium, eds. M. Friedlander, N. Gehrels, D. Macomb, AIP Conf. Proc. #280; 828; (1992)

8. B.E. Schaefer; S.D. Barthelmy; D. Palmer, et al.; ApJLet, In press; (1993)

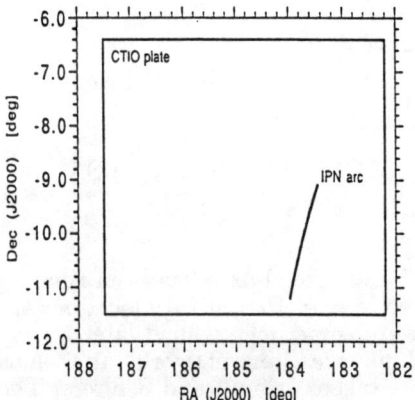

Fig 1. A diagram of the CTIO plate exposure coverage and the IPN arc for GRB930131. This CTIO plate was part of more than a dozen observations made by many observer[8].

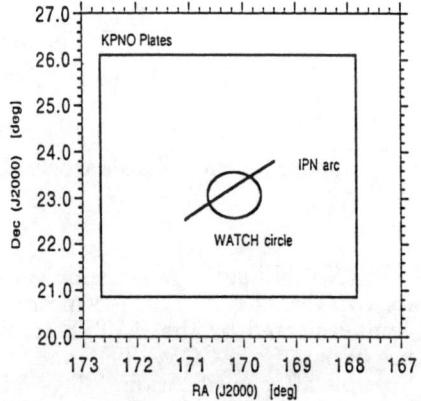

Fig 2. A diagram of the KPNO plates exposure coverage, the Watch error circle, and the IPN arc segment for GRB930614.

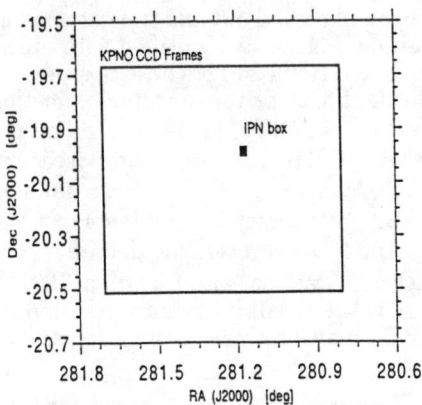

Fig 3. A diagram of the KPNO CCD frames and the IPN error box for GRB930706.

THE ESO - SCHMIDT SURVEY OF GAMMA-RAY BURSTS*

Michel Boër

Centre d'Etude Spatiale des Rayonnements (CNRS/UPS)
BP 4346, F31029 Toulouse Cedex, France

Holger Pedersen

Copenhagen University Observatory
Øster Voldgade 3, DK 1350 Copenhagen K, Denmark

A. Smette

European Southern Observatory, La Silla, Chile

J. Fishman, C. Kouveliotou

NASA Marshall Space Flight Center, Huntsville, Alabama, USA

K. Hurley

Space Sciences Laboratory, University of California, Berkeley, USA

ABSTRACT

A rapid survey of cosmic gamma-ray burst error boxes has been running for two years at the ESO – 1 metre Schmidt telescope. Preliminary locations are communicated by the BATSE team, which are sometimes refined later by the use of data from COMPTEL and/or the 3rd interplanetary network. As soon as possible after notification, $5° \times 5°$ images are taken, using B and R filters. The boxes are then blinked for a new or highly variable object. Preliminary results and future developments of the project are presented and upper limits to the visible radiation from a possible decaying counterpart to a gamma-ray burst are derived.

INTRODUCTION

No Gamma-ray burst source has yet been observed outside the gamma-ray and X-ray ranges. None of them have been unambiguously identified with quiescent sources, with some possible exceptions, like GBS 0526-66 whose position coincides with the N 49 supernova remnant in the LMC, or the marginal detection of an EINSTEIN X-ray source in the error box of GRB 781119[1,2,3]. Some optical transient sources, possibly associated with GRBs have been proposed[4,5,6], but no concensus has been reached as to their reality[7,8], The BATSE results on the distribution of gamma-ray bursts sources[9] give forceful arguments in favor of cosmological models (see e.g. Paczyǹsky, 1992[10]); however, the debate on the origin of GRB sources is still contentious (see e.g. Paciesas and Fishman, 1992[11], and these conference proceedings), and there is the possibility that GRB sources do not have a single origin, e.g. some GRBs may originate from the galactic

* Based on observations done at the European Southern Observatory, La Silla, Chile

disk, while the majority of them come from the galactic halo or are at cosmological distances[12]. Last, but not least, although the soft gamma-ray repeaters are in some respects different from the "classical" GRBs, their origin is "more unknown", if possible, than that of GRBs, and very few data are available on them, except that the SGR 1806-20 was repeating shortly before (and may be during) this workshop. However, their repetitive nature, as well as their softer spectra, make them objects of choice for a multiwavelength study.

Ideally, any multi-wavelength study of GRBs should be done at the same time than as that of the gamma-ray event, in the same direction, and with some high sensitivity device. Though this may seem obvious, it is not easy to achieve in practice. Ground based experiments suffer from low sensitivity, insufficient duty cycles, and/or lack of angular coverage. In space, the situation may not be better, since the optical/UV monitor aboard the HETE satellite, has a field of view of 1.7 sr, and will reach magnitude 9^{13}. Current high energy experiments such as SIGMA and WATCH aboard the GRANAT satellite, BATSE and COMPTEL aboard CGRO, and the gamma-ray burst detector aboard the ULYSSES spacecraft give the possibility of deriving timely, more or less precise GRB positions. This new possibility allows a sensitive search to be initiated as soon as the GRB position is known. The diversity of detetors and/or methods used to derive this localization results in various accuracies and times needed to derive the burst position. For instance, the BACODINE project (Barthelmy, S., these proceedings) provides localizations as fast as 4.5 seconds, but with a 15° error radius, while the accuracy reached by the triangulation may be an arcminute or less, but sometimes with a delay of 1 week. Intermediate cases are the fast positions provided by the BATSE team at MSFC (1h - 1day; accuracy 5°), the SIGMA coded mask and sidelobes positions (1 day; accuracy 3arcmin, fov to 30 arcmin, sidelobes), and the Watch rotating collimator camera(1day; accuracy 1°). These accuracies may differ significantly from the final values, and also from the accuracies reached in the case of X-ray transients. This is due to the rapid fluctuations in bursters (as far as accuracy is concerned, X-ray transients are not different from permanent sources, while gamma-ray bursts often display only one second or so of data at a high enough S/N ratio to enable position derivation), and to the fact that for the present study, quick look data and analysis are used, with the advantage of fast position computations, at the expense of the accuracy and of the stability of the position. For instance, in the case of triangulation, though the relative timing accuracy of quick look data may be close to final, it may happen that the absolute datation differs in the final tape by an amount of 50ms or so from that on the quick look data. This will not reduce the error box size significantly, but will shift it, in the case of a triangulation using Ulysses, by several arcminutes.

OBSERVATIONS

Given the above constraints it is obvious that the Schmidt telescope is the instrument of choice, with a good trade-off between sensitivity and short delay exposures after the burst. These cameras have a large field of view, are usually running in service mode, so that they are more rapidly available for target of opportunity observations than standard telescopes. The present study is based on a program running since 1991 at the ESO 1m Schmidt telescope. The field of view of a 30 x 30 cm plate is 5.5 × 5.5°, which matches the BATSE uncertainty for bright bursts well. For these bursts, a position is usually obtained later using the triangulation method. Even if the position is preliminary, the final one is always contained in the plate, giving confidence that no time is wasted in imaging the wrong field. Among other advantages, La Silla and Huntsville have a local time difference of only 1-2 hours, so that BATSE positions usually reach ESO in the late afternoon, permitting optical observations with minimum delay. In few cases, other experiments, like COMPTEL or WATCH may provide additional data. As soon as the plate has been developed, a blink study is initiated to search for a decaying optical transient. Later the plates are sent to Paris were they are presently digitized and examined in greater detail.

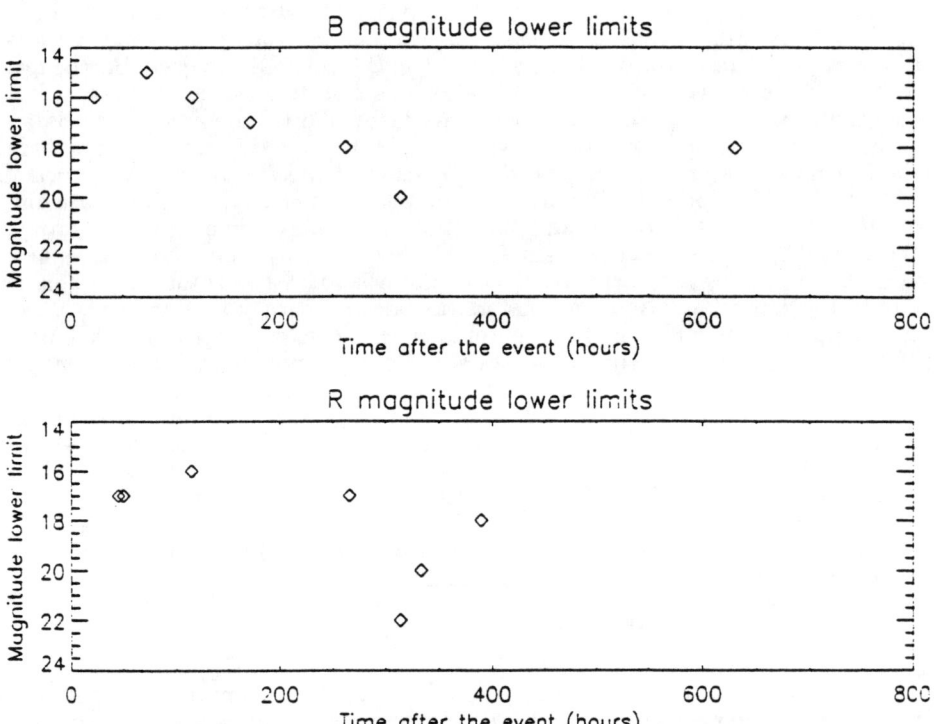

Figure 1. Magnitude lower limits.

Table I. Observing log							
GRB		Plate		Time lag	Mag. lower limit		
date	time	date	time	(hours)	B	V	R
920311	02:20	920313	04:30	50.2			17
		920422	00:46	262.4	18		
		920424	23:32	333.2			20
920406	02:45	920422	08:14	389.5			18
		920502	08:30	629.8	18		
		920509	07:04	796.3			
		920510	08:03	821.3			
920501[1]	21:18	920512	23:21	266.1			17
		920514	23:10	313.9	20		22
		920521	23:07	481.8			
920517	03:18	920519	00:33	45.3			17
		920520	03:33	72.3	15		
		920420	04:01	72.7			
920525	03:27	920602	07:09	171.7	17		
920830	01:45	920901	01:32	23:47			16
921123		921127	01:17	115.0	16		
930118	17:44	930120	07:16	37.5		18	
930131[2]	17:44	930217		419.4			
		930222	05:40	538.7		19	

Table 1 gives the observing log and the lower limits to the magnitude which we derived from the comparison of the plate with the corresponding survey region of the ESO SRC and QBS surveys. Unfortunately the optimum offset, a few hours after the event has not been reached, either because of the presence of the moon, bad weather conditions, position of the burst close to the sun, or technical problems. We are still hoping that in the future this delay will

[1] CCD exposure; observer AS
[2] Old plate, hence very dark: no limit derived

be reduced in some cases. Present SGR activity may provide this opportunity. On one occasion, one of us (AS) was able to take deep CCD exposure of a triangulated position (GRB 920501). No new optical source was detected at the position of the gamma-ray burst source. Lower magnitude limits for the B and R photometric bands are displayed in figure 1, as a function of the time elapsed between the gamma-ray burst and the optical exposure. The best limit obtained for classical GRBs is 20 in B and 22 in R. The shortest time offset is 24 hours. For the SGR, this delay is considerably reduced to few hours, because of its repetitive nature. Typical integration times are about 30 min. in R and 1 hour in B.

When SGR 1806-20 became active again, a campaign was started at ESO to acquire plates each night, for a period still to be decided. So far no object has been detected. But in this case the delay between a recurrence and the exposure is considerably reduced, and plates are taken both before and after the burst. A detailed anaysis of the plates will be performed later.

ACKNOWLEDGEMENTS

This project is a CGRO Guest Investigator program. The Ulysses GRB experiment was built at CESR (CNRS/UPS) in France with support from the Centre National d'Etudes Spatiale, and in Germany by DARA FRG contracts 01 ON 088 ZA/WRK 275/4 7.12 and 01 ON 88014. We thank Dr. Bo Reipurth, Mrs O. and G. Pissaro, at ESO La Silla, for their continuous support to our work. KH ackowledges support for this project under NASA grant NAG5 - 1560 and JPL contract 958056.

REFERENCES

1. Grindlay, J. et al., Nature **300**, 730 (1983).
2. Pizzichini, G. et al., Ap. J. **301**, 641 (1986).
3. Boër, M. et al., A.&A. **202**, 117 (1988).
4. Schaefer, B.E., Ap. J. **364**, 641 (1990).
5. Hudec, R., Peresty, R. and Motch C., A.&A. **235**, 174 (1990).
6. Greiner, J., Wenzel, W., and Degel, W., A.&A. **235**, 174 (1990).
7. Zytkow, A.N., Ap. J. **359**, 138 (1990).
8. Greiner, J., A.&A. **264**, 121 (1992).
9. Meegan, C.A., et al., Nature **355**, 143 (1992).
10. Paczyński, B., Acta Astronomica **41**, 257 (1992).
11. Paciesas, W.S., Fishman, G.J., edt., Proceedings of the 1991 Huntsville Workshop on Gamma-ray Bursts (AIP Conf. Proc. 265, N. Y., 1992).
12. Lingenfelter, R.E., and Higdon, J.C., Nature **356**, 132 (1992).
13. Ricker, G.R., et al., Gamma-Ray Bursts, Observations, Analyses and Theory; Cheng Ho, R.I. Epstein, and E.E. Fenimore eds. (Cambridge University Press, 1991), p. 288.
14. Eichler, D., and Cheng, A.F., Ap. J. **336**, 360 (1989).

A REAL-TIME SEARCH FOR OPTICAL COUNTERPARTS TO GAMMA-RAY BURSTS

J. T. Bonnell*, J. P. Norris*
Code 668.1, NASA/GSFC Greenbelt, MD 20771

S. D. Barthelmy, T. L. Cline, N. Gehrels
Code 661.0, NASA/GSFC Greenbelt, MD 20771

G. J. Fishman, C. Kouveliotou, C.A. Meegan
NASA/MSFC, ES-62, Huntsville, AL 35812

ABSTRACT

We report on the setup and operation of an instrument to search for optical counterparts of gamma-ray bursts. This instrument consists of a wide angle camera mounted on an automated, rapidly slewing telescope located at the NASA/GSFC Optical Test Site. The observations are triggered by a phone line data transmission of burst coordinates from the BACODINE[1] network. The camera used is a 1-inch format SIT video camera with a zoom lens providing a 5-20 degree adjustable field of view. Observations of the burst field are recorded on video tape and subsequently digitized and analyzed.

INTRODUCTION

This is a status report on the set up and operation of an instrument to search for optical counterparts of gamma-ray bursts. This instrument is presently operational and located at the 1.2 meter Satellite Tracking Facility at the Goddard Space Flight Center (GSFC) Optical Test Site in Greenbelt, Maryland. However, as the title implies, no candidates for optical counterparts to gamma-ray bursts have been detected; the search continues.

Our goal has been to establish a program which is capable of a rapid response to gamma-ray burst localizations, achieves a useful sensitivity and resolution in optical imaging, and makes as much use as possible of existing equipment and facilities available at GSFC. We have achieved these goals by utilizing, phone line/modem data transmission of BATSE derived burst coordinates from the BACODINE[1] network, a Silicon Intensified Target (SIT) low light surveillance video camera with a wide field zoom lens, and an automated telescope facility normally used for laser ranging and satellite tracking purposes at GSFC[2]. This program operates in coordination with laser ranging and satellite tracking programs at GSFC and is presently referred to as the Transient Optical And Satellite Tracking (TOAST) system.

INSTRUMENT SUMMARY

The instrument used for optical imaging is a 1-inch format SIT video camera with an f/1.8, 15-150 mm (10x) zoom lens sold commercially for low

---
*Compton GRO Science Support Center

light level applications. The video signal from the camera is recorded with a time stamp on VHS format video tape. For astronomical purposes, the characteristics of this system are;
- sensitive to approximately 8th visual magnitude star images;
- 20 x 30 degree maximum field of view;
- digitized pixel size of 2.8 arcminutes at maximum field;
- 30 video frames per second data rate.

This instrument is mounted coaxially with a 1.2 meter automated, rapidly slewing telescope. The computer controlled slew rate is 6 degrees per second with better than 20 arcsecond pointing accuracy[2].

OPERATIONS SUMMARY

This program is run simultaneously with experimental satellite tracking programs using the 1.2 meter telescope at the GSFC optical test site. Normal operation of these tracking programs typically occurs for an 8 hour shift each night, contingent on local weather conditions.

An onsite pc and dedicated phone line are used to monitor the BACODINE network for a burst notification and coordinate transmission. The pc is set up to sound an alarm, display burst coordinates on screen, and log activity to a file. The response to a gamma-ray burst notification takes priority over tracking programs and is initiated by the telescope operator. The operator then goes through the following check list.

0 - Determine if burst coordinates are visible from local site.
1 - Type in the burst coordinates and start the mount slewing.
2 - Start the video tape with time stamp included.
3 - Zoom the SIT camera out to its maximum field-of-view.
4 - Verify that the dome lights are out.
5 - On the audio channel, record comments about "seeing, etc."
6 - Continue observing/recording for 10-15 minutes.
7 - Remove and label the tape with approximate date/time and field.
8 - Resume normal satellite tracking activities.

Test runs have been conducted, producing data for the development of analysis procedures, and indicate that a 30 second to 1 minute response time is achievable.

Operational runs began on October 5, 1993, however, no BATSE / BACODINE triggers have produced observable burst fields for this system to date (November 30, 1993).

DATA REDUCTION

For the initial identification of an optical transient, the beginning of the video tape of the burst field is manually scanned and compared to a computer generated star field to identify a likely event. An approximate position and time range is noted from the video tape and the appropriate video frames are digitized at a resolution of 640 x 480 pixels.

The digitized video frames are then converted to primary array FITS files with basic header information. Two or three field stars are identified and a preliminary astrometry solution is computed and stored in the FITS header. The astrometry solution is refined using an interactive procedure and the transient event position is converted to epoch 2000 right ascension and declination from the image pixel coordinates. The astrometry software tools used are ASTRON[3] library routines based on the Interactive Data Language (IDL). A sample FITS header with an initial astrometry solution is given below.

```
**************************************
sample fits header created for digitized
video image.  Initial astrometry plate
solution parameters are stored as keyword
values.
**************************************
SIMPLE  =                    T /Written by IDL: 15-Oct-1993 14:40:22.00
BITPIX  =                   16 /Integer*2 (short integer)
NAXIS   =                    2 /
NAXIS1  =                  646 /
NAXIS2  =                  486 /
EQUINOX =                 2000 /EQUINOX OF REF. COORD
CD1_1   =          0.020158272 /DEGREES/PIXEL
CD2_1   =         -0.0087456611 /DEGREES/PIXEL
CD1_2   =         -0.041416891 /DEGREES/PIXEL
CD2_2   =         -0.016401049 /DEGREES/PIXEL
CTYPE1  = 'RA—TAN'              /COORDINATE TYPE
CRPIX1  =              319.000 /REFERENCE PIXEL IN LINES
CRPIX2  =              147.000 /REFERENCE PIXEL IN SAMPLES
CRVAL1  =             304.64999 /R.A. (DEGREES)
CRVAL2  =             62.583000 /DEC (DEGREES)
HISTORY PUTAST:15-Oct-1993 15:01 CD, CRPIX AND CRVAL
HISTORY PARAMETERS WRITTEN
***** END OF FILE *****
```

The sensitivity limit for the present video system corresponds roughly to 8th visual magnitude star images. Photometry procedures to reconstruct a transient event light curve from a series of digitized frames are under development.

FUTURE PLANS

We are still exploring the use of SIT video cameras for astronomical observations of transient events. A near term improvement to our present set up will likely be a more sensitive version of the current camera system. The platform used for mounting small instruments coaxially with the 1.2 meter satellite tracker presently contains additional room which will be used to mount an automated film camera along with other instruments as they become available.

REFERENCES

1. Barthelmy, S. D. et. al., these conference proceedings , ().
2. McGarry, J., Zagwodzki, T., Degnan, J., SPIE *Acquisition, Tracking, and Pointing* **641**, 77 (1986).
3. Landsman, W. B., ASP Conference Series *Astronomical Data Analysis Software and Systems II* **52**, 246 (1993).

OPTICAL FOLLOW-UP OF GAMMA-RAY BURSTS OBSERVED BY WATCH

Alberto J. Castro-Tirado *, Soren Brandt, Niels Lund
Danish Space Research Institute, Gl. Ludtoftevej 7, DK-2800 Lyngby, Denmark

Sergei S. Guziy
Nikolaev State University Observatory, Nikolaev, Ukraine

ABSTRACT

44 Gamma-Ray Bursts have been localized by the WATCH experiments on GRANAT and EURECA. For some of them, Schmidt plates were taken within days after the burst. In other cases, time-correlated plates were found in some of the main astronomical archives. No obvious optical counterpart has been found in any of the investigated plates.

INTRODUCTION

An important clue for resolving the Gamma-Ray Burst puzzle would be the detection of transient optical emission associated with the Gamma-Ray Bursts (GRBs). Optical emission has been predicted by different theories[1-5]. The first search for optical emission arising simultaneously with the GRB was performed[6] by Grindlay et al. in 1974. Recently, five other events were found for which time-correlated plates were available[7,8]. The latest result concerns GRBs detected by the BATSE experiment on GRO: time-correlated plates were found[9] for more than 20 bursts, but none of the plates revealed any interesting brightening in the error box. Thus, in no case there has been a positive detection. For GRB 910814, a rather deep plate (photographic limiting magnitude of 4.5 for a 1 s flash) was found in Dushambe. It implied a limit for $F_\gamma/F_{opt} \geq 42$, with F_γ and F_{opt} being the peak gamma-ray and optical fluxes.

It has been customary in optical searches to consider limits for detection of flashes of 1 s duration, but here we would like to suggest that a better estimation of the upper limit to any optical emission could be obtained by taking the ratio between the optical fluence to the gamma-ray fluence, S_{opt}/S_γ, rather than considering only peak fluxes under the assumption of 1-second optical flashes. GRB time profiles are very different and GRBs with identical peak fluxes can have fluences that differ by several orders of magnitude.

We present here new optical searches for 10 GRBs detected by WATCH on GRANAT[10,11] and 1 GRB detected by WATCH on EURECA[12]. We found simultaneous plates for some of them; in other cases we used plates taken shortly afterwards.

* current address: Laboratorio de Astrofisica Espacial y Fisica Fundamental; P.O. box 50727, 28080, Madrid, Spain

OBSERVATIONS

A collaboration with several observatories around the world has been established in order to secure a rapid follow-up response to some of the GRBs located by the WATCH wide field monitors on board GRANAT (3 units) and the ESA EURECA satellite (1 unit). The WATCH detector is unique, in the sense that bright bursts can be rapidly localized within a 1 degree radius error box, making feasible quick follow-up by wide field optical telescopes at major observatories. In several cases Schmidt plates were taken, and in many cases a corresponding plate could be found in some of the larger plates archives, like Sonneberg in Germany or Ondrejov in the Czhec Republic. As GRANAT was launched on Dec 1, 1989, and the last Harvard plates dated from 1989, no plates could be checked for simultaneous optical emission in the Harvard Collection. Almost all the plates were test-blinked; one of them in Bamberg, and the rest in Sonneberg.

DISCUSSION

Table I shows the GRBs for which Schmidt plates were taken within a few days after the event. Table II shows the GRBs for which plates near to the time of the events were found in one of the plates archives investigated. In all the cases the results of a search for a counterpart for the burst were negative. We discuss the most relevant searches:

1) GRB 901116. The negative result implies a $S_{opt}/S_{6-100 \text{ keV}} \leq 330$ for simultaneous emission.

2) GRB 910219. An optical transient, OT 1905, was reported to be located in this error box[13,14]. However, there were discrepancies between the OT position and the GRB position determined from the third interplanetary network. Nevertheless, based on the Schmidt plate taken at our request at Calar Alto, no optical emission brighter than B = 19 was detected 66 hours after the burst neither at the OT nor at the GRB positions.

3) GRB 911016. This is was one of the most rapidly optical follow-up's to a GRB[15], and possibly the first one covering the whole error box up to a limiting magnitude B = 17. The Schmidt plate was taken only 12 hours after the main event, but nothing was found in our preliminary analysis. We intend to study this plate further.

4) GRB 920723b. A photographic plate at Ondrejov was exposed just 32 minutes after the event. The peak flux for GRB 920723b is the highest one for the GRBs observed by WATCH[16]. A suspicious object was found in the plate, but it turned out to be an airplane flash (R. Hudec, private comunication). $S_{opt}/S_{6-100 \text{ keV}} \leq 2$ for any emission 32 minutes after the event.

5) GRB 920902. No candidate has been found in the crowded field where the burst originated (towards the center of the Galaxy).

6) GRB 920903a. A limit of $S_{opt}/S_{6-100 \text{ keV}} \leq 33$ is derived for simultaneous emission.

7) GRB 920925c. This is one of the longest GRBs detected by WATCH so far. It lasted for 300 s. Its total fluence was $\sim 10^{-5}$ erg cm^{-2}. A simultaneous

plate was found in the Ondrejov plate collection. The plate was exposed during the first five hours of the night and trailed for most of the time. No candidate was found (this result was also confirmed by Hudec, private communication). It implies $S_{opt}/S_{6-100 \, keV} \leq 0.85$.

Table I. Available Schmidt Plates for WATCH bursts.			
WATCH GRB	Observatory	hours after burst	Phot. lim. magnitude
910219	Calar Alto	66	18
911016	ESO	12	16
"	ESO	36	18
920714	ESO	63	18
"	ESO	79	20
920718a	ESO	18	17
920902	ESO	14	17
"	ESO	38	19

Table II. Archival Plates for WATCH bursts.			
WATCH GRB	Observatory	hours after burst	Phot. lim. magnitude
911116	Ondrejov	0	6
"	Odessa	24	13
910310	Sonneberg	83	16
920723b	Ondrejov	0.5	6
920903a	Ondrejov	0	6
920925c	Ondrejov	0	6
"	Sonneberg	24	14
930410	Barcelona	24	11

CONCLUSIONS

No candidate has been found for any of the GRBs here reported as observed by the WATCH experiments. In the future, a more rapid response is necessary in order to set deeper limits to the $S_{opt}/S_{6-100 \text{ keV}}$ ratio.

Acknowledgements. We are grateful to M. Boer, A. Smette, G. Pizarro, R. West (ESO), and K. Birkle (Calar Alto), for obtaining the Schmidt plates on our request. To I. Bues and O. von Ranke (Bamberg), J. Soldan, M. Novak and S. Spurny (Ondrejov) for their collaboration. Thanks to W. Wenzel and H.J. Brauer for the hospitality help at Sonneberg Observatory. We are in debt to P. Arranz, S. Escudero, J. Ganan, J. Garcia, S. Garcia, M. Gil, Juanjo, J. L. Ortiz, J. Prat, F. Reyes-Andres, I. Rivas, J. Ruiz, J. Tortosa, J.M. Trigo and B. Troughton for their valuable help in the goal of achieving a simultaneous plate to a GRB. Conversations with R. Hudec and H. Pedersen were very fruitful.

REFERENCES

1. London, R.A., High energy Transients in Astrophysics (S. Woosley (ed:), 1984), p. 151.
2. Woosley, S., High energy Transients in Astrophysics (S. Woosley (ed:), 1984), p. 485.
3. Melia, F., Rappaport. S. and Joss, C., ApJ **21**, 30 (1986).
4. Hartmann, D., ApJ **336**, 889 (1989).
5. Meszaros, P. and Rees. M. J., ApJ letters, submitted , (1993).
6. Grindlay, J.E., Wright. E.L. and McCrowsky, R.E., ApJ **192**, L113 (1974).
7. Hudec, R. et al., A&A **75**, 71 (1987).
8. Hudec, R., Astr. Lett. and Communications **28**, 359 (1993).
9. Greiner, J. et al., Second Compton Observatory Symposium
10. Castro-Tirado, A. J., Brandt, S. and Lund, N. et al., These proceedings
11. Castro-Tirado, A. J.. Ph.D. Thesis, University of Copenhagen , (1993).
12. Brandt, S., Castro-Tirado, A. J., and Lund, N., These proceedings
13. Hudec, R. et al., Czech Astronomical Institute, preprint 131 , (1992).
14. Vrba, F. J. et al., to be published in ApJ , (1994).
15. Schaefer, B. E., These Proceedings
16. Terekhov, O., Lobachov, V.A., Danisenko. D. V., Lapshov, I., Sunyaev, R. A., Lund, N., Brandt. S. and Castro-Tirado, A. J., in preparation , (1993).

SIMULTANEOUS OPTICAL/GAMMA-RAY OBSERVATIONS OF GRBs

J. Greiner,[1] W. Wenzel,[2] R. Hudec,[3] E.I. Moskalenko,[4] V. Metlov,[5]
N.S. Chernych,[6] V.S. Getman,[7] R. Ziener,[8] K. Birkle,[9] N. Bade,[10]
S.B. Tritton,[11] G.J. Fishman,[12] C. Kouveliotou,[12] C.A. Meegan,[12]
W.S. Paciesas,[12,13] R.B. Wilson[12]

[1] Max-Planck-Institut für Extraterrestrische Physik, 85740 Garching, FRG
[2] Sternwarte Sonneberg, 96515 Sonneberg, FRG
[3] Astronomical Institute Ondřejov, 25165 Ondřejov, CR
[4] Sternberg Astronomical Institute, 119899 Moscow, Russia
[5] Sternberg Astronomical Inst., Crimean Observatory, 334413 Crimea, Ukraine
[6] Crimean Astrophysical Observatory, 334413 Crimea, Ukraine
[7] Inst. of Astrophysics, Academy of Sciences, 743042 Dushanbe, Tadshikistan
[8] Thüringer Landessternwarte, 07778 Tautenburg, FRG
[9] Calar Alto, German-Spanish Astronomical Centre, 04080 Almeria, Spain
[10] Hamburger Sternwarte, Gojenbergsweg 112, 21029 Hamburg, FRG
[11] Royal Observatory, Blackford Hill, Edinburgh EH9 3HJ, U.K.
[12] Marshall Space Flight Center, Huntsville, AL 35812, U.S.A.
[13] University of Alabama, Huntsville, AL, U.S.A.

ABSTRACT

This status report presents details on the project to search for serendipitous time-correlated optical photographic observations of GRBs. The ongoing photography at nine observatories is used to look for plates which have been exposed simultaneously with a GRB detected by BATSE and contain the burst position. The results for the first two years of BATSE operation are presented.

STRATEGY AND INSTRUMENTS

We are correlating two independent surveys, namely the photographic plates (optical) of the observatories involved and the BATSE (γ-ray) burst data. At present the photographic sky patrols of the observatories Sonneberg, Tautenburg and Hamburg (all Germany), Calar Alto (Spain), Ondřejov (CR), Odessa and Crimea (Ukraine), and Dushanbe (Tadshikistan) are used for the northern hemisphere, and the UK (Siding Spring, Coonabarabran, Australia) and the ESO (La Silla, Chile) Schmidt plates for the southern hemisphere GRBs. The instruments are described elsewhere[1], and relevant details are given in Table 1.

With BATSE detecting and localizing $\approx$ 1 GRB per day, and a spatial and temporal optical coverage with the observatories involved of about 80%, we are presently checking 280–300 GRBs per year. In average, we identify
- $\approx$10–15 simultaneous plates per year with 150°×150° field of view (FOV) and limiting magnitude of 1–3 mag.
- $\lesssim$ 1 simultaneous plate per year with 20°×20° FOV and limiting magnitude of 4–7 mag.
- $\approx$10–15 near-simultaneous plates ($\pm$12 hours) per year with 5°×5° FOV and limiting magnitude of 6–15 mag.

The limiting magnitudes refer to the sensitivity for an assumed 1 sec duration flash within the exposure time of typically 0.5–4 hours.

RESULTS

Within almost two years of BATSE observations (1991 April 23 — 1993 Feb. 28) we have identified simultaneous photographic observations for 29 GRBs (Tab. 2). For most of these hits, several simultaneously exposed plates from different observational stations are available. Most of these plates are from the Czech meteor patrol which uses a red sensitive combination of objective and panchromatic emulsion (almost no sensitivity below 400 nm). All plates have been investigated by blink comparison. No optical flash or any brightening of an object was found which could be undoubtedly attributed to a GRB. Thus, any optical emission accompanying a GRB must have been less than the minimum detectable optical flux (erg/cm^2/s) of the corresponding plate. Using the peak γ-ray flux in the 50–300 keV range reached on the 1024ms timescale we derive a lower limit of typically $F_\gamma/F_{opt} > 0.5$–40.

Furthermore, near-simultaneous deep exposures (24 hours before or after GRB) are also registered (Tab. 3). Since these plates have smaller FOV, the large GRB error boxes are not always covered fully. The last column in Tab. 3 gives the percentage of the error box covered by the plate. This has been calculated by adding in quadrature a systematic error of 4° to the statistical error of each burst. In addition to the partial coverage, some of these plates are taken under programmes which are not related to variability studies; i.e. there is no plate of the same area (and same filter/emulsion combination) which could be used for blink comparison. Thus, only part of the near-simultaneous plates could be investigated for optical activity. The blink comparison of those deep plates having suitable comparison plates revealed a number of known as well as new variable stars (Wenzel[3,4]). However, this optical activity is presumably not related to the GRB phenomenon. We have found no optical flash event. Counting only plates which contain 100% of the error box and which have been checked by blink comparison, we constrain the optical brightness within 1–3 hours before or after the GRB to be less than 5–7th mag for a 1 sec flash.

With an increasing number of simultaneous optical observations of GRB locations we can conclude with higher confidence that the optical emission of typical GRBs is (1) at a level at least below $(F_\gamma/F_{opt})^{-1} \approx 2$ at the time of the burst and (2) lower by a factor of 10-200 than the simultaneous emission a few hours after the burst. In the future (with improving statistics) we expect to find more simultaneous plates with deeper limiting magnitudes and possibly also for brighter GRBs.

ACKNOWLEDGEMENT

JG and WW are partly supported by DARA contracts 50 OR 9104 3 and 50 OR 9201, and by DESY-PH contract 05-5S0414. RH is supported by grant 303103 of the Academy of Sciences of the Czech Republic.

REFERENCES

1. Greiner J., et al. , Gamma-ray bursts (AIP 265, 1992), p. 327.
2. Schaefer B.E., Nature **294**, 722 (1981).
3. Wenzel W., Inf. Bull. Variable Stars , No. 3883 (1993).
4. Wenzel W., Mitteil. Veränderl. Sterne, Sonneberg **12**, 161 (1993).

Table 1: Instrument capabilities at the observatories involved

Observatory	Characteristics of the instruments	Field of view (degrees)	Plate scale ("/mm)	Exposure time[1] (hours)	Limiting magnitude for 1 s flash[1] (mag)	No. of plates taken simultaneously	Probability for a time-correlated plate[2]
Sonneberg	sky patrol[3] 7.1/25	22×22	600	0.7	6 (pg)	2	2×10^{-4}
Sonneberg	Astrographs 40/160, 40/200	4×4	50	0.5–1	9 (pg)	1	2×10^{-6}
Tautenburg	Schmidt 1340/2000/4000	3.3×3.3	49.5	0.5–1	14 (pg)	1	7×10^{-7}
Calar Alto[4]	Schmidt	5.5×5.5	82	0.75–1	15 (pg)	1	6×10^{-6}
Ondřejov[5]	fish-eye 1:3.5	150×150	6800	4	3 (pv)	1–10	5×10^{-2}
Odessa	sky patrol 7.5/25–7.5/72	90×40	300–830	0.5	6 (pg)	1–3	1×10^{-4}
Dushanbe	meteor patrol[6]	40×40	830	1	6 (pv)	1–3	2×10^{-3}
Crimea[7]	Astrographs 40/160	10×10	120	0.5	9 (pv)	1/2	7×10^{-7}
La Silla	ESO Schmidt	5×5	75	0.25–2	15 (pg)	1	9×10^{-6}
Siding Spring	UK Schmidt 1240/1830/3070	6.5×6.5	67	1–1.5	15 (pg)	1	9×10^{-6}

[1] Typical values.
[2] Including multiple cameras for Sonneberg, Ondřejov, Odessa and Dushanbe but excluding multi-colour exposures.
[3] Two mountings with 7 cameras (for pg and pv plates) each; cameras have adjacent FOV with small overlap.
[4] The Schmidt telescope of Hamburg Observatory is operated at Calar Alto. However, the programmes of Hamburg Observatory and Calar Alto are performed and cataloged separately.
[5] The Astronomical Institute Ondřejov operates a total of 10 stations throughout Czech and Slovakia.
[6] Three mountings with 6 cameras each; cameras have adjacent FOV with small overlap.
[7] The Crimean site hosts two different institutions: The Crimean Astrophysical Observatory (double astrograph) and the Observatory of the Sternberg Astronomical Institute Moscow. The instruments are identical.

Table 2: Simultaneous plates for BATSE GRBs of the first two years

GRB	Time (UT)	Location (RA, DEC)	No. of simultaneous plates[1]	limiting magnitude mag_{pg} (for 1s flash)	Peak flux F_γ (erg/cm^2/s) 50-300 keV 1024ms	Limit[2] for F_γ/F_{opt}
910505	20:15:18	179°,33°	O, 4	2-3	2.2e-07	0.8
910809C	00:58:27	236°,38°	O, 3	2.5	1.8e-07	0.4
910814	19:14:35	344°,29°	D, 1	4.5	1.6e-06	24
910829	22:39:44	61°,23°	O, 1	1.0	4.8e-08	0.03
910902	22:55:37	298°,-8°	O, 8	2-3	1.7e-07	0.6
911004	01:27:11	352°,59°	O, 8	3-4	4.4e-07	4.0
911025	21:25:16	337°,31°	O, 8	0.5	2.7e-07	0.1
911027	01:45:13	261°,78°	O, 8	1.0	8.1e-08	0.05
911127	04:22:10	271°,50°	O, 2	-0.5	1.8e-06	0.3
911129	17:59:53	44°,-3°	O, 4	2-3	7.4e-08	0.3
911209B	00:56:50	86°,10°	O, 3	1.5	1.0e-07	0.1
911228	17:22:08	279°,46°	O, 3	2.0	1.6e-07	0.2
920209B	21:40:46	287°,60°	O, 5	2-3	9.5e-08	0.3
920224B	21:38:02	201°,9°	O, 9	3.0	7.1e-08	0.3
920227B	20:50:02	200°,51°	O, 7	3-4	4.6e-07	4.2
920305B	01:00:44	268°,60°	O, 5	1-2	2.5e-07	0.4
920505	21:54:51	42°,74°	O, 6	1-4	1.9e-07	1.7
920525	00:41:46	224°,33°	O, 5	3.0	[3]	
920530	22:59:55	165°,21°	O, 5	0.0	1.0e-07	0.02
920701	19:38:19	309°,30°	O, 6	2.0	1.7e-06	2.5
920730	21:31:33	242°,-7°	Sp, 1	4.5	6.1e-07	9.1
920806	18:9:17	286°,-10°	O, 4	0.5	2.1e-07	0.08
920925	21:45:16	155°,69°	O, 8	2.5	5.4e-07	1.2
920928	17:51:18	324°,20°	O, 5	2.0	[3]	
921101	18:3:55	329°,48°	O, 5	2.0	2.0e-06	3.0
921217	19:34:35	90°,19°	O, 7	2.0	1.1e-07	0.2
930131	21:11:11	77°,-3°	O, 4	-2.0	8.3e-08	0.003
930203	01:02:51	100°,87°	O, 6	2.5	7.6e-08	1.8
930214	04:08:17	260°,19°	O, 6	1.0	3.9e-07	0.2

[1] The letters indicate the observatory: D=Dushanbe, O=Ondřejov, Sp=Sonneberg patrol

[2] In the case of several simultaneous plates the ratio F_γ/F_{opt} is calculated for the deepest plate. Note that F_γ/F_{opt} is a true flux ratio (peak γ-ray flux in erg/cm^2/s divided by the optical flux limit in the same units) rather than the usually given^2 "flux" ratio L_γ/L_{opt}, which is really a fluence ratio, i.e. it depends on the burst duration (T_{90}) and the exposure time of the plate. F_γ/F_{opt} is always lower than L_γ/L_{opt} if the burst duration is larger than 1s (cf. for example $L_\gamma/L_{opt} = 1600$ for GB 910814).

[3] F_γ not available because events cannot be processed due to some missing data types.

Table 3: Selected near-simultaneous (± 1 day) plates for BATSE GRBs

GRB	Time (UT)	Location RA, DEC	Observatory[1]; Plate center (RA, DEC)	Time between GRB onset and exposure[2]	mag_{lim} (1s flash)	percentage of error box covered
910518	03:25:33	218°,58°	H; 210°,62°	-2.6 h	13.0	15
*910601	19:22:16	309°,32°	Sp; 300°,40°	+2.5 h	6.0	100
				+2.5 h	6.5	100
910718D	11: 9:17	305°,-45°	U; 295°,-45°	+25.1 h	11.5	25
*910902	22:55:37	298°,-8°	Sp; 300°,0°	-2.2 h	5.5	95
910905	23:48:55	0°,-83°	U; 326°,-84°	-11.0 h	14.5	40
910916	16:31:50	23°,-33°	U; 28°,-25°	-24.6 h	10.0	10
				-23.4 h	13.0	10
*911007	15:32:11	312°,44°	Sp; 315°,40°	+ 5.6 h	7.0	100
911027	01:45:13	261°,78°	Sp; 300°,80°	+16.0 h	5.5	60
*911129	17:59:53	44°,-3°	Sp; 45°,0°	+3.4 h	5.0	100
			Sp; 30°,0°	+2.8 h	5.0	20
911204	17:21:42	69°,-24°	U; 74°,-20°	-1.8 h	11.5	40
				-2.9 h	13.0	40
*911209B	00:56:50	86°,10°	Sp; 75°,20°	+19.2 h	6.5	40
			Sp; 90°,20°	+21.5 h	7.0	50
			U; 95°,15°	-8.8 h	16.0	10
*920121	21:56:51	43°,32°	Sp; 30°,40°	-3.4 h	6.0	30
920128	03:11:29	104°,11°	Sp; 105°,20°	16.8 h	5.0	50
				18.9 h	5.0	50
*920227C	17:46:18	239°,12°	Sp; 240°,20°	+7.4 h	5.5	95
*920227B	20:50:02	200°,51°	Sp; 195°,40°	+3.2 h	6.0	60
			Sp; 210°,40°	+ 2.1 h	6.0	60
920302B	00:14:44	122°,-64°	U; 130°,-70°	-12.3 h	15.0	5
*920305B	01: 0:44	268°,60°	Sp; 270°,60°	+1.6 h	5.5	100
*920429	18:59:34	189°,-20°	Od; 187°,-23°	+58 min	7.0	100
*920525	00:41:46	224°,33°	Sp; 210°,40°	-2.1 h	6.0	50
920701	19:38:19	309°,30°	Sp; 300°,20°	-19.2 h	6.0	60
			Sp; 300°,40°	-19.2 h	6.0	60
*920804	19:53:21	284°,48°	Sp; 270°,60°	-21.9 h	5.5	40
			Sa; 284°,54°	+3.2 h	10.0	30
920806	18: 9:17	286°,-10°	Sp; 285°,0°	-17.5 h	5.0	60
920925	21:45:16	155°,69°	Sp; 150°,80°	+22.9 h	6.0	60
*920928	17:51:18	324°,20°	Sp; 330°,20°	+2.2 h	6.0	60
*921217	19:34:35	90°,19°	Sp; 90°,20°	+2.9 h	3.0	100
*930131	18:57:11	181°,-7°	Od; 189°,3°	+6.5 h	6.5	100

Note: Plates marked with an asterisk have been investigated by blink comparison.
[1] The letters indicate the observatory: H=Hamburg, Od=Odessa, Sa=Sonneberg astrograph, Sp=Sonneberg patrol, U=UK Schmidt
[2] The shortest time is given corresponding to the start of exposure for a plate taken after a GRB, and corresponding to the end of exposure for a plate taken before a GRB. Typical exposure times are 30–60 min.

GAMMA RAY BURSTS AT OPTICAL WAVELENGTHS: FIRST OPTICALLY IDENTIFIED GRB?

R. Hudec and J. Soldán
Astronomical Institute, 251 65 Ondřejov, Czech Republic

ABSTRACT

We outline the recent activities in the analyses of GRB in optical wavelengths at the Astronomical Institute of the Czech Academy of Sciences. We summarize the recent results in the four main directions of these investigations, namely (i) archival searches, (ii) simultaneous searches using the BATSE GRO data, (iii) immediately follow-up searches as well as (iv) CCD optical transient monitoring.

INTRODUCTION

The physical nature of the gamma ray bursts of cosmic origin remains to be unclear. Their detection at optical wavelengths could significantly help to solve the question of their nature: (1) The detected energy ratios between gamma rays and optical light (or limits) are important parameters necessary to select among the theoretical models available. (2) The optical observations are usually able to provide a much better accuracy of the position as is possible from the observation in gamma rays (usually not better than 5 degrees). (3) This fine position should allows very deep and detailed surveys of the position in order to look for quiescent counterparts and, consequently, to understand the physical nature of the objects which are responsible for the bursts.

ARCHIVAL SEARCHES : TWO CONFIRMED OPTICAL TRANSIENTS, FIRST TWO OPTICALLY IDENTIFIED GRB?

We present the evidence of the reality of two bright optical transient (OT) images found on archival astronomical plates taken in 1905 and 1928 at the Harvard College Observatory (HCO). The OT1928 image has been found by B. Schaefer and suggested to be a flashing optical counterpart to GRB781119[4] but has been heavily debated in previous years.[3,5,6,7] For these reasons, the plate was carefully investigated in our Institute with the results as given here. The OT1905 is a recent new discovery.[2] We conclude that the reality of OTs is confirmed now. However, more identifications are needed to confirm first possible identifications of OTs with GRB definitely. This is due to still relatively weak positional coincidence resulting mainly from (still) relatively inaccurate GRB positions.

OT1905 AND GRB910219

This is the first object detected in a very extended study in which archival plates from world's largest plate collections were analysed for particular position. Altogether 9104 photographic plates from Harvard, Sonneberg, Ondřejov and Bamberg Observatories representing 2.06 yr (18 046 hr) of monitoring time were analysed for one particular GRB position (of GRB910219 detected by WATCH experiment and also by the IPN network). This is the optically best analysed error box of any particular GRB. The OT image has been found inside of the WATCH GRB error box, on the plate taken on May 10, 1905. The magnitude of the object was determined to be 6.4 mag for the full plate exposure time i.e. 5.5 mag above the plate limit. This brightness has allowed a very detailed analysis by different methods. First time among archival searches, we were able to confirm the object's reality and to exclude definitely plate fault origin: (1) The object cannot be a plate fault due to the presence of the image distortion and a foggy halo like other bright stars on the plate. Further, the analysis using reflected light microscopy confirms the similarity to real star images. (2) The object cannot be a satellite glint or aircraft flash–the plate was taken in 1905 (3) The object cannot be a double exposure–no similar object on the plate (4) The object cannot be a minor planet–no trailing, no image on the plate taken one day before, position more than 60 deg from the ecliptic (5) The object cannot be a variable star, novae, dwarf novae or flare star–the light amplitude greater than 15 mag on a very brief time scale (6) The possibility that the image was caused by a head-on meteor is highly improbable. Recent estimations[12] indicate the probability of meteor event analogous to observed OT below 10^{-8} inside of searched area.

The parameters of the OT image are as follows: (1) The magnitude at least 5.8 (or brighter: estimated 1-s magnitude is –0.8) (2) The duration below 35 min.– image not trailed while real stars are trailed by inaccurate guiding (3) The extremely colored light (either blue or red)-the image distortion is analogous to those of stars more distant from the plate centre (4) The light amplitude more than 15 mag. (5) The position inside of WATCH GRB error box (45 arcmin radius, 90 % confidence.) (6) The OT error box contains a faint blue candidate (B=21.01, B-V=0.40, U-B=-0.77). (7) This faint quiet candidate is QSO. Are the QSO sources of OT (and GRB) ? For more details see paper by Vrba et al. presented in this volume.

OT1928 AND GRB781119

We have analysed this image detected by Schaefer[4] carefully and have proved that the image is affected by the image distortion caused by the objective, hence definitely cannot be classified as an emulsion defect. The parameters of the OT image are as follows:[1,4] (1) The magnitude 8.7 or brighter (eg. 3.0 for the duration of 1 sec) (2) The duration below 10 min. (3) The extremely colored light (very blue or very red) (4) The position inside of the error box of GRB 781119 ($\sim$ 10 arcmin^2 area, 90 % confidence). (5) The error box contains a faint blue object[4] with B=25.9, B-V $\leq$ 0.0. (6) Is this object a QSO?

SIMULTANEOUS SEARCHES USING GRO BATSE DATA

The Ondřejov meteor network (12 fish–eye cameras at 10 stations) represents the only possibility to get really simultaneous optical data so far (the probability to get simultaneous record is $\sim 10^{-2}$ while only $\sim 10^{-3}$ to $\sim 10^{-7}$ for other programs, see Greiner et al. in this volume). We have analysed 30 GRB positions on simultaneously taken plates so far. The typical limiting magnitudes were between 7 and 11 for stars and between 0 and 3.5 for an 1–sec flash. No simultaneous optical triggers were recorded so far. However: (1) The device is red–sensitive with a very limited sensitivity for light below 400 nm (2) All investigated GRB were faint (3) The photographic emulsion is not suitable for brief flashes of 1–sec or even less (4) The physical processes of GRB phenomena are completely unknown, hence significant delays between gamma rays and optical emissions cannot be entirely ruled out. Significant delays were already observed in the brightenings of the "steady" gamma ray and optical emission of QSO0836+710.[11] The physical processes may be related taking the recent possible optical identification of OT (and possibly GRB) with QSO into account.

Nevertheless, OT images have been already detected which may be related to GRB other than recorded by GRO–BATSE, e.g. OT890802 correlated in time with GRB890802 detected by the GINGA satellite or OT910114 detected by two independent observations and methods.[8]

IMMEDIATELY FOLLOW–UP OPTICAL SEARCHES

Three observatories in the Czech Republic (Ondřejov, Klet' and Úpice) and one in Germany (Sonneberg) participate in the GRO–BATSE and COMPTEL burst alert program. The CCD and photographic telescopes are sensitive enough to detect objects up to 18 mag over a field of view of up to 5 deg radius. No real triggers were detected so far, however the program is just at the beginning with four events followed (SGR 1900+14, GRB930720, GRB931031 and GRB931103) and minimum delay between gamma ray event and deep optical data of 28 hours.

OPTICAL TRANSIENT MONITOR AND OPTICAL BURST NETWORK

It is evident that very strong evidence exists that there are optical transients, perhabs very colored and very brief, on the sky. It is however also evident, that the currently avaliable devices and networks are not suitable to detect , to classify and to verify these OT phenomena by a reliable way. This is why we develop an inexpensive but reliable system which should be able to provide the monitoring necessary. The idea is to concentrate on brigher triggers but to prove them definitely.

A double CCD camera with wide field lenses will monitor 13 % of the full sky hemisphere for optical transient phenomena. Two such systems are planed, separated by 100 km to be able to eliminate all effects in the earth atmosphere and near space. If succesfull, then more such stations could be operated (Worldwide Optical Burst Network) in order to get better sky and time coverage. The device represents a CCD wide field camera with lens of brief focal length and with sophisticated system to exclude all false triggers.[9,10] **The previous experiments in this field have shown that the main problem of these investigations is the high noise to signal ratio and hence it is rather important to be able to eliminate reasonably all background and false triggers.**

We suggest to develop and to run **a double CCD camera with two independent paths equipped with different filters and to develop a special software excluding false triggers and triggers caused by satellites.** The basic parameters of the proposed device are summarized as follows: (1) Lens: wide–field f = 6 mm, 1:1.6 (2) Focal detector : CCD camera ST-6 SBIG. (3) Area of the focal detector : 8.63 x 6.53 mm, 375 x 242 pixels Pixel size: 23 x 27 microns (4) Field of view: 47 x 55 deg^2 or 2606 deg^2 (12.6 % of the full sky hemisphere) (5) Angular resolution: 0.1 deg. (6) No. of optical paths in one device: 2. (7) Limiting magnitude: 5 (for 1s–flashes, 10 for 100s–exposures)

Alternatively, other lenses and/or CCD arrays may be used, resulting in different parameters. **The goal is to develop a very wide–field monitoring of the sky with still reasonable quality and parameters.**

The experiment assumes the development of a new automatic working system for short optical light burst detections using the Texas Instruments Digital Signal Processor (DSP) **TMS320C40** and based on the clasical PC computer as a host computer. Two or (better) four CCD cameras will be used in one station with minimum of two stations separated by at least 100 km to eliminate atmospheric and near space events.

The expected basic features of the system are as follows: (1) The background and the false triggers elimination (2) The real time data acquisition (3) The detector electronics false triggers elimination (4) The aircraft and satellite elimination.

References

[1] Hudec R., Pravec P., Borovička J. and Dědoch A.,1993, A&A, accepted.

[2] Hudec R., Pravec P., Borovička J., 1993, A&A, accepted.

[3] Greiner J.,1992,A&A 264, 121.

[4] Schaefer B., 1981, Nature 294, 722.

[5] Schaefer B., 1990, ApJ 364, 590.

[6] Schaefer B., 1992, in GRB: Observations, Analyses and Theories, Eds. C. Ho, R. Epstein and E. Fenimore, Cambridge University Press, p. 133.

[7] Zytkow A., 1990, A&A 264, 121.

[8] Hudec R., 1993, A&AS 97, 49.

[9] Hudec R. and Soldán J., ApJ, in press (paper presented at the INTEGRAL Workshop, Les Diablerets, Feb 1993).

[10] Hudec R. and Soldán J., 1993b, paper presented at the IAU Symposium No. 161 Astronomy from Wide Field Imaging, Potsdam, Aug 1993.

[11] von Linde J. et al., 1993, A&A 267, L23.

[12] Varady M., 1994, in preparation.

FIRST RESULTS OF THE BATSE/COMPTEL/NMSU RAPID BURST RESPONSE CAMPAIGN

R. M. Kippen, A. Connors, J. Macri, M. McConnell, J. Ryan
Space Science Center, University of New Hampshire, Durham, NH

W. Collmar, J. Greiner, V. Schönfelder, M. Varendorff
Max-Planck Institut für Extraterrische Physik, Garching, Germany

G. J. Fishman, C. Meegan
Space Sciences Laboratory, NASA/Marshall Space Flight Center, Huntsville, AL

C. Kouveliotou
Universities Space Research Association, Huntsville, AL

B. McNamara, T. Harrison
Astronomy Department, New Mexico State University, Las Cruces, NM

W. Hermsen, L. Kuiper
SRON-Leiden, Leiden, The Netherlands

K. Bennett, L. Hanlon, C. Winkler
Astrophysics Division, ESTEC, Noordwijk, The Netherlands

ABSTRACT

The Imaging Compton Telescope (COMPTEL) onboard the Compton Gamma Ray Observatory regularly observes gamma-ray bursts which occur inside the instrument's ∼1 sr field-of-view. COMPTEL images bursts in the 0.75-30 MeV energy range with a typical location accuracy of 1-3 degrees, depending on burst strength, position, duration and spectrum. COMPTEL's imaging capability has been exploited in order to search for fading gamma-ray burst counterparts at other wavelengths through the establishment of a BATSE/COMPTEL/NMSU rapid burst response campaign. This campaign utilizes near real-time identification and preliminary burst location by BATSE, accelerated COMPTEL imaging, and a world-wide network of observers to search COMPTEL error boxes as quickly as possible. Timely, deep searches for lingering counterpart emission of several bursts per year are the realized goal of this campaign. During its first year of operation, the rapid response program has been successfully applied to two strong bursts: GRB 930131 and GRB 930309. These bursts were imaged in record time only hours after their occurrence. Subsequently, several observations were made at radio and optical observatories world-wide.

RAPID BURST RESPONSE

In response to the lack of identification of any gamma-ray burst (GRB) counterparts at other wavelengths, a unique search campaign has been implemented by a collaboration of optical and gamma-ray observers. This "Rapid

Burst Response Campaign" searches for fading GRB counterpart emission by performing deep scans of gamma-ray error boxes as soon after the burst occurrence as possible using existing instrumentation. This is realized only through the highly coordinated use of two instruments on the Compton Gamma Ray Observatory (BATSE and COMPTEL), and a world-wide network of wide-field, ground-based multiwavelength observers. By using BATSE to quickly identify bursts which COMPTEL can image, we are able to perform deep optical/radio scans of degree-sized GRB error boxes within hours. The Rapid Burst Response plan schematically illustrated in figure 1 has been explained in more detail elsewhere.[1]

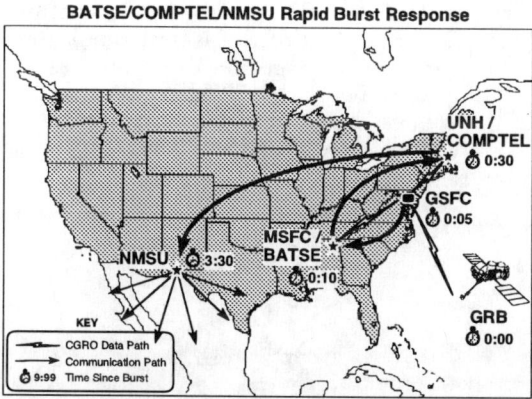

Figure 1. Schematic view of Rapid Burst Response operations.

Rapid response action is initiated when special BATSE count-rate thresholds are triggered at the onset of a burst. These thresholds have been set to indicate only those bursts which are of sufficient intensity to be detectable by COMPTEL. BATSE threshold data are continuously monitored in real-time at Goddard Space Flight Center (GSFC) and triggers are promptly communicated to the BATSE team at Marshall Space Flight Center (MSFC). The BATSE team provides a rough burst direction used to determine if the burst is in the ~1 sr COMPTEL field of view and if it is of cosmic origin. Burst triggers satisfying these criteria are communicated to the University of New Hampshire (UNH) where readily available COMPTEL data are processed and analyzed in an accelerated manner. The result of this analysis is a COMPTEL image (0.75-30 MeV) of the burst. The burst direction information from the COMPTEL image (i.e. GRB error box) is then communicated to New Mexico State University (NMSU) where the counterpart search is coordinated using a world-wide network of wide-field radio and optical observatories. Under optimum circumstances, this entire plan can be executed in ~3.5 hours from the time of the burst occurrence.

The BATSE Rapid Response thresholds were initiated in July 1992, however it was not until several months later that reliable procedures and communications pathways were defined and established between all the network participants. Since the Rapid Response Campaign was initiated only 5 bursts were intense enough to exceed the BATSE thresholds and trigger Rapid Response action, although subsequently an additional 6 bursts have been imaged from this same period. It was not until GRB 930131 that a fully reliable response plan was in place and could be exercised. Two months later, the plan was again successfully applied to GRB 930309. These two successful applications have resulted in degree-sized GRB locations in record time.

GRB 930131

The gamma-ray burst on 31 January 1993 (alias "The Super Bowl Burst") was the most intense burst yet observed by BATSE.[2] It was a unique event in terms of its intensity, short time-scale variation and hardness.[3,4,5] It was also the first successful application of the Rapid Response procedure. Upon notification of the burst by BATSE, COMPTEL data were used to determine the source of emission to an accuracy better than $2°$ (1σ error radius) within 6.5 hours of the burst onset. Being the first real trial of the Rapid Burst Response campaign, several minor problems were encountered, resulting in a less than optimum response time. Observations of the burst locale where made by several optical and radio observatories, the soonest being only 11 hours after the burst. Although no obvious fading counterparts were identified, several interesting objects were observed, prompting further study.[6]

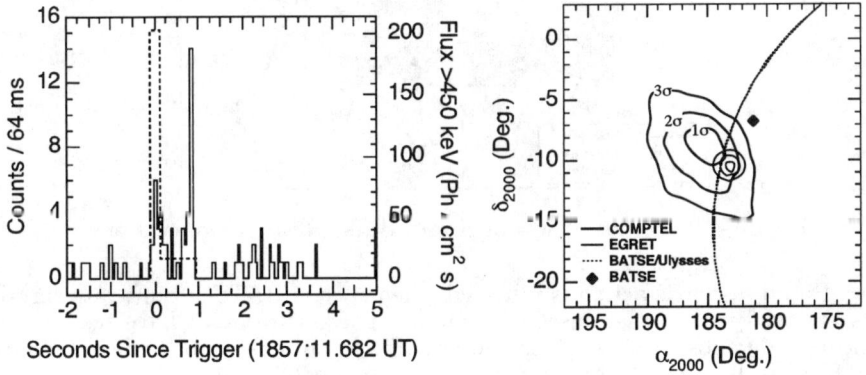

Figure 2. COMPTEL telescope lightcurve and image of GRB 930131.

In measuring bursts, COMPTEL uses two independent modes of operation: a single detector "burst" mode and a double scatter "telescope" mode.[7] Here, we are concerned primarily with the imaging "telescope" data. In the COMPTEL telescope data (0.75-30 MeV), GRB 930131 was intense and short-lived with the majority of significant emission occurring in a ~ 1 s interval. The history of time-tagged telescope events in 64 ms bins is shown in figure 2. Because this burst was intense, the COMPTEL data suffers from severe deadtime effects ($\sim 12\%$ live-time over the ~ 1 s of intense emission). The complex evolution of the burst with time can be seen in the two-bin histogram (dashed line) in figure 2. This represents the burst emission as measured by a single COMPTEL detector, largely unaffected by deadtime effects. The single detector measurement shows that the telescope does not register much of the burst flux due to instrument deadtime.

To determine the location of GRB 930131, a maximum-likelihood technique was employed which places quantitative statistical constraints on the burst source position.[8] The maximum-likelihood ratio skymap for GRB 930131 (figure 2) was generated using only 28 events selected from the ~ 1 s of significant burst emission. The high instrument deadtime and relatively short burst duration result in few telescope events for imaging, thus the statistical uncertainty in the location (±1.5° at the 1σ confidence level) is greater that of other strong bursts measured by COMPTEL (typically ~1°). From figure 2, it is clear that the COMPTEL-derived burst position is consistent with the BATSE[2] and EGRET[5] localizations as well as the BATSE/Ulysses IPN triangulation annulus.[9] The best COMPTEL position along this annulus is α_{2000}= 12h14m, δ_{2000}= -9°41' (±48' (1σ) along the annulus).

GRB 930309

The second successful application of the Rapid Response campaign occurred on 9 March 1993. This burst was weaker, softer and of longer duration than the Super Bowl Burst, however it was of sufficient intensity to trigger the BATSE thresholds initiating Rapid Response. Following prompt notification from BATSE, a COMPTEL image was obtained less than 4.5 hours after the start of the burst. Unfortunately, the burst location was close to the sun, making the field poorly observable to optical instruments. Radio observations were made over the next several days by Westerbork, Owens Valley and the VLA observatories.[10,11]

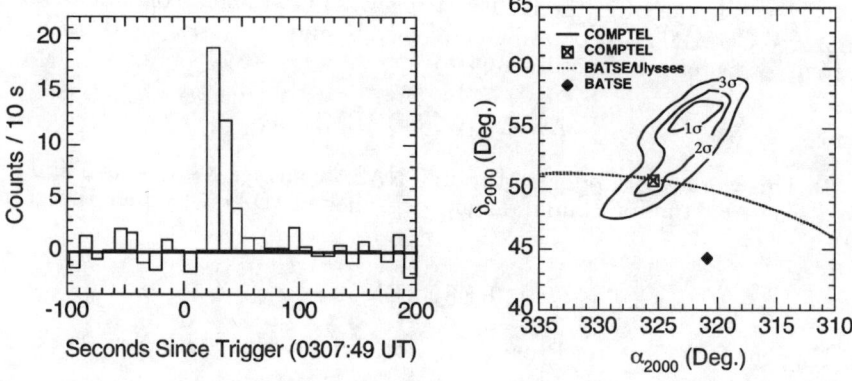

Figure 3. COMPTEL telescope lightcurve and image of GRB 930309.

An interesting candidate object was identified by Westerbork very close to the most likely COMPTEL-derived position[12]. It exhibited a declining flux over time as might be expected from a fading GRB counterpart. Unfortunately, the position of this radio source was inconsistent with the BATSE/Ulysses IPN triangulation annulus[13] as revealed several days later. The source candidate was later determined to be a variable double lobed radio galaxy.

Since GRB 930309 was of moderate intensity, the COMPTEL data does not suffer from significant deadtime effects (~95% live-time for the duration of the burst), however the background during this burst cannot be neglected. To improve the signal to noise ratio (S/N), the intensity-time profile of telescope data shown in figure 3 includes only events from within 6° of the derived burst position. This data selection effectively removes background while retaining source events. To correct data for the remaining background, binned events from 15 orbits prior to the burst were used as a background estimate. COMPTEL measures significant emission starting well after the BATSE trigger and lasting for ~25 seconds before returning to background level.

The maximum-likelihood ratio skymap of GRB 930309 (figure 3) was calculated using 69 events (source plus background) from the most intense 22 seconds of the burst. Since this burst was quite far from the COMPTEL pointing axis (~30°), the location contours are azimuthally elongated. This is an artifact of COMPTEL's off-axis response. Thus, the systematic and statistical location uncertainties combine to yield a burst position accurate only to about 5° RMS. The BATSE/Ulysses IPN triangulation annulus[13] falls within the 2σ confidence contour of the COMPTEL skymap. The best COMPTEL position along this annulus is $\alpha_{2000}=$ 21h42m, $\delta_{2000}=$ 50°47' ($\pm$30' (1σ) along the annulus).

CONCLUSION

GRB 930131 and GRB 930309 have been imaged by COMPTEL in record time, prompting rapid optical and radio follow-up observations. With two successful applications of the Rapid Burst Response Campaign, we have shown that deep, time-correlated searches for gamma-ray burst counterparts are possible using existing instrumentation. It is our intent to continue to improve this campaign, resulting in burst locations distributed to the observer network as quickly as ~1 hour after the BATSE trigger. This will be accomplished through improved COMPTEL processing and analysis, and the use of the BACODINE[14] system for faster burst notification.

ACKNOWLEDGEMENTS

This work was supported through NASA contract NAS5-26645 and by the Deutsche Agentur für Raumfahrtangelegenheiten (DARA) under the grant 50 QV 90968.

REFERENCES

1. Kippen, R. M., et al., Proc. Compton Symposium (AIP, New York, 1993), p. 823.
2. Kouveliotou, et al., Ap. J. (Letters), in press (1993).
3. Ryan, J. et al., Ap. J. (Letters), in press (1993).
4. Kippen, R. M. et al., Proc. 23rd ICRC, Calgary **1**, 85 (1993).
5. Sommer, M. et al., Ap. J. (Letters), in press (1993).
6. Schaefer, B. E. et al., Ap. J. (Letters), in press (1993).
7. Schönfelder, V. et al., Ap. J. Suppl. **86**, 57 (1993).
8. de Boer, H. et al., Data Analysis in Astronomy IV (Plenum Press, 1992).
9. Cline, T. L., Barthelmy, S. and Palmer, D., IAU Circ. **5703**, (1993).
10. Bennett, K. et al., IAU Circ. **5749**, (1993).
11. Harrison, T. E. and McNamara, B., IAU Circ. **5755**, (1993).
12. Hanlon, L. et al., Proc. 27th ESLAB Symposium (Kluwer Academic Publishers, 1993).
13. Hurley, K., private communication, (1993).
14. Barthelmy, S. D., et al., these proceedings (AIP, New York, 1993).

SEARCHES FOR OPTICAL COUNTERPARTS OF BATSE GAMMA-RAY BURSTS

Hans A. Krimm, Roland K. Vanderspek, and George R. Ricker
Center for Space Research and Department of Physics,
Massachusetts Institute of Technology, Cambridge, MA 02139 U.S.A.

ABSTRACT

The Explosive Transient Camera (ETC) is a wide-field CCD camera system capable of detecting short (1-10 s) celestial optical flashes as faint as m $\sim$ 10 over a field-of-view of 0.75 steradians between -15° and +45° declination. This sensitivity would allow observations down to $L_\gamma/L_{opt} > 10^5$ of the optical counterpart to a gamma-ray burst as intense as that of 19 Nov 1978, for which a counterpart was reported by Shaefer[1]. The ETC has been operating automatically under computer control since January 1991. Since the launch of the Compton Gamma Ray Observatory, it has been capable of observing an optical flash coincident with a gamma-ray burst (GRB) detected by the Burst and Transient Spectroscopy Experiment (BATSE). All GRBs which occur during an ETC observation are examined for spatial overlap between the BATSE 68% confidence positional error box and the ETC field of view. Between April 1991 and May 1993, there were four cases of at least partial BATSE/ETC spatial overlap during an ETC observation. In each case upper limits are placed on the optical-to-gamma-ray flux ratio. The expected rate of future BATSE/ETC coincident events is also discussed.

INTRODUCTION

The Explosive Transient Camera (ETC) has been developed primarily to establish the existence of optical counterparts for gamma-ray bursts (GRBs). An optical flash detected by the ETC in coincidence with a BATSE gamma-ray burst detection could produce a localization of the burst position to $\sim$10-20 arc seconds, as opposed to the $\sim$2.5 degree localization possible with BATSE alone[2]. The ETC is able to search for variable sources on short time scales (1-10 s) as well as on time scales ranging from hours to years.

The ETC is housed in a dedicated building on the summit ridge at Kitt Peak National Observatory, U.S.A. and is described in a number of references[3,4,5,6]. It is a fully automated instrument, and normally operates entirely by computer control without human intervention. Data collected each night is automatically transferred to Massachusetts Institute of Technology (MIT) for further analysis.

The ETC operates every clear, dark night and is sensitive to optical transients which may or may not be correlated with detected GRBs. It does not require a trigger from BATSE or any other experiment, but searches for rapid optical flashes by comparing successive exposures of duration t, where t is adjustable and normally 4 - 5 seconds. This procedure is described in detail elsewhere[3].

SEARCH FOR ETC/BATSE OVERLAPS

The ETC observations are correlated with BATSE bursts *retrospectively* in the following way:
1. GRB times are matched with ETC observation times.
2. Coordinates of temporally coincident GRB events are compared with ETC fields-of-view.
3. ETC flash events during relevant observations are studied in detail to determine their origin. As seen in Figure 1, the primary source of false ETC triggers is sunlight reflected from earth-orbiting satellites.

If no true astronomical transient events are found in coincidence with a GRB:
1. The limiting detectable optical fluence, $L_{opt}(min)$ is calculated for the observation.
2. $L_{opt}(min)$ is compared with burst fluence L_γ to derive a limit for L_{opt}/L_γ.
3. Images of the event region before and after the burst are compared to search for slowly rising transients.

If a coincident astronomical transient event is found:
1. The time and location of the optical flash are compared with the GRB.
2. The brightness of the flash L_{opt} is calculated.
3. The ratio L_{opt}/L_γ is calculated from L_{opt} and the GRB fluence.

The probability that the ETC is observing the same part of the sky in which BATSE detects a GRB is proportional to the number of GRBs in the entire sky N_{GRB}, the observing efficiencies ϵ_{BATSE} and ϵ_{ETC} of BATSE and the ETC respectively, and the solid angle coverage Ω_{BATSE} and Ω_{ETC} of the two instruments:

$$N_{ETC-BATSE} = N_{GRB} \cdot (\Omega_{BATSE}/4\pi) \, \epsilon_{BATSE} \cdot (\Omega_{ETC}/4\pi) \, \epsilon_{ETC}. \qquad (1)$$

The BATSE observation rate is:

$$\{ N_{GRB} \cdot (\Omega_{BATSE}/4\pi) \, \epsilon_{BATSE} \} \simeq 300 \text{ yr}^{-1}. \qquad (2)$$

The effective solid angle θ_{BATSE} for ETC/BATSE overlaps depends on the physical solid angle coverage of the ETC, 0.73 steradians in its current configuration and on the size of the BATSE positional error circles. If a GRB is localized to a point outside the ETC field-of-view, an associated optical flash may still be detected if the BATSE localization error circle overlaps the ETC field-of-view to a significant extent. The increase to the effective solid angle is estimated by adding in quadrature the average BATSE systematic error, $4°$ [2] (68% confidence interval) and the average BATSE statistical error, $5.7°$ [7]. This leads to $\Omega_{ETC} \approx 1.22$ ster.

The observing efficiency of the ETC is the ratio of observing time to elapsed time. Between Apr. 1991 and Aug. 1993, the ETC observed for 1043 hr out of 20736 hr elapsed time (day and night), yielding an efficiency $\epsilon_{ETC} = 0.05$. Using the values for Ω_{ETC} and ϵ_{ETC} and Eq. (2) in Eq. (1) gives the estimate:

$$N_{ETC-CGRO} \text{ (predicted)} \simeq 1.5 \text{ yr}^{-1}, \qquad (3)$$

comparable to the rate at which overlaps were observed: $4/(2.33 \text{ yr}) = 1.5 \text{ yr}^{-1}$.

ANALYSIS OF ETC/BATSE OVERLAPS

Between April 1991 and August 1993, there were four cases in which a BATSE GRB occurred during an ETC observation and within or near an ETC field-of-view. No optical transients were detected during any of these observations. These cases are outlined in Table I along with the associated limits to the ratio of optical to gamma-ray fluence.

Table I: ETC-BATSE Overlaps from April 1991 - August 1993

Burst	GRB910616	GRB910702	GRB921108	GRB930203
Duration (T_{90})	1.54 sec	31.1 sec	0.4 sec	0.1 sec
Integrated prob. of ETC detection	0.17	0.72	0.20*	0.43
Flash $m_{lim}(V)$	8.0	6.6	5.9	5.8
Flash $L_{opt}(min)$ 12σ detection	2.3×10^{-9}	8.5×10^{-9}	1.6×10^{-8}	1.8×10^{-8}
GRB fluence L_γ (full burst) (50 - 300 KeV)	2.12×10^{-7}	7.74×10^{-7}	3.53×10^{-8}	3.25×10^{-8}
L_γ/L_{opt} (12σ)	> ~120	> ~125	> ~2	> ~2
GRB fluence L_γ (5 sec. exposure) (50 - 300 KeV)	2.12×10^{-7}	1.55×10^{-7}	3.53×10^{-8}	3.25×10^{-8}
L_γ/L_{opt} (12σ)	> ~120	> ~25	> ~2	> ~2

* Based on estimated size of BATSE error circle.
All fluences in erg/cm^2. The table entries are explained in the text.

The flash durations and fluences in Table I are from the BATSE catalog[7] for GRB910616 and GRB910702 and supplied by the BATSE team[8] for GRB921108 and GRB930203. The integrated probability of detection is an estimate of the likelihood that each burst actually occurred within the ETC field-of-view. It is calculated by convolving the BATSE error circle for each burst (approximated by a two-dimensional Gaussian distribution) with the ETC sky coverage at the time of the burst. Figure 1 shows this relationship graphically for each burst.

The limiting magnitude $m_{lim}(V)$ for a detectable one second flash during an exposure is determined for each ETC camera, based on the magnitude of the faintest observable field stars. The minimum detectable fluence $L_{opt}(min)$ is calculated from $m_{lim}(V)$. If the entire optical fluence of a burst $L_{opt} > L_{opt}(min)$

falls within a single t sec. ETC exposure, it can be detected. If, however, the burst lasts longer than t seconds, only that part of the burst fluence which rises during a single exposure would be detectable as a transient source. If the time

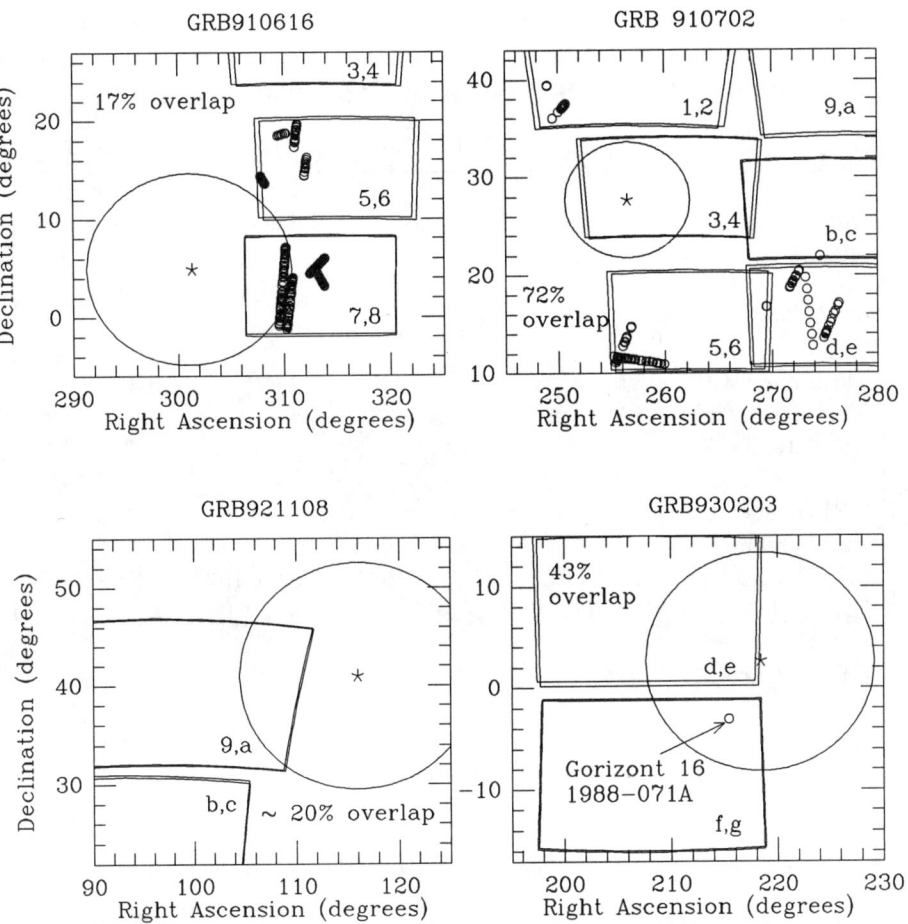

Figure 1. Views of the ETC fields-of-view during the four ETC/BATSE overlaps. The large circles represent the BATSE 68% error circle for each burst, with a star symbol at the best fit GRB location. The large trapezoids represent the fields-of-view of the numbered ETC cameras which are arranged in overlapping pairs. The small circles represent optical flashes within ~40 minutes of each burst. The colinearity of the flashes gives evidence that they can all be attributed to sunlight reflected from earth-orbiting satellites. The single flash seen ~ 35 minutes before GRB930203 is a reflection from a Russian satellite.

profile of the optical light from GRB910702 follows the gamma-ray light curve, an estimated $\sim 20\%$ of the fluence would fall during a single ETC exposure. This reduces the limit L_γ/L_{opt} for GRB910702 in the final row of Table I.

CONCLUSION AND FUTURE PLANS

To date, the ETC has detected no optical counterparts to BATSE GRBs, despite a sensitivity such that a counterpart to the 19 Nov. 1978 GRB for which Schaefer[1] reported an optical counterpart would be detected with $L_\gamma/L_{opt} > 10^5$. Limits for L_γ/L_{opt} are established for the 4 BATSE coincidences listed in Table I.

In addition to optical flash data, the ETC stores several full field images each night. Images taken before and after a burst event in the field-of-view are compared to search for transients rising too slowly to be detected during standard operations. A search carried out for the four burst overlaps gave negative results.

The ETC will continue to observe ~ 2.5 BATSE GRB error circles per year, rising to ~ 7.5 after a planned increase in the ETC field-of-view. The ETC will also receive BATSE GRB coordinates from the BACODINE network[9]. The ETC will be able to slew its cameras to any part of the observable sky within ~ 10 sec. This will increase the expected rate of ETC/BATSE overlaps to ~ 5 - 10 per year.

ACKNOWLEDGEMENTS

The ETC team is grateful for the assistance of C. Kouveliotou (USRA), G.J. Fishman (NASA/MSFC) and the BATSE science team. This work was supported by a NASA Compton GRO Guest Investigation under Grant NAGW-2089.

REFERENCES

1. B.E. Schaefer, Nature, **294**, 722, 1981.
2. M.N. Brock et al., in "Compton Gamma-Ray Observatory," St. Louis, Missouri, 1992, ed. M. Friedlander, N. Gehrels and D.J. Macomb, (AIP Conf. Proc.; **280**; New York, AIP), 709, 1993.
3. R.K. Vanderspek, H.A. Krimm, and G.R. Ricker, This Conference.
4. G.R. Ricker et al., in "High Energy Transients in Astrophysics," ed. S.E. Woosley (AIP Conf. Proc. **115**; New York, AIP), 669, 1984.
5. R.K. Vanderspek, G.R. Ricker, and J.P. Doty, in "Robotic Telescopes in the 1990's," ed. A.V. Filippenko, (Astronomical Society of the Pacific Conf. Proc. **34**; San Francisco), 123, 1992.
6. R.K. Vanderspek, J.P. Doty, and G.R. Ricker, in "Gamma-Ray Bursts," Huntsville, Alabama, 1991, ed. W. Paciesas and G.J. Fishman, (AIP Conf. Proc.; **265**; New York, AIP), 404, 1992.
7. Derived from the "The First BATSE Gamma-Ray Burst Catalog," G.J. Fishman et al., to be pub. in The Astrophysical Journal Supplement Series, 1994.
8. C. Kouveliotou, private communication.
9. S. Barthelmy et al., This Conference.

STELLAR FLARES AND GAMMA-RAY BURSTS

P. Li, K. Hurley
Space Sciences Laboratory
University of California, Berkeley
Berkeley, CA 94720

G. J. Fishman, C. Kouveliotou
ES-62, Marshall Space Flight Center-NASA
Huntsville, AL 35812

ABSTRACT

We have searched for gamma-ray bursts during stellar flares using Ulysses and BATSE/CGRO data. A total of five stellar flares were identified (two from AD Leo and three from AU Mic), but neither BATSE nor Ulysses observed any gamma-ray bursts which could be attributed to them. Using the BATSE trigger threshold, and the known distances to these flare stars, upper limits to the gamma-ray luminosity were obtained. The conditions under which stellar flare gamma-ray bursts could be detected by BATSE as weak events were studied. In particular, we found that if weak events (10^{-7} ergs cm^{-2} s^{-1}) are to be explained by flares as recently suggested, the stellar X-ray emission (>25 keV) must be comparable to its optical emission, and the ratio of L_x/L_{opt} must be higher than that for solar flares by at least 4 orders of magnitude. The stellar flare logN-logS and spatial distributions are studied and their implications for gamma-ray bursts are discussed.

INTRODUCTION

Recent observations from BATSE show that the gamma-ray burst spatial distribution is isotropic, with no concentration towards the galactic plane or center, and that the number of weak events is smaller than that expected from a homogeneous source distribution in a static Euclidean space[1,2]. These results suggest that we are at the center of the gamma-ray burst distribution. The source distance is unknown and a contentious issue at the present time.

Although we do not know what kind of objects cosmic gamma-ray bursts are associated with, we do know that the nearest star to us—the Sun—produces gamma-ray bursts when it flares. Impulsive solar flare gamma-ray bursts last from 10^{-3} s to 10^3 s and clearly have nonthermal spectra above 30 keV[3]. The maximum total energy released during the flare reaches $\sim 10^{32}$ ergs[4]. The energy released in the form of gamma-rays is insignificant in the total energy budget, with luminosity ratios[5] of $L_x/L_{tot} \sim 10^{-6}$, and $L_x/L_{opt} \sim L_x/L_{EUV} \sim 10^{-5}$. Since some temporal and spectral properties of cosmic gamma-ray bursts are similar to those of solar gamma-ray bursts[6,7,8,3], it is natural to inquire whether some cosmic gamma-ray bursts originate from flares on stars other than the sun.

The association between cosmic gamma-ray bursts and stellar flares has been studied on and off since the discovery of gamma-ray bursts[9,10]. It has been proposed that cosmic gamma-ray bursts might be flares on main sequence stars[11] (one of the very first theories published). More recently, based on observations of flares on dK/M stars, it has been proposed that some weak gamma-ray bursts observed by BATSE could be stellar flares[12]. However, most of these studies were theoretical and contained *ad hoc* assumptions. Although stellar flares have been observed in the radio, optical, EUV, and soft X-ray ranges, they have not been observed in gamma-rays > 25 keV[13]. Most of the observations have been in the optical, and, to date, only a few low sensitivity searches for time-and/or position- coincident events have been made[14,15].

SEARCHING FOR GAMMA-RAY EMISSION IN STELLAR FLARES

A search for stellar flares which occurred between the launch of Ulysses (November 1990) and present time has turned up a total of five events (Table I). The search is continuing and more are expected. The first three stellar flares were in the literature[16,17,18], while the last two were kindly provided by Dr. S. Hawley and her team of EUVE Guest Observers[19]. At the times around these flares, no gamma-ray bursts were observed either by BATSE or Ulysses from the directions of these stars, indicating that the gamma-ray emission from these flares must be below the trigger threshold for BATSE[20] (10^{-7} ergs cm^{-2} s^{-1}) and/or for Ulysses[21] (normally 10^{-6} ergs cm^{-2} s^{-1}, but less for short events). Since the flares may last for many Compton Observatory orbits, the BATSE duty cycle for observing a burst from them is ~50%. Therefore we used the Ulysses data, with its 4π sr, and >95% time coverage, to complement the BATSE data. The distances to AD Leo and AU Mic are 4.9 and 9.3 pc respectively. The gamma-ray flux F_x for these stellar flares must satisfy $F_x < F_c = 10^{-7}$ ergs cm^{-2} s^{-1}, where F_c is the trigger threshold. The upper limits to the luminosities L_c are 2.7×10^{32} ergs s^{-1} for the AD Leo flare and 9.7×10^{32} ergs s^{-1} for the AU Mic flare, assuming isotropic emission.

Table I Stellar Flare Observations

DATE	START TIME	DURATION	STAR	SP. TYPE	D (pc)	WAVE-LENGTH	INSTRU-MENT	TOTAL ENERGY (ergs)	L (ergs s^{-1})
91/09/03	4:56:10	3 s	AU Mic	dM1e	9.3	UV	HST	10^{30}	3.3×10^{29}
92/07/15	12:38:00	2 hr	AU Mic	dM1e	9.3	EUV	EUVE	3×10^{34}	4.2×10^{30}
92/09/09	12:46	1 hr	AU Mic	dM1e	9.3	OPT	AAT	10^{34}	2.5×10^{30}
93/03/02	~09:00	4 hr	AD Leo	dM3.5e	4.9	OPT/EUV	Lick/EUVE	6×10^{32}	4×10^{28}
93/03/03	~10:00	3 hr	AD Leo	dM3.5e	4.9	EUV	EUVE	10^{32}	9×10^{27}

The optical flares observed on dK/M dwarf stars (such as dM AD Leo and AU Mic) have many properties similar to those of solar flares[22,23,24]. If we assume the same luminosity ratio $L_x/L_{EUV} \sim 10^{-5}$ as for a solar flare, then the gamma-ray luminosity, estimated from EUV observations, is $\sim 10^{25}$ ergs s^{-1} for the very large AU Mic event of July 15 1992. This value is far below the critical luminosity L_c required for that flare to be observed by BATSE or Ulysses. Therefore we should not expect BATSE to observe these stellar flares if we believe that the properties of AD Leo or AU Mic flares are similar to those of solar flares. In other words, scaled-up solar-type flares on dK/M stars cannot be observed by BATSE or Ulyssses.

CONDITIONS REQUIRED TO TRIGGER BATSE

The results of the previous section indicate that the stellar flares in Table I were not powerful enough at gamma-ray energies to be observed by BATSE or Ulysses. However, we may be limited by the available observations. There are about 10^6 flare stars within 300 pc[13]. The actual number of stellar flares which occurred over the period of this study was probably much larger than the number recorded due to lack of coverage. The maximum optical luminosity is $\sim 10^{32}$ erg s^{-1} for late-type dK/M stellar flares (<25 pc)[13,23].

If $\eta = L_x/L_{opt}$ is the gamma-ray to optical luminosity ratio for stellar flares, then for stellar flare emission to be powerful enough to be observed by BATSE, η needs to satisfy

$$\eta > \left(\frac{F_c}{L_{opt}}\right) 4\pi d^2. \quad (1)$$

The nearest dK/M flare stars are at ~1 pc. Using the lowest distance and highest optical luminosity, a lower limit to η for dK/M stellar flares is $\eta_{min} > 0.1$, which is significantly higher than that of solar flares.

Late-type dK/M stars flare more frequently and their size distributions are well determined; they are power laws with $\alpha \sim 0.8^{23}$. If we take $\eta = 1$, the maximum luminosity in the gamma-ray range will be $(L_x)_{max} = 10^{32}$ erg s^{-1}. Their size distribution can be represented by

$$N(>L_x) = N_0 \left(\frac{L_x}{L_{x0}}\right)^{-\alpha} \quad (L_x)_{min} \le L_x \le (L_x)_{max}, \quad (2)$$

where N is the number of flares per year per star with luminosity greater that L_x. It has been found[23], from studies of $>10^4$ dK/M flares, that $N_0 \sim 20$ star^{-1}yr^{-1} for $L_{x0} = 10^{31}$ erg s^{-1} (equivalent to $L_{opt} = 10^{31}$ ergs s^{-1} assuming $\eta = 1$). The all-sky flare rate with flux larger than F_c can be calculated by differentiating equation (2) and integrating it over the range of allowable luminosities and flare star distribution volumes (assuming a homogeneous distribution); it is:

$$N(>F_c) \sim 400 \left(yr^{-1}\right) \left(\frac{n}{1pc^{-3}}\right). \tag{3}$$

The estimated flare star density n is[13] 0.01-0.1 pc^{-3}. Using an upper limit n=0.1 pc^{-3}, equation (3) gives N(>F$_c$)=40 yr^{-1}, which is 5% of the total gamma-ray burst rate inferred from BATSE[2] (~800 yr^{-1}). These flares must be within 3 pc to be observable with BATSE. As there are only a few known flare stars within 3 pc, identifying a gamma-ray burst from one of them would be relatively simple; however, these results are based on the uncertain assumption that η=1.

SUMMARY AND DISCUSSION

We have studied gamma-ray emission in stellar flares and its possible association with the gamma-ray bursts observed by BATSE and Ulysses. The major results are:
1) A total of five stellar flares have been identified to date from dM stars and no gamma-ray bursts were observed by BATSE or Ulysses during these flares.
2) Upper limits of $L_x < 2.7 \times 10^{32}$ ergs s^{-1} were found for AD Leo and $L_x < 9.7 \times 10^{32}$ ergs s^{-1} for AU Mic.
3) For a large dK/M flare to radiate enough power in gamma-rays to be observed by BATSE, the luminosity ratio L_x/L_{opt} (or L_x/L_{EUV}) must be greater than 0.1.
4) Stellar flares could account at most for 5% of the BATSE bursts if $L_x/L_{opt} =1$.

In order to be observed by BATSE or Ulysses, stellar flares must not only be very large (i.e., reach their maximum luminosities), but also satisfy $L_x/L_{opt}>0.1$. Such a constraint on η is extreme, since it is more than four orders of magnitude higher than that of solar flares. This condition implies that a simple scaling-up of solar flares to dG/K/M stellar flares cannot account for BATSE or Ulysses events. Moreover, if we take η=1, $L_x=(L_x)_{max}=10^{32}$ ergs s^{-1}, and assume that > 25 keV energetic photons are produced by energetic electrons through nonthermal bremsstrahlung, generally believed to be the production mechanism for >25 keV solar energetic emission[24], then the power in energetic electrons reaches 10^{37} ergs s^{-1}. Taking this number as an input to well-developed stellar flare models[25,26], the expected soft X-ray, EUV, and optical flares would be several orders of magnitude higher than those observed, contradicting the existing observations. However, since η is unknown, we cannot completely rule out the possibility of $L_x/L_{opt}=1$ for some unusual stellar flares.

ACKNOWLEDGEMENTS

We would like to thank Dr. S. Hawley and her team members for providing the unpublished AD Leo flare observational data. The Ulysses GRB experiment was constructed at the CESR with assistance from a grant from CNES, and at MPE under FRG Contracts 01 ON ZA/WRK 275/4-7.12 and 01 ON 88014. This project was supported by JPL contract 958056 and NASA grant NAG5-1560.

REFERENCES

1. G. J. Fishman et al., in Proc. of Gamma-Ray Obs. Sci Workshop, ed. W. N. Johnson, Ch. 2, 39 (1991).
2. C. A. Meegan et al., Nature, 355, 143 (1992).
3. B. R. Dennis, Sol. Phys., 118, 49 (1988).
4. S. T. Wu et al., in Energetic Phenomena on the Sun, eds. M. Kundu & B. Woodgate, NASA CP-2439, p. 5-i (1986).
5. H. S. Hudson, Sol. Phys., 133, 357 (1991).
6. K. Hurley, in Cosmic Gamma Rays, Neutrinos, and Related Astrophysics, eds. M. Shapiro & E. Wefel, pp. 337, 80 (1989).
7. K. Hurley, Ann. NY Acad. Sci., 357, 442 (1989)
8. K. Hurley, in X-ray Binaries, eds. W. Lewin & J. Van Paradijs, Cambridge Univerity Press (1993).
9. M. Ruderman, Ann. NY Acad. Sci., 262, 164 (1975).
10. J. C. Higdon and R. E. Lingenfelter, Ann. Rev. A&A, 28, 401 (1990).
11. F. W. Stecker and K. J. Frost, Nature Physical Science, 245, 70 (1973).
12. E. P. Liang and H. Li, A&A, 273, L53 (1993).
13. B. Haisch, K. T. Strong, and M. Rodonò, Ann. Rev. A&A, 29, 275 (1991).
14. H. Hudson and V. Tsikoudi, Nature Physical Science, 245, 88 (1973).
15. J. Pye and I. McHardy, MNRAS, 205, 875 (1983).
16. B. E. Woodgate et al., ApJ, 409, L49 (1992).
17. S. Cully et al., ApJ, 414, L49 (1993).
18. R. D. Robinson et al., ApJ, 414, 872 (1993).
19. S. Hawley et al. in preparation, Abstract in Vol. 25 BAAS, 182nd AAS Meeting, Berkeley (1993).
20. G. J. Fishman et al., in Gamma Ray Bursts, eds. W. S. Paciesas & G. J. Fishman, AIP Conf. Proc. 265, New York (1991)
21. K. Hurley et al., A&AS, 92, 401 (1992).
22. B. R. Pettersen, Sol. Phys. 121, 299 (1989).
23. N. I. Shakhovskaya, Sol. Phys. 121, 375 (1989).
24. E. Tandberg-Hanssen and A. G. Emslie, The Physics of Solar Flares, Cambridge University Press, Cambridge (1988).
25. G. H. Fisher and S. Hawley, ApJ, 357, 243 (1990).
26. C. C. Cheng and R. Pallavicini, ApJ, 357, 234 (1991).

X-RAY AND OPTICAL OBSERVATIONS OF THE COMPTEL ERROR BOX FOR GRB910601

Bernard McNamara, Thomas Harrison, Christina Williams
New Mexico State University, Las Cruces, NM 88003

ABSTRACT

In this paper we present a preliminary investigation of the COMPTEL and IPN error box of GRB910601. Our data consists of a 20 ksec PSPC ROSAT image and follow-up optical B and V CCD images. We find two X-ray sources located within a few arcminutes of the intersection of the IPN track and the COMPTEL error box. One of these objects, source #29, is the strongest X-ray source located within the two degree diameter ROSAT field. It is possibly associated with the star SAO 70303 or a bluer (?) object located slightly to the east of it. No obvious candidate was found for the other ROSAT detection.

INTRODUCTION

One method of searching for the origin of a cosmic gamma-ray burst (GRB) is to perform deep imaging of its error box at X-ray wavelengths. This method assumes that a GRB is related to the outburst phase of an object which emits energy more frequently at lower (i.e. X-ray) wavelengths. In this investigation our strategy was to deeply image GRB error boxes using ROSAT to identify X-ray sources within them. We then attempted to optically identify these sources based upon positional coincidences with blue objects or sources which are known X-ray emitters. Once optical candidates have been identified, further observations will be needed to determine the cause of the high energy emission. This paper reports on the first phase of our investigation of the error box of GRB910601, the identification of X-ray sources.

The COMPTEL 1.5 sigma error box and maximum likelihood position (x) along with the IPN are shown in Figure 1. The ROSAT X-ray sources are depicted as circles. Since typical ROSAT localizations are accurate to about 30 arcseconds the circles are much larger than the positional uncertainties. The six strongest ROSAT sources are represented as solid circles. These six sources are listed in Table 1. In Figure 2 we present optical images of the two X-ray fields (#29 and #30) which are located closest to the IPN track. The ROSAT position is marked along with its 30 arcsecond error circle.

DATA

The CCD images shown in this paper were taken with the New Mexico State University Blue Mesa Observatory 24 inch telescope using a 1024 × 1024 Tektronix chip. They reach a limiting magnitude of V near 20 and B near 19.5. North is at the top of each image and East is to the left. The circles have a radius of 30 arcseconds and reflect the positional accuracy of a moderate strength ROSAT detection. The ROSAT data consists of a 20.8 ksec image obtained with the PSPC. It was obtained UT 12–14 Nov 1992. The exposure times vary for each source from 9273 seconds (#29) to 20521 seconds (#14). The optical data was acquired during June 1993.

DISCUSSION

ROSAT Field #29: The bluest object in this ROSAT error circle is star 4. Its measured value of B-V = -0.12 is however very uncertain due to possible contamination from SAO 70303. The center of the ROSAT error circle is located within 20 arcseconds of SAO 70303. This star may also be responsible for the ROSAT detection. We plan to obtain spectra of objects 1 and 4 when they are next accessible to ground based observations. ROSAT data for this field consists of a total integration time of 9273 seconds. The X-ray flux was flagged as possibly variable and was the strongest of any source located within the field of the PSPC. As seen in Figure 1, this field is located within a few arcminutes of the IPN track.

ROSAT Field #30: This field is also located fairly close to the IPN track. Optically it contains three faint red objects, none of which appear particularly unusual. Our investigation does not reveal any obvious X-ray candidates for the ROSAT source. We intend to optically image this field to a deeper limiting magnitude when it next becomes accessible to ground based observations.

SUMMARY

This investigation has resulted in two ROSAT detections (sources #29 and #30) located close to the intersection of the IPN track and COMPTEL localization of GRB910601. These objects are the first X-ray sources detected in a relatively small gamma-ray burst error box (Boer et al. 1993, Astron Astrophy 277, 503). Field #29 contains the strongest X-ray signal within the entire ROSAT PSPC image. It is possibly associated with SAO 70303 or a bluer(?) object located approximately 15 arcseconds to the east of it. We do not have an identification for the object associated with ROSAT Field #30. We intend to more deeply image this field when it is next observable. Our current images extend down to about V=20. In addition to the above, we also find that the Barium star HD 196673 is located within 3 arcseconds of ROSAT source #4 on our PSPC frame but is far from the IPN track.

Table 1. Strong ROSAT Sources in the Field of GRB910601

Object	Cnts/100sec	Exp (sec)	S/N
29	2.17	9273	14
1	1.88	12861	15
2	1.59	16539	16
30	1.27	11275	11
24	1.17	12246	11
14	1.09	20521	14

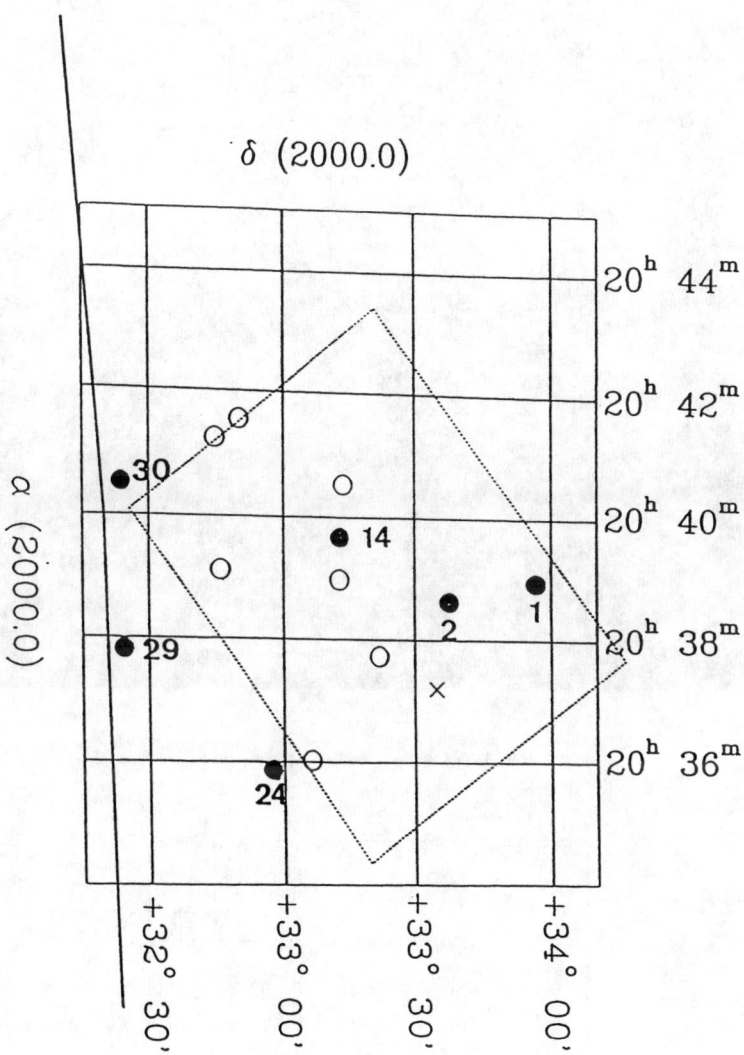

Figure 1

436 COMPTEL Error Box for GRB910601

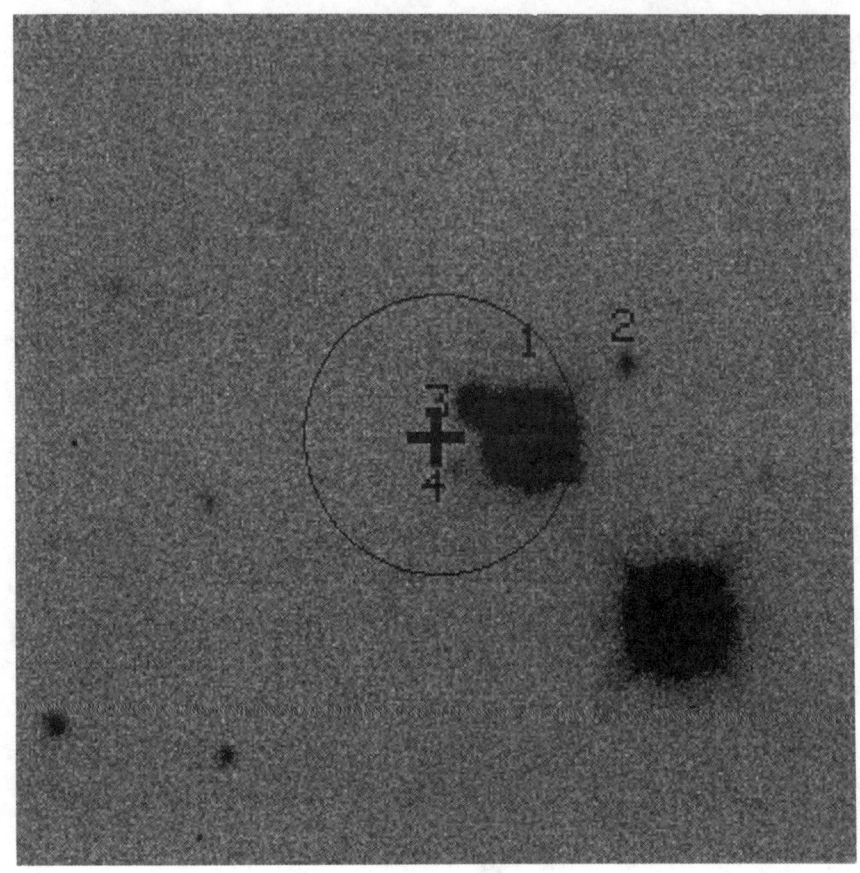

BV Photometry of GRB 910601
Field 29

star id	V	$B - V$
1	8.648 ±0.002	0.703 ±0.001
2	16.251 ±0.083	0.274 ±0.053
3	13.469 ±0.015	1.315 ±0.028
4	16.330 ±0.092	-0.122 ±0.049

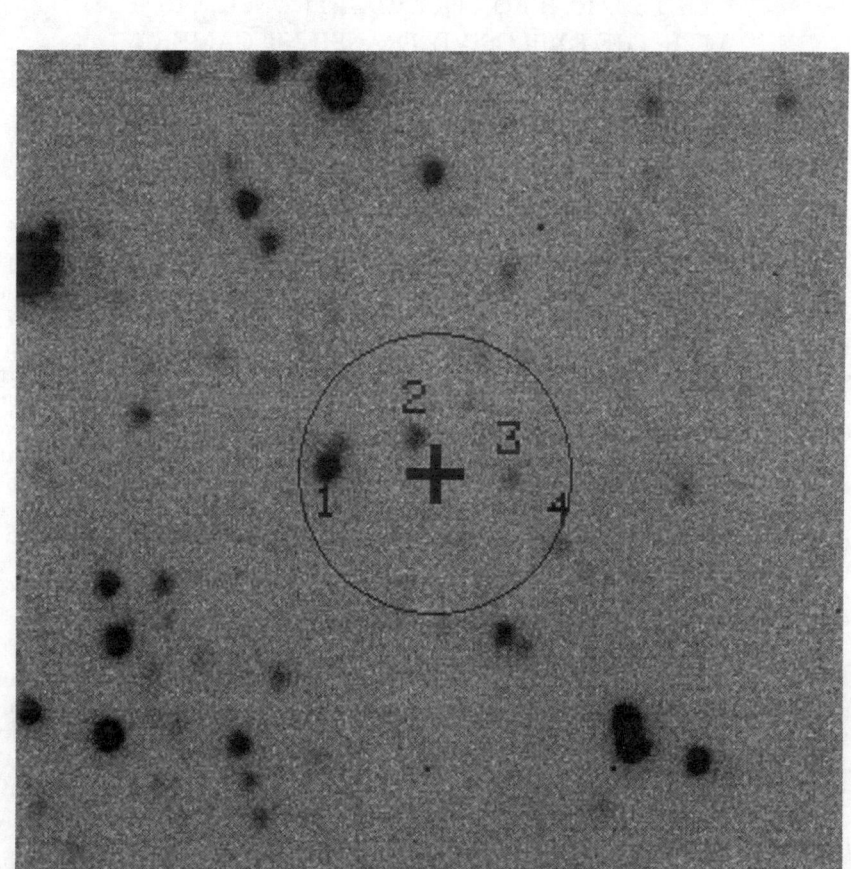

BV Photometry of GRB 910601
Field 30

star id	V	B − V
1	14.036 ±0.022	2.376 ±0.036
2	15.329 ±0.053	2.678 ±0.164
3	16.264 ±0.098	
4	16.931 ±0.173	

THE SEARCH FOR OPTICAL TRANSIENTS WITH THE EXPLOSIVE TRANSIENT CAMERA

Roland Vanderspek, Hans A. Krimm, and George R. Ricker
Department of Physics and Center for Space Research
Massachusetts Institute of Technology, Cambridge, MA 02139-4307

ABSTRACT

For the past 2.5 years, the Explosive Transient Camera (ETC), located on the summit ridge of Kitt Peak, has been conducting an automated wide-field sky search for optical transients. The ETC's CCD cameras, with a total field-of-view of 0.75 steradians, are capable of detecting an optical transient as faint as V=10. To date, the ETC has surveyed over 600 steradian-hours of the night sky between -15° and +45° declination. These observations have resulted in the detection of over 100,000 optical flashes: most, if not all, of these flashes were created by known terrestrial sources, such as moonlit clouds and artificial Earth satellites. After further analysis roughly 500 of these events remain unidentified: the locations and times of a subset of these events are presently being compared with the locations of known satellites, to test the hypothesis that all of these events were created by satellites.

INTRODUCTION

The Explosive Transient Camera (ETC) has been conducting fully automatic observations of the night sky at Kitt Peak since early 1991. The goal of these observations is the detection of short, fast optical transients: in a five-second exposure, the ETC is sensitive to an optical transient as faint as V=10. The ETC's large field-of-view (0.75 steradians), combined with its fully automated operation, makes it a powerful tool for the detection of optical counterparts of gamma-ray bursts.

In this paper, we report preliminary results of the analysis of data taken since the launch of BATSE. The focus of this paper is the extent of ETC observations in that period and the analysis of optical transients detected: Krimm, Vanderspek, and Ricker (1993) contains a discussion of simultaneous observations of fields in which BATSE has detected a GRB.

ETC INSTRUMENTATION

The ETC consists of sixteen wide-field CCD cameras, each with a field-of-view of 20° x 15°. A summary of the ETC configuration is given in Table 1. The CCD cameras are mounted to two sidereal drives in groups of eight: the drives allow the cameras to be slewed to different parts of the sky and to track the sky during observations. Each ETC field-of-view is monitored by two cameras: no optical transient is considered real unless detected by both cameras, so that sources, such as cosmic rays, which can mimic optical transients in a single CCD camera, do not create false optical transients. Custom CCD control electronics allow all sixteen CCDs to be read out simultaneously: the image data from each CCD flow into a dedicated 68000-based single-board computer, known as a Trigger Processor, which is responsible for the bulk of real-time data analysis.

The sixteen Trigger Processors report to a single Overseer Computer, which coordinates the operations of the ETC.

Table 1. ETC Characteristics	
Number of Cameras	16
Field-of-View per Camera	20°x 15°
Total Field-of-View	0.75 steradians
Pixel Size	3.2 arc-minutes
Exposure Time	5.0 seconds
Field Star Sensitivity	V≈11 (5s exposure, 4σ detection)
Sensitivity to Optical Flashes	V≈8.5 (5s flash, 12σ detection threshold)

ETC OPERATIONS

The ETC Overseer Computer has full control of all ETC instrumentation, including the roll-off roof covering the ETC building. The Overseer Computer also has access to precipitation, windspeed, and temperature sensors. This control allows the Overseer Computer to run the ETC completely automatically, with no human interaction required for weeks or months at a time.

Because it has complete control of all of the ETC instrumentation, the Overseer Computer is capable of directing the ETC to perform observations in the manner a human observer on site would. These observations are performed entirely automatically, without nightly commands from MIT. Data collected during the night are stored on hard disk and sent, via Internet, to MIT each morning. Members of the ETC team also receive a daily summary of the each night's observations by electronic mail.

During the day, the ETC waits for astronomical twilight. At twilight, the Overseer Computer reinitializes the hardware and then checks the weather conditions: if the conditions are good, it opens the roof and quickly determines whether the sky is cloudy. If clouds are present, the roof is closed and the Overseer Computer waits fifteen minutes before repeating the check. If the skies are clear, the Overseer Computer initiates real observations. First, the photometric and astrometric observational parameters are determined, then observations begin. The observations run in thirty-minute cycles, after which data are dumped to disk and observational parameters are recalibrated. The cameras begin an observation cycle 1-3 hours east of the meridian: when they have tracked to 1-3 hours west of the meridian, they are slewed back to the east to keep them as close to the zenith as possible.

ETC OBSERVATIONS TO DATE

Since April, 1991, the ETC has monitored over 600 steradian-hours of the night sky for short-timescale optical transients. These observations covered the full range of right ascension in the declination range of -15°to +45°. An exposure map of the observations is shown in Figure 1: the values represented by the gray scale range from 40 to over 200 hours of observation per 1° x 1° element. The sensitivity of these observations can be seen in Figure 2. The histogram in Figure 2 shows the number of steradian-hours of observations as a function of limiting magnitude for a one-second optical flash.

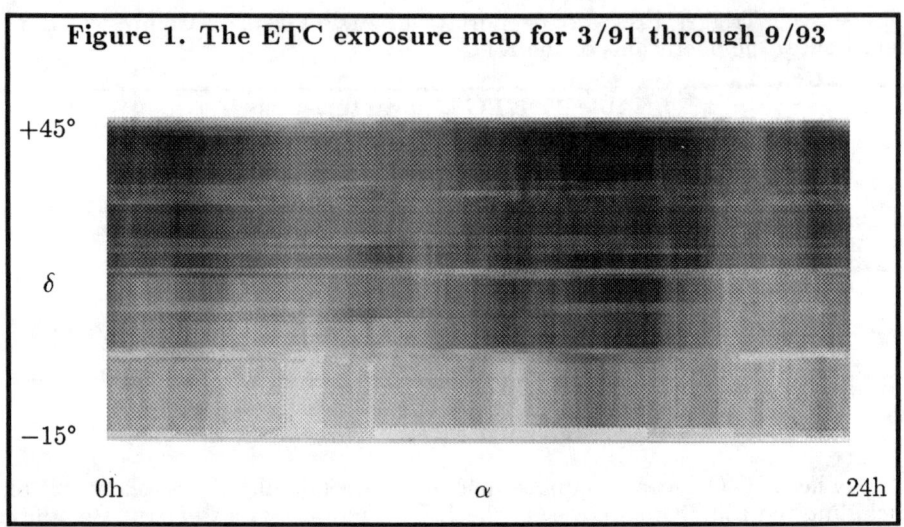

Figure 1. The ETC exposure map for 3/91 through 9/93

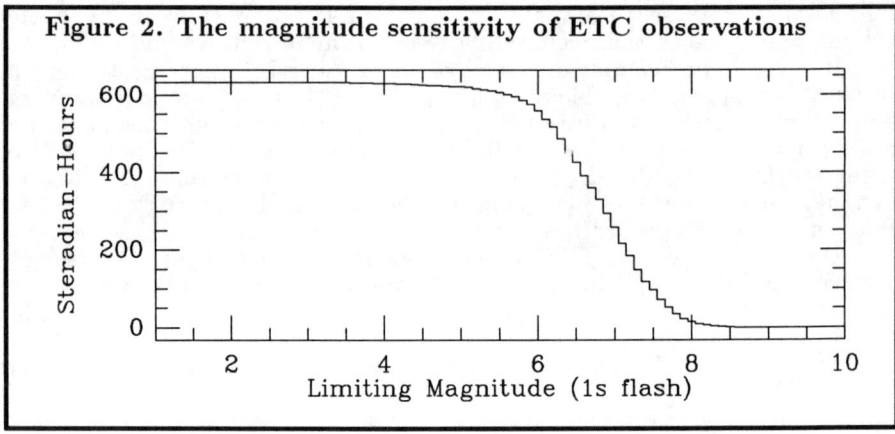

Figure 2. The magnitude sensitivity of ETC observations

IDENTITY OF OPTICAL TRANSIENT EVENTS

In the course of these observations, the ETC has detected over 100,000 optical transients. A large majority of these transients can be recognized as coming from terrestrial sources, such as moonlit clouds and artificial Earth satellites. The different types of event detected by the ETC, and their frequency of occurrence, are tabulated in Table 2.

The post-processing of the events detected by the ETC has identified all but roughly 500 events: the majority of these unidentified events are of type O or type T. A short glint from an orbiting satellite can create both type O and type T events: the type T events can be seen as type O events which occurred on the border between contiguous ETC exposures. Modelling of satellite orbits and populations and glint durations indicates that roughly 1% of all point-like glints should straddle two ETC exposures: the ratio of the number of type T

Table 2. Types of Events Detected by the ETC

Type	Description	Frequency
Type S:	Streak-like satellite glint, recognized by the streak created by the passage of the sunlit satellite.	57.9%
Types G,V:	Point-like satellite glint, recognized by the fact that three or more such events are lined up in right ascension, declination, and time.	10.1%
Type B:	Trigger is at the location of a bright (m<6) star. The source of the trigger could be the star itself, or a satellite glint on top of the star, or the passage of a cloud from in front of the star.	8.1%
Type C:	Trigger was caused by a cloud, either because it was moonlit, or because it passed from in front of a star (there can be some overlap with type B).	21.8%
Type M:	Trigger was caused by moonlight (as stray light or moonlit clouds - - may overlap type C).	1.1%
Type F:	Trigger was caused some hardware failure, which could include things such as a tracking drive failure or the bleeding of charge from a bad column	0.5%
Type O:	The event is seen as a point-like object in a single exposure by two CCD cameras.	0.5%
Type T:	The event is seen as a point-like object in two consecutive exposures by two CCD cameras, with insufficient evidence of motion between the two.	0.1%

events to the sum of the types G, V, and O events is consistent with this result.

A subset of the 86 type T events are presently being compared to the locations of known satellites to determine whether a satellite was the source of the event: this analysis is being performed by R. Rast of the Center for Analysis of Satellite Interference with Astronomy (CASIA). If all of the type T events are seen to have been created by glints from orbiting satellites or debris, then one can conclude that the majority of type O events were creating by glints from orbiting satellites or debris. The only way to conclude that all type O events are terrestrial in nature is to subject them to the same analysis the type T events are undergoing.

A subset of the results of this analysis are shown in Table 3. In that table, events for which columns 2-4 are empty have not yet been correlated with a known Earth-orbiting satellite: further, more comprehensive tests of these events are pending.

PRELIMINARY CONCLUSIONS

In its 30 months of automatic observations, the ETC has detected over 100,000 optical flashes. After a small amount of post-processing, most of these

Table 3. Sample of Attempted Identification of Type T Events.

UT Date and Time of Event (yymmdd hh mm ss)	Catalog Number	Designation	International Identification
910220 10 53 00			
910304 06 35 39			
910501 10 59 39	17296	1987-032B	SL-14 3rd Stage
910701 07 18 23			
910708 07 42 24			
911001 04 21 19	7902	1975-042B	Centaur Stage
911101 11 45 14	16393	1985-117A	Molniya 3-27
911102 04 30 05	12556	1981-060A	Molniya 1-50
911105 07 16 44			
911108 05 51 23	9941	1977-032A	Molniya 3-7
911123 12 47 25	5587	1971-095A	DSCS 2-1
920125 08 40 00	12133	1981-002A	Molniya 3-14
920324 04 36 15	15738	1985-040A	Molniya 3-24

events can be associated with terrestrial sources of optical transients: moonlit clouds, clouds passing from in front of field stars, satellites, meteors, etc. Roughly 500 events remain unidentified: these events are point-like images which are seen in one or two ETC exposures and which cannot be associated, by collinearity in space and time, with any other ETC events. Because these events are very similar to events which were created by satellites, the probability is high that a large fraction of the remaining events were created by satellites as well. Analysis of the type T events, in which a point-like object is seen in two consecutive ETC exposures, is continuing: this analysis will reveal how many of the type T events were created by satellites. A simple model of the probability of a satellite glint being a type T event (rather than a type O event) shows that, if all of the type T events were created by satellites, then the number of type O events is consistent with all type O events having been created by satellites.

An important result of this analysis to date is that the satellites which created the events are large, of order one meter in size, a result which is consistent with simple models of satellite glints. We may tentatively conclude, therefore, that if no known satellite (NORAD tracks all orbiting particles larger than 10 cm) can be found to be the source of an optical transient, the transient is indeed celestial in nature. Of course, if all type O and T events can be correlated with satellites, then none of these events is celestial in nature, and we will be able to set an upper limit on the rate of celestial optical flashes of $10^{-2} sr^{-1} hr^{-1} (3\sigma)$.

We gratefully acknowledge the assistance of the members of MIT's CCD Laboratory. In particular, John Doty, Andrew Kraft, and Peter Goldstein contributed significantly to the analysis of these data. This work is supported under NASA Grant NAGW-2089.

REFERENCES

1. H. A. Krimm, R. Vanderspek, and G. R. Ricker, Proceedings, this conference (AIP Press, N. Y., 1993).

RESULTS FROM THE USNO GAMMA-RAY BURST OPTICAL COUNTERPART SEARCH

Frederick J. Vrba

U.S. Naval Observatory, P.O. Box 1149, Flagstaff, AZ 86002

Dieter H. Hartmann

Dept. of Physics and Astronomy, Clemson University, Clemson, SC 29634

Mark C. Jennings

P.O. Box 66, Corona del Mar, CA 92625

ABSTRACT

The U.S. Naval Observatory 1.0-m telescope was used to obtain deep CCD imaging of eight small Gamma-ray burst error boxes derived from multi-satellite wavefront triangulation and one optical transient field. Over a period of five years a total of 282 hours of open shutter time was used to obtain multiple UBVI-filtered frames covering a total area approximately twice that of the 99% confidence localizations of the Gamma-ray bursts. For 2070 objects in these fields above the survey detection limit of $V = 24.2$, approximately 79,000 calibrated photometric measurements were made. The objectives of this program are to survey the error boxes for objects of unusual colors or light variability and to compare the results of this object census to those obtained for immediately surrounding field regions.

INTRODUCTION

Despite years of effort, employing a number of techniques, no convincing optical counterpart of a classical Gamma-ray burst (GRB) has yet been identified[1]. The increasing degree of isotropy of GRB localizations as found by the BATSE experiment aboard CGRO[2] has brought into question Galactic neutron star models for GRBs[3] and, if anything, has made the quest for a GRB paradigm even more intriguing. However, whether GRB events come from Galactic or cosmological distances, the identification of an underlying source at longer wavelengths would be an important breakthrough in understanding the nature of the GRB phenomenon. Strategies for detecting GRB counterparts at optical wavelengths include all sky[4,5] or rapid responses[6] to current GRB events, searches for archival optical transients within or near GRB localizations[7], and deep searches in the fields of well-localized GRB events[8,9,10]. The logic of the latter strategy is to discover the quiescent counterpart of the GRB via distinctive light variability, photometric colors, or (in the case of Galactic models) proper motion.

In 1987 we initiated a long term program to undertake a truly comprehensive deep search survey of small GRB localizations available at that time from the Interplanetary Network operating in the late 1970s and observable from the U.S. Naval Observatory in Flagstaff, Arizona. This survey employed the USNO 1.0-m telescope with a UV-flooded TI 800^2 CCD providing a 5.7 x 5.7 arcmin field. The survey was designed to obtain multiple observations with typically

60-100 minute exposure times in the UBVI filter system over a time baseline of five years. Thus, not only would accurate color information be obtained for relatively faint objects in or near GRB error boxes, but the survey would provide an unprecedented opportunity to explore the light variability and proper motions of these same objects. In 1992 we presented a report[11] on the observations obtained as of June 1990. We refer the reader to that paper for further details regarding the survey objectives, observational strategies, and photometric reductions. Observations for this survey were terminated in late 1992. Since that time photometric reductions have been completed for 8 of 9 fields surveyed. We present here an interim report on our findings and plans for future work.

SURVEY DATABASE

Table I presents a list of the GRB localizations observed along with the areas of the 99% confidence error boxes, the area of the sky actually observed in this survey, and the areal oversampling ratio. The GRB error boxes observed were obtained either from the literature[12,13,14] or by private communication from K. Hurley. In addition to eight GRB fields, observations were also obtained of the OT310846/OTS1810+31 field[15] although it is unlikely to be associated with the nearby GRB250379b event[16]. For the GRB fields, the total area surveyed is approximately twice that of the 99% confidence error boxes, thus assuring that the underlying GRB sources are typically within the areas surveyed, even if they have significant proper motions.

Table I. Gamma-Ray Localizations Surveyed			
Event	Localization Area	Area Surveyed	Oversampling Ratio
	arcmin^2	arcmin^2	
GB161179/GBS0010-16	4.5	32.5	7.2
GB180479/GBS0552-08	2.9	32.5	11.2
GB290379/GBS1028+46	41	62.0	1.5
GB241178/GBS1205+23	62	63.1	1.0
GB130679/GBS1412+79	0.7	32.5	46.4
OT310846/OTS1810+31 (GB250379b)	NA	32.5	-
GB310379/GBS1925+04	20	32.5	1.6
GB041178b/GBS2006-22	14	32.5	2.3
GB051179b/GBS2252-03	41	81.9	2.0
Σ	186	370	2.0

Fig. 1 summarizes the mean 5% photometric, 20% photometric, and detection limits of a single frame for the UBVI observations. The error bars represent the standard deviations due to the differences in typical seeing between fields with large and small transit zenith distances. Additionally, the mean limit for spatial resolution is given for the V filter based upon the faintest object which was clearly non-pointlike for each field.

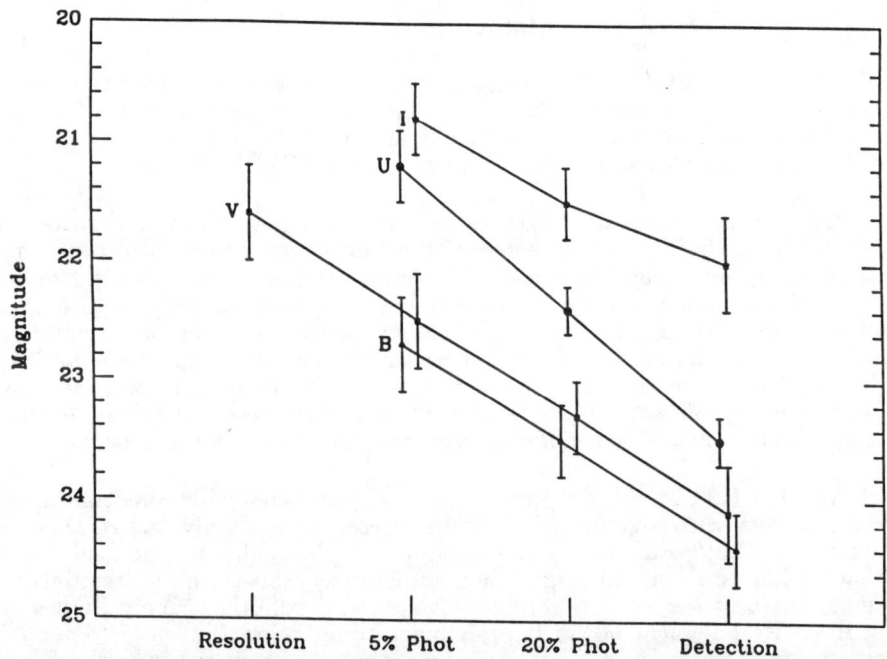

Figure 1. Mean values of single frame photometric limits.

Table II summarizes the total number of frames and amount of open shutter time, in hours, contained in the database. GB310379/GBS1925+04 has fewer hours as observations were terminated in 1990 due to extreme crowding in the field.

Table II. Summary of Observations					
Field	U	B	V	I	Σ
	No. frames (No. hrs)				
GB161179/GBS0010-16	2(3.4)	6(5.3)	12(9.0)	10(6.8)	30(24.5)
GB180479/GBS0552-08	2(3.4)	5(4.3)	17(13.8)	13(10.3)	37(31.8)
GB290379/GBS1028+46	4(6.7)	9(8.4)	21(17.8)	12(9.3)	46(42.2)
GB241178/GBS1205+23	4(6.7)	10(9.0)	16(12.3)	10(7.0)	40(35.0)
GB130679/GBS1412+79	2(3.3)	6(5.2)	16(13.2)	12(9.7)	36(31.4)
OT310846/OTS1810+31 (GB250379b)	2(3.3)	6(5.1)	26(24.2)	14(11.4)	48(44.0)
GB310379/GBS1925+04	1(0.5)	3(1.7)	8(3.2)	8(2.7)	20(8.1)
GB041178b/GBS2006-22	4(6.7)	5(4.2)	9(6.7)	7(5.3)	25(22.9)
GB051179b/GBS2252-03	6(10.0)	9(6.8)	17(15.2)	12(9.8)	44(41.8)
Σ	26(44.0)	56(50.0)	142(115.4)	98(72.3)	ΣΣ = 322(281.7)

PRELIMINARY RESULTS

Although analysis of the database is far from complete, some preliminary results can now be reported. Excluding the data for the GRB1925+04 field, which has yet to be reduced, the following applies to the total areas covered in the survey for the remaining GRB fields and the OTS1810+31 field.

GALAXIES: A total of 120 resolved galaxies were discovered, which is consistent with the number expected at the spatial resolution magnitude limit based upon galaxy count surveys[17]. We estimate that approximately 525 additional objects are likely to be galaxies based upon their colors. The total number of galaxies discovered in each field was found to be consistent with the number expected based upon both the limiting photometric magnitude and the estimated Galactic extinction for the field. None of the objects deemed to be galaxies were significantly variable in brightness. One resolved galaxy, located just outside a formal 99% confidence error box has a large UV excess.

VARIABILITY: No violently variable objects were discovered, although several objects with significant variability were found. Only one of these is contained in a 99% confidence GRB error box. Additionally, the 'fiducial' frame for each field containing the largest number of objects (always an I-filtered frame taken in the best seeing) was carefully compared to all other survey frames for that field. No quiescent object lying N magnitudes below the survey detection limit in any filter band and having a flux increase integrated over the frame exposure time corresponding to $\geq$N magnitudes was thusly detected.

COLORS: Several objects lying in or near GRB 99% confidence error boxes were found with spectral energy distributions (SEDs) inconsistent with dwarf or giant stellar colors or galaxy colors. Most of these are very blue objects whose colors are consistent with being either a white dwarf or QSO. One object was found to have an extraordinary SED, narrowly peaked in the V passband with flux rapidly decreasing toward both longer and shorter wavelengths. This object was observed numerous times during the course of the survey, each time with the same SED and flux levels.

PROPER MOTIONS: Due to considerations of observing efficiency, many of the CCD frames in this survey were taken at large zenith distances and with poor seeing. Since differential color refraction errors will be large in this case, we have have not carried out formal global proper motion solutions to the database. However, we did use the point spread function fitting routine in DAOPHOT[18] to obtain the photometry for this survey. Starting from an initial fiducial coordinate list determined at one epoch, its ability to recover images taken on frames throughout the five year survey places limits on proper motions. It is unlikely that any object of V $\leq$23 mag had a proper motion significantly larger than $\approx$ 1 arcsec/yr. Additionally, the precautions taken to discover transient objects, described in the variability section above, would have revealed fainter stars with large proper motions. Finally, a formal proper motion analysis might indeed reveal objects of V $\leq$23 mag with significant proper motions of $\leq$1 arcsec/yr, but this is of little interest since many stars at these brightness levels will have such motions.

DISCUSSION

Some guidelines for future photometric surveys of GRB fields seem apparent, based on the results from this search. First, long-term variability studies are probably not productive unless such searches are conducted with much greater open shutter time to at least comparable light levels. Second, photometric studies with large color baselines are good at identifying objects with unusual SEDs, but do not necessarily reveal what these objects are. Unfortunately, interesting objects in the fields we searched begin at V >21 mag, thus follow-up studies at higher spectral resolution will involve significant amounts of large aperture telescope time.

FUTURE WORK

We intend to aggressively pursue those objects discovered with unusual SEDs by follow-up optical spectroscopy from collaborators at the MMT facility. As soon as the current analysis phase is complete the entire database will be published. We have obtained deep UBVI imaging of many of the new small localizations from the Third Interplanetary Network which are available from Flagstaff and are currently analyzing these data. It is our intent to continue to provide deep photometric surveys of future small localizations as they become available. Finally, we will continue deep searches of the small targets provided by archival optical transients[7]. Our discovery of a QSO[19,20] likely associated with an optical transient[21] near GRB910219 indicates this to be a promising avenue for further research whether or not related to the GRB phenomenon.

REFERENCES

1. B.E. Schaefer, these proceedings.
2. C.A. Meegan, et al., these proceedings.
3. D.H. Hartmann, these proceedings.
4. H. Krimm, R. Vanderspek, and G. Ricker, these proceedings.
5. J.T. Bonnell et al., these proceedings.
6. S.D. Barthelmy et al., these proceedings.
7. R. Hudec, A&A Suppl. **97**, 49 (1993).
8. H. Pedersen et al., ApJ **270**, L43 (1983).
9. B.E. Schaefer et al., ApJ **270**, L49 (1983).
10. T.E. Harrison, B.J. McNamara, and A.R. Klemola, AJ, in press.
11. F.J. Vrba, D.H. Hartmann, and M.C. Jennings, Gamma-Ray Bursts: Observations, Analyses, and Theories (Cambridge University Press, 1992), p. 138.
12. C. Barat et al., ApJ **280**, 150 (1984).
13. T.L. Cline et al., ApJ **286**, L15 (1984).
14. J.G. Laros et al., ApJ **290**, 728 (1985).
15. C. Motch, R. Hudec, and C. Christian, A&A **235**, 185 (1990).
16. D.H. Hartmann et al., ApJ **336**, 889 (1989).
17. J.A. Tyson, and J.F. Jarvis, ApJ **230**, L153 (1979).
18. P.B. Stetson, PASP **99**, 191 (1987).
19. F.J. Vrba et al., these proceedings.
20. F.J. Vrba et al., ApJ, in press.
21. R. Hudec et al., A&A, in press.

IS A QSO THE SOURCE OF OT050510, AND IS GRB910219 RELATED?

F.J. Vrba, C.B. Luginbuhl
U.S. Naval Observatory Flagstaff Station, P.O. Box 1149, Flagstaff, AZ 86002

D.H. Hartmann
Department of Physics and Astronomy, Clemson University, Clemson, SC 29634-1911

R. Hudec
Astronomical Institute Ondrejov, 25165 Ondrejov, Czech Republic

F.H. Chaffee, C.B. Foltz
Multiple Mirror Telescope Observatory, University of Arizona, Tucson, AZ 85721

K.C. Hurley
University of California, Space Sciences Laboratory, Berkeley, CA 94720

ABSTRACT

A search for Optical Transients near early localizations of the GRB event of 19 February 1991 by Hudec et al.[1] yielded a single convincing candidate on a Harvard Patrol Camera Plate taken 10 May 1905. A deep optical search at the OT position has found a QSO with V=20.6 and z=1.78. Further refinements in the GRB position using the Third Interplanetary Network have however moved the GRB position about 13 arcmin away from the OT/QSO.

THE GRB910219 EVENT

GRB910219 was detected by WATCH[2], PVO and Ulysses[3]. The localization determined from WATCH was $\alpha, \delta(2000) = 14^h08^m, +58°40'$ with a 99% confidence radius estimated at $0°.3$. Based on new pointing information for the GRANAT spacecraft, a revised position of $\alpha, \delta(2000) = 14^h14^m.4, +58°55'$ was determined[4] with an estimated confidence radius of 46'. Preliminary IPN[3] positions, determined after Hudec's plate searches and our CCD observations for this paper were made, lie near $\alpha, \delta(2000) = 14^h14^m.8, +58°42'$ with an error of at most $1'.5$.

THE OT050510 EVENT

A search for optical transients (OT) associated with GRB910219 was carried out by Hudec et al.[1] based on the original WATCH position. The search included over 9,000 plates covering a total exposure exceeding 18,000 hours. A single bright (mag. 6.4) optical transient was discovered just outside the $0°.3$ radius error circle on a Harvard Patrol Camera plate taken 10 May 1905. The position of the OT is given as $\alpha, \delta(2000) = 14^h13^m09^s.8, +58° 36' 30"$, with a 38"×62" error ellipse oriented NE–SW. This position lies within the larger revised WATCH error circle, but is not compatible with the most recent IPN[3] position. The following sections summarize the results of a deep CCD search of the area surrounding the OT position made with the USNO 1.0m telescope at optical wavelengths. A more complete discussion of this study can be found in Vrba, et al.[5].

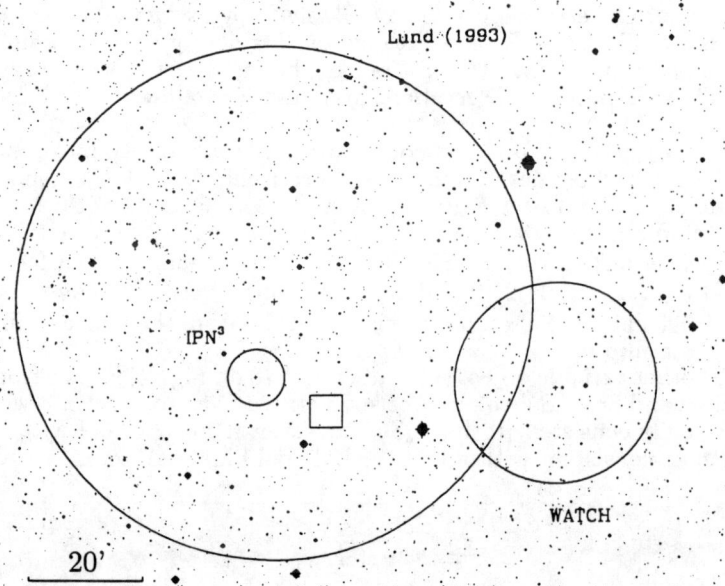

Figure 1. GRB910219 Localizations. Square is CCD Field.

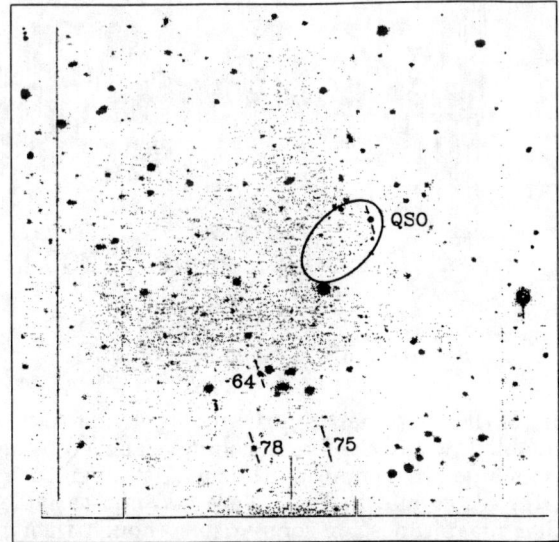

Figure 2. CCD Field.

OPTICAL OBSERVATIONS AT THE OT POSITION

The position of the CCD field in relation to the several localizations for the GRB910219 event is shown in Fig. 1. Fig. 2 shows the 5.′7×5.′7 CCD field and the Hudec OT error ellipse. The CCD field covers over 60 times the area of the OT error ellipse. CCD observations totaling nearly sixteen hours exposure were obtained in the UBVI (Kron-Cousins) bandpasses from February through June 1992, with multiple observations at V and B to allow a search for variable objects.

Two objects with peculiar colors and some variability in V and B were detected, one of which lies within the OT error ellipse. Follow up spectroscopic observations of this object (Fig. 3) show it to be a QSO with z=1.78. The chance of finding a QSO with V≤21.0 in a random high Galactic latitude field is small. Boyle et al.[6] and Boyle, Shanks and Peterson[7] find QSO surface densities of 16 and 37/deg^2 at B≤20.0 and B≤20.9, respectively. These densities imply only a ~0.5% liklihood of an unassociated QSO within the 0.51 arcmin^2 area of the OT error ellipse.

The other variable object (indicated as #78 on Fig. 2) lies far from the OT error ellipse and has UBV colors consistent with a hot subdwarf or a white dwarf and marginally consistent with a QSO. However, it's excess at I with respect to UBV is most consistent with a QSO or hot-cool binary system.

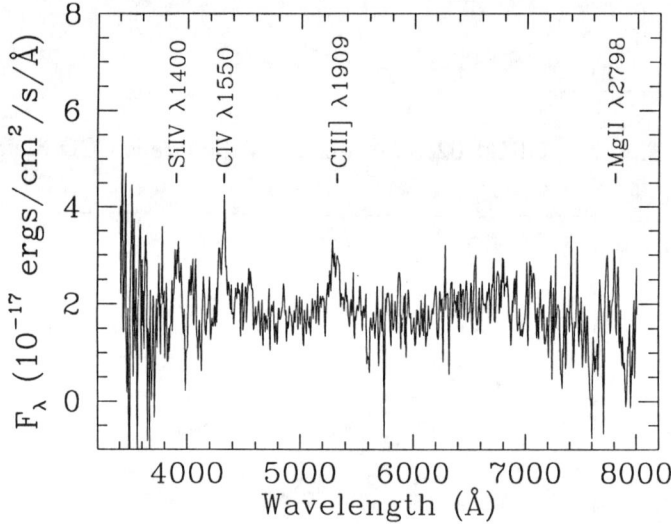

Figure 3. MMT Spectrum of QSO.

A summary of the photometric measures obtained for the QSO and Object #78 is given in Table I, where the fourth column (labelled n) indicates the number of measures included in computing the mean V magnitude listed in column 1. Included in the Table and in Fig. 4 are two non-variable stars (#64 and #75) with similar magnitudes for comparison, though their magnitudes have been offset in the figure by −0.20 and +0.20 mag, respectively, for clarity.

Table I. PHOTOMETRY OF OBJECTS IN CCD FIELD									
Star	V	σ	n	B–V	σ	U–B	σ	V–I	σ
QSO	20.61	0.05	9	0.40	0.04	−0.77	0.08	0.84	0.09
64	20.50	0.03	11	0.55	0.04	−0.19	0.10	0.71	0.09
75	21.43	0.07	11	1.58	0.17			1.82	0.11
78	21.21	0.13	11	−0.09	0.06	−0.81	0.09	0.51	0.14

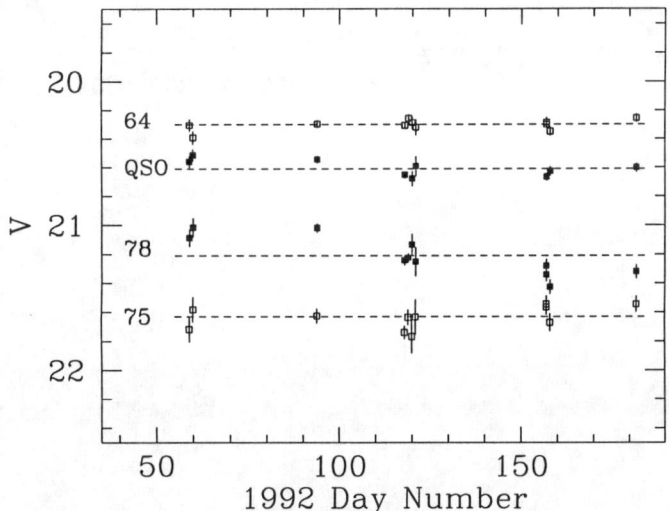

Figure 4. Variability of QSO and Object 78.

DISCUSSION

In this paper we have presented evidence for a likely QSO origin of a highly certain OT located near the GRB910219 event, but separated from it in time by 86 years. Vrba et al.[5] further discuss the implications of the various likely and less-likely associations of the QSO, OT and GRB. Although there may not be any connection between the OT/QSO and the GRB event in this case, such associations will be tested in the future by further well-localized GRBs from timing data. If more QSOs (or other active galaxies) emerge in association with OT and/or GRB positions or a single unambiguous identification of a QSO GRB counterpart is found, the answer to the Galactic/cosmological question will be clear. If OT/QSO/GRB associations are verified by further investigations, solutions to the GRB and OT puzzles would be much closer.

Even if the QSO association with GRBs does not stand up to further scrutiny, the probable discovery of a large-amplitude flare (OT) from a QSO is significant in its own right. Rapid optical variability is a well-known phenomenon for some AGNs, but never has such a large flare been observed on such short time scales.

REFERENCES

1. R. Hudec et al., A&A, (submitted), (1993).
2. N. Lund, S. Brandt, & A.J. Castro-Tirado, Gamma Ray Bursts, AIP 265, ed. W.S. Paciesas & G.J. Fischman (New York, AIP, 1992), p. 53.
3. K. Hurley et al., (in preparation), (1993).
4. N. Lund, private communication, (1993).
5. F.J. Vrba et al., ApJ (in press), (1994).
6. B.J. Boyle et al., MNRAS **227**, 717 (1987).
7. B.J. Boyle, T. Shanks, & B.A. Peterson, MNRAS **235**, 935 (1988).

THE X-RAY SURVEY OF THE SECOND CATALOG GAMMA-RAY BURST ERROR BOXES

Michel Boër[1,2], J. Greiner, P. Kahabka[2], C. Motch[2,3], W. Voges[2]

[1] Centre d'Etude Spatiale des Rayonnements (CNRS/UPS)
BP 4346, F31029 Toulouse Cedex, France

[2] Max Planck Institut für Extraterrestrishe Physik
Postfach 1603, 85740 Garching, Germany

[3] Observatoire de Strasbourg, 11, rue de l'Université
67000 Strasbourg, France

ABSTRACT

Using the ROSAT all-sky survey data, we examined every small GRB error box, mainly from the second catalog of GRBs[1]. We present here preliminary results from this study, the list of X-ray sources in, or near, GRB error boxes, as well as upper limits on the quiescent X-ray emission from GRB sources.

INTRODUCTION

Many attempts have been made to identify Gamma-Ray Bursts at X-ray wavelengths[2,3,4]. The observed objects were choosen primarily on the basis of their good localization, or on their spectral properties. Studies in X-rays were first justified by the galactic neutron star pradigm. Although this hypothesis seems not compatible with the available data ([5,6]), systematic observations may complement the BATSE results on the GRB source distribution and quiescent emission. In addition, it has been suggested GRBs may originate from the galactic arms[7], or that some of the sources are in the galactic disk, while the majority of them are in the halo or at cosmological distances[8]. If this is the case, a systematic survey of many error boxes may constrain this hypothesis, or precise the fraction of near by sources. We used the data of the ROSAT all-sky survey to search in every small error box of the second catalog[1], for quiescent soft X-ray (0.1 - 2.4 keV) counterparts of GRBs.

OBSERVATIONS

The ROSAT all-sky survey was performed during the period July 1990 - January 1991[9]. The exposure times varied between 150s and more than 10ks. The point source sensitivity was about 10^{-13} erg·s^{-1}·cm^{-2} for a 500 seconds exposure. We selected all error boxes of size less than 1 square deg., with the exception of GRB 790101, GRB 791105a, and GRB 791109 which are too extended in one dimension to allow proper processing. Table I summarizes the sources investigated, with the number of ROSAT sources found in or near the error box, the exposure time and the galactic hydrogen column density[10]. In order to make provision for larger uncertainties in the GRB source localisations, we defined an enlarged search radius for each error box, listed table I. These radii, are to be taken as an extension of the region searched, from the edges of the error boxes. Optical investigations are underway to confirm the possible association of these candidates with the GRB sources, and will be reported in a future paper.

DISCUSSION AND CONCLUSIONS

They are very few cases in which a source has been found in a small error box. The probability to find by chance at least a source in a given (extended) error box is quoted in table II. These quantities were computed using the local density of ROSAT sources detected above a maximum likelhood threshold of 8 (approximately 4 gaussian sigmas), in regions of at least 16 sq. deg., surrounding the actual GRB source position (and eventually more according to the error box extent). With these probabilities we give the 3σ upper limits to the source count rate in 3 energy ranges, namely 0.1 - 2.4 keV (whole ROSAT energy band), 0.1 - 0.4 keV, and 0.5 - 2.0 keV. These limits are the same as used for the ROSAT standard analysis.

Three of the soft X-ray sources inside GRB error boxes are known active late type star, at distances ranging from 15 to 40 pc. If we suppose that GRBs originate from these stars, the gamma-ray burst energy is then about 10^{35-36}erg. These energies have been observed exceptionally at X-ray wavelengths[11]: in these cases, the luminosity was always less than 10^{-9} erg $\cdot$ s^{-1} $\cdot$ cm^{-2}, though very little is known on giant flares and even less on the ration L_X/L_γ. However, active stars represent 20% of the ROSAT all-sky survey sources at high galactic latitude, and as much as 50% in the galactic plane[12,13], and it is not suprising to find them in or near some error boxes. However, if some flare stars are GRB sources[14], then it is probable that these sources repeat, and they may be detected by BATSE, or even by another less sensitive experiment, though the reccurence rate of giant flares may be very low.

Let us now suppose that GRB sources are radiating as blackbodies at a temperature of 10^6K. Then ROSAT should detect all sources up to a distance of d_{det} = 400pc (no absorption) or 200pc[15], if we take as mean hydrogen column density 5×10^{20} cm^{-2}. If we make the hypothesis that none of the sources listed table II is related to the GRB phenomenom, then the data presented here allows the derivation of a distance scale for bursters:

$$\left(\frac{n}{n_{tot}}\right) = \left(\frac{d_{det}}{d_{max}}\right)^3 \qquad (1)$$

Since 35 localizations have been studied, equation 1 gives a lower limit for the distance scale of 1.3kpc if no absorption is present between the source and the observer, and 700pc if the hydrogen column density is 5×10^{20} cm^{-2}. These limits, which will be increased by the use of more bursts, are comparable to those obtained with the deepest pointed exposures[16]. Of course, if a source is identified as a GRB source, then this lower limit will be converted to a distribution radius, according to the amount of X-ray sources connected to GRB sources. If one suppose that not all the sources investigated, but only a fraction of them may be galactic, then this limits moves as the inverse of the cubic root of this fraction.

ACKNOWLEDGEMENTS

MB and CM thank the Max Planck Institut für Extraterrestrische Physik and Prof. J. Trümper, for their kind hospitality under the MPG/CNRS exchange agreement. The ROSAT project is supported by the BMFT in Germany. This research has made use of the Simbad database, operated at CDS, Strasbourg, France.

REFERENCES

1. Atteia, J.L. et al., Ap. J. S. **64**, 305 (1987).
2. Grindlay, J. et al., Nature **300**, 730 (1983).
3. Pizzichini, G. et al., Ap. J. **301**, 641 (1986).
4. Boër, M. et al., A.&A. **202**, 117 (1988).
5. Meegan, C.A., et al., Nature **355**, 143 (1992).
6. Hartmann, D., these proceedings
7. Lamb, D.Q., these proceedings
8. Lingenfelter, R.E., and Higdon, J.C., Nature **356**, 132 (1992).
9. Trümper, J.E., Traces of the Primordial Structure of the Universe, eds. H. Böhringer and R.A. Treumann (MPE Report 227, 1991).
10. Dickey, J.M., and Lockman, F.J., A.R.A.A. **28**, 215 (1990).
11. Preibish, Th., et al., A.&A. , *in press* (1994).
12. Hasinger, G., et al., A.&A. **275**, 1 (1993).
13. Motch, C., et al., A.&A. **246**, L24 (1991).
14. Liang, E.P., and Li, H., A.&A. **273**, L53 (1993).
15. Boër, M., et al., A.&A.S. **97**, 69 (1993).
16. Boër, M., et al., A.&A. **277**, 503 (1993).

Table I. Summary of "2nd catalog" observations

GRB name	Size arcmn2	RA 2000.0	DEC 2000.0	Extended search rad arcmin	ROSAT sources in	ROSAT sources out	Exposure time (s)	Nh (10e20)
GRB 780921	890.0	133.4	+34.3	6	1	0	243	2.9
GRB 781016B	3200.0	3.0	+13.3	12	1	1	100	4.27
GRB 781104B	14.0	302.3	-21.5	1	0	0	320	7.77
GRB 781115A	800.0	211.25	+51.94	6	0	1	680	1.38
GRB 781119	8.0	19.71	-28.62	1	0	1	377	1.77
GRB 781121A	140.0	256.41	0.52	2.6	0	1	508	7.9
GRB 781124	48.0	181.93	+23.65	1.5	0	1	410	2.06
GRB 790113	78.0	248.77	-76.61	2	0	0	160	7.17
GRB 790211	1600.0	142.47	5.28	9	3	4	410	3.54
GRB 790305B	0.05	81.50	-66.08	1	1	1	1063	5.56
GRB 790307	10.0	210.67	-46.99	1	0	0	127	11.9
GRB 790313	24.0	94.10	-46.15	1	1	0	979	7.07
GRB 790325B	2.0	272.94	+31.40	1	0	0	670	6.15
GRB 790329	41.0	157.78	+45.64	1.4	0	0	526	1.2
GRB 790331	20.0	292.00	3.68	1	0	0	440	26.35
GRB 790406	0.26	348.51	-49.66	1	0	0	200	1.45
GRB 790412B	1200.0	93.81	-5.27	7	0	1	520	23.7
GRB 790418	2.9	88.59	-6.99	1	0	0	528	19.43
GRB 790419	220.0	335.50	-41.75	3.3	0	0	338	1.29
GRB 790504	58.0	348.40	+32.18	1.7	1	0	520	6.26
GRB 790514	640.0	38.61	+60.97	5.7	0	1	629	6.29
GRB 790613	0.76	212.98	+78.68	1.0	0	0	807	2.93
GRB 790622	110.0	326.34	-41.17	2.3	0	0	385	2.18
GRB 791014	82.0	96.72	-34.65	2.0	0	0	670	5.57
GRB 791018/1	13.5	221.09	-32.20	1.1	0	0	70	5.09
GRB 791018/2	13.5	230.70	-3.41	1.2	0	0	173	7.75
GRB 791031A	620.0	257.59	-82.38	5.6	0	1	215	6.72
GRB 791101	220.0	295.05	+38.27	3.0	0	0	580	1.96
GRB 791105B	35.0	343.57	-2.31	1.3	0	0	339	5.55
GRB 791111/1	350.0	214.70	-33.96	6.0	0	0	348	5.3
GRB 791111/2	350.0	229.67	+3.20	6.0	0	0	329	3.82
GRB 791116	3.7	3.23	-15.70	1.0	0	0	323	2.02
GRB 791215	1700.0	52.41	+51.66	9.2	1	1	538	51.8
GRB 791220/1	140.0	74.77	-60.84	3.7	0	1	661	2.43
GRB 791220/2	140.0	299.30	+66.77	3.7	1	1	1334	10.1
GRB 800103/1	550.0	30.37	-34.19	7.4	0	0	320	1.53
GRB 800103/2	550.0	350.14	-35.14	7.4	1	1	738	8.39
GRB 800105	2100.0	13.79	+7.91	10	0	2	431	5.4
SGR 1806-20	250.0	272.15	-20.73	3.5	0	1	285	181.5

Table II. Probability that a given box contains at least 1 source and count rate upper limits in GRB error boxes

GRB name	Probability		3 sigma upper limits (10^{-3} s^{-1})		
	Inside GRB box	Extended box	0.1-2.4 keV	0.1-0.4 keV	0.5-2.0 keV
GRB 780921	0.2017	0.5246	36.7	26.6	22.1
GRB 781016B	0.6858	0.9870	21.6	9.64	29.0
GRB 781104B	0.0001	0.0054	20.1	15.9	22.4
GRB 781115A	0.2765	0.5206	12.9	12.9	8.8
GRB 781119	0.0050	0.0199	34.8	16.1	27.8
GRB 781121A	0.0139	0.0277	75.7	73.5	24.4
GRB 781124	0.0202	0.0659	27.7	18.6	17.3
GRB 790113	0.0058	0.0157	31.2	32.5	31.3
GRB 790211	0.3624	0.8070	14.1	17.4	15.8
GRB 790305B	0.0038	0.0038	4006.	132.	4776.
GRB 790307	0.0034	0.0216	41.7	44.0	37.6
GRB 790313	0.0072	0.0178	21.2	10.53	15.3
GRB 790325B	0.0005	0.0014	7.07	10.2	4.52
GRB 790329	0.0231	0.0553	16.5	16.6	10.8
GRB 790331	0.0030	0.0088	14.1	9.23	9.73
GRB 790406	0.0001	0.0012	18.4	12.8	8.93
GRB 790412B	0.2166	0.5134	13.1	11.5	11.4
GRB 790418	0.0016	0.0031	9.58	10.9	9.50
GRB 790419	0.1240	0.2434	20.3	20.1	14.7
GRB 790504	0.087	0.0522	13.9	12.5	11.7
GRB 790514	0.2276	0.4448	10.3	8.46	9.80
GRB 790613	0.0016	0.0022	5.86	6.84	5.86
GRB 790622	0.0136	0.275	31.8	36.4	13.9
GRB 791014	0.0263	0.0631	11.0	12.2	12.3
GRB 791018	0.0012	0.0043	82.9	74.3	72.2
GRB 791031A	0.1037	0.2516	23.7	26.1	23.9
GRB 791101	0.0692	0.1586	10.6	11.7	13.6
GRB 791105B	0.0078	0.0294	20.5	17.5	13.3
GRB 791111	0.1074	0.1431	26.6	27.3	16.0
GRB 791116	0.0009	0.0063	16.2	12.9	8.91
GRB 791215	0.2627	0.5113	13.9	13.6	9.10
GRB 791220	0.1292	0.2191	12.2	14.1	11.2
GRB 800103	0.6331	0.7689	20.3	16.4	22.0
GRB 800105	0.3354	0.3862	17.4	14.6	12.0
SGR 1806-20	0.0289	0.0489	25.5	22.2	25.3

RECENT SMALL GAMMA-RAY BURST ERROR BOXES IN THE ROSAT ALL-SKY SURVEY

Michel Boër[1,2], J. Greiner, P. Kahabka[2], C. Motch[3,2], W. Voges[2]
M. Sommer[2], K. Hurley[4], M. Niel[1], J. Laros, R. Klebesadel[5]
C. Kouveliotou, G. Fishman[6], T. Cline[7]

[1] Centre d'Etude Spatiale des Rayonnements (CNRS/UPS)
BP 4346, F31029 Toulouse Cedex, France

[2] Max Planck Institut für Extraterrestrishe Physik
Postfach 1603, 85740 Garching, Germany

[3] Observatoire de Strasbourg, 11, rue de l'Univesité
67000 Strasbourg, France

[4] University of California, Space Sciences Laboratory
Berkeley, CA 94720, USA

[5] Los Alamos National Laboratory, NM 87545, USA

[6] NASA Marshall Space Flight Center, Huntsville, AL 35812, USA

[7] NASA Goddard Space Flight Center, Greenbelt, MD 20771, USA

ABSTRACT

Using the ROSAT all-sky survey data, we examined recent small GRB error boxes. In one case, a source has been found inside the GRB error box of size 13 arcmin^2. Upper limits to the source count rates are derived and discussed.

INTRODUCTION

Until now, all observations of GRB error boxes have been performed after the GRB event itself, even when the sources were seremdipitously observed as part of another program[1]. This suppose that the activity of the GRB sources does not change significantly before and after the event, or that the decay time of the GRB soft X-ray counterparts is on the order of some months or years. However, many objects were localized by the third interplanetary network[2], with accuracies on the order of some tenth arcmin^2, during the period 1991 - 1992. These positions may be compared with the ROSAT all-sky survey data, which was performed in the course of the 2nd semester 1990, allowing to search for an X-ray counterpart before the burst occurence. Moreover, this additional data enlarge the sample of the sources observed by this instrument[3] or other experiments and provides additional or tighter constraints.

OBSERVATIONS

The X-ray observations and data processing are basically the same than the work done for the survey of historical error boxes[3]. We present here the main characteristics and differences. The ROSAT all-sky survey was performed during the period July 1990 - January 1991[4]. The exposure times varied between 150s and more than 10ks. The point source sensitivity was about 10^{-13} erg·s^{-1}·cm^{-2} for a 500 second reference exposure. We selected 16 error boxes derived from the 3rd interplanetary network data[2]. The sizes in all cases were less than 150 arcmin^2. Table I summarizes the sources investigated, with the number of ROSAT sources found in or near the error box, the exposure time and the galactic hydrogen column density[5], as well as cross correlation results with the SIMBAD database. In order to make provision for larger uncertainties in the GRB source localizations, we defined an enlarged search radius for each error box. Optical investigations of the ROSAT sources are underway to confirm a possible association of these candidates with the GRB sources, and study the nature of the other X-ray sources. Table II gives the probability that at least one source is found by chance in a given error box; this probability is computed by taking into account the local source density over a large area surrounding the error box. The last three columns are 3σ upper limits to the count rate in 3 energy ranges, namely 0.1 - 2.4 keV, 0.1 - 0.4 keV, and 0.5 - 2.0 keV.

DISCUSSION AND CONCLUSIONS

There is only one case of a positive detection inside a GRB error box, namely GRB 920711. Preliminary investigations of this source at optical wavelengths do not show any obvious counterpart, like a coronal source or an AGN. However, deeper searches are underway and will be reported later. The ROSAT data for this source shows that 20 photons were detected, almost all of them around 1 keV. No time variability (e.g. a clustering in time) has been noted, and the light curve is consistent with that of a persistent source, as far as the statistics on the few numbers of counts allow it. Note that because of the proximity of this source to the North Ecliptic Pole, 2500s of data were acquired. Assuming that a 16km radius spherical source is radiating as a 1 keV blackbody spectrum, this gives a flux of 5×10^{-14} erg·s^{-1}·cm^{-2} which places the source at 3 kpc. If we suppose that only 1/1000 of this area is emitting in the ROSAT energy range, the distance is further limited to 100pc. Figure 1 gives the temperature of the source as a function of distance, or the accretion rate, assuming that the source is heated by accretion, in two cases, 1) the entire surface of a 16 km radius neutron star radiates (solid line), and 2) only 1/1000 of this surface is heated (dashed line), for instance by accretion funneled along the polar caps by a 10^{12}G magnetic field.

One feature of the present work is that all the observations reported here were performed 1 - 2 years prior to the GRB. If bursts originates from, e.g., neutron star mergers at cosmological distances ($z \approx 0.2$ for the bright sources studied here), there is no obvious reason to detect an enhanced level of X-ray activity, at our sensitivity limit, before the gamma-ray event. On the other side, if at least some of the GRB sources are local (see e.g. Colgate and Leonard; Li, Duncan and Thompson; Lamb and Quashnock, these proceedings), or are the result of giant stellar flares[6], we might expect either some X-ray activity before the burst, or to find a stellar coronal source inside the GRB error box. This is clearly not the case, with the possible exception of the source of GRB 920711.

These data may be used also to constrain further the GRB distance scale. Here we suppose, as in the previous paper that GRB sources are radiating as blackbodies at a temperature of 10^6K. Then ROSAT should detect all sources up to a distance of d_{det} = 400pc (no absorption) or 200pc^3, if we assume a mean hydrogen column density $5 \times 10^{20}\,cm^{-2}$. Including the small archival sources observed also by ROSAT, 50 GRB error boxes have been studied in detail at X-ray wavelengths. If we make the hypothesis that none of the X-ray source detected is related to the GRB phenomenom, then the data presented here allows to further constraint the source distance scale, as done for archival boxes. Equation 9 of Boër et al.[7] gives a lower limit for the distance scale of 2kpc if no absorption is present between the source and the observer, and 1 kpc if the hydrogen column density is $5 \times 10^{20}\,cm^{-2}$. These limits compare favorably with those obtained with the deepest pointed exposures[7]. Of course, if a source is identified as a GRB source, e.g. GRB 920711, then this lower limit will be converted to a distribution radius, according to the amount of X-ray sources associated to GRB sources. This limits is connected also to the fraction of galactic GRB sources.

ACKNOWLEDGEMENTS

MB and CM thank the Max Planck Institut für Extraterrestrische Physik and Prof. J. Trümper, for their kind hospitality under the MPG/CNRS exchange agreement. The ROSAT project is supported by the BMFT in Germany. This research has made use of the Simbad database operated at the CDS, Strasbourg, France.

REFERENCES

1. Boër, M., et al., A.&A. **249**, 118 (1991).
2. Cline, T., et al., Proceedings of the 1991 Huntsville Workshop on Gamma-ray Bursts, Paciesas, W.S., Fishman, G.J., edt. (AIP Conf. Proc. 265, N. Y., 1992), p. 72.
3. Boër et al., these proceedings
4. Trümper, J.E., Traces of the Primordial Structure of the Universe, eds. H. Böhringer and R.A. Treumann (MPE Report 227, 1991).
5. Dickey, J.M., and Lockman, F.J., A.R.A.A. **28**, 215 (1990).
6. Liang, E.P., and Li, H., A.&A. **273**, L53 (1993).
7. Boër, M., et al., A.&A. **277**, 503 (1993).

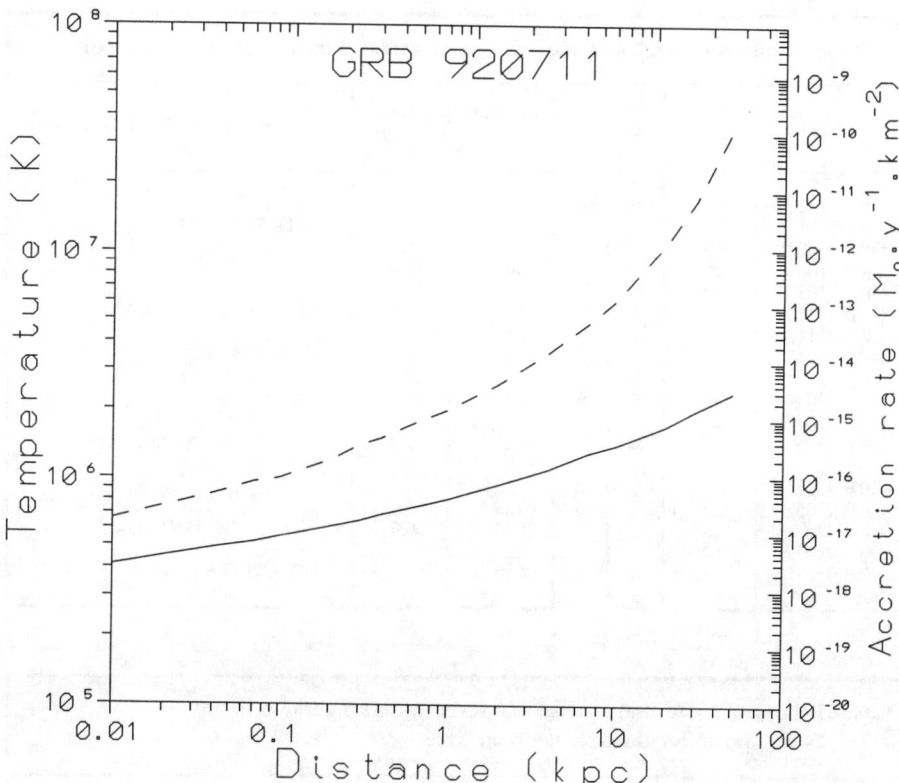

Figure 1. Temperature (left hand scale), and accretion rate (right hand scale) as a function of the distance for the X-ray source inside the GRB 920711 error box. Solid line: accretion over the whole 3200 km² surface of the neutron star. Dashed line: only a surface of 3.2 km² radiates. These curves have been computed using the data taken in the 0.5 - 2.0 keV energy band. The influence of the galactic absorption is negligible.

Table I. Summary of X-ray observations of recent bursts localized by 3 satellites

GRB name	Size arcmn2	ROSAT sources in	ROSAT sources out	Exposure time (s)	Nh (10e20)	Comments and SIMBAD correlation
GRB 910122	40	0	0	147	7.0	CPD-70 2702 A5 CPD-70 2703 K0
GRB 910219b	15	0	0	684	1.3	
GRB 910522	25	0	0	546	135.4	
GRB 910717	150	0	0	298	24.4	
GRB 911104	40	0	0	520	1.0	Close to galaxy z 1407.2-3522
GRB 911109	30	1	0	457	41.4	
GRB 911118	15	0	0	362	5.2	HD 96334 (A0)
GRB 920311	35	0	1	502	32.5	= HD 75653 (F7V)
GRB 920325	35	0	0	445	4.3	SAO 108569 (F8)
GRB 920406	20	0	0	154	5.5	ESO 141-56 (gal. S)
GRB 920501	12	0	0	491	60.25	Compatible with Hurley et al. TOO source
GRB 920517	45	0	0	342	6.56	
GRB 920525	20	0	0	323	5.11	
GRB 920711	13	1	0	2500	6.7	<> from HRI TOO source
GRB 920720	10	0	0	585	0.98	
GRB 920723	13	0	1	616	17.2	HD 337674 (G0)

Table II. Probability to find at least one source in GRB error box and 3 σ upper limits to the count rate

GRB name	Probability	Count rates upper limits (10^{-3} c/s) (kev)		
		0.1 - 2.4	0.1 - 0.4	0.5 - 2.0
GRB 910122	0.0062	42.0	35.2	33.3
GRB 910219b	0.0065	10.8	10.6	8.25
GRB 910522	0.0082	14.4	13.0	12.3
GRB 910717	0.0403	21.9	26.7	20.0
GRB 911104	0.0165	14.3	14.6	11.7
GRB 911109	0.0109	34.5	15.8	20.8
GRB 911118	0.0021	18.3	18.5	14.4
GRB 920311	0.0074	12.2	11.4	10.3
GRB 920325	0.0091	15.0	14.0	11.8
GRB 920406	0.0022	45.7	40.1	36.5
GRB 920501	0.0017	16.2	14.8	11.3
GRB 920517	0.0062	22.2	17.3	15.3
GRB 920525	0.0042	23.4	23.5	18.6
GRB 920711	0.0140	5.17	4.33	3.72
GRB 920720	0.0059	12.6	11.4	10.5
GRB 920723	0.0047	29.7	25.4	10.5

X-RAY BURST RATES FROM ROSAT

Andrew Kahn and Hakkı Ögelman
University of Wisconsin, Dept. of Physics, Madison, WI 53706

ABSTRACT

We searched for x-ray bursts in the background of three of the pointings of the ROSAT PSPC detector totaling 77076 seconds of exposure time. We searched each field for bursts at three levels of duration of $\Delta t \leq 1$, 10, and 100 seconds with corresponding fluences around $(3\text{-}7) \times 10^{-11}$ ergs cm^{-2} We found no significant bursts in any of the searches. We determined the 95% upper limit to the burst rate to be 0.45 bursts sec^{-1} sr^{-1}. This rate was calculated utilizing less than 0.3% of the available ROSAT data and could be reduced by more than two orders of magnitude if the remaining data were to be used.

INTRODUCTION

Our search for X-ray bursts uses the ROSAT position sensitive proportional counter (PSPC) data, which has a circular field with a radius of 1 degree and an energy range of 0.1 - 2.4 keV. But in order To shrink the amount of random bursts, we used only the central region out to about 20 arcmin; past that the point spread function radius begins to get very large. This also conveniently aviods having to deal with the effects of the ribs of the detector interfering with the data. Then from this data we removed all the sources found from a source detection algorithm so that our search could be conducted on the remaining "background". We searched three different images, which totals to 77076 seconds of exposure time, 117111 accepted events (after removing the outer regions and sources), and an average solid angle of 8.7×10^{-5} sr per pointing.

WHAT IS A BURST?

The ROSAT photon events table (PET) consists of all the accepted events which have been time, position, and amplitude tagged. The search begins with the first photon, from a chronologically ordered list, and then searches the remainder of the table for additional events that have occured within the specified time limit of the first one ($\Delta t \leq 1$, 10, or 100 seconds). Those photons are then checked to find ones that have occured within the point spread function radius of the initial event. If the number of events found to fit these qualifications is greater than a preset number of minimum events per burst, then the group of events is declared a "burst". Then the search moves on to the second event and attempts to find a burst begining with that event and so on down the events table. Each burst is also checked to make sure it is an entirely new burst and not part of a previous burst already found. If some of the events within a new burst were found to exist in a previous burst then it is no longer declared a new burst and the remaining events which were not originally part of the old burst are then added to the old burst. Once the entire list is finished the result is some number of bursts found.

ARE THOSE BURSTS SIGNIFICANT?

There is a certain amount of bursts we would would expect to find due to random coincidence, but it is difficult to calculate that amount exactly. What we did instead was to perform a Monte Carlo simulation. The location of an event in the PET is changed by randomly choosing another event in the table and reassigning its location as the location of the initial event. Once this is done for the entire table what results is a new list that still contains the exact positions and times of all the events but the positions and times are no longer correlated. This new shuffled list can then be searched for bursts to get an idea of how many random bursts we would expect to see in this particular field of view. Each time this procedure is done the PET is shuffled in a new and completely random manor. Thus after repeating it 1000 times we can get an average number of random bursts found for each set of initial parameters.

For each image used, we searched for burst durations of $\Delta t \leq$ 1, 10, and 100 seconds. And for each Δt we set the minimum number of events per burst at various levels. In every case, the number of bursts found from the real events table was approximately equal to the number found from the shuffled events tables. Thus, we declared none of these bursts to be significant.

CALCULATING AN UPPER LIMIT

We kept on searching, each time increasing the minimum events per burst requirement, until we found zero bursts (the level at which no bursts were found in the real data also corresponded to the same level in the shuffled tables). Since no bursts were found at this level, statistics tells us that to 95% confidence a maximum of 3 bursts could have existed. The upper limit of the burst rate is then equal to 3 bursts divided by the total exposure time and the average solid angle of the field. This rate corresponds to the specific Δt at which the search was set at and a particular fluence. That fluence was detemined by the average energy per count, the value of the minimum number of counts per burst which produced zero bursts, and the effective area of the detector. In the absence of specific spectral distributions of burst photons, the effective area was averaged over all energies and off axis angles to obtain the value of 86 cm^2. This results in an upper limit to the burst rate for each image (see table 1). The results from each image were combined by summing the total exposure time and finding the average solid angle. Since there were no bursts found in any field, an upper limit of three bursts still exists. Thus we have a final value for the burst rate of 0.45 bursts sec^{-1} sr^{-1}. This rate corresponds to three levels: a $\Delta t \leq$ 1 second and a fluence of 3×10^{-11} ergs cm^{-2}, a $\Delta t \leq$ 10 seconds and a fluence of 4×10^{-11} ergs cm^{-2}, and a $\Delta t \leq$ 100 seconds and a fluence of 7×10^{-11} ergs cm^{-2}. These fluences were determined by averaging the individual fluences of each image.

Table 1: Individual image and final results

center of data set ra	dec	fluence ($10^{-11} ergs\ cm^{-2}$) $\Delta t \leq 1 sec$	$\Delta t \leq 10 sec$	$\Delta t \leq 100 sec$	burst rate ($sec^{-1} sr^{-1}$)
$5^h 31^m$	$-65°55'$	3.96	4.95	10.9	0.899
$6^h 32^m$	$-23°25'$	2.69	2.69	4.49	2.13
$10^h 58^m$	$-52°27'$	2.59	3.45	4.31	2.01
total		3.08	3.70	6.57	0.450

CONCLUSION

In order to extrapolate this result to place a constraint on the gamma ray burst rate, we must estimate the X-ray to gamma ray luminosity ratios of gamma ray bursts. For a value of 0.02 (*Laros et al.*, 1984) our value of 0.45 bursts sec^{-1} sr^{-1}, which is 1.8×10^8 bursts yr^{-1} over the whole sky, would correspond to a gamma ray fluence on the order of 10^{-9} ergs cm^{-2}. Similar searches for X-ray bursts, but at a slightly higher fluence than ours ($\sim 10^{-10}$ ergs cm^{-2}), have previously been conducted using Einstien data (*Helfand and Vrtilek*, 1983). Although their result of $<10^5$ yr^{-1} places a better constraint on the log N-log S distribution of gamma ray bursts than ours does, our result is actually only a preliminary estimate. Our rate was calculated utilizing three pointings totaling ~ 7000 seconds; that is less than 0.3% of the available ROSAT data. If this search were to be conducted on the remaining data, this rate could be reduced by no less than two orders of magnitude.

REFERENCES

Laros, J., Evans, W., Fenimore, E., Klebesadel, R., Shulman, W., and Fritz, G., *Ap. J. 286*, 681, 1984.

Helfand, David J., and Vrtilek, Saeqa Dil, *Nature 304*, 41, 1983

AN ASCA ATTEMPT OF GRB OBSERVATION IN X-RAY RANGE

Atsumasa Yoshida

The Institute of Physical and Chemical Research (RIKEN)

Yasushi Ogasaka, Toshio Murakami

Institute of Space and Astronautical Science

ASCA team

ABSTRACT

We are attempting to search GRBs from the extended halo around nearby galaxies in X-ray band with the Gas Imaging Spectrometer on board the ASCA satellite which was launched in February, 1993. GIS/ASCA has the imaging capability and can detect X-rays in 0.5 keV to 10 keV band in which GRBs commonly emit the radiation. GIS is sensible to bursts with $S_X \approx 10^{-10}$ erg cm^{-2} which correspond to $\sim 10^{42}$ erg GRBs occurring in M31. We have performed a search on the ASCA PV phase data of M31, M33, LMC and SMC region, however we could not get any positive detection so far.

INTRODUCTION

From the extensive observations with BATSE/CGRO, we have learned that the angular distribution of gamma-ray bursts (GRBs) is completely isotropic while V/V_{max} implies spatial inhomogeneity[1]. This suggests that GRBs must be very local, cosmological, or originating in a Galactic halo extended largely enough to make our offset from the Galactic center result in no detectable anisotropy. If an extended Galactic halo is the site of GRBs, we may observe GRBs from halos around nearby galaxies such as M31.

Highly sensitive low-background X-ray telescopes may have benefits for a search of bursts from nearby galaxies even though GRBs emit only several percent of gamma-ray radiations in X-ray band[2]. From this viewpoint, Liang and Li proposed a search of X-ray flashes associated with GRBs from M31 using the ROSAT Position Sensitive Proportional Counter[3,4]. In February, 1993, the ASCA satellite was launched from Kagoshima in Japan. It carries two CCD X-ray spectrometers (Solidstate Imaging Spectrometer; SIS) and two imaging gas-scintillation proportional counters (Gas Imaging Spectrometer; GIS) in the focal plane of the X-ray mirrors, and can detect X-rays in a fairly large energy band from ~ 0.5 keV to 10 keV. From the observations with GBD/*Ginga*, we have learned that X-rays in this band are commonly emitted from GRBs[2], hence we can detect X-ray part of GRBs with ASCA satellite.

In this paper, we describe our attempt to search GRBs with ASCA, and mention the preliminary results using the PV phase observations on some nearby galaxies.

DETECTABILITY

We use the data from GIS which has larger field-of-view (FOV) than SIS. Background rate in GIS has a radial dependence in the counter because there is a relatively large background near the counter side wall. Therefore, we use data from the region of radius < 17 arcsec from the center of each GIS sensor for the search. This region sees $2.4 \times 10^{-5}\pi$ str of the sky.

First we evaluate the rate of that isotropically distributed GRBs come into the above solid angle by chance. We apply PVO logN-logP distribution[5]:

$$N(> P_\gamma) = 21 \left(\frac{P_\gamma}{2 \times 10^{-5} \text{ erg sec}^{-1} \text{ cm}^{-2}} \right)^{-1.5}, \quad (1)$$

for $P_\gamma > 2 \times 10^{-5}$ erg sec^{-1} cm^{-2}, and assume a slope of -0.8 below $P_\gamma = 2 \times 10^{-5}$ erg sec^{-1} cm^{-2}. To convert P_γ to the X-ray fluence, we assume that a GRB peak duration (FWHM) in X-ray band is ~ 5 sec; hence, an expected rate of GRBs for which GIS can detect more than 10 counts, $R_{>10}$, is evaluated by

$$R_{>10} \sim 1 \times 10^4 \left(\frac{\delta}{0.01} \right)^{0.8} \text{ yr}^{-1} (4\pi)^{-1}, \quad (2)$$

where $\delta = S_X/S_\gamma$ is the X/γ fluence ratio. Multiplying the effective GIS FOV and a duty cycle of ~ 0.5, we can evaluate the chance rate to be $\lesssim 0.03$ yr^{-1}. Therefore, the chance GRB incidence is negligible for searching bursts in a certain pointing data to a nearby galaxy during a observing period of $\lesssim$ Msec.

Secondly, we estimate expected GIS counts from GRB occurred in nearby galaxies. An expected X-ray fluence may be given by the following formula:

$$S_X \approx 1.9 \times 10^{-10} \left(\frac{\delta}{0.01} \right) \left(\frac{Q_\gamma}{10^{42} \text{ erg}} \right) \left(\frac{D}{670 \text{ kpc}} \right)^{-2} \text{ erg cm}^{-2}, \quad (3)$$

where D and Q_γ are the distance from the source and the total energy emitted in gamma-ray band during a burst. Assuming a power-law X-ray spectra of index α, expected counts for each GIS sensor, C_{GIS}, can be evaluated by taking account of the detector/mirror responses:

$$C_{\text{GIS}} \approx C_0 \left(\frac{\delta}{0.01} \right) \left(\frac{Q_\gamma}{10^{42} \text{ erg}} \right) \left(\frac{D}{670 \text{ kpc}} \right)^{-2} \text{ counts}, \quad (4)$$

where a column depth of 1×10^{21} is assumed, and C_0 is 2.8, 3.6, and 5.7 for $\alpha = 0.5, 1.0$, and 2.0 respectively. Since δ is several percent[2], we may expect > 5 GIS counts for GRBs with $Q_\gamma = 10^{42}$ erg.

Thirdly, we estimate GRB rate expected from M31. We assume "dark halo distribution"[6];

$$n = \frac{n_0}{1 + (R/R_C)^2}. \quad (5)$$

Following BATSE result, we choose $n_0 = 94.6$ yr^{-1}, and $R_C = 28$ kpc. Applying this distribution to M31, we get an expected GRB rate inside the effective GIS FOV pointing to the center of M31; ~ 0.03 events day^{-1}. This number is about 400 times larger than the chance rate mentioned above. Furthermore, M31 has the luminous mass about three times more than the Galaxy. Hence we may expect three times higher rate, ~ 0.09 events day^{-1}, requiring observations of 20–30 days.

OBSERVATIONS AND ANALYSIS

Table 1 summarizes the ASCA PV phase observations of nearby galaxies for which we are searching GRBs.

Since photons from a real GRB event come onto a certain portion of the detector surface, we can get a good S/N ratio for the event using the position information of photons. We are performing a search in the following ways;

(1) Divide each GIS sensor surface into 25 cells. Each cell has the size of $\sim 6.8 \times 6.8$ arcmin^2 which is larger than the mirror resolution since Half-Power-Diameter of X-ray mirror is about 3 arcmin.
(2) Data of each cell are accumulated every 8 seconds into a bin.
(3) Since typical GRB duration is about a few ten seconds, we perform a running integration every 8-second-bin. We use an integration time of 64 seconds (i.e. 8 bins) for this analysis.
(4) A background level is calculated for each cell of each sensor. A background rate follows the Poisson distribution with a mean of 0.1 - 2, which depends on whether steady sources are in the FOV. If an excess is found with a certain significance, data of corresponding cells of another GIS sensor are studied. We record an event as a candidate when there is an excess in both data from two identical sensors.

To test the logic, we have applied the above method on the data of the blank sky and of X-ray bursting source. Applying the 20 ksec data of NEP region, we did not find any excess with the confidence of 1×10^{-4}. From the data on EXO 0748-676, we have detected an X-ray burst from the above method.

target	start time	exposure	distance
M31	93/07/28 15:00	20ksec×6 pointing	670kpc
M33	93/07/22 21:40	20ksec×2 pointing	730kpc
LMC	93/08/20 22:45	40ksec	47kpc
SMC	93/07/26 22:08	40ksec	58kpc

Table 1. Summary of nearby galaxies that we are searching GRBs.

RESULTS

M31 / M33

From a search performed on the PV phase data listed in Table 1., which have an effective observation time of 134 ksec, we could not find any burst-like event with the confidence of $1 - 10^{-4}$. The upper limit on the burst occurrence rate is $1/1.5$ events day^{-1} for events with X-ray fluence of > 5 GIS counts.

Since an expected rate is ~ 0.09 events day^{-1}, we need about at least 1 Msec observation data for the positive detection. We are planning to propose about 2 week observation of M31.

LMC / SMC

We have the PV phase data of 35 ksec effective time for LMC and SMC region. Analysis is now in progress.

REFERENCES

1. C. A. Meegan et al., Nature **355**, 143 (1992).
2. A. Yoshida, et al., Publ. Astron. Soc. Japan **41**, 509 (1989).
3. E. P. Liang, Astrophys. J. **380**, L55 (1991).
4. H. Li and E. P. Liang, Astrophys. J. **400**, L59 (1992).
5. E. Fenimore et al., Proc. Compton Symp. (St. Louis) (AIP press, 1992), p. 744.
6. S. Mao and B. Paczyński, Astrophys. J. **389**, L13 (1992).

SEARCHES FOR BURSTS OF TEV GAMMA RAYS ON TIME-SCALES OF SECONDS

V.Connaughton,M.Chantell,A.C.Rovero,T.Whitaker,T.C.Weekes
Whipple Observatory, S.A.O., P.O.Box 97, Amado, AZ 85645

C.W.Akerlof,D.I.Meyer,M.S.Schubnell
University of Michigan, U.S.A

D.J.Fegan,S.Fennell,J.Hagan,N.A.Porter,M.Punch
University College, Dublin, Ireland

J.Gaidos,G.Sembroski,C.Wilson
Purdue University, U.S.A

A.M.Hillas,J.Rose,M.West
University of Leeds, U.K

A.D.Kerrick,P.Kwok,D.A.Lewis,R.C.Lamb,G.Mohanty
Iowa State University, U.S.A

ABSTRACT

The Whipple Observatory gamma-ray telescope has a high sensitivity to sources of gamma rays in the 0.4 to 4 TeV energy range. Although this sensitivity is used primarily to search for discrete sources of gamma-rays the instrument also has sensitivity to gamma-ray bursts on time-scales from milliseconds to seconds. The field of view is limited but the source location capability is good. Such bursts could radiate with peak luminosity at TeV energies and could originate from (a) primordial black holes or (b) cosmic strings; they could also be the high energy counterparts of BATSE-type bursts and hence of unknown origin. The search of the Whipple data-base for statistically unlikely consecutive events on time-scales of seconds will be described and compared with the theoretical predictions.

INSTRUMENT

The 10 m Optical Reflector at the Whipple Observatory on Mount Hopkins in southern Arizona (elevation 2.3 km) has been used in studies of very high energy gamma rays and cosmic rays since 1968. It consists of 248 hexagonal mirrors mounted on a steel frame whose radius of curvature is equal to the focal length of the ground-glass, front- aluminized, mirrors. The point spread function has an angular width (FWHM) of less than 0.13°. The alt-azimuth mount can be programmed to track sidereally with an accuracy of $< 0.1°$. Since April, 1988, the reflector has been equipped with a fast High Resolution Camera consisting of 109 pmt's. The diameter of the inner 91 pixels is 0.25° and the diameter of the overall field of view is 3.5°. The camera is triggered when two tubes exceed a preset threshold; the trigger rate is between 3 and 10 Hz.

IMAGING TECHNIQUE.

Gamma-ray showers are identified by shape and orientation of the recorded Cherenkov light image; more than 99.7% of the background can be thus rejected with loss of less than 50% of the signal.

The telescope is routinely used on all clear dark nights to search for discrete sources of very high energy gamma rays. To date two sources have been detected with some certainty: the Crab Nebula (Vacanti et al. 1991) and Markarian 421 (Punch et al. 1992). Because of the high statistical significance of these detections the imaging technique can be used in a search for gamma-ray burst phenomena.

The unique advantage of this instrument is that its response to gamma-ray showers is well-known and hence the signature of a gamma-ray burst can be identified with confidence. At the same time the recorded background events can be used as an effective check against the occurrence of false bursts i.e. non-statistical clustering of triggers in time. This is particularly important for atmospheric Cherenkov detectors which are exposed to the night-sky and possible spurious sources of pulsed radiation.

Since 1977 the Whipple telescopes have been used in a variety of configurations to search for the gamma-ray emission from the final stages of evaporation of primordial black holes (Porter and Weekes, 1979). Most recently the High Resolution Camera on the 10 m optical reflector has been used in these searches (Nolan et al. 1990; Connaughton et al. 1991).

For the purposes of this search we define a candidate gamma-ray burst as an occurrence of n events with the shape expected of a gamma-ray shower in a time t. In addition the axes of the gamma-ray candidate images should intersect; the point of intersection can be outside the geometrical field of view.

OBSERVATIONS.

A pilot search for bursts of TeV gamma rays was undertaken using data from the 1991-1992 season of data-taking with the 10m High Resolution Imaging Camera.

A total of 45 hours taken between April and June 1992 covering a variety of sources comprised the final database. The data files were parameterized and a reduced data set containing gamma-ray like events was defined by selecting events on the basis of the shape of its resulting image. No event selection was attempted using the orientation parameter of the image since the potential burst sources could lie anywhere within the field of view. For the same reason, the traditional method of comparing the 'ON' source region with a run taken in a suitable 'OFF' source region is not a valid way of defining the background data set in this search.

The background files were produced by taking each complete parameterized data file, scrambling all the events while maintaining the original time sequence, and applying the image cuts to the new scrambled file. The events surviving the cuts would be identical to those in the real reduced data set but they would have a random distribution in time. This scrambling process was repeated so that each file had 25 corresponding background files.

METHOD.

To search for bursts in the above data set, a 1s moving window was applied to each file. The window was positioned on each event in the file and its time was compared with the times of the previous events. In this manner, bursts of photon-like events were isolated and the occurrences of two to five events per second counted.

The numbers obtained were compared to random expectation by applying the same technique to each set of 25 scrambled files and calculating the average occurrence of 2 to 5-fold coincidences.

A routine was developed to display the bursts as they appear on the PMT array of the detector by superimposing each reconstructed image on a $3° \times 3°$ grid along with the other images comprising the burst and their major axes.

RESULTS.

The observed number of bursts (3 candidate gamma-ray events in 1 sec) was 7; the predicted rate was 8.51. The number can be further limited by consideration of the orientation of three images which must intersect; the tolerance for intersection is the cone defined by the angle, alpha $< 15°$. In addition the displacement from the centroid of the image to the point of origin is a function of the ellipticity of the image. Only one of the seven candidate bursts satisfies these criteria; hence we derive an upper limit (at the 99% level) of 6.1 events in 1.64×10^5 sec or 1116 yr^{-1}.

THE PRIMORDIAL BLACK HOLES PHENOMENON.

In the early Universe primordial black holes (PBH) of arbitrarily small masses may have formed; the simplest mechanism is the gravitational collapse of overdense regions. Such PBH's in the Universe would go largely unnoticed were it not for the Hawking radiation; in particular the final stages of black hole evaporation may produce a burst of very high energy gamma-ray radiation that can be detected in ground-based telescopes.

A NEW LIMIT ON THE PBH DENSITY.

The gamma-ray burst limit derived above can be used to set a limit to the PBH density. The PBH burst luminosity used is that given in eqn. (16) in Halzen et al. (1991); for dt = 1 sec and $E_D = 0.5$ TeV we get $N_{\text{gamma}} = 8 \times 10^{28}$. We can thus determine the maximum distance that a PBH could be from the detector and still be detected; this is equal to 0.5 pc. For a collection area of 1×10^9 cm^2 and solid angle of 3×10^{-3} ster, we find the sensitive volume, V=1.2×10^{-4}pc^3. The upper limit to the PBH density,

$$N_{\text{pbh}} = R/V = 1116/1.2 \times 10^{-4} = 9.3 \times 10^6 \quad \text{PBH's per pc}^3 \text{ per yr.}$$

It should be noted that the limit from this pilot search is based on the analysis of less than 10% of the data-base from just one of the five years of observations in the Whipple Observatory High Resolution Camera data-base A more significant limit based on observations at a much higher energy threshold (> 50 TeV) over a 3.3 year observing period has recently been published (Alexandreas et al. 1993).

PREVIOUS PBH LIMITS REVISITED.

The original prediction of the expected burst power was that there would be 10^{30} gamma rays of energy 5 TeV emitted in 0.1 sec. No significant evidence was found for bursts with these characteristics and upper limits were derived based on this prediction. A re-evaluation of the final stages of pbh evaporation (Halzen et al. 1991) reduces the predicted emission and broadens both the spectrum and duration of the expected gamma-ray bursts. It is therefore necessary to revise the previously derived upper limits; these are listed below.

Experiment	Reference	N_{pbh}	New limit $\text{pc}^{-3} \text{ yr}^{-1}$
10m	Porter+Weekes,1979	7.1×10^{28}	6.0×10^8
S.L.'s	" "	7.7×10^{28}	4.9×10^7
10m (A)	Nolan et al,1990	3.0×10^{28}	9.3×10^6
10m (B)	" "	3.0×10^{28}	1.8×10^6
10m,high z	Connaughton et al, 1991	2.6×10^{27}	8.7×10^7

More recent work (Cline et al. 1993) suggests that uncertainties in the particle physics of the density of states of the final stages of PBH evaporation mean that a search based on a 1s time-scale is too restrictive and that future searches should involve windows ranging from 10^{-7} s to a few seconds.

GAMMA-RAY BURSTS FROM SUPERCONDUCTING COSMIC STRINGS.

The detection of a burst of TeV photons or neutrinos offers one of the few possibilities of directly detecting a cosmic string. Because of the greater sensitivity of TeV gamma-ray telescopes the photon detection appears more likely. Two possibilities are considered:

(a) Cosmic Strings

Ordinary (non-superconducting) Cosmic Strings (CS's) release extremely energetic particles in the cusp region on very short times-scales; these particles decay down to gamma rays and neutrinos which are emitted in narrow beams. MacGibbon and Brandenberg (1993) have calculated the maximum photon cusp radiation from individual CS's and compared it with the detection capabilities of existing ground-based gamma-ray telescopes. The duration of the signal at the detector is determined by the spreading out of the photon arrival times due to the decay process; this time is difficult to predict.

(b) Superconducting Cosmic Strings

Some kinds of cosmic strings behave like superconductors. In a recent study

(Samura and Kobayakawa, 1993), the gamma-ray emission from Superconducting Cosmic Strings (SCS's) was evaluated; the SCS emits heavy fermions in a jet when it reaches a saturated current. If the mass of fermions is greater than a few hundred TeV, the resultant burst from a SCS out to a distance of 1 kpc was shown to be detectable to a telescope with the sensitivity of the Whipple HRC; the time-constant of the bursts was 1 sec.

For a telescope with the sensitivity of the Whipple telescope they show that the absence of any detected bursts in a year of operation can set an upper limit to the number of strings of 10^8 out to a distance of 10 kpc.

BATSE COUNTERPART SEARCH.

To extend the spectra of GRB's to higher energies it is desirable to have simultaneous coverage of a burst by BATSE and an atmospheric Cherenkov telescope. Because of the limited field of view and low duty-cycle of the latter, this is not easy to achieve.

As shown above the Whipple atmospheric Cherenkov telescope has an energy threshold of about 0.5 TeV and a burst sensitivity of about 3 photons for a burst of duration 1 sec. A burst of 3 events corresponds to about 2.4 ergs, a fluence of $(2.4/10^9) = 3 \times 10^{-9}$ ergs-cm^{-2}. Non-imaging Cherenkov telescopes might have a factor of 2 lower sensitivity. This is better than BATSE and suggests that all the BATSE bursts with flat spectra (differential power index = -2) would be detectable by existing Cherenkov telescopes if their fields of view overlapped.

The probability of overlap in a single year with a single atmospheric Cherenkov telescope is small; however given the anticipated length of the CGRO BATSE mission (> 8 years) and the number of atmospheric Cherenkov telescopes now in operation ($\sim$ 10), the probability of overlap is > 1. We have begun a search for the overlap with ground-based telescopes in the first few years of BATSE operation. For this search we are using the observational data-bases of the following atmospheric Cherenkov observatories: Whipple Observatory, U.S.A. (Mt. Hopkins); CANGAROO, Adelaide-Tokyo (Woomera, Australia); Crimean Astrophysical Observatory, Ukraine; Themistocle, France (Themis); Institute of High Energy Physics, Beijing: Tata Institute, India (Pachmari); Potchefstroom, South Africa; Durham, U.K. (Narrabri, Australia); Wisconsin-Purdue (South Pole).

REFERENCES

Alexandreas, D.E. et al., 1993, Proc. 23rd ICRC (Calgary), 1, 428. Chantell, M. et al. 1992; Proc. CGRO Symposium (St. Louis), 833. Cline, D., Hong, W., 1993, Proc. of this workshop. Connaughton, V., et al. 1991; Proc. 22nd ICRC (Dublin), 1, 69. Connaughton, V. et al., 1993, Proc. 23rd ICRC (Calgary), 1, 112. Halzen, F. et al. 1991; Nature, 353, 807. MacGibbon, J.H., Brandenberger, R.H., 1993: Phys. Rev. D. 47, 2283. Nolan, K. et al., 1990, Proc. 21st ICRC (Adelaide), 2, 150. Samura,T. and Kobayakawa, K., 1993, Proc. 23rd ICRC (Calgary), 1, 128. Porter, N.A. and Weekes, T.C., 1979, Nature, 277, 199. Punch, M. et al., 1992, Nature, 358, 477. Reynolds, P.T. et al., 1993, Ap. J. 404, 206. Vacanti, G. et al., 1991, Ap.J. 377, 467.

SOUDAN 2 MUONS IN COINCIDENCE WITH BATSE BURSTS

David M. DeMuth, Marvin L. Marshak, Greg L. Wagner

Physics & Astronomy, University of Minnesota, MN 55455

ABSTRACT

We explore the possibilities of statistically significant temporal and spatial coincidences between underground muons at Soudan 2 and Gamma Ray Bursts at the GRO–BATSE detector. Our search uses data from the April 91 to March 92 BATSE burst catalog[1] to seek correlations within a 100 second window of coincidence. Sixteen of 180 BATSE triggers have temporally and spatially coincident muons in the Soudan 2 detector. We estimate the chance probability of each coincidence assuming the null hypothesis on the basis of a study of the multiplicities of spatially coincident muons observed over a two day period centered on the time of burst.

INTRODUCTION

The origin and mechanism of Gamma–Ray Bursts (GRB), discovered[2] in the seventies, continues to be a source of debate. For example, there is no agreement on whether the origin of this phenomenon is galactic or cosmological. Suggested models for the GRB mechanism include expanding super Eddington colliding shells and merging neutron stars.

The BATSE instrument, one of four on the Compton Gamma-Ray Observatory triggers on bursts at ~0.8 per day and has since its launch in April 1991 recorded in excess of 800 bursts. Simultaneously, the Soudan 2 underground proton decay detector continues to record ~1 TeV and greater cosmic ray muon events at a rate of ~0.5 Hz. By conventional mechanisms, there should be no correlation among these two data streams because (1) the energy range is different and (2) TeV photons do not efficiently produce muons. Nonetheless, past reports have suggested a controversial, but possible connection between high-energy astrophysical sources and underground muons. For this reason, we have examined the possibility of spatial and temporal coincidences between GRBs detected by BATSE and underground muons detected by Soudan 2.

THE SOUDAN 2 DETECTOR

The Soudan 2 Detector is located in northern Minnesota at a latitude of 47° 49" 13' N and longitude of 92° 14" 31' W under 2090 m.w.e. of rock composed mostly of silicon, oxygen and iron. Its primary purpose is to search for proton decay in iron sheets located in 224 independent modules stacked in an array of 8m by 14m by 5m high. Each module has dimension of 1m by 1m by 2.5m meters high and has mass of ~4.3 metric tons. Readout of the ionization created when the charged particles traverse the detector occurs at 14 mm spacings and result in a ~1° tracking resolution. The main detector is surrounded by yet another planar detector constructed of ~1500 extruded aluminum proportional tubes. This outer shield allows for the efficient vetoing of spurious proton decay events as well as the effective tagging of high energy muons. Muons generally traverse straight tracks through the detector as represented in Figure 1.

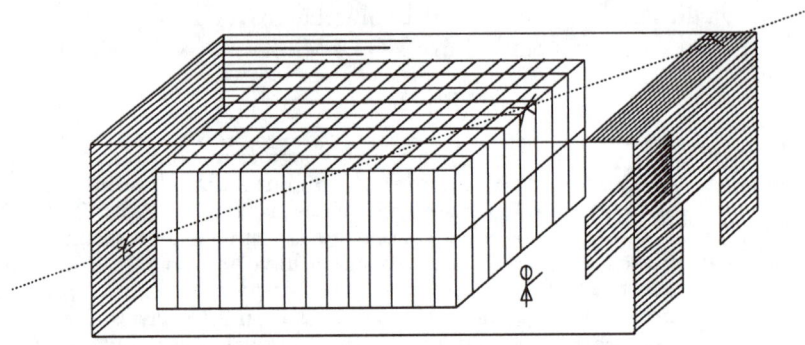

Figure 1. Soudan 2 Proton Decay Detector

Most muons observed underground result from meson decay originating from interactions of cosmic ray primaries with atmospheric nuclei. On average the initial interaction occurs at $\sim$75 gm cm^{-2} from the top of the atmosphere with primary average energy† of $\sim$20 TeV/nucleus. The differential muon flux at the surface has spectral index of -3.7. Because of the small muon interaction cross section, they are able to penetrate deep into the Earth. To reach Soudan 2, a muon surface energy of 0.8 TeV is necessary while the average surface energy of a detected muon is $\sim$1.3 TeV. Muons are detected at Soudan at a rate of $\sim$0.5 Hz. Since these atmospheric muons in Soudan 2 are well understood, a search for a signal above the atmospheric background is feasible.

THE SEARCH

We seek to assess the possible existence of statistically significant coincidences between Soudan 2 muons and BATSE detected photons. We consider 180 BATSE bursts, all of which have nonzero T_{90} duration times and a 10° or less spatial error. A muon database time coincident with the first BATSE catalog contains $\sim$5x10^6 single and multiple muons.[3] Our search technique is as follows. We make the necessary conversions of the BATSE burst coordinates in right ascension and declination to local detector coordinates. Then we make temporal and spatial comparisons with the Soudan 2 data. Temporal coincidence between the muons and photons are defined as when $UT_S \epsilon (UT_B - 50, UT_B + D_B + 50)$ where UT_S and UT_B are the universal time of the respective muon event and the burst and D_B is the BATSE T_{90} duration. The margin of 50 seconds before and after the BATSE defined time was chosen arbitrarily to allow for possible time jitter between photon emission and the emission of muon-parent quanta of the GRB. Figure 2 displays the number of muons satisfying the above temporal condition of coincidence as a function of BATSE trigger number.

† Energy, in lab frame of reference, and depth determined by HEMAS and GEANT monte carlo calculation.

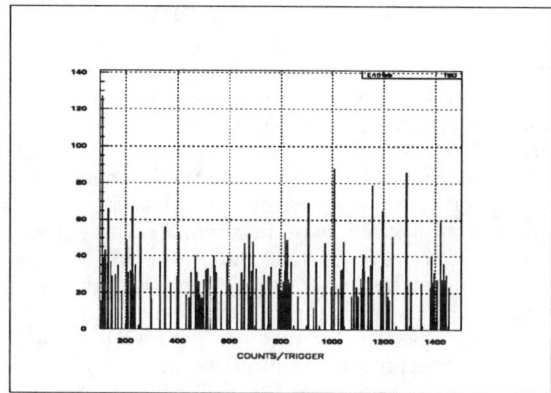

Figure 2. Temporal Coincidence

Spatial coincidence is achieved when the space angle between the BATSE direction and the Soudan 2 direction falls within the error cone defined by the quadrature summed systematic of 4° and statistical errors as reported in the BATSE catalog. Because Soudan 2 does not measure track direction, we consider coincidences under both the assumptions of a down–going or up–going track. We also require that the muon tracks pass standard Soudan 2 event and track cuts designed to assure event reliability. This search produced 24 candidate muon events that correspond to 16 trigger concidences. These events are listed in Table I. Figure 3 shows the locations of the 24 coincident muons events on an Aitoff projection of the celestial sphere. The largest muon multiplicity found was four muon events coincident with BATSE trigger 229.

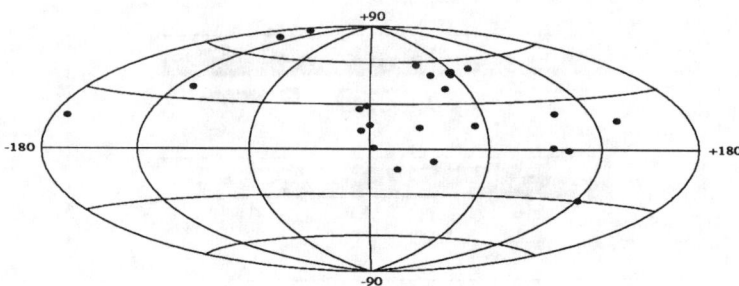

Figure 3. Coincident Muons in Galactic Coordinates

THE BACKGROUND

The existence of spatial and temporal coincidences between BATSE-detected bursts and Soudan 2-detected muons is not in itself surprising. The interesting question is whether the number of coincidences observed is more than expected by chance. To assess this probability, we determined the background–the number of chance coincidences in the absence of a real effect–by the following method. For each trigger, we opened an initial time window beginning the previous day, and searched over two days contiguous intervals of length of $I_B = 100 + D_B$ seconds for events that have spatial coincidence within the same spatial tolerance described above. We correct the results of this search for detector on-time. Table II indicates the chance probability of observing the actual number of muons observed or more muons within the described spatial cone and time interval. For example, 4 muon events were observed in coincidence with BATSE trigger 229. There is a 4% probability of observing 4 or more muons in Soudan 2 in spatial and temporal coincidence with that trigger. Work is in progress towards assessing the overall chance probability of BATSE–Soudan 2 coincidences listed in Table II. As an example of the background calculation, Figure 4 shows the background multiplicity distribution, the spatially coincident muon multiplicities and the omnidirectional muon multiplicities corresponding to the 2120 observation trials for length I_B of trigger 229 and the distribution of angular dot products between all spatially coincident muon events respectively.

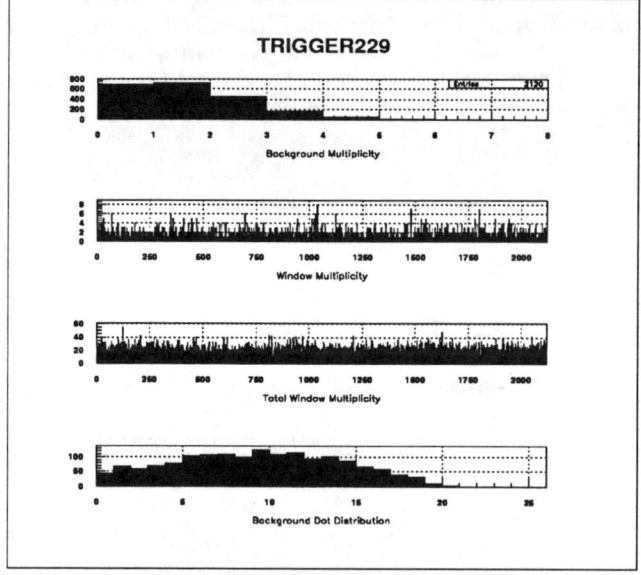

Figure 4. Trigger 229 Background

NEUTRINO INDUCED MUONS

It has been suggested[4] that the central engine of the GRB may produce a few dozen ~30 GeV– 1 TeV neutrinos per squared meters in conjunction with the high energy γ ejecta, (> Tev). Neutrinos induce muons when undergoing the charge-current interaction $\nu_\mu + N \rightarrow \mu + \pi's + K's + X$. A cross section[5] of $\sigma_\nu \sim 0.8 \times 10^{-38} \text{cm}^2 \times E_\nu$ for $E_\nu <$ 3600 GeV along with the expectation of ~1 ν/kton-year interacting within the confines of the Soudan 2 detector implies a search for neutrinos from GRBs is not feasible. However, the volume of an underground detector is effectively expanded because of its sensitivity to neutrino-induced muons from the rock, which makes the detection of neutrinos more likely. Muons with zenith angle of ~80° traverse ~10 km.w.e. of rock and are very likely to be induced by a ν_μ. Learned et al. estimate[6] ~1/8 neutrino induced interactions ton^{-1} year^{-1} for $E_\nu >$ 300 MeV. We expect to do further analysis of such large angle muon events to search for possible coincidences with BATSE-measured GRBs.

ACKNOWLEDGEMENTS

The collaborating institutions for the Soudan 2 experiment include Argonne National Laboratory, the University of Minnesota, Oxford University, the Rutherford–Appleton Laboratory, Tufts University and Western Washington University. This work was undertaken with the support of the the U.S. Department of Energy and the State of Minnesota. We wish to thank the Minnesota Department of Natural Resources for allowing us to use the facilites of the Tower-Soudan State Park and the dedicated laboratory workers, W. Miller, J. Beaty, J. Meier, G. Benson, D. Carlson, B. Anderson, J. Proton, and B. Dahlin.

REFERENCES

1. G.J. Fishman et al., The First BATSE Gamma–Ray Burst Catalog, September 1993.
2. R.W. Klebesadel, I.B. Strongby, and R.A. Olson, APJ **182**, L85 (1973).
3. P.J. Litchfield, R.A.L., SEARCH, Soudan Collaboration Track Fitting Software.
4. Bohdan Paczyński, These Proceedings..
5. T.K. Gaiser, Cosmic Rays and Particle Physics (Cambridge, 1990), p. 111.
6. J.G. Learned, F. Reines, and A. Soni, Phys. Rev. Lett. **43**, 907 (1979).

APPENDIX

Table I. BATSE – SOUDAN Coincidence

Trig #	Run	Event	UT	RA	Dec
110	27006	757	20364.11	174.52	-22.00
121	27109	1266	11468.98	177.45	15.84
121	27109	1283	11502.93	181.12	22.39
207	27541	108	12342.38	229.88	55.13
207	27541	112	12345.72	211.96	61.26
229	27727	1145	16731.71	227.67	55.82
229	27727	1156	16752.94	232.36	55.80
229	27727	1168	16781.21	251.18	51.24
229	27727	1176	16791.99	237.18	46.86
237	27763	499	63381.52	41.30	25.65
373	28201	758	38383.88	300.33	5.04
685	29720	589	43936.84	33.98	60.04
829	30761	413	84414.06	228.04	41.94
840	30847	1603	26627.75	95.31	49.27
1046	31750	95	80449.42	305.54	23.08
1086	31918	21	85162.37	267.63	6.64
1086	31918	21	85162.37	255.59	6.72
1086	31918	37	85194.49	268.89	1.03
1086	31918	37	85194.49	255.13	10.75
1102	31960	228	80878.02	279.90	28.22
1096	31967	419	28387.62	17.54	63.28
1112	32050	30	15879.29	67.39	80.18
1123	32109	16	25269.01	120.52	0.57
1382	33927	105	78021.92	293.92	54.05

Table II. Statistics

Trig #	Mult.	Prob.
110	1	0.09
121	2	0.09
207	2	0.11
229	4	0.04
237	1	0.37
373	1	0.42
685	1	0.63
829	1	0.12
840	1	0.18
1046	1	0.38
1086	2	0.02
1102	1	0.41
1096	1	0.38
1112	1	0.22
1123	1	0.22
1382	1	0.34

SEARCH FOR ULTRA HIGH ENERGY RADIATION FROM GAMMA-RAY BURSTS

Richard Schnee

Physics Board, University of California, Santa Cruz, CA 95064

representing

The CYGNUS Collaboration

D.E. Alexandreas,[a,b] G.E. Allen,[c] D. Berley,[c,d] S. Biller,[a] R.L. Burman,[e]
M. Cavalli-Sforza,[f] C.Y. Chang,[c] M.L. Chen,[c] P. Chumney,[a] D. Coyne,[f]
C. Dion,[c] G.M. Dion,[a,g] D. Dorfan,[f] R.W. Ellsworth,[h] J.A. Goodman,[c]
T.J. Haines,[c] M. Harmon,[a] C.M. Hoffman,[e] L. Kelley,[f] S. Klein,[f]
D.E. Nagle,[e] S.C. Schaller,[e] D.M. Schmidt,[e] R. Schnee,[f] C. Sinnis,[e]
A. Shoup,[a] M.J. Stark,[c] D.D. Weeks,[e] D.A. Williams,[f] J.-P. Wu,[i]
T. Yang,[f] G.B. Yodh,[a] and W. Zhang[e,j]

[a] The University of California, Irvine, CA 92717.
[b] Now at Istituto Nazionale di Fisica Nucleare, Padova, Italy.
[c] The University of Maryland, College Park, MD 20742.
[d] Permanent address: The National Science Foundation, Washington, DC 20550.
[e] Los Alamos National Laboratory, Los Alamos, NM 87545.
[f] The University of California, Santa Cruz, CA 95064.
[g] Now at ICRR, University of Tokyo, Tokyo, Japan.
[h] George Mason University, Fairfax, VA 22030.
[i] The University of California, Riverside, CA 92521.
[j] Now at NASA Goddard Space Flight Center, Greenbelt, MD 20771.

ABSTRACT

A recent model suggests that some gamma-ray bursts may have photon energies as high as 10 TeV. Extensive air shower arrays could detect strong bursts with hard spectra extending above such energies. Using data from the CYGNUS array, we have searched for evidence of emission of ultra high energy ($\gtrsim$ 100 TeV) radiation coincident with gamma-ray bursts observed either by BATSE or by the third Interplanetary Network. No statistically significant excess was found for the six bursts whose locations were accurately determined by the IPN, or for any point in the sky within 2σ of the BATSE location coordinates of an additional 52 bursts. Flux upper limits depend greatly on the zenith angle of the burst, but typical limits above 100 TeV are $\sim 4 \times 10^{-10}$ cm^{-2} s^{-1}. The flux upper limits for three bursts imply that the observed spectrum softens between 2 MeV and $\sim$100 TeV.

INTRODUCTION

The origin of gamma-ray bursts (GRB's) remains a mystery in part due to the lack of observed counterparts at other wavelengths. Although detecting lower energy counterparts seems more likely, searches at ultra high energies are also important. At least one recent theory[1] suggests that some GRB's may have photon energies as high as 10 TeV. Furthermore, the EGRET instrument[2] has detected bursts with hard power-law spectra, no evidence of softening, and photons with energies as high as 10 GeV.

The CYGNUS array could detect the hardest observed GRB's if the bursts' energy spectra continue to ultra high energies. For example, the differential flux measured with EGRET[3] at ~10 MeV during the 7-second burst 1B910503 has a differential spectral index of about $\gamma = -2.2$. Extrapolating this flux to ultra high energies results in about 7 detected source events expected in a bin with 0.1 expected background events for a source overhead; for a zenith angle $\theta \approx 30°$, about 2.5 detected source events and 0.02 background events are expected. Unfortunately, this GRB and the very strong "superbowl" burst[4-7] detected on 31 January 1993 were not in the CYGNUS field of view.

Many bursts detected at lower energies (20 keV to 2 MeV) by BATSE[8] were in the CYGNUS field of view. Some of these bursts also would be detectable if their spectra continued to ultra high energies. Extrapolating the $\gamma = -1.7$ spectrum[9] measured by BATSE for burst 1B911104 results in about 2500 expected source events in a bin with fewer than 0.01 expected background events.

We have examined the CYGNUS data set for evidence of ultra high energy (UHE) emission coincident with bursts detected by BATSE and/or the third Interplanetary Network[10] (IPN). BATSE has been detecting GRB's at a rate of about one per day since it began operation in April 1991.[11] BATSE has a flux threshold of $\sim 10^{-7}$ erg cm^{-2} s^{-1} and is sensitive to γ-rays in the energy range of 20 keV to 2 MeV.[11] The IPN, which has consisted of Ulysses, the Pioneer Venus Orbiter, the Compton Gamma-Ray Observatory, and other satellites, accurately localized about 30 GRB's by triangulation from three spacecraft.

We have considered the 260 bursts in the first BATSE catalog,[12] as well as 18 bursts whose locations were well determined by the IPN[13] between April 1991 and July 1992. Six of the IPN bursts and 52 of the BATSE bursts (according to BATSE's best location coordinates) had $\theta \leq 60°$ and so were in the CYGNUS field of view.

THE CYGNUS EXPERIMENT

The CYGNUS-I extensive air shower array,[14] located in Los Alamos, New Mexico, has been operating since April 1986. The trigger rate has been about 3.5 s^{-1} since the summer of 1989. The angular resolution is 0.7°.[15] The energy threshold depends strongly on the zenith angle of the incident particle. The median energy of detected γ-rays (E_{med}) is about 50 TeV for showers from zenith, about 90 TeV for showers with zenith angle $\theta \approx 30°$, and about 1000 TeV for showers with $\theta \approx 60°$.

METHOD OF SEARCH

We search for UHE emission during T_{90}, the time BATSE received 90% of the burst flux, rounded up to at least a second (burst durations vary from hundredths to hundreds of seconds). The observed number of events from the direction of the burst is compared to the expected number of background events, as calculated using data within a few hours of the burst. For each of five well located IPN bursts a single circular bin of 1.5° is used. This bin size maximizes the sensitivity of the search.[16] For one IPN burst, the source locus is a short arc, so three overlapping such bins are used.

The remaining 52 bursts from the BATSE catalog have 1σ location errors ranging from 4° to 19° (much larger than the angular resolution of the CYGNUS array). There is a 92% probability that a burst is within 2σ of the reported location.[12] Therefore, an angular area with radius twice that of the quoted BATSE 1σ error around each burst and with $\theta < 60°$ is searched by dividing it into a non-overlapping grid of (44 to 665) square bins with dimensions 2.6° in declination (δ) by $2.6°/\cos\delta$ in right ascension (α). The number of observed events in each of the bins is compared with the number expected from the cosmic ray background. To ensure sensitivity to a source near the edge of a bin on this grid, the process is repeated with the grid shifted 1/2 of a bin width in α, then 1/2 of a bin width in δ, then 1/2 of a bin width in both.

For each of the five well localized IPN bursts, the probability P_f of background fluctuations yielding at least the number of observed events is determined, including the uncertainties in the background calculation. For these bursts, the burst probability P_{burst} is simply P_f. For each BATSE burst, and for the remaining IPN burst, P_f is found for each bin, and the smallest probability is called P_{min}. Then P_{burst} is the probability of background fluctuations yielding at least one P_f as small as P_{min}. Figure 1 shows the integral distribution of the burst probabilities P_{burst}.

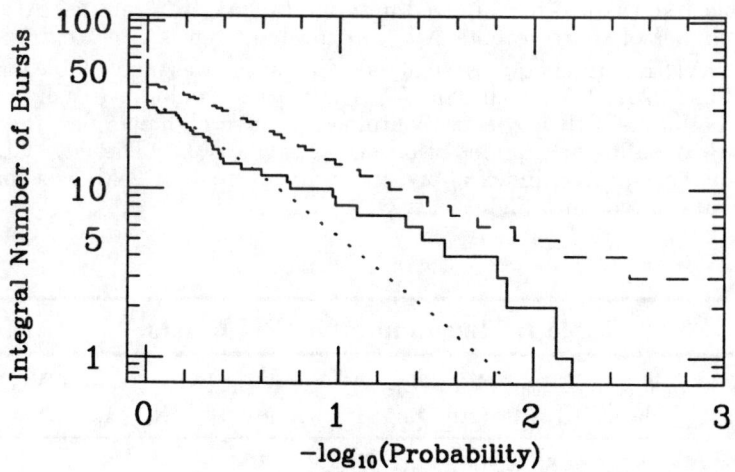

Figure 1. Integral number distribution of the 58 bursts as a function of post-trials burst probability P_{burst}. The dotted line shows the expected distribution, which exhibits a discontinuous jump at $-\log_{10} P_{burst} = 0.0$ due to the finite number of bursts expected to have no events in the entire area searched. The dashed line indicates the excess needed to yield a 1% confidence level for the sample.

To test this distribution, we determine whether the most significant excess is consistent with no signal. Final burst probabilities P_{burst} are sorted into ascending order $P_1, P_2, \ldots, P_{58}$, where P_i is the ith smallest probability of any burst in the sample. For each P_i we calculate the probability that i or more bursts have probabilities less than or equal to P_i.

RESULTS

The most significant excess in the distribution corresponds to four bursts with $P_{\text{burst}} \leq 0.015$. The probability of background fluctuations causing so unlikely an excess at some point in the distribution is 10%, consistent with the hypothesis of no signal.

The 90% CL upper limit on the flux of photons above an energy E is obtained by normalizing the upper limit on the number of excess events to the background cosmic ray flux (see ref. 15 for details). The upper limits depend on the source zenith angle, the source spectrum at earth, and the assumed cosmic-ray composition. For extragalactic sources, absorption[17] by intergalactic radiation fields[18-20] may drastically reduce the flux above the detector's energy threshold (for quantitative results, see ref. 21). Here we assume no absorption and a source spectrum matching the $E^{-2.7}$ cosmic ray spectrum, but the upper limits change by $\lesssim 10\%$ for spectra as hard as E^{-2}. Flux upper limits for sources with small zenith angles ($\sim 30°$) and moderate durations (~ 10 s) are typically

$$F_{\text{ul}}(> 100 \text{ TeV}) \sim 4 \times 10^{-10} cm^{-2} s^{-1}. \quad (1)$$

Table I shows the flux upper limits for the six IPN bursts. Also given are the number of source events N_{extrap} expected from a direct extrapolation using the BATSE high-energy spectral index β,[9] which was unavailable for bursts 1B910421 and 920720. Although the CYGNUS array would have easily detected three of the bursts if their spectra continued to ultra high energies, there is no significant excess in the number of observed events N_{obs}. The 90% CL upper limit β_{ul} on the spectral index above an assumed break at 2 MeV is smaller if the spectrum softens at a higher energy.

Table I. Flux Limits for IPN bursts.

Burst	θ_{zen} (deg.)	E_{med} (TeV)	$F_{\text{ul}}(> E_{\text{med}})$ (cm^{-2} s^{-1})	BATSE index β	N_{extrap}	CYGNUS β_{ul}	N_{obs}
1B 910421	27	80	9.5×10^{-10}				0
1B 910627	51	600	3.6×10^{-10}	-2.03 ± 0.02	1		0
1B 911104	40	100	3.4×10^{-9}	-1.67 ± 0.03	2500	-2.03	1
1B 911126	31	90	1.0×10^{-10}	-2.02 ± 0.05	29	-2.16	0
920325	26	70	2.2×10^{-9}	-2.04 ± 0.02	8	-2.12	0
920720	30	90	5.8×10^{-10}				0

CONCLUSIONS

For the 58 visible bursts, we saw no evidence for emission of UHE radiation. The results for three bursts clearly indicate either a softening of the production spectrum at high energies or the presence of UHE γ-ray absorption. Such absorption would imply a cosmological origin for the bursts.

ACKNOWLEDGMENTS

We thank the BATSE collaboration, G. Fishman spokesman, for sharing of preliminary data on γ-ray bursts. We are indebted to C. Kouveliotou of the BATSE Collaboration for transmission of the burst data and for extended communications concerning the operation of BATSE and interpretation of the data. We thank M. Brock for information on BATSE location errors. We are grateful to T. Cline, K. Hurley, and their collaborators in the third Interplanetary Network, for detailed information on burst locations. We thank R.S. Delay for his vital assistance in maintaining and operating the experiment and B. Dingus for very helpful discussions. This work is supported in part by the National Science Foundation, Los Alamos National Laboratory, the U.S. Department of Energy, and the Institute of Geophysics and Planetary Physics of the University of California.

REFERENCES

1. P. Meszaros, P. Laguna, and M. J. Rees, Astrophys. J. **415**, 181 (1993).
2. B. L. Dingus *et al.*, these proceedings (1993).
3. E. J. Schneid *et al.*, Astron. Astrophys. **255**, L13 (1992).
4. C. Kouveliotou *et al.*, Astrophys. J. Lett. (in press).
5. J. Ryan *et al.*, Astrophys. J. Lett. (in press).
6. M. Sommer *et al.*, Astrophys. J. Lett. (in press).
7. K. Hurley *et al.*, these proceedings (1993).
8. G. J. Fishman *et al.*, in Proc. Gamma Ray Observatory Workshop, edited by W. N. Johnson (NASA/GSFC, Greenbelt, 1989), p. 39.
9. D. Band *et al.*, Astrophys J. **413**, 281 (1993).
10. K. Hurley *et al.*, Astron. Astrophys. Supp. **97**, 39 (1993).
11. C. A. Meegan *et al.*, Nature **355**, 143 (1992).
12. G. J. Fishman *et al.*, Astrophys. J. Supp. (in press).
13. K. Hurley and T. Cline (private communications).
14. D. E. Alexandreas *et al.*, Nucl. Instrum. Methods **A311**, 350 (1992).
15. D. E. Alexandreas *et al.*, Astrophys. J. **405**, 353 (1993).
16. D. E. Alexandreas *et al.*, Nucl. Instrum. Methods **A328**, 570 (1993).
17. A. I. Nikishov, Zh. Eksp. i Teor. Fiz. **41**, 549 (1961) [Sov. Phys. JETP **14**, 393 (1962)].
18. R. J. Gould and G. Schreder Phys. Rev. Lett.**16,** 252 (1966); Phys. Rev. **155,** 1404 (1967); Phys. Rev. **155,** 1408 (1967).
19. J. V. Jelley, Phys. Rev. Lett. **16**, 479 (1966).
20. F. W. Stecker, O. C. De Jager, and M. H. Salamon, Astrophys. J. Lett. **390**, L49 (1992).
21. D. E. Alexandreas *et al.*, Astrophys. J. (in press).

THE IDENTIFICATION OF A SUPERNOVA REMNANT WITH A SOFT GAMMA RAY REPEATER

D. A. Frail

National Radio Astronomy Observatory, Socorro, NM 87801

S. R. Kulkarni

Division of Physics, Mathematics and Astronomy, 105-24, Pasadena, CA 91125

ABSTRACT

Earlier we had shown that the well-localized SGR 1806−20 was coincident with the radio supernova remnant G 10.0−0.3, with a low chance probability. Recent high energy observations have left little doubt as to the veracity of this association. We present new radio observations of G 10.0−0.3 revealing a hierarchy of nested, amorphous regions, in which a distinctive point-like peak is seen. We consider two alternative models for SGR 1806−20. Either it is a pulsar like the Crab or it is an accreting X-ray binary like Cir X-1. The radio and X-ray data suggest that some type of hybrid model is required.

INTRODUCTION

G 10.0−0.3 is listed as a supernova remnant in the catalog of Green (1991). This classification is based on data taken at 5 GHz and 408 MHz by Shaver and Goss (1970), who identified G 10.0−0.3 as an extended, non-thermal radio source in the Galactic plane. These low resolution images left the true morphology of G 10.0−0.3 uncertain. To resolve this issue we acquired some VLA data at 20 and 90 cm (Kulkarni and Frail 1993). These images suggest that G 10.0−0.3 is an evolved, amorphous supernova remnant, consistent with an old ($t_{age} < 10^4$ yrs) pulsar-powered nebula or "plerion".

Kulkarni and Frail (1993) argued that G 10.0−0.3 was coincident with the soft gamma ray repeater SGR 1806−20, with a small chance probability (3×10^{-3}). This prediction received a timely confirmation when a hard X-ray burst was detected simultaneously by BATSE and the ASCA satellite at the position of G 10.0−0.3 (Kouveliotou, et al. 1993, Tanaka et al. 1993). Kulkarni & Frail concluded that all SGRs were associated with supernova remnants and that if SGRs were neutron stars, as is commonly thought (e.g. Norris et al. 1991), then they must be *young* neutron stars. Since SGRs appear to be rare we further speculated that the SGR phenomena takes place in only a small subset of the young pulsar population (1-2%) or when *all* pulsars undergo brief (∼ 500 yr) but major structural changes in their youth.

NEW OBSERVATIONS

Despite what had been learned from the radio images in Kulkarni & Frail (1993) the quality of the VLA data was still sufficiently poor to leave open several questions about the source. On Sept 14.0 1993 a multifrequency observing campaign was conducted at 0.3, 1.4, 4.8 and 8.4 GHz to understand the true nature of the source via the usual diagnostics of continuum radio astronomy: morphology, and α, the spectral index. Here the flux density at frequency ν is

$S_\nu \propto \nu^{-\alpha}$.

At the lowest frequencies an amorphous nebula of size 9×6 arcmin^2 is apparent (Figure 1). The total flux in these essentially fully sampled images is 3.3 Jy and 1.3 Jy at 0.3 GHz and 1.4 GHz, respectively, giving a spectral index $\alpha = -0.6$. At 1.4 GHz a plateau of emission with a full radius of 40″ can be seen located slightly northeast of the nominal center of G 10.0−0.3. At the higher resolution of the 4.8 and 8.4 GHz images this plateau is resolved into a compact core region with size of few arcseconds and complex filamentary structure (Figure 1). This central plateau region also has $\alpha = -0.6$, as measured between 1.4 GHz and 4.8 GHz.

The morphology of the radio images - the emission from the central regions and the hierarchical structures culminating in a peak - is most easily understood in a model in which a compact source located at the core is the source of a relativistic flow. This realization led us to assert (Kulkarni, Frail, Kassim, Murakami & Vasisht 1993a) that SGR 1806−20 (now more appropriately called SGR 1805−20) was located at the radio core.

Kulkarni, Frail, Kassim, Murakami & Vasisht (1993b) consider two alternative models for the radio nebula G 10.0−0.3. Either it is (1) a plerion, a pulsar powered synchrotron nebula like the Crab Nebula, or (2) a nebula powered by jets emanating from an accreting binary such as the radio nebula associated with Cir X-1 (Stewart, Caswell, Haynes & Nelson 1993) or the nebula W50 associated with SS 433. Plerions are powered by a relativistic wind flowing out from a spinning down pulsar, while in Cir X-1 and other such rare binaries, some fraction of the accretion power comes out in the form of collimated relativistic outflows which energize the nebula.

The radio and X-ray observations are used to place constraints and to make predictions for the two models. To summarize the results of Kulkarni *et al.* (1993b), the radio spectral index and the nested morphology of G 10.0−0.3 favor a Cir X-1 type of model, whereas the plerion model is favored by the X-ray data and the absence of radio variability (see below). Clearly a hybrid model is required to satisfy both these observations. We propose that the persistent luminosity is due to a Vela-like pulsar which, for some unspecified reason, undergoes super-Eddington bursts of particles. The particles are not injected in a spherical wind, as in a pulsar, but in collimated outflows or jets, as in sources like SS433 or Cir X-1. Indeed a hint of such a flow is visible in the 8.4 GHz image (Figure 1). We note that the time averaged burst luminosity is comparable to the persistent luminosity (from X-rays), suggesting that the burst mechanism is not an insignificant drain on the energy budget of the pulsar.

CONTINUING OBSERVATIONS

As part of an ongoing effort to better to understand the SGR phenomena we have initiated a number of observing efforts. Some of these are summarized here.

The model described above can be tested with future high resolution radio imaging and deep IR observations of the core region. Radio monitor of G 10.0−0.3 has been ongoing since the SGR recently became active again at high energies (Kouveliotou, *et al.* 1993). Weekly or bi-monthly observations have been made at both 20 cm and 3.6 cm. As of this date (November 29, 1993) there had been no variations in the total radio flux of G 10.0−0.3 or the peak flux of the radio core, above 10% at either frequency.

In the hope of finding a unified model for SGRs we have searched for a suitable SNR candidate for the third SGR, SGR 1900+14. The error ellipse for SGR 1900+14 is large and intersects the Galactic plane and therefore there are many candidate objects at radio wavelengths. However, there appears to be only one interesting object that has the required properties to contain an SGR (Vasisht, Kulkarni & Frail 1993).

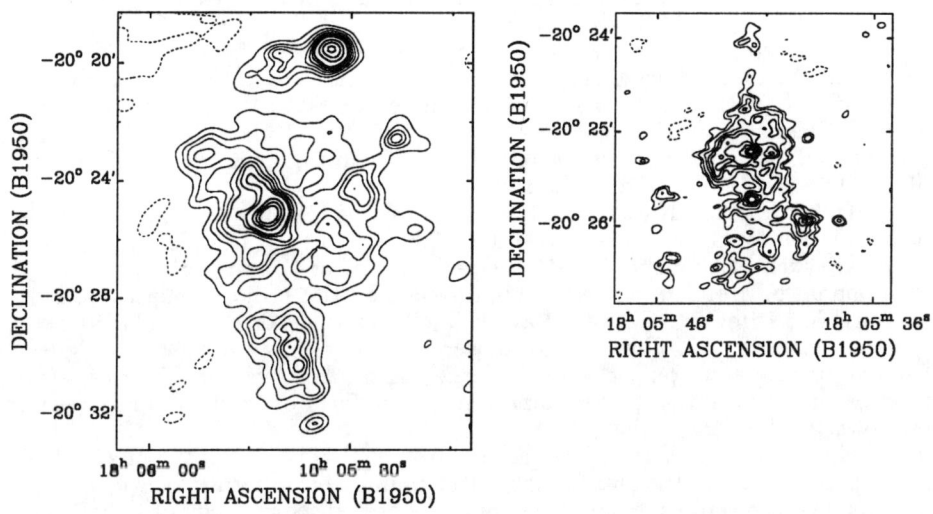

Figure 1. Radio images of the supernova remnant G 10.0−0.3 at 1.4 GHz (left) and 8.4 GHz (right).

REFERENCES

1. Green, D. A., Pub. Astron. Soc. Pacific **103**, 209-220 (1991).
2. Kulkarni, S. R. & Frail, D. A., Nature **365**, 33-35 (1993).
3. Kulkarni, S. R., Frail, D. A., Kassim, N. E., Murakami, T. & Vasisht, G., I.A.U. Circ. (5879, 1993a).
4. Kulkarni, S. R., Frail, D. A., Kassim, N. E., Murakami, T. & Vasisht, G., Nature (submitted, 1993b).
5. Kouveliotou, C. et al., Nature (submitted, 1993).
6. Norris, J. P, Hertz, P, Wood, K. S. & Kouveliotou, C., Astrophys. J. **366**, 240-252 (1991).
7. Shaver, P. A. & Goss, W. M., Austr. J. Phys. Astrophys. Suppl. **14**, 133-196 (1970).
8. Stewart, R. T., Caswell, J. L., Haynes, R. F. & Nelson, G. J., Mon. Not. Roy. Astr. Soc. **261**, 593-598 (1993).
9. Tanaka, Y. et al., Nature (submitted, 1993).
10. Vasisht, G., Kulkarni, S. R. & Frail, D. A., Nature (submitted, 1993).

X-RAY IDENTIFICATION OF SGR1806-20

T. Murakami, T. Sonobe, Y. Ogasaka, T. Aoki
Inst. of Space and Astronautical Sci. Sagamihara, Kanagawa 229 Japan

A. Yoshida
Inst. of Physical and Chemical Res. Wako, Saitama 351 Japan

S. R. Kulkarni
Cal-Tech , 105-24, Pasadena, California 91125, USA

ABSTRACT

Among gamma-ray burst sources (GRBs) three clearly repeat (SGR) and produce bursts that distinguish themselves from "classical" gamma-ray bursts by having a soft spectrum and a short duration.[1-6] However, no conclusive identification of an SGR burst with a compact source has yet been made. We report here the identification of SGR1806-20 with AX1805.7-2025, a new, persistent X-ray source, by the imaging of a burst with the Japanese X-ray satellite ASCA.[7] The burst is coincident in time with that detected by BATSE[8,9] on board the CGRO. The position of the X-ray source is coincident with the point source in the supernova remnant (SNR) G10.0-0.3, which was identified as the radio counterpart by Kulkarni et al.[10,11] This result provides strong evidence for a neutron star origin of SGRs.

OBSERVATIONS

On September 29, 1993, the BATSE team discovered that SGR1806-20 was actively bursting again[8], almost a decade after the last activity was observed in 1983.[4] This, happily, forced us to conduct observations with ASCA for two days between October 9.67 and 11.60 UT, 1993. Two days of TOO observation was extraordinary for ASCA. The revised location of SGR1806-20 is a narrow diamond-shaped region approximately 4 arcmin wide and 1° 20′ long (see Fig. 1 of Kulkarni and Frail[12]). The observations were performed at four consecutive positions A to D from north to south so as to cover the entire error region with the four focal plane detectors; two CCD cameras (SIS) with a 20 × 20 arcmin square field of view (FOV) and two gas scintillation imaging spectrometers (GIS) with a 50 arcminute diameter circular FOV. The effective integration time was 15 ksec each for positions A, C, and D, and 30 ksec for position B—two days in total. Prior to our analysis, the BATSE team communicated to us (private communication, C. Kouveliotou) that they had detected a burst of 28 msec duration from SGR1806-20 at 22h24m58s UT, October 9. At this time, ASCA was pointed to position A, and we were fortunate to have detected this burst[7] as shown in Fig. 1. The monitor of GIS showed a sharp increase of about 170 counts for the sum of two GIS.

Each GIS event is processed on board and the position on the focal plane (x-y coordinates) and the energy are determined for each qualified event. The observation mode employed during this pointing was a medium bit-rate mode, so readout was performed every 31.25 ms. Limited by this readout frequency

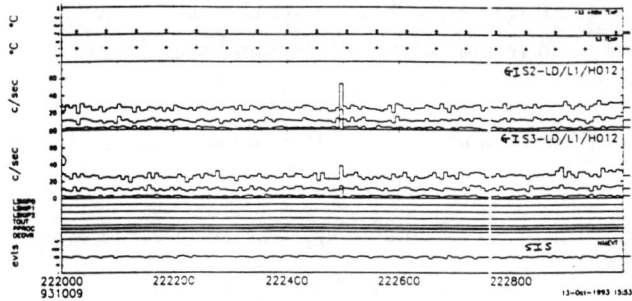

Figure 1. Counting rate histories of the monitors with 4 sec bin. At 24m58s there are sharp increases of counts only in GIS counters. For SIS, the event is outside the FOV.

and the memory capacity, only ten photons with position and energy data out of the total about ∼170 incident photons were recorded. As ASCA has been designed to focus on imaging and spectroscopy observations, there is no capability to handle such a high counting rate. In the same observation, a previously uncatalogued persistent X-ray source was discovered. It is clearly visible in the GIS image shown in Fig. 2. The individual positions of ten burst photons are plotted in the same GIS image. The image suffers a considerable distortion (oval-shaped) near the rim of the focal plane at position A. This is a consequence of the off-axis aberration in the ASCA telescope optics and the image non-linearity of GIS for objects close to the edge of the FOV. These effects have been incorporated in determining the burst location by employing a measured off-axis point response function, which is from an observation of the Crab Nebula at about the same off-axis angle. The 90% confidence error region, as derived from statistical analysis of ten photons, is approximately 2×4 arcmin (Fig. 2), and coincides with the newly discovered X-ray source AX1805.7−2025.

This source was observed again in a subsequent pointing nearly at the center of the FOV (position B) of both SIS and GIS, which allowed us a more precise determination of its location. The celestial coordinates of the source are R.A. = 18h05m41s, Decl. = $-20°25'07''$ (equinox 1950.0) with a systematic error circle radius of 1 arcmin. The X-ray spectrum can be expressed by a power-law of a photon index 2.2±0.2 (90% confidence limit) with a heavy absorption by a column of $\sim 6 \times 10^{22}$ H atoms cm^{-2}. The flux in the range 2–10 keV is approximately 6×10^{-12} erg cm^{-2} s^{-1} as observed. The intensity remained constant over two days, and no change was observed before and immediately after the burst.

The region containing AX1805.7−2025 was previously surveyed by many satellites, and the source was undetected.[13] The lowest upper limits is 1×10^{-11} erg cm^{-2} s^{-1} (September, 1988) from the Ginga observation.[14]

DISCUSSION

Recently, Kulkarni et al. conducted a multi-band radio observation of

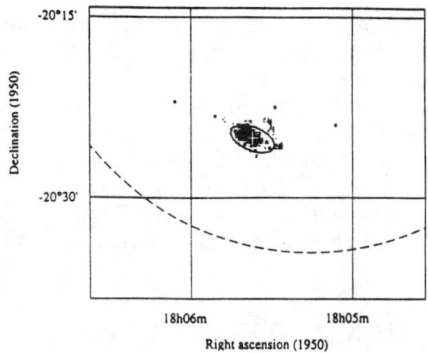

Figure 2. The GIS image obtained in position A. A persistent X-ray source, AX1805.7−2025, appears near the edge of the 50 arcmin diameter FOV (outer dashed circle). The individual positions of ten photons from the burst are shown with filled squares. The 90% confidence error region for the burst is also indicated.

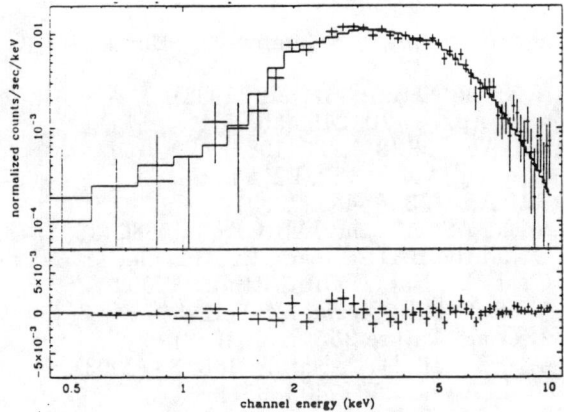

Figure 3. Spectrum of the persistent X-ray source AX1805.7−2025. The best fit model is a power-law of 2.2±0.2 with a heavy absorption of 6×10^{22} H atoms cm^{-2}

SNR G10.0–0.3 at the Very Large Array (VLA). In the images obtained on September 14, 1993, they found a compact nebula with the centroid at R.A. = 18h05m41s.76, Decl. = −20°25′13″ (equinox 1950.0), superposed on an extended plateau of emission.[10,11] They argue that the morphology is indicative of a "plerion," a centrally brightened pulsar-powered synchrotron nebula. Remarkably, AX1805.7−2025 is coincident with the centroid of the compact radio nebula. We may conclude that AX1805.7−2025 is the burst source, not only from the positional coincidence but also from other unique features. First, the X-ray spectrum of AX1805.7−2025 resembles that from a synchrotron source, and second, the flux ratio between X-ray and radio are consistent with a synchrotron source such as the Crab Nebula. The persistent luminosity of AX1805.7−2025 estimated at 15 kpc in distance is roughly 10^{35} erg cm^{-2} sec^{-1} also support a synchrotron source. Thus, the pulsar model provides the most straightforward interpretation of our data. Clearly, the most decisive evidence for a pulsar is the detection of pulsations. We are in the process of searching for pulsation in the much longer observation of Oct. 20, which is the follow-up of the Oct. 9 observation. The present result is the first identification of a SGR with an X-ray point source and also provides strong evidence for a neutron star origin of the SGR's. Together with the earlier association of the SNR N49 in the Large Magellanic Cloud with SGR0526-66,[6] we claimed that these two associations argue in favor of a neutron star.

ACKNOWLEDGEMENTS

The authors are grateful to C. Kouveliotou and the BATSE team for sending a quick notice and providing prompt information of bursts from SGR1806-20. They are indebted to Prof. Y. Tanaka and the ASCA team member for the strong support in this observation.

REFERENCES

1. Liang, E.P., and Petrosian, V., Gamma-Ray Burst (AIP Conference Proc. 141, 1984).
2. Kouveliotou, C. et al., Nature **362**, 728 (1993).
3. Atteia, J-L. et al., Ap. J. **320**, L105 (1987).
4. Laros, J.P. et al., Ap. J. **320**, L111 (1987).
5. Kouveliotou, C. et al., Ap. J. **322**, L21 (1987).
6. Cline, T.L. et al., Ap. J. **348**, 485 (1990).
7. Tanaka, Y. and the ASCA team, IAU **Circle**, 5880 (1993).
8. Kouveliotou, C and the BATSE team, IAU **Circle**, 5875 (1993).
9. Kouveliotou, C, et al., Nature **submitted**, (1993).
10. Kulkarni, S.A. et al., IAU **Circle**, 5879 (1993).
11. Kulkarni, S.A. et al., Nature **255**, L45 (1982).
12. Kulkarni, S.A. and Frail, D.A., Nature **365**, 33 (1993).
13. Norris, J.P. et al., Ap. J. **366**, 240 (1991).
14. Murakami, T. et al., Astron. & Astrophys **227**, 451 (1990).
15. Stewart, R.T. et al., MNRAS **261**, 593 (1993).

BURST ORIGINS AND EMISSION PROCESSES (THEORY)

Cosmological Models
Galactic Models
Soft Gamma-Ray Repeaters—Theory

FIREBALLS

Tsvi Piran

Racah Institute for Physics, The Hebrew University, Jerusalem 91904, Israel

ABSTRACT

The sudden release of copious γ-ray photons into a compact region creates an opaque photon–lepton fireball due to the prolific production of electron–positron pairs. The photons that we observe in the bursts emerge only at the end of the fireball phase after it expanded sufficiently to become optically thin or after it converted its energy to the kinetic energy of relativistic baryons which convert it, in turn, to electromagnetic pulse via the interaction with interstellar matter. It is essential, therefore, to analyze the evolution of a fireball in order to comprehend the observed features of γ-ray bursts. We discuss various aspects of fireball hydrodynamics and the resulting emitted spectra.

1. Introduction - The Inevitability of Fireballs

The recent observation of the BATSE experiment on the COMPTON-GRO observatory have demonstrated, quite convincingly that γ-ray bursts (grbs) originate from cosmological sources[1,2]. Preliminary evidence for the predicted[3,4] correlations between the duration the hardness, the strength and the hardness of the bursts[5,6] supports this conclusion. The correlation suggests, in agreement with an analysis cosmological C/C_{min} distribution[4] that the weakest bursts originate from distances of $z \approx 1$, corresponding to a release of $E \approx 10^{51}$ergs (if the emission is isotropic).

The rapid rise time observed in some of the bursts implies that the sources are compact and that in some cases the size of the source R_i is as small as 100km. The copious release of energy within such a small volume results in an initially optically thick system of photons, electrons and positrons which we call a "fireball". The term "fireball" refers here to an opaque radiation - plasma whose initial energy is larger than its rest mass. The initial optical depth in cosmological grbs for $\gamma\gamma \to e^+e^-$ is[8]:

$$\tau_{\gamma\gamma} = f_g E \sigma_T / R^2 m_e c^2 \approx 10^{19} f_g E_{i,51} R_{i,7}^{-2}, \qquad (e1)$$

where $E_{i,51}$ is the initial energy of the burst in units of 10^{51}ergs, $R_{i,7}$ is the radius into which the energy is injected in units of 10^7cm and f_γ is the fraction of primary photons with energy larger than $2m_e c^2$. Since $\tau_{\gamma\gamma} \gg 1$ the system reaches rapidly thermal equilibrium (regardless of the initial energy injection mechanism) with a temperature: $T = 6.4 E_{i,51}^{1/4} R_{i,7}^{-3/4}$MeV. At this temperature there is a copious number of e^+-e^- pairs which in turn contribute to the opacity via Compton scattering.

The huge initial optical depth prevent us from observing directly the radiation released by the source regardless of the specific nature or the source. The observed radiation emerges only after the fireball has expanded significantly and became optically thin. We should divide, therefore, the discussion of cosmological grbs to two parts: the nature of the energy source (in another paper in this volume[8] we discuss the binary neutron star merger model) and the evolution of a fireball (which we address here). The the fireball phase determines the observational features of grbs. This can be compared to the situation in stars in which energy is generated in the core but it leaks out to through an optically thick envelope and the observed spectrum is determined by the conditions at the photosphere. Similarly the observed grb spectra is determined by the way that fireballs evolve and release their energy.

Before turning to a discussion of the fireball evolution we discuss two recent proposals to avoid fireballs in grbs. The first idea resembles the fireball to some extend as it is based on a relativistic motion of the source. The observed photons are blue shifted to γ-rays because of the relativistic motion of the source. The local temperature is much lower and the fraction f_g of high energy photons at the source is sufficiently small that $\tau_{\gamma\gamma}$ would be less than one and the photons would escape freely (this is the case, for example, in the pure radiation fireball that we discuss latter). The opacity of a source with a spectral index α moving toward the observer with a relativistic factor γ is[9]:

$$\tau_{\gamma\gamma} \approx 6.5 \times 10^6 (2\gamma)^{-(2+\alpha)} R_6 A_{12}^{-2} F_{-7} D_{100}^2 , \qquad (2)$$

where $F = F_{-7} \times 10^{-7}$ ergsec^{-1}cm^{-2} is the observed flux, $A = 10^{12} \times A_{12}$ cm^2, is the area of the emitting region and $D = 100 \times D_{100}$ kpc is the distance to the source. Even at the galactic halo one requires $\gamma > 30$ for $\alpha = 2$ or $\gamma > 100$ for $\alpha = 1$. This solution raises, therefore, several other problems which are as serious as the problem that it solves: What is the accelerating mechanism that accelerates the grb sources to such a high relativistic velocities? What is the source of the huge kinetic energy required for the bulk motion of such sources?

Alternatively, a fireball will not appear if the energy is releases non-electromagnetically and it is converted to photons at significantly larger distance in which eq. 1 yields $\tau_{\gamma\gamma} \ll 1$. This can happen, for example, if the source emits weakly interacting particles which are somehow converted in route to photons. Such a model based on emission of axions by supernova has been recently suggested[10]. It does not explain, however, how to reconciles the supernova rate and the grb rate (which differ by a factor of a thousand) ? and how can GeV photons[11] emerge from such sources?

2. Fireball Evolution - an Overview

Consider, first, a pure radiation fireball. Initially, when the local temperature T is large, the opacity is large due to e^+e^- pairs[12] and the radiation cannot escape freely. The fireball expands and cools and this opacity, τ_p, decreases exponentially with decreasing temperature. At $T_p \approx 20$ KeV, $\tau_p \approx 1$, the fireball becomes transparent, the photons escape freely and the fireball phase ends. While the local temperature is T_p the photons are blue shifted roughly to the original temperature due to the relativistic motion of the fireball at that stage. Preliminary calculations[12,13,14] show that unlike the spectra observed in grbs (see however [15]) the spectra emitted from a pure radiation fireball is a blended thermal spectrum.

In addition to radiation and e^+e^- pairs, astrophysical fireballs may also include some baryonic matter which may be injected with the original radiation or may be present in an atmosphere surrounding the initial explosion[13,16,17]. This affect the fireball in two ways: The electrons associated with this matter increase the opacity, delaying the escape of radiation. More importantly, the baryons are accelerated with the rest of the fireball and convert part of the radiation energy into bulk kinetic energy.

As a loaded fireball with a baryonic mass, M, evolves two important transitions take place. One transition corresponds to the change from optically thick to optically thin conditions. The opacity itself has a contribution from electron-positron pairs as well as electrons associated with the baryons. Initially, when the local temperature T is large, the opacity is dominated by τ_p. However, the matter opacity, τ_b, decreases only as R^{-2}, where R is the radius of the fireball. Generally, at the point where $\tau_p = 1$, τ_b is still > 1 and the final transition to $\tau = 1$ is delayed and occurs at a cooler temperature. The photons escape freely at this stages. The electrons and the baryons are however still coupled to the photons until the mean free path for a Compton scattering of an electron on a photons drop to unity. This happens at a slightly larger radius.

The second transition corresponds to the switch from radiation dominated to matter dominated conditions, i.e from $\eta > 1$ to $\eta < 1$, where $\eta \equiv E/Mc^2$, the ratio of the radiation energy E to the rest energy M. In the early radiation dominated stages when $\eta > 1$, the fluid accelerates in the process of expansion, reaching relativistic velocities and large Lorentz factors. The kinetic energy too increases proportionately. However, later when $\eta < 1$, the fireball becomes matter dominated and the kinetic energy is comparable to the total initial energy. The fluid therefore coasts with a constant radial speed. The overall outcome of the evolution of a fireball then depends critically on the value of η when τ reaches unity (or equivalently on whether $R_\eta > R_\tau$ or vice versa). If $\eta > 1$ when $\tau = 1$ most of the energy comes out as high energy radiation, whereas if $\eta < 1$ at this stage most of the energy has already been converted into kinetic energy of the baryons and we have to examine the fate of those extreme relativistic baryons.

The initial ratio of radiation energy to mass, η_i, determines in what order the above transitions take place. Shemi and Piran[13] identified four regimes:

(i) $\eta_i > \eta_{pair} = (3\sigma_T^2 E_i \sigma T_p^4 / 4\pi m_p^2 c^4 R_i)^{1/2} \approx 10^{10} E_{i,51}^{1/2} R_{i,7}^{-1/2}$ (corresponding to $M < M_{pair} = 5 \times 10^{-13} m_\odot) E_{i,51}^{1/2} R_{i,7}^{1/2}$): In this regime the effect of the baryons is negligible and the evolution is of a pure photon-lepton fireball. When the temperature reaches T_p, the pair opacity τ_p drops to 1 and $\tau_b \ll 1$. At this point the fireball is radiation dominated ($\eta > 1$) and so most of the energy escapes as radiation.

(ii) $\eta_{pair} > \eta_i > \eta_b = (3\sigma_T E_i / 8\pi m_p c^2 R_i^2)^{1/3} \approx 10^5 E_{i,51}^{1/3} R_{i,7}^{-2/3}$ (corresponding to $M_{pair} < M < m_b = 5 \times 10^{-8} m_\odot E_{i,51}^{2/3} R_{i,7}^{2/3}$): Here, in the late stages, the opacity is dominated by free electrons associated with the baryons. The comoving temperature therefore decreases far below T_p before τ reaches unity. However, the fireball continues to be radiation dominated as in the previous case, and most of the energy still escapes as radiation.

(iii) $\eta_b > \eta_i > 1$ (corresponding to $M_b < M < 5 \times 10^{-4} m_\odot E_{i,51}$): The fireball becomes matter dominated before it becomes optically thin. Therefore, most of the initial energy is converted into bulk kinetic energy of the baryons, with a final Lorentz factor $\gamma_f \approx \eta_i$.

(iv) $\eta_i < 1$: This is the Newtonian regime. The rest energy exceeds the radiation energy and the expansion never becomes relativistic. This is the situation, for example in supernova explosions in which the energy is deposited into a massive envelope.

3. Extreme Relativistic Scaling Laws

After an initial acceleration phase in which the fireball reaches relativistic velocities and $\gamma \gtrsim$ few each shell of an extreme relativistic fireball satisfies to order $o(\gamma^{-2})$ the following conservation laws[14]:

$$r^2 n \gamma = \text{const.}, \qquad r^2 e^{3/4} \gamma = \text{const.}, \qquad r^2 (n + 4e/3)^2 = \text{const.}, \qquad (3)$$

where n and e are the local baryon and energy densities. These scalings were derived for a homogeneous radiation dominated fireball[13,12] by noting the analogy with an expanding universe. The same relations are valid, however, for each individual radial shell in the fireball even in the more general inhomogeneous case. These scaling laws also apply to Paczyński's[18] solution for a steady state relativistic wind. They are valid even for fractions of individual shells provided that some general conditions on the angular motion are satisfied.

Eqs. 3 yields a scaling solution which is valid everywhere provided that $\gamma \gtrsim$ few. Let t_0 be the time and r_0 be the radius at which a fluid shell in the fireball first becomes ultra-relativistic, with $\gamma \gtrsim$ few. Label various properties of the shell at this time by a subscript 0, e.g. γ_0, n_0, e_0, and $\eta_0 = e_0/n_0$. Defining the auxiliary quantity D, where

$$\frac{1}{D} \equiv \frac{\gamma_0}{\gamma} + \frac{3\gamma_0}{4\eta_0 \gamma} - \frac{3}{4\eta_0}, \qquad (4)$$

we find that

$$r = r_0 D^{3/2} (\gamma_0/\gamma)^{1/2}, \qquad n = n_0 D^{-3}, \qquad e = e_0 D^{-4}, \qquad \eta = \eta_0 D^{-1}. \qquad (5)$$

These are parametric relations which give r, n, e, and η of each fluid shell at any time in terms of the γ of the shell at that time. The relation for r in terms of γ is a cubic equation. This can in principle be inverted to yield $\gamma(r)$, and thereby n, e, and η may also be expressed in terms of r.

The parametric solution 5 describes both the radiation-dominated and matter-dominated phases of the fireball within the frozen pulse approximation. For $\gamma \ll \eta_0 \gamma_0$, the first term in eq. 4 dominates and we find $D \propto r$, $\gamma \propto r$, which yields the radiation-dominated scalings of eqs. e7. This regime extends out to a radius $r \sim \eta_0 r_0$. At larger radii, the first and last terms in eq. 4 become comparable and γ tends to its asymptotic value of $\gamma_f = (4\eta_0/3 + 1)\gamma_0$. This is the matter dominated regime. (The transition occurs when $4e/3 = n$, which happens when $\gamma = \gamma_f/2$.) In this regime, $D \propto r^{2/3}$, leading to the matter dominated scalings laws (eqs. 10).

It is unlikely that a realistic fireball will be spherically symmetric. In fact strong deviation from spherical symmetry are expected in the most promising neutron star merger model, in which the radiation is expected to emerge through funnels along the rotation axis[8]. The initial motion of the fireball might be fairly complex but once $\gamma \gg 1$ and provided that some some simple conditions are satisfied then the motion of each fluid element decouples from the motion of its neighbors and it can be described by the same asymptotic solution, as if it is a part of a spherical shell. We define the spread angle α as $u^r \equiv u \cos \alpha$ and the angular range over which different quantities vary as $\Delta \theta$. Eqs. 3 hold locally if:

$$\alpha < \Delta\theta \quad \text{and} \quad (\alpha < 1/\gamma \quad \text{or} \quad 1/\gamma < \alpha \ll 1). \qquad (6)$$

4. Physical Conditions in the Fireball

4.i Radiation-Dominated Phase

The fireball is initially radiation-dominated. During this phase ($e \gg n$) and:

$$\gamma \propto r, \qquad n \propto r^{-3}, \qquad e \propto r^{-4}, \qquad T_{obs} \sim \text{constant}, \qquad (7)$$

where $T_{obs} \propto \gamma e^{1/4}$ is the temperature of the radiation as seen by an observer at infinity. (Strictly, the radiation temperature depends on e_r, the energy density of the photon field alone; for $T \ll m_e c^2$, $e_r = e$, but for $T > m_e c^2$, e contains an additional contribution from the electron position pairs[13] we neglect this complication for simplicity). The scalings of n and e given in eqs. 7 correspond to those of a fluid expanding uniformly in the comoving frame. Although the fluid is approximately homogeneous in its own frame, because of Lorentz contraction it appears as a narrow shell in the observer frame, with a radial width given by:

$$\Delta R \sim R/\gamma \sim \text{constant} \sim R_i \quad . \qquad (8)$$

We interpret eq. 7 and the constancy of the radial width Δr in the observer frame to mean that the fireball behaves like a pulse of energy with a frozen radial profile, accelerating outward at almost the speed of light.

If there are no baryons this phase last until the local temperature drops to T_p and the fireball becomes optically thin. Preliminary calculations[12,13,14] show that unlike the spectra observed in grbs (see however [15]) the spectra emitted at this stage is a blended thermal spectrum.

If baryons are present the radiation dominated phase lasts from the initial size, R_i, until $\gamma = \eta$ at R_η

$$R_\eta = 2R_i\eta = 2 \times 10^{11} \text{cm } R_{i,7}\eta_4, \qquad (9)$$

where the initial thermal energy is converted to the kinetic energy of the baryons:

4.ii Matter-Dominated Phase

In the alternate matter-dominated regime ($e \ll n$), we obtain from eq. 3 the following different set of scalings,

$$\gamma \to \text{constant}, \qquad n \propto r^{-2}, \qquad e \propto r^{-8/3}, \qquad T_{obs} \propto r^{-2/3}. \qquad (10)$$

The modified scalings of n and e arise because the fireball now moves with a constant radial width in the comoving frame. (The steeper fall-off of e with r is because of the work done by the radiation through tangential expansion.) Moreover, since $e \ll n$, the radiation has no important dynamical effect on the motion and produces no significant radial acceleration. Therefore, γ remains constant on streamlines and the fluid coasts with a constant asymptotic radial velocity. The width of the fireball remains constant with:

$$\Delta R = R_i. \qquad (11)$$

Eventually, as the particle density decreases this phase ends when the electrons decouple from the photons at R_c:

$$R_c = (a^{1/8}/k^{1/2})\sigma_T^{1/2}(4\pi/3)^{-3/8}E_i^{3/8}R_i^{3/8} = 1.6 \times 10^{15} \text{cm } E_{i,51}^{3/8} R_{i,7}^{3/8}. \qquad (12)$$

4.iii Free Coasting

At very late times in the matter-dominated phase the frozen pulse approximation breaks down. At $R \approx R_c$ the electrons decouple from the photons and at $R > R_c$ the baryons, electrons and photons coast freely. The spread in the Lorentz factor of the baryons leads to a spreading of the fireball whose width becomes:

$$\Delta R = R_i + R/\gamma^2 = 10^7(R_{i,7} + R_{15}/\eta_4^2) \quad (13)$$

The second term, that expresses the additional spreading, is comparable to the original width at

$$R_w \approx \eta^2 R_i \approx 10^{15}\text{cm } \eta_4^2 R_{i,7} \quad (14)$$

However if $R_w < R_c$ the spreading does not begin until $R = R_c$.

$$\Delta R \approx \begin{cases} R/\eta^2 \approx 10^7\text{cm } R_{15}\eta_4^{-2} & \text{for } R > R_w \text{ and } R > R_c \\ R_i = 10^7\text{cm } R_{i,7} & \text{otherwise} \end{cases} \quad (15)$$

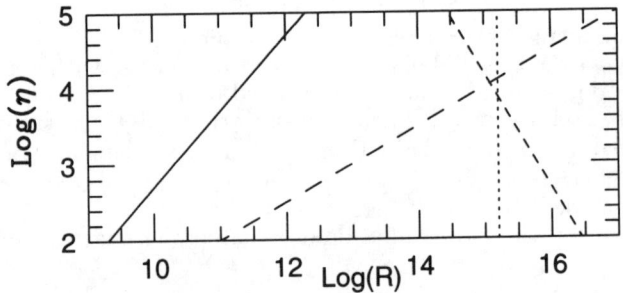

Figure 1. R_η (solid line), R_w (long dashed line), R_c (dotted line), and R_γ (short dashed line) as a function of η for $E_i = 10^{51}$ergs and $R_i = 10^7$cm.

5. Interaction of the Fireball with the ISM

Mészaros, and Rees[19,20] suggested that the interaction between the ultra-relativistic baryons and the interstellar matter (ISM) provides a way to convert back the kinetic energy of the baryons to electromagnetic energy. The situation is similar to the one in supernova remnants (SNRs) in which the kinetic energy of the ejecta is converted to radio emission due to interaction with the ISM. The mean free path of a relativistic baryon in the ISM is $\gtrsim 10^{26}$cm, hence the interaction between the baryons and the ISM cannot be collisional. However, from the existence of SNRs we can infer that a collisionless shock can form (possibly via magnetic interaction).

The interaction becomes significant at R_γ where the fireball sweeps an external mass of $M_0/\gamma_F = (E_i/\eta c^2)/\gamma_F$ and looses half of its initial momentum:

$$R_\gamma = \left[\frac{M_0}{(4\pi/3)n\gamma_F}\right]^{1/3} = 1.3 \times 10^{15}\text{cm } E_{i,51}^{1/3}\eta_4^{-2/3}n^{-1/3} \quad (16)$$

Because of a numerical coincidence $R_w \approx R_\gamma \approx R_c$ for our canonical parameters. R_c is independent of η while R_w increases with η and R_γ decreases with η. Therefore, $R_\gamma < R_c < R_w$ for $\eta > 10^4$ and $R_w < R_c < R_\gamma$ for $\eta < 10^4$.

Just like in SNRs the interaction between the fireball and the ISM produces a forward moving shock, which propagates into the interstellar matter, and a reversed shock propagating into the fireball (see Fig. 2). Following Katz[21] we denote as region 1 the interstellar matter, $n_1 = n$ and $e_1 \ll n_1 m c^2$. Regions 2 and 3 describe the shocked material. Pressure equilibrium along the contact discontinuity between 2 and 3 requires $e = e_2 = e_3$. Region 4 denotes the fireball where $n_4 m c^2 \gg e_4$.

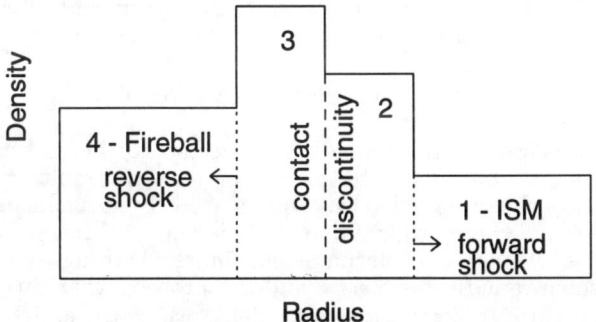

Figure 2. Schematic density profile across the shocks. Region 1 is the interstellar matter. Regions 2 and 3 describe the shocked material, with a contact discontinuity between 2 and 3. Region 4 is the unshocked material of the fireball.

A critical parameter is f, the ratio of densities (in the local fluid's frame) between region 4 (the fireball's material) and region 1 (the external matter):

$$f = \frac{n_4}{n_1} = \frac{M}{[(4\pi)R^2 \Delta R \gamma] n_1} = \frac{1}{3}\left(\frac{R_\gamma}{R}\right)^2 \left(\frac{R_\gamma}{\Delta R}\right) \approx \quad (17)$$

$$\approx \begin{cases} 5 \times 10^7 E_{i,51} \eta_4 n^{-1} R_{15}^{-3} & \text{for } R > R_w \text{ and } R > R_c \\ 5 \times 10^7 E_{i,51} \eta_4^{-1} n^{-1} R_{i,7}^{-1} R_{15}^{-2} & \text{otherwise} \end{cases}$$

where R_{15} is the radius in units of 10^{15} cm.

The shock conditions between 1 and 2 yield[22,23]:

$$\gamma_{1,2} = 0.5 \sqrt{e/n_1 m_p c^2} \; ; \; n_2 = 4\gamma_{1,2} n_1, \; ; \; e \equiv e_2 = \gamma_{1,2} n_2 m c^2 \quad (18)$$

where $\gamma_{1,2}$ is the Lorentz factor of the motion of the shocked fluid relative to the rest frame of an external observer.

The Lorentz factor of the shock front itself is $\sqrt{2}\gamma_{1,2}$. Similar relations hold for the reverse shock (with 3,4 replacing 1,2). The definition of f yields: $\gamma_{3,4} = f^{-1/2} \gamma_{1,2}$ and $n_3 = f^{1/2} n_2$. Using this we can express $\gamma_{1,2}$ and $\gamma_{3,4}$ in terms of γ_F:

$$\gamma_{1,2} = f^{1/4} \gamma_F^{1/2}/\sqrt{2} \; ; \; \gamma_{3,4} = f^{-1/4} \gamma_F^{1/2}/\sqrt{2}. \quad (19)$$

This holds if $f < \gamma_F^2 \approx \eta^2$. Otherwise the reverse shock is not relativistic and:

$$\gamma_{1,2} \approx \gamma_F \quad ; \quad \gamma_{3,4} \approx 1. \tag{20}$$

Since f decreases with R the reverse shock is initially non relativistic. Using eq. 17 we find that $f(R_\gamma) < \eta^2$ only if $R_w > R_\gamma$. In this case $f(R_\gamma) \approx (\eta^2/3)(R_\gamma/R_w)$ and a mildly relativistic reverse shock develops, with $\gamma_{3,4}(R_\gamma) \approx (R_w/R_\gamma)^{1/4}$. If $R_\gamma > R_w$ and $R_\gamma > R_c$, $f = \eta^2/3$ at R_γ. f decreased with R and one might expect that a relativistic reverse shock will develop latter. The reverse shock reaches, however, the inner boundary of the fireball shell when the fireball reaches R_γ and a rarefraction wave begins to move forwards from the back of the fireball before a relativistic reverse shock develops.

6. Energy Generation via Synchrotron Cooling

The kinetic energy of the fireball is converted to thermal energy at the shocks. This happens in an optically thin region and the resulting photons can escape and produce the observed grbs. The most likely mechanism for the conversion of the thermal energy to kinetic energy is via synchrotron cooling of the ultra-relativistic electrons. This mechanisms requires a strong coupling between the electrons, which radiate the energy, and the protons, that carry the kinetic energy. It also requires a strong magnetic field. Mészaros and Rees[24] discuss various emission mechanisms from the shocks. We discuss here an example of the simplest model.

We assume equipartition between the magnetic and the thermal energies. Using eq. 18 we find:

$$B = .5 \text{Gauss}\gamma_{1,2} n_1^{1/2} \quad \text{and} \quad \epsilon_L = 10^{-8} eV \gamma_{1,2} n_1^{1/2}, \tag{21}$$

where ϵ_L is the corresponding Larmour energy. Assuming equipartition between the kinetic energy of the shocked electrons and the shocked protons the typical Lorentz factor of the electrons is larger by (m_p/m_e) then the Lorentz factor of the protons. The typical energy of an emitted photon is $\epsilon_{synch} = (m_p/m_e)^2 \gamma_{1,2}^2 \epsilon_l$. This is blue shifted by another factor of $\gamma_{1,2}$ for an observer an rest. Thus:

$$\epsilon_o \approx 4 \times 10^{-2} eV \gamma_{1,2}^4 n_1^{1/2} \approx \begin{cases} 2 \times 10^4 \text{MeV} f_4 \eta_4^2 n_1^{1/2} & \text{relativistic reverse shock} \\ 2 \times 10^8 \text{MeV} \eta_4^4 n_1^{1/2} & \text{otherwise} \end{cases} \tag{22}$$

Similarly, the typical energy of a synchrotron photon emitted by the reverse shock is:

$$\epsilon_{o,rs} \approx 4 \times 10^{-2} eV \gamma_{1,2}^2 \gamma_{3,4}^2 n_1^{1/2} \approx 2 \text{MeV} \eta_4^2 n_1^{1/2} \tag{23}$$

Interestingly enough this is in the right energy range regardless of the question whether the reverse shock is relativistic or not. This suggests that the observed radiation might come from the reverse shock and demonstrates the potential of this mechanism. However it also shows the difficulty that this mechanism poses. Eqs. 22 and 23 depend on a relatively high power of η. For our canonical parameters the radiation form the reverse shock is in the γ-ray range. The load parameter, η can, however, vary easily by orders of magnitude from one burst to another. It is possible that similar processes produce x-ray bursts, uv-bursts as well as bursts with much harder γ-rays Alternatively, it is possible that the situation is much more subtle and either η is relatively constant or other mechanisms control the emission. In either it is not clear yet why does the energy emerge in soft γ-rays and the resolution of this puzzle might provide the clue to the enigma of grbs.

7. Beaming and Timing

The emitting source is moving relativistically towards the observer and the observed photons are blue shifted both in a pure radiation fireball, that releases its photons when it becomes optically thin, or in a loaded fireball that emits a grb when it interacts with the ISM. Thus, each observer detects blue shifted radiation from a narrow angle $\approx 1/\gamma$. This does not mean, however, that the overall grb is beamed in such a narrow angle. The overall angular spread depends on the width of the emitting region which depends on the source model (in principle the emission could be over 4π if the fireball is spherically symmetric) and is independent of γ.

The duration of the burst depends on several factors. The original duration of the pulse is $R_i/c \approx 10^{-3} R_{i,7}$. This time scale increases due to spreading of the pulse (which takes place in the free coasting phase) and could be as long as:

$$\Delta T_1 \approx 1.5 \times 10^{-3} \sec E_{i,51}^{1/3} \eta_4^{-8/3} n^{-1/3} \qquad (24)$$

for a loaded fireball with $R > R_w$ and $R > R_c$. The duration also increases due to the small angular spread of the signal A given observer will detect radiation from an angular scale $1/\gamma$ around his line of sight. This will lead to a typical duration of[21]:

$$\Delta T_2 \approx R_\gamma/\gamma_F c = 5 \sec E_{i,51}^{1/3} \eta_4^{-5/3} n^{-1/3}. \qquad (25)$$

The relatively strong dependence of η is an advantage here, as it provides a possible explanation to the large variability in durations of grbs. Clearly ΔT is the longer of ΔT_1 and ΔT_2.

8. Conclusions

We have shown that fireballs with a large initial ratio η_i of radiation energy to rest mass energy show certain common global features during their expansion and evolution. After a short initial acceleration phase, the fluid reaches relativistic velocities, and the energy and mass become concentrated in a radial pulse whose shape remains frozen in the subsequent expansion. The motion is then described by an asymptotic solution (eqs 7, 10), which gives for each individual shell scaling laws similar to those of a homogeneous sphere.

The expanding fireball has two basic phases: a radiation dominated phase and a matter dominated phase. Initially, during the radiation dominated phase the fluid accelerates with $\gamma \propto r$ for each Lagrangian shell. The fireball is roughly homogeneous in its local rest frame but due to the Lorentz contraction its width in the observer frame is $\Delta r \approx R_i$, the initial size of the fireball. Ultimately, a transition takes place to the matter dominated phase and all the energy becomes concentrated in the kinetic energy of the matter, and the matter coasts asymptotically with a final Lorentz factor $\gamma_F \approx \eta$. The matter dominated phase is itself further divided into two sub-phases. At first, there is a frozen-coasting phase in which the fireball expands as a shell of fixed radial width in its own local frame, with a width $\sim \gamma_F R_i \sim \eta_i R_i$. Because of Lorentz contraction the pulse appears to an observer with a width $\Delta r \approx R_i$. Eventually, the spread in γ_F as a function of radius within the fireball results in a spreading of the pulse and the fireball enters the coasting-expanding phase. In this final phase, $\Delta r \approx R_i/\gamma_F^2$, and the observed pulse width increases linearly with the radius from which the radiation is emitted.

The fireball can become optically thin in any of the above phases. Once this happen the system ceases to behave like a fluid, and the radiation moves as a pulse with a constant width, while the baryons enter a coasting phase like the one described above.

For most realistic loads the fireball becomes matter dominated before it becomes optically thin (unless there is some unknown yet mechanism that separates the baryons from the e^+e^- pairs). In this case the observed grb is produces at $R_\gamma \approx 10^{15}$cm from the source where the kinetic energy of the baryons is converted to thermal energy and γ-rays due to the interaction of the relativistic baryons with the ISM. We have seen that this process leads to the right time scales for grbs and could potentially lead to the right spectrum. However, at present the simple estimates of the spectrum do not necessarily yield signals at the γ-ray range. An explanations of this feature might provide the key to our understanding of fireballs and grbs.

This work was supported in part by a BRF grant to the Hebrew University and NASA grant NAGS-1904 to the CFA.

Reference

1. Meegan, C.A., et. al., *Nature*, **355** 143.
2. Meegan, C.A., et. al., this volume.
3. Paczyński, B. 1992, *Nature*, 355, 521.
4. Piran, T., 1992, *Ap. J. L.***389**, L45.
5. Norris et. al., 1993, this volume.
6. Davis et. al., 1993, this volume.
7. Piran, T. and Shemi, A., 1993, *Ap. J. L.*, **403**, L67.
8. Piran, T., 1993, This volume.
9. Krolik,J.H. and Pier, E.A., 1991, *Ap. J.*, **373**, 277.
10. Leob, A., 1993, this volume.
11. Dingus, B., et. al., 1993, this volume.
12. Goodman, J., 1986, *Ap. J. L.*, **308** L47.
13. Shemi, A. and Piran, T. 1990, *Ap. J. L.***365**, L55.
14. Piran, T., Shemi, A. and Narayan, R., 1993, *MNRAS.*, **263**, 861.
15. Palmer, D., M., et. al., 1993, this volume.
16. Paczyński, B., 1990. *Ap. J.*, **363**, 218.
17. Cavallo, G., and Rees, M.J. 1978, *MNRAS.*, **183**, 359.
18. Paczyński, B., 1986, *Ap. J. L.*, **308**, L51.
19. Mészaros, P. & Rees, M. J., 1992. *MNRAS.*, **258**, 41p.
20. Mészaros, P. & Rees, M. J., 1992, *Ap. J. L.*, in press.
21. Katz, J. 1993, *Ap. J.*in press.
22. Blandford, R., D., and McKee, C. F., 1976, *Phys. of Fluids*, **19**, 1130.
23. Blandford, R., D., and McKee, C. F., 1976, *MNRAS.*, **180**, 343.
24. Mészaros, P. & Rees, M. J., 1993, this volume.

SHOCK MODELS AND O, X, γ SIGNATURES OF GAMMA-RAY BURST SOURCES

P. Mészáros [1] and M.J. Rees [2]

[1] Pennsylvania State University, 525 Davey Lab, University Park, PA 16803
[2] Institute of Astronomy, Madingley Road, Cambridge CB3 0HA, England

Abstract

The 0.1-100 MeV spectrum of Gamma-ray bursts, together with the rare X-ray detections and optical upper limits, provides strong constraints on possible emission mechanisms and physical models. In almost any scenario, the initial energy density in a burst is so large that the resulting relativistic plasma expands with $v \sim c$ producing a blast wave ahead of it and a reverse shock moving into the ejecta, as it plows into the external medium. We evaluate the spectrum from such shocks in both cosmological and galactic bursts, over the range from the IR to > GeV, and compare them with the spectral behavior reported by BATSE, and with X-ray and optical constraints. For burst durations $\gtrsim 1$ s acceptable γ-ray spectra and L_x/L_γ ratios are obtained in the cosmological case for models where turbulent field growth is inefficient and the ejecta has frozen-in fields, either being matter dominated or else magnetically dominated but with enough matter that gets mixed in to accelerate electrons. Some cosmological models with turbulent field growth and radiation in the blast wave only may also be acceptable. Galactic disk models can produce bursts of similar gamma-ray fluence and duration, but they violate the X-ray paucity constraint, except for the shorter bursts ($\lesssim 1$ s).

1. Introduction

The detection of gamma-ray burst sources (GRB) at other energies, e.g. in the X-ray band or the optical/UV band, would be of great importance for identifying possible counterparts and for understanding the physics of these objects. Whatever the distance at which GRB occur, the energy density is expected to so large that it leads to a relativistic outflow. The expanding relativistic flow is eventually decelerated by the external medium, leading to a blast wave moving ahead and a reverse shock propagating back into the ejecta (Rees and Mészáros , 1992). The emission from such shocks satisfies the energetic and temporal average properties of GRBs, and produces optically thin non-thermal radiation (Mészáros and Rees, 1993). For relativistic expansion, the apparent diameter of the emitting shock regions is larger by at least a factor Γ (the bulk Lorentz factor) than in the corresponding nonrelativistic case. There is therefore much less likelihood of strong suppression of the radiation down to optical frequencies

© 1994 American Institute of Physics

due to self-absorption. For this reason, and also because the radiation mechanisms are more straightforward than in some alternative models (no pulsar-type or stronger magnetic fields are required, etc.), predictions can be made about γ-ray spectral properties as well as for other energy bands. Measurements of γ-ray spectral breaks, indices and fluences, as well as any O-UV/ X-ray measurements or upper limits would have relatively straightforward implications for this promising class of models.

2. Model Shock Structure and GRB Spectra

Even when the magnetic fields present in the ejecta or in the external medium are weak, they may be amplified up to equipartition values in the shocks (as seems to happen in supernova remnants and radio sources). The shock radiative efficiency will be a function of the bulk Lorentz factor $\Gamma \simeq \eta = E_o/M_o c^2$, where E_o is the initial energy release and M_o is the baryon loading. It will depend also on the magnetic field strength, the particle acceleration mechanism and the density of the external medium (Mészáros and Rees, 1993, Katz, 1993). The minimum Lorentz factor of electrons accelerated in the shocks will be $\gamma_m = \kappa \Gamma$, where $1 \lesssim \kappa \lesssim m_p/m_e$, and the typical electron energy slope expected is -2 to -3. The blast wave consists of external gas that has been shock heated, and the sound speed is relatively high. Because of the very high Γ, the shock will be strong, so electron (and proton) diffusive acceleration is expected. For the reverse shock, in the case of a baryon-loaded fireball the sound speed in the ejecta may be low, and the reverse shock will be strong and at least mildly relativistic. If the fireball is magnetically dominated (almost completely unloaded, e.g. Narayan, Paczyński and Piran, 1992), it will have a large outward Lorentz factor even relative to the frame of the contact discontinuity. The reverse shock is then strong, the field in the shocked region providing most of the pressure that drives the blast wave.

Three models are considered. The first model (F), assumes there is a frozen-in magnetic field in the ejecta, and there is shock acceleration in the reverse shock. This shock contributes a synchrotron component, and a higher energy component due to inverse Compton (IC) scattering of the synchrotron photons by the same electrons, (e.g. Mészáros, Laguna and Rees, 1993). In the blast wave, it is assumed that there is relativistic particle acceleration, but turbulent magnetic field growth is not efficient. The blast wave electrons, however, can IC scatter the synchrotron photons from the reverse shock, leading to three spectral components (double IC scattering is neglected, since it occurs in the Klein-Nishina regime). Another model (T) assumes that frozen-in fields are negligible but that turbulent field growth and particle acceleration are efficient in both the blast wave and reverse shock. This produces five spectral components: blast synchrotron and IC, reverse synchrotron and IC, and combined IC of reverse synchrotron photons on blast wave electrons. In principle, different efficiencies of field growth and acceleration may occur in each shock, but for simplicity these are taken approximately equal. The simplest model, (P), assumes that turbulent field generation and acceleration occur only in the strong blast wave, and that the ejecta acts only as a piston, the reverse shock being devoid of frozen-in or turbulent magnetic fields as well as being an inefficient accelerator. This model produces only two spectral components, synchrotron and IC from the blast wave. These components are not, however, of the same strength as the corresponding ones in (T) above, since the electrons only lose energy to synchrotron and IC from the blast, and not to the reverse shock photons.

The energy loss timescales of the electrons can be estimated by assuming the electron spectral index p, the minimum electron Lorentz factor γ_m, the field strength, and the physical conditions in the ejecta and blast wave, e.g. Mészáros, Laguna and Rees, 1993. The total energy available in each shock is a fraction of the kinetic energy of the ejected material ($\sim E_o$, the initial energy liberated). This is shared approximately equally between both shocks, since they are in pressure equilibrium. The total fluence is then determined by multiplying this fraction by the efficiency and by the total bolometric fluence potentially available, $S_b = E_o/(4\pi\theta^2 D^2)$ ergs cm^{-2}, where θ is the possible beaming angle and D is the distance.

The synchrotron spectrum has an energy slope $-(p-1)/2$ above frequencies $\nu_m \sim 10^6 B\gamma_m^2$, where $-p$ is the electron power-law index above γ_m. We took p to be around -3, giving a photon energy index about -1 above $\nu_m(\gamma_m)$, or power index $\alpha = 0$. For a single value of γ_m, the photon energy spectrum below $\nu_m(\gamma_m)$ would be expected to have a slope $+1/3$, or power slope $+4/3$. Synchrotron self-absorption leads to an energy slope 2 or 2.5 (power slope 3 or 3.5), but it usually occurs below the UV range for the parameters of these models. The power per decade slopes 4/3 and 0 are close to the observed "fiducial" values 1 and 0 deduced for those objects where a break is required (Schaefer, et al., 1993; c.f. Band et al. 1993)). However, γ_m may not be uniform throughout different regions of the shock, which could lead to a rounding off of the breaks. Thus the power per decade composite spectrum of GRBs, whether galactic or extragalactic, is made up of between two to five subcomponents with slopes roughly 4/3 (0) below (above) break energies determined by the comoving mean values of B, γ_m.

3. Results and Discussion

The spectra are discussed in more detail elsewhere (Mészáros and Rees, 1993b, 1994). Figure 1 shows a turbulent cosmological model which is typical, e.g., for a 5 s duration the spectrum has a break at about 1 MeV and for 30 s at about 0.5 MeV. The X-ray fluence is a factor $\lesssim 10^2$ below this. The O/UV fluence predicted by this model is comparable to the X-ray fluence. Other models (e.g. frozen-in, piston) differ in the γ/X and γ/UV predictions, e.g. the piston model predicts smaller values than Fig. 1. The X-ray and O/UV fluences shown in Fig.

1 are compatible with current X-ray paucity constraints (e.g. Hartmann, 1993, Hurley, 1989), and with existing optical upper limits. The HETE threshold (Ricker, 1992) at X and UV is of order 10^{-10} ergs cm^{-2} and is shown as the last "floor" at low frequencies in Fig. 1. The simultaneous emission at IR and radio is self-absorbed, and below the threshold shown.

The break energies are not standard, depending on η, κ, the field strength, the external density and the shock in question (reverse or blast). Longer duration bursts tend on average to have somewhat softer break energies (duration and break depend as different inverse powers of η). Galactic disk burst models could also produce acceptable γ-ray spectra in the impulsive regime, but they produce significantly more X-rays than the typical X-ray paucity constraint allows, except for durations shorter than about 1 s. They also predict larger O/UV fluences than the cosmological models. Galactic models could satisfy the X-ray paucity constraint only if the input is non-impulsive, in which case the duration is an additional free parameter, determined by the energy input mechanism. This could, for instance (e.g. Begelman, Mészáros and Rees, 1993) be given by a magnetic flare in a galactic neutron star. Blast waves in an extended galactic halo would also need to be nonimpulsive. We note also that all models predict emission in the GeV range and above.

Acknowledgements: We are grateful to NASA (NAGW-1522, NAG5-2362), to NCSA and to the Royal Society for support, and to H. Papathanassiou for computational assistance.

References

Band, D., *et al.* , 1993, Ap.J., 413, 281
Begelman, M.C., Mészáros , P. and Rees, M.J., 1993, M.N.R.A.S., 265, L13.
Hartmann, D., *et al.* , 1993, in *High Energy Astrophysics*, (World Scientific).
Hurley, K., 1989, in *14th Texas Symp.*, *Ann.N.Y.Acad.Sci.*, 571, 442.
Katz, J.I., 1993, Ap.J., in press
Mészáros , P. and Rees, M.J., 1993, Ap.J., 405, 278
Mészáros , P. and Rees, M.J., 1993b, Ap.J.(Letters), 418, L59
Mészáros , P., Laguna, P. and Rees, M.J., 1993, Ap.J., 415, 181
Narayan, R., Paczyński , B. and Piran, T., 1992, Ap.J.(Lett.), 395, L83
Rees, M.J. and Mészáros , P., 1992, M.N.R.A.S., 258, 41P
Ricker, G., *et al.* , 1992, in *Gamma-ray Bursts*, C. Ho, *et al.* , eds. (Cambridge U.P.), p. 288
Schaefer, B., *et al.* , 1992, Ap.J.(Lett.), 393, L51

Figure 1: An example of the spectrum as function of burst duration (in seconds) for a cosmological turbulent model with fields 10^{-6} of the shock equipartition value, external density $n_o = 1$ cm^{-3}, $\theta = 10^{-1}$, $\kappa = 10^3$. The flat portion at fluence 10^{-10} ergs cm^{-2} represents an arbitrary detection threshold, below which the spectrum is not plotted.

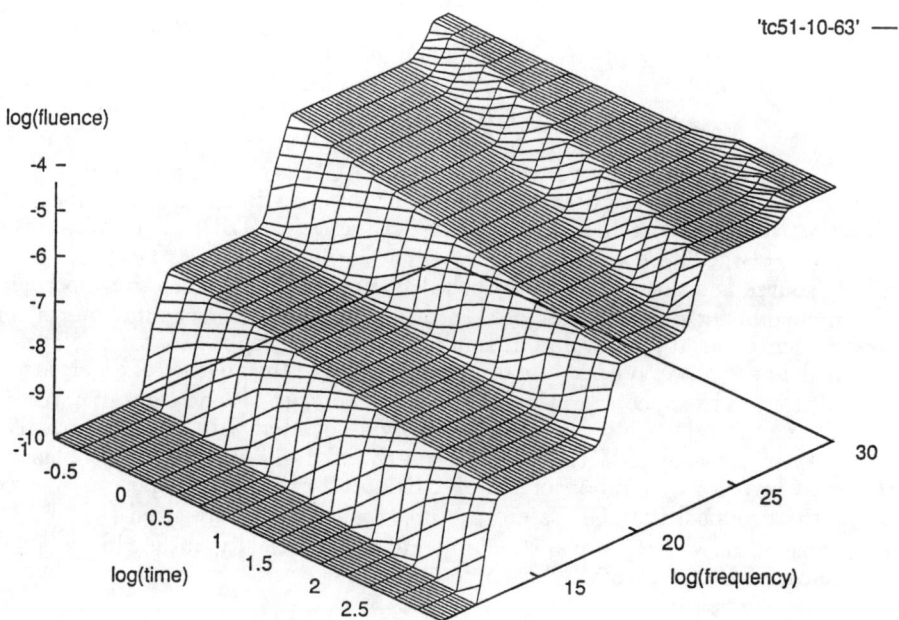

GAMMA-RAY BURSTS AND GAMMA-RAY BLAZARS

Charles D. Dermer
Naval Research Laboratory, Code 7653, Washington, DC 20375-5352, USA

Reinhard Schlickeiser
MPI für Radioastronomie, Auf dem Hügel 69, D-53121 Bonn, Germany

ABSTRACT

We propose that cosmological gamma-ray bursts are produced when modestly sized black holes tidally disrupt stars, energizing relativistically outflowing jets and producing bursts for favorably oriented observers. The crucial test of this model is the detection of soft X-ray flares with durations of $\sim 10^2$ s and fluences $\sim 10^{-10} - 10^{-6}$ ergs cm^{-2} at rates of several to several hundred per day, reflecting burst events from misaligned jets.

INTRODUCTION

The hypothesis that gamma-ray bursts (GRBs) originate from sources at cosmological distances has received strong support from observations made with the Burst and Transient Source Experiment (BATSE) on the Compton Observatory[1,2]. The large diversity of time histories[3] suggests that the burst energy source is associated with a turbulent nonstationary accretion flow onto a cosmic compact object. Lack of association with quiescent counterparts[4] indicates that catastrophic events in small galaxies are responsible for the bursts. The peak in the power output at gamma-ray energies, the absence of a pair attenuation cutoff in the spectra in the MeV regime, and sub-second time structure[5,6] seem to require relativistic beaming, possibly associated with jet emission. The similarity of broadband GRB spectra with the high-energy spectra of blazars[7] argues in favor of a common origin of GRBs and gamma-ray blazars. We examine the proposal that GRBs are produced when modestly sized ($\approx 10^6 M_\odot$) black holes tidally disrupt stars[8], energizing relativistically outflowing jets and producing GRBs for favorably oriented observers.

GRBs FROM STELLAR DISRUPTION IN GALAXY CENTERS

Black holes of mass $10^6 M_6$ Solar masses, with $M_6 \sim 1$, may reside in the centers of many galaxies, including our own[9,10]. The frequency of stellar capture within the tidal radius r_t of black holes with masses of this magnitude has been estimated to be $\sim 10^{-4} M_6^{4/3}$ yr^{-1} for likely conditions found in the inner regions of galaxies[11,12]. For the capture of a star similar to the Sun, $r_t \approx 5 \times 10^{12} M_6^{1/3}$ cm. Thus $r_t \approx 30 r_g$, where the gravitational radius $r_g = GM/c^2 = 1.5 \times 10^{11} M_6$ cm. Since tidal disruption is only important when r_t is larger than the radius of the event horizon $2r_g$, this limits the hole mass to less than $\sim 6 \cdot 10^7 M_\odot$ [13], so that this process is not relevant for the most luminous quasars or active galactic nuclei, as suggested by Usov and Chibisov[14]. The resulting disruption of a star captured within the tidal radius would generate a flash of emission persisting for times exceeding the light travel time $t_f \sim r_t/c = 5\xi M_6$ s, where we scale $r_t = \xi r_g$. The entire duration of the capture event depends sensitively on the nature of the star involved in the capture, the orbital trajectory of the star, and details of stellar disruption[15], but we can expect a significant fraction of the energy to be emitted during this brief flare phase.

Much of the captured star's gravitational potential energy Mc^2 may also be lost through expulsion or direct capture by the central black hole, or radiated at later times. Letting $\eta = 10^{-3}\eta_{-3} \sim 1$ represent the typical radiative efficiency, we can obtain the characteristic temperature T of the flare by assuming that the energy ηMc^2 is radiated quasi-thermally within the tidal radius. Thus $2\pi r_t^2 \sigma_{SB} T^4 \simeq \eta Mc^2/t_f$, implying $kT(\text{keV}) \simeq 0.55 \eta_{-3}^{1/4} m^{1/4} M_6^{-3/4} \xi_{30}^{-3/4}$, where $m = M/M_\odot$ and $\xi = 30\xi_{30}$. We therefore expect that stellar capture will be accompanied by a transient soft X-ray flare with characteristic temperature $\lesssim 1$ keV. At cosmological distances, the bolometric fluence $S = \eta Mc^2/4\pi d_L^2$, where d_L is the luminosity distance corresponding to a particular cosmological model. Defining $d_{28} = d_L/(10^{28}$ cm$)$, we find that $S \simeq 1.4 \times 10^{-6} m \eta_{-3} d_{28}^{-2}$ ergs cm^{-2}.

Observations made with the Ginga satellite[16,17] have detected X-ray precursors and tails associated with GRBs which have characteristic temperatures between ~ 1 and 2 keV. The ratio of the fluxes in the 1.5-10 keV and the 1.5-375 keV bands are typically 5% at the peak of the burst. Because the duration of the X-ray emission lasts longer than the gamma rays, the X-ray fluence is $\approx 20\%$ of the gamma-ray fluence. The bright bursts observed by Ginga have gamma-ray fluences between $10^{-5} - 10^{-4}$ ergs cm^{-2}, implying that the X-ray fluences are an order of magnitude smaller. The good agreement between the measured values and the previous fluence and temperature estimate suggests that X-ray precursors and tails represent the emission from stellar disruption by $\sim 10^6 M_\odot$ black holes.

Accretion of matter onto black holes in the centers of galaxies is also thought to be responsible for high-energy emission from extragalactic blazars at cosmological distances[13]. Recent observations[7,18] made with the Compton Observatory show that the bulk of the luminosity from blazars is often radiated at gamma-ray energies, and that the rapid variability and large apparent gamma-ray luminosity are a consequence of relativistic outflow and beaming of plasma jets[19]. Several recent papers[20,21,22] have proposed a connection between the GRB phenomenon and beaming and emission processes in active galaxies. However, the lack of association[4] between GRB source positions and directions to known active galaxies suggests that bursts take place only under extraordinary conditions in obscure locales, such as through stellar capture by black holes in faint galaxies, for example, dwarf elliptical galaxies.

We propose a spectral emission model based on inverse Compton scattering of the soft X-ray photons produced in the flash phase, assuming that the jet is energized during the stellar capture event. The disrupted turbulent stellar material emits soft X-ray photons which traverse the jet and are scattered to high energies by relativistic electrons which may be accelerated by shocks in the outflowing plasma of the jet. For simplicity, we assume that the energetic electrons are isotropically distributed in the comoving frame of the fluid, and that the fluid moves outward along the jet axis with relativistic bulk Lorentz factor Γ. If ϕ is the opening angle of the jet, the emission will be scattered within solid angle $\Delta\Omega \propto \Gamma^{-2}$ when $\Gamma^{-1} > \phi$, and $\Delta\Omega \propto \phi^2$ when $\Gamma^{-1} < \phi$. For extragalactic radio sources, $\phi \sim 0.1$. If GRB sources have similar opening angles, then we expect that for two-sided jets, the beaming fraction $f_b = 2 \times \Delta\Omega/4\pi \approx 1/200$ when $\Gamma \gtrsim 10$.

The ratio of fluxes Φ of the beamed and isotropic soft X-ray components is determined by f_b, the luminosity L_{iso} in X-rays, and the power L_{jet} injected into the jet. When viewing within the opening angle of the jet, $\Phi_{jet}/\Phi_{iso} \approx f_b^{-1} L_{jet}/L_{iso}$. Depending on the overall shape of the beamed spectral component (see below), the X-ray paucity constraint on GRB spectra is satisfied when $L_{jet} \sim 0.1 L_{iso}$, noting that X-ray observations of GRBs at energies ~2-10 keV may also detect a substantial beamed component. Because the duration of the gamma-ray burst phase is $\approx 25\%$ of the soft X-ray phase, a beaming fraction $< 1\%$ implies fluences of bright GRBs $\approx 10^{-4}$ ergs cm^{-2}, in agreement with observations, and consistent with our previous estimate based on emission from stellar disruption at cosmological distances. A beaming factor of 10% is also consistent with the observations if $L_{jet} \approx L_{iso}$.

Calculations of GRB spectra depend on the details of electron energization and interaction processes within the jet. The beamed high-energy emission of blazars steepens between X-ray and gamma-ray energies[23], which can be explained if electrons are injected with a power-law distribution and cool through inverse Compton interactions[24]. GRBs display characteristic spectra with photon number index $\alpha \approx 1$ ($\Phi \propto \epsilon^{-\alpha}$) at photon energies $\epsilon < 100$ keV, which steepen to $\alpha \approx 2$ at $\epsilon \gtrsim 500$ keV[5,6]. This spectral shape can be reproduced if electrons are accelerated by strong shocks in relativistic fluids with injection spectrum $\propto \gamma^{-r}$, where $1.5 < r < 2$, and subsequently evolve to an equilibrium spectrum $\propto \gamma^{-p}$, where $p \approx 1.5 - 2$ at Lorentz factors $\gamma < \gamma_c$ where incomplete Compton cooling occurs, and $p = 2.5 - 3$ at higher energies due to Compton losses. A low-energy cutoff to the power-law electron injection function will also produce a hard X-ray spectrum.

The position of the spectral break, which we define as the energy where the photon spectral index equals 2, varies by several orders of magnitudes in gamma-ray blazars[7], whereas in GRBs it is typically between 100 keV and several MeV, although a larger range may be involved due to selection effects. The characteristic spectral shape and the dominance of the gamma-ray luminosity is produced when nonthermal electrons are injected into a relativistic jet moving outward[23]. The spectral break arises from incomplete Compton cooling of the outflowing electrons. Note that even though intense broadband radiation is produced in blazars at radio, IR, and optical wavelengths, we do not expect comparable emission in GRBs during the bursting event because the lower-energy radiation will be dissipated over a much longer timescale.

Electron beaming relieves the $\gamma-\gamma$ opacity problem which plagues isotropic models[26,27]. The threshold for pair production is given by $\epsilon_1 \epsilon_2 (1 - \cos\theta) \geq 2(m_e c^2)^2$. A hard gamma-ray photon has energy $\epsilon_h \sim \gamma^2 kT$. Taking $\cos\theta \sim 1/2$ as typical for the cosine of the scattering angle between hard and soft photons, we see that the threshold is met when $\gamma \gtrsim m_e c^2/kT$, which means that only photons with $\epsilon \gtrsim (m_e c^2)^2/kT \sim 100\text{-}200$ MeV are subject to pair attenuation. If we consider instead interactions between two hard photons, we must determine their relative scattering angles. For $\Gamma \gg 1$, $1 - \cos\theta \sim \Gamma^{-2}$, so that the threshold condition now implies $\gamma^2/\Gamma \gtrsim m_e c^2/kT$. Thus only photons with $\epsilon > \Gamma m_e c^2$ are subject to pair attenuation, again representing energies of $\sim 10^2$ MeV for $\Gamma \sim 10^2$.

MODEL PREDICTIONS

Although quantitative details have yet to be worked out we note that this model can offer explanations for various features of GRB phenomenology. The general hard-to-soft spectral evolution within individual bursts[28] may reflect the temporal variation of black-hole fueling and jet energization processes following stellar capture and disruption. The luminosity-temperature correlation of burst spectra at different time intervals within a burst[29] could result from optical depth effects within the jet: the shielding of soft photons by a larger electron current impedes the cooling so that the apparent burst temperature increases while, at the same time, the overall luminosity rises. Material within the jet can produce transient blue-shifted absorption features due to K shell photoionization[30] which could mimic absorption features reported in some GRB spectra. The appearance of second harmonics[31,32] is, however, difficult to understand in this model other than through coincidences in line energies. There could also be considerable pair production when Γ is small, which might produce transient annihilation features which have been reported[5] in burst spectra.

BATSE observations imply $\sim 10^3$ GRBs per year with fluxes greater than the BATSE threshold[1]. If 1% of the $\sim 10^{10}$ galaxies in the universe have a $10^6 M_\odot$ black hole, then one capture event every 10^4 yrs easily satisfies the observed rate with a beaming fraction $f_b \sim 1 - 10\%$. Observations of soft X-ray flares associated with the unbeamed population of GRBs, occuring $\sim 10 - 10^3$ as often as GRBs with fluences in the range $10^{-10} - 10^{-6}$ ergs cm^{-2}, is the outstanding prediction of our model. Such events may already have been detected with the Einstein Observatory.[33] Upper limits obtained with the Wide Field Camera on Rosat[34] are inconclusive. Because of its low energy range (62-206 eV), photoelectric absorption by interstellar matter in our Galaxy would attenuate the signal by at least an order of magnitude. The probability of detection of X-ray flares by the Rosat PSPC ranges from $\sim$0.2% to 20% per day, but in our case pointed observations are not necessary (compare with Ref. [35]). We strongly encourage analysis of the Rosat PSPC data for X-ray transients with the characteristics described above. Observations of X-ray flares and measurements of the magnitude of photoelectric absorption in their spectra can help establish or refute the cosmological scenario in general[36], and this model in particular.

We thank Kevin Hurley, Edison Liang, and Brad Schaefer for useful discussions. RS acknowledges support by the DARA (50 OR 9301 1) of his GRO guest investigator program GRO-90-44.

REFERENCES

1. C. A. Meegan, et al., Nature **355**, 143 (1992).
2. B. Paczyński, Huntsville Workshop on Gamma-Ray Bursts, ed. W. Paciesas & G. Fishman (AIP, New York, 1992), p. 144.
3. G. J. Fishman, et al., A&AS **97**, 17 (1993).
4. B. E. Schaefer, Gamma-Ray Bursts, ed. C. Ho, R. I. Epstein, & E. E. Fenimore (Cambridge University Press, New York, 1992), p. 107.
5. J. C. Higdon & R. E. Lingenfelter, ARA&A **28**, 401 (1990).
6. A. K. Harding, Phys. Rep. **206**, 327 (1991).
7. C. E. Fichtel, et al., A&AS **97**, 13 (1993).

8. B. Carter, ApJ **391**, L67 (1992).
9. M. J. Rees, Nature **333**, 523 (1988).
10. R. Genzel & C. H. Townes, ARA&A **25**, 377 (1987).
11. J. Frank & M. J. Rees, MNRAS **176**, 633 (1976).
12. M. J. Rees, Science **247**, 817 (1990).
13. R. D. Blandford, Active Galactic Nuclei, ed. T. J.-L. Courvoisier & M. Mayor (Springer, New York, 1990), p. 161.
14. V. V. Usov & G. V. Chibisov, Sov. Astron. **19**, 115 (1975).
15. B. Carter & J. P. Luminet, Nature **296**, 211 (1982).
16. A. Yoshida, et al., PASJ **41**, 509 (1989).
17. T. Murakami, et al., Nature **350**, 592 (1991).
18. R. C. Hartman, et al., ApJ **385**, L1 (1992).
19. C. D. Dermer & R. Schlickeiser, Science **257**, 1642 (1992).
20. J. Krolik & E. A. Pier, ApJ **373**, 277 (1991).
21. B. McBreen, S. Plunkett, & L. Metcalfe, A&AS **97**, 81 (1993).
22. J. J. Brainerd, ApJ **394**, L33 (1992).
23. W. Hermsen, et al., A&AS **97**, 97 (1993).
24. C. D. Dermer & R. Schlickeiser, ApJ **416**, 458 (1993).
25. P. N. Bhat, et al., Compton Gamma-Ray Observatory, ed. M. Friedlander, N. Gehrels, & D. J. Macomb (AIP, New York, 1993), p. 912.
26. W. K. H. Schmidt, Nature **271**, 525 (1978).
27. M. G. Baring, ApJ **418**, 391 (1993).
28. D. Band, et al., Huntsville Workshop on Gamma-Ray Bursts, ed. W. Paciesas & G. Fishman (AIP, New York, 1992), p. 169.
29. V. E. Kargatis, E. P. Liang, & K. Hurley, Huntsville Workshop on Gamma-Ray Bursts, ed. W. Paciesas & G. Fishman (AIP, New York, 1991), p. 201.
30. G. S. Bisnovatyi-Kogan, A&AS **97**, 65 (1993).
31. T. Murakami, et al., Nature **335**, 234 (1988).
32. E. E. Fenimore, et al., ApJ **335**, L71 (1988).
33. D. J. Helfand, E. Gotthelf, & T. T. Hamilton, Compton Observatory Science Workshop, ed. C. R. Shrader, N. Gehrels, & B. Dennis (NASA Conf. Publication 3137, Greenbelt, 1992), p. 317.
34. A. Owens, et al., MNRAS **260**, L25 (1993).
35. E. P. Liang, ApJ **380**, L55 (1991).
36. B. E. Schaefer, Compton Gamma-Ray Observatory, ed. M. Friedlander, N. Gehrels, & D. J. Macomb (AIP, New York, 1993), p. 803.

FOCUSING OF ALFVÉNIC POWER IN NEUTRON STAR MAGNETOSPHERES

Marco Fatuzzo *
Dept. of Physics, University of Michigan, Ann Arbor, MI 48109

Fulvio Melia **
Dept. of Physics and Steward Observatory, U. of Arizona, Tucson, AZ 85721

ABSTRACT

Highly dynamic magnetospheric perturbations in neutron star environments can naturally account for the features observed in Gamma-ray Burst spectra. However, if GRB's have an extragalactic origin, as is implied by the uniform yet spatially truncated distribution observed by the BATSE experiment, then noncatastrophic isotropic emission mechanisms may be ruled out on energetic and timing arguments. As such, we consider MHD processes which can produce strongly anisotropic γ-rays with an observable flux out to distances of $\sim 1-2$ Gpc. In particular, we show that sheared Alfvén waves propagating along open magnetospheric field lines at the poles of magnetized neutron stars transfer their energy dissipationally to the charges generating the current which sustains the field misalignment, and thereby focus their power into a spatial region (i.e., the shear) that can be many times smaller than that of the crustal disturbance. This produces a strong (observable) flux enhancement along certain directions.

INTRODUCTION

The lack of a precise determination of a distance scale to Gamma-ray burst (GRB) sources has greatly hindered our theoretical understanding of these objects. Much of what we know about these bursts is based on inferences drawn from clues provided by their spectra, including their rapid variability, their complex temporal structure, and the emission of a substantial fraction of their power at energies in excess of ~ 1 MeV.

While it is possible to account for these observations by invoking a model in which the bursts originate within the magnetosphere of strongly magnetized neutron stars, this paradigm must be reconciled with the uniform, yet spatially truncated GRB distribution observed by the BATSE experiment on CGRO[1]. These observations seem to rule out nearby (i.e. Galactic) single population models, and have therefore led to renewed speculation that GRB's originate at cosmological redshifts. But a naive estimate of the burst energy required for such distant sources yields a value that is significantly larger than that which a neutron star could reasonably produce unless the event was catastrophic (e.g., the coalescence of a neutron-star binary[2]), which does not seem to be borne out by the time history of typical bursts.

A resolution to this apparent conflict was proposed by Melia and Fatuzzo[3], in which sheared Alfvén waves generated near the polar cap of strongly magnetized neutron stars produce streams of relativistic particles that are focused by

* Compton GRO Fellow
** Alfred P. Sloan Fellow

the underlying magnetospheric structure. These energetic charges upscatter the radio-frequency photons (emitted at larger radii) into γ^{-1} cones aligned with the underlying magnetic field lines, resulting in an enhanced γ-ray flux along preferred lines of sight. This anisotropic emission is such that a pulsar glitch releasing $\sim 10^{45}$ ergs of energy could be viewed as a GRB out to a distance of $\gtrsim 1$ Gpc. A key assumption of this scenario is that the Alfvénic power can indeed be emitted anisotropically. We show here that the required focusing is a natural consequence of the dissipational properties of sheared Alfvén waves whose shear lengthscales ($s \lesssim 10$ cm) are much smaller than the size of the region ($\gtrsim 10^4$ cm) encompassing the overall Alfvén wave fluctuation.

SHEARED ALFVÉN WAVE DISSIPATION

Sheared Alfvén waves may be described by magnetic perturbations of the form

$$\mathbf{B}_A = B_a(y) \exp(ikz - i\omega t)\hat{x} , \qquad (1)$$

where $B_a(y)$ is an odd function that characterizes the shear geometry. It is clear from the form of $B_a(y)$ and Ampére's Law that an electric field E_{Az} must exist inside the sheared region until a sufficiently strong current J_s is produced parallel to the underlying magnetic field $\mathbf{B}_0 = B_0\hat{z}$. If the equilibrium Goldreich-Julian particle density n_0 is insufficient to support a current large enough to short out E_{Az}, charges must be copiously stripped off the stellar surface, thereby inducing a charged particle flow to give the required J_s. Since the Alfvén speed in these environments is $v_\alpha \gg c$, the waves travel with a phase velocity $u_\alpha = \omega/k \approx c$. In order for J_s and the encompassing magnetic shear to remain in phase, the particle flow must be relativistic, and thus, have an average density

$$n_s \approx \frac{B_{a0}}{4\pi e s} \approx 10^{19} \text{cm}^{-3} \left(\frac{B_{a0}}{10^{12} \text{ G}}\right) \left(\frac{s}{10 \text{ cm}}\right)^{-1} , \qquad (2)$$

where B_{a0} is the magnitude of the magnetic perturbation and s is the lengthscale of the shear (which is assumed to be much smaller than the lengthscale S of the encompassing plane wave regions). For convenience, we define $\eta \equiv S/s$. We note that the stripped particles escape from the system by flowing out along the open magnetospheric field lines.

In standard pulsar theory, radio emission results from the coherent motion of "bunches" of electrons streaming along open field lines with Lorentz factors $\gamma \sim 10^{4-5}$. As such, strong transient radio emission is expected to be a natural byproduct of sheared Alfvén waves if similar particle energies are reached, and if this emission is produced with front-back symmetry along the local field-line direction, a large fraction of the overall radio flux will naturally be funneled back onto the polar cap. Taking into account the coherent nature of the processes responsible for pulsar emission, we parametrize the flux impinging onto the stellar surface by $F_r = \xi \eta^{-2}(n_s/n_C)^2 (L_C/\pi R_{pc}^2)$, where L_C and n_C are the Crab pulsar luminosity and magnetospheric number density, respectively, and where R_{pc} is the radius of the open field line polar cap. Assuming a stellar radius of $R_* = 10^6$ cm, R_{pc} can be related to the pulsar period P via $R_{pc} = 1.4 \times 10^4$ cm $(P/1s)^{-1/2}$. With $L_C = 10^{32}$ ergs s^{-1} and $n_C = 10^{13}$ cm^{-3}, this yields

$$F_r \approx 10^{30} \text{ergs cm}^{-2} \text{ s}^{-1} \xi \left(\frac{\eta}{10^3}\right)^{-2} \left(\frac{B_{a0}}{10^{12} \text{ G}}\right)^2 \left(\frac{s}{10 \text{ cm}}\right)^{-2} \left(\frac{P}{5 \text{ s}}\right), \quad (3)$$

where $\eta^{-1} = (S/s)^{-1}$ is the sheared flow "filling factor". The parameter ξ encompasses both geometric and emission uncertainties, and as such, is poorly known. We note that if ξ becomes too small ($\lesssim 0.1$ for the range of parameters considered here), the wave dissipation lengthscale due to field line annhilation becomes much larger than R_*, and the SAW mechanism becomes inefficient at producing γ-rays. However, since the Crab pulsar is itself very inefficient at converting spin-down energy into radio emission compared to typical pulsars, and since we have made the conservative assumption that the (coherent) radio flux scales as η^{-2} (i.e., the square of the total number of particles), it is reasonable to assume that $\xi \gg 0.1$ (see also the discussion below).

The presence of F_r results in a radiative drag on the relativistic current-carrying charges. By analogy with MHD phenomena, the current driving electric field ($\mathbf{E} = E_{Az}\hat{z}$) must be generated within the shear at the expense of the magnetic wave energy. However, the sheared waves are distinguished from pure MHD fluctuations for two important reasons. First, the charges which generate J_s are constrained to always move along the same $\mathbf{B_0}$ field lines, so that SAWs cannot easily change their initial structure. Second, the simple concept of Ohm's law is not valid for the relativistic flow inside the shear. Indeed, once the particles become relativistic, the current quickly decouples from the driving electric field, and since the radiative drag increases rapidly with γ (the particle Lorentz factor), one might expect that a mildly relativistic flow will be favored by the system. Though E_{Az} depends on the microphysics of the shear (including all the annihilation processes, such as the tearing mode instability), its value may be estimated with a relatively simple argument under the assumption that the annihilation time scale within the shear is the shortest of the relevant time scales. The strength of the electric field is limited by the rate at which the oppositely-directed magnetic fluctuations are driven together by the large magnetic pressure gradients associated with SAWs. Since the Alfvénic field lines are strongly coupled to B_0 via flux freezing with the charged medium, this transfer of Alfvénic power into the shear is dictated by the diffusion rate within the resistive plasma in the region $|y| > s$, where the resistance is provided primarily by e^-/radio photon scatterings. As long as most of the wave energy is channeled into the shear before the waves break, we may equate the Alfvénic luminosity generated at the stellar surface with the power dissipated by the current as it converts magnetospheric energy into upscattered radiation. This yields an average electric field strength

$$E_{Az} \approx \min\left[5 \times 10^{11} \text{sV cm}^{-1} \left(\frac{B_{a0}}{10^{12} \text{ G}}\right),\right.$$

$$\left. 10^{10} \text{sV cm}^{-1} \xi \left(\frac{B_{a0}}{10^{12} \text{ G}}\right)^3 \left(\frac{P}{5 \text{ s}}\right)^2 \left(\frac{B_0}{10^{12} \text{ G}}\right)^{-1} \left(\frac{S}{10^4 \text{ cm}}\right)^{-3}\right], \quad (4)$$

where we have used $\eta = S/s$. The first expression on the RHS of eq. (4) represents the maximum attainable value of E_{Az} and occurs when the diffusion velocity approaches c.

We assume a typical pulsar spectrum specified as a steep power law with (flux density) index μ above a break at frequency $\epsilon_0/h \approx 500$ MHz. With $\gamma \gg 1$, a lab frame photon with energy ϵ will be blue-shifted to $\sim 2\gamma\epsilon$ in the electron rest frame, which is well below the resonant energy $\epsilon_B \equiv (B_0/44.14 \times 10^{12}\text{G})m_e c^2$, and its angle of propagation relative to the particle direction (and hence $\mathbf{B_0}$) is $\sim \gamma^{-1}$. As such, $\epsilon_B/\epsilon \approx 6 \times 10^9 (B_0/10^{12}\,\text{G})(\epsilon/h\,500\,\text{Mhz})^{-1} < \gamma^2$, and $\sigma_{\rm MC} \approx 4\sigma_T (\gamma\epsilon/\epsilon_B)^2$ [4,5]. Balancing the accelerating force eE_{Az} by the radiative drag $\gamma^2 F_r \sigma_{\rm MC}/c$, one therefore obtains

$$\gamma_s \approx 10^6\, f(\xi) \left(\frac{B_{a0}}{10^{12}\,\text{G}}\right)^{1/4} \left(\frac{B_0}{10^{12}\,\text{G}}\right)^{1/4}$$
$$\left(\frac{P}{5\,\text{s}}\right)^{1/4} \left(\frac{S}{10^4\,\text{cm}}\right)^{-1/4} \left(\frac{\dot\epsilon_0}{h\,500\,\text{MHz}}\right)^{-1/2}, \quad (5)$$

where $f(\xi) = \min[1, (\xi/\xi_0)^{-1/4}]$ and the parameter ξ_0 is defined as the smallest value of ξ for which the diffusion velocity equals c (e.g., $\xi_0 = 35$ for the represented parameter space). This result is consistent with the assumptions discussed above (e.g., that the particle motion is relativistic and sufficiently energetic to produce the required radio luminosity and that $\sigma_{\rm MC}$ have an ϵ^2 dependence). Ultimately, the current J_s must decay in concert with the Alfvénic magnetic field, even though the particles remain relativistic. Evidently, the initially fully charge separated regions must merge together and neutralize. This behavior is expected since the power transferred from the wave to the particles is reduced as the magnetic field decays. As such, an increasing number of charges undergoing collisions will not be energized quickly enough to remain in phase with the wave, and are therefore swept up by the lagging oppositely charged wave region.

The scaling of the magnetic fields B_0 and B_{a0} in the above equations was chosen for convenience and not to suggest that $B_{a0} \sim B_0$. We note, however, that the presence of significant reconnection in the nonlinear regime would have the desirable effect of enhancing the annihilation rate within the shear (see the paragraph preceeding Eq. 4).

DISCUSSION

In applying the above discussion to the cosmological gamma-ray burst model, we must now generalize to a more realistic magnetospheric geometry in which the field lines are more or less radial close to the stellar surface. It is evident from §2 that SAWs focus their energy into the internal current flow. Thus, as long as $\gamma^{-1} \ll s/R_*$, the flux is enhanced by a factor η in certain directions, for which a source at a distance D will have an observable γ-ray flux

$$F_\gamma \approx 10^{-7}\,\text{ergs cm}^{-2}\,\text{s}^{-1} \left(\frac{\eta}{10^3}\right) \left(\frac{B_{a0}}{10^{12}\,\text{G}}\right)^2 \left(\frac{D}{1\,\text{Gpc}}\right)^{-2}. \quad (6)$$

The observability of these bursts at cosmological distances imposes strict (but not unrealistic) conditions on the model parameters, such as the required burst power ($L_{\rm burst} \sim 10^{44} - 10^{46}$ ergs s^{-1}), whose magnitude depends on whether the SAWs are generated only near the polar cap (where the field lines are most strongly coupled to toroidal crustal activity) or are generated throughout the entire stellar surface.

Since the particle flow remains optically thin to the radio photons impinging upon the star, the ϵ^2 dependence of the cross-section (see above) implies that the incipient radio spectrum is upscattered to a γ-ray spectrum with (power) index $\mu + 2 + 1$, and very importantly, that the spectral radio break at ϵ_0 is translated to the corresponding γ-ray break at

$$\epsilon_{\text{break}} \sim 2\gamma_s^2 \epsilon_0 \approx 4.4 \text{ MeV } g(\xi)$$
$$\left(\frac{B_{a0}}{10^{12}\text{ G}}\right)^{1/2} \left(\frac{B_0}{10^{12}\text{ G}}\right)^{1/2} \left(\frac{P}{5\text{ s}}\right)^{1/2} \left(\frac{S}{10^4\text{ cm}}\right)^{-1/2}, \qquad (7)$$

independent of ϵ_0. Here, $g(\xi) = \min[1, (\xi/\xi_0)^{-1/2}]$. This result compares favorably with the observed value of ϵ_{break} (which after redshift is taken into account is seen to fall within the range ~ 100 keV -3 MeV [6]), and suggests that ξ/ξ_0 may be as large as 100. A more detailed description of the resulting γ-ray spectrum is given in Reference 3 (see, for example, Figure 2 therein). For completeness, we note that a cylindrical shear would correspond to $\eta \sim (S/s)^2$ and the parameter S in Equation (7) should be replaced by s. Such a strictly confined shear region would thus appear to be unlikely, though it cannot be ruled out without a more detailed calculation.

REFERENCES

1. C. A. Meegan et al., Nature **355**, 143 (1992).
2. R. Narayan, B. Paczyński, and T. Piran, Ap. J. **395**, L83 (1992).
3. F. Melia and M. Fatuzzo, Ap. J. **398**, L85 (1992).
4. F. Melia and M. Fatuzzo, Ap. J. **346**, 378 (1989).
5. C. D. Dermer, Ap. J. **360**, 197 (1990).
6. B. E. Schaefer et al., Ap. J **393**, L51 (1992).

ESCAPE OF HIGH-ENERGY PHOTONS FROM RELATIVISTICALLY EXPANDING GAMMA-RAY BURST SOURCES

Alice K. Harding and Matthew G. Baring *
NASA Goddard Space Flight Center, Code 665, Greenbelt, MD 20771

ABSTRACT

Four bright gamma-ray bursts detected by BATSE have also been detected at higher energies by EGRET. All are consistent with power-law spectra extending to energies as high as, in the case of GRB930131, 1 GeV. The optical depth to photon-photon pair production in these sources is extremely large for distances more than a few pc away if the radiation is emitted isotropically in the observer's frame. While it has been shown that the pair production optical depth can be dramatically reduced if the source is moving with a relativistic bulk Lorentz factor Γ, calculations have been limited to cases of a beam with opening angle $1/\Gamma$. The beaming angles required for optically thin sources are so small for Galactic halo or cosmological distances that the implied number of non-repeating sources is unreasonably high. Spherical expansion has also been considered but only in the case of an infinitely thin shell. We have investigated the pair production optical depth in relativistically expanding sources for more general cases, including shells of finite thickness and arbitrary opening angle. The new limits on required velocity for given beaming angles and thickness will place realistic constraints on gamma-ray burst source models.

INTRODUCTION

The discovery by BATSE that the spatial distribution of gamma-ray bursts (GRBs) is isotropic and non-homogeneous[1] suggests that the sources are either in an extended halo or at cosmological distances. The observed fluxes of GRBs at earth imply high luminosities for isotropically emitting sources: $L \simeq 10^{42-43}\,\mathrm{erg\,s^{-1}}$ at a distance of $d = 100$ kpc and $L \simeq 10^{50-51}\,\mathrm{erg\,s^{-1}}$ when $d = 1$ Gpc. Such luminosities combined with the compact source size implied by the observed rapid time variability gives photon densities high enough to make GRBs optically thick to photon-photon pair production by many orders of magnitude. Such a situation is in direct conflict with observed spectra of GRBs which show no attenuation at energies as high as 1 GeV.

One solution to the problem is to allow beaming of the radiation, which raises the pair production threshold to energies above those observed. Relativistic motion of the source is a natural way to achieve beaming of the emission and avoid attenuation in GRBs. The radiation from a source that is isotropically emitting in the comoving frame will be beamed roughly within an angle $1/\Gamma$ in the observers frame, where Γ is the bulk Lorentz factor. Calculations for the case where the source opening angle is of order $1/\Gamma$ show that the the pair production optical depth has the dependence $\tau_{\gamma\gamma}(\varepsilon) \propto \Gamma^{-(1+2\alpha)}\varepsilon^{\alpha-1}$, where α

---

* NAS–NRC Research Associate

is the photon spectral index and ε is the photon energy in units of mc^2, and therefore is reduced by a factor $\Gamma^{-(1+2\alpha)}$ below the optical depth for isotropic radiation.[2,3] The minimum bulk Lorentz factors required to make $\tau_{\gamma\gamma} < 1$ in the bright "superbowl" burst (GRB930131) detected by EGRET[4] up to an energy of 1 GeV are $\Gamma \gtrsim 10^3$ at a distance of 1 Gpc and $\Gamma \gtrsim 10$ at 30 kpc (ref. 5).

An advantage of relativistic beaming is a smaller required luminosity at the source because the observed flux, $\phi \sim \Gamma^2 L/4\pi d^2$, is enhanced by a solid angle factor, Γ^2 (ref. 2). However, the number of sources must be a factor Γ^2 higher in order to account for the observed number of GRBs. In the case of cosmological GRBs, this factor could be as high as 10^6 for the above limits on Γ, which is unacceptably large for many models. Source geometries with beaming angles larger than $1/\Gamma$ could ease this problem if the high energy photons were able to escape. Fenimore, Epstein & Ho[6] have shown that a relativistically expanding, spherical shell will allow escape of high energy gamma-rays, but their calculation was limited to the case of an infinitely thin shell. In this study, we have extended the calculation of pair production optical depth to the full range of intermediate source geometries between opening angles of $1/\Gamma$ and a spherical expansion, and allowing shells of arbitrary thickness. We find that thick sources with large opening angles will also allow the high energy photons to escape with only moderately larger minimum required Lorentz factors.

PAIR PRODUCTION OPTICAL DEPTH

We consider a source expanding with constant and homogeneous bulk Lorentz factor Γ, and opening angle $2\Theta_B$ and thickness ΔR in the observers frame. Suppose that a test photon, with energy ε_t and angle cosine $\mu_t = \cos \Theta_t$ with respect to the axis bisecting $2\Theta_B$, is emitted at time $t = 0$ from the inner radius R_0 of the shell and moves radially through the source to eventually escape and reach the observer. Then the radial distance r_t of the test photon at any time t from the center of the expansion is

$$r_t = R_0 + ct \quad (1)$$

The γ-γ pair production optical depth of the source for this photon is[7]

$$\tau_{\gamma\gamma}(\varepsilon_t, \mu_t) = \frac{1}{2\pi^2 c} \int dr_t \, dr_{ti} \, d\mu_{ti} \, d\phi \, d\varepsilon_i \, \sigma_{\gamma\gamma}(\chi)(1 - \mu_{ti}) \dot{n}(\varepsilon_i, \mu_{ti}, r_{ti}) \quad . \quad (2)$$

Here, subscript i denotes the interacting source photons, $\mu_{ti} = \cos \theta_{ti}$ is the cosine of the angle between the directions of the test and interacting photons, and $\chi = \sqrt{\varepsilon_t \varepsilon_i (1 - \mu_{ti})/2}$ is the center-of-mass (CM) frame energy. The threshold for pair production is at $\chi = 1$. If the source generates isotropic emission in the comoving frame with a power-law spectrum $\dot{n}_c(\varepsilon_c, \mu_c)\varepsilon_c^{-\alpha}$, then the emission rate in the observers frame, using the fact that the number of photons is Lorentz invariant, will be

$$\dot{n}(\varepsilon_i, \mu_i) = \frac{d^2\Omega}{V_o} \alpha \, \Phi_{obs} \, \varepsilon_i^{-\alpha} \frac{(1-\beta)^\alpha}{(1-\beta\mu_i)^{1+\alpha}}, \quad -\infty \le \varepsilon_i \le \infty \, , \quad (3)$$

where μ_i is the cosine of the angle between the interacting photon and the expansion direction at its position, and $\beta = (1 - 1/\Gamma^2)^{1/2}$. Here, Φ_{obs} is the number of photons from a source observed (at a distance d) at energy mc^2 on earth per second per unit area per mc^2, and V_o is the initial source volume

$$V_o = \frac{\Omega}{3}\left\{(R_0 + \Delta R)^3 - R_0^3\right\} , \qquad (4)$$

where $\Omega = 2\pi(1 - \cos\Theta_B)$ is the solid angle of the expansion.

If the test photon is emitted from the back of the shell at time $t = 0$ it remains within the expanding volume only when

$$R = R_0 + \beta ct \leq r_t \leq R_0 + \Delta R + \beta ct . \qquad (5)$$

The test photon can interact with photons at positions within some look-back volume; for the high-Γ expansions considered here, this volume is relativistically distorted[6,8] into a highly eccentric ellipsoidal geometry. Denote the radius of a typical interacting photon by r_i. Such an interacting photon was emitted at time t_i ($<t$) and at a distance $r_{ti} = c(t-t_i)$ from the test photon. Restrictions to the integration limits in Eqn (2) arise because the interacting photons can only be emitted from the region the expanding plasma occupied at the time of emission. In the radial direction, this volume is specified simply by

$$R_i \leq r_i \leq R_i + \Delta R , \quad R_i = R_0 + \beta ct_i , \qquad (6)$$

and the angular constraint is independent of time:

$$0 \leq \Theta_i \leq \Theta_B , \qquad (7)$$

where Θ_i is the angle between the radial vector to the interacting photon and the expansion axis. The values of r_{ti} that are achievable are further constrained by the causality condition

$$0 \leq r_{ti} \leq ct = r_t - R_0 . \qquad (8)$$

We have been able to reduce the calculation of $\tau_{\gamma\gamma}$ in Eqn (2) to just two numerical integrations by performing the integrations over r_t, ϕ and ε_i analytically (see ref. 7). Assuming that the plasma emits uniformly at any one time, the angular constraint results in an analytic determination of the azimuthal integration, since at a given radius, each azimuthal angle ϕ within the cone of expansion contributes equally. In the case considered here, where the test photon moves along the axis of the expansion, this azimuthal integration is trivial.[7] The ε_i integration can be performed by changing variables to the (CM) frame energy $\chi = \sqrt{\varepsilon_t \varepsilon_i (1 - \mu_{ti})/2}$, which enables the integration over the cross-section to be separated and approximated analytically.[3] The remaining triple integration can be reduced[7] to a double integral by observing that R_0 effectively scales the spatial dimensions and that a change of integration variables to r_t/R_0 and r_{ti}/r_t enables analytic determination of the r_t integration. We will assume throughout that the emission from the source is constant in time.

MINIMUM LORENTZ FACTOR

The pair production optical depth as described in Eqn (2) depends on the free parameters: Γ, $\Delta R/R_0$, Θ_B, d, (we have assumed $\mu_t = 1$ here, with the test photon on the axis of the expansion) and on the observed parameters: Φ_{obs}, α, ε_t, Δt. The burst variability timescale Δt gives an upper limit on the source size which we take as $R_0 = \Gamma c \Delta t = \Gamma\, 10^7$ cm, based on the apparent size of the expanding shell perpendicular to the light of sight, as seen by a stationary observer[8]. From Eqn (2), we have explored the minimum bulk Lorentz factor $\Gamma_{\rm MIN}$ at which $\tau_{\gamma\gamma} = 1$ as a function of the above free parameters.

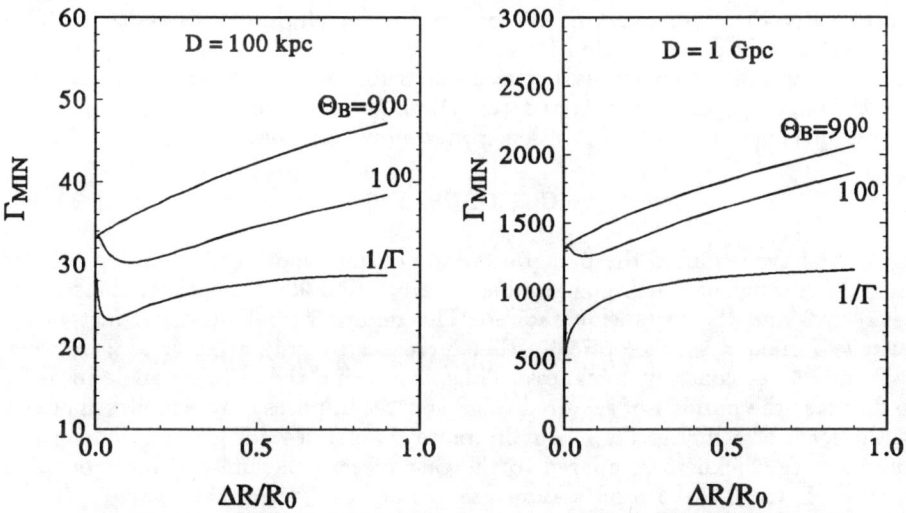

Figure 1. The minimum bulk Lorentz factor $\Gamma_{\rm MIN}$ for GRB930131, as obtained from the pair production condition $\tau_{\gamma\gamma}(\varepsilon_t) = 1$ in Eq. (2) for $\varepsilon_t = \varepsilon_{\rm MAX} = 1\,{\rm GeV}$. Results are shown for two different source distances, (a) $D = 100\,{\rm kpc}$ and (b) $D = 1\,{\rm Gpc}$, with three expansion opening half-angles Θ_B for each, as labelled.

Figure 1 illustrates the dependence of $\Gamma_{\rm MIN}$, on the geometry of the expanding region for the burst source GRB930131 at two different sample distances. This burst had photons observed out to $\varepsilon_{\rm MAX} = 1$ GeV (see ref. 4; we take $\varepsilon_t = \varepsilon_{\rm MAX}$), a power-law spectral index of $\alpha = 2$ and flux $\phi_{\rm obs}(1\,{\rm MeV}) = 2\,{\rm ph\,cm^{-2}\,s^{-1}\,MeV^{-1}}$. The $1/\Gamma$ curves in Fig. 1 correspond to the cases where the opening angle of the expanding emission region was set to $\Theta_B = 1/\Gamma$, which approximates the special case of Baring & Harding[9] where $\tau_{\gamma\gamma} \propto \Gamma^{-(1+2\alpha)}$ when $\Delta R/R_0 \sim 1/\Gamma$. The $\Theta_B = 90°$ curves yield $\Gamma_{\rm MIN}$ as an increasing function of $\Delta R/R_0$, a consequence of the increase in the volume that is causally connected to the escaping test photon. As Θ_B decreases, the available phase space and therefore the optical depth drop, so that $\Gamma_{\rm MIN}$ decreases roughly proportional to the solid angle Ω of the expansion. This decrease below the $\Theta_B = 90°$ case is not observed in Fig. 1 for the thin-shell limit $\Delta R/R_0 \lesssim \Omega/2\pi$, since then the causally-connected volume is more severely restricted by the radial dimensions of the expansion rather than its angular size. In fact, the thin-shell limit gives Γ_{MIN} independent of Θ_B, and is amenable to analytic approximation[7]. Therefore all the curves in Fig. 1 with identical d converge to one point on the Γ_{MIN} axis at $\Delta R/R_0 = 0$; note that extremely low $\Delta R/R_0$ points are not depicted in Fig. 1. The minima that appear in the $1/\Gamma$ curves follow from competition between the angular and radial restrictions on available volume. The increase in Γ_{MIN} toward $\Delta R/R_0 = 0$ is due to the increase in photon density required to produce the observed flux (cf. Eqn 3).

While the thin-shell limit is an elucidating special case, it is unlikely to be physically appropriate to gamma-ray bursts since it is probably difficult to generate relativistically-expanding emission regions much thinner than R_0/Γ. This has been clearly demonstrated for relativistic fireballs[10]. The results presented here assume that the power-law source spectrum extends beyond the maximum energy ε_{MAX} of photons detected (cf. Eqn 3): this seems reasonable as the source spectrum shows no deviation from a power-law out to ε_{MAX}.

CONCLUSIONS

We have explored the pair production optical depth of expanding sources for the full range of opening angles between $1/\Gamma$ and 90^0, and source thicknesses between 0 and the expansion radius. The minimum bulk Lorentz factors required to make a source optically thin increases by only a factor of 2 between $1/\Gamma$ and 90^0 at constant thickness. Thus, increasing the opening angle in order to decrease the number of required GRB sources imposes only a modest penalty in the form of a higher Γ_{MIN}. Furthermore, Γ_{MIN} does not greatly increase for finite source thickness compared to the case of zero thickness at large opening angles[6]. Both of these results allow greater flexibility in constructing optically thin GRB source models, i.e. those which avoid fireballs, especially at cosmological distances. These calculations will also be applicable to fireball models[11] where the GRB is produced by interaction with an external medium[12], in determining when the fireball and the swept up material become optically thin.

REFERENCES

1. C. A. Meegan et al., Nature **335**, 143 (1991).
2. J. H. Krolik & E. A. Pier, ApJ **373**, 277 (1991).
3. M. G. Baring, ApJ **418**, 391 (1993).
4. M. Sommer et al., ApJ Lett., in press (1993).
5. A. K. Harding, Proc. of Second Compton Symposium (ed. N. Gehrels & J. Norris, in press (1994)).
6. E. E. Fenimore, R. I. Epstein & C. Ho, A & A Supp **97**, 59 (1993).
7. M. G. Baring & A. K. Harding, in prep., (1994).
8. M. J. Rees, Nature **211**, 468 (1966).
9. M. G. Baring & A. K. Harding, Proc. 23rd Int. Cos. Ray Conf. **1**, 53 (1993).
10. T. Piran, A. Shemi & R. Narayan, MNRAS **263**, 861 (1993).
11. B. Paczyński, ApJ **308**, L43 (1986).
12. P. Mészáros & M. J. Rees, ApJ **405**, 278 (1993).

GAMMA-RAY BURSTS FROM BLACK HOLE-NEUTRON STAR MERGERS

Jordi Isern and Margarita Hernanz
Centre d'Estudis Avançats de Blanes, CSIC
Camí de Sta. Bàrbara s/n, 17300 Blanes (Girona), Spain
and
Laboratori d'Astrofísica, IEC, Diagonal, 645, 08028 Barcelona, Spain

Robert Mochkovitch and Xavier Martin
Institut d'Astrophysique de Paris, 98 bis Boulevard Arago, 75014 Paris, France

ABSTRACT

We present a scenario for the production of cosmological γ-ray bursts in terms of an expanding fireball, originated by the merging of a neutron star and a stellar mass black hole. The disrupted neutron star releases a great amount of energy in the form of neutrino-antineutrino pairs, which convert their energy into γ-rays by annihilation. The formation of an optically thick wind, preventing γ-rays from escaping and converting their energy into kinetic energy of the ejected material, is avoided in our model, because angular momentum conservation means that a region not polluted with baryons forms along the rotation axis of the system. The annihilation of neutrino-antineutrino pairs in this matter-free funnel can originate the γ-rays which will escape freely.

INTRODUCTION

The Burst and Transient Source Experiment (BATSE) on the Compton Gamma-Ray Observatory has shown that the distribution of 153 γ-ray bursts detected has an isotropic angular distribution but is not radially uniform[1]. These results are inconsistent with the distribution of galactic objects and suggest that the bursts may be at cosmological distances. Therefore, they must be very energetic events, releasing $\sim 10^{50}$ erg sr^{-1} on a timescale of seconds. Possible scenarios for such phenomena could be the coalescence of two neutron stars[2-4] or the accretion-induced collapse of a white dwarf[5], because both can release up to 10^{53} erg as neutrino-antineutrino pairs. The conversion of less than 1% of $\nu\bar{\nu}$ pairs in γ-rays by annihilation[6] could, thus, generate γ- ray bursts. But in all these models, an optically thick wind forms[7,8] which prevents the escape of γ-rays. A possible solution to this problem is a scenario where a neutron star is disrupted by a stellar mass black hole. The thick disk formed emits pairs of neutrinos and antineutrinos which expel a wind from the disk.

The transfer of energy by the neutrinos emitted from a neutrinosphere to matter occurs essentially through interaction with nucleons and through creation of e^+e^- pairs. If the wind driven by the deposited energy implies a mass loss rate $\dot{M} < 10^{-2}(\dot{E}/c^2)$, with $\dot{E}$ the rate of energy deposition, γ-rays will escape[7]. Then, the wind, dominated by e^+e^- pairs and photons, becomes very relativistic, with Lorentz factors higher than 100, and the blueshifted radiation falls in the γ-ray range[2,9]. But models which take into account neutrino-matter interaction show that the mass loss rate is fairly higher than the above limit[8,10]. Typically :

$$\dot{M} \approx \dot{E}\left(\frac{GM}{r_\nu}\right)^{-1} = \left(\frac{2r_\nu}{r_g}\right)\left(\frac{\dot{E}}{c^2}\right) \geq 10\left(\frac{\dot{E}}{c^2}\right) \qquad (1)$$

where M is the mass of the compact object, r_ν the radius of its neutrinosphere and $r_g = 2GM/c^2$ its Schwarzschild radius. Thus, an optically thick wind is formed which prevents the γ-rays from escaping, because their energy is invested in kinetic energy of the wind. A possible way to solve this problem is to consider a scenario where a neutron star is disrupted by a black hole. The neutron star forms a thick disk which emits $\nu\bar{\nu}$ pairs. Although these neutrinos expel a wind from the disk, angular momentum conservation avoids baryon contamination along the rotation axis. Thus, $\nu\bar{\nu}$ annihilation in this clear funnel can be at the origin of the γ-ray burst.

MODEL

The scenario we propose consists of a binary system formed by a neutron star and a black hole. The neutron star, with a mass of $1.4M_\odot$, is tidally disrupted by a black hole of a few solar masses ($5M_\odot$ in our model). For an orbital separation of $\sim 5r_g$, the neutron star overflows its Roche lobe[11], being the mass loss either dynamical, with a timescale of a few milliseconds, or driven by gravitational radiation, with a timescale of a few tens of milliseconds[12]. We assume that in the resulting disk the neutrinosphere has its inner boundary at $3r_g$ and a circular section of radius $2r_g$. The material from the wind, driven by neutrino emission, has a large angular momentum and flows away from the system vertical axis, leaving there a matter-free region[13]. Therefore, neutrinos can reach this baryon-free region along the rotation axis of the system, annihilate and create a fireball, from which γ-rays will emerge.

The energy released by $\nu\bar{\nu}$ annihilation on the rotation axis of the system, $q_{\nu\bar{\nu}}(r)$ has been calculated[6] assuming that neutrinos move on straight lines (so neglecting aberration, Doppler shift and general relativity effects):

$$q_{\nu\bar{\nu}}(r) \approx 3 \times 10^{23} T_\nu^8 \int (1 - \mathbf{\Omega}_\nu \cdot \mathbf{\Omega}_{\bar{\nu}})^2 d\Omega_\nu d\Omega_{\bar{\nu}}$$
$$= 3 \times 10^{23} T_\nu^8 \Phi(r) \text{ erg cm}^{-3} \text{ s}^{-1} \qquad (2)$$

where T_ν is the temperature at the neutrinosphere in MeV and $\mathbf{\Omega}_\nu$ and $\mathbf{\Omega}_{\bar{\nu}}$ are unit vectors along the direction of propagation of neutrinos and antineutrinos interacting at a radius r. The deflection of the neutrino trajectories due to general relativistic effects implies that the resulting momentum of the annihilating neutrinos will be directed towards the black hole below some radius $r_0 \approx 3r_g$. Therefore, the energy deposited at $r < r_0$ will fall into the black hole and the flow of e^+e^- pairs and photons moving outwards will start at r_0.

The temperature of the neutrinosphere must satisfy a constraint, because it has to generate a certain amount of energy: for a typical cosmological γ-ray burst, the luminosity is of the order of $(10^{51}/4\pi)$ erg s^{-1} sr^{-1}. So, we can write

$$\int_{r_0}^{\infty} q_{\nu\bar{\nu}} r^2 dr = 3 \times 10^{23} T_\nu^8 \int_{r_0}^{\infty} \Phi(r) r^2 dr = \frac{10^{51}}{4\pi} \qquad (3)$$

on the assumption that the fireball is confined along the axis. This is the case if the wind originated in the disk can confine the flow of e^+e^- pairs and photons until it reaches the ultra-relativistic regime. After that, relativistic beaming (with an opening angle $\tan^{-1}(1/\Gamma) < 30°$, with $\Gamma = (1-\beta^2)^{-1/2}$, where $\beta = v/c$, the Lorentz factor) will limit the lateral expansion of the flow. For the adopted geometry of the system, we have calculated the integral $\Phi(r)$ in equation (2). The neutrinosphere temperature obtained is $T_\nu = 4.76\text{MeV}$. The corresponding total neutrino luminosity $L_\nu = \Sigma_\nu \times 3 \times \frac{3}{8}\sigma T_\nu^4 = 1.2 \times 10^{54}$ erg s^{-1}, with $\Sigma_\nu = 4\pi r_g^2$ the surface of the neutrinosphere. Therefore, only 0.1% of the neutrino energy is converted into γ-rays. This large value of L_ν means that all the energy available from the disruption of the neutron star, several 10^{53} erg, will be released in less than one second. The duration of the observed burst is (1+z) times longer, being z the redshift.

If we assume that the relativistic flow is confined along the vertical axis within a solid angle $\Delta\Omega$, the flow equations are the following[2,14]

$$(U + P)\Delta\Omega r^2 v Y^2 = L_{\nu\bar{\nu}} \quad (4)$$

$$TY = T_0 \quad (5)$$

$$\frac{dL_{\nu\bar{\nu}}}{dr} = \Delta\Omega r^2 \left(1 - \frac{r_g}{r}\right)^{1/2} q_{\nu\bar{\nu}} \quad (6)$$

where T_0 is a constant, $Y = (1 - r_g/r)^{1/2}(1 - v^2/c^2)^{-1/2}$. The substitution of equation (5) into (4) gives

$$\Delta\Omega \frac{16}{3}(1 + \eta(T))\sigma T_0^4 \beta(1-\beta^2) = \frac{L_{\nu\bar{\nu}}}{r^2}\left(1 - \frac{r_g}{r}\right) = F(r) \quad (7)$$

where $\eta(T)$ varies between 7/4, for ultra-relativistic electrons, and 0 after annihilation. The condition $(dF/dr)_{r_s} = 0$ gives the sonic radius r_s, where $\beta = \beta_s = 1/\sqrt{3}$. We obtain $r_s/r_g = 5.12$ and $T_0 = 555(1+\eta)^{-1/4}$keV. The results for β and the Lorentz factor Γ as a function of r can be found in our previous paper[15]. At large distances, Γ is proportional to r and T to r^{-1}. We assume that the wind from the disk is able to confine the fireball as long as its temperature (and pressure) is large. If this is true until Γ reaches the value $\Gamma \approx 2$, then relativistic beaming will prevent the lateral expansion of the wind. The Lorentz factor attains the value $\Gamma = 2$ at $r \simeq 150$ km, and the corresponding temperature is $T \approx 200 - 300$keV. At this radius, the wind temperature T_w can be estimated to be $T_w \approx 500 - 600$keV (for $T_\nu \approx 4 - 5$MeV).

DISCUSSION

The next step is to compare the results with the observations. First of all, we will consider the burst statistics. If we assume that the wind confines the fireball within an angle $\theta \approx 30 - 50°$ away from the vertical axis of the system, the fraction of detectable bursts is $f = 1 - \cos\theta = 0.15 - 0.40$. The rate of black hole-neutron star mergers has been estimated as $10^{-5} - 10^{-6}$ yr^{-1} per galaxy[16]. The product of both quantities gives an estimation of the rate of γ-ray bursts which is not far from the one deduced from the observations, $\sim 10^{-6}$ yr^{-1} per galaxy[17].

Concerning the spectrum, one would expect a thermal one with a temperature $T \approx 500/(1 + z)$keV, but this is in direct conflict with the observed spectra[18,19]. All the models of γ-ray bursts based on an expanding fireball have this same problem. However, the interaction between the fireball and an external medium could perhaps produce a non-thermal spectrum[13,20]. Concerning the peak fluxes, those predicted by our model are rather low ($\sim 10^{-6} h^2$ erg cm^{-2} s^{-1}keV^{-1}, for z=1, with h the Hubble constant in units of 100 km s^{-1} Mpc^{-1}), specially if h=0.5.

Other observed properties are the time profile and the duration. γ-ray bursts show a large variety of these aspects. Our model can explain short bursts with smooth profiles. Models for the bursts with a complex time structure and lasting up to 100 s have been proposed by Narayan et al[21]. In some cases, the neutrino annihilation model can also lead to a variability at the millisecond timescale. This could correpond to highly turbulent disks, where induced shocks would lead to a succession of short pulses of neutrino emission at the neutrinosphere. These pulses could produce, after $\nu \bar{\nu}$ annihilation, the characteristic temporal behaviour of complex bursts.

ACKNOWLEDGEMENTS

This work has been partially funded by the DGICYT grant PB91-060, the Spanish-French Action "Physics of White Dwarfs and Brown Dwarfs" and the CESCA grants "Structure and Evolution of Galaxies" and "Accretion onto White Dwarfs".

REFERENCES

1. C. A. Meegan et al, Nature **355**, 143 (1992).
2. B. Paczyński, Ap. J **308**, L43 (1986).
3. D. Eichler, M. Livio, T. Piran, D. N. Schramm, Nature **340**, 126 (1989).
4. A. Shemi, T. Piran, Ap. J **365**, L55 (1990).
5. R. Ramaty, A. Dar, Proc. COSPAR Meeting (in press, 1993).
6. J. Goodman, A. Dar, S. Nussimov, Ap. J **314**, L7 (1987).
7. B. Paczyński, Ap. J **363**, 218 (1990).
8. S. E. Woosley, E. Baron, Ap. J **391**, 228 (1992).
9. J. Goodman, Ap. J **308**, L47 (1986).
10. R. C. Duncan, S. L. Shapiro, I. Wasserman, Ap. J **309**, 141 (1986).
11. J. M. Lattimer, D. N. Schramm, Ap. J **192**, L145 (1974).
12. J. P A. Clark, D. M. Eardley, Ap. J **215**, 311 (1977).
13. P. Mészáros, M. J. Rees, Ap. J **397**, 570 (1992).
14. R. A. Flammang, Mon. Not. R. Astr. Soc. **199**, 833 (1982).
15. R. Mochkovitch, M. Hernanz, J. Isern, X. Martin, Nature **361**, 236 (1993).
16. R. Narayan, T. Piran, A. Shemi, Ap. J **379**, L17 (1991).
17. S. Mao, B. Paczyński, Ap. J **388**, L45 (1992).
18. J. C. Higdon, R. G. Lingenfelter, Ann. Rev. Astron. Astrophys. **28**, 401 (1990).
19. B. E. Schaefer et al, Ap. J **393**, L51 (1992).
20. P. Mészáros, M. J. Rees, Ap. J **405**, 278 (1993).
21. R. Narayan, B. Paczyński, T. Piran, Ap. J **395**, L83 (1992).

RADIO AND OPTICAL EMISSION, SPECTRAL SHAPES AND BREAKS IN GRB

J. I. Katz

Dept. of Physics, Washington University, St. Louis, Mo. 63130

ABSTRACT

Relativistic blast wave models of GRB predict the spectrum of the emitted synchrotron radiation. The electrons in the shocked region are heated to a Wien distribution whose "temperature" is 1/3 of the mean electron energy. This energy determines a characteristic (break) frequency of synchrotron radiation. At much lower frequencies a spectrum $F_\nu \propto \nu^{1/3}$ is predicted independently of the details of the emitting region. This is consistent with the observed soft X-ray emission of GRB. It implies low visible and radio intensities, unless there are collective emission processes.

INTRODUCTION

Most of the controversies surrounding GRB involve, directly or indirectly, the shapes of their continuum spectra. Different theoretical models predict different spectral characteristics. For example, if the radiation emerges from a stationary region optically thick to gamma-gamma pair production, there will be a spectral break at an energy $O(m_e c^2)$[1]. If the source emits as a black body the spectrum will resemble a Planck function, as may be the case for SGR (but not for classical GRB). Electrons with a power-law distribution of energies produce optically thin synchrotron radiation with a power-law spectrum and no characteristic energies or spectral breaks.

The low-frequency extension of the gamma-ray spectrum determines the observability of GRB outside the gamma-ray band[2]. Observations at visible frequencies[3] are widely believed to hold the key to identifying the quiescent counterparts, astronomical sites, and physical mechanisms of GRB. In addition, observations at soft X-ray frequencies[4] may provide a direct measure of the intervening column density of (chiefly) oxygen, while observations of radio dispersion similarly measure[5,6,7] the intervening column density of free electrons. These observations may settle the question of Galactic vs. cosmological distances for GRB, as well as measure properties of the intergalactic medium if the distances are cosmological.

HYPOTHESIS

A model of GRB has been developed which unambiguously predicts the shape of the low-frequency part of their spectra. This model involves debris[8] accelerated by a relativistic fireball interacting with a clumpy surrounding interstellar medium[7,9,10]; the observed gamma-rays are produced in relativistically shock-heated interstellar matter and fireball debris.

In the more familiar case of relativistic particle acceleration at a nonrelativistic shock (such as in supernova remnants) only a small fraction of the particles are accelerated. The acceleration process provides no characteristic energy scale for the accelerated particles, so their spectrum is a power law, broken only at the energy at which their gyroradii carry them out of the region of acceleration. This conventional model of shock acceleration is inapplicable to relativistic shocks in GRB.

In the present model of GRB all the charged particles in the shocked matter are accelerated, and the internal energy per particle sets an energy scale. This internal energy is determined by the hydrodynamic jump conditions at the shock, which in turn are set by the velocity of the debris and the densities of the debris and interstellar medium. Because there is a characteristic energy scale (enforced by conservation of energy and number, which are inapplicable to particles accelerated from a thermal reservoir by an imposed flow field) there is no reason to expect a power-law energy distribution of the relativistic particles. They rapidly interact with each other by means of plasma waves (which mediate the collisionless shock), and come to an equilibrium Wien distribution

$$N_e(E) \propto E^2 \exp(-E/k_B T), \qquad (1)$$

where the temperature parameter $k_B T$ is 1/3 of the mean energy $\mathcal{E}$ per particle. To obtain an equilibrium distribution it is sufficient that wave-wave and wave-particle interactions be rapid; in this manner all particles are coupled to each other despite the absence of collisions. Estimates show that the plasma wave interaction and acceleration times, typically $O(\omega_g^{-1})$ in strong turbulence, where ω_g is the gyrofrequency, are very much shorter than the other characteristic times in the problem, the hydrodynamic rarefaction and the synchrotron radiation times, so that these latter processes affect $\mathcal{E}$ but not the form (1).

HIGH ENERGY SPECTRA

The observed GRB spectra at high photon energies do not show the exponential cutoff implied by the Wien particle spectrum (1). There are two possible explanations:

1. The radiating electrons interact with each other by means of plasma waves which have a very high brightness temperature (far in excess of the individual particle energies), and which therefore do not constitute a genuine heat bath. As a result, the form (1) is not thermodynamically required, and a power-law spectrum (rather than an exponential cutoff) for $E > \mathcal{E}$ is possible. In order that this high energy tail not dominate the energy content (which would be inconsistent with the definition of $\mathcal{E}$) the electron energy distribution $N_e(E) \propto E^{-p}$ must have an index $p > 2$ and the spectral index of its synchrotron radiation (defined by $F_\nu \propto \nu^{-s}$) $s = (p-1)/2 > 1/2$, consistent with the observed[2] $s \approx 1$ at high energies.

2. At any time (and even more so in time-average) the observed radiation is integrated over radiating volumes with a distribution, probably very broad, of values of $\mathcal{E}$, magnetic field, and Doppler shift. As a result, the inferred distribution of energies of radiating particles only shows an exponential cutoff at energies higher than the greatest $\mathcal{E}$ found anywhere in the radiating volume. Observed breaks in the spectrum[11,12] reflect a characteristic $\mathcal{E}$ in the radiating region, and their evolution through a burst reflects the evolution of $\mathcal{E}$ as the blast wave progresses through the interstellar medium.

LOW ENERGY SPECTRUM

For any electron energy distribution with power law exponent $p < 1/3$ synchrotron radiation at frequencies below the spectral peak is dominated by the highest energy electrons because the power radiated at a given frequency[13] is $\propto E^{1/3}$. This condition is met by the Wien distribution, for which $p \to -2$ for $E \ll \mathcal{E}$, and by most plausible distributions below their characteristic energy $\mathcal{E}$. The integrated spectrum then has the index $s = -1/3$:

$$F_\nu \propto \nu^{1/3}, \quad (2)$$

characteristic of low-frequency synchrotron emission below the spectral peak[13]. This result survives averaging over an emission region with a range of electron energy distributions, $\mathcal{E}$, magnetic field, and Doppler shift, as long as the frequency of observation is everywhere below the characteristic synchrotron frequency (Doppler-shifted to the observer's frame) for electrons of energy $\mathcal{E}$, and is therefore a robust prediction of relativistic blast wave models.

The predicted spectrum (2) is consistent with data[2] on GRB at X-ray energies below 10 KeV, where the observed photon count rate per unit energy $N_\gamma \propto (h\nu)^{-0.7}$ is equivalent to $s = -0.3$, indistinguishable from $s = -1/3$. This supports the applicability of relativistic blast wave models to GRB. The form (2) also resolves the X-ray paucity problem[14] which arises in models of GRB emission close to neutron stars.

It is possible to extrapolate (2) to lower frequencies with confidence, once the basic model is accepted. An intense GRB with a flux 10^{-5} erg/cm^{2}sec in a soft gamma-ray bandwidth of 400 KeV has a gamma-ray flux of 10 mJy (1 Jy $\equiv 10^{-23}$ erg/cm^2 sec Hz) and a visible flux of 0.2 mJy. The total visible power corresponds to a $\approx$ 18th magnitude star, difficult to detect as an optical transient.

For the same bright GRB extrapolation to 1 GHz leads to a predicted flux of about 2 μJy. The effective flux of a brief transient measured by a broad-band receiver may be further reduced by dispersion. In addition, self-absorption in an incoherent source leads to an independent upper bound[7]

$$F_\nu < 2\pi\nu^2 m_p \frac{r_s^2}{D^2} \approx 0.5 \; \mu\text{Jy} \left(\frac{\nu}{10^9 \text{ Hz}}\right)^2 \left(\frac{r_s}{2 \times 10^{15} \text{ cm}}\right)^2 \left(\frac{1 \text{ Gpc}}{D}\right)^2, \quad (3)$$

where r_s is the radius of emission and D is the distance to the GRB. The self-absorption bound (3) will exceed the μJy level for the brightest GRB, which may be much closer than 1 Gpc, and after the observable gamma-ray emission, when the blast wave expands to $r_s \gg 2 \times 10^{15}$ cm.

As the blast wave expands the number of radiating electrons increases $\propto r^3$. At a given frequency ν the radiated power density per electron (in the co-moving frame in which the electron and field distributions are assumed isotropic) is $\propto (\nu\gamma_e/B)^{1/3}$; γ_e and B each vary[7] $\propto r^{-3/2}$. Applying a Lorentz factor γ to transform to the observer's frame multiplies the spectral density by $\gamma^{1+s} = \gamma^{2/3} \approx \gamma_F^{1/3} \propto r^{-1}$, where γ_F is the fireball Lorentz factor. Hence the spectral brightness increases $\propto r^2 \propto t^{4/5}$ until a time $t_c \propto \nu^{-5/12}$ characteristic of the frequency of observation[7] is reached, after which the brightness may fall exponentially; t_c is about 3×10^4 times longer at 1 GHz than at 300 KeV, and may be days to weeks. The peak brightness at 1 GHz may be $\sim 10^4$ times brighter ($\sim$ 20 mJy) than it was when the spectral peak was at 300 KeV. These numerical results are necessarily rough.

The preceding arguments would have predicted that pulsars should be unobservable at radio frequencies! Fortunately, collective emission processes occur which are not bound by single electron radiation rates or by limits on brightness temperatures. We may hope (with faint reason) that collective processes are similarly effective in GRB.

I thank M. Davis, P. Horowitz, and B. E. Schaefer for discussions and NASA NAGW 2918 for support.

REFERENCES

1. B. J. Carrigan and J. I. Katz, Ap. J. **399**, 100 (1992).
2. B. E. Schaefer, these proceedings (AIP, New York, 1994).
3. B. E. Schaefer, Gamma-Ray Bursts, Observations, Analyses and Theories (Cambridge University Press, Cambridge, 1992), p. 107.
4. B. E. Schaefer, Compton Gamma-Ray Observatory (AIP, New York, 1993), p. 803.
5. V. L. Ginzburg, Nature **246**, 415 (1973).
6. D. M. Palmer, Ap. J. (Lett.) **417**, L25 (1993).
7. J. I. Katz, Ap. J. **422**, in press (1994).
8. A. Shemi and T. Piran, Ap. J. (Lett.) **365**, L55 (1990).
9. M. J. Rees and P. Mészáros, MNRAS **258**, 41p (1992).
10. P. Mészáros and M. J. Rees, Ap. J. **405**, 278 (1993).
11. B. E. Schaefer, et al., Ap. J. (Lett.) **393**, L51 (1992).
12. R. D. Preece, these proceedings (AIP, New York, 1994).
13. J. D. Jackson, Classical Electrodynamics (Wiley, New York, 1975).
14. J. N. Imamura and R. I. Epstein, Ap. J. **313**, 711 (1987).

AXION BURSTS FROM SUPERNOVAE AT COSMOLOGICAL DISTANCES

Abraham Loeb

Astronomy Dept., Harvard University, 60 Garden St., Cambridge MA 02138

ABSTRACT

The observed spatial distribution of γ−ray bursts indicates that they probably originate at cosmological distances. At this distance scale their variability timescale and flux above MeV, combined with their highly non-thermal spectra, require that the energy release will not be contaminated by more than $\sim 10^{-5} M_\odot$ in baryons. This constraint is difficult to satisfy in realistic astrophysical environments. We show that the baryonic contamination constraint is avoided if axion bursts from supernovae are converted to γ−rays over cosmological distances. Nonthermal bursts with the relevant flux, duration, variability and spectra are obtained just for the range of axion masses of $10^{-5} - 10^{-4}$eV that accounts for the cold dark matter in the universe. The observed rate of γ−ray bursts implies that axions should be converted efficiently to photons in only one out of $\sim 10^4$ supernovae.

INTRODUCTION

The recent results from the BATSE experiment on board the Compton Gamma Ray Observatory indicate that the source distribution of γ−ray bursts (GRB) is non–Euclidean at low fluxes and isotropic. These observations rule out a galactic disk population, and suggest that GRB most likely[1,2] originate at a cosmological distance D_H. The characteristic GRB flux F_γ then translates to a source luminosity (assuming isotropic emission),

$$L_\gamma \approx 10^{50} \frac{\text{erg}}{\text{sec}} \left(\frac{F_\gamma}{10^{-7}\text{erg cm}^2\,\text{s}^{-1}}\right) \left(\frac{D_H}{3\text{Gpc}}\right)^2, \qquad (1)$$

surprisingly close but smaller than the binding energy output of a hot neutron star. Moreover, the typical duration of bursts is similar to the cooling time of a hot neutron star. These coincidences, combined with the observed frequency of bursts ($\sim$ 2day^{-1}), led to the suggestion that coalescing neutron star binaries are the energy sources of GRB[1]. Nevertheless, a merger hydrodynamics results in fluid velocities that are subsonic relative to the sound speed in the neutron star core ($\sim c/\sqrt{3}$). Without viscous shock dissipation, it is unclear how the binding energy of the binary can be released in a sufficiently bright neutrino burst ($L_\nu \sim 10^{53}$erg s^{-1}) that would eventually energize a γ−ray burst.

Even without a reference to the origin of their energy supply, cosmological GRB face theoretical difficulties. Many of the bursts show variability on timescales shorter than a fraction of a second and must therefore originate from a source size $< 10^{10}$cm. With the characteristic photon flux above MeV, the optical depth for pair creation by photon–photon collisions is therefore $> 10^{10}(D_H/3\text{Gpc})^2$ at the source. The resulting burst spectrum would therefore be thermalized unless the γ−rays are emitted from a relativistic photosphere that moves with a Lorentz factor $> 10^2$ towards the observer[3]. Since the total energy released is of order 10^{51}ergs, the maximum amount of baryonic contamination allowed is $\sim 10^{-5} M_\odot$. If the energy release was loaded with more baryons the emerging photons would have been degraded to low energies and the burst duration would have been lengthened beyond the observed bounds. This constraint appears to be rather acute for realistic astrophysical environments, especially near the surface of a neutron star where neutrino heating can easily ablate $> 10^{-3} M_\odot$ of baryons[4]. It seems unlikely that the environment near the rotation axis of a neutron-star merger would remain evacuated from baryons to a level of $10^{-5} M_\odot$ for more than a few dynamical times. In view of this difficulty it seems worthwhile to consider novel physical processes that would allow energy to be released in a weakly-interacting form and later be transformed to an electromagnetic form over cosmological distances, far away from the source. A variety of astrophysical constraints exclude the possibility that the radiative decay of supernova neutrinos with a nonzero magnetic moment would result in γ-ray bursts. In this paper we examine the alternative option that axions, produced in supernova events at cosmological distances, mediate the energy release in γ-ray bursts.

SUPERNOVA AXIONS

The axion was discussed extensively in the literature as the most attractive solution to the strong CP problem and as a potential cold dark matter candidate[5]. The misalignment production of axions results in a cosmological density parameter,

$$\Omega_a = 0.2 \times 10^{\pm 0.4} h_{75}^{-2} g(\Theta_1^2) \Theta_1^2 \left(\frac{m_a}{10^{-5} \text{eV}} \right)^{-1.18}, \qquad (2)$$

where m_a is the axion mass, h_{75} is the Hubble constant over $75 \,\text{km s}^{-1} \,\text{Mpc}^{-1}$, Θ_1 is the initial misalignment angle, $g(\Theta_1^2)$ is a function of order unity, and the error bars reflect theoretical uncertainties. Due to this highly non-thermal production, the universe is filled with a Bose-Einstein condensate of axions at almost zero peculiar momentum. This process allows the relatively light axion to serve as a potential cold dark matter candidate.

The emission of axions by supernovae was discussed recently in detail[6] in an attempt to constrain the axion properties after a neutrino burst was detected from SN1987A. The axion luminosity was found to be dominated by nucleon-nucleon bremsstrahlung. For $m_a \ll 10^{-2}$eV the axions stream freely through the neutron star with a total luminosity,

$$L_a \approx (2 \times 10^{50 \pm 1.5}) \frac{\text{erg}}{\text{sec}} \left(\frac{m_a}{10^{-4} \text{eV}} \right)^2 \left(\frac{T_{ns}}{30 \text{MeV}} \right)^{3.5}, \qquad (3)$$

where T_{ns} is the mass averaged temperature of the neutron star. For the axion masses which account for the dark matter in the universe [Eq. 2] the value of L_a coincides with the inferred luminosity L_γ of cosmological GRB [Eq. 1]. Axions in the mass range of $10^{-5} - 10^{-4}$eV affect only weakly the neutrino cooling rate of a hot neutron star, since $L_a \ll 10^{53}$erg/sec. Therefore, the duration ($\sim 0.1 - 10^2$ sec) and variability timescales ($\sim 0.1 - 10^4$msec) of the resulting axion burst would be comparable to those of GRB. Short bursts (~ 0.1sec) could be associated with a collapse to a black hole. On the other hand, long bursts ($\sim 10^2$ sec) could result from a significant rotation that slows down the collapse until the angular momentum is transported away from the core. (The presence of rotation in core collapse events is indicated by young pulsars.) Since the neutron star is transparent to axions, their integrated bremsstrahlung emissivity would differ appreciably from a blackbody shape. The volume emissivity of axions makes the specific properties of their burst highly sensitive to the details of the collapse. In particular, significant non–sphericity, angular momentum, fragmentation, or magnetic fields are just a few of the available degrees of freedom that could control the specific properties of axion bursts and make them different from each other just like GRB. Although these ingredients are likely to affect the properties of the neutron star core in realistic circumstances, they are often ignored for computational convenience in stellar collapse codes. Motivated by the similarity of axion bursts and GRB, the remaining question is whether supernova axions can be converted to γ–rays due to their electromagnetic coupling as they travel across the universe. Radiative decay is irrelevant since the axion lifetime is $\sim 10^{45}(m_a/10^{-4}\text{eV})^{-5}$ sec. Scattering of supernova axions on the cosmic background radiation can be induced by the cosmological axion condensate (to a final axion state that coincides with the condensate and a scattered photon that carries most of the initial axion energy); however the total rate for this process is still small. A mechanism of potential importance is axion–photon conversion in the magnetic fields that surround the progenitor star or that are embedded in the interstellar or intergalactic media[7]. Although this mechanism involves a variety of astrophysical degrees of freedom, there are difficulties in applying it using plausible physical conditions.

The frequency of GRB (~ 2 day^{-1}) implies that only $\sim 10^{-4}$ of all supernova events in the universe get efficient conversion of their axions to photons. It is therefore not surprising that SN1987A was not one of these rare events[8]. The scarcity of bursts relative to supernovae may be linked either to the axion properties or to the external conditions responsible for the axion–photon conversion outside the supernova envelope. It could result, for example, from the fact that the axion is relatively light ($10^{-6} - 10^{-5}$ eV) and has weak couplings, or that only a small fraction of all lines of sight are associated with magnetic environments that allow efficient axion–photon conversion. It is also possible that only rare phenomena, like binary coalescence (of neutron star–neutron star or neutron star–black hole systems), manage to provide the conditions necessary for bright bursts. If Type II supernovae follow each of the GRB, it would be possible to detect their light curve even at cosmological distances through dedicated optical searches. Their association with specific bursts requires that the positional error boxes of the bursts be $< (2°)^2$, much smaller in area than the error boxes provided by the Gamma Ray Observatory.

Axion bursts are expected to be deficient of low energy axions because of the effects of degeneracy blocking and collisional decoherence on the nucleons that produce them by bremsstrahlung. This is consistent with the lack of x–rays in GRB ($< 2\%$ of the total energy). The spread in travel times across the universe of axions with different energies above E should be small, < 2 msec $h_{80}^{-1}(m_a/10^{-4}\text{eV})^2(E/\text{MeV})^{-2}$, and difficult to detect.

SUMMARY

In conclusion, axion bursts from cosmological supernovae share the qualitative temporal and spectral characteristics of γ−ray bursts, provided axions are the cold dark matter in the universe. The specific process for converting supernova axions to γ−rays in one out of $\sim 10^4$ supernovae is still a missing link that must be identified in order that this association be viable. It would be remarkable indeed if the "invisible axion" revealed its existence in γ−rays. In principle, the radiative decay of other types of particles (excluding neutrinos) may also explain the origin of cosmological γ−ray bursts.

REFERENCES

1. T. Piran, this conference   (1993).
2. S. Mao and B. Paczyński, Ap. J. **388**, L45 (1992).
3. P. Meszaros and M. J. Rees, this conference   (1993).
4. S. E. Woosely, Ap. J. **405**, 273 (1993).
5. E. W. Kolb and M. S. Turner, The Early Universe (Addison-Wesley, N. Y., 1990), p. 401.
6. A. Burrows, M. S. Turner, and R. P. Brinkmann, Phys. Rev. D **39**, 1020 (1989).
7. G. Raffelt and L. Stodolsky, Phys. Rev. D **37**, 1237 (1988).
8. E. L. Chupp, W. T. Vestrand, and C. Reppin, Phys. Rev. Lett. **62**, 505 (1989).

GAMMA-RAY BURSTS FROM RELATIVISTIC BEAMS IN NEUTRON STAR MERGERS

Robert Mochkovitch, Sacha Loiseau
Institut d'Astrophysique de Paris, 75014 Paris, France

Margarita Hernanz, Jordi Isern
C.E.A.B. 17300 Blanes, Girona, Spain

ABSTRACT

The coalescence of two neutron stars produces a merged configuration consisting of a dense core surrounded by a thick rotating disk. Two opposite relativistic beams can form along the system axis, fed in energy by the annihilation of neutrinos and antineutrinos emitted from the hot disk. We show that the relativistic factor of the beams can reach $10^2 - 10^3$ because the core polar cap remains cooler than the disk and a large fraction of the energy is released beyond the sonic point. The interaction of the ejected shell with the interstellar medium may eventually lead to a gamma-ray burst if the kinetic energy of the shell can be efficiently radiated through non thermal processes in shocks [1].

INTRODUCTION

The merging of two neutron stars has been proposed as a possible origin for cosmological gamma-ray bursts [2,7]. The formation rate of compact binary neutron stars is very uncertain [5,8–10] but appears to be about one order of magnitude larger than the burst rate. The energy released during and after the coalescence is several times 10^{53} erg so that the conversion of only 1% of that energy into γ-rays can power a burst. The compactness of the source also explains the variability of some bursts on timescales as short as one millisecond.

The mechanism generally invoked for the transfer of energy from neutrinos to ordinary matter is $\nu\bar{\nu}$ annihilation into e^+e^- pairs. The absorption and emission of neutrinos by nucleons is also important close to the neutrinosphere. The deposition of energy drives a wind from the merger surface, the final relativistic factor Γ of this wind being of the order of $\eta = \dot{E}/\dot{M}c^2$ for $\eta \gg 1$ ($\dot{E}$ and $\dot{M}$ are respectively the energy injection rate and the mass loss rate). The Rees and Mészáros scenario [1] where the kinetic energy of the ejecta is dissipated in shocks to produce the γ's requires $\eta \sim 10^2 - 10^3$ and therefore a very small baryonic pollution. The zone along the system axis, which is preserved from massive contamination by centrifugal forces [11,12], appears as a natural place to generate a relativistic wind. However, the polar cap of the merger can also inject baryons into the wind and the final amount of pollution will critically depend on its temperature as shown below.

STRUCTURE AND EVOLUTION OF THE MERGER

Following the coalescence of the two neutron stars, the merger rearranges itself in a few milliseconds. The resulting structure consists of a central core surrounded by a rapidly rotating thick disk where most of the angular momentum of the original binary is stored (see Fig. 1). The subsequent evolution of the merger will depend on the time scale τ_j for angular momentum transport in the disk. Accretion onto the central core occurs at a rate $\sim M_{disk}/\tau_j \sim 1\, M_\odot/\tau_j$ and the neutrino luminosity can then be expected to be

$$L_\nu \sim \alpha \frac{GM_c}{R_c}\frac{M_{disk}}{\tau_j}$$
$$\sim 1.5\, 10^{54}\left(\frac{\alpha}{0.5}\right)\left(\frac{M_c}{1\, M_\odot}\right)\left(\frac{M_{disk}}{1\, M_\odot}\right)\left(\frac{10\, km}{R_c}\right)\left(\frac{0.1\, s}{\tau_j}\right)\, erg.s^{-1}, \quad (1)$$

where M_c and R_c are respectively the mass and radius of the central core. The factor $\alpha < 1$ accounts for the fact that the disk material is already deep into the potential well of the core when accretion begins.

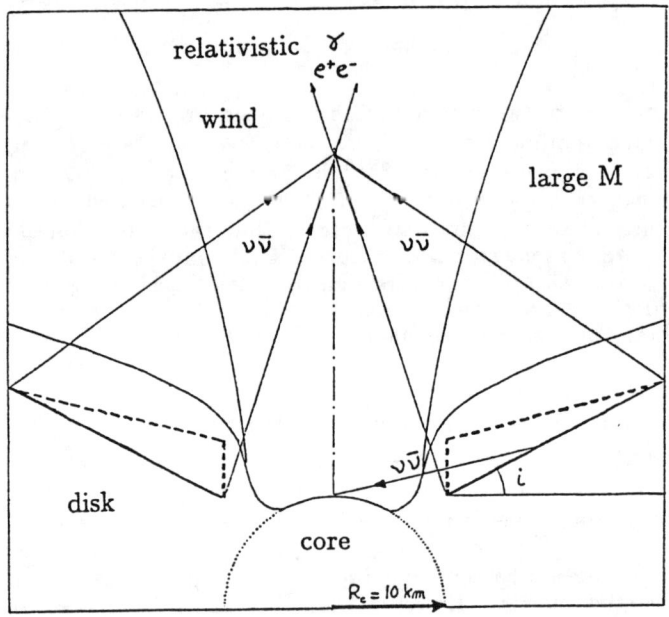

Figure 1. Simplified view of the system geometry. The disk neutrinosphere is represented by the thick full and dashed lines; full line: simple geometry with a unique temperature $T_\nu = 10$ MeV and an inclination angle i; dashed line: modified geometry with a "wall" at temperature $T_\nu^W < 10$ MeV. $\nu\bar\nu$ annihilation along the system axis produces e^+e^- pairs and γ-rays. The polar cap of the merger is heated by the neutrino flux coming from the disk. The baryon free funnel where the relativistic wind can form is shown.

The time scale for the transport of angular momentum is very difficult to estimate. If the disk becomes turbulent and one adopts an α-prescription for the turbulent viscosity (Shakura and Sunyaev, 1973)

$$\tau_j \sim \frac{R^2}{\alpha v_s H_d} \sim 0.1 \left(\frac{0.1}{\alpha}\right) \left(\frac{R}{50\ km}\right)^2 \left(\frac{30\ km}{H_d}\right) \left(\frac{10000\ km.s^{-1}}{v_s}\right)\ s\ , \qquad (2)$$

where R and H_d are the typical radius and thickness of the disk and v_s is the sound velocity. In the following we shall adopt $\tau_j \sim 0.1$ s as typical but shall keep in mind that this value of τ_j remains extremely uncertain.

We assume that most of the neutrino luminosity is radiated by the inner part of the disk, up to a maximum radius R_M so that

$$L_\nu \sim 2\pi R_M^2 \times \frac{7}{8}\eta\sigma T_\nu^4\ , \qquad (3)$$

where T_ν is the temperature at the neutrinosphere and $\eta = 3$ if all three neutrino flavors contribute to the luminosity. Relation (3) then gives

$$T_\nu \sim 10 \left(\frac{L_\nu/\eta}{5\ 10^{53}\ erg.s^{-1}}\right)^{\frac{1}{4}} \left(\frac{30\ km}{R_M}\right)^{\frac{1}{2}}\ MeV\ . \qquad (4)$$

The luminosity produced by the annihilation of neutrinos close to the axis is proportional to the ninth power of the neutrinosphere temperature

$$L_{\nu\bar{\nu}} = \int_{R_c}^\infty r^2 q_{\nu\bar{\nu}}(r) dr \sim \frac{10^{53}}{4\pi} \left(\frac{T_\nu}{10\ MeV}\right)^9\ erg.s^{-1}.st^{-1}\ , \qquad (5)$$

where $q_{\nu\bar{\nu}}$, the power released per unit volume, has been taken from [13-15]. For a neutrino emission lasting $\tau_j \sim 0.1$ s and for $T_\nu \approx 10$ MeV the available energy is therefore large enough to account for a cosmological γ-ray burst.

We have computed the mass loss rate and the related value of η by solving the relativistic wind equations [16,17]

$$r^2 \rho v Y = \dot{M} = const, \qquad (6)$$

$$H\dot{M}Y + \mathcal{L}_\nu = \dot{E} = const, \qquad (7)$$

$$\frac{dP}{dr} = \rho H \frac{d\ln Y}{dr}, \qquad (8)$$

where all symbols have their usual meaning and $Y = (1 - \frac{r_g}{r})^{1/2}(1 - \frac{v^2}{c^2})^{1/2}$, $r_g = \frac{2GM_c}{c^2}$ and $H = c^2 + \frac{P+U}{\rho}$; $\dot{M}$ is the mass loss rate per steradian and $\mathcal{L}_\nu$ satisfies the relation

$$\frac{d\mathcal{L}_\nu}{dr} = -r^2(q_{\nu n} + q_{\nu\bar{\nu}}), \qquad (9)$$

$q_{\nu n}$ and $q_{\nu\bar{\nu}}$ being the power per unit volume generated respectively by neutrino absorption and emission processes and $\nu\bar{\nu}$ annihilation.

The neutrinosphere temperature at the polar cap, T_ν^{cap}, which serves as an inner boundary condition has been obtained from the energy balance between heating by disk neutrinos and local reemission. Finally, the integration of (6 - 9) has been performed from the sonic point inward following a procedure described in [18].

RESULTS AND CONCLUSION

We first computed η assuming the neutrinosphere geometry represented by a thick full line in Fig. 1. For an inclination angle $i = 30°$ we get $\eta \approx 8$ and $T_\nu^{cap} \approx 5.6$ MeV. We see that the wind along the axis is already relativistic, contrary to the slower wind coming from the disk. This is essentially due to the comparatively lower temperature of the neutrinosphere, $\dot{M}$ being extremely sensitive to T_ν. However, the values of η larger than 100 needed in the Rees and Mészáros scenario [1] require a more favourable geometry such as the one represented by a thick dashed line in Fig. 1. Around the polar cap, the disk neutrinosphere now starts with a "wall" of temperature $T_\nu^W < 10$ MeV (the rest of the neutrinosphere still being at $T_\nu = 10$ MeV). The energy coming from $\nu\bar{\nu}$ annihilation is therefore much reduced just above the polar cap (due to the T_ν^9 dependence of $q_{\nu\bar{\nu}}$) and most of it is now released beyond the sonic point. We find that η strongly depends on T_ν^W, covering the whole range of interest $(10^2 - 10^3)$ for $5 < T_\nu^W < 7$ MeV. The behavior of the wind temperature and velocity are shown for a typical model ($T_\nu^W = 6$ MeV) in Fig. 2. The sonic point is located at $r = 14.8$ km, $\eta = 175$ and $T_\nu^{cap} = 4$ MeV. The total energy from $\nu\bar{\nu}$ annihilation is $E_{\nu\bar{\nu}} = 9 \; 10^{51}/4\pi$ erg.st^{-1} (only $3 \; 10^{-3}$ of it being released below the sonic point) with a total ejected mass $M_{ej} = 5.7 \; 10^{28}/4\pi$ g.st^{-1} (for a neutrino emission lasting 0.1 s).

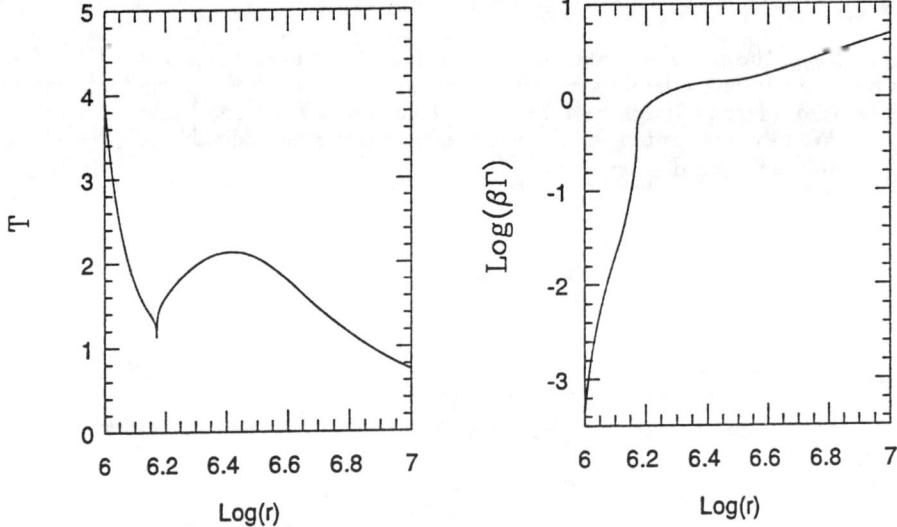

Figure 2. Behavior of the wind temperature (in MeV) and velocity (through the $\beta\Gamma$ factor) for a "wall" temperature $T_\nu^W = 6$ MeV.

Our results (see [19] for details) show that, at least for a specific neutrinosphere geometry, the ejection of matter with relativistic factors $\Gamma = 10^2 - 10^3$ is possible in neutron star mergers. High Γ values are obtained due to the relatively low temperature of the polar cap neutrinosphere ($\lesssim 4$ MeV) and because the major part of the energy is injected beyond the sonic point.

The geometry we have considered for the neutrinosphere certainly needs to be confirmed but we believe it is not unrealistic in view of existing SPH merging calculations [20,21]. Another critical parameter is the temperature of the disk neutrinosphere which is related to the very uncertain time scale for angular momentum transport in the disk through Eq. (1) and (4). Due to the T_ν^9 dependence of $E_{\nu\bar\nu}$ a choice of T_ν only 25% smaller than our adopted value of 10 MeV would have reduced the burst energy by a factor of 10. It is clear that this sensitivity of the model to some parameters can be a problem, since burst statistics probably requires that a fraction $\gtrsim 10\%$ of all merging events must lead to an observable burst.

REFERENCES

1. M.J. Rees and P. Mészáros, M.N.R.A.S. **258**, 41p (1992).
2. B. Paczyński, Ap. J. **308**, L51 (1986).
3. J. Goodman, Ap. J. **308**, L47 (1986).
4. D. Eichler, M. Livio, T. Piran and D.N. Schramm, Nature **340**, 126 (1989).
5. R. Narayan, T. Piran and A. Shemi, Ap. J. **379**, L17 (1991).
6. B. Paczyński, Acta Astron. **41**, 217 (1991).
7. T. Piran, R. Narayan and A. Shemi, Proc. Huntsville GRO Meeting (W.S. Paciesas and G.J. Fishman, 1992).
8. J.P.A. Clark, E.P.J. van den Heuvel and W. Sutantyo, Astron. Astrophys. **72**, 120 (1979).
9. E.S. Phinney, Ap. J. **380**, L17 (1991).
10. A.V. Tutukov and L.R. Yungelson, M.N.R.A.S. **260**, 675 (1993).
11. P. Mészáros and M.J. Rees, M.N.R.A.S. **257**, 29p (1992).
12. R. Mochkovitch, M. Hernanz, J. Isern and X. Martin, Nature **361**, 236 (1993).
13. J. Cooperstein, L.J. van den Horn and E. Baron, Ap. J. **309**, 653 (1986).
14. J. Cooperstein, L.J. van den Horn and E. Baron, Ap. J. **321**, L29 (1987).
15. J. Goodman, A. Dar and S. Nussimov, Ap. J. **314**, L7 (1987).
16. R.A. Flammang, M.N.R.A.S. **199**, 833 (1982).
17. B. Paczyński, Ap. J. **318**, 363 (1990).
18. R.C. Duncan, S.L. Shapiro, and I. Wasserman, Ap. J. **309**, 141 (1986).
19. R. Mochkovitch, M. Hernanz, J. Isern and A. Loiseau, A&A (to be published, 1994).
20. F. Rasio and S.L. Shapiro, Ap. J. **401**, 226 (1993).
21. M.B. Davies, W. Benz, T. Piran and F.K. Thielemann, Ap. J. (to be published, 1994).

RADIO AND NEUTRINO EMISSION FROM THEORETICAL GAMMA-RAY BURSTERS

Bohdan Paczyński, James Rhoads, and Guohong Xu

Princeton University Observatory, Princeton, NJ 08544-1001

bp@astro.princeton.edu, rhoads@astro.princeton.edu, xu@astro.princeton.edu

Gamma-ray bursts are highly super-Eddington events which almost certainly eject some matter at ultra-relativistic speeds. When the ejecta interact with the interstellar or intergalactic matter a strong synchrotron radio emission is likely to be generated, as is the case with supernovae remnants and radio galaxies. The strongest gamma-ray bursts should be followed by radio transients with peak fluxes as high as 20 mJy. The time scale depends on the distance scale: it is less than a minute for the bursters in the galactic halo, and about a week for the bursters at cosmological distances. These should be detectable with the VLA (Paczyński & Rhoads 1993).

The rapid time variability of many gamma-ray bursts implies that the Lorentz factors of the ultra-relativistic ejecta are likely to vary as well. Therefore, there are likely to be collisions between the various faster and somewhat slower ejecta, converting some kinetic energy of expansion into quasi-thermal energy, and leading to nucleon – nucleon collisions energetic enough to create pions. There are at least two different types of processes that can generate high energy gamma rays: those which are driven by the electron interactions with photons or the magnetic fields, and those following the decay of pions created by nucleon – nucleon collisions. The main difference between the two is that the latter generate high energy neutrinos and anti-neutrinos. We estimate that in our scenario the neutrinos may have energies $\sim$ 30 GeV if the colliding nucleons have a Maxwellian energy distribution, and up to $\sim$ TeV if the nucleons have a power law distribution. The strongest gamma-ray bursts are observed to deliver $\sim 10^{-4}$ erg cm^{-2} in 100 – 2,000 keV photons. In our scenario even more energy may be delivered in a high energy neutrino burst, which might be detectable with DUMAND and other future experiments (Paczyński & Xu 1994).

Paczyński, B., & Rhoads, J. 1993, ApJ, 418, L5.

Paczyński, B., & G. Xu. 1994, ApJ, in press.

GAMMA-RAY BURSTS FROM NEUTRON STAR MERGERS

Tsvi Piran

Racah Institute for Physics, The Hebrew University, Jerusalem 91904, Israel

ABSTRACT

Binary neutron stars merger (NS^2M) at cosmological distances is probably the only γ-ray bursts model based on an independently observed phenomenon which is known to be taking place at a comparable rate. We describe this model, its predictions and some open questions.

Cosmological γ-Ray Bursts and Fireballs

Compton-GRO has demonstrated, quite convincingly, that γ-ray bursts (grbs) originate from cosmological sources[1,2]. Evidence for the predicted[3,4] correlations between the duration, the strength and the hardness of the bursts begins to emerge[5,6]. Preliminary analysis suggests that the weakest bursts originate from $z \approx 1$, in agreement with fits to a cosmological C/C_{min} distribution. This corresponds to a local rate[4] of[4] $\approx 10^{-6}$/year/galaxy (depending on the cosmological model and on other factors). The energy released in each burst depends also on the cosmological model $10^{50} \gtrsim E \lesssim 10^{51}$ ergs if the energy emission is isotropic.

The intense energy released in a small volume (evident by the rapid rise time of some of the pulses) implies that any cosmological grb source is initially optically thick[7] to $\gamma\gamma \to e^+e^-$. The large initial optical depth prevent us from observing directly the photons released by the source regardless of the specific nature or the source. The sources produce an optically thick radiation-electron-positrons plasma "fireball", which behaves like a fluid, expands and reaches relativistic velocities[8,9]. The observed radiation emerges only after the fireball has expanded significantly and became optically thin.

One should divide, therefore, the discussion of cosmological grbs to a discussion of the nature of the energy source (for which we present a model here) and a discussion of the fireball phase (which we address elsewhere in this volume). For the paper it is sufficient to recall that the fireball must reach ultra-relativistic velocities with a Lorentz factor $\gamma \gtrsim 10^2$ to produce a grb. Since $\gamma \approx E/Mc^2$ (where E is the total energy of the fireball and M is the mass of the baryons in the fireball) the condition $\gamma > 10^2$ sets a strong upper limit on the amount of baryons: $M < E/\gamma c^2 \approx .510^{-5} M_\odot (E/10^{51} ergs)(\gamma/10^2)^{-1}$. This condition poses a strong constraint on grb models.

NS^2M and GRBs - Agreement at a Glance

Neutron star binaries, such as the one observed in the famous binary pulsar PSR 1916+13, end their life in a catastrophic merge event (denoted here NS^2M). Using the three observed binary pulsars we can estimate the expected rate of NS^2M events[10,11] as $\approx 10^{-5.5\pm.5}$/year/galaxy. An energy comparable to a neutron star binding energy ($\gtrsim 5 \times 10^{53}$ ergs) is released in NS^2Ms mostly as neutrinos and gravitational radiation. The neutrino signal is comparable in its

© 1994 American Institute of Physics

signature to supernova neutrino signals which are thousand times more frequent. It is unlikely that it will ever be detected. The gravitational radiation pulses, have however, a unique signature and Two gravitational radiation detectors, LIGO and VIRGO are currently constructed to detect them.

Several years ago Eichler, Livio Piran and Schramm[13], (see also [14,15,16,17]) suggested that grbs originate at NS^2Ms. Between 10^{-2} to 10^{-3} of the total energy released in NS^2Ms is sufficient to power a grb at a cosmological distance. The required energy could be converted to electromagnetic energy either via[13,15] $\nu\bar{\nu} \to e^+e^-$ or via magnetic processes in an accretion disk that forms in the merger[18]. The rates of NS^2Ms estimated from binary pulsars and the observed rate of grbs measured by BATSE and estimated from cosmological fits are within half an order of magnitude from each other. A remarkable agreements in view of the large uncertainties involved in both estimate.

Numerical Simulations of NS^2M - Some Answers to Further Questions

It worthwhile, therefore, to explore whether the mergers can produce clean enough fireballs (i.e. fireballs with sufficiently low baryonic load) as required from the fireball analysis and to ask whether enough energy can be converted to electromagnetic energy in this events. To address these iissues we[19] developed a numerical code that follows neutron star binary mergers and calculates the thermodynamic conditions of the coalesced binary. The process of coalescence, from initial contact to the formation of an axially symmetric object, takes only a few orbital periods. Some of the material from the two neutron stars is shed, forming a thick disk around the central, coalesced object. The mass of this disk depends on the initial neutron star spins; higher spin rates resulting in greater mass loss, and thus more massive disks. For spin rates that are most likely to be applicable to real systems, the central coalesced object has a mass of $2.4M_\odot$, which is tantalizingly close to the maximum mass allowed by any neutron star equation of state for an object that is supported in part by rotation. Using a realistic nuclear equation of state we estimate the temperatures after the coalescence: the central object is at a temperature of $\sim 10MeV$, whilst the disk is heated by shocks to a temperature of 2-4MeV.

A typical density cut perpendicular to the equatorial plan is shown in Fig. 1. The disk is thick, almost toroidal; the material having expanded on heating through shocks. This disk surrounds a central object that is somewhat flattened due to its rapid rotation. An almost empty centrifugal funnel forms around the rotating axis and there is practically no material above the polar caps. This funnel provides a region in which a baryon free radiation-electron-position plasma could form[20]. Neutrinos and antineutrinos from the disk and form the polar caps would collide and annihilate preferentially in the funnel (the energy in the c.m. frame is larger when the colliding ν and $\bar{\nu}$ approach at obtuse angle, a condition that easily holds in the funnel). The numerical computations do not show any baryons in the funnels. The resolution of our computation is insufficient, howeer, to show that the baryonic load in the funnel is as low as needed. The neutrinos radiation pressure on polar cap baryons can generate a baryonic wind that will load the flow. Estimates of this effect[21,22] show that it is negligible if the temperature on the polar caps is sufficiently low. The estimated temperature from our computations is $\approx 2MeV$, which is marginal. Our temperature estimate is, however, least certain in low temperature regions like this.

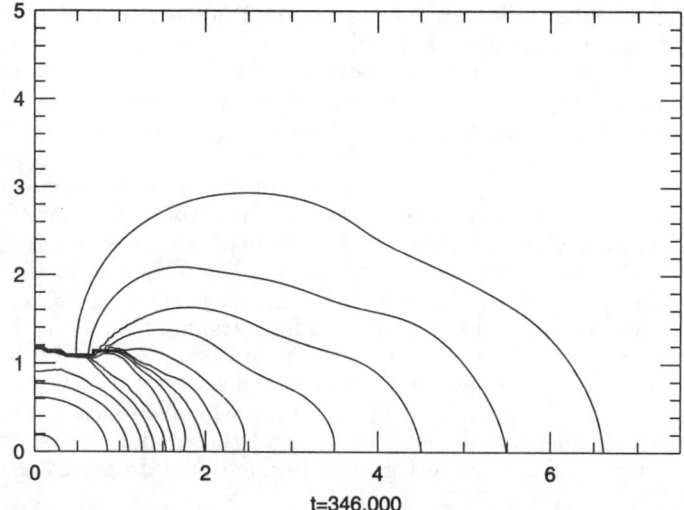

Figure 1. Logarithmic density contour lines at the end of the computation of the merger. The contours are logarithmic, at intervals of 0.25 dex (from [19]).

If the core does not collapse directly to a black hole it will emit its thermal energy as neutrinos. The neutrino flux is sufficiently large that $\approx 10^{-2}$ to 10^{-3} of it could be converted to electron-positron pairs via $\nu\bar{\nu} \to e^+e^-$ and produce a grb. The time scale for the neutrino burst is short enough to accommodate even the shortest rise times observed. An additional energy source that could power a grb is the accretion of the disk surrounding the central object. This energy source can operate on a longer time scale and it takes place regardless of the question of whether the central object collapse directly to a black hole or not.

Open Questions and Predictions

The numerical calculations support earlier suggestions[17] that the energy release in anisotropic and that an empty funnel forms around the rotating axis of the binary system. The fireball is highly non spherical and it expands along the polar axis and forms a jet. This poses an immediate constraint on the model. If the width of the jet is θ than we observe grbs only from a fraction $2\theta^{-2}$ of NS²Ms. The rates of grbs and NS²Ms agree only if $\theta \gtrsim 0.2$ (unless the rate of NS²Ms is much higher than the current estimates). A condition which at first glance is satisfied by the funnel seen in Fig. 1.

The duration and spectra of grbs vary greatly from one burst to another. Both are determined by the fireball phase but the source might contribute in producing fireballs with different Lorentz factors and different initial durations. Within the funnel the baryonic load will vary as a function of the angular position leading to varying final Lorentz factors which, in turn, produce bursts with different durations and spectra. Another source of variability could arise from the interplay between the two energy sources in NS²Ms: Neutrino annihilation and accretion energy of the disk. These mechanisms would operate on different time scale and produce different looking bursts. An additional source of diversity[19] is the distinction between systems that collapse directly to a black hole and those that undergo a longer rotating core phase. Finally, black hole-neutron star binaries are predicted to be as common as neutron star binaries[10]. A black hole neutron star merger[14] would produces grbs with different characteristics than NS^2M.

NS^2M events can take place in a variety of host systems including dwarf galaxies, or even in the intergalactic space if the neutron star binary is ejected from the host galaxy when it forms[18]. Hence, unlike other cosmological models it is not essential that an optical counter part will be observed in the location of grbs[23]. A unique prediction of the NS^2M model is that grbs should be accompanied by gravitational radiation signals from the final stages of the merger and vice versa (the latter is true only up to the anisotropic emission factor discussed earlier). This coincidence could prove or disprove this model. It could also serve to increase the sensitivity of the gravitational radiation detectors[12]. Hopefully, this coincidence will be detected and the model will be confirmed when gravitational radiation detectors will become operational at the turn of the century.

I would like to thank Ramesh Narayan for many helpful discussions. This work was supported in part by a BRF grant to the Hebrew University and NASA grant NAG5-1904 to the CFA.

Reference

1. Meegan, C.A., et. al., *1992, Nature*, **355** *143*.
2. Meegan, C.A., et. al., *1993, this volume*.
3. Paczyński, B. *1992, Nature, 355, 521*.
4. Piran, T., *1992, Ap. J. L.***389**, *L45*.
5. Norris et. al., *1993, this volume*.
6. Davis et. al., *1993, this volume*.
7. Piran, T. and Shemi, A., *1993, Ap. J. L.*, **403**, *L67*.
8. Goodman, J., *1986, Ap. J. L.*, **308** *L47*.
9. Paczyński, B., *1986, Ap. J. L.*, **308**, *L51*.
10. Narayan, R., Piran, T. and Shemi, A., *1991, Ap. J. L.*, **379**, *L17*.
11. Phinney, E. S., *1991, Ap. J. L.*, **380**, *L17*.
12. Kochaneck C. and Piran, T., *1993, Ap. J. L., in press*.
13. Eichler, D., Livio, M., Piran, T., and Schramm, D. N. *1989, Nature*, **340**, *126*.
14. Paczyński, B., *1991, Acta Astronomica*, **41**, *257*.
15. Goodman, J., Dar, A. and Nussinov, S. *1987, Ap. J. L.*, **314**, *L7*.
16. Piran, T., *1990, in Wheeler, J. C., Piran, T. and Weinberg, S. Supernovae World Scientific Publications*.
17. Piran, T., Narayan, R. and Shemi, A., *1992, in Paciesas W. S. and Fishman, G. J. eds. Gamma-Ray Burst, Huntsville, 1991, AIP press*, *149*.
18. Narayan, R., Paczyński, B., and Piran, T., *1992, Ap. J. L.*, **395**, *L83*.
19. Davies, M. B., Benz, W., Piran, T., and Thielemann, F. K. *1993, submitted to Ap. J..*
20. Mochkovich, R. et. al., *1993, this volume*.
21. Duncan, R., Shapiro, S. L., and Wasserman, I., *1986, Ap. J.*, **340**, *126*.
22. Woosley, S. E., and Baron, E., *1992, Ap. J.*, **391**, *228*.
23. Schaffer, B. et. al., *1993, this volume*.

GRBs FROM COMPTON DRAG OF RELATIVISTIC FLOWS

Amotz Shemi
Wise Observatory & School of Physics and Astronomy
Tel Aviv University, Tel Aviv, 69978, Israel.

ABSTRACT

Kinetic energy of relativistic fireballs is reconverted to GRBs by upscattering optical-UV photons, presumably in dense globular clusters or in galactic nuclei.

INTERACTION OF COSMIC FIREBALLS WITH RADIATION FIELDS

In cosmological optically thick GRB models the explosion energy is converted into kinetic energy of a relativistically expanding fireball[2,3,4]. A generic problem is the reconversion of this energy to γ-ray photons. I propose a model[5] in which the relativistic flow upscatter ambient interstellar photons of local radiation fields. For a Lorentz factor $\Gamma > 100$ and dense optical - UV radiation fields the emergent signal is a typical GRB. In optimal conditions almost 100% of the energy is converted to high energy photons; this can explain the nonthermal spectra and the rather long duration of many bursts. It resolves the inevitable large compactness of cosmological GRBs and their sensitivity to baryonic load. The time variability is naturally explained if the radiation fields are inhomogeneous. The characteristic energy $\sim \Gamma^2 h\nu$ is $300-500$KeV, where $h\nu \sim 1-10$eV. Also, if the flow becomes optically thin when surrounded by a dense (even thermal) radiation field, the emergent spectra is roughly a power law only due to the dynamical effects on the radiation diffusion.

Presumably the explosions occur in dense globular clusters at cosmological distances. The average radiation intensity in dense cores is probably sufficiently dense, and the ISM density is sufficiently low, so that the flow is braked predominantly by inverse Compton energy loss (e.g., M15 core radius is 0.08 pc and the stellar population is larger than a few 10^5 stars). Upscattering of X-ray photons should also be considered if the explosion occurs near an X-ray source in the core. Another possibility is explosions that cause intense fireballs in galactic nuclei. In the first case mergers of neutron star binaries can be the sources of the fireballs. In the second case fireballs can originate during disruption of normal stars by a central, massive black hole[1].

I thank I. Goldman, J. Katz, D. Maoz, M.J. Rees and B. Yanny for discussions.

REFERENCES

1. Carter, B. , Ap. J. **391**, L67 (1992).
2. Paczynski,B., Ap. J. **363**, 218 (1990).
3. Shemi, A. & Piran, T., Ap. J. **365**, L55 (1990).
4. Shemi, A., Huntsville Gamma-Ray Bursts Workshop (This proceedings, 1993).
5. Shemi, A., Preprint (submitted to MNRAS, 1993).

COSMIC FIREBALLS AND GAMMA RAY BURSTS

Amotz Shemi

Wise Observatory & School of Physics and Astronomy
Tel Aviv University, Tel Aviv, 69978, Israel.

ABSTRACT

The evolution of a cosmic fireball, a very hot and optically thick pair-plasma bubble contaminated by baryonic matter, was studied analytically and numerically. If the initial energy exceeds the rest mass ($\eta \equiv E_0/Mc^2 \gg 1$) the fireball passes an early rearrangement phase, where matter and energy are concentrated towards the front, and propagates at nearly the speed of light ($\Gamma \gg 1$) with a frozen radial profile of a shell. The shell width is narrower as η increases and it preserves the initial scale as $\eta \gg 1$. The relativistic expansion of the shell is best described by a set of simple scaling laws. A remote observer will detect a blend of blackbody spectra of different temperatures but the signal will still resemble a Planckian curve. The spectrum will be harder at early times and softer later on. Newtonian fireballs ($\eta \lesssim 1$) become broad and roughly homogeneous, never become ultrarelativistic ($v \sim \eta^{1/2}$) and never acquire a frozen shape. Gamma ray burst sources, if more distant than a few parsecs, are likely to involve a fireball phase and ultrarelativistic motions.

HYDRODYNAMICS OF COSMIC FIREBALLS

The injection of a large amount of radiation energy into a compact region inevitably leads to a cosmic fireball - an extremely hot and opaque pair-plasma object[1]. The fireball can be contaminated by a baryonic mass, but is dominated by radiation pressure, thus expands adiabaticaly and reaches an ultra-relativistic Lorentz factor $\Gamma \gg 1$. Analytical calculations of homogeneous fireballs[5] show two classes, radiation - dominated ($\eta > 1$) and matter - dominated ($\eta < 1$) fireballs. The local temperature decreases as $T \propto R^{-1}$, where R is the radius of the fireball, but if $\eta \gg 1$, the observed, blue-shifted temperature ΓT can be comparable to the initial temperature T_0. In the later stages of the expansion, unless η is extremely large, the matter dominates the opacity and the dynamics, and the final Γ roughly equals η.

The opacity has contributions from electron-positron pairs and from the electrons associated with the baryons. Initially, when the local temperature T is high, the opacity is dominated by e^+e^- pairs, but it decreases exponentially with decreasing temperature, and falls to unity when $T = T_p \approx 20$ KeV. The matter opacity decreases only as R^{-2}. If at the point where the pair opacity is vanished the matter opacity is still large, the final transition to an optically thin phase is delayed and occurs at a cooler temperature. The value of η determines in what order the above transitions take place. One can distinguish between four regimes[5,6]: (i) $\eta > \eta_{pair} = [3\sigma_T^2 E_0 \sigma T_p^4/4\pi m_p^2 c^4 R_0]^{1/2}$: In this regime the effect of the baryons is negligible and the evolution is of a pure photon-lepton fireball. When the temperature reaches T_p, the pair opacity τ_{pair} drops to 1 and $\tau_{baryons} \ll 1$. At this point the fireball is still radiation dominated and so most of the energy escapes as radiation. (ii) $\eta_{pair} > \eta > \eta_{baryons} =$

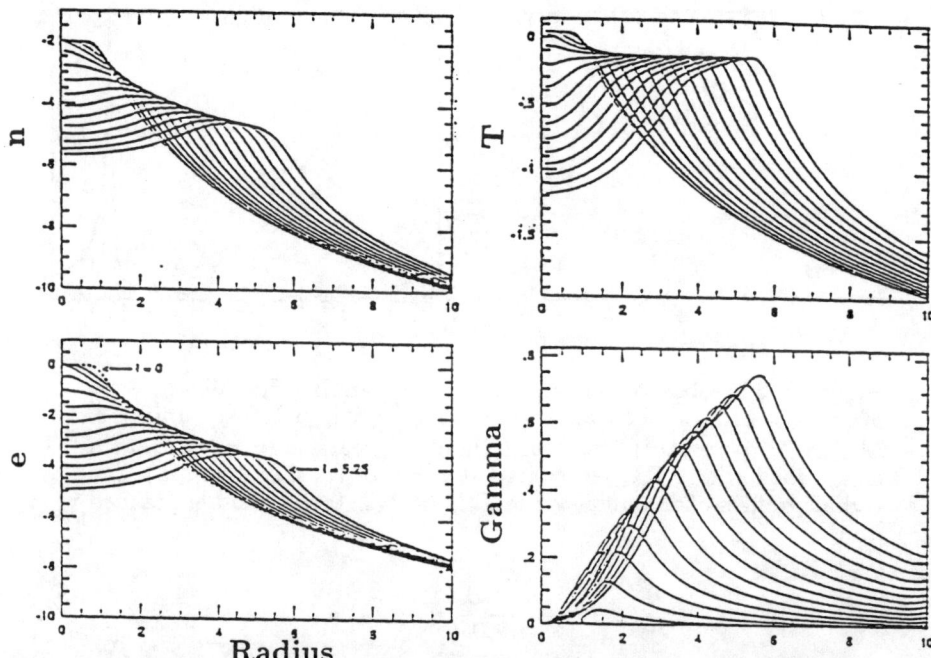

Figure 1. The first 5.246 time units of $\eta = 100$ fireball. The radial profiles of the energy density e, the Lorentz factor γ, the baryon density n and the observed temperature $\propto \gamma e^{1/4}$ (vertical axis in a logarithmic scale); each figure shows 15 slices of equal time lapse ($0.35 t_0$). The e and n profiles are rearranged within $\approx 3 t_0$ into a concentrated pulse of width $\approx 2 R_0$ at the fireball front.

$(3\sigma_T E_0 / 8\pi m_p c^2 R_0^2)^{1/3}$: In the late stages the opacity is dominated by free electrons associated with the baryons. The comoving temperature therefore decreases far below T_p before τ reaches unity. However, the fireball continues to be radiation dominated as in the previous case, and most of the energy still escapes as radiation. (iii) $\eta_{baryons} > \eta > 1$: The fireball becomes matter dominated before it becomes optically thin. Therefore, most of the initial energy is converted into bulk kinetic energy of the baryons, with a final Lorentz factor $\Gamma_f = \eta + 1$. (iv) $\eta < 1$: The Newtonian regime, where the rest energy exceeds the radiation energy and the expansion never becomes relativistic.

An Eulerian, spherically-symmetric, time-dependent code, which calculates the relativistic conservation equations, was developed. Assuming a perfect fluid and a relativistic equation of state $p = e/3$, these equations read[4,6]

$$\frac{\partial}{\partial t}(n\gamma) + \frac{1}{r^2}\frac{\partial}{\partial r}(r^2 n u) = 0 \qquad (1)$$

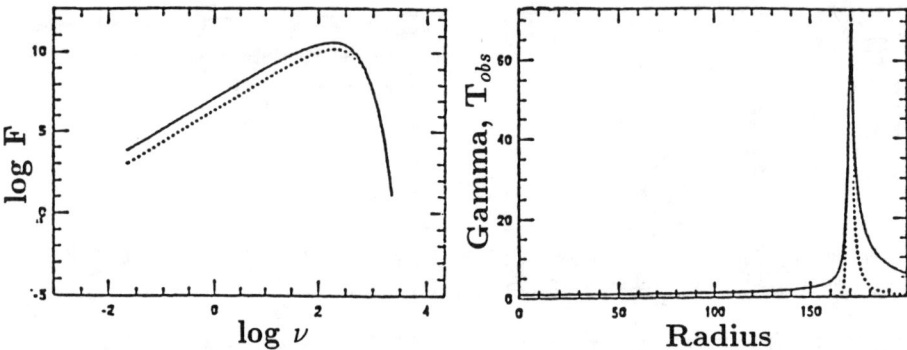

Figure 2. The observed spectrum of an expanding fireball with $\eta = 50$ at $t = 175$. a) The photon flux $\mathrm{Log}(F_\nu)$ in arbitrary units vs. $\mathrm{Log}(h\nu)$ ($h\nu$ in mc^2) (solid line) and the blackbody curve from a single shell where T_{obs} is peaked, at $r = 170$ (dashed line). b) The radial profiles of maximal value $T_{obs} = 2\gamma T$ (solid) and γ (dashed line).

$$\frac{\partial}{\partial t}(e^{3/4}\gamma) + \frac{1}{r^2}\frac{\partial}{\partial r}(r^2 e^{3/4} u) = 0 \qquad (2)$$

$$\frac{\partial}{\partial t}((n + 4/3e)\gamma u) + \frac{1}{r^2}\frac{\partial}{\partial r}(r^2(n + 4/3e)u^2) = -\frac{1}{3}\frac{\partial e}{\partial r} \qquad (3)$$

where n and e are the mass and energy density, respectively, p is the pressure, $\gamma = (1 - v^2)^{-1/2}$, and $u = v\gamma$ is the four velocity. For $\eta > 1$ the flow passes an initial short and strong acceleration phase, in which the fireball rearranges into a narrow front (figure 1).

The acceleration continues until locally the energy density falls considerably below the rest mass, and the fireball coasts with a constant comoving velocity profile. In an intermediate stage, depending on η, the fireball becomes optically thin, and the photons escape in about one crossing time. The global characters of the homogeneous fireball appear also in the narrow front configuration. The spectrum, a blend of thermal spectra from different shells, is modified blackbody, whose typical temperature is comparable to T_0 if $\eta \gg 1$, but it is $\ll T_0$ if $\eta \lesssim 1$. Such a blend of spectra does not seem sufficient to explain nonthermal GRB spectra (figure 2.).

Newtonian $\eta < 1$ fireballs essentially differ from $\eta > 1$ cases. They show a homogeneous shape, rather than being rearranged into a narrow front, never become ultra-relativistic, and their final temperature is very low (figure 3).

After the initial phase the solution satisfies simple scaling laws[4], which are valid rather a long time after the fireball onset. Denote variable values at the first phase where the flow becomes relativistic by '$_1$', these scaling laws take the form

$$r = r_1 \frac{\gamma_1^{1/2} D^{3/2}}{\gamma^{1/2}}, \qquad n = \frac{n_1}{D^3}, \qquad e = \frac{e_1}{D^4}, \qquad \eta = \frac{\eta_1}{D}, \qquad (4)$$

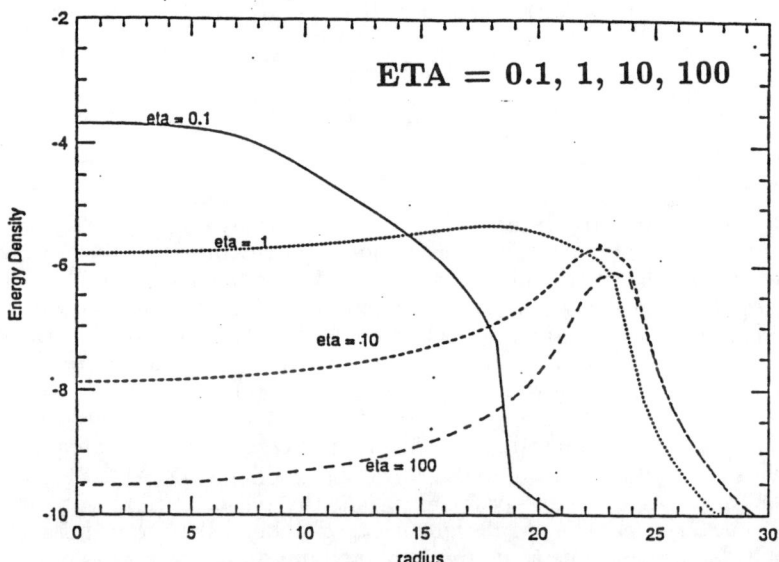

Figure 3. Energy density profiles at $t = 23.33$ for $\eta = 0.1$, 1, 10 and 100, corresponding to solid, dotted, short-dashed and long-dashed lines, respectively. The width in $10^{-0.5}$ of the maximal height is $\simeq$ 25, 2.5 and 1.5, for $\eta = 1$, 10 and 100, respectively. In the Newtonian $\eta = 0.1$ case the pulse width scales roughly as $vt \leq \sqrt{2\eta t} \approx 10$.

where:

$$\frac{1}{D} \equiv \frac{\gamma_1}{\gamma} + \frac{3\gamma_1}{4\eta_1\gamma} - \frac{3}{4\eta_1}$$

The good agreement between numerical and analytical solutions at $t = 3\eta t_0$ is demonstrated elsewhere[4].

Primary application of cosmic fireballs are cosmological GRBs, emitting typically $\lesssim 10^{51}$ erg. Galactic halo sources of $\sim 10^{41}$ erg, unless highly collimated, are optically thick like the cosmological ones, thus also must involve a fireball phase. However, the cosmological sources are less sensitive to mass load, and involve higher T_0 and extreme relativistic expansion with $\gamma \approx 10^3$. Consequently, models of cosmological GRBs are preferable[3].

I thank R. Narayan and T. Piran for discussions.

REFERENCES

1. Cavallo, G. & Rees, M. J., MNRAS **183**, 359 (1978).
2. Piran, T., Narayan, R. & Shemi, A., Gamma-Ray Bursts, Huntsville Al, 1991, eds. W. S. Paciesas & G. J. Fishman, AIP press (1992, 149).
3. Piran, T. & Shemi, A., Ap. J **403**, L67 (1993).
4. Piran, T., Shemi, A. & Narayan, R., MNRAS **263**, 861 (1993).
5. Shemi, A. & Piran, T., Ap. J. **365**, L55 (1990).
6. Shemi, A., Ph.D. Thesis (Tel - Aviv University, 1993).

ON THE NATURE OF NONTHERMAL RADIATION FROM COSMOLOGICAL γ-RAY BURSTERS

Vladimir V. Usov

Dept. of Physics, Weizmann Institute, Rehovot 76100, Israel

ABSTRACT

Relativistic electron-positron winds with strong magnetic fields are considered as a source of radiation for cosmological γ-ray bursters. Such a wind is generated by a millisecond pulsar with a very strong magnetic field. It is shown that at the distance of $\sim 10^{13}$ cm from the pulsar the magnetohydrodynamic approximation for the pulsar wind is broken, and intense electromagnetic waves may be generated. The frequency of these waves is equal to the frequency of the pulsar rotation. Outflowing particles are accelerated in the field of intense electromagnetic waves to Lorentz factors of the order of 10^6 and generate nonthermal synchro-Compton radiation. The typical energy of nonthermal photons is ~ 1 MeV. A high-energy tail of the γ-ray spectrum may be up to $\sim 10^4$ MeV. Baryonic matter is ejected occasionally from the pulsar magnetosphere. The baryonic matter ejection and subsequent suppression of the γ-ray emission may be responsible for the time structure of γ-ray bursts.

INTRODUCTION

The BATSE experiment on board the *Compton Gamma Ray Observatory* found that the γ-ray burst sources are distributed isotropically in the sky, and it sees the edge of the distribution[1]. These two facts can be naturally explained if γ-ray bursters are at cosmological distances[2-8]. Besides, there are other indications of a cosmological origin for γ-ray bursts (see ref. 7 and references therein).

A generic problem with a cosmological model is that the huge initial energy density implies a very large optical depth to electron-positron pairs, thermalization of the electron-positron plasma, and a blackbody spectrum with small modifications[9,10]. This is a clear conflict with the observed spectra of γ-ray bursts, which are well fit either by power laws or by broken power laws[11,12]. It was suggested[13,14] that a very strong magnetic field may be in the electron-positron plasma which flows away from the γ-ray burster. Here it is shown that outflowing particles may be accelerated outside the γ-ray photospheres of relativistic electron-positron winds with strong magnetic fields. These particles may be responsible for the nonthermal radiation of γ-ray bursts.

PARTICLE ACCELERATION AND γ-RAY EMISSION

Below, I assume that millisecond pulsars with extremely strong magnetic fields, $B_s \simeq$ a few $\times 10^{15}$ G, are a cosmological source of γ-ray bursts[13]. In such a model the neutron star rotation is a source of energy for γ-ray bursts. The rotation of the magnetic neutron star decelerates because of the electromagnetic torque. The rate of kinetic energy loss for a supposedly nearly orthogonal magnetic dipole is[15]

$$-\frac{dE_{\rm kin}}{dt} \simeq L_{\rm md} = \frac{2}{3}\frac{B_s^2 R^6 \Omega^4}{c^3} \simeq 2\times 10^{51}\left(\frac{B_s}{3\times 10^{15}\,\rm G}\right)^2 \left(\frac{\Omega}{10^4\,\rm s^{-1}}\right)^4 \rm erg\,s^{-1}, \quad (1)$$

where $E_{\rm kin}$ is the rotational kinetic energy, $R \simeq 10^6$ cm is the radius of the neutron star and Ω is the angular velocity.

Like ordinary known pulsars, electron-positron pairs are created in the magnetosphere of a millisecond pulsar with a very strong magnetic field. The electron-positron plasma and radiation are in quasi-thermodynamical equilibrium in the environment of such a pulsar[13].

The electron-positron plasma with strong magnetic fields flows away from the pulsar at relativistic speeds. During outflow, the electron-positron plasma accelerates and its density decreases. At a distance $r_{\rm ph}$ from the pulsar where the optical depth for the main part of photons is $\tau_{\rm ph} \sim 1$, the radiation propagates freely. If we don't take into account the magnetic field, the radius of the γ-ray photosphere for a spherical optically thick electron-positron wind is[9]

$$r_{\rm ph} \simeq \left(\frac{L_p}{4\pi\sigma T_0^4 \Gamma_{\rm ph}^2}\right)^{1/2}, \quad (2)$$

where T_0 is the temperature of electron-positron plasma at $r \simeq r_{\rm ph}$ in the co-moving frame, $\sigma = 5.67 \times 10^{-5}$ erg cm^{-2} K^{-4} s^{-1} is the Stefan-Boltzmann constant and $\Gamma_{\rm ph}$ is the mean Lorentz factor of plasma particles at $r \simeq r_{\rm ph}$.

Since $\Gamma_{\rm ph} \simeq r_{\rm ph}/r_{lc}$ and $T_0 \simeq 2 \times 10^8$ K (ref. 3), we have

$$r_{\rm ph} \simeq \left(\frac{L_p r_{lc}^2}{4\pi\sigma T_0^4}\right)^{1/4} \simeq 3.5\times 10^8 \alpha^{1/4}\left(\frac{B_s}{3\times 10^{15}\,\rm G}\right)^{1/2}\left(\frac{\Omega}{10^4\,\rm s^{-1}}\right)^{1/2} \rm cm, \quad (3)$$

$$\Gamma_{\rm ph} \simeq 10^2 \alpha^{1/4}\left(\frac{B_s}{3\times 10^{15}\,\rm G}\right)^{1/2}\left(\frac{\Omega}{10^4\,\rm s^{-1}}\right)^{3/2}, \quad (4)$$

where $r_{lc} = c/\Omega$ is the radius of the pulsar light cylinder.

At $r < r_{\rm ph}$ the optical depth increases sharply with decreasing r. Therefore, any energy which is inherited by particles and radiation at a distance to the pulsar a few times smaller than $r_{\rm ph}$ will be thermalized before it is radiated at $r \simeq r_{\rm ph}$.

The luminosity of the γ-ray photosphere is $\alpha L_{\rm md} \sim (0.01-0.1)L_{\rm md}$. The temperature which corresponds to the blackbody-like radiation from the γ-ray photosphere is $\sim 2\Gamma_{\rm ph} T_0 \sim 10^{10}$ K, and the typical energy of γ-rays is ~ 1 MeV.

Kinetic energy which is released in the process of deceleration of the neutron star rotation transforms mainly to the magnetic field energy but not to the energy of particles[16,17]. The pulsar luminosity in magnetic fields is $(1-\alpha)L_{\rm md} \simeq L_{\rm md}$. The strength of the magnetic field which is generated outside the pulsar light cylinder because of the neutron star rotation is

$$B \simeq B_s \left(\frac{R}{r_{lc}}\right)^3 \left(\frac{r_{lc}}{r}\right) \simeq 3.3\times 10^{14}\frac{R}{r}\left(\frac{B_s}{3\times 10^{15}\,\rm G}\right)\left(\frac{\Omega}{10^4\,\rm s^{-1}}\right)^2 \rm G. \quad (5)$$

The magnetic field is frozen in the outflowing electron-positron plasma if the distance to the pulsar is not too large (see below). This field can transfer the energy from the pulsar environment to the region outside the γ-ray photosphere, $r > r_{\rm ph}$, without its thermalization.

The magnetic field does not change qualitatively the motion of relativistic electron-positron wind inside the γ-ray photosphere. At $r \simeq r_{\rm ph}$, the density of electrons and positrons drops sharply, and the particle acceleration because of the magnetic field may be essential.

Acceleration of particles in the pulsar wind at $r > r_{lc}$ is characterized by the following dimensionless parameter[18]

$$\eta = \frac{\Omega^2 \Phi^2}{4\pi f c^3}, \qquad (6)$$

where

$$f = \rho v_r r^2, \qquad \Phi = r^2 B_r, \qquad \rho = n_\pm m, \qquad (7)$$

$n_\pm$ is the laboratory frame number density, m is the mass of electron, v_r is the radial velocity and B_r is the radial component of the magnetic field.

Continuity of the magnetic flux gives $\Phi = $ constant. At the pulsar light cylinder, $r = r_{lc}$, we have $B_r \simeq B_s(R/r_{lc})^3$ and $\Phi \simeq B_s R^2 (\Omega R/c)$. Using this value of Φ and taking into account that $v_r \simeq c$ for relativistic flow, from equations (1), (6) and (7) we obtain

$$\eta \simeq \frac{L_{\rm md}}{mc^2 \dot{N}_\pm}, \qquad (8)$$

where $\dot{N}_\pm$ is the flux of electrons and positrons from the pulsar.

For the electron-positron wind the flux of particles at $r > r_{\rm ph}$ is[9]

$$\dot{N}_\pm \simeq \frac{4\pi c r_{\rm ph} \Gamma_{\rm ph}^2}{\sigma_T} \simeq 2 \times 10^{48} \alpha^{3/4} \left(\frac{B_s}{3 \times 10^{15} \,\rm G}\right)^{3/2} \left(\frac{\Omega}{10^4 \,\rm s^{-1}}\right)^{7/2} \,\rm s^{-1}, \qquad (9)$$

where $\sigma_T = 6.65 \times 10^{-25}$ cm^2 is the Thomson cross section.

Equations (1), (8) and (9) yields:

$$\eta \simeq 10^9 \alpha^{-3/4} \left(\frac{B_s}{3 \times 10^{15} \,\rm G}\right)^{1/2} \left(\frac{\Omega}{10^4 \,\rm s^{-1}}\right)^{1/2}. \qquad (10)$$

The density of electron-positron plasma at $r \simeq r_{\rm ph}$

$$n_\pm = \frac{\dot{N}_\pm}{4\pi c r_{\rm ph}^2} \simeq 4 \times 10^{19} \alpha^{1/4} \left(\frac{B_s}{3 \times 10^{15} \,\rm G}\right)^{1/2} \left(\frac{\Omega}{10^4 \,\rm s^{-1}}\right)^{5/2} \,\rm cm^{-3} \qquad (11)$$

is essentially higher than the critical value[18]

$$n_{cr} = \frac{\Omega B}{4\pi c e} \simeq 4 \times 10^{16} \frac{R}{r} \left(\frac{B_s}{3 \times 10^{15} \,\rm G}\right) \left(\frac{\Omega}{10^4 \,\rm s^{-1}}\right)^3 \,\rm cm^{-3}. \qquad (12)$$

Therefore, the magnetic field is frozen in the plasma, and the magnetohydrodynamic (MHD) approximation can be used to describe the wind motion[18]. In this approximation, particles may be accelerated to Lorentz factors of the order of $\eta^{1/3}$ (ref. 18), which is an order of magnitude more than $\Gamma_{\rm ph}$. However, the $\eta^{1/3}$ estimate of Γ is valid only if either the interaction between particles and photons is negligible or the mass density of radiation, $\rho_\gamma = aT^4/c^2$, is smaller than the mass density of particles, $\rho_\pm = n_\pm m$ (here $a = 7.56 \times 10^{-15}$ erg cm^{-3} K^{-4} is radiation density constant). Near the γ-ray photosphere the density of radiation is very high, and the interaction between particles and radiation is strong. This interaction results in the increase of the mass density of accelerated matter. Substituting ρ_γ for ρ in equation (6), we have the following estimate for the Lorentz factor of accelerated particles near the γ-ray photosphere: $\Gamma_m \simeq (\eta \rho_\pm/\rho_\gamma)^{1/3}$. For $B_s \simeq 3 \times 10^{15}$ G, $\Omega \simeq 10^4$ s^{-1}, $T = T_0$ and $\alpha \simeq 0.01-0.1$, we have $\Gamma_m \sim \Gamma_{\rm ph}$. Hence, there is no essential acceleration of particles because of the magnetic field near the γ-ray photosphere in the MHD approximation. Therefore, the region of the pulsar wind near the γ-ray photosphere is not promising for a generation of strong nonthermal radiation.

With the distance from the pulsar the density of particles decreases in proportion to r^{-2}. The critical density decreases somewhat slower (see equation (12)). At the distance

$$r_{\rm nth} \simeq 1.3 \times 10^{14} \alpha^{3/4} \left(\frac{B_s}{3 \times 10^{15} \text{ G}}\right)^{1/2} \left(\frac{\Omega}{10^4 \text{ s}^{-1}}\right)^{1/2} \text{ cm} \qquad (13)$$

the plasma density, $n_\pm$, is equal to the critical one, n_{cr}. At $r > r_{\rm nth}$ the MHD approximation is broken for the pulsar wind with the magnetic field which alternates in polarity on the scale length of $\sim \pi(c/\Omega) \sim 10^7$ cm (refs 18,19), and intense electromagnetic waves with the frequency of Ω can propagate outside[19,20]. In this case the process of particle acceleration changes qualitatively, namely: particles can be accelerated in this wind zone to Lorentz factors of the order of $\eta^{2/3} \sim 10^6$ (refs 21-23) in contrast to the $\eta^{1/3}$ estimate given by the MHD theory. Particles are accelerated on the scale length of $\sim r_{\rm nth}$ and generate synchro-Compton radiation. The radiative damping length for intense electromagnetic waves with the wavelength $\lambda = 2\pi(c/\Omega)$ is[23]

$$l \simeq \frac{6}{\pi^4} \frac{\Omega^3}{c^3 r_0^3 n_\pm^2} \left(\frac{n_{cr}}{n_\pm}\right) \text{ wavelengths}, \qquad (14)$$

where $r_0 = e^2/mc^2 = 2.8 \times 10^{-13}$ cm is classical electron radius.

For a millisecond pulsar with a very strong magnetic field, $B_s \simeq 3 \times 10^{15}$ G, $\Omega \simeq 10^4$ s^{-1} and $\alpha \simeq 0.1$, from equation (14) we have $l \simeq 10^2 \lambda \simeq 2 \times 10^9$ cm $\ll r_{\rm nth}$ at $r \simeq 2 r_{\rm nth}$. Therefore, low-frequency intense electromagnetic waves have to be reradiated into nonthermal high-frequency emission with the luminosity up to $(1-\alpha)L_{\rm md} \simeq L_{\rm md}$. The typical energy of nonthermal photons, $\epsilon_\gamma \simeq (\pi/4)\hbar\Omega\eta^2 \sim 1$ MeV (ref. 23), is suitable to be identified with the energy of breaks which are observed in the spectra of γ-ray bursts[11,12]. A long high-energy tail of the γ-ray spectrum may be up to $\sim \hbar(eB/mc)\eta^{4/3} \sim 10^4$ MeV.

Baryonic matter may be ejected occasionally from the neutron star magnetosphere because of some kind of plasma instabilities (Usov, in preparation), and the γ-ray emission may be suppressed for some time[9,24]. This process may be responsible for the time structure of γ-ray bursts. It is expected that the flux variations in nonthermal γ-rays is as short as

$$\tau_0 \simeq \frac{r_{nth}}{2c\Gamma^2} \simeq \frac{r_{nth}}{2c\eta^{2/3}}, \qquad (15)$$

where Γ, the Lorentz factor of the outflowing plasma particles, is $\sim \eta^{1/3}$ at $r_{ph} \ll r < r_{nth}$. For a pulsar with $B_s \simeq 3 \times 10^{15}$ G, $\Omega \simeq 10^4$ s^{-1} and $\alpha \simeq 0.1$, from equations (10), (13) and (15) we have $\tau_0 \simeq 10^{-4}$ s.

The expansion energy of baryonic matter can be reconverted into nonthermal radiation when it interacts with an external medium[25,26]. Another component of nonthermal radiation which may be observed in the burst spectra is annihilation lines[27]. Positrons which are responsible for these lines can be produced by burst photons interacting with a medium surrounding a γ-ray burster.

REFERENCES

1. C.A. Meegan et al., Ap. J. **355**, 143 (1992).
2. V.V. Usov and G.V. Chibisov, Soviet Astr. **19**, 155 (1975).
3. B. Paczyński, Ap. J. **308**, L43 (1986).
4. B. Paczyński, Acta Astr. **41**, 257 (1991).
5. T. Piran, Ap. J. **389**, L45 (1992).
6. C.D. Dermer, Phys. Rev. Lett. **68**, 1799 (1992).
7. W.A.D.T. Wickramasinghe et al, Ap. J. **411**, L55 (1993).
8. P. Tamblyn and F. Melia, Ap. J. **417**, L21 (1993).
9. B. Paczyński, Ap. J. **363**, 218 (1990).
10. R. Narayan, B. Paczyński and T. Piran, Ap. J. **395**, L83 (1992).
11. B.E. Schaefer et al., Ap. J. **393**, L51 (1992).
12. D. Band et al., Ap. J. **413**, 281 (1993).
13. V.V. Usov, Nature **357**, 472 (1992).
14. C. Thompson and R.C. Dancan, Ap. J. **408**, 194 (1993).
15. J.P. Ostriker and J.E. Gunn, Ap. J. **157**, 1395 (1969).
16. M.A. Ruderman and P.G. Sutherland, Ap. J. **196**, 51 (1975).
17. J. Arons, in Proc. Workshop "*Plasma Astrophysics*" (Varenna, Italy, 1981), p. 273.
18. F.C. Michel, Ap. J. **158**, 727 (1969).
19. V.V. Usov, Astrophys. Space Sci. **32**, 375 (1975).
20. E. Asseo, F.C. Kennel and R. Pellat, Astr. Astrophys **44**, 31 (1975).
21. J.E. Gunn and J.P. Ostriker, Ap. J. **165**, 523 (1971).
22. F.C. Michel, Ap. J. **284**, 384 (1984).
23. E. Asseo, F.C. Kennel and R. Pellat, Astr. Astrophys **65**, 401 (1978).
24. A. Shemi and T. Piran, Ap. J. **365**, L55 (1990).
25. M.J. Rees and P. Mészáros, MNRAS **258**, 41P (1992).
26. P. Mészáros and M.J. Rees, Ap. J. **405**, 278 (1993).
27. H.S. Fencl, R.N. Boyd and D.H. Hartmann, Ap. J. **407**, L21 (1993).

RELATIVISTIC BULK MOTION AND STATISTICS OF BEAMED GAMMA-RAY BURSTS

Insu Yi [*]

Harvard-Smithsonian Center for Astrophysics, 60 Garden St., Cambridge, MA 02138

ABSTRACT

Gamma-ray bursts may be from jet-like sources in relativistic bulk motion. In this case, emission of gamma-rays is strongly beamed along the direction of the bulk motion. We analyze effects of the relativistic bulk motion and the resulting beamed emission on statistics of the gamma-ray bursts.

INTRODUCTION

We assume that sources are described either by a 'standard candle' (a standard luminosity with a spectral index) or by a 'standard source' (a standard Lorentz factor γ, a standard intrinsic luminosity in the source frame with a spectral index). Diverse observed types of GRBs may indicate deviation from our simplifying assumption. The photon spectrum is described by the photon index α, $\phi(E) \propto E^{-\alpha}$ being the number of photons emitted per unit time per unit photon energy (E) interval. We take $\alpha = 1, 2$. The observed flux (in the band of energies E_1 and E_2) from a source at redshift z with the luminosity $L = \int_{E_1}^{E_2} dE E \phi(E)$ is then given by

$$F(z; L) = \left(\frac{H_0^2 L}{16\pi c^2}\right)(1+z)^{1-\alpha}(\sqrt{(1+z)}-1)^{-2} \tag{1}$$

for a flat universe (with zero cosmological constant) and c is the speed of light. In $\alpha = 1$ case, with

$$x_F(L) \equiv 1 + z_F(L) = \left(1 + \frac{H_o}{4\sqrt{\pi}c}\left(\frac{L}{F}\right)^{1/2}\right)^2 \tag{2}$$

we get the cumulative number of bursts with observed fluxes larger than F

$$N(>F) \propto \int dL \Phi(L) \int_1^{x_F(L)} (x^{-5/2} - 2x^{-3} + x^{-7/2}) dx \tag{3}$$

and $\langle V/V_{max} \rangle$

$$\left\langle \frac{V}{V_{max}} \right\rangle \equiv \left\langle \left(\frac{F_{min}}{F}\right)^{3/2} \right\rangle$$

$$= \left(\frac{16\pi c^2 F_{min}}{H_o^2}\right)^{3/2} \frac{\int dL \Phi(L) L^{-3/2} \int_1^{x(L)} dx x^{-7/2}(x^{1/2}-1)^5}{\int dL \Phi(L) \int_1^{x(L)} dx x^{-7/2}(x^{1/2}-1)^2} \tag{4}$$

[*] E-mail: Internet: yi@cfa.harvard.edu

where $x(L) = x_F(L)$ with $F = F_{min}$ and $\Phi(L)$ is the luminosity distribution of GRBs (see below). Similar results are derived for $\alpha = 2$.

The observed apparent luminosity from a source in relativistic bulk motion is boosted along the direction of the motion[1]. For standard sources with an intrinsic luminosity in the comoving source frame,

$$L_{int} \equiv L_o = \int_{E_1}^{E_2} E\phi(E)dE, \qquad (5)$$

the apparent distribution of observed luminosities depend on the angle between the bulk motion and the observer's line of sight. Large γ's are required in order for the emission region to be transparent to pair-producing photon-photon interactions. For typical bursts, $\gamma \sim O(100)$ can make a source sufficiently transparent[1]. For our analysis, we do not constrain either γ or specific emission mechanisms. We take γ as a model parameter.

RESULTS AND DISCUSSIONS

Using the apparent luminosity distributions, we plot $\langle V/V_{max} \rangle$ as a function of L_o for $\gamma \leq 10^3$ with $\alpha = 1, 2$.

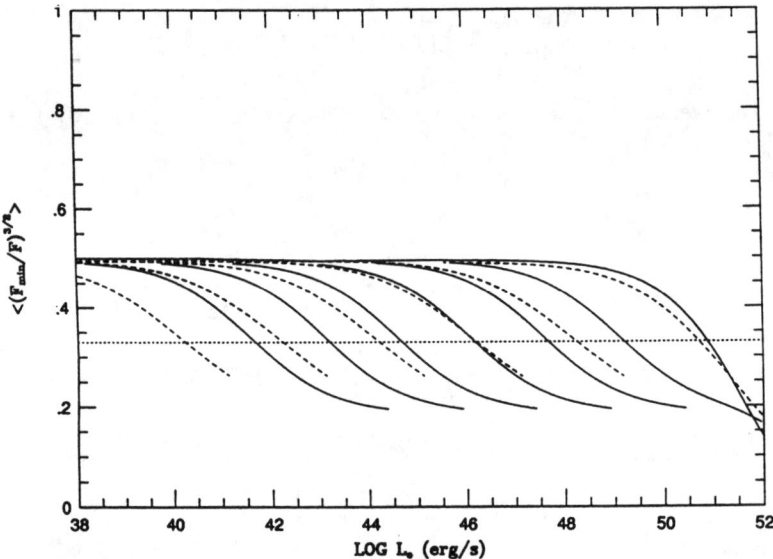

Figure 1. $\langle V/V_{max} \rangle$ vs. L_o for Lorentz factors, $\gamma = 10^{0.5-3.0}$. The solid lines are for $\alpha = 1$ (from left to right, $\gamma = 10^{3,2.5,2,1.5,1,0.5}$ and isotropic case) and the dashed lines correspond to $\alpha = 2$ (from left to right, $\gamma = 10^{2.5,2,1.5,1,0.5}$ and isotropic case). The dotted horizotal line corresponds to the approximate observed value $<V/V_{max}> = 0.33$ which is satisfied in all cases with different L_o for the γ's adopted. The results are almost unaffected by a change in the observed $\langle V/V_{max} \rangle$ from 0.33 to 0.32.

In the beamed cases, the approximate observed value $\langle V/V_{max}\rangle \approx 0.33$ is easily satisfied as in the non-beamed cases. Obviously, larger γ's require smaller L_o's. As in the non-beamed case, $\alpha = 1$ requires a larger L_o (i.e. a larger distance to the most distant sources) than that in $\alpha = 2$ case. In $\alpha = 1$ non-beamed case, the most distant bursts have a redshift $z_{max} \approx 1.5$ in order to satisfy the observed $\langle V/V_{max}\rangle$. However, in the beamed case, $z_{max} \approx 5.5$, which means that if the observed $\langle V/V_{max}\rangle \approx 0.33$ has to be satisfied, the most distant bursts could be much farther than those in the non-beamed case. Similarly, in $\alpha = 2$ case, we get $z_{max} \approx 1$ (non-beamed) and $z_{max} \approx 2$ (beamed). In either case, z_{max} increases by a factor ≥ 2 by beaming. (z_{max} could be slightly affected by $h \approx 0.5 - 0.75$.)

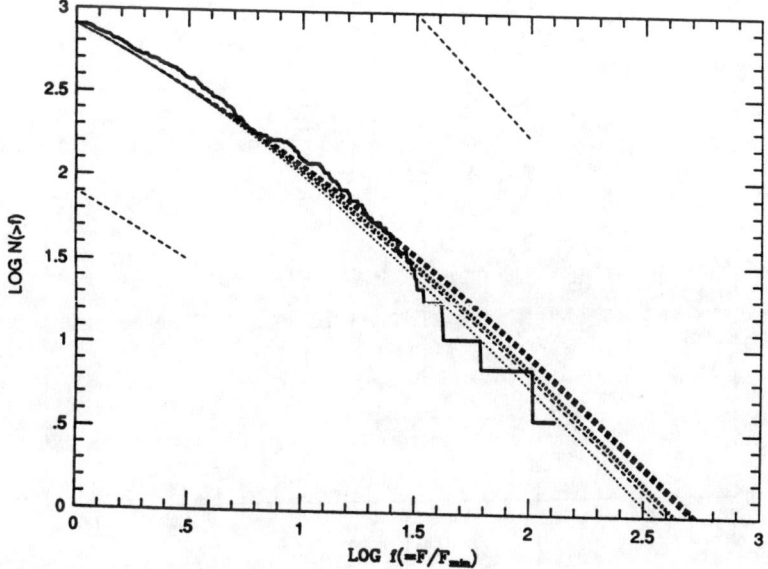

Figure 2. log N - log f for various Lorentz factors for $\alpha = 1$ (solid lines; the bottom line corresponds to the isotropic case) and $\alpha = 2$ (dotted lines; the bottom line corresponds to the isotropic case). Two dashed straight lines correspond to slopes of -0.8 (left) and -1.5 (right). Although the deviations from the isotropic cases are rather small, the effect of beaming tends to erase the change of slope from -0.8 to -1.5 in the cosmological model. For $\alpha = 2$, the effects are more important in the observed flux range $\log f < 2.5$.

In Fig. 2, we plot log N - log f($= F/F_{min}$) for the same parameters as in Fig. 1 together with the observed distribution. We normalize the total number of bursts at $800_{yr}{}^{-1}$. The effect of the luminosity distributions tends to smooth out transition in the slope, roughly from $N \propto f^{-1.0}$ to $f^{-1.5}$, in the observed range $\log f \leq 2.5$. The effects of beaming are rather small (slightly larger in $\alpha = 2$ case than in $\alpha = 1$). More noticeable deviation from the non-beamed (or standard candle) case requires either much wider or a far steeper luminosity distributions than Eq. (10). Currently, the beamed emission is not ruled out based on these simple statistics[2]. Steeper or wider distributions would almost erase the transition in slope from $\propto f^{-1.5}$ to $\propto f^{-1.0}$. In the strong burst limit, $H_0(L/F)^{1/2}/4\sqrt{\pi}c \ll 1$, the luminosity distribution does not affect $F^{-3/2}$ dependence regardless of α. The effects of luminosity distributions due to relativistic bulk motion may be similar to some source evolution (more bursts at lower redshifts than at higher redshifts).

Large γ's provide wide distributions of observed time scales such as GRB durations even if the corresponding intrinsic time scale in the comoving frame is a standard value for all sources.

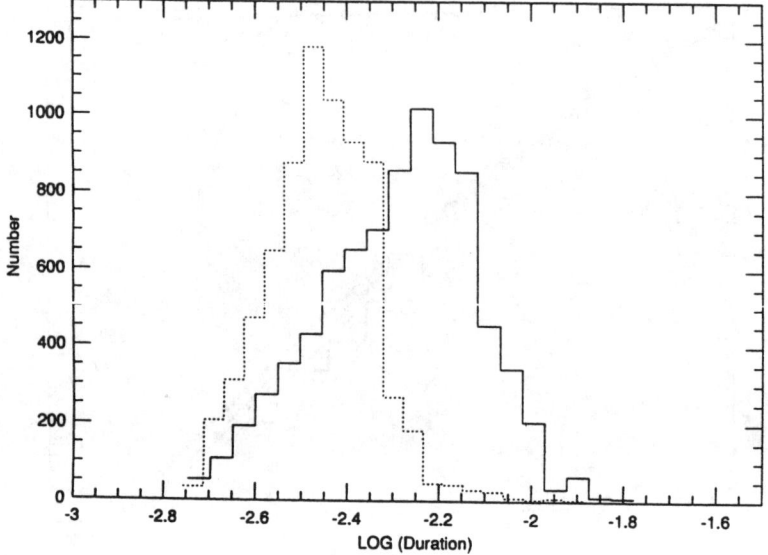

Figure 3. Distributions of durations for cosmological gamma-ray bursts for $\gamma = 10^{2.5}$. The duration is in (arbitrary) unit of the intrinsic duration in the jet frame. The solid histogram is for $\alpha = 1$ and the dotted histogram for $\alpha = 2$. The width of the distribution in the standard source case gives the ratio between the maximum duration and the minimum duration ~ 10.

In Fig. 3, using Monte-Carlo realizations, we show the expected distributions of the apparent observed durations for $\alpha = 1$ and $\alpha = 2$ with $\gamma = 10^{2.5}$. Since the duration is in arbitrary units of intrinsic duration (a single value) which is not currently constrained, not the absolute values but the widths of the distributions are meaningful. The distributions are negatively skewed (more shorter durations) despite rather extended tails toward long durations. Longer durations are rare since these bursts are observable only at smaller distances from an observer (corresponding to much smaller comoving volumes). For $\alpha = 1$ and $\gamma = 10^{2.5}$, the mean duration is shorter than the intrinsic one by a factor

$\sim 6.2 \times 10^{-3}$ and for $\alpha = 2$, by a factor $\sim 4.2 \times 10^{-3}$.

In Fig. 4, we plot the relation between the duration and the flux using a large number of randomly selected bursts. The flux − duration relation gives a definite prediction in the beamed cases. Fig. 4 is a unique prediction of the beamed bursts as long as they are described by a standard source. Bursts with longer durations will have relatively smaller scatter around smaller fluxes than those with shorter durations. However, if the bursts are composed of several types of standard bursts, the observed flux − duration relation should be a superposition of the types with different spectral indices and Lorentz factors. A large scatter in the flux for bursts with long durations would imply that the bursts may not be described by a single standard burst type. A similar argument may also be applied to the non-beamed cosmological bursts. Two spectral indices show qualitatively similar relations. Any direct comparison with observations does not seem meaningful as intrinsic durations are likely to be distributed with some finite (unspecified) widths.

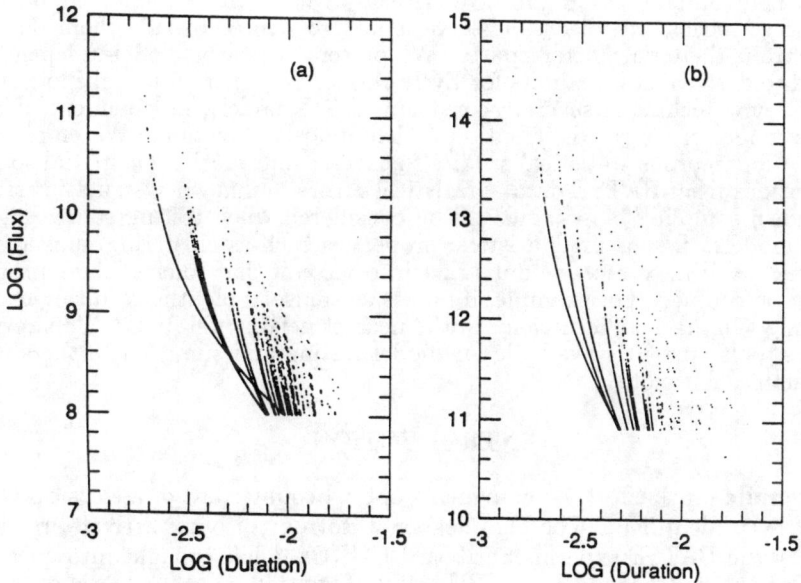

Figure 4. Duration vs. flux relation for $\gamma = 10^{2.5}$. (a) for $\alpha = 1$ (b) for $\alpha = 2$. In both cases, GRBs with longer durations are expected to show smaller scatter in their fluxes than those with shorter durations. This trend is unique and valid only if GRBs may be described by standard sources. Two spectral indices show qualitatively similar relations.

REFERENCES

1. I. Yi, Phys. Rev. **D**, in press (1993).
2. I, Yi, Ap. J., submitted (1993).

SEARCHING FOR A GALACTIC ORIGIN OF GAMMA-RAY BURSTS

Dieter H. Hartmann
Department of Physics and Astronomy
Clemson University, Clemson, SC 29634

ABSTRACT

Distances to Gamma-ray bursts (GRBs) are not known. The isotropic angular distribution on the sky and the non-uniform distribution of burst brightnesses observed with BATSE argue for a cosmological origin of GRBs. Small, brightness-dependent, deviations from isotropy are observed, however, and arguments have been made in favor of a Galactic origin of some bursts, if not all. Models using several burst populations with different distance scales have also been suggested in the literature. The arguments for and against various burst sources are reviewed. Galactic components under consideration include the Oort cloud, local spiral arm segments, companion galaxies of the Milky Way, and a hypothetical extended Galactic halo (EGH). Most of the local models retain the old neutron star paradigm. We critically review the evidence supporting Galactic scenarios and discuss observational procedures to rule them out, or to constrain their parameter space. We present a nearly model-independent method to derive distance limits for EGH bursts. Together with multipole analysis this new technique severely constrains EGH models, independent of any specific model of source distribution and luminosity function. We emphasize the need for rigorous statistical studies to extract information on spatial source distributions from BATSE data. Statistical errors, unknown systematic errors, and known sampling biases must all be considered when testing source distribution models. Discussing statistical properties such as clustering, anisotropy, recurrence, V/Vmax, etc, one must take into account that some of these properties can be coupled. For example, if bursts are anisotropic, the resulting excess clustering will mask as recurrence in the nearest neighbor statistic. This review presents tools currently available for model testing and summarizes the status of Galactic GRB models.

INTRODUCTION

Despite much effort, no convincing optical counterpart of a classical GRB has yet been identified[1]. The high degree of isotropy of burst arrival directions found by the BATSE experiment aboard CGRO[2-4] has brought into question Galactic neutron star models for GRBs and, if anything, has made the quest for a GRB paradigm even more intriguing. Perhaps the most natural explanation is through cosmological models. The observed high degree of isotropy certainly argues in favor of cosmology and the reduced frequency of fainter bursts could be explained through the geometry of the expanding universe. To develop a feel for the largest conceivable distances consider total conversion of rest mass energy into electromagnetic energy. Assume that the time scale is the gravitational crossing time, τ_g, i.e. the Schwarzschild radius divided by the speed of light. The resulting luminosity is then

$$L_{\max} = \frac{mc^2}{\tau_g} = \frac{mc^3}{R_s} = \frac{mc^3}{2Gm/c^2} = \frac{c^5}{2G} \sim 2 \, 10^{59} \text{ ergs},$$

about 10^{21} Eddington luminosities! For typical burst fluxes of $\sim 10^{-6}$ ergs cm^{-2} s^{-1} the corresponding luminosity distance would be in excess of 10^7 Mpc. The Hubble size of the universe is about 3,000 h_{100}^{-1} Mpc, so that the corresponding redshift clearly exceeds $\sim 10^3$. This redshift seems too large for photons to survive on their path to the Earth. Indeed, at the observed photon energies relevant for bursts (10 keV to 1 GeV) the total optical depth of the universe exceeds unity for redshifts larger than ~ 80.[5] The energy associated with a typical bursts at a much smaller redshift, say $z = 0.1$, is of order 10^{50} ergs, assuming a duration of 10 s, which is only a small fraction of a neutron star's binding energy. Beaming would reduce this energy scale further. Although this amount of energy is impressive, scenarios are conceivable that divert $\sim 0.1\%$ of stellar binding energy into the gamma-ray band. Other objections to cosmological models were voiced, but essentially all obstacles can be overcome by taking relativistic effects into account. The physics of cosmological fireball models of GRBs is reviewed elsewhere in these proceedings.

There are no unsurmountable objections to cosmological models and one feels encouraged to abandon the old paradigm of Galactic neutron stars. Wastebaskets have been filled before in the course of science history, but there is a natural tendency to resist paradigm shifts, unless the data indeed require it. Attempts to provide life-support to the Galactic neutron star paradigm continue. Attempts were made to achieve consistency with the data through extended halo models, with an associated scale of ~ 100 kpc. Other solutions assume bursters to be much closer to home (Oort cloud scale), or mix disk and halo populations to satisfy BATSE's boundary conditions. We discuss the search for evidence supporting the notion that some bursts, or perhaps all, do originate in the general vicinity of our Galaxy. We work our way up from the smallest to the largest, non-cosmological, scale. It seems clear that a minimum distance of ~ 100 AU is established for at least some bursts through the requirement that burst triangulation and direct positioning with BATSE, COMPTEL, or EGRET agree to within their errors. If bursts were much nearer than 100 AU one expects the plane wave assumption to break down, which would cause triangulated positions, derived from different detector pairs, to diverge. This is not observed and we thus assume that bursts are more distant than ~ 100 AU. This rules out possible models associated with the solar sphere of influence, here identified with the location of the heliopause at a heliocentric distance of ~ 100 AU. Models involving planets obviously do not work because of their confinement to the ecliptic, but even spherically symmetric models confined to the solar system seem to be ruled out by the above argument Table 1 lists the various distance scales considered in this review and comments on their relevance for GRB models. This review is limited in space and scope so that we recommend several complementary reviews[6-18] to readers interested in observational and theoretical aspects of GRBs that are not limited to the Galactic paradigm.

GALACTIC ORIGINS ?

Oort Cloud Models

Comets have long played a role in modeling bursts,[19] but they were generally considered as a means to extract energy from the gravitational field of neutron stars. BATSE data reopened the possibility that comets could play a role much closer to home, i.e., that bursts might originate within the Oort cloud. Short-period comets are not likely to be related to GRBs, because of their highly anisotropic distribution,[20] although it is hard to imagine that bursts should only involve the small fraction of comets residing in the outer Oort cloud.[21] Comet-comet collisions at typical relative velocities of ~ 1 km s^{-1} might produce γ-rays by electrostatic processes.[22] While the microscopic physics in these collisions, such as particle acceleration, is poorly understood, at least global energy arguments can be made to support this picture.

Collisions of Primordial Black Holes (PBHs) with comets in the Oort cloud could occur frequently enough if the number density of small-mass PBHs is preferentially enhanced in stellar comet clouds.[23] PBHs passing a comet cause rapid collapse of comet material along the PBH path, which could cause a temperature increase to more than ~ 10 MeV via Coulomb heating. The mini-fireball propagating through the comet material completely vaporizes and ionizes the gas, but might only accrete a small fraction of the comet's total mass. Transit times are of order comet size / PBH velocity $\sim 10^{0-9}$ m / 10^{4-5} m s^{-1} $\sim \mu$s – days. This range is somewhat too large in comparison to the observed duration distribution. Direct evaporation of nearby PBHs as a source for some bursts has recently been reconsidered,[24] but these events are expected to be very short (ms) with emission extending into the GeV range.

Maoz[25] searched for a correlation between the burst angular distribution and long-period comets' aphelia directions, and argued that the association is unlikely, but not ruled out, based on angular data alone. The highly non-uniform radial profile of comet density[26] (n $\propto$ r$^{-3.5}$) also argues against the Oort cloud hypothesis because the observed V/V$_{max}$ distribution implies a relatively uniform source distribution for the closer, brighter subset of bursts. This argument is even stronger for comet-comet collision models, where the burst rate is proportional to the square of the comet density. Clark et al.[21] use statistical arguments to emphasize that bursts should originate in the inner Oort cloud, if comets are involved at all, and that their properties do not satisfy the BATSE constraints.[21,27] To match the observational constraints White[28] considered the possibility of a spherical comet cloud between 40 and 400 AU, marginally consistent with the minimum distances mentioned above.

Table I. Gamma-Ray Burst Distance Scales				
Scale	Scale	Association	Models	Constraints
10^{14} cm	10^1 AU	solar system	No	parallax/isotropy
10^{15} cm	10^2 AU	heliopause	No	parallax
10^{17} cm	10^4 AU	Oort cloud	Yes	isotropy
10^{18} cm	10^5 AU	stellar halo	No	nearby stars/α Cen
10^{19} cm	10^0 pc	local fluff	No	energetics
10^{20} cm	10^1 pc	LISM	No	energetics/isotropy
10^{21} cm	10^2 pc	spiral arms	Yes	isotropy/V/Vmax
10^{22} cm	10^3 pc	local disk	Yes	isotropy/V/Vmax
10^{23} cm	10^1 kpc	halo	No	isotropy
10^{24} cm	10^2 kpc	extended halo	Yes	isotropy/M31

Stellar halos and the LISM

The extent of the Oort cloud might approach $\sim 10^5$ AU, which is a significant fraction of the distance to nearby stars. If the Sun is surrounded by such an extended cloud, composed of comets or other objects considered to be the cause of GRBs, then it is reasonable to assume that other stars will be surrounded by similar structures. In that case we can use the lack of excess events from nearby stars,[27] such as α Cen, to constrain model parameters through Monte Carlo simulations. An alternative approach presented at this meeting[29] assumes that every bursts that originates inside our local structure has an associated "cousin" originating in other structures. This is of course only true in a statistical sense, but one can then derive an ensemble limit to the burster distance scale without making any assumptions about the details of the local radial profiles of burst sources or the luminosity function. Arguing that the local stellar halo must be confined to a size small enough to avoid excess burst activity from nearby stars Brown et al.[29] derive a limiting size of $\sim$ 65,000 AU from the 1B data. While this limit begins to constrain outer Oort cloud models, it is not yet strong enough to rule out the traditional Oort cloud. However, any model that invokes a star-centered halo is constrained to this size.

If not directly related to individual stars, the sun's position within structures of the local interstellar medium might be such that bursts appear nearly isotropic on the sky and exhibit a spatial edge. Perhaps the mystery of GRBs is related to another mystery of galactic astronomy; the existence of a local cloudlet (the "local fluff") in which the Sun appears to be embedded.[30] Recent photometric observations[31] of hot white dwarfs in the solar neighborhood with EUVE have measured the density of neutral hydrogen in the fluff as n(HI) $\sim$ 0.1 cm^{-3}. The characteristic column density of the fluff, $\Sigma = 10^{18}$ cm^{-2}, implies a radial extent of $\sim$ 2–3 pc. At these distances typical burst fluxes and durations correspond to a total energy release of $\sim 10^{31}$ ergs. If variability time scales of $\sim$ 1 ms suggest that the accessible energy reservoir is limited to a radius of 3 10^7 cm, the fluff must have an energy density of $\sim 10^8$ ergs cm^{-3}. Cosmic rays, thermal, and magnetic energy densities fall short by about 20 orders of magnitude. So, while the geometry of the fluff may be consistent with the BATSE observations, there seems to be no diffuse energy source in the interstellar medium sufficiently strong to provide enough power for GRBs. Similar energy arguments also hold for the local ISM (LISM), defined by a scale of $\sim$ 10 pc. In addition, the solar neighborhood within $\sim$ 200 pc does not appear to by isotropic.

Local Disk Models

Based on simulated neutron star kinematics,[32,33] isotropy,[34] and lack of angular correlations[35] of bursts in the IPN catalog,[36] local Pop I neutron star models require sampling distances of less than a few hundred parsec, which implies a mean $V/V_{max} \sim$ 0.5 for these source. However, IPN data require detection by multiple detectors and thus sample only brighter events. The KONUS experiment provided localizations for fainter events directly from relative count rates (like BATSE) and analysis of KONUS data prior to the launch of BATSE suggested that fainer bursts tended to cluster towards the Galactic plane.[37] A possible Galactic association of bursts showing cyclotron lines was also argued[38] for fifteen KONUS and one HEAO bursts with accurately measure positions.

BATSE does not confirm this picture,[2] but a recent analysis[39] of brightness data for 241 BATSE bursts as a function of C_p/C_{lim} suggested a slight concentration to the Galactic disk. Quashnock & Lamb[40] analyzed the 1B data based on their Type I/II classification. They found that the 54 Type I bursts of medium brightness exhibit a Galactic dipole moment of $D = \langle\cos\theta\rangle = 0.204 \pm 0.079$, and a Galactic quadrupole moment of $Q = \langle\sin^2(b)\rangle - 1/3 = -0.104 \pm 0.041$. Using Monte Carlo simulations that include uneven sampling of the sky with BATSE these authors find that the probability to find such deviations from isotropy by chance is $\sim 7 \; 10^{-5}$, and conclude that GRBs are Galactic in origin. Analyzing the data further, they identify spiral arm segments as source regions. Briggs et al.[41] analyzed post-1B data to confirm this anisotropy, but find for the corresponding set of 51 GRBs a dipole moment of $D(\text{post} - 1B) = -0.134 \pm 0.081$, and a quadrupole moment of $Q(\text{post} - 1B) = -0.056 \pm 0.042$. The significance of the deviations is lower and, more importantly, the direction of the dipole is reversed. These results strongly argue against the Galactic disk origin hypothesis. Although isotropy and non-uniformity argue against Galactic disk scenarios,[35,42] it is not impossible to construct solutions with beamed emission.[43,44]

Two- or multi-component solutions could involve populations at very different distance scales or with very different luminosity functions. However, if bright (nearby) bursts do show some anisotropy, the high degree of isotropy of the combined populations then requires an even more isotropic distribution of faint (far) sources, almost certainly forcing them to be cosmological. A local NS — local NS model (with two distinct luminosity functions) was suggested by Lingenfelter and Higdon,[45] but was shown to be inconsistent with the observed brightness statistics.[46] A disk plus halo model was considered by Smith and Lamb,[47] who showed that perhaps up to 70% of all BATSE bursts could be local disk neutron stars ($\leq$ 1 kpc), while the rest could reside in a dark matter halo with core radius of $R_c \sim 22$ kpc. A model combining Galactic and cosmological distributions was suggested by Katz.[48] The basic problems with multi-component models is the uniformity of burst appearence; there are no established burst classes with distinct spectral, temporal, flux, or angular patterns. There is also the possibility that stellar flares from the local population of disk stars contaminate the data.[49]

Extended Halo Models

Burst distributions on halo scales ($\sim 10-100$ kpc) were considered long before the launch of BATSE,[50–54] but the new data showed that standard dark matter halos do not have the required statistical properties. However, it was argued[55–59] that neutron stars in an extended galactic halo (EGH) could satisfy all observational constraints. An extended halo created by injection of high velocity neutron stars from the Galactic plane[60] causes large anisotropies, unless early bursting activity is supressed.[61] Injection from a halo distribution of parent bodies alleviates some of the problems associated with disk injection.[59] The recent discovery[62] of high velocity radio pulsars well above the Galactic plane lends some support to the notion that neutron star formation in the halo could provide some GRB sources.

The radial extent of such a halo population is constrained by the BATSE statistic of 743 events[63] showing a dipole moment of $D = \langle\cos\theta\rangle = 0.018 \pm 0.003$, where θ is the angle between the burst and galactic center directions and the error is statistical only. The measured quadrupole moment is $Q = \langle \sin^2(b) \rangle - 1/3 = -0.012 \pm 0.002$, where b is the galactic latitude of the burst. The V/V_{max} value for this sample is 0.320 ± 0.013. The multipole values are very close to isotropy, $(D,Q)_{iso} = (0,0)$, so that small effects due to uneven sampling of the sky may become important. Therefore, the BATSE team carefully derived a sky exposure map.[64] The dipole- and quadrupole moments induced by uneven sampling of the sky are $D = -0.013 \pm 0.021$, and $Q = -0.005 \pm 0.011$. Errors in (D,Q) due to localization uncertainties are expected to be small.[65]

The high degree of isotropy requires a very extended halo of bursting objects. The crucial question is how to populate such an extended region and how to avoid the highly anisotropic contributions from disk objects. A clue to the answer might come from a recent pulsar proper motion survey,[62] which suggests larger space velocities than previously observed. A significant fraction of pulsars in the high velocity tail are not bound to the Galaxy. An interesting aspect of these observations is the presence of several high velocity pulsars that appear to be moving toward the Galactic plane, while most pulsars are known to migrate away from the plane. Some of these pulsars could be due to runaway OB-stars and some could be returning pulsars that were born in the plane with high velocities. However, some of the pulsars have too small characteristic ages for this to be the case. Although there are less than a handful of these pulsars, their observed properties cast some doubt on the traditionally held views about pulsar birthplaces. A possible interpretation of these observations is that neutron star formation is an ongoing process high above the Galactic plane.[56-59] The formation of neutron stars in an already extended halo could produce a spatial distribution that satisfies the BATSE isotropy constraint. If neutron star formation in the halo is due to the merger events of pre-galactic white dwarf binaries, nucleosynthesis during these events and the emission of neutrino bursts constrains the merger rate.[58] The number ratio of neutron stars formed in the halo to that formed in the disk should be less than $\sim 10^{-3}$. We therefore have less than 10^5 to 10^6 sources available to produce the observed rate of $\sim 10^3$ bursts per year. This implies a recurrence time of less than 100 to 1,000 years, which is consistent with current estimates of ~ 10 years. However, one of the major questions is why this small population produces such prolific bursters, while disk neutron stars appear to be γ-ray quiet. One possible solution could be a correlation between the velocity of neutron stars and their emission directionality.[66,67] If burst emission is beamed preferentially along the direction of motion, as suggested by the "magnetars" model,[67,68] the detection probability increases with source distance from the galactic center, and the contamination problem is reduced.[66]

Isotropic injection at some location r_0 with velocity v_0 leads to expanding galactocentric shells. The brightness distribution is a function of many (unknown) properties; radial distribution of birthplaces and birthdates, initial velocities, burst rate as a function of age, and luminosity function. It is thus not too hard to artificially match the observed V/V_{max} statistics, but one has less freedom with the angular distribution. The observed anisotropies depend on the sampling depth from the Earth and the radial "occupation probabilities" of the shells sampled. Ignore effects due to varying detector thresholds and the dependence on luminosity function, and consider the appearance of bursts uniformly located on shells of radius $x = R/R_\odot$. A single shell contributes a dipole moment $D = \frac{2}{3} x^{-1}$ and a quadrupole moment

$$Q = \frac{1}{8}x^{-1} \left\{ x^3 + x - \frac{1}{2}(1-x^2)^2 \ln\left(\frac{1+x}{|x-1|}\right) \right\} - \frac{1}{3}.$$

The dipole provides the most stringent constraints for this class of models, but because of the relationship between the two moments ($Q \propto D^2$ for large x), it is advantageous to consider both of them simultaneously. The multipole limits from 743 events severely constrain EGH models. One way to constrain EGH models is to integrate over shells by means of Monte Carlo simulations using a parametrized density profile and luminosity function.[69-73] Alternatively, one can derive model-independent constraints exclusively from geometric considerations. The argument is similar to the one we made in the section on stellar halos. If the Galaxy has a GRB generating halo, then it is reasonable to assume that other galaxies have them as well. The nearest structure we can use to constrain models is of course the halo of M31. No excess burst activity from that direction has been observed. In a statistical sense, each burst in our Galaxy has an associated cousin burst in M31. The absence of bursts from M31 thus leads to a statistical upper limit on the galactocentric distance of GRBs. Unlike for stellar halos (that were assumed to be centered on the star) this upper limit yields lower limits on the moments D,Q because a smaller shell radius corresponds to larger deviations from isotropy due to the Sun's offset from the galactic center. Summing up all 1B bursts, one finds consistency. However, the limits on deviations from isotropy are now greatly reduced with 743 bursts (see above), while the GRB brightness distribution has not changed shape. Still, no M31 excess of faint bursts has been observed, which implies that EGH models are now inconsistent with the observations,[74] independent on how one wishes to arrange the bursts on various shells and what luminosity function one choses.

Of course, the previous argument does not rule out the possibility that some bursts, or even a significant fraction of all bursts, originates in the Galactic halo. If the total event rate from the halo is $\sim 10^2$ per year, the recurrence time of less than ~ 1000 years for an estimated 10^5 halo objects. If a typical burst fluence is 10^{-6} ergs cm^{-2}, this implies a lifetime storage requirement of $\sim 10^{48}$ ergs. If this energy is provided by accretion or thermonuclear explosions, less than $\sim 10^{-5}$ $M_\odot$ have to be stored in orbit around the neutron star. The low metallicity of population II objects could be responsible for a much higher efficiency of fall-back of material onto the neutron star,[75] leading to the build-up of a planetesimal accretion disk[76] with an estimated total mass of $\sim 10^{-5}$ $M_\odot$ and the mass range of planetesimals is 10^{-12} $M_\odot$ to 10^{-6} $M_\odot$. Accreting neutron stars in the halo undergoing rare Rayleigh-Taylor instabilities due to pycnonuclear reactions do not satisfy the energy requirements.[77] The accretion rate onto high-velocity neutron stars could be significantly enhanced by Kelvin-Helmholtz instabilities in the magnetopause, but the enhancement are not large enough to account for the observed burst frequency.[78]

A typical γ-ray burst flux of 10^{-6} ergs cm^{-2} s^{-1} and duration of ~ 10 s imply, assuming isotropic emission, an energy release of $\sim 10^{36}$ ergs if the distance scale is 100 pc. EGH models increase that scale by a factor 10^3, and cosmological models by more than a factor 10^6. Thus, the energies for EGH or cosmological models are about 10^{42} ergs and 10^{48} ergs, respectively. The observed variability of γ-ray burst light curves suggests a small volume, comparable to the size of neutron stars, for the delivery (impulsive or steady over the short duration of the burst) of this energy. Even if this amount of energy is delivered as pure radiation, two photon pair production ($\gamma\gamma \to e^-e^+$) will create a pair plasma on time scales that are short in comparison with burst durations. If no baryonic matter is present in the deposition region a pair fireball will be created at temperatures of ~ 100 keV, for EGH distances, or ~ 100 MeV for cosmological distances.[79] For EGH models the Compton optical depth exceeds 10^9, so that regardless of the energy injection mechanism a thermal pair fireball is created. The resulting bulk motion of these expanding pair fluids lead to Lorentz factors $\Gamma \sim 10$ for EGH and much larger in cosmological models.[79] While these factors are sufficient to Doppler boost the spectrum at ~ 100 keV into the γ band, they do not allow a significant reduction in the $\gamma\gamma$ opacity,[80] so that the spectrum would still be approximately thermal, in conflict with the observations. Fireballs at EGH distances are likely to face the problem of "baryonic pollution," which removes burst energy from the photons and transfers it to bulk motion, thus quenching the γ-ray signal. It is thus questionable whether one can construct an EGH burst model that yields not only the right fluxes, time scales, and angular distribution, but also the correct spectra. While particle acceleration mechanisms and spectrum formation processes for bursts from nearby neutron stars are well studied much work needs to be done to see whether it will be possible to explain at least some γ-ray bursts with pair fireballs in the halo.

OUTLOOK

Whether GRBs turn out to be cosmological, located in a large halo, or members of an extended solar system, it is clear that either case would teach us something new. The properties of the solar comet population are poorly known because the small number of planet orbit crossing comets are an insufficient tracer of the outer solar system. If the burster origin turns out to be cosmological, a new probe of large scale structure may become available. If some of the events are indeed lensed, information on the lensing systems could be obtained. And if bursts originate in galactic halos, their required properties suggest that some new class of objects or some new formation mechanisms must be at work; again we would learn much.

At this meeting several crucial questions were raised and debated: Is there more than one population of bursters? Soft repeaters and classical bursts are already separated, but are there different classes of classical bursts and/or other "pollutants"? Just how isotropic is the burst distribution? Is there absolutely no angular clustering? Do classical bursts repeat? Is there evidence for a galactic concentration of at least some bursts? What fraction of bursts show line features, and are the properties of those that do different? How do photons with energies above 1 GeV get away from the source? And many more. There appear to be more questions than answers, but we remain optimistic that the GRB phenomenon will not elude us much longer. Part of that optimism is related to the recent identification of a counterpart to the SGR 1806−20. Kulkarni and Frail[81] identified this source with the Galactic supernova remnant G10.0−03 based on a radio survey of the IPN error box. After BATSE discovered renewed activity from this source[82] X-ray observations with the ASCA satellite[83] confirmed this identification. This breakthrough highlights the importance of determining precise burst positions so that multiwavelengths studies can be carried out. More precisely localized bursts using multi spacecraft triangulation[84,85] as well as more rapid localizations (even during the burst) now seem possible,[86] providing hope for identification of quiesent counterparts in the near future. Low energy transients could provide a breakthrough if simultaneous detection of "optical" and γ-ray emission is accomplished. Hopes are on 1) the Explosive Transient Camera (ETC) at Kitt Peak, 2) the GSFC/MSFC rapid response network[86] (BACODINE) 3) operation of the High Energy Transient Experiment (HETE),[87] expected to be launched in 1994, and 4) radio afterglows.[88] Dispersed radio emission appears to be particularly promising for determining the burster distance scale.[89] This work was supported by NASA grant NAG 5-1578.

REFERENCES

1. B. E. Schaefer, these proceedings.
2. C. A. Meegan, et al., Nature **355**, 143 (1992).
3. G. J. Fishman, et al., ApJS , in press (1993).
4. C.A. Meegan, et al., these proceedings.
5. A. A. Zdziarski, & R. Svensson, ApJ **344**, 551 (1989).
6. E. P. Liang & V. Petrosian, Gamma-Ray Bursts (AIP 141, 1986).
7. D. H. Hartmann & S. E. Woosley, Multiwavelength Astrophysics (Cambridge, 1988), p. 189.
8. K. Hurley, Ann. N. Y. Acad. Sci. **571**, 442 (1989).
9. K. Hurley, Nucl. Phys. **10B**, 21 (1989).
10. J. C. Higdon, & R. E. Lingenfelter, ARAA **28**, 1990 (401).
11. K. Hurley, High-Energy Astrophysics (Nat. Acad. Press, 1991), p. 204.
12. C. Ho, R. I. Epstein, & E. E. Fenimore, Gamma-Ray Bursts (Cambridge, 1992).
13. W. S. Paciesas & G. J. Fishman, Gamma-Ray Bursts (AIP 265, 1991).
14. A. K. Harding, Phys. Reports **206**, 327 (1991).
15. D. Q. Lamb, Neutron Stars: Theory and Observation (Kluwer, 1993), p. 545.
16. K. Hurley, X-ray Binaries (Cambridge Univ. Press, 1993), p. in press.
17. D. H. Hartmann, The Lives of Neutron Stars (Kluwer, 1994), p. in press.
18. K. Hurley 1993, ApJ Suppl. , in press (1993).
19. M. Ruderman, Ann. N. Y. Acad. Sci. **262**, 165 (1975).
20. G. S. Watson, Statistics on the Sphere , Wiley Publ.: New York (1983).
21. T. E. Clarke, O. Blaes, & S. Tremaine, ApJ , in press (1993).
22. J. I. Katz, AIP: The COMPTON Observatory **280**, 1090 (1993).
23. K. F. Bickert & J. Greiner, AIP **280**, 1059 (1993).
24. D. B. Cline & W. Hong, ApJL **401**, L57 (1992).
25. E. Maoz, ApJ **414**, 877 (1993).
26. M. Duncan, T. Quinn, & S. Tremaine, AJ **94**, 1330 (1987).
27. J. Horack et al., ApJ , in press (1993).
28. R. S. White, ApSS , in press (1993).
29. L. E. Brown, D. H. Hartmann, & L.-S. The, these proceedings.
30. D. P. Cox, & R. J. Reynolds, ARAA **25**, 303 (1987).
31. S. Vennes, et al., ApJL , in press (1993).
32. D. H. Hartmann, R. I. Epstein, & S. E. Woosley, ApJ **348**, 625 (1990).
33. B. Paczynski, ApJ **348**, 485 (1990).
34. D. H. Hartmann & R. I. Epstein, ApJ **346**, 960 (1989).
35. D. H. Hartmann & G. R. Blumenthal, ApJ **342**, 521 (1989).
36. J.-L. Atteia, et al., ApJS **64**, 305 (1987).
37. J.-L. Atteia, et al., Nature **351**, 296 (1991).

38. O. C. De Jager & B. E. Schaefer, AIP: Gamma-Ray Bursts 265, 226 (1993).
39. J.-L. Atteia & J.-P. Dezalay, A&A , in press (1993).
40. J. M. Quashnock & D. Q. Lamb, MNRAS , in press (1993).
41. M. Briggs, et al., these proceedings.
42. S. Mao & B. Paczynski, ApJL 389, L13 (1992).
43. W. Kundt & H.-K. Chang, ApSS 200, 151 (1993).
44. I. G. Mitrofanov, ApSS 165, 137 (1990).
45. R. E. Lingenfelter & J. C. Higdon, Nature 356, 132 (1992).
46. B. Paczynski, Acta Astr. 42, 1 (1992).
47. I. A. Smith & D. Q. Lamb, ApJL 410, L23 (1993).
48. J. I. Katz, ApJ , in press (1993).
49. E. P. Liang, A&A 273, L53 (1993).
50. G. Fishman, et al., ApJL 223, L13 (1978).
51. M. Jennings, & R. S. White, ApJ 238, 110 (1980).
52. M. Jennings, AIP: High Energy Transients in Astrophysics 115, 412 (1984).
53. J.-L. Atteia, & K. Hurley, Adv. Sp. Res. 6, 39 (1986).
54. T. Yamagami, & J. Nishimura, ApSS 121, 241 (1986).
55. J. J. Brainerd, Nature 355, 522 (1992).
56. D. Eichler & J. Silk, J., Science 257, 937 (1992).
57. D. H. Hartmann, et al., AIP 265: Gamma-Ray Bursts , 120 (1992).
58. D. H. Hartmann, Comm. in Astrophys. 16, 231 (1992).
59. D. H. Hartmann, et al., ApJS , in press (1993).
60. I. S. Shklovskii & I. G. Mitrofanov, MNRAS 212, 545 (1985).
61. H. Li & C. D. Dermer, Nature 359, 514 (1992).
62. P. A. Harrison, A. G. Lyne, & B. Anderson, MNRAS 261, 113 (1993).
63. C. A. Meegan, et al., these proceedings.
64. M. N. Brock, et al., AIP: Gamma-Ray Bursts 265, 399 (1992).
65. J. M. Horack, et al., ApJ 413, 293 (1993).
66. H. Li, et al., these proceedings.
67. R. C. Duncan & C. Thompson, ApJL 392, L9 (1992).
68. R. C. Duncan, H. Li, & C. Thompson, C., AIP 280, 1074 (1993).
69. I. Wasserman, ApJ 394, 565 (1992).
70. J. Hakkila, et al., AIP: Gamma-Ray Bursts 265, 70 (1992).
71. J. Hakkila, et al., AIP: Compton Gamma-Ray Observatory 280, 704 (1993).
72. J. Hakkila, et al., ApJ , in press (1993).
73. J. Hakkila, et al., AIP: Gamma-Ray Bursts , these proceedings (1994).
74. D. H. Hartmann, et al., ApJ , in preparation (1994).
75. S. E. Woosley, Astr. Soc. Pacific , in press (1993).
76. D. N. C. Lin, S. E. Woosley, & P. Bodenheimer, P., Nature 353, 827 (1991).
77. O. Blaes, R. Blandford, P. Madau, & S. Koonin, ApJ 363, 612 (1990).
78. A. K. Harding, & M. Leventhal, Nature 357, 388 (1992).
79. T. Piran & A. Shemi, ApJL 403, L67 (1993).
80. J. H. Krolik & E. A. Pier, ApJ 373, 277 (1991).
81. S. R. Kulkarni, & D. A. Frail, Nature 365, 33 (1993).
82. C. Kouveliotou, et al, IAU Circ. 5875.
83. N. Murakami, et al., these proceedings.
84. K. Hurley, et al., A&A Suppl. 97, 39 (1993).
85. K. Hurley, et al., these proceedings.
86. S. D. Barthelmy, et al., these proceedings.
87. G. Ricker, et al., Gamma-Ray Bursts (Cambridge Univ. Press, 1992), p. 288.
88. B. Paczynski, & J. Rhoads, ApJL 418, L5 (1993).
89. D. M. Palmer, ApJL 417, L25 (1993).

GAMMA-RAY BURST CONTINUUM SPECTRA FROM MAGNETIC INVERSE COMPTON SCATTERING

Matthew G. Baring *

NASA Goddard Space Flight Center, Code 665, Greenbelt, MD 20771

ABSTRACT

The Thomson limit of inverse Compton scattering in the strong magnetic fields of neutron stars is considered as a model for producing gamma-ray burst continuum spectra. Emission spectra are obtained as solutions of a set of photon and electron kinetic equations for various assumptions about the collision cross-section; comparison of these spectra indicates that use of the full magnetic Thomson cross-section in these upscattering (CUSP) models is necessary.

INTRODUCTION

In gamma-ray burst (GRB) studies prior to the launch of the Compton Gamma-Ray Observatory, neutron star models for burst emission locales were extremely popular. At that stage, a popular model was where the gamma-rays from these sources were generated by inverse Compton scattering in strong fields (which is resonant at the cyclotron energy), what has commonly been called the cyclotron upscattering process (CUSP). Ho and Epstein[1] first pointed out that even non-resonant (classical) Thomson upscattering of X-rays from a neutron star surface by relativistic electrons beamed away from the star could simultaneously produce flat GRB spectra and satisfy the constraint imposed by the paucity of X-rays from bursts, spawning a variety of treatments of the upscattering mechanism that incorporated the effects of the strong field. These included both kinetic equation analyses[2,3] and Monte Carlo simulations,[4,5,6] under a variety of assumptions about the resonant cross-section. All predicted flat γ-ray spectra and a beaming of the high-energy radiation along the field that would help avoid absorption by magnetic single-photon pair production.

These analyses are currently being generalized[7] to treat the case of multiple scattering of photons in the emission region, a situation that is appropriate for bursts located in a galactic halo. Simplifying approximations to the resonant cross-section that were made in some of these earlier treatments are then no longer applicable.[7] This paper presents GRB emission spectra in the case of single photon scatterings; it is found that some cross-section approximations are quite inaccurate even for low source optical depths. The spectra are produced as equilibrium solutions of a system of photon and electron kinetic equations, with relativistic electrons scattering quasi-thermal X-ray photons that are injected into the emission region from, for example, a neutron star surface. As with the earlier treatments, the emission spectra are strongly dependent on the observing angle, with the highest energy γ-rays beamed closest to the field lines. A criterion for the applicability of the single scattering case is presented.

* NAS–NRC Research Associate

MAGNETIC INVERSE COMPTON SCATTERING SPECTRA

Consider the situation where monoenergetic soft photons collide with relativistic, beamed electrons. Assume that the soft photons of energy ε_s (typically around 1 keV) are injected continuously into the GRB emission region at a rate $\dot{n}_s(\varepsilon) \propto \delta(\varepsilon - \varepsilon_s)$. This crudely mimics a thermal component from a neutron star surface, though other X-ray sources are conceivable. The relativistic electron population will be assumed to be injected into the emission region with a one-dimensional power-law energy distribution with momenta along the magnetic field lines. Such an anisotropic distribution is easily generated in strong neutron star fields because the timescale for de-excitation of Landau levels, the synchrotron cooling timescale, is extremely short ($\sim 10^{-16}$ sec for $B = 10^{12}$ Gauss). For simplicity, the magnetic field will be assumed to be uniform.

Emission spectra can be obtained by determining solutions to the simplest kinetic equation for photons subject to inverse magnetic Thomson scattering:

$$0 = \dot{n}_s(\varepsilon) + \frac{s_T(\varepsilon, \mu)}{t_{esc}} - \frac{n_\gamma(\varepsilon, \mu)}{t_{esc}} \quad . \tag{1}$$

Here $s_T(\varepsilon,\mu)/t_{esc}$ is the rate for photon production, for given energy ε and angle cosine μ relative to the magnetic field lines) in Compton collisions with the non-thermal electrons. Pair production is neglected, and steady-state conditions are assumed. The $n_\gamma(\varepsilon,\mu)/t_{esc}$ term defines electron escape with a timescale of $t_{esc} = R/c$. Hereafter all photon and particle densities will be scaled by the characteristic density $1/(\sigma_T R)$, where σ_T is the Thomson cross-section and R is the size of the emission region. All energies will be scaled by $m_e c^2$.

The steady-state kinetic equation for the non-thermal electrons under the influence of Thomson scattering takes the form of a Fokker-Planck equation,[7] since for $\varepsilon_s \ll 1$ the scattering yields only differential changes in the electron energy. It solves easily[7] to yield the steady-state electron distribution:

$$n_e(\gamma) = -\frac{1}{\dot{\gamma}_T} \int_\gamma^\infty Q(\gamma') \, d\gamma' \quad . \tag{2}$$

Here $Q(\gamma)$ is the rate of electron injection into the emission region and is taken to be a power-law; $\dot{\gamma}_T$ is the electron cooling rate due to resonant Thomson scattering. Note that this solution is valid only when $\dot{\gamma}_T < 0$; an alternative form for the domain where $\dot{\gamma}_T > 0$ (i.e. for electron heating, which occurs when $\gamma \sim 1$) is required.[7,8] For the scenario considered here, where photons are not repeatedly scattered, the optical depth must be much less than unity and therefore is omitted in Eq. (1). A criterion for this case is given in Eq. (6).

When integrated over the electron distribution $n_e(\gamma)$, expressions for the spectrum of photon production[2,9] $s_T(\varepsilon_f)$, for general Thomson scattering differential cross-sections, yield

$$s_T(\varepsilon_f) = \int_1^\infty \frac{d\gamma}{\gamma^2 \beta} n_e(\gamma) \int_0^\infty \frac{d\varepsilon_i}{\varepsilon_i} \int_{\mu_-}^{\mu_+} d\mu_i \, n_s(\varepsilon_i, \mu_i) \frac{1}{\sigma_T} \frac{d\sigma_T(\omega_i)}{d(\cos\theta_f)} \quad , \tag{3}$$

where the the subscripts i and f (initial and final) denote before and after scattering. Here ω_i is the photon energy in the electron rest frame (ERF), and the negligible energy change in the ERF that occurs in the Thomson limit is defined by the equality $\omega_i = \gamma \varepsilon_i (1 - \beta \mu_i) = \gamma \varepsilon_f (1 - \beta \mu_f)$.

The full differential magnetic Thomson cross-section in the frame where the electron is at rest (ERF), for arbitrary angles of the initial and final (scattered) photons, can be obtained from Eqs. (14) and (15) of Herold[10] (see also Eq. 2 of ref. 2):

$$\frac{d\sigma_T(\omega_i)}{d(\cos\theta_f)} = \frac{3\sigma_T}{8} \left\{ \sin^2\theta_i \sin^2\theta_f + \frac{1}{4}[1 + \cos^2\theta_i][1 + \cos^2\theta_f] \Sigma_\kappa\left(\frac{\omega_i}{B}\right) \right\} \quad (4a)$$

where

$$\Sigma_\kappa(\psi) = \frac{\psi^2}{(\psi-1)^2 + \kappa^2} + \frac{\psi^2}{(\psi+1)^2} \quad , \quad \kappa = \frac{2\alpha_f B}{3} \quad (4b)$$

defines the scattering resonance. Here κ is a scaling of the natural line width $\Gamma = 4\alpha_f B^2/3$ of the cyclotron harmonic, the only resonance that appears in the Thomson limit. Also, the field strength B is (hereafter) scaled in units of the quantum critical field $B_c = m_e^2 c^3/(e\hbar) = 4.413 \times 10^{13}$ Gauss.

A particularly useful approximation is to assume that the scattering is isotropic in the ERF, and also that the photons travel almost parallel to the field lines in the ERF. The assumption of quasi-isotropic scattering is generally quite good, however the initial photons will only travel along the field lines in the electron rest frame when they collide with extremely relativistic electrons. These approximations dramatically simplify the cross-section, which can be obtained from Eq. (4) by the substitutions $\theta_i \to 0$ and $(1 + \cos^2\theta_f) \to 4/3$. A principal objective of this paper is to assess how accurately such a simplification (referred to hereafter as the I-|| cross-section) can model emission spectra.

For soft photons that are injected isotropically over a restricted range of angles, the equilibrium distribution of low-energy photons satisfying Eq. (1) takes the form

$$n_s(\varepsilon_i, \mu_i) = \frac{n_s(\varepsilon_i)}{\mu_+ - \mu_-} \quad , \quad \mu_- \leq \mu_i \leq \mu_+ \quad . \quad (5)$$

It is then a simple matter to derive the resulting photon production spectrum, integrated over a range of emission angles $\mu_l \leq \mu_f \leq \mu_u$; analytic expressions for the spectrum have been obtained both for the simple case of quasi-isotropic scattering of photons that move along the field lines in the ERF,[2,9] and also for the full cross-section[9] in Eq. (4).

A comparison of the spectra yielded by Eq. (3) using these two forms for the scattering cross-section is made in Fig. 1a, for power-law electrons $n_e(\gamma) \propto \gamma^{-p}$, indicating that I-|| cross-section underestimates the spectrum generally by about a factor of two. Asymptotic approximations[2,9] to Eq. (3) give spectral indices of $-(p+2)$, 0 and $p+2$, respectively, for low, intermediate and high emitted photon energies ε_f. Fig. 1b depicts the comparison of electron cooling rates[9] $\dot{\gamma}_T$ used in Eq. (2) for these two forms of the cross-section; the I-|| cross-section clearly underestimates the cooling rate for low γ. This difference is very

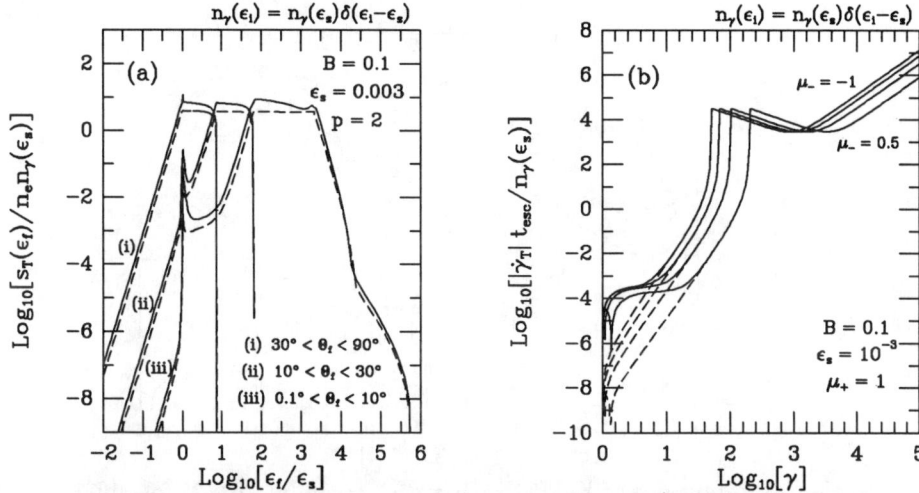

Figure 1. Photon production spectra from Eq. (3) [Fig. 1a] for γ^{-p} electrons, and electron cooling rates $\dot{\gamma}_T$ [Fig. 1b] appearing in Eq. (2) and calculated in Baring;[9] in both figures the solid curves represent evaluation using the full cross-section in Eq. (4) while the dashed curves represent the I–|| cross-section. The soft photons have the angular distribution in Eq. (5), with Fig. 1a having $\mu_+ = 1$ and $\mu_- = 0$. Fig. 1b displays the cases $\mu_- = -1$, -0.5, 0 and 0.5.

important for the determination of the cooling e^- distribution in Eq. (2), and consequently for the resulting photon emission spectrum. The cusps in Fig. 1b are where $\dot{\gamma}_T \to 0$, and define the onset of electron heating at low γ.

Figure 2 displays the solutions of the system of photon and electron kinetic equations, i.e. Eqs. (1) and (2), comparing the spectra obtained from the full cross-section in Eq. (4) and the I–|| cross-section; clearly the latter produces grossly erroneous results at low (X-ray) energies. The spectra for the full cross-section when viewed relatively close to the field lines are still quite flat at soft X-ray energies and show a prominent break in the soft gamma-ray range (at $\varepsilon \sim 10^{-1} \approx 50\,\text{keV}$). Also apparent are sharp drops in the hard gamma-ray spectrum due to rapid variations in the electron cooling rate depicted in Fig. 1b; a detailed analytic discussion of the spectral indices and features is given in Baring,[7] where the differences between the spectra in Fig. 2 and the results of Dermer[2] are discussed. Unphysical divergences at low energies in the spectra in Fig. 2, due to the $\dot{\gamma}_T \to 0$ singularity in Eq. (2), are not displayed.

The solutions presented here are for when photons are scattered at most about once before leaving the emission region, i.e. the scattering cascade is first-order in the photon distribution. This corresponds to the ratio of electron (l_e) to soft photon (l_s) injection luminosities being low[7]:

$$\frac{\pi}{\alpha_f B} \frac{l_e}{l_s} \ll 1 \ . \tag{6}$$

If the soft X-ray photons are assumed to come from a neutron star surface,

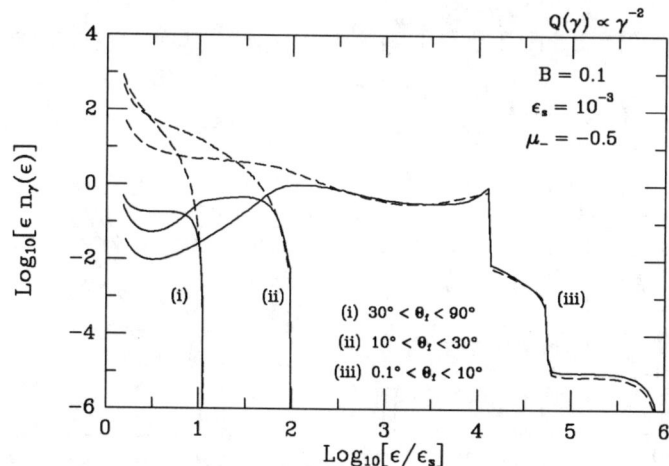

Figure 2. Solutions of Eq. (1) with $n_e(\gamma)$ given by Eq. (2) for different ranges of emission angles Θ_f relative to the field lines. The solid curves represent solutions using the full cross-section in Eq. (4) while the dashed curves represent the I-∥ cross-section. The soft photons have the angular distribution in Eq. (5), with $\mu_+ = 1$.

then their thermal nature will largely fix the value of l_s. At the same time, the observed gamma-rays determine l_e (since $n_\gamma(\epsilon)$ obtained from Eq. (2) are proportional to l_e), which obviously depends on the source distance. It is clear that for large enough distances, Eq. (6) is violated and first-order cascades are no longer appropriate; generally this occurs for bursts located outside the galactic disk.[7] Work is in progress[7] to generalize the solutions of the Thomson cascade model presented here to include the case of multiple photon scatterings. Clearly then, as indicated by the results here, it will be necessary to include the full magnetic Thomson cross-section in Eq. (4).

REFERENCES

1. Ho, C. and Epstein, R. I., Ap. J. **343**, 227 (1989).
2. Dermer, C. D., Ap. J. **360**, 197 (1990).
3. Vitello, P. and Dermer, C. D., Ap. J. **374**, 668 (1991).
4. Daugherty, J. K. and Harding, A. K., Ap. J. **336**, 861 (1989).
5. Fenimore, E. E., et al., in Gamma-Ray Bursts, Observations, Analyses and Theories, eds. C. Ho et al. (CUP, Cambridge, 1991), p. 305.
6. Brainerd, J. J., Ap. J. **384**, 545 (1992).
7. Baring, M. G., Ap. J. to be submitted, (1994a).
8. Preece, R. D. and Harding, A. K., in Gamma-Ray Bursts, Observations, Analyses and Theories, eds. C. Ho et al. (CUP, Cambridge, 1991), p. 365.
9. Baring, M. G., Ap. J. to be submitted, (1994b).
10. Herold, H., Phys. Rev. D **19**, 2868 (1979).

THE OBSERVATION OF UNUSUAL GAMMA RAY BURSTS FROM PRIMORDIAL BLACK HOLE EVAPORATION

David B. Cline and Woopyo Hong

Department of Physics and Astronomy
University of California at Los Angeles
405 Hilgard Avenue, Los Angeles, California 90024

ABSTRACT

We discuss the possibility that a small fraction of the observed Gamma Ray Bursts(GRBs) originates from the Primordial Black Hole(PBH) final state evaporation. We review a class of very short time span GRBs with hard photon spectra that could be candidates for PBH evaporations.

Ever since the invention of "Primordial Black Holes" there have been suggestions for the experimental detection of such objects[1]. However, the real time detection depends on the final state evolution of the PBH; as the PBH's temperature is rising as it sheds mass, into the region of hadronic interactions and hadronic final states[1](See Table 1 for details). The most extreme model used to simulate this final state was the Hagedorn model that predicted an explosion lasting $\sim 10^{-7}$ sec (and very heavy particle luminosity), whereas other QCD inspired calculations suggested a final state collapse time of the order of seconds[3]. We questioned the validity of the QCD inspired calculations by pointing out that final state interactions and non-perturbation effects could increase the low energy particle luminosity and even the collapse time[4].

A reasonable model of the emitted particles degrees of freedom, $\alpha(M)$, is shown in Fig. 1, where we indicate the regions of uncertainty where there could be a rapid increase in the effective degrees of freedom due to the quark-gluon phase transition, leading to a rapid burst in the PBH evaporation or at very high energy, where there could be many new particle types that would also increase the rate of evaporation. We also show, in Fig. 1, the regions in PBH temperature where short γ bursts may occur when the PBH mass is 10^{14} or 10^9 grams! Unfortunately, Experimental Physicists or Astrophysicists don't have the time or patience (or perhaps ability) to understand in detail quantum gravity and also build detectors. However, we take these suggestions as indicating there is a possibility that PBH have a violent ending that might be experimentally detectable[4].

We have studied available GRBs data and noticed that there is a class of fairly short GRBs (< 200 msec) that in many cases seem to have a fairly hard γ spectrum[5]. Fig. 2 shows the profile of some of these events. However, we must be careful that the harder spectrum is not somehow involved in the detection process for these very short GRBs. It occurs to be that one way to demonstrate that a GRB event was coming from an exotic source like a PBH, would be to have observed even a shorter final structure. For example, if nanosecond structure were observed this would presumably limit the size of the progenitor to

$$\begin{bmatrix} \text{Size of} \\ \text{Object} \end{bmatrix} \lesssim [\text{nanosecond}] \times [c] \leq 30\,\text{cm}$$

In this case there would be no known astrophysical object, i.e. neutron star, several solar mass black hole, etc., that could be the origin of the burst. Unfortunately, there are no detectors in space now, or being planned, that would have nanosecond time resolution and sufficient area to collect adequate statistics to observe such structure.

Table 1. Primordial Black Hole Phenomenology

1) Photon fluxes emitted by a PBH:

$$\frac{d^2 N_\gamma}{dt\, dE_\gamma} = \frac{\Gamma(M, E_\gamma)}{2\pi\hbar} \left[\exp\left(\frac{8\pi GME_\gamma}{\hbar c^3}\right) - 1 \right]^{-1}$$

2) PBH temperature vs. mass:

$$T \simeq 10^{16} \left[\frac{1g}{M}\right] \text{MeV}$$

3) PBH mass loss rate with particle degrees of freedom $\alpha(M)$:

$$\frac{dM}{dt} = -\frac{\alpha(M)}{M^2}$$

$$\alpha(M) \approx \begin{cases} [7.8 d_{s=1/2} + 3.1 d_{s=1}] \times 10^{24} \text{g}^3\text{s}^{-1} & ;\text{Standard Model} \\ m^{-5/2} \exp(m/\Lambda_H) & ;\text{Hagedorn Model} \end{cases}$$

4) PBH masses significant for observation[2]:

$$4.3 \times 10^{14} < M_* = (3\alpha(M_*)t_o)^{1/3} < 7.0 \times 10^{14}\text{g} \quad \text{for SM}$$

The hardness ratio in Fig. 2 is defined as the fluence in channel 3 ($\sim$ 100 to $\sim$ 300 keV) to the fluence in channel 2 ($\sim$ 50 to $\sim$ 100 keV) of BATSE detector. As we see in Fig 2. there is a strong tendency for harder spectra with shorter GRB durations. This may indicate that the GRBs originate from two difference

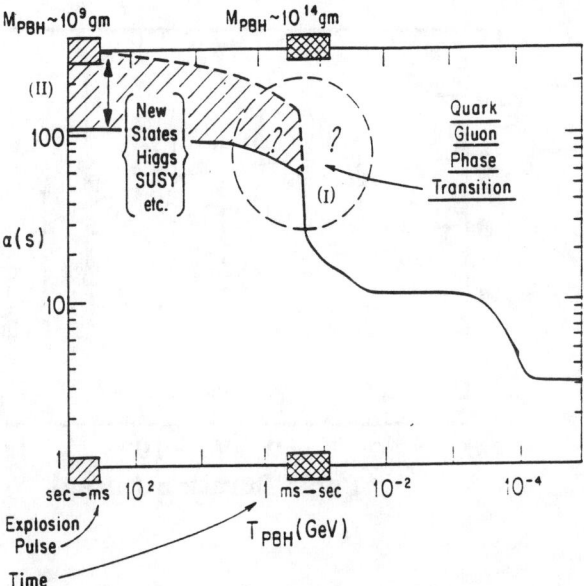

Figure 1: The running coupling or density of states factor α showing regions of uncertainty due to the quark-gluon phase transitions (I) or the increase in the number of new elementary particles (II). It is possible that intense short γ bursts could occur at either of these temperatures.

sources[6]; shorter bursters could be contributed by PBHs explosion. Probably, the most direct test for this argument could be to calculate the hardness ratios for PBH evaporations at several different temporal stages[7]. This hardness ratio can be expressed as

$$\text{Hardness } (\delta\tau) = \frac{\int_0^{\delta\tau} dt \int_{100 \text{ keV}}^{300 \text{ keV}} E_\gamma \frac{d^2 N_\gamma}{dt\, dE_\gamma} dE_\gamma}{\int_0^{\delta\tau} dt \int_{50 \text{ keV}}^{100 \text{ keV}} E_\gamma \frac{d^2 N_\gamma}{dt\, dE_\gamma} dE_\gamma}.$$

In conclusion, we believe it is essential to study unusual cosmic events, such as shorter GRBs, to possibly identify unusual behavior that could be characteristic of PBH explosions. We have described a class of GRB that are intriguing from this stand point.

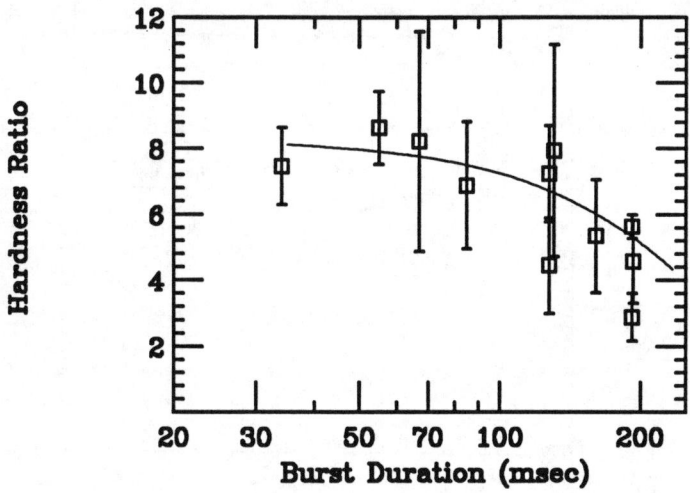

Figure 2: Hardness ratio for some of the γ bursts reported in the literature. Note that the short time bursts have a much harder spectrum, a trend that would be expected if some of the short bursts came from PBH evaporation. A simple fitting of these data indicates an anti-correlation of hardness vs. burst duration.

REFERENCES

1. D.N. Page and S.W. Hawking, Ap. J. **206**, 1 (1976).
2. F. Halzen, et al., Nature **353**, 807 (1991).
3. M.J. Rees, 1977, Nature **266**, 333-334.
4. D.B. Cline and W.P. Hong, 1992, Ap. J. Letts. **401**, L57.
5. G.J. Fishman, et al., Ap. J. Suppl. (1993).
6. C. Kouveliotou, et al., Ap. J. **413**, L101.
7. D.B. Cline and W.P. Hong, work in progress.

GAMMA-RAY BURSTS FROM THE ACCRETION OF SOLID BODIES ONTO HIGH-VELOCITY GALACTIC NEUTRON STARS

Stirling A. Colgate and Peter J. T. Leonard

T-6, MS B275, Los Alamos National Laboratory, Los Alamos, NM 87545

ABSTRACT

We propose a simple model for the gamma-ray bursts based on high-velocity Galactic neutron stars that have accretion disks. The latter are formed from a mixture of material from the supernova shell and that ablated from a pre-supernova binary companion. Accretion onto the neutron star from this disk when the disk is still largely gaseous may result in a soft gamma-ray repeater phase. Much later, after the neutron star has moved away from its birthplace, solid bodies form in the disk, and some are perturbed into hitting the neutron star to create gamma-ray bursts. This model makes several predictions that are consistent with the observations. The observed combination of a high degree of isotropy on the sky coupled with the observed value of $< V/V_{max} >$ is not, at first glance, predicted, but is not impossible to attain in our model.

INTRODUCTION

The main argument used against a Galactic origin for the gamma-ray bursts (GRBs) is their apparent isotropy on the sky coupled with the observed value of $< V/V_{max} >$ (Meegan et al. 1992). However, it is dangerous to rule out all Galactic GRB models based on only one combination of two observations, especially when the physics involved in the Galactic models is much easier to believe than that invoked in cosmological GRB models. Indeed, the recent evidence for repeat GRBs (Wang & Lingenfelter 1993; Quashnock & Lamb 1993a) and a trace of a correlation of GRBs with the Galactic disk (Atteia & Dezalay 1993; Quashnock & Lamb 1993b) suggest the necessity of a Galactic GRB model. Hence, the issue here is the likelihood that the popular neutron star (NS) merger model can produce the observed GRBs at cosmological distances compared with a near conspiracy of the distribution and properties of Galactic sources necessary to produce the observed isotropy and $< V/V_{max} >$.

In this paper, we consider a GRB model based on the accretion of solid bodies onto high-velocity Galactic NSs. This theory predicts several of the observed properties of GRBs, but the high degree of isotropy of GRBs on the sky coupled with the observed value of $< V/V_{max} >$ is not naturally predicted. However, is not impossible to obtain the observed combination in Galactic GRB models. Two existing ideas on how to accomplish this are 1) a local model involving the nearby spiral structure of the Galaxy plus repeat bursts (Quashnock & Lamb 1993a, 1993b), and 2) an extended halo model involving high-velocity NSs with a delayed turn on in bursting (Li & Dermer 1992).

Before we go into the details of our model, we would like to point out some problems with the physics involved in the cosmological GRB model based on mergers of pairs of NSs. This model faces the extreme difficulty of obtaining a hard GRB spectrum, and the relatively long duration, but episodic, emission of GRBs. The neutrino emission, necessary to produce a neutrino fireball, is less efficient than the process within a supernova, where the leptons are initially trapped and compressed to a high Fermi level before they escape by diffusion with an increase in energy of up to a factor of $\sim 10^2$. This process will not take place in the NS merger case because the matter is already de-leptonized or transparent. The accretion from the disk formed during the merger onto the central merged object could indeed be episodic and delayed, but the re-leptonization by initial expansion would not then lead to trapping by the relatively slow and turbulent process of alpha disk accretion. If a neutrino fireball were formed, the escape of the thermal radiation, up shifted to gamma rays by the relativistic blue shift of the expanding fireball, is strongly inhibited by the pair opacity and ion dynamical friction. The emission of gamma rays from the relativistic shock (Lorentz factor $\sim 10^3$) of compact supernovae has proven to be extremely small. Here the total energy of $\sim 10^{49}$ ergs is comparable with that required for cosmological GRBs, and the required Lorentz factor is similar (Colgate 1968; Colgate & McKee 1973; Colgate & Petschek 1979). After expansion of the relativistic fireball and the conversion of its initial energy entirely into ultra highly-relativistic nucleons (Lorentz factor of $\sim 10^6$), the reconversion of this nucleon kinetic energy back into hard photons in the ISM, below ~ 100 Mev in photon energy, is extremely inefficient because of the lack of coupling processes. If this energy has to flow through electron processes, then the resulting bremsstrahlung spectrum will be extremely soft, since it must occur in the transparent limit. These difficulties justify a continuing investigation of Galactic GRB models.

THE GALACTIC MODEL

The Galactic model depends upon the formation of Type II or Ib core collapse supernova (SN) in pre-SN binary systems. We select a typical pre-SN binary of a 10 $M_\odot$ companion and a 5 $M_\odot$ pre-SN star where the masses were reversed before mass transfer, and are now in a 10-day orbit with $a = 0.2$ AU. We expect that the NS formed in the collapse and SN explosion of the 5 $M_\odot$ star will be ejected at a much higher than average velocity.

High-velocity NSs are created by asymmetric accretion during the SN process from slowly rotating pre-SN stars. Slow rotation occurs because of tidal locking caused by meridional circulation (Tassoul & Tassoul 1990), as observed by Griffin et al. (1993) at an evolution phase long before mass transfer. Herant, Benz & Colgate (1992) have shown that up to ~ 0.1 $M_\odot$ may be accreted asymmetrically creating a single large plume in the opposite direction and leading to a recoil velocity greater than $\sim 10^3$ km s^{-1}. This accretion is $\sim 10^2$ times more sensitive to rotation than the NS star itself, and thus rapidly-rotating NSs are not likely to attain high velocities, while slowly-rotating ones are likely to reach the escape velocity from the Galaxy.

The mass ejected from a 10 $M_\odot$ secondary star by the impact of the ejecta from the SN is of the order of ~ 1 $M_\odot$ (Wheeler, Lecar & McKee 1975), and $\sim 10^{-5}$ of this mass will be captured into a disk around the NS, based on Nakamura & Piran (1991) and our own calculations. This mass of $\sim 2 \times 10^{28}$ g is more than enough to make the planetoids necessary to explain the observed GRB rate. We estimate one such SN event in the Galaxy per $\sim 10^2$ years, since binary stars are as common as non binaries. Since there are roughly $\sim 10^3$ GRBs per year, then each high-velocity NS+disk system must produce $\sim 10^5$ events in total. Each Galactic halo burst requires $\sim 2 \times 10^{41}$ ergs (less for local disk models) and so the mass accreted in the lifetime of the system must be $\sim 10^{26}$ g. Since the assumed bursts here are $\sim 10^3$ times the Eddington luminosity and the mass required even then is $\sim 0.5\%$ of what is available, this leaves ample mass for the soft gamma-ray repeater phase and later the formation of a dynamically-dominant body required to gravitationally scatter the smaller bodies into the NS, and (mostly) very large orbits.

Soft gamma-ray repeaters (SGR) will occur when the disk is still largely in a gaseous state. The gas in the accretion disk moves inwards due to the alpha viscosity process, and accumulates at the Alfven radius of the NS. Once the pressure in the "thin" disk exceeds the magnetic pressure from the NS, the gas streams rather suddenly onto the poles of the NS along the magnetic field lines, producing a soft GRB. Such events occur before the NS has had a chance to move very far from its SN remnant, which would explain the observed correlation between SGRs and SN remnants (e.g., Kulkarni & Frail 1993). The SGR phase terminates due to the thinning of the disk by mass accretion. This is $\sim 10^3$ years for a $\sim 10^{28}$ g disk and the Eddington limit. It will be longer depending upon the fraction of "on time" of the repeater and the extent to which the emission exceeds the Eddington limit. The shut-off condition for this alpha phase of accretion is when the disk becomes thin enough so that convection and frictional heating, necessary to drive the alpha viscosity turbulence, can no longer take place. This corresponds to a thickness of radiative cooling in one rotation period, and hence predicts a threshold thickness of ~ 3 to 10 g cm^{-2}. The total mass at ~ 1 AU is then $\sim 10^{27}$ g, about $\sim 10\%$ of the original mass, and comfortably close to one earth mass. The accretion disk then cools rapidly and grains condense, the alpha process stops, and the SGR phase ends. There is a long delay before the classical GRB phase begins, due to the need to build up large solid bodies from the grains.

Once cooling has reduced the temperature below the point where molecules and grains can form, a prediction of the time for the formation of a single dynamically-dominant solid body can be made. We assume that the growth of molecules, grains, solid bodies, etc... is proportional to the cross sectional area and a constant virial velocity. For constant average density, the rate of accretion growth for a body of mass M is $dM/dt \propto r^2$, where r is the radius of the body. Since $M \propto r^3$, the total time to accrete to a radius r becomes $t_o \times r/r_o$, where t_o is the initial sticking time of molecules of radius r_o. The value of t_o for molecules at the density of the disk at ~ 1 AU (i.e., ~ 10 g cm^{-2}), and a velocity of $\sim 10^4$ cm s^{-1} is roughly ~ 0.1 s. The solid bodies must grow until they reach the size where the gravitational scattering cross section exceeds the collisional cross section. This occurs at $r \sim 30$ km, or the start of runaway accretion leading to a single dynamically-dominant solid body. The ratio in size of a molecule to ~ 30 km is $\sim 3 \times 10^{15}$, and so $\sim 10^7$ years is required. (This is comparable with the time to form the solar system.) The average mass of the solid bodies not accreted into the dominant one is $\sim 10^{20}$ g. This is the starting point of gamma-ray bursts.

The accretion of solid bodies by NSs (Harwit and Salpeter 1973) was considered in detail by Colgate and Petschek (1981; henceforth CP81) including the distortion, and compression of the body due to the gravity and magnetic field of the NS. In this case, where we have assumed a strong magnetic field for the NS (i.e., $\sim 3 \times 10^{12}$ g), the magnetic pressure, $B^2/8\pi$, is $\gtrsim \rho v^2$ of the accreting matter (where ρ and v are mass density and velocity, respectively) even for a direct impact and including the expected tidal compression ratio of ~ 30 (CP81). Thus the cold degenerate matter will enter the field as unstable sheets (CP81), thus transferring the kinetic energy of impact to both the distortion energy of the magnetic field as well as the thermal energy of the plasma. The distortion energy of the field will be given up by reconnection, producing the very high energy part of the spectrum by $E_{parallel}$ acceleration (Colgate 1992). The thermalized matter will settle along the field lines supported by the NS as a constant entropy atmosphere, maintained by convection and field interpenetration. As pointed out in CP81, the maximum emission rate when the plasma pressure is confined by magnetic field is not the Eddington limit, but instead of the order of $(c/4)(B^2/8\pi) \sim 3 \times 10^{33}$ ergs cm^{-2} s^{-1}, or a luminosity $\sim 3 \times 10^{46}$ ergs s^{-1}. A small fraction of this, $\sim 10^{-5}$, is sufficient to satisfy a Galactic halo model.

OBSERVATIONAL PREDICTIONS

Our GRB model makes several observational predictions. Due to the high velocities of the NSs and the delay before GRBs are produced, the correlation with star formation regions in our Galaxy should be weak. No correlation of GRBs with pulsars is expected, since few of the slowly-rotating high-velocity NSs are observable as pulsars, simply because they are rotating so slowly. A correlation of GRBs with the Galactic disk should eventually become obvious after more GRB events are observed. An instrument that is more sensitive than BATSE may help detect such a correlation, and also GRBs from M31.

The maximum GRB luminosity is $\sim 10^{38}$ erg s^{-1} for local Galactic models, and perhaps $\sim 10^3$ times larger for extended halo models.

The accretion rate of solid bodies by a given NS (and thus the resulting GRB rate) increases rapidly once a dynamically-dominant solid body is formed, probably via a phase of runaway accretion. Afterwards, the GRB rate falls off exponentially as the smaller solid bodies are depleted via being scattered into the NS, and (mostly) very large orbits.

The growth of the solid bodies continues with time as the NS moves away from the Galactic plane. Since, as pointed out above, the radii of the solid bodies grow in proportion to time, the GRB events that occur farther out in the Galactic halo involve, on average, more massive solid bodies. Assuming that the maximum luminosity of GRBs is less than proportional to the mass, because of either an upper-limiting maximum emission rate (e.g., the Eddington limit) or diffusion processes, then GRBs that originate farther out in the Galactic halo have, on average, a longer duration. Since the flux received from distant GRB events is smaller, we predict that longer bursts are, on average, fainter. Such bursts should also be softer, because larger mass implies greater thermalization.

The spectrum of total GRB energies from a given GRB source is like the mass spectrum of small solid bodies that form via accretion. The decrease in the number of bodies as accretion occurs is $dN/dt \propto -Nr^2$, or $dN/dt \propto -N\, dM/dt$. Since $N \propto M^{-1}$, we find $dN/dM \propto M^{-1}$. Since the total energy of a GRB, E, is proportional to M, then $dN/dE \propto E^{-1}$.

Repeat bursts are expected as a result of the tidal fracturing of the small solid bodies during close encounters with the dynamically-dominant body. Repeat bursts are not likely to be of the same total energy, since the masses of the solid bodies depend so strongly on their radii (i.e., $M \propto r^3$). There should be a broad spectrum of repeat times, and the fainter bursts (small solid bodies without enough "oomph" to reach the maximum luminosity) may both precede and/or follow the main burst.

In summary, we believe this model makes many exciting predictions concerning supernova, neutron stars, planetary systems, and gamma-ray bursts.

REFERENCES

1. J. -L. Atteia and J. -P. Dezalay, A&A **274**, L1 (1993).
2. S. A. Colgate, Canadian Journal of Physics **46**, 476 (1968).
3. S. A. Colgate, Proceedings of the Los Alamos Workshop on Gamma-Ray Bursts, edited by C. Ho et al. (Cambridge University Press, 1992), p. 75.
4. S. A. Colgate and A. G. Petschek, ApJ **229**, 682 (1979).
5. S. A. Colgate and A. G. Petschek, ApJ **248**, 771 (1981).
6. S. A. Colgate and C. R. McKee, ApJ **181**, 903 (1973).
7. R. E. M. Griffin et al., A&A **274**, 225 (1993).
8. M. Harwit and E. E. Salpeter, ApJ **186**, L37 (1973).
9. M. Herant, W. Benz and S. A. Colgate, ApJ **395**, 642 (1992).
10. S. R. Kulkarni and D. A. Frail, Nature **365**, 33 (1993).
11. H. Li and C. D. Dermer, Nature **359**, 514 (1992).
12. C. A. Meegan et al., Nature **355**, 143 (1992).
13. T. Nakamura and T. Piran, ApJ **382**, L81 (1991).
14. J. M. Quashnock and D. Q. Lamb, MNRAS , in press (1993a).
15. J. M. Quashnock and D. Q. Lamb, MNRAS , in press (1993b).
16. J. -L. Tassoul and M. Tassoul, ApJ **359**, 155 (1990).
17. V. C. Wang and R. E. Lingenfelter, ApJ **416**, L13 (1993).
18. J. C. Wheeler, M. Lecar and C. F. McKee, ApJ **200**, 145 (1975).

DUAL POPULATION, GALACTIC NEUTRON STAR MODELS OF GAMMA-RAY BURSTS REVISITED

J. C. Higdon* and R. E. Lingenfelter
CASS, University of California, San Diego, La Jolla, CA 92093
*Visiting from Claremont McKenna College, Claremont, CA 91711

ABSTRACT

We investigate in more detail the properties of our recent[1,2] two-population model for gamma-ray bursts. We calculate the gamma-ray burst statistical properties, $\langle V/V_{max}\rangle$, $\langle \cos\Theta\rangle$, and $\langle \sin^2 b\rangle$, as a function of the detection flux threshold for bursts coming from both the galactic disk and massive halo populations. We employ a range of halo models[3,4] infered from the main observational constraints on the large scale galactic structure. We compare our values of $\langle V/V_{max}\rangle$, $\langle \cos\Theta\rangle$, and $\langle \sin^2 b\rangle$, with those measured by BATSE and other detectors. We find that the measured values are consistent with a range of halo distributions, mixed with local disk distributions, which can account for as much as $\sim 20\%$ of the observed BATSE bursts. We also demonstrate, contrary to recent arguments[5,6], that the size-frequency distributions of our dual population models are quite consistent with the BATSE observations.

INTRODUCTION

The gamma-ray burst measurements by the BATSE experiment on the Compton Gamma-Ray Observatory show[7] that their distribution is spatially non-uniform, and isotropic on the sky. Various calculations of models constructed solely from single populations all suggest (e.g. Ref. 8) that such a distribution is consistent either with burst sources at cosmological distances, or with sources in an extended galactic halo population with implausibly large core radii. But such a distribution is not consistent with that expected solely from a galactic disk population of neutron star sources. Nonetheless, a local (< 1 kpc) galactic neutron star origin of at least a fraction of the bursts is strongly suggested[9-11] by observations of absorption lines at 20 to 50 keV, attributed to cyclotron scattering and absorption in 10^{12} G magnetic fields, and by the black body limits on the soft x-ray emission from bursts.

We suggested[1,2], however, that all of the observations can be reconciled, if there are at least two distinct populations of burst sources, consisting of galactic disk neutron stars and galactic halo neutron stars, which have different emission mechanisms, differing by a factor of 10^5, or more, in mean luminosity.

Here we make detailed calculations of the variation of the gamma-ray burst statistical properties, $\langle V/V_{max}\rangle$, a measure of spatial uniformity, versus $\langle \cos\Theta\rangle$ and $\langle \sin^2 b\rangle$, measures of angular isotropy, and cummulative distribution of peak count rates, for combinations of disk and halo populations of burst sources, as a function of increasing detector sensitivity. These calculations show that such combined populations are not only consistent with the BATSE measurements of these properties, but also with those of less sensitive burst detectors, such as KONUS or PVO. These models also allow much smaller and more plausible halo core radii.

DISK & HALO SOURCE POPULATIONS

We consider two-population models consisting of galactic disk and halo populations of gamma-ray burst sources, and we explore the statistical properties of the bursts as a function of the detection flux threshold. We calculate three examples to show the range of possible distributions.

To model the density distribution of the galactic disk sources, we assume for these examples, $n(z,\rho) = n_o \exp^{-\rho/h-|z|/z_o}$, where ρ is galactocentric distance projected along the plane, h of 3.5 kpc is the radial scale along the galactic plane, and z_o of 1 kpc is the scale height transverse to the plane. To model the density distribution of the massive halo sources, we assume, $n(r) = exp(-r/R)^2 n_c[1 + (r/r_c)^2]^{-1}$, where r is the galactocentric distance, r_c is the radius of the uniform core of the halo, and $R = 250$ kpc is an exponential cutoff introduced for purposes of integration rather than the abrupt cutoff assumed in the halo models of Ostriker and Caldwell[3,4]. We consider their two halo core radii[3,4], r_c: 7.5, and 15 kpc, infered[3,4] from the constraints on galactic structure, and also consider a larger radius of 30 kpc to compare with previous halo models. A modest (~ 10) variation in luminosity is used, assuming a power-law luminosity distribution, $dn/dL \propto L^{-2}$ for $L_{min} \leq L \leq L_{max}$, and a halo-to-disk luminosity ratio of 10^5.

Cummulative size-frequency distributions of peak counts, are shown in Figure 1, as a function of the peak count rate calculated for two-population, disk and halo, gamma-ray burst models with the different halo core radii of 7.5 (dotted), 15 (dashed) and 30 (dot-dashed) kpc. These are compared with the BATSE distribution for the complete sample of 191 bursts, from the 1024 ms trigger, excluding overwrites and writeovers. As can be seen, contrary to recent arguments[5,6], the size-frequency distributions of our dual population models are quite consistent with the BATSE observations. Our results also agree well with Smith & Lamb's[12] dual population model with an extended ($r_c = 22.5$ kpc) halo.

The resultant variations of $\langle V/V_{max} \rangle$ and $\langle \cos\Theta \rangle$, as a function of the detection flux threshold, for the dual-population models of the combined galactic disk and halo models of gamma-ray burst sources, are shown in Figure 2 for these different halo core radii. We compare these model calculations with the BATSE[7], KONUS[13] and PVO[14] values of $\langle V/V_{max} \rangle$ of 0.324±0.016, 0.43±0.024 (including the < 1 sec bursts[1]), and 0.46±0.07, and the corresponding 1σ limits[15] on $\langle \cos\Theta \rangle$ of 0.048±0.027, $\sim 0 \pm 0.09$, and $\sim 0 \pm 0.08$. Since positions are known[16] for only 51 of the PVO bursts and only 42 of the KONUS bursts, the 1σ uncertainties on $\langle \cos\Theta \rangle$ are 0.08 and 0.09. We also compare (Fig. 3) the calculated values of

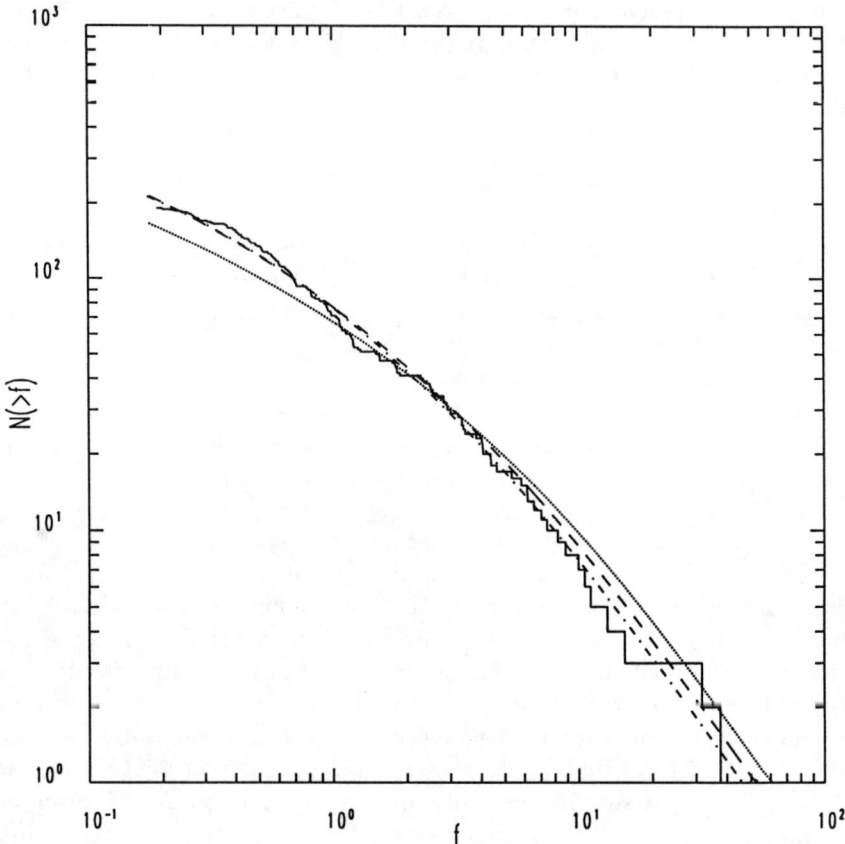

Fig. 1. Cummulative size frequency distributions of peak counts as a function of the peak count rate calulated for two-population, disk and halo, gamma-ray burst models with a disk scale height of 1 kpc and different halo core radii of 7.5 (dotted), 15 (dashed) and 30 (dot-dashed) kpc. These are compared with the BATSE distribution for the complete sample of 191 bursts, from the 1024 ms trigger, excluding overwrites and writeovers.

$\langle \sin^2 b \rangle$, with the BATSE value[15] of 0.329±0.014, ignoring the much larger KONUS and PVO limits of 0.16 to 0.18.

We see from in Figures 2 and 3, the values of $\langle V/V_{max} \rangle$ and $\langle \cos\Theta \rangle$ and $\langle \sin^2 b \rangle$, calculated for two-population models and varying detector sensitivity are consistent with those measured by both the BATSE and the PVO burst detectors for a wide range halo distributions with observationally reasonable[3,4] core radii $\sim$ 15 kpc, mixed with local disk distributions, which account for $\sim$ 20% of the observed bursts. This is contrary to the results of Hakkila et al.[6], who suggest that only negligible disk components are allowable in a dual population models.

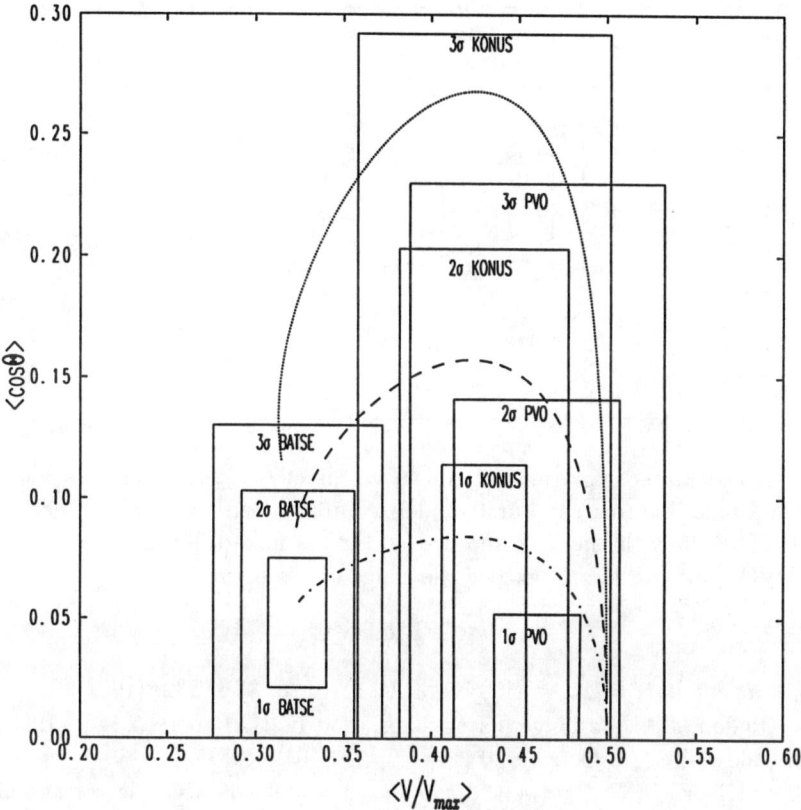

Fig. 2. Calculated variations of $\langle \cos\Theta \rangle$ versus $\langle V/V_{max} \rangle$ for two-population, disk and halo, gamma-ray burst models with a disk scale height of 1 kpc and different halo core radii of 7.5 (dotted), 15 (dashed) and 30 (dot-dashed) kpc. Also shown for comparison are the values measured by the BATSE[7,15], as well as KONUS[13] and PVO[14] burst detectors, whose detection thresholds differ from that of BATSE by factors of the order of 10 and 100, respectively.

SUMMARY

We show that a wide range halo distributions including observationally reasonable[3,4] core radii between 7.5 and 30 kpc, mixed with local disk distributions, which account for $\sim 20\%$ of the observed bursts, are consistent with the spatial and intensity distributions measured by the BATSE, KONUS and PVO burst detectors, whose detection thresholds span a range of the order of 100. Such models can therefore account for the absorption features seen in a fraction of the burst spectra that require nearby (< 1 kpc) galactic neutron star sources.

We thank NASA for support under grants NAG-5 2010 (JCH) and NAGW 1970 (REL) and also the Irvine Foundation (JCH).

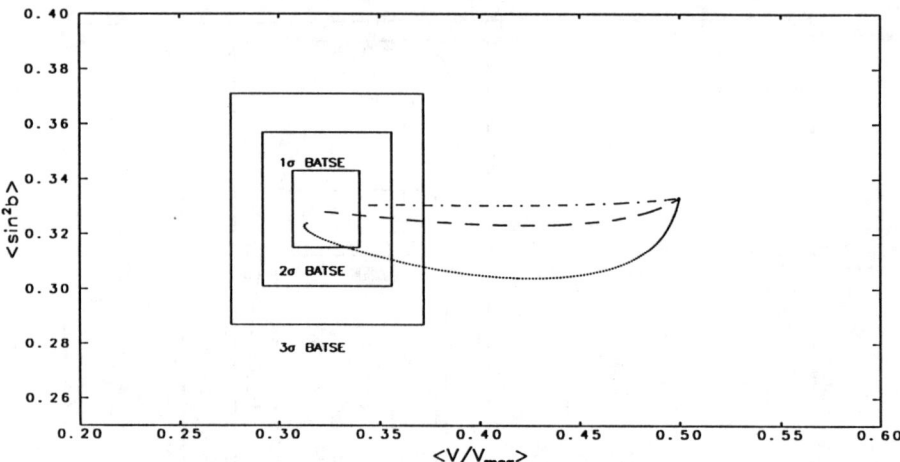

Fig. 3. Calculated variations of $\langle \sin^2 b \rangle$ versus $\langle V/V_{max} \rangle$ for the two-population, disk and halo, gamma-ray burst models compared with the values measured by the BATSE[7,15] with the 1, 2 and 3 σ statistical uncertainties. The 1σ KONUS and PVO limits on $\langle \sin^2 b \rangle$ exceed the range of the figure.

REFERENCES

1. Lingenfelter, R. E., & Higdon, J. C., Nature, **356**, 132, (1992).
2. Higdon, J. C., & Lingenfelter, R. E., 23d ICRC Papers, **1**, 45, (1993).
3. Caldwell, J., & Ostriker, J., Ap. J., **251**, 61, (1981).
4. Ostriker, J., & Caldwell, J., in Large Scale Charateristics of the Galaxy, (Dordrecht: Reidel) p. 441, (1979).
5. Paczynski, B., Acta A., **32**, 1, (1992).
6. Hakkila, J., et al. in Compton Gamma-Ray Observatory, (New York: Am.Inst.Phys.), p. 704, (1993).
7. Meegan, C. A., et al. in Compton Gamma-Ray Observatory, (New York: Am.Inst.Phys.), p. 681, (1993).
8. Paciesas, W. S., & Fishman, G. J., eds. Gamma-Ray Bursts, (New York: Am.Inst.Phys.), 427 pp, (1992).
9. Higdon, J. C., & Lingenfelter, R. E., Ann. Rev. A. & Ap., **28**, 401, (1990).
10. Lamb, D. Q., Wang, J. C. L., Wasserman. I. M., Ap.J., **363**, 670, (1990).
11. Murakami, T., et al. Nature, **350**, 592, (1991).
12. Smith, I. A., & Lamb, D. Q., Ap. J., **410**, L23, (1993).
13. Mazets, E. P., et al., Ap. Space Sci., **80**, (1980).
14. Hartmann, D., et al., in Gamma-Ray Bursts, eds. C. Ho, R. Epstein & E.E. Fenimore, (Cambridge Univ. Press), p. 45, (1992).
15. Briggs, M. S., et al., in Compton Gamma-Ray Observatory, (New York: Am.Inst.Phys.), p. 686, (1993).
16. Atteia, J. L., et al., Ap. J. Supp., **64**, 305, (1987).

HALO POPULATION OF QUARK NUGGETS AS γ-BURSTERS : THE GLOW OF DARK MATTER ?

J.E.Horvath

Dept. of Space Phys.& Astron., Rice U., P.O.Box 1892, Houston, TX 77251

and Instituto Astronômico e Geofísico, Universidade de São Paulo

Av. M.Stéfano 4200, Agua Funda (04301-904) São Paulo SP, Brazil

ABSTRACT

New data from the BATSE experiment onboard the Compton Observatory pose a challenging situation for model-building of the gamma-ray burst sources. Although not strictly required, it is possible that "exotic" objects like quark matter nuggets provide a substantial fraction of the observed events. It is shown in this work that an extended halo nugget population may produce gamma-ray bursts consistent with a subset, and perhaps all, of the observed sample. The mechanism involves a breaking of chiral symmetry of the nugget matter with an energy release of few MeV/baryon which generates a highly relativistic dense e^+e^- plasma from $\nu\bar{\nu}$ interactions. The model attempts to relate the gamma-ray burst sources to a specific cold dark matter candidate. Thus, gamma-ray bursts could be interpreted as the signature (glow) of the dark matter.

INTRODUCTION

The recent results [1] on gamma-ray bursts distribution as measured by the BATSE experiment onboard the Compton Observatory are exciting news for high-energy astrophysics. As discussed elsewhere (see for example Ref.2) the data support an isotropic but inhomogeneous distribution of the sources; contrary to earlier expectations of a concentration towards the galactic disk. The latter idea arose mainly from the strong consensus that neutron stars must be involved in the production of such energetic phenomena. Although there is some evidence that a subset of bursts show spectral features consistent with that hypothesis, it has been also recognised[3] that at least $\sim 75\%$ of the observed bursts do not need to be related to compact stars. Furthermore, it has been suggested that it is unlikely that a single class of bursts can explain the BATSE data[4], thus we are left with the possibility that at least a fraction of the bursts originate from truly "exotic" sources, which could be located in a galactic halo[5] or at extragalactic distances[6,7]. We shall sketch below the hypothesis of primordial quark nuggets being sources of gamma-ray bursts.

Primordial quark nuggets are hypothetical substellar-mass dark matter candidates, presumably produced at the QCD energy scale. They are a physical consequece of the intriguing "strange matter" scenario[8] (hereafter SM ; see also Ref.9 for a pioneer work along these lines), in which a cold form of the quark gluon plasma (with strangeness/baryon~ 1) may be more stable than the ordinary nucleons we are made of. While the nuggets could be indeed produced at high temperatures in the early universe, considerable controversy[10,11,12] arose about their survival against evaporation and/or boiling at temperatures below the QCD scale $T_{QCD} \sim 100$ MeV. The most recent results[13,14] including detailed physics have yielded that nuggets containing more than $\sim 10^{39-40}$ baryons (but obviously less than the causal limit $A_C = 10^{49}(100\,MeV/T_{QCD})^2$ at the epoch of formation) as present relics. We stress that there has been considerable work to identify and explore other forms of "strange matter" which should play an important role in the theory[15,16,17,18,19]. The crucial point is that the SM picture is based on a Fermi gas model, and therefore few-quark correlations, which may hide a rich phase structure, cannot be revealed in this approach. From the physical point of view it is more likely that if SM exists, its phase structure in the $T - \mu$ diagram must show a diversity of structures in complete analogy with the ordinary nuclear physics. However, the decay rate SM → any phase is not a simple matter to address. There may be physically forbidden or may require, for example, substantial compression (the primordial nuggets have not been formed under external pressure). Indeed, we shall see below that we are most likely dealing with a collective phenomenon and this complicates the estimation of the decay rate.

Provided some lower energy state (phase) than SM exists, primordial halo nuggets at typical distances of tens of kpc should, from time to time, undergo spontaneous transitions to the latter which render a huge amount of energy. This expectation follows from the binding energy per baryon unit, which is in the ballpark of nuclear binding energies $\sim$ several MeV. Note that from the observational requirement of having ~ 1 burst/day as suggested by the existing data, and assuming that the nuggets constitute the massive dark halo (which gives the total number of them), we can calculate the probability of decay in a Hubble time for a given nugget which turns out to be $P_d \simeq 10^{-8}$, consistent with a strongly suppressed decay process. Hereafter we shall discuss the possible generation of gamma-ray bursts from these events.

PHASE TRANSITIONS IN SM

As a specific example of unexpected phases, the possibility of stable "strange hadronic matter" (SHM) has been recently suggested[19] in which bulk aggregates of $(n, p, \Lambda, \Xi^0, \Xi^-)$ can be bound by $\simeq 20$ MeV/baryon. Its existence is not at odds with SM nuggets, on the contrary, it can be viewed as the hadronic analogue of SM in which the chiral symmetry is broken. However, for such a transition SM → SHM to occur, the quarks must arrange themselves in the right quantum number combinations, giving an unknown strong suppression. Yet another possibilities are quark matter crystallisation[20] which may happen whenever a periodic lattice is energetically preferred to the disordered state, or the existence of a self-bound correlated diquark state[21]. More work needs to be done before we can say something more quantitative on these hypothetic phases and we shall not refer to them further.

Let us address the specific case of a nugget decay into bound SHM. For the sake of definiteness we consider a nugget of $A_N = 10^{49}$ and $\rho_N = 2\,\rho_o = 5.4 \times 10^{14}$ g cm^{-3}, implying a radius $R_N = 2 \times 10^3$ cm. Initially, the nugget may be taken as being in equilibrium with the photon background at a temperature T_o. When the transition takes place, releasing an energy $\Delta\varepsilon = \alpha$ MeV/baryon; the nugget temperature increase can be found from the expression

$$\Delta\varepsilon\,\eta = \int_{T_o}^{T_i} c_V\,dT. \qquad (1)$$

where η is the inverse of the volume filled by one baryon number unit in the SHM, assumed to be $(1\, fm)^3$. Since the specific heat c_V is largely dominated by the degenerate fermions, we shall approximate it by the free Fermi gas expression $c_V \sim 10^{22}(T/10^{11}K)$ erg cm^{-3} K and integrate equation (1) to give after inverting it

$$T_i = 1.7\, \alpha^{1/2} \times 10^{11}\, K. \qquad (2)$$

such a high temperature is, of course the natural result of the release of an energy amount of the order of nuclear binding energies.

For the given conditions of density and temperature, the energy loss (and thus the cooling history of the nugget) is expected to be dominated by neutrino emission. However, an important difference between this model and the cooling of a neutron or strange star[22,23] is that the mean free paths of all neutrino flavours at temperatures like those of equation (2) are $\geq R_N$, so rather than diffusing out the neutrinos escape promptly. The cooling equation is

$$L(T) = -\frac{4}{3}\pi R_N^3 \times c_V(T) \frac{dT}{dt} \qquad (3)$$

where $L(T) = \xi \times 10^{36}\, (T/10^{11}K)^8 \times \frac{4}{3}\pi R_N^3$ erg s^{-1} parametrises the neutrino luminosity in terms of typical parameters of a modified URCA value[24]. A simple integration gives the analytic cooling law

$$T_{11}(t) = \left(\frac{t}{\tau} + C\right)^{-1/6} \qquad (4)$$

where $\tau = 1.6\, \xi^{-1} 10^{-4}\, s$, $C = T_{i11}^{-6}$ and the temperatures have been scaled to the initial value $T_{11} \equiv (T/10^{11}\, K)$.

Due to the enormous neutrino density; a fraction of ν and $\bar{\nu}$ can annihilate and produce an ultrarrelativistic e^+e^- plasma[25]. The luminosity going into e^+e^- scales as a high power of the temperature as inferred from phase space considerations. Inside the nugget $\nu\bar{\nu} \to e^+e^-$ is strongly inhibited by the e^- degeneracy and the efficiency of this process drops very quickly far away from the nugget. Appropiate expressions have been calculated in Ref.23 for the strange matter case and Ref.25 for the nuclear matter case. Since $L_{e^+e^-}$ depends very strongly on T_{11} the precise value of the numerical coeficcient which sets the luminosity scale is quite irrelevant as long as it does not change dramatically in the strange hadronic matter case. We assume

$$L_{e^+e^-} = 2 \times 10^{41}\, T_{11}^9 \left(\frac{R_N}{10^3\, cm}\right)^3\, erg\, s^{-1}. \qquad (5)$$

Combining equations (4) and (5) we get the total energy injected in pairs outside the nugget

$$E_{e^+e^-} = \int_0^\infty L_{e^+e^-}\, dt = 4 \times 10^{38} \left(\frac{R_N}{10^3\, cm}\right)^3 \left(\frac{\tau}{10^{-3}\, s}\right) T_{i11}^3\, erg. \qquad (6)$$

which shows explicitely the dependence on the relevant parameters. To produce bursts stronger than, say, $\sim 10^{-5}$ erg cm^{-2} at a few tens of kpc distances we should have $E_{burst} \sim E_{e^+e^-} \geq 10^{41}$ erg, which is easily met for $T_{i11} \geq 3$ (or equivalently $\alpha \sim 5$). The injection region[23] is essentially a spherical shell of thickness $\sim 0.3 R_N$ and the time scales for the injection of 90% and 99% of the total energy $E_{e^+e^-}$ are $t(90) = 16\, T_{i11}^{-6}$ ms and $t(99) = 1.6\, T_{i11}^{-6}$ s respectively. As expected, these are quite short and sensitive to the initial value T_{i11}. Had we assumed that the nugget rearranges without substantial modification of the SM thermal and emission properties, the figures would have been $T_i = 0.5\,\alpha^{1/2} \times 10^{11}\, K$; $E_{e^+e^-} = 1.5 \times 10^{38}\, (R/10^3\, cm)^3\, (\tau/10^{-3}\, s)\, T_{i11}^5$; $t(90) = 24\, T_{i11}^{-4}$ ms and $t(99) = 0.18\, T_{i11}^{-4}$ s ; which gives an idea of the range of conditions for the scenario to work.

HALO FIREBALLS AND GAMMA-RAY BURSTS

The fate of a pure e^+e^- fireball as the one described above has been addressed by many authors[6,26,27,28]. An adiabatic expansion occurs in which random thermal energy is converted into bulk kinetic energy. Gamma radiation can only escape only when the fireball becomes optically thin at $T_{esc} \sim 20$ KeV. Since the radiation will be blueshifted to the observer, the typical energy of the photons will be of the order of the initial plasma temperature. The latter can be estimated by knowing the energy density $\epsilon = E_{e^+e^-}/V$; we find that since the injection region is small, the initial plasma temperature T_o is a significant fraction (about 0.2) of T_i. This in turn implies a highly relativistic flow characterised by a Lorentz factor $\Gamma \sim T_o/T_{esc} \sim$ few $\times 10^3$. A more detailed analysis shows[28] that the emerging quasi-blackbody spectrum is indeed a blend of several shells at slightly different temperatures, supporting the idea of a possible hard-to-soft evolution which may be detected in actual bursts.

Our results suggest that if quark nuggets are responsible for gamma-ray bursts, featureless, short events are to be expected as the most "pure" manifestation of their presence (i.e. coming from nuggets not embedded in a significantly clumped region of interstellar material). However, and due to the spatially compact nature of these objects, larger plasma temperatures T_o, and thus higher Lorentz factors than the previously discussed strange-star case must be expected, opening the possibility that the interaction with the interstellar material[29,30,31] is the ultimate cause for the complex profiles and longer timescales hiterto observed. Note that the energy injection time scale will not determine the duration of the burst if this is the case[31] in which case the values of $t(90)$ and $t(99)$, are less important than the Γ factor of the expanding e^+e^- plasma. However, one should remember that halo fireballs might not satisfy the X-ray paucity constraint as shown by Mészáros in these Proceedings.

It would be interesting to isolate and characterise a "pure" burst sub-sample (which may be actually quite small) to see to what extent the model is supported. Another important test would be provided by a detected burst excess in the direction of M31, a possibility that may be near the present experimental sensitivities. Pure dark matter halos behaving asymptotically as r^{-2} may conflict with the burst statistics, but a faster decay is easily compatible with the data[32] and in any case it should be remarked that r^{-2} laws are not required to explain the dynamics of our galaxy for distances $\gg 50$ kpc, so there is plenty of space for a dark halo cutoff well before distances which merge it with the one of M31.

Perhaps the most exciting possibility would be the detection of the nuggets by the MACHO search experiments[33,34] but this goal will require a careful multi-exposure procedure[35] which has not been yet attempted, although it is within the present technical capability.

ACKNOWLEDGEMENTS

It is a pleasure to acknowledge F.C.Michel and I.Smith for useful discussions and remarks on a draft version of this work. This work has been done during a visit to Rice University made possible through a PosDoctoral Fellowship of the Brazilian Foundation FAPESP (São Paulo, SP).

REFERENCES

1. C.Meegan et al., Nat **355**, 143 (1992).
2. K.Hurley, Nat **357**, 112 (1992).
3. K.Hurley, Nuc.Phys. **10B**, 21 (1989).
4. R.E.Lingelfelter and J.C.Higdon, Nat **356**, 132 (1992).
5. M.C.Jennings, High Energy Trans. in Ap. (ed.S.Woosley (AIP,NY), 1984).
6. B.Paczyński, ApJLett **308**, L43 (1986).
7. T.Piran, ApJLett **389**, L45 (1992).
8. E.Witten, PRD **32**, 242 (1984).
9. A.Bodmer, PRD **4**, 1601 (1971).
10. C.Alcock and E.Farhi, PRD **32**, 1273 (1985).
11. J.Madsen, H.Heiselberg and K.Riisager, PRD **34**, 2947 (1986).
12. C.Alcock and A.V.Olinto, PRD **39**, 1233 (1989).
13. K.Sumiyoshi and T.Kajino, Nuc.Phys. **24B**, 80 (1991).
14. J.Madsen, Nuc.Phys. **24B**, 84 (1991).
15. R.L.Jaffe, PRL **38**, 195 (1977).
16. F.C.Michel, PRL **60**, 677 (1988).
17. F.C.Michel, ApJLett **327**, L81 (1988).
18. E.P.Gilson and R.L.Jaffe, PRL **71**, 332 (1993).
19. J.Schaeffner et al., PRL **71**, 1328 (1993).
20. F.Iachello,W.D.Langer and A.Lande, unpublished (1975).
21. J.E.Horvath, Phys.Lett. **242**, 412 (1992).
22. R.F.Sawyer and A.Soni, ApJ **230**, 859 (1979).
23. P.Haensel, B.Paczyński and P.Amsterdamski, ApJ **375**, 209 (1991).
24. J.Goodman,A.Dar and S.Nussinov, ApJLett **314**, L7 (1987).
25. B.L.Friman and O.V.Maxwell, ApJ **232**, 541 (1979).
26. J.Goodman, ApJLett **308**, L247 (1986).
27. A.Shemi and T.Piran, ApJLett **365**, L55 (1990).
28. T.Piran,A.Shemi and R.Narayan, unpublished (1993).
29. G.Cavallo and M.Rees, MNRAS **183**, 359 (1978).
30. P.Mészáros and M.Rees, unpublished (1993).
31. M.Rees and P.Mészáros, MNRAS **258**, 41 (1992).
32. I.A.Smith and D.Q.Lamb, ApJLett **410**, L23 (1993).
33. R.Aubourg et al., talk given at 2nd DEAC Meeting, Obs. de Meudon, France (1991).
34. R.Bennett et al., talk given at 12th TEXAS/PASCOS Symposium (1992).
35. A.Gould, ApJ **392**, 442 (1992).

NEARBY NEUTRON STARS AS THE SOURCES OF THE GAMMA-RAY BURSTS

Wolfgang Kundt and Hsiang-Kuang Chang
Institut für Astrophysik der Universität Bonn, Germany

ABSTRACT

We interpret the puzzling γ-ray bursts as emitted by cooling sparks from near the surface of spasmodically accreting, old neutron stars[1]. Their spiky, anisotropic radiation is oriented w. r. t. the Galactic disk via interstellar accretion, whose orbital angular momentum tends to counteralign with the Galactic spin; in this way, larger source numbers in directions of the Galactic disk are compensated by smaller beaming probabilities, resulting in a near-isotropic arrival distribution, as observed by BATSE. The source distances range between 10 pc and 500 pc. Their radiated energies are of order 10^{35} erg, corresponding to accreted clumps (blades) of typical mass 10^{15} g per burst. Magnetic surface field strengths range between 10^{10} G and 10^{13} G, perhaps somewhat weaker than those of newborn neutron stars. In this communication, we explain why we are still convinced that the soft repeaters have distances $\lesssim 50$ pc, and how our model can be updated to encorporate the recently observed hard tails ($E \lesssim 10\,\text{GeV}$ for $t \lesssim 10^2\,\text{s}$).

PLEA FOR A GALACTIC SOURCE DISTRIBUTION

At this (Huntsville) workshop, a Galactic origin of the γ-ray bursts was dismissed for no other reason than the distribution of their ($\gtrsim 700$) arrival directions and fluxes: it looks like a shell distribution, not like that of a (warped) disk. In our recent communication[1] we have argued, and numerically simulated, that such a reasoning is inconclusive: there are plausible radiation patterns, oriented w.r.t. the Galactic disk, for which both the subset of strong bursts, and the set of all bursts appear isotropic: a shell distribution can be mimiced by a beaming fraction that decreases with decreasing (Galactic) latitude. Our model assumes that the bursts are emitted during spasmodic, clumpy accretion by (old) neutron stars whose accretion disks are collected in transit from Galactic clouds. It predicts isotropic arrival directions at the level of $|(dN/d\cos\theta)/\langle dN/d\cos\theta\rangle - 1| \gtrsim 10\%$; anisotropy should get noticeable for observed burst numbers N in excess of $10^{3.3\pm0.3}$.

Galactic neutron stars have been the favoured model for many years, because neutron stars are sufficient in (i) expected number and (ii) accretion flux and total power, (iii) just small enough to make their appearance as thermal X-ray emitters, (iv) compact and hard enough to allow for hard radiation, and short variability timescales ($\lesssim 10^{-3.7}$ s), because (v) spectral peaks (of $E^2 \dot N_E$) at $E = 10^{\pm 0.5}$ MeV are reminiscent of transient $e^{\pm}$-pair production, as are (vi) occasionally reported emission lines at (420±70) keV, and because (vii) occasional absorption lines and their first harmonics, above 20 keV, look like cyclotron-scattering features in magnetic fields of strength $\gtrsim 10^{12}$ G. Moreover, (viii) the long-duration bursts look like superpositions of short-duration events, (each rising faster than decaying, whilst the spectrum softens), *i.e.* they look like caused by swarms of accreted clumps; and (ix) the hard tails discussed in the last section look like delayed magnetospheric reactions.

If the γ-ray bursts came from beyond the Galactic disk, we would have to worry why we had no signature yet of this expected Galactic accretion phenomenon. Why do we not detect its radiation?

THE SOFT REPEATERS

Only three out of several hundred bursters are known with certainty to repeat; do they form a different class? Their dynamic range nearly exhausts the surveyed range (in energy): their brightest events lead the distribution, whereas their (soft) repetitions can be some 10^3 times fainter. These properties are expected of the nearest sources of a uniform class (of distance $\lesssim 50$ pc): of more distant events, we are too far to see the fainter (and softer) repetitions.

This interpretation is at variance with reported extended X-ray and radio detections, which claim associations with distant ($\gtrsim 17$ kpc) "middle-aged plerions"[2,3], though not at all at variance with associations with X-ray point sources[4,5,6] – which may be nearby neutron stars – and with nearby pulsar nebulae[3,7,8]. For SGR 1806-20, the recently reported associations suggest a large (> 10 kpc) $\Sigma - d$-distance but a small (< 1 kpc) X-ray distance (if due to thermal radiation from a neutron star), and IR distance (if the two reddened stars are of low mass, consistent with a PSR nebula). And SGR 1900+14 projects onto a radio-loud area. Both detections owe their convictive power to the first 'detection', the coincidence of the 5 March 79 event 0526-66 with SNR N49 in the LMC, whose importance was denied, e.g., by Schmidt[9], Colgate and Petschek[10], and by Zdziarski[11]: At the distance of LMC, the burst's brightness (of $S \leq 10^{-2.7}$ erg cm^{-2} s^{-1}) would have corresponded to an (isotropic) power of $L \leq 10^{44.8}$ erg s^{-1}, brighter than the brightest stellar sources by some 10^6. Even if this enormous luminosity were simulated by (excessive) beaming, we should ask ourselves why there have not occurred some 30 similar (and similarly beamed) events in the Galaxy (with its 30 times larger mass). Moreover, at the distance of LMC, SGR 0526-66 would have had a pair-production opacity in excess of 10^6, inconsistent with its highly nonthermal spectrum. A distance of $\lesssim 50$ pc would revise the opacity downward towards tolerable values.

We judge a coincidence in sky projection less convincing than the luminosity-function argument, in conjunction with fundamental physics which would have to be evaded by extreme (and implausible) relativistic beaming. Once we consider the reported 2.5 projections as either accidental, or of ill-determined distance, the SGRs offer themselves as the expected inner edge of the spatial source distribution.

HARD TAILS

Our model (of spasmodically accreting neutron stars) predicts spectra which cut off at photon energies of 0.5 GeV, because they are controlled by gravitational infall velocities. This prediction is (so far) in accord with the observed burst spetra, which even soften with time as expected, but is at variance with two or three recent EGRET detections of photon energies reaching, or exceeding, 10 GeV after a delay time of $\lesssim 10^2$ s, (starting with 910503, [B.E. Schaefer, private communication]). These 'hard tails' correspond to flat burst-energy spectra, $E^2 \dot{N}_E$ = constant, i.e. involve comparable energies to their triggering events; see figure 1.

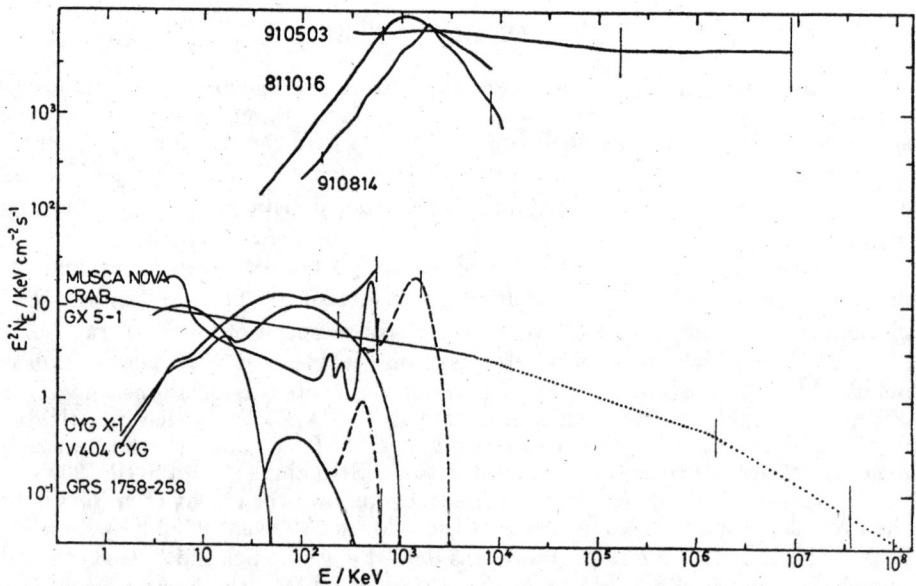

Figure 1. Spectra of three γ-ray bursts (named after their date of occurrence), $E^2 \dot{N}_E$ versus E, together with the spectra of a few other hard and soft X-ray sources, some of them at outburst (broken) and at a quiet epoch (full). Hardest are the Crab pulsar (dotted) and its 'nebula' (full). Cyg X-1, Nova Muscae 1991, and V404 Cyg are black-hole candidates, whose properties are often reminiscent of neutron-star sources[12]. The figure is reproduced from Kundt and Chang[1], but extended to include the recent EGRET data at $E \leq 9.6$ GeV.

Apparently, a mechanism is at work in the bursters that causes high-energy emission to (escort and) follow transrelativistic emission, at comparable total energy (if similarly beamed), with a delay time between 0 and 10^2 sec which can exceed the onset time (of $\lesssim$ msec) by more than a factor of 10^5.

In our model, the magnetically confined 'blades' scratching the neutron star's surface will eventually come to rest, braked both by crashes and by the magnetosphere, and expand to neutron-star surface densities whilst the ditches they have carved fill in from the sides, forced by the star's gravity. These interactions and rearrangements will temporarily distort the magnetosphere near the orbit plane(s), which wants to relax to its pre-burst configuration. We estimate the (diffusive) magnetospheric relaxation time to be of order

10^2 sec, and expect it to involve flares of significant electric voltage, reaching $\phi \approx (\mu\Omega/Rc)(\triangle R/R) = 10^{12}$ V $\mu_{30}\Omega_0(\triangle R/R)_{-4}$ for stars of magnetic moment $\mu = 10^{30}$ G cm^3, spin angular frequency $\Omega = 10^0$ s^{-1}, and ditch width $\triangle R$ of order 1 m [e.g., Kundt and Schaaf[13]]. Such flares give rise to relativistic pair formation, preferentially along fieldlines in the accretion plane, and the pairs collide with the accretional X-ray photons, resulting in γ-rays of energy $e\phi \lesssim 10^{12}$ eV for $\mu_{30}\Omega_0(\triangle R/R)_{-4} \lesssim 1$. Such hard γ-rays are (transiently) reconverted to pairs if their energy E exceeds GeV/$B_{9.3}$, where $B_{9.3} := B_\perp/10^{9.3}$G is the transverse magnetic field component (encountered by the photon) in units of 2×10^9 G. 10 GeV photons want $B_\perp < 10^{8.3}$ G, a condition which should be satisfied at distances beyond a few stellar radii, depending on the star's magnetization.

Our model therefore allows for energetic hard tails, of photon energy less than some 10^3 GeV, and more abundantly less than some 10 GeV, because hard photons require strong magnetic fields for the pair-formation voltage but weak magnetic fields for escape. Responsible for the delayed energy release – and for the upgraded photon energy – is the transiently distorted magnetosphere.

ACKNOWLEDGEMENTS

We thank Brad Schaefer and Jochen Greiner for informative news, and Ravi Subrahmanyan for the manuscript.

REFERENCES

1. W. Kundt, H.-K. Chang, Ap&SS **200**, 151 (1993).
2. S.R. Kulkarni, D.A. Frail, Nature **365**, 33 (1993).
3. S.R. Kulkarni, D.A. Frail, N. Kassim, T. Murakami, G. Vasisht, IAU Circ., No. 5879 (1993).
4. R.E. Rothschild, R.E. Lingenfelter, F.D. Sewart, O. Vancura, in: 1st Compton γ-ray Observatory, AIP Proceedings **280**, 808 (1993).
5. Y. Tanaka, IAU Circ., No. 5880 (1993).
6. B.A. Cooke, IAU Circ., No. 5883 (1993).
7. W. Kundt, H.-K. Chang, Ap&SS **193**, 145 (1992).
8. S.R. Kulkarni, K. Matthews, G. Neugebauer, N. Reid, T. Soifer, G. Vasisht, IAU Circ., No. 5883 (1993).
9. W.K.H. Schmidt, Nature **271**, 525 (1978).
10. S.A. Colgate, A.G. Petschek, ApJ **248**, 771 (1981).
11. A.A. Zdziarski, A&A **134**, 301 (1984).
12. W. Kundt, D. Fischer, J. Ap. Astr. **10**, 119 (1989).
13. W. Kundt, R. Schaaf, Ap&SS **200**, 251 (1993).

BEAMED GAMMA-RAY BURSTS FROM THE GALACTIC HALO: MODEL COMPARISONS WITH BATSE DATA

Hui Li*, Robert Duncan† & Christopher Thompson‡

*Dept. Space Physics & Astronomy, Rice University, Houston, TX 77251
†Dept. Astronomy, University of Texas at Austin, Austin, TX 78712
‡CITA, 60 St. George Street, Toronto, Ontario M5S 1A1, Canada

ABSTRACT

We consider scenarios for gamma-ray bursts (GRBs) from high-velocity neutron stars which are born in the galactic disk and escape into an extended galactic halo.[1,2] We found previously[3] that the beaming of bursts along magnetic field lines, as expected on physical grounds, has a dramatic effect on the observable GRB distribution if the beaming (magnetic) axis is approximately aligned with the recoil velocity of the star. In this paper we give new, detailed comparisons of model results with BATSE observations. Beaming greatly reduces the number of bursts observable from sources born in neighboring galaxies such as M31. We describe some testable predictions from the model.

INTRODUCTION

The BATSE experiment has revealed a nearly isotropic but inhomogeneous distribution of gamma ray bursts (GRBs).[4] One possible location for the burst sources is in an extended galactic halo. Here we consider the idea[1,2] that GRBs are emitted by a population of high-velocity, bursting stars which are born in the galactic disk with recoil velocities $\mathbf{V_r} > 800$ km s^{-1}, large enough to propel them them out of the Galaxy. The simplest such models do not satisfy the Pioneer Venus Observer (PVO) constraint that the bright burst counts have a distribution consistent with a uniform density of sources; furthermore, they predict a large dipole anisotropy $\langle \cos\theta \rangle$ toward the galactic center. One possible way to solve these problems is for the bursting stars to exhibit a "delayed turn-on" in their bursting rate,[2] effectively reducing the density of nearby sources. An alternative solution,[3] that we will concentrate on here, involves burst beaming.

Figure 1 illustrates the basic idea: gamma-rays are produced only within indicated cones (of angular radius ϕ) about the star's magnetic axis $\pm\vec{\mu}$, which we require in this model to be nearly aligned with the peculiar stellar recoil velocity, $\mathbf{V_r}$. (Reasons for this alignment are discussed below.) This recoil is acquired in a random direction when the burster is born in the disk of the Galaxy (e.g., at point **A** in Fig. 1). However, as the stars sails into the halo, (e.g., reaching point **B** and beyond) the line of sight from Earth (**E**) to the star becomes increasingly aligned with $\mathbf{V_r}$. Thus more and more escaping bursters become visible at larger distances r from the galactic center. The fraction of all bursters which are detectable increases with distance as the transverse area of the beaming cone in the Galaxy, namely as $\sim r^2$. This tends to cancel the radial dependence of the density of escaping stars, $n \propto r^{-2}$, out to a radius $\sim R_o/\phi$, where $R_o \sim 8.5$ kpc is a galactic disk dimension. At distances larger than this core radius, all bursters are detectable at Earth, and the $n \propto r^{-2}$ free-streaming density trend prevails, accounting for the "boundedness" (i.e. $\langle V/V_{max}\rangle < 0.5$) found by BATSE.

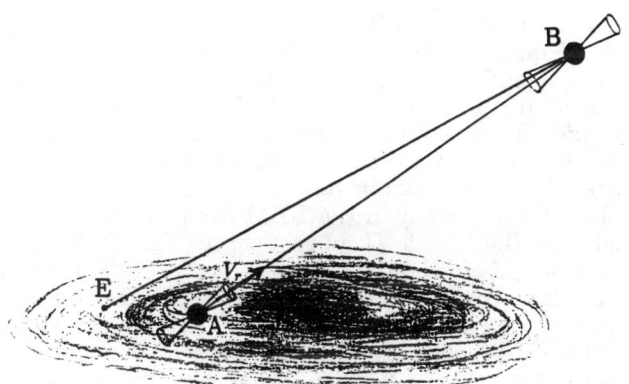

Figure 1 A schematic plot of a burst source leaving the Galaxy (shaded region) with GRB emissions that are beamed within a cone along its recoil velocity. Point **E** is the location of Earth.

PHYSICS OF BURST BEAMING AND ALIGNMENT

Before discussing numerical models, we briefly describe physical mechanisms which could give GRB beaming aligned with the recoil velocity. Hard gamma rays which emerge from high-B environments tend to beamed along field lines because of the transverse pair-creation opacity, and because gamma rays produced by such mechanisms as curvature radiation and Compton upscattering are strongly beamed. Thus the GRB beam axis is likely to be parallel to the stellar magnetic moment $\pm\overrightarrow{\mu}$. We then require that $\overrightarrow{\mu} \parallel \pm\mathbf{V_r}$.

One clear motivation for this kind of alignment comes from the "magnetar" model, although it might also occur in other physical contexts. Magnetars are discussed in ref. 3, 5, 6, and 7; with applications to soft gamma repeaters detailed in ref. 8. Here we simply note that any recoil mechanism which imparts impulses to the stellar surface with coherence time longer that the rotation period of the star yields $\mathbf{V}_r \parallel \pm\mathbf{\Omega}$. In magnetars this mechanism is probably magnetically-induced anisotropic neutrino emission during the first ~ 30 s of the star's life.[5,6] (Several alternative recoil mechanisms for magnetars [cf. ref. 5] also would give $\mathbf{V}_r \parallel \pm\mathbf{\Omega}$.) If we furthermore have $\mathbf{\Omega} \parallel \pm\overrightarrow{\mu}$ to within the burst beaming angle $\phi \sim 20°$, then the alignment requirement of the HBM is satisfied. In fact, near-alignment of the rotation and magnetic axes naturally occurs in large-scale dynamos, as in the familiar cases of stars and planets. Such dynamos produce magnetar fields. For more details about beaming and alignment, see ref. 3 and 9.

MODEL COMPARISONS WITH BATSE

The particular version of "halo beaming model" (HBM) we analyze here has the following simple properties: (1) bursters are born at positions distributed like young Pop. I stars in the galactic disk, (2) with randomly directed recoils $V_r = 1000$ km/s; (3) they emit GRBs at a constant rate, with (4) constant luminosity, and (5) the gamma ray emission is beamed parallel and anti-parallel to $\mathbf{V_r}$, within an angular radius $\phi = 20°$.

We ran a Monte Carlo code which simulated the trajectories of more than 4000 neutron stars in the potential of our Galaxy (within 400 kpc).[2,3] Model results for angular statistics and $\langle V/V_{max}\rangle$ are shown in Fig. 1a̅ of ref. 3. These curves give an adequate fit to the total body of BATSE data for a significant range of BATSE sampling depths $D \geq 100$ kpc. (We will show that external galaxies put essentially no constraints on D in the HBM.)

We now extend the analysis of ref. 3 in two ways. Rather than use the $\langle V/V_{max}\rangle$ statistic, we make a detailed comparison between HBM and BATSE burst brightness (i.e., peak photon flux F) distributions. Secondly, since the largest deviations from isotropy in the HBM occur within subsets of GRBs selected for brightness (i.e., small-D subgroups; see Fig. 1(a) in ref. 3), we investigate whether analysis of brightness subsets in the BATSE catalog constrains the model more strongly than the total body of BATSE data does.

With these objectives in mind, we filtered the Monte Carlo HBM results to simulate BATSE sky coverage and detection incompleteness, using tables provided by BATSE.[10] Figure 2 shows cumulative log $\mathcal{N}$–log F_{1024} data from the first BATSE catalog (thick solid curve) compared with the best-fit HBM result (thin curve). We plot only long ($T_{90} > 2$ s) bursts here (cf. ref. 11), but we found equally good fits when just short bursts, or all the bursts, were used (see ref. 9 for graphs). The peak photon emission rate $\dot{N}$ was adjusted to make the fit of Fig. 2, giving a best-fit "standard candle" value $\dot{N} = (4.0 \pm 1.1) \times 10^{46}$ photons s^{-1}. This implies a BATSE sampling depth $D \approx 140$ kpc and a peak luminosity $\dot{E} \approx 6 \times 10^{39}$ erg s^{-1}. Note that the values of $\dot{N}$ and $\dot{E}$ quoted here include the beaming reduction factor; equivalent isotropic values would be larger by $(1 - \cos\phi)^{-1} = 17$.

In Fig. 3, we plot measures of angular (an)isotropy in galactic-based coordinates (vertical axis) for subsets of the BATSE catalog in which the peak photon flux is greater than or equal to a given value, shown on the horizontal axis. The upper graph in Fig. 3 gives the dipole anisotropy toward the galactic center, $\langle \cos\theta \rangle$, where θ is the angle between an observed burst and the galactic center. The middle graph shows the galactic disklike quadrupole, $\langle \sin^2 b \rangle$, where b is galactic latitude. The bottom graph, gives the galactocentric quadrupole, $\langle \cos^2\theta \rangle$. The HBM makes the distinctive prediction[3] that this quantity exceeds 1/3.

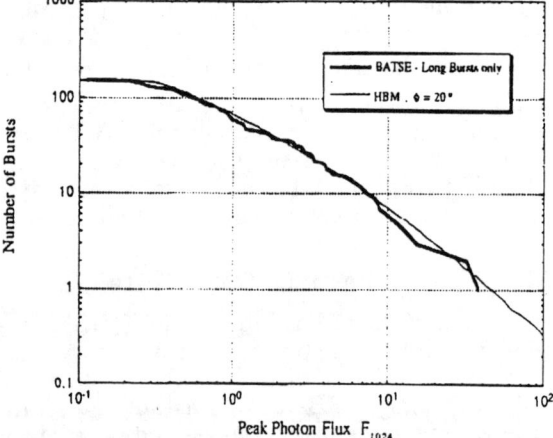

Figure 2 The number of GRBs with peak photon flux on 1024 ms time intervals greater than a given value, $\mathcal{N}(>F_{1024})$ plotted versus F_{1024} for long ($T_{90} > 2$ s) bursts in the BATSE catalog [thick line], and for the HBM [thin line].

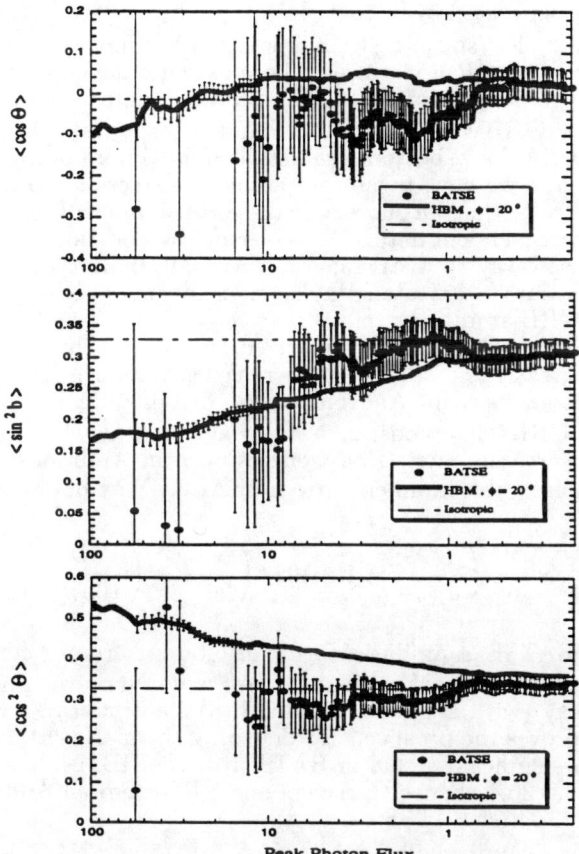

Figure 3 Cumulative plots of angular (an)isotropy measures versus peak photon flux, for *all* GRBs in the BATSE catalog (long and short), as explained in the text. Note that data points in these plots are *not* statistically independent.

Thick solid lines in Fig. 3 are the HBM results for $\phi = 20°$. Small error bars centered on these lines show 1-σ numerical (Monte Carlo) uncertainties. These error bars are only evident near the right-hand side of the curves, since they are smaller than the width of the line elsewhere. Thin dashed lines show the prediction if gamma-ray bursts are isotropic on the sky at all intensity levels, as expected (to within an excellent approximation) in the cosmological burster hypothesis. All calculations were corrected for the anisotropy of BATSE sky coverage. Large error bars represent the BATSE catalog data (with a cumulative point for each burst) when compared with the isotropy hypothesis. (Slightly larger error bars [not shown] are appropriate when comparing the BATSE data with the HBM. See ref. 9 for more details.)

CONCLUSIONS

Fig. 3 shows that the first BATSE catalog is statistically consistent with both isotropy *and* the halo beaming model. Because the two model $\langle\cos\theta\rangle$ curves (upper graph) are separated by $\sim 1\sigma$ or less at all F, this statistic is not a very good discriminant. The $\langle \sin^2 b \rangle$ test (middle graph) is more sensitive; here the HBM fits slightly better than isotropy. The maximum deviation from isotropy (i.e. maximum for any value of F) is 2.0-σ, but the data never deviates more than 1.2-σ from the HBM. On the other hand, isotropy fits the $\langle \cos^2 \theta \rangle$ data better than the HBM does (bottom graph). The maximum deviation from the HBM is 2.4-σ, where we have taken into account corrections to the error bars mentioned above. Note that we quote only the most significant result from many tests on partially independent data sets. Nevertheless, the plots themselves offer a sensitive, unbiased way to distinguish between alternative models. We give more details about our statistical methods, and some perspectives on what these results mean for GRB studies, in ref. 9.

We have not included bursters born in Andromeda in this analysis. This is because beaming greatly reduces the number of such stars which can be observed at Earth. If bursters born in Andromeda (at distance $d_A = 670$ kpc away) continue to emit GRBs at a constant rate at ages $t > d_A/V_r \sim 10^9$ yrs then the number of observable bursts from sources born in Andromeda, $\mathcal{N}_A$ is still smaller than the observable number $\mathcal{N}_{MW}$ from Milky Way bursters by a factor[9]

$$\frac{\mathcal{N}_A}{\mathcal{N}_{MW}} \approx \frac{\phi^2}{6}\left(\frac{D}{d_A}\right)^2 = 4 \times 10^{-4} \left(\frac{\phi}{20°}\right)^2 \left(\frac{D}{100\,\mathrm{kpc}}\right)^2,$$

where D is the BATSE sampling depth. All bursts from Andromeda-born sources are concentrated within solid angle $\pi\phi^2$ centered on the position of Andromeda on the sky (with a smaller number in the antipodes), a circumstance which greatly improves the prospects for detecting them. Nevertheless they are unlikely to give a significant signal in BATSE. If the HBM is a good idealization, there is only a 20% chance that even one GRB from an Andromeda-born star is present in the first BATSE catalog.

This work was supported by grants NAG 5-1515 and NAG 5-2045 at Rice University, by grant NAGW-2418 at the University of Texas, and by the NSERC of Canada.

REFERENCES

1. Shklovski, I.S. & Mitrofanov, I.G., *MNRAS*, **212**, 545 (1985)
2. Li, H. & Dermer, C.D., *Nature*, **359**, 514 (1992)
3. Duncan, R.C., Li, H. & Thompson, C., *Compton Gamma Ray Observatory*, eds. M. Friedlander et al. (NY: AIP), 1074 (1993)
4. Meegan, C.A. et al., *Nature*, **355**, 143 (1992)
5. Duncan, R.C. & Thompson, C., *Astrophys. J.*, **392**, L9 (1992)
6. Thompson, C. & Duncan, R.C., *Astrophys. J.*, **408**, 194 (1993)
7. Thompson, C. & Duncan, R.C., *Compton Gamma Ray Observatory*, eds. M. Friedlander et al. (NY: AIP), 1085 (1993)
8. Thompson, C. & Duncan, R.C., in preparation (1993)
9. Duncan, R.C., Li, H. & Thompson, C. in preparation (1993)
10. Fishman, G.J. et al., *Astrophys. J. Supp.*, in press (1994)
11. Kouveliotou, C. et al., *Astrophys. J.*, **413**, L101 (1993)

WHY 'GALACTIC' GAMMA-RAY BURSTS MIGHT DEPEND ON ENVIRONMENT: BLAST WAVES AROUND NEUTRON STARS

Martin J. Rees[1], Peter Mészáros[2], and Mitchell C. Begelman[3],[4]

ABSTRACT

Although galactic models for gamma-ray bursts are hard to reconcile with the isotropy data, the issue is still sufficiently open that both options should be explored. The most likely 'triggers' for bursts in our Galaxy would be violent disturbances in the magnetospheres of neutron stars. Any event of this kind is likely to expel magnetic flux and plasma at relativistic speed. Such ejecta would be braked by the interstellar medium (ISM), and a gamma-ray flash may result from this interaction. The radiative efficiency, of this mechanism would depend on the density of the circumstellar ISM. Therefore, even if neutron stars were uniformly distributed in space (at least within 1-2 kpc of the Sun), the observed locations of bursts would correlate with regions of above-average ISM density.

1. INTRODUCTION

Before there was any firm evidence on the isotropy of classical gamma-ray bursts (GRBs), the most plausible interpretations involved magnetospheres of neutron stars within our Galaxy. Indeed, on the basis of general theoretical plausibility, many people would have bet strongly against a cosmological interpretation. The remarkable isotropy discovered by the BATSE experiment (together with the 'flatter than Newtonian' counts) clearly shifts the odds substantially. If one lays aside theoretical preconceptions, the cosmological interpretation may now seem strongly favoured. However, Bayesians who allow their assessment to be influenced by some prior view of the relative plausibility of the alternative hypotheses may now find the arguments quite evenly balanced. For example, we may think the isotropy is 100 times easier to account for in a cosmological than in a galactic model; however, if we previously would have bet 100 to 1 in favour of a galactic origin, we end up betting 'evens'. This is our rationale for continuing to consider both options.

If they are not cosmological, GRBs would most likely populate a relatively nearby region of the Galactic disk, at distances $\lesssim (1-2)$ kpc. [As noted by other speakers, 'halo' models entail (at least in a mild form) many of the same

---

[1] Institute of Astronomy, Madingley Road, Cambridge CB3 0HA, England.
[2] Pennsylvania State University, 525 Davey Lab, University Park, PA 16803.
[3] Joint Institute for Laboratory Astrophysics, University of Colorado and National Institute of Standards and Technology, Boulder, CO 80309
[4] Department of Astrophysical, Planetary and Atmospheric Sciences, University of Colorado, Boulder, CO 80309.

theoretical problems as cosmological models]. The bursts could then be due to violent disturbances in the magnetospheres of neutron stars.[1,2] Any disturbance of the kind proposed is also likely to expel magnetic flux and plasma into the interstellar medium (ISM) surrounding the neutron star, possibly at relativistic speed. The new point we wish to discuss here is that the blast wave driven into the ISM by a magnetospheric disturbance could also produce a flash of gamma-rays with the characteristics observed to be typical of GRBs. We briefly summarise here the physics of the interaction between relativistic ejecta from a neutron star and the ambient interstellar medium. Fuller details are in reference (3).

2. RADIATIVE PROPERTIES OF BLAST WAVES

Suppose an amount of energy $E_0 \sim 10^{39} E_{39}$ ergs is impulsively released from a neutron star magnetosphere into a medium of number density $n(r)$ cm^{-3}, where r is the distance from the source of the energy. (The time structure and duration of bursts may be partly due to a more complicated pattern of energy release.) The initial energy produces a highly relativistic fluid, with Lorentz factor η, if the mass M_0 initially released along with the energy satisfies $E_0/M_0 c^2 \equiv \eta \gg 1$. After an amount $\eta^{-1} M_0$ of external mass has been swept up a blast wave forms ahead of the ejecta, which starts to decelerate. In this decelerating regime, if radiation were *in*efficient the bulk Lorentz factor of the blast wave, after having reached the value $\Gamma \simeq \eta$, would vary with radius according to

$$\Gamma \sim \left(\frac{3E_0}{4\pi m_p c^2 n r^3}\right)^{1/2}. \qquad (1)$$

The blast wave, however, may radiate away enough of its energy in a sufficiently short time scale to be of interest for explaining GRBs. This can occur, for example, if the magnetic energy density is amplified behind the shock front (due to turbulent shear, etc.) to a significant fraction (λ) of equipartition with respect to the shocked ambient gas, or in a reverse shock;[4,5] Compton losses can also be important.

In the comoving frame, the magnetic field is given by

$$B' \sim 0.3 \lambda^{1/2} n^{1/2} \Gamma \text{ G}. \qquad (2)$$

The highest efficiency is obtained if the gamma rays are synchrotron radiation. To produce synchrotron photons of observed (Doppler-boosted) energy ε_{MeV} MeV requires that electrons be accelerated to random Lorentz factors (in the fluid frame) γ such that

$$\gamma \Gamma \sim 2.6 \times 10^7 (\lambda n)^{-1/4} \varepsilon_{MeV}^{1/2}. \qquad (3)$$

This can be accomplished, in principle, by Fermi acceleration at the strong shock front.[6] For blast wave radii $\gtrsim 10^{13}$ cm and typical interstellar conditions $n \sim 1$ cm^{-3}, synchrotron radiation at energies above 1 MeV can be highly efficient.[3]

If we assume that shock acceleration to a Lorentz factor γ requires $100\zeta_2$ gyro-orbital times, then the maximum synchrotron photon energy coming from the blast wave varies according to $\varepsilon_{max} \sim 0.4\zeta_2^{-1}\Gamma$ MeV. Note that ε_{max} depends on the highly uncertain shock acceleration rate through ζ. Photons of energy ε will come predominantly from inside the radius at which $\varepsilon \sim \varepsilon_{max}$, i.e., where the blast wave has slowed to $\Gamma \sim 2.5\zeta_2\varepsilon_{MeV}$. If $E \sim 10^{39}E_{39}$ erg, this radius is given by $r_{max} \sim 3 \times 10^{13}(E_{39}/n\zeta_2^2\varepsilon_{MeV}^2)^{1/3}$ cm, corresponding to a maximum burst duration of $\Delta t_{max} \sim r/c\Gamma^2 \sim 160(E_{39}/n)^{1/3}(\zeta_2\varepsilon_{MeV})^{-8/3}$ s. This estimate suggests that the maximum burst duration might be anticorrelated with observing frequency. Note, however, that the extreme sensitivity to ζ makes it difficult to extract useful numerical estimates from this formula.

A necessary condition for the blast wave to radiate efficiently at energy ε is that $r_{rad} < r_{max}$, which is equivalent to the condition

$$n > 0.03 E_{39}^{-4/5} \lambda^{-9/5} \zeta^{8/5} \varepsilon_{MeV}^{2/5} \text{ cm}^{-3}. \quad (4)$$

While the numerical values of the parameters in eq. (4) are very uncertain, the condition suggests a correlation between burst efficiency (and therefore detectability) and the density of the ambient ISM.

3. INFERENCES FROM BURST STATISTICS

If bursts repeat on a timescale of order t_r years, then the local population of bursters comprises of order $10^3 t_r$ neutron stars. Given a Galactic pulsar birthrate[7] of $\sim 10^{-11}$ pc^{-2} yr^{-1}, the mean age of a bursting neutron star is $t_{burst} \sim 10^7 R_{kpc}^3 (t_r/f)$ yr $\equiv 10^{10} t_{10}$ yr, where R_{kpc} is the mean distance to bursts in kpc and f is the fraction of the time during which the deposition of burst energy in the ISM would lead to a detectable burst. Since the dipole spindown time of a pulsar is $\sim 10^9 P^2 B_{11}^{-2}$ yr, the typical spin period of neutron stars responsible for the local bursts would be $\sim 3 B_{11} t_{10}$ s. If $t_{10} < 3(v_{100}/B_{11})^{1/2} n_\infty^{-1/2}$, these pulsars would still be producing wind bow shocks in the ISM, and would not be accreting interstellar gas.

The contact discontinuity between the shocked pulsar wind and the ISM is then located at $r_W \sim 10^{13} B_{11}^{-1} t_{10}^{-1} v_{100}^{-1} n_\infty^{-1/2}$ cm. This number is smaller than r_{max} for 1 MeV photons provided that $B_{11} t_{10} n^{1/2} > 0.3$, suggesting that detectable bursts from blast waves would come primarily from a relatively old population of pulsars, $t_{burst} \gtrsim 10^9$ yr, and/or from neutron stars passing through denser regions of the ISM. In either case, we estimate $t_r/f \gtrsim 100$. Note that, in the simplest interpretation, f would be the volume filling factor of ISM with high enough density to make the blast wave readily detectable.

4. DISCUSSION

We have extended previous ideas about plausible radiation mechanisms for Galactic GRBs, pointing out that relativistic blast waves driven into the ISM by magnetospheric disturbances around neutron stars can yield bursts of gamma-rays with roughly the observed range of timescales and fluences. Our extremely simple conjectures about the radiative properties of synchrotron emitting blast waves do not reveal the expected spectral properties of such bursts, but they do suggest a plausible correlation between the radiative efficiency at MeV energies and the density of the ambient medium.

The question of what might trigger gamma-ray bursts in this picture is unresolved. A neutron starquake model or other impulsive events that violently disturb the magnetosphere seem attractive on energetic grounds. The bursts may be due to neutron stars with unusually strong dipole fields, which have been advocated by Duncan and Thompson (these proceedings) in their interpretation of the soft repeaters. Rotational or gravitational energy would be adequate to power numerous bursts per neutron star. As already explained, the stars would not be accreting from the ISM; however impact of comets or asteroids[8] are further possibilities.

A relatively robust conclusion is that, for any kind of trigger mechanism which involves violently shaking a neutron star magnetosphere, a strong gamma-ray burst can be generated by interaction of the expanding energy flow (whatever its form) with the ISM. This does not exclude a gamma-ray burst from the magnetosphere itself, but in the light of the evidence for two classes of classical GRBs,[9,10] one might perhaps attribute the short bursts to the latter mechanism and the long ($t \gtrsim 2$ s) to the blast wave. The efficiency of this mechanism depends on the ISM (which may introduce longer timescale variability), so the blast wave component would be specially dominant for bursts occurring in regions where the ISM has reasonably large density (e.g. clouds, not necessarily molecular).

[We are discussing primarily the "classical" bursts rather than the soft repeaters. However it is worth noting parenthetically that the one feature of the latter which causes problems for Duncan and Thompson's model (these proceedings) is the sharp initial spike on the famous "March 5th" event. If the event released some relativistic plasma, this intense and short-lived precursor burst would be accounted for if $\Gamma \gtrsim 500$. If the ejection were directional, its energy could of course be well below the value of 10^{44} ergs inferred on the basis of isotropic emission.]

We recall that old pulsars would be expected to have a smooth distribution in the galaxy, constituting a halo population or a disc population with a large scale height. If the bursts came from $\gtrsim 10$ kpc distances, one would expect a strong systematic concentration towards the Galactic Centre; on the other hand, if the burst distances are only 1 kpc or less and their distribution directly traced that of the old neutron stars, the non-uniformities revealed by V/V_m would be perplexing. A $\log N - \log S$ slope flatter than 3/2 at low fluences can be easily understood in terms of a dropoff in the number of sources beyond a local high density excess associated with our immediate neighbourhood. However, it would seem a bit of a coincidence that the anisotropy should be so small relative to the deficit from Euclidean counts – this would imply that we were relatively near the centre of a kpc-scale region where the mean ISM density was higher than outside.

This model, based on a local burst population made conspicuous by a denser gaseous environment, would predict that the spatial distribution would be modulated by the highly irregular and structured distribution of the ISM. We could even account for the spiral-arm effects discussed by Quashnock & Lamb.[11] If this effect indeed exists, our explanation seems more plausible than attributing all bursts to neutron stars just a few million years old which still remember the spiral arm they came from, since the latter would require a much higher repetition rate.

We acknowledge partial support from NSF grant AST91-20599 and NASA grant NAG5-2026 (MCB), from NASA grant NAGW-1522 (PM) and from the Royal Society (MJR).

REFERENCES

1. Blaes, O., Blandford, R.D., Goldreich, P., & Madau, P. 1989, ApJ, 343, 839
2. Ramaty, R., Bonazzola, S., Cline, T.L., Kazanas, D. & Mészáros, P., 1991, Nature, 287, 122
3. Begelman, M.C., Mészáros, P., & Rees, M.J. 1993, MNRAS, 265, L16
4. Mészáros, P., Laguna, P , & Rees, M.J. 1993, ApJ, 415, 181
5. Mészáros, P., Rees, M.J. & Papathanassiou, H., 1994, ApJ, submitted
6. Ellison, D.C. and Reynolds, S.P., 1991, ApJ, 382, 242
7. Narayan, R. and Ostriker, J.P., 1990, ApJ, 352, 222
8. Harwit, M. and Salpeter, E.E., 1973, ApJ, 186, L37.
9. Kouveliotou, C., et al, 1993, ApJ, 413, L101
10. Lamb, D.Q., Graziani, C. & Smith, I., 1993, ApJ, 413, L11
11. Quashnock, J.M. and Lamb, D.Q. 1993, MNRAS (in press).

GALACTIC ARM AND DISK PLUS HALO MODELS OF GAMMA-RAY BURST SOURCES

I. A. Smith

Dept. Space Physics & Astronomy, Rice University, Houston, TX 77251-1892

ABSTRACT

Evidence was presented recently that at least some of the gamma-ray burst sources are in our Galactic spiral arms[1]. A large range of possible spiral arm spatial distributions for the gamma-ray burst sources is investigated here, and it is shown that none of these are able to simultaneously fit the BATSE $<V/V_{\max}>$, $<\cos\theta>$, and $<\sin^2 b>$ observations if all the sources are in the spiral arms. The current observations made by BATSE and by previous experiments are still consistent with a combined Galactic disk (or arm) plus extended Galactic halo model, provided the halo has a finite extent. A prediction of the disk plus halo model is that the fraction of the bursts observed to be in the "disk" population rises as the detector sensitivity improves. A careful re-examination of the numbers of bursts in the two populations for the pre-BATSE databases could rule out these models.

GALACTIC ARM MODELS

For a uniform distribution of sources in n_D dimensions, with an arbitrary luminosity law, one finds $<V/V_{\max}> = n_D/(n_D + 3)$. Thus a 2-D Galactic disk cannot have $<V/V_{\max}> < 0.4$, making it impossible to fit the observed BATSE results[2] ($<V/V_{\max}> = 0.324\pm0.016$ for 336 bursts, $<\cos\theta> = 0.048\pm 0.027$ and $<\sin^2 b> = 0.320 \pm 0.014$ for 447 bursts) using only disk sources[3,4]. However, it is potentially possible to have a small enough $<V/V_{\max}>$ if all the sources are in a 1-D Galactic arm. It is therefore necessary to perform a careful study of Galactic arm models for the burst sources to see if they can give the observed BATSE results[5].

A cylindrical geometry for the Galactic arm sources is used, as shown in Figure 1. The sources are assumed to be standard candles that are distributed axially symmetrically; the source number density is assumed to be independent of z (to make the distribution 1-D). The Sun is taken to be a distance ρ_0 from the axis of the arm. Because ρ_0 is small compared to the distance of the Sun from the galactic center ($R_0 = 8.5$ kpc), the curvature of the arm is ignored. For the Galactic arm models, it is necessary to use $\cos\psi$ rather than the usual $\cos\theta$ (see Figure 1); if the axis of the arm is between the Sun and the Galactic Center, then $\cos\psi = \cos\theta$, otherwise $\cos\psi = -\cos\theta$ (as in Figure 1).

A wide range of source number density distributions $n(\rho)$ have been considered. Figure 2 plots $<\sin^2 b>$ versus $<V/V_{\max}>$ for the case when $\rho_0 = 0$ (i.e. the Sun is on the axis of the arm) for different $n(\rho)$. (Note that for $\rho_0 = 0$, the curves do not depend on the particular values of ρ_{cd}, $\bar{\rho}$, or ρ_c that are chosen.) It can be seen that none of these curves come close to fitting the BATSE data point (for which 1σ error bars are shown). The arms that come closest to the BATSE point are the thin Gaussian shell arms, and an arm that has a constant density out to a sharp cut-off[4]; in both these cases, a luminosity spread will broaden the effective source distribution, making the real curves lie

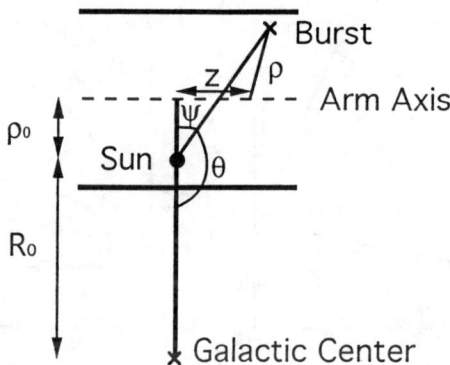

Figure 1. Cylindrical geometry for Galactic arm sources.

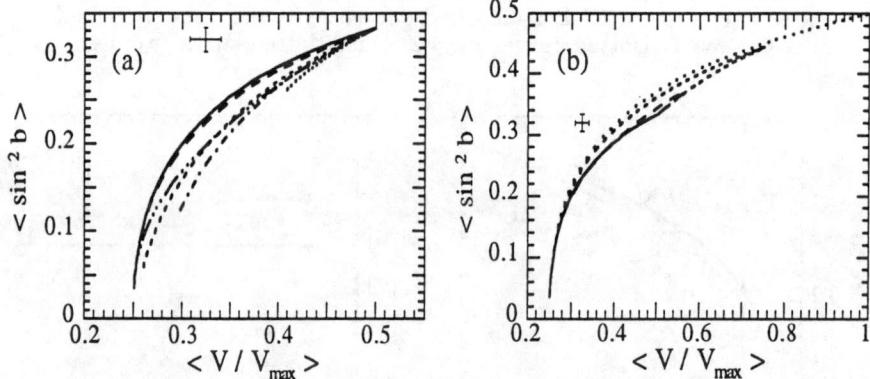

Figure 2. $< \sin^2 b >$ versus $< V/V_{\max} >$ for Galactic arm models with different source number density distributions with $\rho_0 = 0$. (a) Solid curve: $n(\rho) = n_a$ if $\rho \leq \rho_{cd}$, $n(\rho) = 0$ if $\rho > \rho_{cd}$, n_a is a constant. Long dash dotted curve: $n(\rho) = n_a e^{-\rho/\bar{\rho}}$. Other curves: $n(\rho) = n_a/[1 + (\rho/\rho_c)^\alpha]$. Long dashed curve: $\alpha = 10$. Short dashed curve: $\alpha = 3$. Short dash dotted curve: $\alpha = 2$. Dotted curve: $\alpha = 1$. (b) Solid curve: $n(\rho) = n_a$ if $\rho \leq \rho_{cd}$, $n(\rho) = 0$ if $\rho > \rho_{cd}$, n_a is a constant. Other curves: $n(\rho) = n_a e^{-(1/2)[(\rho-\rho_s)/\sigma]^2}$. Long dashed: $(\rho_s/\sigma) = 2$. Short dashed: $(\rho_s/\sigma) = 5$. Dotted: $(\rho_s/\sigma) = 50$.

further from the BATSE point. Also note that a less sensitive detector viewing the Gaussian shell arm could see $< V/V_{\max} > > 0.5$; since this has not been observed, it makes this choice of geometry unlikely.

The Earth is offset 1 to 2 kpc from the axis of the spiral arms measured by the HII regions or the free-electron density[6]. Therefore, it is likely that the Earth would also be offset from the axis of an arm of gamma-ray burst sources. Figure 3 shows the effect of different offsets ρ_0 from the axis of an exponential arm. Figure 4 does the same for a Gaussian shell arm. The BATSE data point is given with 1σ error bars. It can be seen that as ρ_0 increases, the

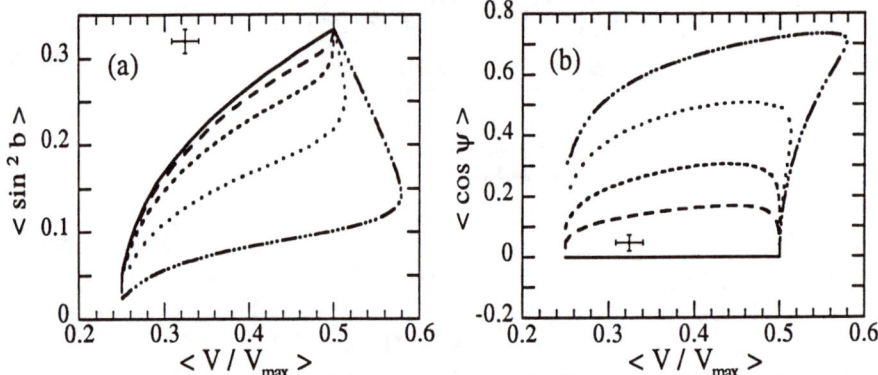

Figure 3. (a) $<\sin^2 b>$ versus $<V/V_{\max}>$, and (b) $<\cos\psi>$ versus $<V/V_{\max}>$ for an exponential arm $n(\rho) = n_a e^{-\rho/\bar{\rho}}$. $\bar{\rho} = 1$. Solid curve: $\rho_0 = 0$. Long dashed curve: $\rho_0 = 0.5$. Short dashed curve: $\rho_0 = 1$. Dotted curve: $\rho_0 = 2$. Dash dotted curve: $\rho_0 = 4$.

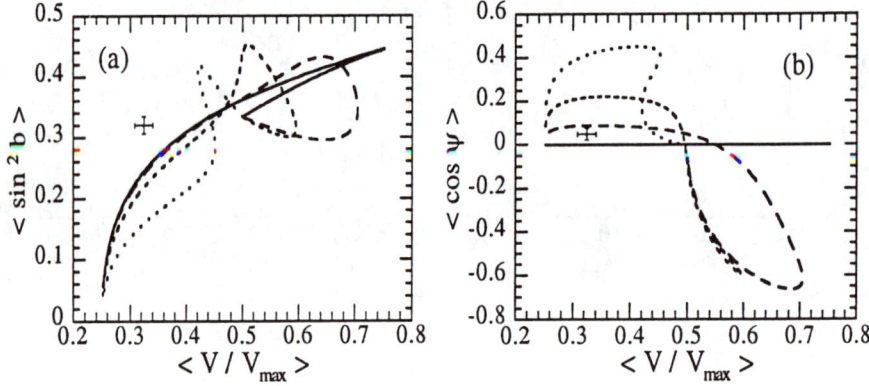

Figure 4. (a) $<\sin^2 b>$ versus $<V/V_{\max}>$, and (b) $<\cos\psi>$ versus $<V/V_{\max}>$ for a Gaussian arm $n(\rho) = n_a e^{-(1/2)[(\rho-\rho_s)/\sigma]^2}$. $\rho_s = 0.5$, $\sigma = 0.1$. Solid curve: $\rho_0 = 0$. Long dashed curve: $\rho_0 = 0.1$. Short dashed curve: $\rho_0 = 0.25$. Dotted curve: $\rho_0 = 0.5$.

$<\sin^2 b> - <V/V_{\max}>$ curves lie increasingly distant from the BATSE data point.

Figures 3 and 4 also show that, even for small offsets ρ_0, the $<\cos\psi> - <V/V_{\max}>$ curves lie above the BATSE data point. It would be possible to keep $<\cos\theta>$ small, and still have a large offset, if there was a cancellation between two arms, one located between the Sun and the Galactic Center and the other beyond the Sun[1]. However, Figures 3 and 4 show that this case would produce a small value of $<\sin^2 b>$. Therefore, it does not appear to be possible to simultaneously fit the BATSE $<V/V_{\max}>$, $<\cos\theta>$, and $<\sin^2 b>$ observations using these simple Galactic arm models.

DISK PLUS HALO MODELS

Although there are constraints on the allowed Galactic halos[4], the Galactic disk (or arm) plus extended Galactic halo model is still consistent with the current observations made by BATSE and by previous experiments, provided the halo has an edge[7,8].

As an example, consider an exponential halo plus disk model. For the exponential halo, the sources are distributed spherically symmetrically about the Galactic Center with number density distribution $n(R) = n_h e^{-R/\bar{r}}$ where R is the distance of the source from the Galactic Center. The sources are assumed to be standard candles, and the solar system is displaced a distance $R_0 = 8.5$ kpc from the Galactic Center. To fit the BATSE data, $\bar{r} = 45$ kpc is chosen, and the distance to the faintest halo source that can be detected $D_h = 165$ kpc.

For the disk, the sources are taken to be standard candles with number density distribution $n(z) = n_d e^{-|z|/z_0}$. The disk has infinite extent in the Galactic plane, z_0 is the disk scale-height, and D_d is the distance to the faintest disk source that can be observed. In Smith & Lamb[7], a disk with $D_d/z_0 = 2/3$ was chosen; here, a thinner disk with $D_d/z_0 = 2$ is used for illustrative purposes.

To fit the BATSE data, $N_d = 39$ disk bursts and $N_h = 154$ halo bursts are used (i.e. 20.2% of the bursts come from disk sources); these are exactly the same numbers used for the "dark matter" halo plus disk example in Smith & Lamb[7], allowing a direct comparison with the results here.

The values of $<V/V_{\max}>$, $<\cos\theta>$, and $<\sin^2 b>$ for the disk, halo, and combined disk plus halo are given in Table 1. The values for the combined model are all easily consistent with the BATSE observations. $<\sin^2 b>_d$ is quite small for this example; it is much smaller than the $<\sin^2 b>_d = 0.2932$ obtained using $D_d/z_0 = 2/3$ in Smith & Lamb[7]. $<\sin^2 b>_d = 0.2242$ would be a 3σ deviation from isotropy for 67 disk bursts, while $<\sin^2 b>_d = 0.2932$ would be a 3σ deviation from isotropy for 497 disk bursts: future BATSE observations could constrain the allowed values of D_d/z_0 for the disk sources.

Table 1. Average quantities for exponential halo plus disk.

	Disk	Halo	Disk + Halo
$<V/V_{\max}>$	0.4566	0.2844	0.319
$<\cos\theta>$	0.0	0.0627	0.050
$<\sin^2 b>$	0.2242	0.3321	0.310

The criteria explained in Smith & Lamb[7] were used to find the $C_{\max}/C_{\min}$ distribution for 193 BATSE bursts. Figure 5 shows this and the $C_{\max}/C_{\min}$ distribution for the combined exponential halo plus disk (solid curve); it can be seen that there is a good fit. There are 14 halo bursts brighter than the brightest disk burst: if it is possible to distinguish between the disk and the halo bursts (for example, because of their duration, or variability), a firm prediction of the disk plus halo model is that the very brightest bursts should have temporal and spectral behaviors characteristic of the halo class of bursts.

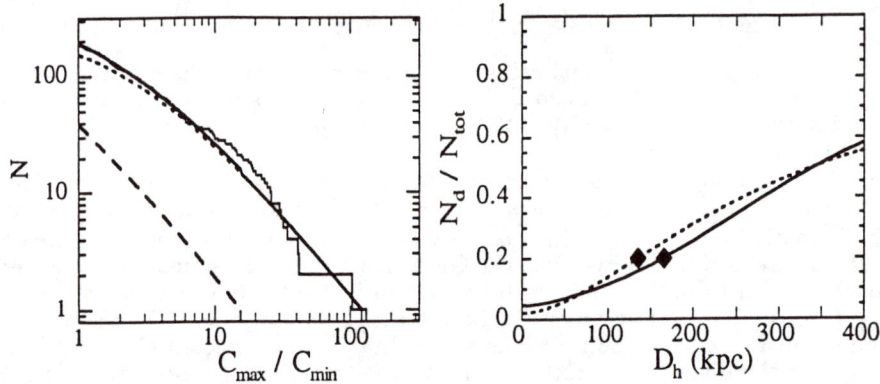

Figure 5. C_{max}/C_{min} distribution for exponential halo plus disk. Short dashed curve: halo bursts only. Long dashed curve: disk bursts only. Solid curve: combined disk plus halo.

Figure 6. N_d/N_{tot} as a function of D_h for the exponential halo plus disk example (solid curve) and "dark matter" halo plus disk example (dashed curve). The black diamonds are at the values of the parameters used to fit the BATSE data.

A prediction of the disk plus halo model is that the fraction of the bursts observed to be in the "disk" population N_d/N_{tot} rises as the detector sensitivity improves[8]. This is illustrated in Figure 6, which plots N_d/N_{tot} as a function of the maximum distance that a detector can see halo sources D_h for the exponential halo plus disk above (solid curve) and the "dark matter" halo plus disk used in Smith & Lamb[7] (dashed curve). The black diamonds mark the halo observing distances used for the BATSE fits in the two examples. By studying the N_d/N_{tot} found by different gamma-ray burst satellites, it may be possible to rule out the disk plus halo models. Note that the relative number of bursts in the two classes will change with the detector sensitivity for any two-population model in which the scale heights of the two populations are different, for example two "dark matter" halos whose core radii are different. The same is true if there is only one spatial distribution for the sources, and the sampling distance is different for the two classes of bursts.

This work was supported at Rice University by grant NAG 5-1515.

REFERENCES

1. J. M. Quashnock and D. Q. Lamb, M.N.R.A.S., in press (1993).
2. C. A. Meegan et al., in Compton Gamma-Ray Observatory (AIP, New York, 1993), p. 681.
3. S. Mao and B. Paczyński, Ap. J. (Letters) **389**, L13 (1992).
4. J. Hakkila et al., Ap. J., in press (1994).
5. I. A. Smith, Ap. J. (Letters), in preparation (1994).
6. J. H. Taylor and J. M. Cordes, Ap. J. **411**, 674 (1993).
7. I. A. Smith and D. Q. Lamb, Ap. J. (Letters) **410**, L23 (1993).
8. I. A. Smith, Ap. J., in preparation (1994).

A POSSIBLE CYCLOTRON LINE SIGNATURE FROM QUIESCENT GAMMA-RAY BURST COUNTERPARTS

John C. L. Wang
Joint Institute for Laboratory Astrophysics, University of Colorado
Campus Box 440, Boulder, CO 80309, USA;

Robert W. Nelson
Canadian Institute for Theoretical Astrophysics, University of Toronto
60 Saint George Street, Toronto, ON M5S 1A7, Canada

ABSTRACT

If γ-ray bursts are associated with isolated, magnetized neutron stars in the Galaxy, they may be accreting directly from the interstellar medium after the burst event. In addition to a soft thermal component ($T_e \sim 100$ eV), we predict that the faint *quiescent* emission spectra from these sources will possess a unique *nonthermal* signature: for magnetic field strengths $B \sim (0.7\text{-}7) \times 10^{12}\,G$, $\sim 0.5\text{-}5\%$ of the total luminosity could be emitted in a narrow ($E/\Delta E \sim 2\text{-}4$) cyclotron emission line which peaks between $\sim 5\text{-}20$ keV below the cyclotron resonance energy ($E_B = 11.6[B/10^{12}\,G]$ keV). Unlike the soft thermal emission, this hard cyclotron component will not be strongly absorbed by the intervening H I gas. For stars with $B \sim 10^{12}\,G$, this nonthermal feature should be detectable by the instruments onboard ASCA and on the next generation of X-ray satellites (e.g., ASTRO-E). We therefore propose that deep X-ray images be taken on γ-ray burst sources with excellent positions (error boxes $\lesssim 1'$) such as is being planned for GRB920501[1] to look for this feature. A positive detection would strongly suggest that at least some of the γ-ray bursters are associated with isolated, magnetized neutron stars in the Galaxy.

1 INTRODUCTION

Over a quarter century after they were first discovered,[2] the origin of the classical γ-ray bursts remains a mystery. One reason for this continuing enigma is that there exist no convincing counterparts (simultaneous, quiescent, or archival) to these sources in any of the energy bands (radio, IR, optical, UV, soft X-rays) for which a search has been carried out.[3,4,5,6,7,8]

If some fraction of classical bursts originate from strongly magnetized Galactic neutron stars,[8,9] then *after* the burst event subsides, the star may accrete directly from the interstellar medium (ISM). We obtain a typical luminosity by assuming that the star undergoes Bondi accretion,[10] $L_{accr}^{typ} = 5 \times 10^{30} n_{10} v_{50}^{-3}$ erg/sec, where $n = 10 n_{10}$ cm^{-3} is the density of the ISM through which the neutron star is moving (supersonically) with speed $v = 50 v_{50}$ km/sec.

If such highly sub-Eddington accretion occurs onto magnetic neutron stars, we predict that the quiescent spectrum should include a prominent cyclotron emission line superposed on the Wien tail of the underlying thermal emission.

In section 2, we describe our model for the cyclotron emission. In section 3, we present the results for the emergent line luminosity and spectra.

2 MAGNETIC ACCRETION AND RADIATIVE TRANSPORT

We consider accretion onto a magnetic neutron star with luminosities satisfying $L_{accr} \ll L_{Edd} = 1.8 \times 10^{35} M_{1.4} R_6^{-2} (A_{cap}/1.3\,\mathrm{km}^2)$ erg/sec. The quantity L_{Edd} is the Eddington limit for magnetic polar cap accretion, A_{cap} is the polar cap area, $M = 1.4 M_{1.4} M_\odot$ and $R = 10^6 R_6$ cm are the neutron star's mass and radius, respectively. We model the polar cap atmosphere as a plane parallel electron-proton plasma slab threaded by a uniform magnetic field of strength $B = 10^{12} B_{12}$ G oriented parallel to the slab normal. We ignore the possibility of collisionless shocks and assume that electrons and ions (mostly protons) enter the polar cap atmosphere at free fall velocity after being channeled along field lines. The bulk of the energy in the accretion stream is carried by the accreting protons which deposit their energy ($\sim$ 200 MeV/nucleon) in the neutron star atmosphere after decelerating through multiple magnetic Coulomb scatterings with atmospheric electrons.

The presence of a strong magnetic field dramatically alters the microphysics of proton stopping and the subsequent conversion of accretion energy to radiation.[11,12] In particular, for field strengths $B_{12} \ll 9 M_{1.4} R_6^{-1}$, accreting protons have sufficient center of mass energy to excite large numbers of electron Landau transitions. In such strong magnetic fields, these excited electrons will then decay to their Landau ground state primarily through single step *radiative* transitions. Thus, a significant fraction of the accretion energy is converted directly to cyclotron photons distributed along the path of the decelerating proton. We emphasize that these cyclotron photons are highly nonthermal with energies

$$E_B = 11.6 B_{12} \text{ keV} \gg k_B T_e = 190 \left(\frac{F_0}{10^{-4}}\right)^{1/4} M_{1.4}^{1/4} R_6^{-1/2} \text{ eV}, \quad (1)$$

where $F_0 \equiv L_{accr}/L_{Edd}$.

Once produced, the cyclotron photons try to escape the atmosphere through multiple magnetic Compton scatters. We have used a Monte Carlo code to compute the polarized radiative transfer of these cyclotron photons through the magnetized plasma.[13] The dominant transfer effects are (1) magnetic Compton scattering which results in both angle and frequency redistribution (Comptonization), (2) polarization mode switching, and (3) absorption via inverse magnetic Bremsstrahlung. As a result of eqn. (1), the atmosphere may be treated as a cold plasma and we use the cold plasma polarization modes in our treatment of polarized transfer.[14,15,16,17]

There are four parameters in our model: The neutron star's mass and radius, the field strength, B, and the dimensionless accretion luminosity, $F_0 = L_{accr}/L_{Edd}$. Without exception, we take $M_{1.4} = 1 = R_6$. We therefore construct a two-parameter family of models in B and F_0. The value of F_0 fixes the plasma temperature (cf. eqn. [1]). The emergent line luminosity and spectra depend most strongly on B.

3 EMERGENT LINE LUMINOSITY AND SPECTRUM

Figure 1 shows the fraction of accretion luminosity that escapes in the cyclotron line, L_{line}/L_{accr}, as a function of B. The curves are labelled by $F_0 =$

10^{-2} (solid), 10^{-4} (dashed), and 10^{-6} (dotted). The dependence of L_{line}/L_{accr} on F_0 enters weakly only through the absorption cross section, $\sigma_{abs} \propto 1/T_e \propto F_0^{-1/4}$. At high values of B, few cyclotron photons are produced so L_{line}/L_{accr} falls off rapidly. The rapid fall-off at low B arises from the transfer microphysics: the energy lost per scatter by photons to electron recoil along field lines is small compared to the thermal Doppler width, so that line photons are trapped inside the line core[17,18] and are destroyed by absorption before they can escape.

From Figure 1, we see that the optimum conditions for line photon escape obtains when $0.7 \lesssim B_{12} \lesssim 7$ (where $0.005 \lesssim L_{line}/L_{accr} \lesssim 0.05$). In this regime, the field is sufficiently strong to avoid line trapping, but not so strong as to quench the initial cyclotron photon production.

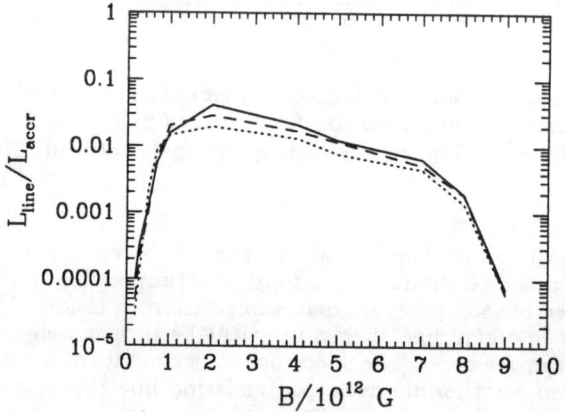

Figure 1. The fraction of the total accretion luminosity in the non-thermal cyclotron emission line as a function of magnetic field strength. The curves correspond to $F_0 = L_{accr}/L_{Edd}$ =(solid) 10^{-2}, (dashed) 10^{-4}, and (dotted) 10^{-6}.

Figure 2 shows the integrated line flux that would be observed for $F_0 = 10^{-2}$ (solid) and 10^{-4} (dashed) for a source placed at a distance of $D = 1$ kpc. We assume the cyclotron line is emitted isotropically from both magnetic poles. To convert F_0 to an absolute luminosity, we take $A_{cap} = 1.3$ km^2, so that $L_{Edd} = 1.8 \times 10^{35}$ erg/sec. Thus, the solid (dashed) curve corresponds to $L_{accr} = 1.8 \times 10^{33}$ (1.8×10^{31}) erg/sec. The curves slide up (down) for smaller (larger) D.

From this figure, we see that faint magnetized neutron stars in moderately dense regions of the ISM could already be observable out to distances ~ 1 kpc through their cyclotron line emission with the present generation of X-ray detectors. For example, an 80 kilosecond integration with the Gas Imaging Spectrometer (GIS) onboard the X-ray satellite ASCA enables one to detect sources down to a flux level of about 10^{-13} erg/cm^2/sec (between 0.5 and 10 keV).[19]

The shape of the emergent photon number spectra (summed over polarization) is shown in Figure 3 for $B = 10^{12}\,G$ (left) and $5 \times 10^{12}\,G$ (right). For both panels, $F_0 = 10^{-4}$; the spectral shape does not depend strongly on F_0. Note

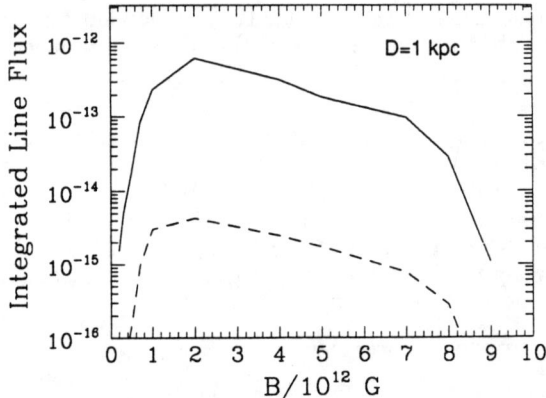

Figure 2. The integrated *line* flux (ergs/cm^2/sec) as a function of field strength assuming isotropic emission for a source placed at $D = 1$ kpc from the observer. The curves correspond to $F_0 =$(solid) 10^{-2}, (dashed) 10^{-4}.

that for $B = 10^{12}\, G$, the entire line lies within the ASCA energy window (0.5 to 10 keV). We obtain the continuum by adopting a (unpolarized) black body spectrum with an effective temperature determined from equation (1). To this spectrum we add the emergent line spectrum, with the proper weighting, computed from the line transfer code. The spectrum thus consists of a soft thermal component plus a hard nonthermal cyclotron emission line superposed on the Wien tail of the underlying thermal continuum. This general form should persist even for harder (e.g., power law) continua as long as the bulk of atmospheric electrons have energies $\ll E_B$.[13,20]

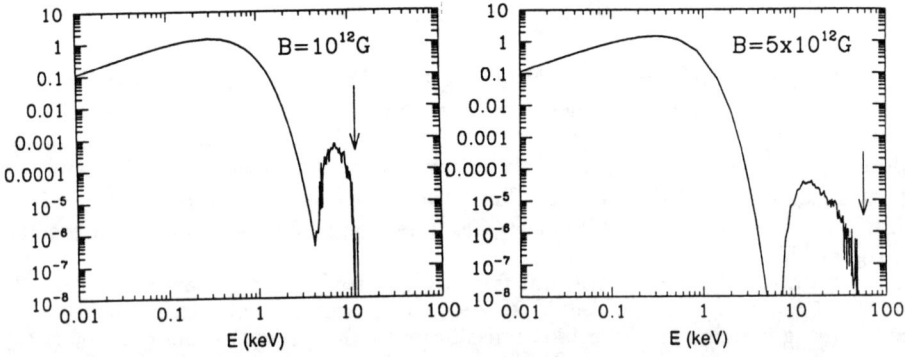

Figure 3. The emergent photon number spectra (summed over polarization and normalized to unit area) from a magnetized polar cap neutron star atmosphere with (left) $B = 10^{12}\, G$, $F_0 = 10^{-4}$, $L_{line} = 0.02 L_{accr}$, and (right) $B = 5 \times 10^{12}\, G$, $F_0 = 10^{-4}$, $L_{line} = 0.01 L_{accr}$. Arrows in figure give location of cyclotron resonance energy, E_B.

The vertical arrows in the figure denote the location of the cyclotron energy, E_B, where line photons are born. The photon energy degradation due to electron recoil (i.e., Comptonization) is clearly evident. In addition, owing to polarization mode switching and the strong energy dependence of the magnetic scattering cross sections,[13] these line features are narrow ($E/\triangle E \sim$ 2-4 for $0.7 \lesssim B_{12} \lesssim 7$).

Thus, if strongly magnetized isolated neutron stars can accrete from the ISM, they will give rise to a prominent cyclotron emission line which peaks between $\sim$ 5-20 keV for $0.7 \lesssim B_{12} \lesssim 7$. This feature should be visible out to distances $\sim$ 1 kpc with the sensitivity of current detectors such as those onboard the ASCA satellite. If deep searches for the quiescent hard X-ray counterparts to γ-ray burst sources (e.g., GRB920501[1] and GRB781119[21]) were to reveal such an emission feature, it would strongly suggest that these sources are associated with magnetized Galactic neutron stars. Furthermore, unless the neutron star is an aligned rotator, the cyclotron emission line should be pulsed. Detection of pulsation would further strengthen the case for an association with Galactic magnetized neutron stars.

This work was supported in part by NASA grants NAGW-666 and NAGW-766, NSF grants AST91-19475 and AST91-20599, and the NSERC of Canada.

REFERENCES

1. Hurley, K. *et al.*, Proc. XXIII International Cosmic Ray Conference **1**, 116 (1993).
2. Klebesadel, R. W., Strong, I. B., and Olson, R. A., Ap. J. Lett. **182**, L85 (1973).
3. Schaefer, B. E., Ap. J., in preparation (1994).
4. Boër, M., these proceedings.
5. Greiner, J. *et al.*, these proceedings.
6. Vrba, F. J., Hartmann, D. H., and Jennings, M. C., these proceedings.
7. Owens, A. *et al.*, *Compton Gamma Ray Observatory Symposium. AIP Conference Proceedings No. 280*, eds. M. Friedlander, N. Gehrels, and D. Macomb (AIP, N.Y., 1993), p. 798.
8. Higdon, J. C. and Lingenfelter, R. E., Ann. Rev. Astron. Astrophys. **28**, 401 (1990), and references therein.
9. Yoshida, A. *et al.*, Pub. Astron. Soc. Japan **43**, L69 (1991).
10. Bondi, H., M.N.R.A.S. **112**, 195 (1952).
11. Nelson, R. W., Salpeter, E. E., and Wasserman, I., *Proc. of Taos Workshop on Physics of Isolated Pulsars*, eds. K. A. Van Riper, R. I. Epstein, and C. Ho (Cambridge University Press, Cambridge, 1992), p. 145.
12. Nelson, R. W., Salpeter, E. E., and Wasserman, I., Ap. J. **418**, 874 (1993).
13. Nelson, R. W., Wang, J. C. L., Salpeter, E. E., Wasserman, I., Ap. J., in preparation (1994).
14. Gnedin, Yu. N. and Pavlov, G. G., Soviet Phys. – JETP **38**, 903 (1974).
15. Ventura, J., Phys. Rev. D **19**, 1684 (1979).
16. Nagel, W. and Ventura, J., A&A **118**, 66 (1983).
17. Wang, J. C. L., Wasserman, I. M., Salpeter, E. E., Ap. J. Suppl. **68**, 735 (1988).
18. Wasserman, I. and Salpeter, E. E., Ap. J. **241**, 1107 (1980).
19. Tanaka, Y., private communication (1993).
20. Wang, J. C. L., Wasserman, I. and Salpeter, E. E., Ap. J. **338**, 343 (1989).
21. Pizzichini, G. *et al.*, Ap. J. **301**, 641 (1986).

BATSE REQUIREMENTS FOR A COLLIDING COMET SOURCE OF GAMMA-RAY BURSTS

R. STEPHEN WHITE
Institute of Geophysics and Planetary Physics
University of California, Riverside, CA 92521.

ABSTRACT

The BATSE[1,2] and Venera[3] results place tight restrictions on possible sources of Gamma Ray Bursts, GRB. Collisions of magnetized comets in the solar system appear to satisfy the conditions of the observed isotropic direction distribution and the measured $N(>P_{max})$ vs P_{max} distributions where X is fluence, flux, maximum flux or ratio of count rate at maximum to the threshold count rate. The distances to the bursts (GRB comet cloud), and density of comets is tailored to the observed data. To satisfy observations of $N(>P_{max})$ vs P_{max} for the maximum gamma ray fluxes, $P_{max} > 10^{-5}$ erg cm^{-2} s^{-1} (about 30 bursts yr^{-1}), the comet density, n, should increase as $n \sim a^1$ from about 40 to 100 AU where a is the comet heliocentric distance. The turnover above 100 AU requires $n \sim a^{-1/2}$ to 200 AU to fit the Venera results and $n \sim a^{1/4}$ to 400 AU to fit the BATSE data. The masses of comets are from: 40-100 AU, about 9 earth masses, m_E; 100-200 AU about 25 m_E; and 100-400 AU, about 900 m_E. The spherical GRB comet cloud is neither the Oort Cloud nor the Kuiper Belt. Current minimum distances to bursts determined from time of arrival of the interplanetary network in combination with Watch on Ulysses or Comptel or EGRET on CGRO neither verify nor prohibit this burst source. The gamma ray burst flux of 10^{-5} erg cm^{-2} s^{-1} corresponds to a luminosity at 100 AU of 3×10^{26} erg s^{-1}. Two colliding Halley's comets at a distance of 100 AU have a combined kinetic energy of 3×10^{28} erg, a factor of about 100 greater than required by the bursts. Betatron acceleration in the compressed magnetic fields between the colliding comets could accelerate electrons to energies sufficient to produce the observed high energy gamma rays by Bremsstrahlung. Many of the observed features of gamma ray bursts can be explained by the solar comet collision source.

ISOTROPY AND UNIFORMITY

The comet collision model gives the spherical distribution of gamma ray bursts determined by BATSE[1,2] and the Log(N>X) vs LogX distributions from BATSE[1,2], Venera[3] and PVO[4] where X is fluence, flux, maximum flux, count rate or ratio of maximum to threshold count rate. At the higher maximum fluxes the integral burst rate plots seem to follow the expected $X^{-1.5}$ curve indicating a uniform distribution of bursts in space, but at lower fluxes, below about 10^{-5} erg cm^{-2} s^{-1}, the distributions flatten toward lower maximum fluxes[1,2,3].

The effect of the decreased burst rate space density with increasing distance on the number of gamma ray bursts yr^{-1} is shown in Figs. 1A,B. In Fig. 1A, the results of Mazets and Golenetskii[3] are plotted as log $N(>P_{max})$ vs Log P_{max} where N is the number of bursts yr^{-1} and P_{max} is the burst maximum flux. For fluxes $P_{max} > 10^{-5}$ erg cm^{-2} s^{-1} the data fit a $P_{max}^{-3/2}$ power law but at lower fluxes drop below that fit. Nor does the data fit the P_{max}^{-1} line expected for a disk shape distribution of burst rates. However, if the space density of burst rates is constant out to a given radius and then decreases with radius it is possible to fit the data. The turnover occurs at about 30 bursts yr^{-1}. Two power laws on the graph start at P_{max} of 10^{-5} erg cm^{-2} s^{-1} and continue to lower fluxes, one as P_{max}^{-1} and the other as $P_{max}^{-0.5}$. These result from burst rates that are independent of angle (isotropic) but decrease differently with radius. A similar plot, Fig. 1B, Meegan et al.[1,2] from the BATSE observations on the Compton Gamma Ray Observatory, give log $N(>C_{max}/C_{min})$ vs log (C_{max}/C_{min}). The C_{max} is the maximum count rate in the burst, and C_{min} is the detector threshold counting rate. The power laws of -1 and -0.5 start at a possible break in the observations and continue to lower ratios. This break occurs at about 50 burst yr^{-1}. The different burst rates of 30 and 50 bursts yr^{-1} are not considered significant for our estimates. The distances to the collisions of the comets that furnish the P_{max} and C_{max}/C_{min} are also given along the upper horizontal axes of the figures. The number of bursts yr^{-1} observed above the Venera threshold is about 100 and above the BATSE threshold about 800.

COMET CLOUD

Theories of the formation of the solar system comet clouds have assumed that comets were formed in the outer planetary region near the ecliptic plane, then evolved to their current orbits through planetary, stellar, galactic tide and giant molecular cloud perturbations, see e.g. Duncan et al.[7] or that they formed in the outer parts of the collapsing protosun at radii $< 5 \times 10^3$ AU at various inclinations to the ecliptic plane, see e.g. Hills[8] and Stagg & Bailey[9]; or were injected into the solar system from the galaxy. Comets with planet orbit crossing trajectories would long ago have been removed. However, the planets are ineffective in disturbing comets with perihelia, q, >40 AU Duncan et al.[7] and passing stars, the galactic tide and giant molecular clouds do not change trajectories that are closer to the sun than about 3×10^3 AU. Consequently, a spherical cloud from 40-400 AU with q>40 AU should be quite stable and last, unperturbed, for the age of the solar system.

The volume collision rate, η (collisions s^{-1} km^{-3}), for a random distribution of comets is

$$\eta = dN/dV = n^2 \sigma v \qquad (1)$$

where N (bursts s^{-1}) is the comet collision rate, n (comets km^{-3}) is the space density of comets, σ (km^2) the comet nucleus cross sectional area, v (km s^{-1}) the comet velocity and V (km^3) the volume of space in which the collisions occur. For circular Keplerian orbits around the sun, $v \sim a^{-1/2}$, = 3 km s^{-1} at 100 AU. The gamma ray burst luminosity, L, about 1% of the kinetic energy of the colliding comets is taken as $0.01 \times 2 \times (1/2\, m_c v^2) \sim 1/a$ where m_c is the mass of each comet. The flux at the earth is then $F = L\,(4\pi\,a^2)^{-1} \sim a^{-3}$. The comet nuclei are taken similar to Halley's comet with densities of 0.5 g cm^{-3}, radii R = 5 km and $\sigma = \pi R^2 = 80$ km^2. We find the relation $N(>X) \sim X^\alpha$ has the values α = -1.5, -1, and -0.5 for comet volume densities varying with radius as a^1, $a^{1/4}$ and $a^{-1/2}$, respectively. Both Venera and BATSE results for $P_{max} > 10^{-5}$ erg cm^{-2} s^{-1} (N $\sim$ 30 bursts yr^{-1}) give α = -1.5 and are satisfied by the colliding comet model with n ~ a^1 for 40<a<100 AU. From $10^{-5} \leq P_{max} \leq 3 \times 10^{-7}$ erg cm^{-2} s^{-1}, ($30 \leq N \leq 100$ bursts yr^{-1}), $\alpha = -0.5$, the Venera results are satisfied by the colliding comet model with n ~ $a^{-1/2}$ from 100<a<200 AU. The BATSE results from $10^{-5} \leq P_{max} \leq 10^{-7}$ erg cm^{-2} s^{-1} ($50 \leq N \leq 800$ bursts yr^{-1}), $\alpha = -1$, are satisfied by n ~ $a^{1/4}$ from 100<a<400 AU. The mass of the model comet is 3×10^{17}g and the masses of comets in the cloud regions of 40-100, 100-200 and 100-400 AU are about 9, 25 and 900 m_E, respectively.

BETATRON ACCELERATION

The kinetic energy in the two colliding comets is 3×10^{28} erg, about 100 times the required luminosity for a burst of maximum flux 10^{-5} erg cm^{-2} s^{-1} lasting one second. A conversion efficiency of about 1% is then required for converting comet kinetic energy into gamma rays through electron bremsstrahlung. Electrons in the comet plasma cloud are accelerated by electric fields generated by the compression of the magnetic fields between the approaching comets.

The solar wind electrons and ions carry magnetic field lines away from the sun that drape around and stream behind the comet. The magnetic field lines are supported by the cloud of electrons and ions surrounding the comet. We use the results of the magnetometer measurements[10] of Halley's Comet realizing they were taken at 1 AU from the sun and not at 100 AU as with our comet example. The magnetometer on the satellite Giotto found the magnetic field in the pile up region around the nose to average about 60 nT over a distance through the nose of 4×10^5 km.

Consider the collision of two Halley's Comets traveling at velocities, v, of 3 km s^{-1} with uniform magnetic fields, B_o, each of 60 nT extending for distances, r_o, of 2×10^5 km around the comet nuclei[10] (Fig. 2). As the comets approach, the magnetic field between them is compressed and betatron acceleration increases the electron momenta, pc, at time to collision, t_c, to

$$(pc) = (pc)_o t_{co}^{1/2} (t_c)^{-1/2} \quad (2)$$

a relativistically good expression where $(pc)_o$ and t_{co} are the initial electron momentum and time to collision.[5,6] If an electron starts with kinetic energy 1 eV at $t_{co} = 10^5$ s (3×10^5 km) its kinetic energies at 1 s (3000 m), 10^{-1} s (300 m), 10^{-2} s (30 m) and 10^{-3} s (3 m), are 0.10, 0.6, 3 and 10 MeV, respectively and its magnetic fields are 0.01, 0.1, 1 and 10 Tesla, respectively. In this model ~ 10% of the comet's kinetic energy is converted into energetic electrons with energies up to 10 MeV and ~ 10% of that energy is converted by bremsstrahlung to gamma rays. Energies of ~ 3×10^{27} erg for electrons and ~ 3×10^{26} erg for gamma rays are required. An electron number density of ~ 3×10^{13}, 10 MeV electrons cm^{-3} in a volume of ~ 10×10^{18} cm^3 gives the required ~ 3×10^{27} erg. A magnetic field of ~ 10 T furnishes the magnetic field density of ~ 4×10^8 erg cm^{-3} in the same volume that balances the electron energy. A slab between the comets, 1400 km × 1400 km × 3 m, illustrative only, satisfies the requirements. This dimension also assures that the GRB rise times and fluctuations can be as short as milleseconds.

The efficiencies could be mitigated with larger colliding comets and higher starting magnetic fields. Suppose $R_{comet} = 50$ km then $m_c = 3 \times 10^{20}$ g. At $t_c = 10^{-3}$ s and $r_c = 3$ m, T_e is 30 MeV and B is 100 T (10^6 G). The required efficiencies for high energy accelerated electrons and gamma rays are down by a factor of 10^3 to 10^{-4} and 10^{-5}, respectively. The required volume of magnetic field of 10^6 G could be a slab 140 km × 140 km × 3m. The 140 km dimension is now near the comet diameter of 100 km. If, in addition, the initial time to collision and therefore the initial magnetic field extended a distance of a factor of 10 greater, the slab would be 45 km × 45 km × 30 m.

The magnetic fields around comets at 100 AU are not known. The interplanetary magnetic field is expected to decrease with distance from the sun to the value of about 0.4 nT, the interstellar magnetic field at the heliopause at about 100 AU[11]. The internal composition of the comet nucleus is not considered sufficiently conducting to significantly delay the outward diffusion of the comet's magnetic field. For the interplanetary magnetic field to drape around Halley's comet, a conducting cloud of ions and electrons is required. Currents are generated in the high conductivity plasma that permit the magnetic field to be compressed, rather than diffuse away.

At 100 AU, temperatures of 50 K, the electron densities around comets are not known. Production of the plasma cloud by other mechanisms appears necessary as comets are visible to observers on the earth usually only when closer than about 5 AU. Rise times of milleseconds and time durations of milliseconds to 300 s are seen. Most are shorter than 30 s with a peak in the distribution near 10s. In the model these burst times are controlled by the changing electromagnetic fields with the accompanying acceleration of electrons.

LOCATING BURSTS

The observed angular distributions[1,2,3] of gamma ray bursts are compatible with isotropy within statistics and uncertainties in angular resolution. To satisfy that constraint our model cloud of comets is isotropic as viewed from the sun. Measured from the earth at 1 AU the distribution will be seasonal and could be measured in large statistical samples if angular resolutions are accurate to fractions of a degree. Measurements from 2 or more satellites separated by distances of at least 1 AU with detectors having angular resolutions of a few minutes of arc could find the distances to the burst sources by triangulation. The Interplanetary Network[12] method of measuring the direction by time of arrival differences at 3 or more satellites widely spread could be extended to 4 or more for triangulation measurements. Angular measurements of large statistical samples accurate to degrees from satellites at distances of many AU from the sun would observe angular distributions with maxima in the direction of the sun. And Bailey[13] pointed out nearly 20 years ago that observations of the comets, themselves, may be possible by occultations of stars.

The measurement of the minimum distance to a burst source by the interplanetary network is limited to the repeater GRB790305 (Tom Cline, private communication), not explained by this model. By using the difference in time of arrival from two widely spaced spacecraft and accurate measurement of direction by an instrument on one of the spacecraft it is possible to measure distances to nearby solar system sources or minimum distances to sources farther out[14]. To date the best measurements are time differences between the spacecraft Ulysses and CGRO and angle measurements by the Watch detector on Ulysses or the COMPTEL or EGRET detectors on CGRO. For the ideal case when the uncertainties in burst arrival times are negligible, the minimum distance to a burst, P, is (Alanna Conners private communication)

$$P \geq S(1-(c\Delta t/S)^2)^{1/2}(2n\Delta\theta)^{-1} \qquad (3)$$

for n standard deviations in the angle uncertainty, $\Delta\theta$, of a given detector. The S is the distance between the two satellites and t the time of arrival differences. For n = 2, reasonably optimistic values of $\Delta\theta = 1$ deg, $S = 5$ AU and $c\Delta t/S = 0.7$ give a minimum source distance of about 50 AU. The distances to the comet collision sources of this model, at this time, can neither be verified nor denied.[15]

Helpful discussions with T. Cline, A. Conners, M. Duncan, K. Hurley, C. Meegan, H. Rahman and J. Ryan are gratefully acknowledged.

REFERENCES

1. C. A. Meegan et al., Bul. Am. Astron. Soc. **23**, 1470 (1992).
2. C. A. Meegan et al., Nature **355**, 143 (1992).
3. E. P. Mazets and S. V. Golenetskii, Astronomia **32**, 16 (1987).
4. K. Chuang et al., Ap. J. **291**, 242 (1992).
5. R. S. White, Proc. 23rd Inter. Cosmic Ray Conf., Calgary **1**, 49 (1993).
6. R. S. White, Accepted in Astrophysics and Space Science, (1993).
7. M. Duncan, T. Quinn, and S. Tremaine, Astron. J. **94**, 1330 (1987).
8. J. Hills, Astron. J **86**, 1730 (1981).
9. C. R. Stagg and M. E. Bailey, Mon. Not. R. Astr. Soc. **241**, 507 (1989).
10. F. M. Neubauer et al., Nature **321**, 352 (1986).
11. S. T. Suess, Reviews of Geophysics **28**, 97 (1990).
12. J. -L Atteia et al., Ap. J. Suppl. **64**, 305 (1987).
13. M. E. Bailey, Nature **259**, 290 (1976).
14. A. Conners et al., Astron. Astrophys. Suppl. Ser. **97**, 75 (1993).
15. D. Hartmann, This Workshop, **Session4**, (1993).

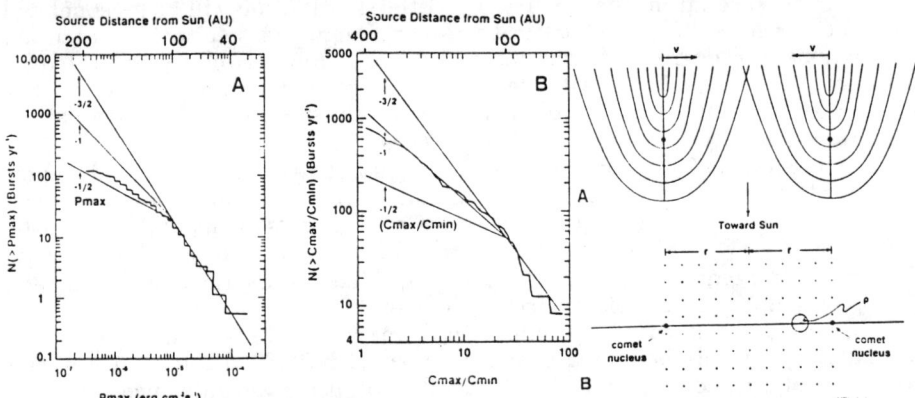

Figure 1. Integral burst frequency plots. (A) Mazets and Golenetskii[3]. (B) Meegan et al.[1,2].

Figure 2. Schematic drawing of the magnetic fields of two comets before collision. (A) Plan view. (B) Front view.

ASTROPHYSICS OF VERY STRONGLY MAGNETIZED NEUTRON STARS: A MODEL FOR THE SOFT GAMMA REPEATERS

R.C. Duncan* & C. Thompson[†]
*Dept. Astronomy, University of Texas at Austin, Austin, TX 78712
[†]CITA, 60 St. George Street, Toronto, Ontario M5S 1A1, Canada

ABSTRACT

We suggest that neutron stars with fields much stronger than $B_{QED} = m_e^2 c^3/e\hbar = 4.4 \times 10^{13}$ G are the source of the March 5 event and the soft gamma repeater (SGR) bursts.[1,2,3] Crustal fractures driven by magnetic stresses in young magnetars (age $\leq 10^4$ yr) release enough energy to power the SGR events. A much more energetic burst, preceded by a hard initial transient (as in the March 5 event) is triggered by a large scale readjustment of the stellar field. The cooling of a pair plasma trapped in the stellar magnetosphere results in hyper-Eddington, quasi-black body emission with a simple light curve and weak spectral evolution, in agreement with observations of both SGR bursts and the soft tail of the March 5th event. We outline five separate lines of reasoning which indicate $B \geq 10^{14}$ G for the March 5 burster.

A rapidly rotating, newborn neutron star supports an efficient α–Ω dynamo during an early post-collapse phase of convection.[1] This suggests the existence of a class of neutron stars with dipole magnetic fields much stronger than those of ordinary radio pulsars, $B_{dipole} \sim 10^{14} - 10^{15}$ G. As the fields of these "magnetars" evolve, catastrophic releases of magnetic energy occur, roughly analogous to stellar flares. We have studied the observational consequences of these events. In this note we outline our results, which will be published in more detail elsewhere.[4]

As the magnetic field diffuses through the stellar interior, it evolves into a configuration that is unstable to a rapid, large scale interchange.[3] By this time, the star will have been spun down by magnetic torques to a long period P (8 seconds in the case of the March 5, 1979 burster). The corotation charge density in the magnetosphere is quite low, $n_c \sim (B/cP)$, and in fact is much too small to provide a current comparable to the displacement current $(1/4\pi)\partial E/\partial t \sim \delta B c/4\pi R_* \gg B/P$ generated by the magnetic fluctuation δB. As a result, the magnetospheric charges are rapidly accelerated *throughout* the region of fluctuating fields, leading to the creation of secondary charges by the usual mechanisms. This process is not reconnection in the usual sense, because particle acceleration is not confined to a thin ohmic layer.

The net result is a pair fireball, part of which expands away from the star along open field lines and part of which is trapped in regions of closed field lines. We identify the initial hard transient of the 1979 March 5 event with the escaping fireball. The energy trapped near the star is converted into an optically thick pair/photon plasma, which is also contaminated by a trace of baryonic matter blown off the surface. We suggest that the soft-spectrum, long-duration (> 200 s) oscillatory tail of the March 5 event was radiated by such a magnetically-confined plasma. The subsequent SGR bursts had spectra and peak fluxes comparable to those of the March 5 soft tail, but with much shorter durations and smaller total energies, $E_{SGR} \sim 10^{41}$ erg. We suggest that these bursts were radiated by the same mechanism, and were triggered when the diffusing crustal magnetic field had built up sufficient stress to crack the crust, abruptly releasing both magnetic and crustal deformation energy in the form of Alfvén radiation.

How does the confined pair plasma release energy? Its surface is covered by a cool, optically thick, pair-depleted skin where the opacity is dominated by the electron-baryon contaminant. This skin is subject to an instability which causes the temperature gradient to increase, the photon pressure to drop, and the heavy electron-baryon component to settle to the stellar surface. As energy is radiated away, the skin layer moves inward. An analytic solution for this "cooling wave" (given below) shows that the emergent radiative flux F is insensitive to the deep interior temperature of the pair plasma, because F is determined at the temperature $T_\star$ where the electron-proton density comes to dominate the pair density. This flux gets as large as 10^4 times the Eddington flux (as indicated for the soft tail of the March 5 event and the ensuing SGR events) only if electron scattering is suppressed by a very strong surface magnetic field, $\sim 10^{14}$ G or larger.[2,4] An ordinary pulsar field of $\sim 10^{12} - 10^{13}$ G is also disfavored for the March 5 source on the grounds that it is too weak to confine the energy released in the soft tail, and for other reasons detailed below.

Solving the coupled energy and mass conservation equations, under the assumption that the baryon-photon fluid is in quasi-hydrostatic equilibrium, one finds (in units where $\hbar = c = 1$),[4]

$$F \simeq (g\Delta R)^{1/4} \left[\frac{g}{\kappa(T_\star, B)} \cdot T_\star^4 \right]^{1/2}. \tag{1}$$

In this equation, ΔR is the radial dimension of the confinement volume, and $T_\star = 34$ keV for $B = 10^{14}$ G (with a weak dependence on B, Y_e and ΔR). The electron-scattering opacity is suppressed below the Thomson value κ_T by a factor $\kappa(T, B) = (2\pi^2)^{-1} (T/\omega_B)^2 \kappa_T$, where ω_B is the electron cyclotron frequency. The emergent spectrum is approximately blackbody at a temperature

$$T_{\text{eff}} = \left(\frac{F}{\sigma_{SB}} \right)^{1/4} = 21 \text{ keV} \left(\frac{B}{10^{14}\text{ G}} \right)^{1/4} \left(\frac{\Delta R}{10\text{ km}} \right)^{1/16}. \tag{2}$$

The light curves of SGR events are given simply by $L(t) = F \cdot A(t)$, where $A(t)$ is the evolving surface area of the cooling, magnetically-confined pair plasma. The spectrum shows little variation with changing luminosity L, due to the very weak dependences in the formula for T_{eff} given above. Non-linear magnetic radiative processes such as stimulated photon splitting will slightly modify the emergent spectrum.[5] Although simplified, this model gives good quantitative fits to the observed flux and durations of SGR events.

We suggest that the same basic radiative mechanism is responsible for both Type II X-ray bursts and the SGR events, although the physical trigger is different.[4] There are a number of remarkable similarities between these two types of X-ray transients. In particular, individual Type II XRB's show very weak spectral evolution, and often display flat-top light curves.[6] We suggest that the sudden accretion of material from a disk onto the neutron star (the standard triggering mechanism[6] for Type II XRB's) excites MHD waves in the magnetosphere which damp and generate a trapped pair plasma. The burst is emitted as the plasma cools. The much lower flux of Type II XRB's ($L \sim L_{edd}$) can be attributed to a weaker confining field, $B \sim 10^{12}$ G, and perhaps a smaller confinement volume. The fact that SGR events do *not* show any correlation between burst energy and time lapsed to the next (or previous) burst,[7] reinforces our conclusion that these events are not triggered by accretion, but by an instability of the stellar magnetic field.

What are some other observational signatures of magnetars, when they are not actively emitting SGR bursts? A strong internal magnetic field can, through Hall drift and ambipolar diffusion, power a large surface X-ray flux, as well as *delay* cooling in non-superfluid regions of the stellar interior.[5] Neutrino cooling does not necessarily limit the surface photon flux when the internal heat source exceeds $\sim 10^{35}$ erg s^{-1}, as it does in the case of canonical pulsars,[8] because field lines can channel heat flow away from the interior of the star to its surface. The temperature in the deep crust adjusts to whatever value is required to maintain the surface X-ray luminosity, which amounts to

$$L_X \simeq 8 \times 10^{34} \left(\frac{B_{crust}}{10^{15} \text{ G}}\right)^{16/5} \left(\frac{t}{10^4 \text{ yr}}\right)^{-4/5} \text{ erg s}^{-1} \qquad (3)$$

from the crust alone.[4] The Hall turbulent cascade downward from a scale ℓ results in an ohmic dissipation rate that is proportional to B^3/ℓ^2 per unit volume.[9] A strong magnetic spot would, as result, produce a significant hemispheric asymmetry in the X-ray flux.

We now outline five different indicators that $B \geq 10^{14}$ G for the 1979 March 5 burst source:

(1) Magnetic torques can spin down the star to $P = 8$ s in the age $t_{SNR} \sim 10^4$ yrs of the surrounding supernova remnant N49 only if the surface dipole field is as strong as[1]

$$B_{dipole} = 6 \times 10^{14} \left(\frac{P}{8 \text{ s}}\right) \left(\frac{t_{SNR}}{10^4 \text{ yr}}\right)^{-1/2} \text{ G}. \qquad (4)$$

The displacement of the star from the SNR center indicates a recoil velocity ~ 800 km s^{-1}, sufficient to disrupt even a tight a binary.[1] This suggests that the star was probably not spun down by accretion.

(2) The soft tail of the March 5 event had a total energy $E_{tail} \approx 3 \times 10^{44}$ erg. Confinement of this energy (in the form of a pair plasma) by a closed magnetic flux loop of size ΔR requires that the field pressure at the outer boundary of the loop exceed $B^2(\Delta R)/8\pi > \frac{1}{3}E_{tail}(\Delta R)^{-3}$. One can also argue that ΔR does not greatly exceed the stellar radius, because otherwise the emissions would not be strongly modulated on the 8 s rotation period. Then the surface field must be stronger than

$$B_\star > 4 \times 10^{14} \left(\frac{\Delta R}{10 \text{ km}}\right)^{-3/2} \text{ G}, \qquad (5)$$

in the rough approximation where $B(\Delta R)/B_\star \sim 2^{-3}$.

(3) Magnetic suppression of Thomson scattering[2] plays an important role in allowing a luminosity as large as $L \sim 10^4 L_{Edd} \sim 10^{42}$ erg s^{-1}, where L_{Edd} is the Eddington luminosity, to escape from near the surface of the neutron star. The cooling wave solution [eqn. (1)] gives

$$B_{photosphere} = 1.2 \times 10^{14} \left(\frac{L}{10^{42} \text{ erg s}^{-1}}\right) \left(\frac{\Delta R}{10 \text{ km}}\right)^{-9/4} \text{ G} \qquad (6)$$

at the photosphere, under the assumption that the emitting area is $2\pi(\Delta R)^2$. Note that eqn. (6) is a weighting of the magnetic field strength over the surface of the confining flux tube.

(4) If the March 5 event was indeed triggered by a large scale rearrangement of the stellar magnetic field, then the total energy of the event, $E_B \approx 5 \times 10^{44}$ erg, must be no more than a fraction of the available magnetic free energy. Approximately the external field as a dipole, we have $\frac{1}{12} B_{dipole}^2 R_\star^3 \gg E_B$ and

$$B_{dipole} \gg 0.8 \times 10^{14} \text{ G}. \qquad (7)$$

(5) The quiescent X-ray emission $L_X \sim 7 \times 10^{35}$ erg s^{-1} detected from the March 5 burst source[10], if powered by a turbulent Hall cascade[4,5], requires a field

$$B_{crust} \sim 3 \times 10^{15} \text{ G} \qquad (8)$$

in the deep crust. If the stellar field arises from an $\alpha - \Omega$ dynamo, then is not surprising that the internal field should be primarily toroidal and somewhat stronger than the external dipole field.

We have noted some striking similarities between the March 5 source and the peculiar X-ray pulsar 1E2259+586.[3,5] This leads us to consider the possibility that 1E2259+586 is an isolated magnetar, as well as the possibility that both 1E2259+586 and the SGRs are accreting neutron stars. 1E2259+586 is an unusual X-ray pulsar in that it has a very soft spectrum, and has been spinning down quite steadily over the time that it has been observed. (We note that a possible recent shift in the spindown rate[11] has a natural explanation as a crustal glitch in the magnetar model.)

The X-ray emission of 1E2259 could be produced by the decay of the crustal field, as we have described above. The dipole field required to spin down the star to its present period of 7 s in the $\sim 1.5 \times 10^4$ yr age of the surrounding SNR is[3] $B \simeq 4 \times 10^{14}$ G. The dissipation of this field energy in the same amount of time correpsonds to $L_X = 3 \times 10^{35}$, within a factor $2-3$ of the observed[11] X-ray luminosity. Accretion is, of course, an alternative explanation for the X-ray emission, which should be taken seriously given the possible detection of a cyclotron line.[11] Perhaps the best alternative model for 1E2259+586 (one which accomodates the lack of an optical companion or binary modulation) is one in which the neutron star has a disk but no binary companion.

Nonetheless, it is very difficult to understand how accretion could power a burst of the enormous energy, fast rise time and (initial) hardness of the March 5 event. The hyper-Eddington flux of the soft tail of that burst, which lasted for at least 200 s, is especially difficult to understand with this approach.

This work was supported by the NSERC of Canada, by NASA grant NAGW-2418, and by the Texas Advanced Research Program.

REFERENCES

1. Duncan, R.C. & Thompson, C., *Astrophys. J.*, **392**, L9 (1992)
2. Paczyński, B., *Acta Astron.*, **42**, 145 (1992)
3. Thompson, C. & Duncan, R.C., *Astrophys. J.*, **408**, 194 (1993)
4. Thompson, C. & Duncan, R.C., in preparation (1993)
5. Thompson, C. & Duncan, R.C., *Compton Gamma Ray Observatory* eds. M. Friedlander *et al.* (NY: AIP) 808 (1993)
6. Lewin, W.H.G., & van Paradijs, J., & Taam, R.E., *Space Sci. Rev.*, **62**, 223 (1992/3)
7. Laros, J.G., *et al.*, *Astrophys. J.*, **320**, L111 (1987)
8. Van Riper, K.A., *Astrophys. J. Supp.*, **75**, 449 (1991)
9. Goldreich, P. & Reisenegger, A., *Astrophys. J.*, **395**, 250 (1992)
10. Rothschild, R.E., *et al. Compton Gamma Ray Observatory*, eds. M. Friedlander *et al.* (NY: AIP) 808 (1993)
11. Iwasawa, K., Koyama, K., & Halpern, J.P., *Publ. Astron. Soc. Japan*, **44**, 9 (1992)

INSTRUMENTATION AND NEW ANALYSIS TECHNIQUES

GAMMA-RAY OPTICAL COUNTERPART SEARCH EXPERIMENT (GROCSE)

Carl Akerlof, Marco Fatuzzo, Brian Lee
University of Michigan, Ann Arbor, MI 48109

Richard Bionta, Arno Ledebuhr, Hye-Sook Park
Lawrence Livermore National Laboratory, Livermore, CA 94550

Scott Barthelmy, Thomas Cline, Neil Gehrels
NASA/Goddard Space Flight Center, Greenbelt, MD 20771

ABSTRACT

The requirements of a gamma-ray burst optical counterpart detector are reviewed. By taking advantage of real-time notification of bursts, new instruments can make sensitive searches while the gamma-ray transient is still in progress. A wide field of view camera at Livermore National Laboratories has recently been adapted for detecting GRB optical counterparts to a limiting magnitude of 8. A more sensitive camera, capable of reaching $m_v = 14$, is under development.

INTRODUCTION

The persistent mystery of the origins of the gamma-ray burst phenomena has escalated the interest in searching for counterparts at other wavelengths. Since follow-up observations with conventional radio and optical telescopes require relatively small error boxes, the BATSE burst locations by themselves have not yet led to detections at these longer wavelengths. Data from various Interplanetary Network satellite configurations has provided much better localization but the time delays to acquire and process the results is at least 8 hours.[1] With a response time of days following the event, no reliable optical counterparts have been found for GRBs down to a limiting magnitude of $m_v = 24$ with the exception of a few soft gamma ray repeaters.[2] We wish to radically reduce this response time to seconds by making use of the BACODINE GRB early warning system[3] developed by Scott Barthelmy at Goddard. Charles Meegan in the introductory talk[4] for this workshop showed a histogram of the durations of BATSE events (see figure 1). The median event durations are approximately 6 seconds for 50% of the flux (τ_{50}) and 20 seconds for 90% (τ_{90}). The goal of the GROCSE experiment is to aim a sensitive wide field of view camera to estimated burst coordinates within these time scales.

No reliable theories have attempted signal detection limits to compute the optical luminosity of GRBs. Since there are no Arabic names for these phenomena, we can immediately infer a limiting visual magnitude of the order of 2 or more. Recent results from the ETC detector[5] push the limit to at least 6. Since

© 1994 American Institute of Physics

the intrinsic detection limit for such an optical system is set by the brightness of the night sky, we can roughly estimate the performance of a detector from the following approximate relations:

Night sky photon rate per pixel:

$$N_{\text{sky}} = 9 \times 10^7 \frac{\pi}{4} \frac{\text{pixel area}}{f^2} \text{ photons/pixel} - \text{sec} \qquad (1)$$

where *pixel area* is given in cm² and f is the focal length/aperture ratio.
Photon flux from a star with magnitude m_v:

$$N(m_v) = 10^6 (2.51189)^{-m_v} \text{ photons/cm}^2 - \text{sec} \qquad (2)$$

If all the starlight falls within a single CCD pixel, the 5 σ detection limit is given by:

$$m_v = 1.086 \log \left(18.7 \cdot f \cdot D^2 \sqrt{\frac{\eta t}{\text{pixel area}}} \right) \qquad (3)$$

where η is the detector efficiency, t is the exposure time in seconds, and D is the lens aperture in cm.

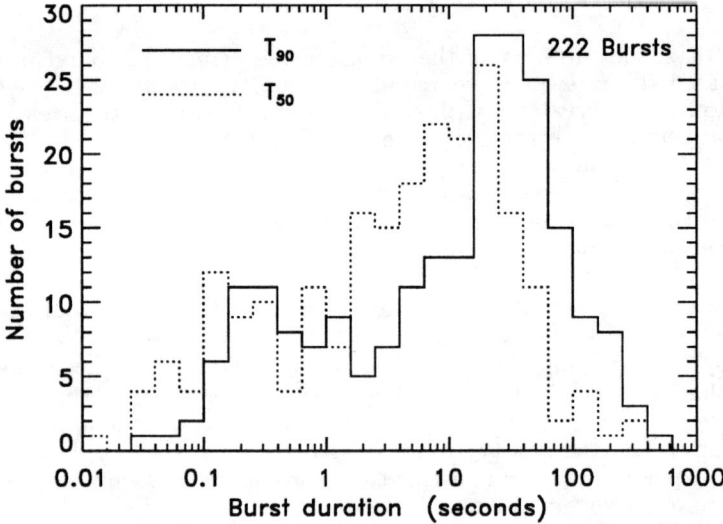

Figure 1. BATSE burst duration distribution (from C. Meegan, et al.)

Certainly this formula is not expected to predict the absolute performance level of a single system but for similar image sensor technology it should enable comparison of two different optical systems. From the known performance of the ETC detector, it appears possible to reach a magnitude $m_v = 14$ sensitivity with a lens aperture of 89 mm and moderate f number.

A limiting visual magnitude of 14 is an interesting benchmark because it defines a power level of about 10^{-11} ergs/cm^2-sec at the Earth. Since the brightest GRBs have been detected with energy fluxes of the order of 10^{-4} ergs/cm^2, $m_v = 14$ corresponds to an optical X-ray/γ-ray power ratio of 10^{-6}, assuming a typical pulse duration of 10 seconds. This is also roughly the ratio of the quantum energies of the photons. In the absence of any guidance from a theory of gamma-ray bursts, this seems like a useful goal to aim towards.

The maximum event rate can be derived directly from the BATSE value of about 0.8 events per day. By using the GRB coordinates supplied by the BACODINE system, we should be able to cover π steradians of sky with an average duty cycle of 10%, averaged over a year. This corresponds to 0.02 events/day or 7 events/year.

Table 1. Livermore WFOV Camera

focal length	250 mm
aperture	89 mm (f 2.8)
imaged FOV	0.621 steradians
image reduction	3.8 : 1
23 image intensifier - CCD sensors	
CCDs	384 × 576 Thomson
pixel size	23 μ × 23μ
pixel coverage	1.2 arc-min
exposure time	0.1 - 1.0 sec
camera mount	Contraves inertial guidance test system
slew rate	100°/sec ; 200°/sec
limiting magnitude	8.0
Response time	BATSE → GSFC 4.5 secs GSFC → LLNL 2.7 secs Camera slewing ~5.0 secs

To begin our search for optical counterparts, the GROCSE collaboration has adapted a wide field of view camera which was originally designed for the Strategic Defense Initiative program (SDI). The essential element of this camera is a wide field lens of a rather unusual design. All surfaces are spherically concentric, leading to a similarly shaped focal surface, as shown in figure 2. The SDI application required rapid frame rates so that the focal surface is divided into 23 rectangular segments, each of which is coupled via a coherent fiber optic bundle to an image intensifier followed by a second fiber optical bundle interfaced to a CCD. The disadvantage of such a system is a fairly poor overall quantum efficiency. The salient characteristics of this camera are listed in Table I. A schematic drawing of the camera is depicted in figure 3.

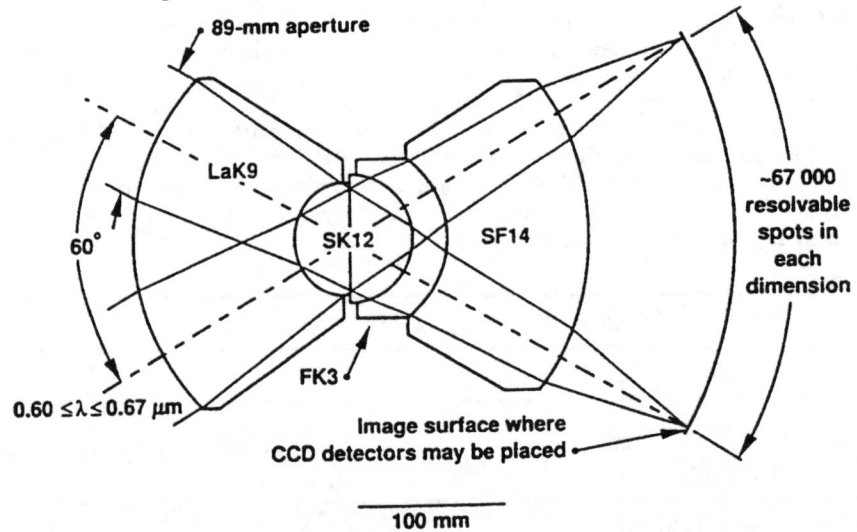

Figure 2. Lens design for the Livermore 89 mm diameter wide-field-of-view camera.

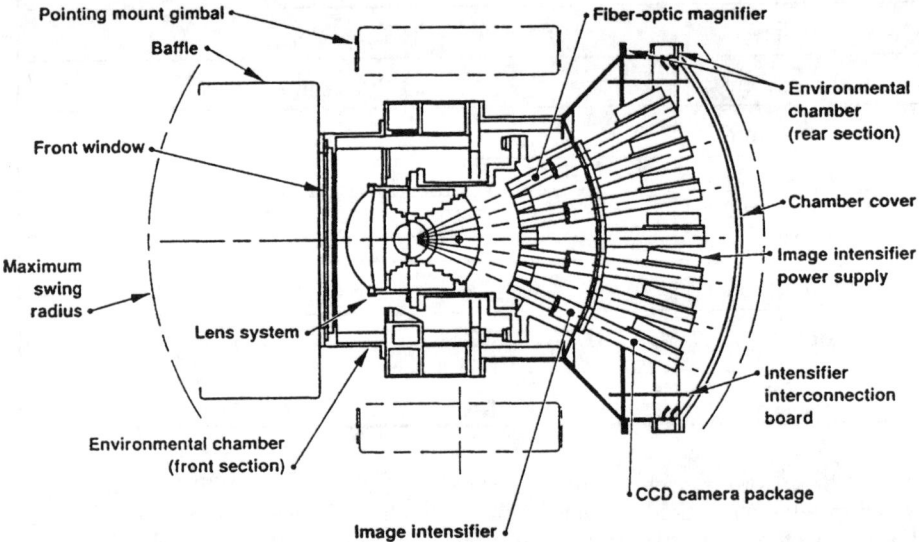

Figure 3. Schematic diagram of the Livermore wide-field-of-view camera imaging system.

The camera is mounted in a Contraves alti-azimuth inertial guidance test system capable of accurate and rapid slewing to a designated celestial coordinate. The total response time to a BATSE burst is given by the sum of three numbers: 4.5 seconds from burst detection onboard GRO to coordinate computation at Goddard, 2.7 seconds to send a burst coordinate packet via Internet to Livermore, and about 5 seconds to slew the camera. This sum of about 12 seconds is approximately half of the τ_{90} burst duration.

Data taking, using the BATSE-generated trigger system, has just begun. The main task we now face is developing the image processing software that can identify transient stellar-like objects against a background of satellite tracks. Comparison of star images with the SAO catalog shows that we are operating with a limiting magnitude of about 8. During the next 12 months, we expect to bring the current wide field camera into routine operation, taking observations whenever weather permits. At the same time, we are also developing a prototype system using similar optics but with considerably improved imaging quantum efficiency. With such a second generation optical detector, we hope to attain the threshold sensitivity required to finally identify GRBs at optical wavelengths.

REFERENCES

1. Kevin Hurley, this proceedings
2. Bradley Schaefer, this proceedings
3. Scott Barthelmy, this proceedings
4. Charles Meegan, this proceedings
5. Hans Krimm, this proceedings

ALBEDO EFFECT ON THE EXPECTED IN-FLIGHT PERFORMANCE OF THE GAMMA-RAY BURST MONITOR ON BOARD THE SAX SATELLITE

F. Alberghini,[1] D. Dal Fiume,[1] F. Frontera,[1,2] G. Pizzichini[1]

[1] Te.S.R.E.-C.N.R.-Via Gobetti 101, Bologna Italy
[2] Physics Dept., University of Ferrara, via Paradiso 12, Ferrara Italy

ABSTRACT

The performance of the gamma–ray burst monitor on board the Italian Dutch X-Ray Astronomy satellite SAX has been investigated. Following previous simulations, we compute the effect of the Earth's atmospheric albedo on the expected observations of gamma–ray bursts.

1. Introduction

The SAX satellite is a spacecraft, dedicated to X-ray astronomy, under construction by the Italian Space Agency (ASI) in collaboration with the Netherlands Agency for Aerospace Programs (NIVR). It will fly on a circular, almost equatorial, orbit at an altitude of 600 km. On board SAX the Phoswich Detector System (PDS) is the high energy instrument, operative in the 15-300 keV energy band [1]. It has a lateral active anticoincidence shield that will be used also as a gamma–ray burst monitor (GRBM). The shield is made of four optically independent CsI(Na) scintillator slabs positioned on the lateral surface of a rectangular parallelepiped. Their dimensions are: 10 mm thickness and 275 mm length; for two detectors the width is 402 mm, while for the other two it is 384 mm. The nominal energy range of GRBM is 60 to 600 keV. We will use GRBM mainly to study detailed temporal intensity profiles. The minimum temporal resolution is 0.48 msec. The PHA spectra of the shield count rates, with 128 s integration time, and the integral count rates for each lateral shield, with 1 s integration time, are continuously transmitted to the ground station as housekeeping.

In a previous paper [2] the results of a simulation of the PDS and GRBM instrumentation on board SAX were reported. It was shown that the different count rates for each slab, after subtracting background and albedo counts, allow a determination of the burst direction. The maximum expected angular resolution is 10 arcmin and the field of view is 2π, due to the spacecraft mass which covers one half of the sky. The sensitivity and the angular resolution are functions of the burst intensity and of the relative burst–instrument orientation.

In this paper we study, at least in a preliminary way, the albedo effect on the count rates.

2. Model atmosphere and burst

Our aim is to evaluate the effect of the burst albedo by the earth's atmosphere both on the burst time history and on the energy spectrum as measured by GRBM. To this end we study the response of the atmosphere to a gamma–ray burst (GRB) by using a Monte Carlo simulation (code EGS4 [3]). For the time being we modelled the atmosphere as a plane slab. We plan to take into account the Earth's curvature in our future work. For the model atmosphere we used two configurations:

1) a single air slab at standard temperature and pressure: in the simulations we used a barometric formula in order to obtain physically meaningful values;
2) an atmosphere made of 20 air slabs: the density and pressure for each slab were derived from the COSPAR model atmosphere [4].

The two configurations give results in good agreement with each other. We plan to use more recent upper atmosphere models [5] in our future work, but we expect to find only minor changes in our results given that the differences between the two model atmospheres [4] and [5] are within 10–15% in the upper layers.

The geometry used in our simulations is shown in Figure 1.

We used different orientations of GRBM relative to the atmospheric slab and to the incoming burst direction. Since the satellite is three–axis–stabilized, θ and ϕ (see figure 1) shall not change during the event unless the satellite is slewing. Because of the flat model atmosphere we used only $\alpha < 10°$ (see figure 1).

We first performed our simulation runs with a uniform energy distribution for the incident gamma-rays in order to obtain statistically meaningful estimates of the important parameters at all energies. We then used the model spectrum of a real GRB (GRB910814 [6]), a power law with an energy break at 1890 keV. Because of the high value of the photon index above the break (-3.70), only the spectrum below the energy break has been used. For test purposes we also simulated some monochromatic lines between 100 keV and 1.5 MeV. The albedo spectrum contributes to the total observed PHA spectrum mainly below 400 keV. Our simulation code has a lower cut at 30 keV, but this is not relevant. It is clear from figure 2 that the albedo is already negligible below 40 keV. We recall that the lower energy threshold of the instrument is 60 keV.

The intensity of the 511 keV annihilation line in the albedo spectrum is always very low.

We simulated 2 cases of temporal intensity profiles:
1) the burst intensity falls uniformly within one temporal bin only (0.48 ms);
2) the intensity profile is modelled by a triangle [7].

By using sequences of *"one bin"* bursts (case 1) with the appropriate normalization we obtain the mean profile (burst and albedo) seen by GRBM in case 2.

3. Results

The zenith angle distribution (figure 3) of the scattered photons has a maximum at angles from the horizon between 45° and 55°.

The arrival times of the albedo photons at the spacecraft are energy independent. We find that the main contribution to the delay of the albedo photons is due to the travel time to the atmosphere and back to SAX. The delay due to the interactions in the atmosphere is within our time resolution of 0.48 ms.

Figure 4 shows the deformation of a triangular intensity profile due to the time spread of the albedo photons. As one can see in detail from the following figures 5 and 6, which show the effect of the albedo scattering on the intensity measured by the GRBM, the albedo intensity never reaches more than 7 % of the original burst intensity and the delay always ranges between 3 and 6 msec.

The instrumental model must be completed by a simulation of the entire satellite mass distribution . We also need to use a more recent model for the atmosphere, which has to be spherical, in order to consider the more general case in which the direction of the GRB is not aligned with SAX and Earth.

References

1. F. Frontera et al., Il Nuovo Cimento **15C**, 867 (1992).
2. M. Pamini et al., Il Nuovo Cimento **13C**, 337 (1990).
3. R.L. Ford, W.R. Nelson, "The EGS code system: computer programs for the Monte Carlo simulation of electromagnetic cascade showers (version 4)" (SLAC-210 UC-32, June 1985).
4. COSPAR Working Group IV, "CIRA 1965" (North-Holland, 1965).
5. A.E. Hedin, J. Geophys. Res. **92**, 4649 (1987).
6. B.E. Schaefer et al., in "Gamma-Ray Bursts", AIP Conference Proceedings 265, W.S. Paciesas and G.J. Fishman eds. (American Institute of Physics, New York, 1992), p. 180.
7. J.C. Lochner, in "Gamma-Ray Bursts", AIP Conference Proceedings 265, W.S. Paciesas and G.J. Fishman eds. (American Institute of Physics, New York, 1992), p. 289.

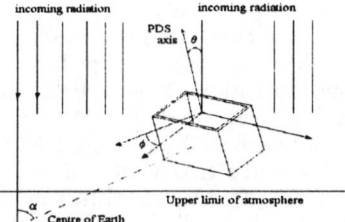

FIGURE 1.- This figure describes the angles used in our simulations. The GRB direction is perpendicular to the air slab. α is the angle at the Earth's center, between the GRB direction and the spacecraft position. φ: the angle by which the GRBM must be rotated around the PDS viewing axis so that the "narrow" slabs are parallel to the GRB-PDS axis plane. θ: rotation angle around the axis perpendicular to the narrow slabs so that the PDS viewing axis becames parallel to the incoming radiation. Note: the order of the two rotations cannot be interchanged.

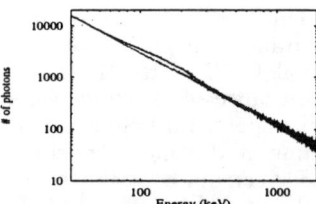

FIGURE 2.- The incident GRB spectrum (lower line) and the sum of incidennt and albedo spectrum. The number of incident photons is $1 * 10^6$.

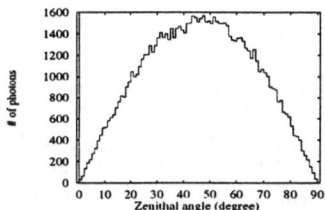

FIGURE 3.- The distribution in zenith angle for albedo photons when they leave the atmosphere. The angles are measured from the horizon. The number of incident photons is $2 * 10^6$.

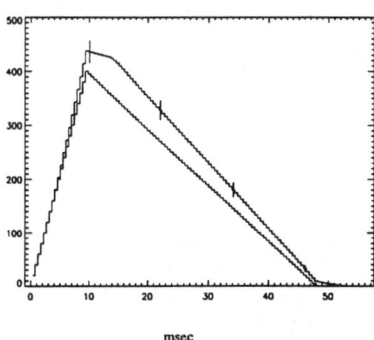

FIGURE 4.- Temporal profile deformation of a GRB with a triangular intensity profile. Lower line: GRB light curve; upper line: apparent GRB light curve, including albedo photons, with 1 sigma errors. In the case shown $\theta = 90°$ and $\phi = 45°$ (see fig. 1).

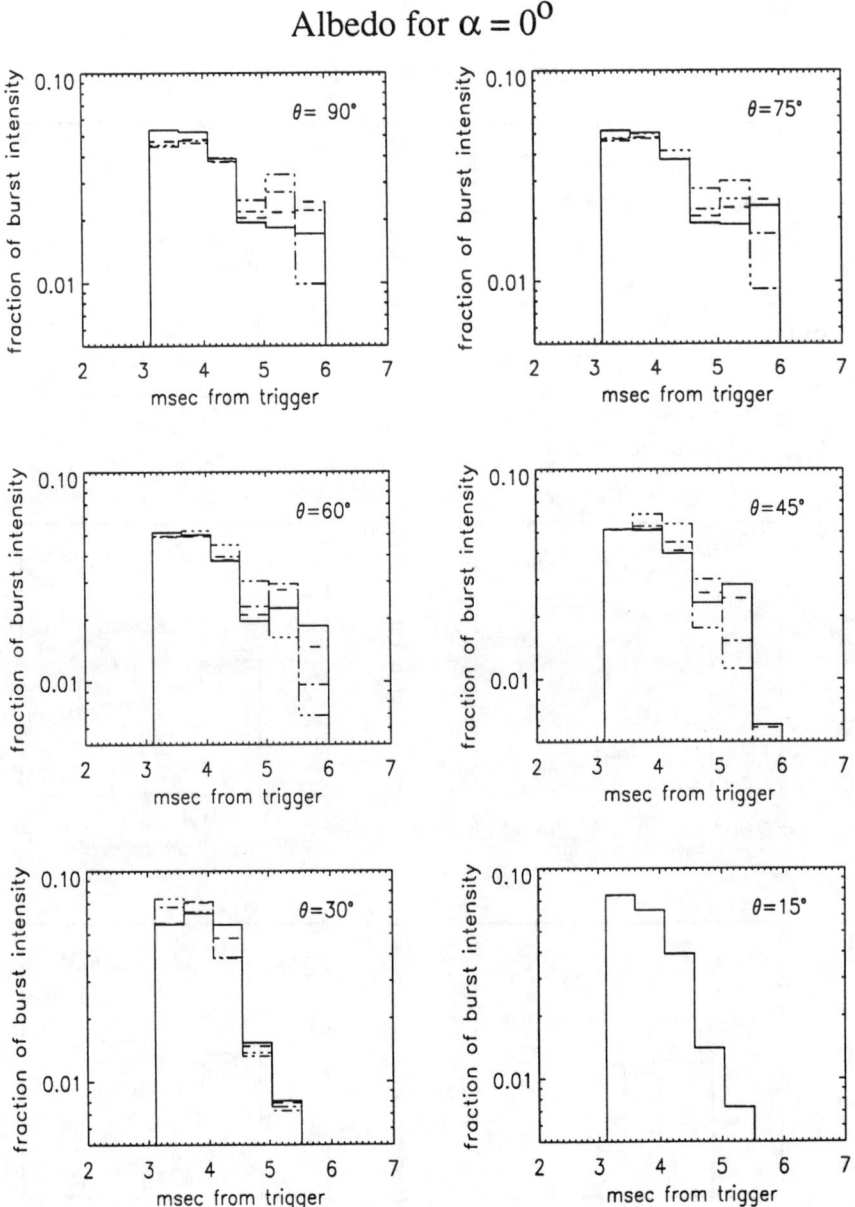

Figure 5. The intensity of albedo relative to the intensity of the incident burst, at different values of θ and ϕ (see Figure 1). In each panel we report 4 albedo time profiles: solid line $\phi=0°$; dashed line $\phi=15°$; dot-dashed line $\phi=30°$; dot-dot-dashed line $\phi=45°$. For small values of θ the albedo does not depend on ϕ.

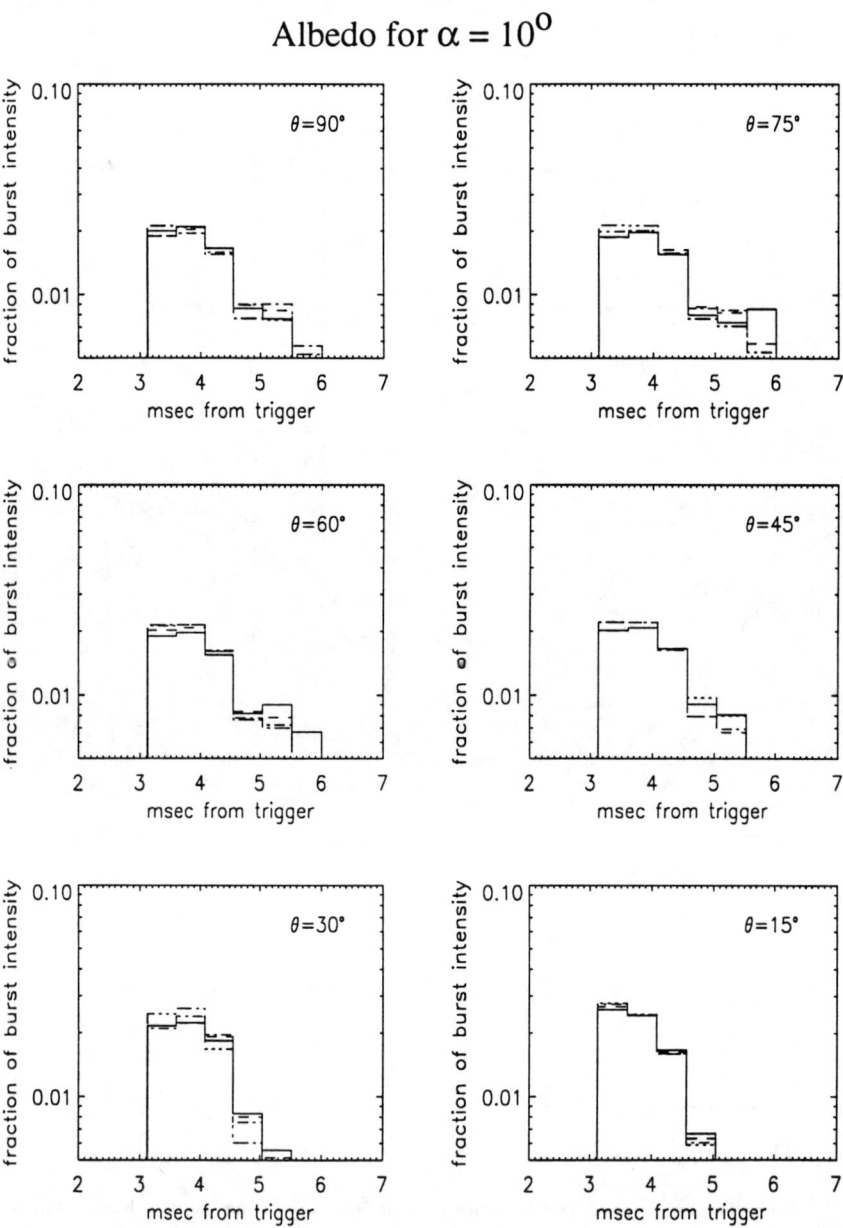

Figure 6. The intensity of albedo relative to the intensity of the incident burst, at different values of θ and ϕ (see Figure 1). In each panel we report four albedo time profiles: solid line $\phi=0°$; dashed line $\phi=15°$; dot-dashed line $\phi=30°$; dot-dot-dashed line $\phi=45°$. For small values of θ the albedo does not depend on ϕ.

BACODINE
The Real-Time BATSE Gamma-Ray Burst Coordinates Distribution Network

S.D. Barthelmy, T.L. Cline, N. Gehrels
Code 660
T.G. Bialas, M.A. Robbins
Code 564
NASA - Goddard Space Flight Center
Greenbelt, MD 20771

J.R. Kuyper
U. of Maryland
College Park, MD 20742

G.J. Fishman, C. Kouveliotou, C.A. Meegan
NASA - Marshall Space Flight Center
Huntsville, AL 35812

ABSTRACT

The real-time transmission of the data from the C-GRO spacecraft allows for access to the BATSE instrument data that can be used to make simultaneous or near-simultaneous multi-band observations of Gamma Ray Bursts (GRBs). We discuss a system that (1) monitors this telemetry stream, (2) extracts the appropriate information from the BATSE portion, (3) detects the occurrence of a GRB, (4) calculates the approximate coordinates for the burst, and (5) distributes those coordinates to instruments, observatories, and other interested parties. This is done with custom hardware plus software located at the Goddard Space Flight Center mission operations center for C-GRO. The maximum time delay between the arrival of the GRB photons at the BATSE detectors and the calculation of the coordinates is 5.5 seconds. The accuracy of the calculated coordinates is ± 10 degrees. The coordinates can be distributed by dedicated phone connections, by direct computer-to-computer socket connections, and by e-mail to any automated, optical telescope systems with wide field-of-view capabilities. This rapid distribution of coordinates (0.3 to 30 sec) allows for a 5 to 6 order of magnitude improvement over previous efforts in the response time of follow-up observations of the optical emission of GRBs. This greatly increases the likelihood of optical flash detection and hence, of the quiescent counterpart identification of GRBs.

INTRODUCTION

Gamma Ray Bursts (GRB) have been studied for more than 20 years with more than a thousand detected since their discovery[1]. Yet, with the notable exception of the transient series called "soft gamma-ray repeaters"[2,3,] which may or may not be related to "classical" GRBs, none have identified source objects. Alternative strategies, therefore, have been initiated to search for emission in other energy regimes.

Two different strategies have been employed to make follow-up observations to identify the GRB source object in the optical and other band passes. The first method is to "stare" at the sky in the hopes that a GRB will occur in the FOV. Because it is impossible to predict when and where a GRB will occur, very large fields-of-view (FOV) are required to get a reasonable probability of "observing" a GRB. Because the large FOVs result in poor spatial resolution and crowded fields at the detector[4], these methods have poor sensitivity -- 4th magnitude at best. The second method improves the sensitivity by limiting the FOV at the expense of making the observation with some time delay after the GRB. By waiting till after the GRB, rough positional information from instruments in orbit (C-GRO-BATSE/COMPTEL, IPN, WATCH-GRANAT/EUREKA, IPN) can be obtained to direct narrower FOV instruments to the GRB error box. While the time delays between the GRBs and the observations have been steadily improving over the years -- currently they are down to 5-to-36-hours range[5-7] -- still no positive identification of a GRB source object has been made. The general conclusion of these observations is that the optical emission from the source objects must have faded below ~20th magnitude within several hours.

Since detector technology is unlikely to develop in the next few years to the point where the first method (stare mode) is likely to yield any new or different results, attempts are being made to improve the second method (follow-up). Clearly, the desired goal is to obtain positional information on the GRB while the burst is still ongoing and get that information to instruments capable of making rapid follow-up observations.

THE SYSTEM

To that end, we have built a system of hardware and software that calculates positions of GRBs detected by the BATSE instrument on C-GRO and distributes those coordinates to instruments around the world within a time short enough that most GRBs are still bursting. The system is called the BATSE COordinates DIstribution NEtwork (BACODINE). It is only now possible to do this because of the failure of the on-board C-GRO tape recorders. Currently, all the data from GRO is transmitted to ground in real-time -- with essentially no time delay.

BATSE detects 0.8 GRBs per day. Folding in multiplicative factors for the fraction of the sky that any ground-based instrument can see (0.35), the fraction when it is night (0.50, for the IR to UV band passes), the fraction of clear-sky weather (0.75), and the "program efficiency (0.40, see discussion below); the rate of making follow-up observations is 0.042 GRBs per day or once every 24 nights. If factors for New Moon (0.50), for the brightest GRBs (0.50), and for the longest GRBs (0.50) are included; then the rate is 0.005 per day or once every 6 months.

BATSE has two types of data from the 8 Large Area Detectors (LAD)[8]. The high time/spectral resolution data are buffered on-board the spacecraft for delayed transmission to the ground, but there is also 1.024-sec resolution rough spectral resolution data which are transmitted in real time. Figure 1 is a timeline of the sequence of steps in the flow of this real-time data from BATSE to the BACODINE processing system. The count rates for the 8 LADs in 4 energy intervals are accumulated for two 1.024-sec intervals and then transmitted to a TDRS satellite over the next 2.048 seconds. The next two 1.024-sec counting rate samples are accumulated while the transmission of the previous 2 samples is in progress, and this process repeats continuously. For those portions of the orbit where C-GRO can get direct line-of-sight transmission of its high-gain antenna to either of the two TDRS satellites, the data is relayed to the NASA White Sands Ground Station in New Mexico where is it retransmitted to DOMSAT and then transmitted back down to the Goddard Space Flight Center Data Capture Facility. There is an additional 1.0 seconds of time delay due to four hops of ground-to-geosync-orbit light-travel time and buffering within the White Sands facility. Once received at GSFC, the entire 2.048 sec of data is processed to yield GRB coordinates within 0.1 sec. The fastest method of coordinates distribution (see below) takes an additional 0.3 seconds. If the GRB started during the beginning of the first 1.024-sec count-rate sample, then the total time delay between when the gamma rays interacted with the BATSE LADs and when the coordinates are available at an instrument to make follow-up observations is 5.50 seconds. If the GRB started at the end of the second 1.024-sec sample, then the time delay is 3.45 seconds. More than half of the GRBs are longer than 5.5 seconds, thus allowing follow-up observations to be made while the burst is still occurring.

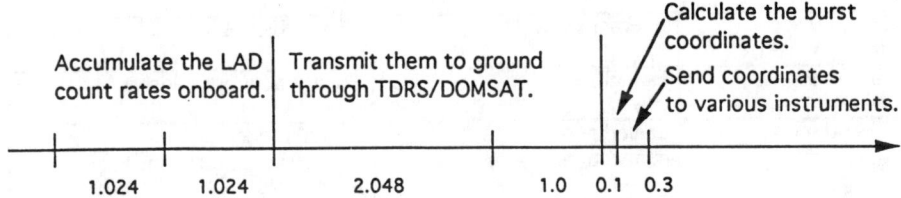

Fig 1: The time delay sequence of events between the GRB and the availability of the coordinates. The 2 asterisks show the time positions of 2 GRBs -- one at the begiining of the first 1.024-sec time sample and another at the end of the second time sample.

While there are several programs running on several computers to do the processing of the BATSE data, the sequence of processing can be broken down into five general steps. (1) The program monitors the telemetry stream continuously extracting the count-rates for the 8 LADs in the 4 energy intervals (25-50, 50-100, 100-300, >300 keV). It uses these rates to determine the current background rates. It also extracts some general purpose housekeeping information (spacecraft clock, RA,Dec orientation of the SC, etc). (2) It monitors the "burst-in-progress" flag generated by the BATSE flight processor and when set true, it (3) takes the current count rates, subtracts the previously accumulated background rates to get the source-only rates, finds the 3 brightest detectors, and (4) solves the set of 3 simultaneous equations of the dot-product of the unknown burst direction and the detector normals of the 3 bright detectors. (5) The burst direction is then sent to a list of instruments that are capable of making follow-up observations.

Currently, the algorithm used to calculate the GRB direction assumes "ideal response" physics for the LADs. This approximation yields an uncertainty for the BACODINE burst position of about a 20° diameter error circle. As mentioned above in the "rate of observations" calculation, the program has an efficiency for bursts of 40%. By this we mean that sometimes the program will get confused by the data and calculate a totally wrong position. See below for a discussion on the improvements in this area.

Table 1 lists the 5 methods that will be available for distributing the BACODINE GRB coordinates. Currently, two methods are implemented. The fastest method is the dedicated phone line. Around sunset at the instrument site (assuming it's an optical instrument), a phone/modem connection is made between the BACODINE computer and the computer at the instrument site. This connection is maintained throughout the night and should a burst occur during this time then the coordinates (RA, Dec, UT) are sent over the connection. At 9600 baud it takes 0.3 seconds. The second fastest method, and much less costly, is the Internet socket connection. Sockets is a technique to connect two computers over a network. Like the dedicated phone method, the socket connection is made at some initial time and maintained for long periods of time. The time delay for the propagation of the coordinates packet varies due to the distance between the two computers, the number of routers and gateways in between, and the amount of other network traffic. However, we have routinely shown that for a connection between Maryland and California (coast to coast US) the propagation time is 2.7 seconds[9].

Table 1: Coordinates Distribution Methods

TIME DELAY	METHOD/MEDIA	COMMENTS
0.3 sec (Implemented)	Modem	Dedicated phone connection. The fastest.
2.5 - 4.0 sec (Implemented)	Internet socket	GROCSE, MD-to-CA 2.7 sec
10 - 30 sec	E-mail	To any network address.
30 - 90 sec	Dialed phone connections	Digital modem, Synthesized voice, and FAX.
60 - 180 sec	Phone "beeper"	RA,Dec,UT in the message displayed on the beeper.

THE FUTURE

Currently, we are operating with routine connections to two instruments; one by each of the two implemented methods. We are looking for more instruments to include in the BACODINE system. Collaborations can be formed with groups who have instruments which are "able" to make appropriate follow-up observations. By able, we mean that the instrument has a FOV comparable to the accuracy of the BACODINE coordinates and that it can make the observations at least within a few minutes of the burst time. The later usually means that the system is automated to the point of being under computer control, fast moving, and with a minimum of humans in the loop. While we previously pointed out the motivation for minimizing delay time to make follow-up observations, we need to caution that since so little is known about GRBs we do not want to discourage any groups with instruments that would fall into the greater-than-10-minutes category. Because the observation of GRBs is a probability game, we would also like to encourage groups with sites at different longitudes and latitudes (in particular the southern hemisphere). Again, given that so little is know about GRBs, we want to encourage any and all band-passes: radio, IR, optical, UV, X and Gamma rays. Significant search strategies can be formulated for each band, and the science returned from each is useful and unique.

The current method for calculating the GRB burst positions from the BATSE data assumes an "ideal physics" response function for the LADs to the gamma rays from the burst. This is only an approximation. There are two major effects which distort the count rates in the LADs for a given burst. They are a non-cosine(theta) response and an earth and spacecraft scattered gamma ray contribution[10]. The non-cos(theta) response results from the LADs not being infinitely thin detectors (i.e. they have some effective area at a burst-to-detector-normal angle of 90°) and that they have some effective area for burst directions from behind the detector. We are currently working on the non-cosine(theta) correction and should have it implemented within a few months. We believe that the GRB position uncertainty will decrease by a few degrees. The earth scattering correction is larger and requires significantly more computations which may not be compatible with the real-time aspects of the BACODINE system. The "program efficiency" mentioned in the GRB rate calculations comes from the program's less than perfect handling of gaps in the telemetry caused by all the Loss and Reacquisition of Signal when C-GRO breaks and remakes TDRS connection during its orbit. Some of this confusion can be eliminated by writing smarter algorithms for handling the background. Also, the GRB follow-up observation rate will be improved by 30% when a TDRS satellite and ground station in Australia becomes fully operational in February 1994.

The e-mail distribution method will be implemented by the end of 1993. Methods 4 and 5 will be available some time in 1994. Currently under development is an automated, wide FOV, rapidly moving, CCD camera system that will be connected to BACODINE. It is fully automated and computer controlled. While its primary goal is be make rapid follow-up observations, it is envisioned that the computer automation can be extended to the near real-time analysis (tens of seconds) of the CCD images such that pixels that changes in intensity by an appropriate amount could be identified and then the much tighter coordinates of that "transient object" could be calculated and distributed to other more traditional, narrow FOV telescopes for much fainter, in-depth study and position determination.

ACKNOWLEDGMENTS

We would like to acknowledge the efforts of Cynthia Lenart & Kevin McGee (GSFC) for their development of the phone/modem control software. We also thank J.F. Ormes and members of the GRO Project office and PACOR for their enthusiastic support. Without the help of all these people, this undertaking would have been immeasurably more difficult, if not impossible. This work is supported under the GRO Phase 3 Guest Investigator program.

REFERENCES

1. R. Klebesadel, Olsen, I.B. Strong; Ap.J.Let.; ??, pp???; 1973

2. S.R. Kulkarni, D.A. Frail; Nature; 33, pp365; 1993

3. Y. Tanaka; I.A.U.Circular #5880; 1993

4. J. Greiner, et al.; Proceedings of the Compton Symposium, eds. M. Friedlander, N. Gehrels, D. Macomb, AIP Conf. Proc. #280; pp828; 1992

5. S.D. Barthelmy, et al.; these proceedings; 1993

6. B.E. Schaefer, S.D. Barthelmy, et al.; Ap.J.Let.; In press; 1993

7. M. Boer; these proceedings; 1993

8. G. Fishman, et al.; GRO Science Workshop Proceeding; pp2-39; 1989

9. C. Akerlof, et al.; these proceedings; 1993

10. M.N. Brock, et al.; Proceedings of the Compton Symposium, eds. M. Friedlander, N. Gehrels, D. Macomb, AIP Conf. Proc. #280; pp709; 1992

BATSE: BURST PERFORMANCE AND EXPERIMENT STATUS

G.J. Fishman[1], C.A. Meegan[1], C. Kouveliotou[2], R. Mallozzi[3], J. Horack[1], T. Koshut[3], G. Pendleton[3], W.S. Paciesas[3], R.B. Wilson[1] and M.N. Brock[1]

1 Code ES66, NASA/Marshall Space Flight Center, Huntsville, AL 35812 USA
2 Universities Space Research Association, Huntsville, AL 35812 USA
3 Physics Department, University of Ala. in Huntsville, Huntsville, AL 35812 USA

ABSTRACT

The current performance of BATSE is described, emphasizing the aspects related to gamma-ray burst observations. Flight software modifications since launch are briefly described which affect burst data, along with the methods of distinguishing true cosmic gamma-ray bursts from other sources of triggers. Finally, we describe a previously unreported, rare type of trigger that may have significance for atmospheric, ionospheric, and lightning research: terrestrial gamma-ray flashes.

BATSE FLIGHT HARDWARE AND SOFTWARE PERFORMANCE

The flight performance of the BATSE experiment flight hardware continues to be excellent, since its initial turn-on in April 1991. All PMT's, power supplies, detectors, and flight electronics continue to operate without failure. There has been no noticeable degradation in the resolution or performance of any detector. Several significant changes have been made in the flight software, primarily to compensate for the loss of the tape recorders on the observatory. Since the implementation of these changes, very little significant gamma-ray burst data have been lost due to telemetry gaps. The on-board gamma-ray burst trigger criteria have remained the same since launch (c.f. Fishman et al. 1992) except for brief periods when other celestial sources (primarily GRO J0422+32 and Cyg X-1) were active and producing large numbers of false triggers. Significant changes to the experiment flight software and operations that have been made since launch are listed in Table 1.

In addition to these operational improvements, there are on-going improvements in the quality and accuracy of burst data. These include: 1) Burst location accuracy improvements through the understanding and elimination of various systematic errors, 2) Improved energy calibration and detector efficiency calibration primarily though the on-orbit observation of known sources and 3) Improved detector response matrices and spectral analysis software. Most of these efforts are being accomplished under the direction of G. Pendleton. Along with this, the BATSE detector background is continuing to be modeled extremely accurately, by two independent semi-empirical methods (Rubin et al. 1993a; Skelton et al. 1993). This modeling will lead to a greater sensitivity in several BATSE research areas, including the detection of un-triggered gamma-ray bursts which do not meet the on-board trigger criteria (Rubin et al. 1993b).

Table 1

Major Flight Software and Operations Changes

Item	Reason/comments
1. Hold burst data in BATSE memory until real-time telemetry coverage	Prevents burst data loss due to failure of tape recorders on the spacecraft
2. Use of 1 kbs data stream for BATSE	Recovers many data functions of continuous data during telemetry gaps
3. Abbreviated burst readout for weak bursts	Many data types not needed for weak bursts; data space can be better utilized
4. Disable trigger in regions of high electron fluxes	Eliminates false triggers; more live-time for real GRB's
5. Operate most spectroscopy detectors at high gains	Emphasize and improve the search for low energy spectral features
6. Quick-Alert burst system	Rapid notification of the location of strong bursts to remote observers
7. Real-time burst system (BACODINE) (under development at GSFC - Barthelmy et al. 1994)	Will provide real-time notification of selected bursts to remote observers (with burst locations)

BURST TRIGGERS - TRUE AND FALSE

The total BATSE trigger rate and its two components are shown in Figures 1. BATSE triggers that are assessed to be caused by classical cosmic gamma-ray bursts are termed "true". All others (including those due to Soft Gamma Repeaters, SGR's) are termed "false". It can be seen from Figure 1 that the false trigger rate has dropped substantially since launch as solar activity is in the declining phase. Along with the reduced solar flare triggers, triggers due to various types of electron precipitation events have also decreased as the magnetosphere is less disturbed during solar-quiet times. These events, described below, are usually the greatest source of false triggers. The classification of triggers from the experiment is made by the BATSE Burst Team, under the direction of C.A. Meegan, taking into account many observable quantities. The methods of separating true bursts from false bursts are described by Mallozzi et al. (1993); Fishman et al. (1992), and Meegan et al. (1993). Independent burst observations and other analyses lead us to believe that the number of "true" gamma-ray bursts which are not real is less than 1% and that the number of "false" triggers which are really gamma-ray bursts is less than 2%. That these percentages are not equal reflects the fact that the assessment of bursts tends to be on the conservative side. The trigger determination for BATSE is facilitated by the eight independent detectors pointed in different directions. This gives BATSE a decisive advantage over gamma-ray burst experiments of only one or two detectors in distinguishing between true and false bursts.

Increased detector counting rates due to precipitating magnetospheric electrons often produce false triggers. The detectors themselves employ charged particle anti-coincidence shields. However, since the electrons have energies less than several hundred keV, most do not penetrate the outer detector coverings. These stopping electrons produce bremsstrahlung photons in the covers and in other nearby detector and spacecraft materials. Because the electrons are traveling in nearly circular orbits around the local magnetic field before impact, the detectors which face opposite directions record similar counting rates. Those detectors with axes that are closely aligned with the local magnetic

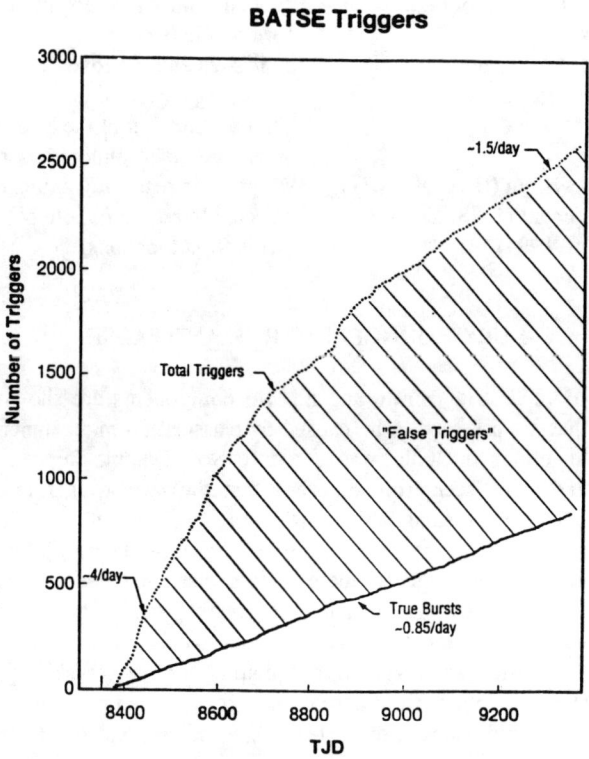

Figure 1. BATSE triggers during the first 2.7 years of the Compton Observatory mission, separated into true gamma-ray bursts and false triggers, of all types. The decreased solar activity after ~TJD 8700 has lead to a dramatic decrease in the false trigger rate. The rate of true bursts has been steady at ~0.85 bursts/day. There are several periods when the apparent gamma-ray burst rate is slightly reduced due to a decreased live-time of the experiment caused by excess false triggers.

field vector have the smallest rate increase. An anisotropy parameter which uses the rates from all eight detectors is a very sensitive measure of these events (Datlowe et al 1994). Early in the mission, it was noted that many of the events occurred in the vicinity of a powerful very low frequency (VLF) transmitter in Australia which was capable of scattering trapped electrons, causing them to precipitate (Horack and Fishman 1991; Horack et al. 1992). The intensity of these events is a sensitive function of the local time at which the spacecraft passes near the transmitter site, since the VLF waves do not propagate upward during the daytime (Datlowe et al. 1994).

A second type of electron precipitation trigger occurs when the BATSE detectors view bremsstrahlung photons from electrons precipitating into the atmosphere, at some distance from the spacecraft. These events are recognized by their location near the horizon and they are always at a relatively high geomagnetic latitude in the spacecraft orbit. Furthermore, these events, like those above, always have a characteristic, soft bremsstrahlung spectrum, never exceeding 300 keV.

ATMOSPHERIC GAMMA-RAY FLASHES

On rare occasions, the BATSE detectors have triggered on brief, intense flashes of hard gamma rays coming from the direction of the Earth's atmosphere. These flashes must originate at altitudes at least 30 km above the surface of the Earth, in order to be observable by the orbiting detectors. At least a dozen events have been detected over the past two years by BATSE. Several of these events are seen to come from the directions of what appear to be large storm systems, although concurrent weather imagery is not available in most cases. The photon spectra from the events are hard and consistent with bremsstrahlung from energetic (MeV) electrons. The most likely origin of these high energy electrons, while speculative at this time, is presumably a rare type of high altitude electrical discharge above thunderstorm regions. The unique features of these events (apart from their terrestrial origin) are their extremely hard spectra and their short duration.

Approximately half of the events consist of two closely-spaced pulses and one event has at least five distinct pulses of similar shape but variable spacing. It is likely that other, weaker events of similar origin go undetected due to the trigger criteria implemented by the experiment. Since the minimum sampling time for triggering the BATSE burst mode is 64ms, over ten times longer than the duration of most of these events, weaker events do not, in most cases, trigger the BATSE system.

It is believed that prior experiments were incapable of detecting the phenomenon reported here for various reasons, or these events were overlooked as being spurious. The time profiles of four events are shown in Figure 1. They consist almost entirely of pre-trigger, time-tagged-event (TTE) data. No events of this type are seen to come from above the local horizon.

The possibility of runaway discharges above thunderstorms was first discussed almost 70 years ago (Wilson 1925; Boys 1926), as it was recognized that ionization from a lightning discharge could create a conducting channel for further current flow and lead to a "runaway" discharge to perhaps very high altitudes. These have been modeled by Cole et al. (1966).

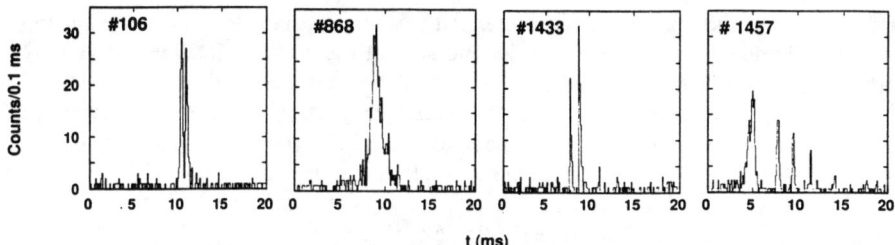

Figure 1. Time profiles of four of the gamma-ray flashes seen to come from the Earth's atmosphere. The time resolution of the plots is 0.1ms per bin. Typical rise and fall times of the peaks are ~0.1ms to 2ms.

SUMMARY AND FUTURE BATSE OPERATIONS

BATSE continues to provide an abundance of gamma-ray burst data. Its detectors and data system are performing well and have shown no degradation or hardware failures since launch. The recent re-boost of the Compton Observatory should assure many additional years of data. The publication of the First BATSE Gamma-ray Burst Catalog (Fishman et al. 1994) has been completed, and work is progressing on the Second Catalog. The inherent redundancy of burst data types, along with the software changes described above, have overcome the data losses which would have otherwise occurred. Other software modifications and improvements are expected to be made in the future.

The burst Quick Alert system has already allowed rapid counterpart searches of bursts as never before possible. The implementation of the BACODINE system will bring a breakthrough in burst research, allowing the possibility of counterpart searches while bursts are in progress.

REFERENCES

Barthelmy, S. et al., these proceedings.
Boys, C.V., Nature 118, 749-750 (1926).
Cole, Jr., R.K., Hill, R.D. and Pierce, E.T. J. Geophys. Res. 71, 959-964 (1966).
Datlowe, D.W. et al. J. Geophys. Res., submitted (1994).
Fishman, G.J. et al., p.13, in Gamma Ray Bursts, AIP Proc.#265 (1992).
Fishman, G.J. et al.. Ap. J. Supp. in press (1994).
Horack, J.M. and Fishman, G.J., NASA TM-103546 (1991).
Horack, J.M. et al., p.373, in Gamma Ray Bursts, AIP Proc.#265 (1992).
Mallozzi, R.S. et al. p. 1122, in Compton Gamma Ray Obs., AIP Proc.#280 (1993).
Meegan, C.A. et al., p1117, in Compton Gamma Ray Obs., AIP Proc.#280 (1993).
Rubin, B.C. et al., p.1127, in Compton Gamma Ray Obs., AIP Proc.#280 (1993a).
Rubin, B.C. et al., p. 719, in Compton Gamma Ray Obs., AIP Proc.#280 (1993b).
Skelton, R.T. et al., p. 719, in Compton Gamma Ray Obs., AIP Proc.#280 (1993).
Wilson, C.T.R. Proc. Phys. Soc., London, 37, 32D-37D (1925).

THE FOURTH INTERPLANETARY NETWORK: ARCSECOND LOCALIZATIONS FROM SPACECRAFT AT 100 AU

K. Hurley
University of California, Space Sciences Laboratory
Berkeley, CA 94720

T. Cline
NASA - Goddard Space Flight Center
Greenbelt, MD 20771

Abstract

The NASA Solar System Exploration Division has been studying a series of "faster, better, cheaper" missions, whose flagship may be the Pluto Fast Flyby at the end of the millenium. The Space Physics Division is studying a Solar Probe, and the Cosmic and Heliospheric Working Group has an Interstellar Probe on its drawing boards. These mission concepts share a number of characteristics. They are highly focussed, with tight schedules, budgets, and mass constraints. They also reach very large distances from Earth, 200 AU in the case of the Interstellar Probe, making them ideal platforms for the ultimate interplanetary network for gamma-ray burst triangulation. We describe a simple, low mass, low cost GRB and particle detector which could be easily replicated and accomodated on these missions, and provide up to arcsecond location capability for bursts.

Introduction

The era of massive, multi-faceted planetary exploration missions has, at least temporarily, come to a close in the United States. In its place is the "faster, better, cheaper" concept. One example is the Pluto Fast Flyby (PFF) mission. A large Mariner Mark II-class Pluto-Charon mission had been endorsed in 1991, and carried a gamma-ray burst detector in its strawman payload[1]. The downsized PFF mission has just three scientific goals, related only to the Pluto-Charon system, and they will be met with four instruments whose total mass is about 5 kg. Another is the Interstellar Probe. In 1990, a study group defined a strawman payload that included 13 instruments (among them a GRB detector) with a total mass of 123 kg. A reduced Interstellar Probe is now being studied, whose baseline allocation is a 20 W, 20 kg payload. A similar fate has befallen the Solar Probe, which now contains a 17 kg payload, including a hard X-ray detector. At the same time, the Solar System Exploration Division has introduced the concept of "Discovery" missions, whose guidelines are 1) development costs <$150M, 2) development time < 3 years, and 3) launch vehicle no larger than a Delta II[2]. The missions under study range from a Mercury Polar Flyby, to a Venus Composition Probe, to Small Missions to Asteroids and Comets. Outer planet Discovery missions are presently not being studied due to the cost and schedule implications of long flight times and the cost impact associated with the need for a Radioisotope Thermoelectric Generator (RTG).

Missions such as these offer a unique opportunity to establish a fourth interplanetary network (IPN) of burst detectors. The distances from Earth range from 1-2 for the Discovery missions, to ~5 for the Solar Probe, which gets a Jupiter gravity assist, to 40 for the PFF, to 200 for the Interstellar Probe. In contrast, the most widespread network to date, the 3rd IPN, is contained within the orbit of Jupiter (6 AU). With this opportunity, however, comes a challenge: it is unlikely that more than 10% of the spacecraft resources would be devoted to a scientific objective such as this, so any GRB detector would have to have a mass of ~500 g, and operate on only about 500 mW.

Instrument Concept

There is always a temptation to propose innovative technologies for new missions. It should probably be resisted in this case. Although semiconductor detectors such as photodiodes, HgI_2, and CdZnTe are promising, they are only available in small sizes, and little is known about their longevity and behavior in space. Missions such as the PFF have an 8 year cruise phase, and the Interstellar Probe must operate for up to 25 years. Phototubes and scintillators have proven to be rugged and extremely reliable for long space missions, and the former have become more compact and sensitive in recent years. For a given effective area and stopping power, the difference in mass between this sensor and one based on semiconductors is negligible. Our design is based on a Hamamatsu R1847 phototube coupled to a plastic/CsI(Tl)/plastic phoswich. The choice of CsI(Tl) is based on 1) our Ulysses experience[3] (the mass of a sensor unit was 260 g, of which 19g was for the PMT), 2) its relatively high stopping power, and 3) the fact that it is not hygroscopic. Its thickness assures good sensitivity to gamma-rays up to ~150 keV, but transparency to higher energy RTG gamma-radiation. The plastic scintillator is included for three reasons. First, the 4th IPN will operate in a period of solar maximum, and discrimination between X-rays and solar particles will be useful; this is why a second layer of plastic has been included between the CsI and the photocathode. Second, the PFF has no particle detectors, and this instrument could answer the question of whether or not Pluto has a magnetosphere. Finally, missions with particle detectors, such as the Interstellar Probe, may find it useful to have some measure of redundancy in their particle measurements. Figure 1 illustrates the sensor. A pulse shape discrimination circuit is included in the analog electronics to separate electron and proton interactions from gamma-rays. If the plastic scintillators are 2 mm thick, the CsI 3 mm, and if a 0.3 mm Al window is used, the detector would be sensitive to electrons and protons with energies ≳1 and 25 MeV, respectively, and gamma rays with energies between ~10 and 150 keV. The surface area is 20 cm^2, and one detector views 2π sr. The geometrical factor, ~60 cm^2 sr, is relatively large and results in excellent sensitivity to particles.

The radiation-hardened digital electronics is based on two Field Programmable Gate Array chips, a simple microprocessor such as the 80C85, and some memory chips. It triggers on an increase in the gamma-ray count rate in one of two energy ranges and one of two time intervals. Count rate data are stored for 10 s prior to the trigger and 100 s after, with 8 ms resolution. 16 channel energy spectra are also stored with 250 ms resolution for 500 s. About 200 kb of memory are required. Particle energy spectra are stored continuously, but with lower time resolution. Gamma-ray data in one or two energy windows will also be read out continuously with 250 ms resolution. Table 1 summarizes the mass, power, and size of the various components.

Table I. Properties of a 4th IPN experiment

Unit	Mass (g)	Power (mW)	Dimensions (cm)
PMT & phoswich	125	150	5 dia. x 9
Analog electronics	125	200	10 x 10 x 1
Digital electronics	150	250	10 x 12.5 x 1

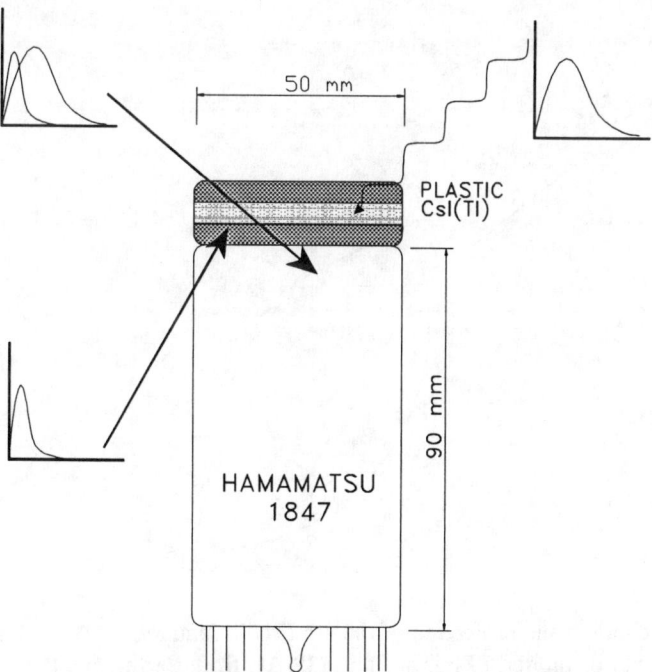

Figure 1. GRB/particle detector for a small mission. The pulse shapes for photons, penetrating, and non-penetrating particles are indicated.

Anticipated Performance

Small missions to outer planets will carry an RTG for power, so the detector background will be similar to that of Ulysses. (More precisely, small missions will probably carry smaller RTGs which are less radioactive, but because the spacecraft sizes are smaller, the two effects will tend to cancel.) Since the detector size is also similar, the sensitivity may be estimated to be the same as that of the Ulysses GRB experiment. Based on the Ulysses rate of burst detection, an experiment such as this will detect a burst with fluence $\gtrsim 10^{-6}$ erg cm^{-2} once every 5-7 days, or 400 events over an 8 year period for PFF. The configuration of the spacecraft in the 4th IPN is shown in Figure 2 for a date in early 2008, when PFF is close to Pluto. We have simulated the arrival of a moderately intense burst on this network from an "average" direction, i.e., one which neither minimizes nor maximizes the location errors. The weaker bursts can be localized to about 1'. For a moderately intense bursts, a cross-correlation uncertainty of ±100 ms can be achieved, giving an error box with a typical dimension of 5". A more unusual event, with a fast rise time, would give uncertainties five times smaller, leading to an arcsecond-size error box. (Examples are Oct 22 1992 and Jan 31 1993, for which current IPN location uncertainties are around 15"[4]).

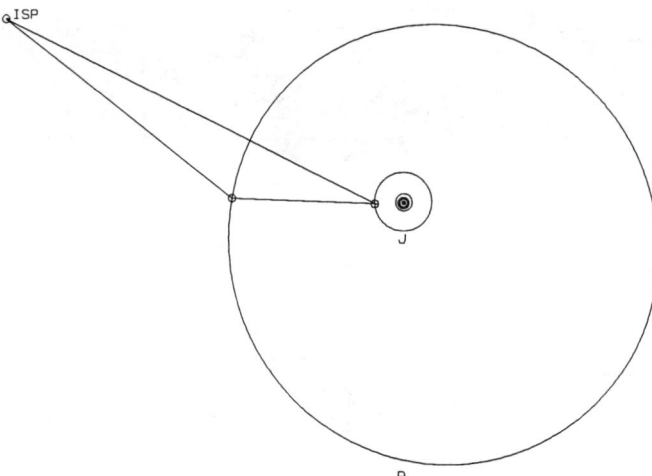

Figure 2. Ecliptic plane projection of the 4th IPN in January, 2008. The Solar Probe is in the vicinity of Jupiter, PFF is at Pluto (32 AU from Earth), and the Interstellar Probe, launched in 2000, is headed for the "nose" of the heliosphere, travelling at 10 AU/year (80 AU from Earth). All previous IPNs were contained within the orbit of Jupiter. The first point of Aries is toward the bottom.

Conclusions

When gamma-ray bursts were first discovered, it seemed that precise localizations would solve the mystery of their origin. "Precise" then meant of the order of 10'. "Deep" searches involved plates whose limiting magnitudes were perhaps 23, with ≤3 m class telescopes. Over the past decade or so, "precise" has come to imply 1', which is roughly compatible with the evolution from plates to CCD cameras, and limiting magnitudes beyond 24. It still appears that counterpart identification is the only hope for discovering the cause of cosmic gamma-ray bursts. New technology and 10 m class telescopes have already pushed magnitude limits to 29, and at high galactic latitudes, 100 objects per square arcminute are imaged[5]. Simply to avoid source confusion, error box sizes of the order of 5" are needed.

A 4th IPN which can produce arcsecond-size error boxes is entirely feasible. It employs current detector and electronics technologies, and the experiment costs are likely to be quite low, particularly if replication is possible, so that economies of scale can be realized. Such a network would also achieve the limiting accuracy of the triangulation technique, at least with small spacecraft. Since it is unlikely, that distances significantly greater than several hundred AU will be reached by future spacecraft in less than a lifetime, and since the experiment size cannot be increased, further reductions in error box sizes cannot be achieved by this technique. If counterparts cannot be identified in the error boxes of the 4th IPN, mother nature is indeed perverse.

Acknowledgments

It is a pleasure to acknowledge conversations with R. Lin, and design work by D. Curtis and N. Madden.

References

1. K. Hurley, Presentation to the Neptune/Pluto Outer Planet Science Working Group, Flagstaff, AZ (1991)
2. C. Pilcher, Discovery (Newsletter of the Solar System Exploration Division) 93-1, 6 (1993)
3. K. Hurley, Astron. Astrophys. Suppl. Ser. 92(2), 401 (1992)
4. K. Hurley, these proceedings (1993)
4. B. Peterson, S. D'Odorico, M. Tarenghi, and E. Wampler, The Messenger 64, 1 (1991)

X-RAY TELESCOPE ARRAY FOR GAMMA-RAY BURST LOCALIZATION

Nobuyuki Kawai

The Institute of Physical and Chemical Research (RIKEN),
2-1 Hirosawa, Wako, Saitama 351-01, Japan

ABSTRACT

X-ray mirror telescope is the most promising technology for GRB localization with 10 arcsec accuracy. We propose an array of small X-ray mirrors covering a wide field of view. We discuss the feasibility and compare it to the detection probablity of GRB in the ROSAT field of view.

INTRODUCTION

The counterparts of gamma-ray burst sources have been searched in optical bands in many instances, but it has not been so successful so far. One of the main difficulty is that the accuracy of the localization with the gamma-ray or hard X-ray photons (degrees to arcminuites) is not sufficient to correlate with the optical objects. For a better localization, search for optical flashes associated with gamma-ray bursts have been planned and conducted. However the ground based observations such as ETC needs coincident detection on spacecraft for confirmation. The background events is also a problem. HETE intends to localize GRB in UV bands to arcseconds accuracy, but it is not proven that majority of GRB is accompanied by optical flashes.

On the other hand, X-ray band is most promising for a fine localization. The observations by Ginga have proven that most of GRBs have X-ray emission in the soft X-ray bands. Recent advances in technologies (X-ray optics, solid-state imaging detectors, and data-processing computers) can be applied to this purpose as well.

FEASIBILITY: A CASE STUDY WITH ROSAT

For a feasibility study of X-ray telescope array, ROSAT PSPC[1] is a good starting point. For our purpose, high throughput, moderately fine position resolution ($\sim 0.5 arcmin$) and a large field of view is required for the telescope. ROSAT PSPC has a good position resolution, moderately wide field of view (f.o.v. diameter = 2°), and low background rate. We can use it as a "measure" for a unit of our telescope array.

Next, we try to estimate the $\log N - \log S$ normalization in the soft X-ray band. The PVO $\log N - \log P$ is given[2] as:

$$N(>P) = 21 \left(\frac{P}{2 \times 10^{-5} \, erg \, s^{-1} cm^{-2}} \right)^{-3/2} yr^{-1} \qquad (1)$$

for bursts with $P > 2.2 \times 10^{-5} \, erg \, s^{-1} cm^{-2}$. For weaker, more frequent bursts, we assume a slope of -0.8, assuming $P_{0\gamma} = 2.2 \times 10^{-5} \, erg \, s^{-1} cm^{-2}$ as the "knee". To express this relation using the X-ray fluence (or integrated counts), we assume the $\gamma(PVO)/X(3-10\mathrm{keV})$ flux ratio to be 2 % [3]. The "knee" flux (corresponding to $P_{0\gamma}$) for the 3 − 10keV X-ray flux is $4 \times 10^{-7}\ erg\ s^{-1}cm^{-2}$. Assuming a power-law spectrum with photon index 1.5 and absorption column $\log N_H = 21.5$, this flux corresponds to 8500 c/s for ROSAT PSPC. We further assume a typical duration of 5 s (FWHM) for GRBs to estimate the integrated counts. We extrapolate the $\log N - \log S$ power-law three orders of magnitude down from the knee, and we obtain a GRB $\log N - \log S$ relation for the all sky expressed in terms of ROSAT PSPC sensitivity:

$$N_{all\ sky}(>S) = 1.7 \times 10^4 \left(\frac{S}{10\ PSPC\ counts}\right)^{-0.8} yr^{-1} \quad (2)$$

By considering the exposure time efficiency of 50 %, field of view with 1°radius with 75 % grid transmission efficiency, we obtain the frequency of GRBs recorded in the field of view of PSPC:

$$N(>S) = 0.36 \left(\frac{S}{10\ PSPC\ counts}\right)^{-0.8} yr^{-1} \quad (3)$$

DESIGN CONSIDERATIONS

We have shown that with ROSAT PSPC one GRB with more than 10 photons is expected to be recorded in 3 years. To obtain a reasonable number of bursts ($>\sim 5$), it is necessary to increase the effective area or the field of view. Assuming a power-law $\log N - \log S$, we estimate the effect of these two parameters. We consider an array of m small telescopes. Each telescope has an effective area r ($r \ll 1$) times that of the ROSAT PSPC, covers a different field of the sky, and has similar observation efficiency as ROSAT. Then for this array we expect

$$N(>S) = 0.36\ m\ r^{0.8} \left(\frac{S}{10\ counts}\right)^{-0.8} yr^{-1} \quad (4)$$

For example, if $r = 0.01$ and $m = 400$ is chosen, factor 10 improvement over ROSAT PSPC can be achieved.

Since the number of telescope is large, compromise in the scale and performance is obviously necessary. For our purpose, we do not require image quality comparable to the ROSAT mirror. In order to achieve $10''$ localization with 10 photons, we only need $30''$ angular resolution for each photon. It should be noted, however, that we require so much angular resolution uniformly over the entire field of view. To obtain $30''$ resolution over a 2 °field of view, the focal plane detector needs to have at least 256×256 pixels. Also, the image needs to be stable within the same requirement. If the telescope changes its direction (as in a spinning satellite), the read-out of the detector must be fast enough to avoid blurring of the image. Another requirement is the background rate. In the case of ASCA and ROSAT, the internal background of the focal plane detectors (CCD or gas counters) is comparable or less than $10^{-3} c\ s^{-1} keV^{-1} cm^{-2}$, negligibly small for GRBs. The diffuse X-ray background is $\sim 10^{-3} c\ s^{-1} arcmin^{-2}$ for ROSAT, which is again negligible. The X-ray emission from the galactic

X-ray sources is not a problem either, except for a few extremely bright objects. Considering these, achieving sufficiently low background is possible.

The main data product is the photon event list, which has the information of position (x, y), telescope ID, pulse height, and arrival time. The photon events will be dominated by the diffuse X-ray background. For the same example with $r = 0.01$ and $m = 400$, the diffuse background rate is $\sim 50c/s$. A source with 1 $Crab$ flux yields $\sim 30c/s$. For most of the cases, pointed at the sky, the data production rate is rather small.

To complement the telescope array, a monitor counter in higher energy range is useful. A copy of Ginga GBD consisting of a proportional counter $(2 - 25\text{keV})$ and a scintillation counter $(10-400\text{keV})$ with a field of view matched with telescopes will be sufficient. These also give timing and spectral information.

TECHNICAL DIFFICULTIES

This project has three major challenges: mirrors, focal plane detectors, and the space craft.

For each of the X-ray mirror, a moderately fine angular resolution ($< 30''$) is required over a wide field of View ($> 2°$ diameter). For ROSAT, a good angular resolution is only achieved on the optical axis. However the Soft X-ray telescope on board the Yohkoh has a uniform $2''$ angular resolution over 0.5 °field of view (the sun). A modified Wolter type I mirror with hyperbola-hyperbola reflection, achieves better resolution off axis at the expense of on-axis resolution. We hope to design a mirror in a similar philosophy, though we require lower resolution for a larger field of view.

Another essential constraint is the cost: the array consists of hundreds of mirrors. However, we require only a modest effective area ($\sim 10\ cm^2$) and a moderately fine resolution, which can be produced at much lower cost than the ROSAT or Yohkoh mirrors. Development of replication and machining technique for the X-ray mirror is actively promoted for number of projects at present. We hope to apply these technique for this array.

For the focal plane detector, an imaging capability with more than 256×256 pixels. A natural choice seems to be CCDs, which now fly on two X-ray astronomy satellite, Yohkoh and ASCA. For our purpose, photon energy must be read out separately (photon counting, not integration). It requires a low-noise read-out, if not to an extreme level like SIS (X-ray CCD camera) on ASCA. On the other hand, the image must be read fast enough to avoid blurring on an unstable or spinning spacecraft. In the case of the space station, which rolls synchronously with the orbital revolution, the image must be read out at a speed of 8 frames/s. A detailed design study is necessary to make the noise level and speed compatible within our requirements.

The spacecraft is a difficult component to obtain. An array of 400 telescopes, each has a diameter of 8 cm (1/10 of the ROSAT mirror), occupies an area of $2 \sim 3 m^2$. A dedicated satellite of this size (somewhat larger than Ginga) with stable pointing capability is ideal, and does not impose fast read out for the focal plane instrument, but such mission is obviously difficult to get approved. A possibile platform is the exposed module of JEM (Japanese experiment module) on the space station (if it ever flies). One exposed module has a size of $4 \times 1.4 \times 2.5 m$, sufficiently large for our payload. The nature of the mission (monitoring GRBs) does not require a fine pointing capability (therefore it can be one of the few sensible astronomical experiments on the space station), but its rotation with orbital revolution and the insufficient attitude stability is still a problem. A good aspect monitor is required to correct for movements of the field of view. A fast read-out is necessary as mentioned above. The interference with the other components of the space station is also a problem. AO for the experiments on the exposed modules of JEM was originally scheduled this year (1993), but was postponed.

SCIENCE PRODUCTS

As calculated in the above, about 10 gamma-ray bursts with > 10 photon detection are expected in a year with this mission. The position can be determined with an accuracy of $10''$ for these bursts. We expect larger number of weaker events for which the positional accuracy is somewhat worse. Since the number of photon is limited, we cannot expect a good spectrum from the telescope alone. The monitor proportional counter and the scintillation counter will provide complementary spectral and time history information.

As a secondary science, this array is extremely useful for monitoring X-ray sources on a long time scale. With a single telescope with an effective area $1/100$ of that of ROSAT, a $1\,mCrab$ source gives $7 \times 10^{-3} c/s$, which is larger than the background rate. It can be detected in $1000\,s$. The flux variation of the source can be monitored on daily basis with an accuracy of 5 %, if the field of view is stationary. Although the sky coverage is limited, long term monitoring of selected AGNs and galactic sources in the X-ray band offers a unique observation not available in other experiments.

REFERENCES

1. ROSAT Call for Proposals, Technical Appendix (MPE, 1990).
2. Fenimore et al., Proc Compton Symp. (St.Louis, 1992), p. 744.
3. Laros et al., Ap. J. **286**, 681 (1984).

X-RAYS FROM GAMMA RAY BURSTS

Edison P. Liang

Department of Space Physics & Astronomy
Rice University, Houston, TX 77251-1892

ABSTRACT

We reconsider the need for the monitoring of GRBs and SGRs in the 0.1- 20 keV range. In addition to good spatial resolution for localization purposes, we propose that such an instrument should have detection sensitivity and spectral resolution much above those to be provided by HETE and XTE in order to sample deeper than BATSE and to detect any possible spectral features in the X-ray range, including atomic and cyclotron lines and absorption edges.

INTRODUCTION AND SUMMARY

The purpose of this paper is to emphasize the renewed urgency of the need for a new window on the mystery of cosmic gamma ray bursts (GRBs). Our thesis is that 0.1-20 keV X-ray is the most logical new window. We propose that a mission devoted to the study of X-rays from GRBs and Soft Gamma Repeaters (SGRs) as well as other X-ray transients should be an all-sky monitor with good spectral resolution (<8% at 7 keV), imaging capability (resolution <few arc minutes) and high sensitivity (< 10^{-10} erg/s.cm^2).

Why emphasize X-rays? Some of the most obvious reasons include:
1. We know that they exist based on detections by Hakucho, Ginga and other satellites.
2. Because X-ray detectors have much higher sensitivities than gamma ray detectors, they can see much deeper.
3. Assuming that the BATSE -0.8 logN-logP slope holds to lower fluxes, this means that the event rates in X-rays should be much higher than the BATSE GRB rate.
4. Low energy cutoffs, atomic absorption edges and emission and absorption lines in the X-ray band are potential distance indicators (e.g. Schaefer 1993) as well as diagnostic tools.
5. Ginga results (Murakami et al 1992) show that the X-ray tails typically last much longer (>100 sec) than the GRB itself, thus having better chance to reveal any underlying long period.
6. Some of the X-ray precursors and tails were found to have blackbody-like spectra (Murakami et al 1992). If the blackbody nature can be confirmed it will rule out many of the cosmological scenarios.
7. X-rays give much better photon statistics than gamma rays, thus having better chance to reveal submillisecond structures and periods.
8. There is a better chance of detecting polarization in the x-rays than gamma rays.
9. X-rays can provide better localization.
10. There is increasing evidence that there exists a continuum of high energy transient phenomena spanning between classical GRBs and X-ray bursts. In addition to the SGRs, there are many "classical-like" GRBs that are also rich in X-rays (see Yoshida 1994, this volume). X-ray experiments will be more adept in detecting such events.

Based on the above reasons, we propose that the top priority in the next phase of the GRB experimental effort is not necessarily a bigger and better BATSE but an X-ray instrument to complement BATSE.

POTENTIAL DIAGNOSTICS WITH X-RAYS

Here we list some examples of GRB diagnostics using 0.1-20 keV X-rays.
1. The amount and frequency of absorption by the ISM of our Galaxy below ~2 keV can distinguish between local versus extended halo or cosmological origins (e.g. Schaefer 1993).
2. The Kα emission and K & L edges of any heavy element (Si - Fe) in the circumburster matter (CBM) can reveal the source environment. If the source is cosmological with a redshift of ~ 1 we expect such emission line and edges to be redshifted by a factor of ~2. In a separate contribution (Liang 1994, this volume) we propose that the X-ray deficiency below ~ 200 keV in most GRBs could be due to absorption by Fe-group-rich CBM, in which case we predict the likely presence of a Kα emission line at ~ 6.7 keV. To detect such lines we need to have spectral resolution better than 8%.
3. Any cyclotron lines of magnetic fields in the range 10^{10} - 10^{12}G will show up in this X-ray range and be likely resolvable at the 8% level.
4. The ~keV blackbody spectrum of any precursor or tail (Murakami et al 1992), if confirmed, will strongly constrain the source distance and emission models.
5. The rapid spectral evolution of the X-rays will provide important diagnostics of the source emission mechanisms and parameters, complementary to the gamma ray spectral evolutions. In particular, if the X-rays are isotropic and the gamma rays are strongly beamed, we expect cross-correlational analyses to shed much light on their differences.
6. The presence or lack of X-ray polarization can help to constrain the beaming model and emission mechanisms of GRBs.
7. If GRBs indeed originate from an extended halo of > 50 kpc radius, then at the X-ray flux level of ~ 10^{-10} erg/s.cm^2 we should start to see some clustering around M31. Similarly if they originate from an extended Oort cloud we should see some clustering around nearby stars for the faintest bursts (cf. discussions below).

ALL-SKY GRB X-RAY FLASH (XRF) RATE ESTIMATES

We estimate the all-sky XRF detection rate based on the following conservative assumptions:
1. X-to-gamma ray peak energy flux ratio ~ 1%
2. XRF detection threshold ~ 10^{-10} erg/s.cm^2 (equivalent to ~2.5 cts/s/pixel for ROSAT-PSPC for a -1 photon spectral index). This is the flux a 2×10^{-6} erg/s.cm^2 Galactic Halo GRB (at 50 kpc) would produce if relocated to M31. Equivalently, it is the flux a 4×10^{-6} erg/s.cm^2 Oort Cloud GRB (at 10^4 AU) would produce if relocated to another star 1 pc away.
3. The BATSE -0.8 logN-logP slope continues to lower fluxes.

If both the X-ray and gamma ray emissions are isotropic then the all-sky XRF rate would be ~ 6.3 times the BATSE GRB rate (assumed threshold ~10^{-7}erg/s.cm^2 in gamma rays) or ~ 5000/yr = 14/day. But if the gamma rays are strongly beamed while the x-rays are isotropic (likely true for a blackbody precursor or tail), the X-ray to gamma ray visibility ratio is $f=4\pi/\Omega_\gamma$ (>>1) where Ω_γ is the gamma beam solid angle. Then the all-sky XRF rate becomes 14f/day. In many cosmological scenarios the required gamma beam angle is very tiny to avoid gamma-gamma pair production catastrophies. In some cases the predicted XRF rate becomes so high that even the existing limits from the Einstein and ROSAT surveys may usefully constrain the beam angle.

COMPARISON OF PROPOSED DETECTOR REQUIREMENTS WITH HETE & XTE

The following table compares the published design parameters of the two pending all-sky X-ray monitors to be launched by NASA in the near future with the desired characteristics proposed here.

	HETE(WFXM)	XTE(ASM)	PROPOSED NEED
FOV	2x(1 str)	3x(6° x 90°)	~ 2π
Ang.Res.	6'	3'x15'	few'
Eng.Range	2-25 keV	2-10 keV	0.1-20 keV
$\Delta E/E$	15%@6keV	3 channels	<8%@6keV
Threshold (erg/s.cm^2)	8×10^{-9}	10^{-8}	10^{-10}
Expected XRF rate/yr	~10f	~10f	~1.5×10^3f

(f is ratio of x-ray to gamma ray beam angle)*

*Note that the x-ray logN-logP may not be the same as the gamma ray logN-logP.

EPL thanks Charles Hailey for valuable discussions. This work was partially supported by NASA grant NAG 5-1515.

REFERENCES

Liang, E.P.1994, contribution in these proceedings.
Murakami, T. et al 1992, Proc. Taos Workshop on Gamma Ray Bursts p.239, ed. C. Ho, R. Epstein and E. Fenimore (Cambridge, UK)
Schaefer, B. 1993, AIP Conf. Proc. No. 280, p.803, ed. M. Friedlander, N. Gehrels and D.J.Macomb (AIP, NY).
Yoshida, A. 1994, contribution in these proceedings.

RULER: AN INSTRUMENT TO MEASURE GAMMA-RAY BURSTER DISTANCES

A. Owens[1], J. Greiner[2], T. Mineo[3], K. Pounds[1], M. Rees[4], B. Sacco[3], L. Scarsi[3], B. Schaefer[4], S. Sembay[1], O. Terekhov[6] and A. Wells[1]

[1]Physics Dept., Leicester University, UK, [2]MPE Garching, Germany, [3]IFCAI Palermo, Italy, [4]IoA, Cambridge University, UK, [5]NASA/GSFC MD 20771, USA, [6]IKI, Moscow, CIS.

ABSTRACT

We describe a satellite-borne soft X-ray burst detector specifically designed to determine the GRB distance scale. This is achieved experimentally by measuring the soft X-ray extinction by interstellar absorption of a representative sample of bursts. A determination of this scale will eliminate whole classes of models by providing the first quantitative description of the 3-dimensional distribution of burst sources.

INTRODUCTION

Until the launch of GRO, GRBs were widely believed to originate on galactic neutron stars (NS) at distances of a kpc or so. The terra-gauss neutron star paradigm, as it became known, was appealing for several reasons, namely; *1)* questions of energetics and confinement were well within current NS models, *2)* cyclotron features at energies representative of terra-gauss fields had been detected in ~ 20% of GRBs by KONUS[1] and strikingly confirmed by GINGA[2], *3)* redshifted annihilation lines at energies appropriate to production at a NS surface had been reported in 15% of KONUS bursts, *4)* weak few-second pulsations and millisecond variability has been reported in a few burst time histories whose periods were consistent with radio pulsars, X-ray pulsars and X-ray binaries[3-5] and *5)* nuclear γ-ray lines have been reported in the spectra of GB791119[6].

The remarkable isotropy observed by BATSE, coupled with strong evidence of an inhomogeneous radial distribution[7], excludes all disk populations and all but the most contrived halo distributions. This includes all known NS distributions. Additionally, BATSE has failed to detect any line features in more than 600 GRB spectra[8] (by far the largest statistical sample to date). When coupled with the lack of an identifiable counterpart at any wavelength (or even a convincing correlation with *any* astronomical population) the data have led to a confused proliferation of models. Whereas models are vague on much of the burst details, they can at least be grouped according to a distance scale. Generally, most models can be grouped into 3 distance scales. These are; extragalactic (3 Gpc), galactic halo (100 kpc) and Solar system (0.001 pc). The RULER experiment is specifically designed to address this scale by accurately measuring the soft X-ray extinction of a representative sample of bursts caused by interstellar absorption[†]. The various inter-model distributions have been explored by Owens & Schaefer[9] and Schaefer[10] and are expected to be quite distinctive, ranging from essentially zero column for solar system models to densities $> 10^{22}$ H cm^{-2} for extragalactic scenarios.

INSTRUMENTAL

RULER is a high spectral and angular resolution all-sky-monitor (ASM) composed of 5 wide-field cameras employing coded aperture imaging techniques. Fig. 1 shows a

---

[†]In principle, line-of-sight column densities could also be determined from Lyman α profiles if burst spectra were found to extend down to the far uv.

cross-sectional view of the ASM. Each camera module consists of a coded mask located 25 cm in front of a 6 × 6 cm plane of silicon detectors. The complete ASM as-

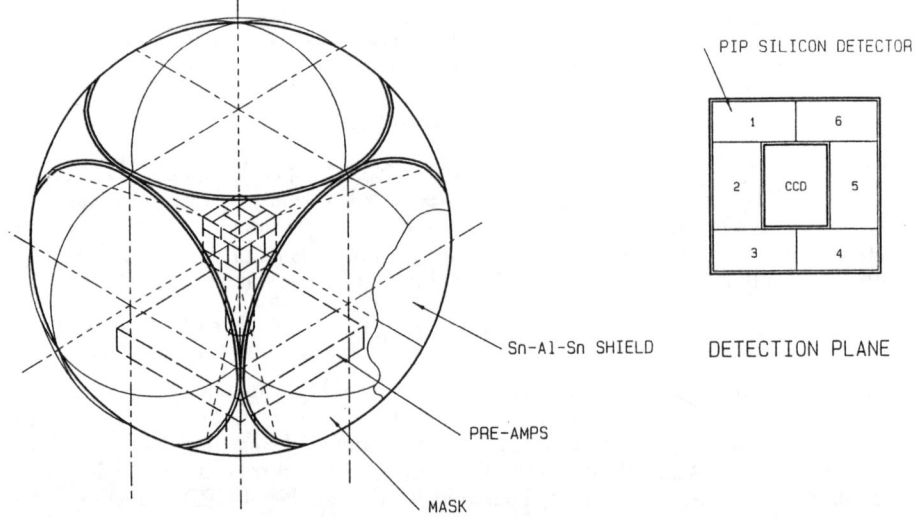

Figure 1. A cross-sectional view of the all-sky-monitor.

sembly is formed by mounting the detection planes of all 5 modules on the sides of an aluminum cube. The 6th face is used to attach to a cooling system. This inner cube is in turn surrounded by a larger structure supporting the coded masks. The FOV of each module (~ 90° FWHM geometric) is defined by composite walls made of Sn-Al-Sn of

Table I. The principal RULER experimental characteristics.

PRINCIPAL DETECTOR: 1 All Sky Monitor (ASM) composed of 5 Wide Field cameras.
WIDE FIELD CAMERAS: Each WFC composed of 1 CCD and 6 PIPS detectors.
ENERGY RANGE: CCDs - 100 eV to 15 keV; PIPS - 100 eV to 50 keV.
ENERGY RESOLUTION: 50 eV @ 500 eV ($E/\Delta E = 10$).
TOTAL FIELD-OF-VIEW: 170° FWFM geometric ~ 3π steradians.
DETECTION SENSITIVITY: ~10^{-5} ergs cm^{-2} for a 1 s burst duration. Rate ~ 120 yr^{-1}.
COLUMN DETERMINATION ACCURACY: ~ 0.3 of a decade (H cm^{-2}) for ~ 26 yr^{-1}.
SUBSYSTEMS: 5 Wide Field Cameras. Field-Of-View ~ 2 ster each.
2 CsI GRB detectors. Energy range 50 keV to 10 MeV.
1 Analog Processing Unit (APU).
1 Digital Processing Unit (DPU).
1 Mass Memory Unit (MMU).
1 ASM assembly; weight 50 kg; dimensions 56 × 56 × 56 cm.
CRYOGENICS: Single stage passive cooler of radiating area ~ 10^4 cm^2; temp 160 K.
TEMPORAL RESOLUTION: Imaging - 0.25 - 1.0 sec ± 1.5 ms; Spectroscopy
- 10 μs intrinsic; ± 1.5 μs after ground processing.
MEMORY SIZE: 125 MB Static CMOS.
EXPECTED BIT RATE: 750 kbit/sec download; 9600 bits/sec upload.
TOTAL WEIGHT: 110 kg; 210 kg including proposed AeroAstro Spacecraft.
TOTAL POWER: 65.5 W quiescent: 10 W de-ice power (low duty cycle).

thicknesses 100 μ, 1 mm and 100 μ, respectively. The total FOV of all 5 modules is ~ 3π sr. The ASM is mounted on a cylindrical tower secured to the upper equipment deck of the spacecraft. To reduce dark current, the detectors are cooled to ~ 160 K by 4 passive radiator panels. Because the response of the silicon detectors does not extend into the γ-ray regime, the GRB nature of the event will be verified by two conventional GRB detectors contained in the spacecraft platform. Two detectors are flown for redundancy and also to ensure that during a burst at least one detector is unobstructed by the tower or ASM.

Each detection plane is comprised of two types of detectors; a large area CCD which is used in conjunction with the mask to produce an image of the sky, and 6 passivated implanted planar silicon (PIPS) detectors of total area ~ 27 cm^2, - used exclusively for spectroscopy. The ubiquitous Al optical filters will be evaporated directly onto the surface thus eliminating the problem of supporting large area thin films and also ensuring the low energy response is not dominated by edges introduced by organic support materials such as Lexan. The PIPs detectors will have a 150 μ thick depletion layer ensuring a usable response up to ~ 50 keV and therefore some over-lap with the conventional GRB detectors. Thus, it should be possible to measure burst spectra from 0.1 keV to 10 MeV.

The masks have dimensions 56 cm × 56 cm and will be constructed from titanium, 100 μ thick. They will be curved to minimize off-axis vignetting and for ease of replication the coding will be based on large uniformly redundant arrays. Assuming a CCD spatial resolution of 27 μ and a unit cell size of 100 μ, then a point source can be located to an accuracy of, $\theta = \theta_m/SNR \sim 1.4/n_\sigma$, where θ_m is the unit cell size in units of arcmins and n_σ is the the the number of standard deviations of the signal above the noise. Thus, setting the minimum trigger signal to 5σ ensures a source location accuracy of 17 arcsecs. The actual precision depends ultimately on the aspect reconstruction error, but is expected to be better than 30 arcsecs. The total weight and power consumption of this experiment are ~ 110 kg and ~ 65 W, respectively, making it suitable for a Pegasus launch. The mission lifetime is expected to be > 3 years for the preferred 550 km equatorial orbit.

EXPECTED PERFORMANCE

The burst detection properties of the instrument have been investigated using the Monte-Carlo method in association with the XSPEC spectral fitting package[11]. The input burst parameters were derived from the following sources; - a) the BATSE log N/log S curve[7], b) the BATSE spatial distribution[7], c) the burst duration distribution given in the review of Hurley[12], and d) the positionally dependent soft X-ray background taken from ROSAT PSPC data[13] and the diffuse hard X-ray background spectrum taken from HEAO-1 measurements[14]. A generic GRB spectrum given by,

$$\frac{dN}{dE} = \begin{cases} \beta(E/E_o)^{-0.7} & 0.1\,\text{keV} \leq E \leq 10\,\text{keV} \\ 2\beta(E/E_o)^{-1.0} & 10\,\text{keV} \leq E \leq 100\,\text{keV} \\ 200\beta(E/E_o)^{-2.0} & E > 100\,\text{keV} \end{cases} \quad (1)$$

was assumed for all bursts. Here $E_o = 1$ keV and β is the normalization in units of photons cm^{-2}s^{-1}keV^{-1} which is related to the burst fluence, S (ergs cm^{-2}s^{-1}), above 30 keV at the Earth by, $\beta = S / 2.5 \times 10^{-6}\Delta T$. The calculation proceeds as follows. GRB's are generated randomly on the celestial sphere and the expected X-ray background in that direction determined. The burst duration and size are then randomly sampled from the appropriate distributions. Next, the burst and background fluxes at the detector are calculated and a simulated detector energy-loss spectrum (complete with Poisson noise) generated using the instrument response matrix in conjunction with the XSPEC FAKEIT facility. The SNR of the burst is calculated for the full burst duration and if it passes some predefined threshold, XSPEC unfolds the data using the detector response function and best-fits it assuming the input spectrum to be an absorbed power-law. The background estimation is assumed to be taken 250 sec either side of the burst. The best-fit column and spectral index of the burst are derived along with their respective errors.

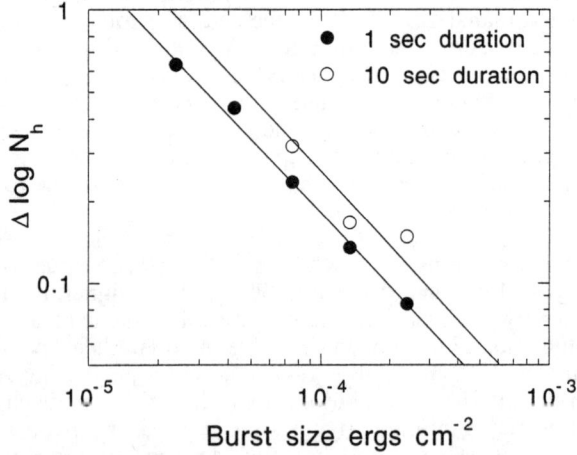

Fig. 2. Figure showing the minimum fluence required to derive a column to a particular precision. Two burst durations are considered, - 1 sec and 10 sec.

Fig. 2 shows the minimum fluence required to derive a column to a particular precision. Two curves are given, - for 1 sec and 10 sec duration bursts. Since the expected range of column densities is a little more than 2 decades of H cm^{-2}, we arbitrarily define the limiting column precision to be 1 decade of H cm^{-2}. From the graph, we see that the corresponding 1 sec and 10 sec limiting fluences are 1.5×10^{-5} ergs cm^{-2} and 2.3×10^{-5} ergs cm^{-2}, respectively. Next we explored the ability of RULER to discriminate between various source models. The results of a simulation in which burst sources were assumed to belong to a nearby extragalactic population are shown in Fig. 3. The initial number of bursts generated was 500 (~1.25 years observation), of which, 25 satisfied the SNR threshold of 50σ in the 2-10 keV band. (We have empirically determined that this SNR is required to determine the column density to within Log N_h ± 0.3 at the 68% confidence level.) The figure shows the best-fit column plotted against the line-of-sight column to the edge of the galaxy determined from the Stark HI radio survey[15]. Obviously for our input model, the data should lie on a straight line of slope unity - which is what is observed. The distribution of data points for extragalactic plus host galaxy absorption models will lie above this line, while galactic models will lie below it. As an example of the latter, the shaded region shows the expected range of data points for a burst population following the galactic supernovae and supernova remnant distributions[16]. It should be noted that, while the background is very high in this experiment due to the wide field-of-view, there is no appreciable systematic error on the estimate of the background during the burst since burst durations are quite short.

Lastly, we point out that since RULER will spend a relatively small fraction of its time carrying out its prime objective, the instrument will also undertake a comprehensive secondary science program. Most importantly, RULER can provide an all-sky-monitor capability in the critical 0.1 to 2 keV range for the brightest sources.

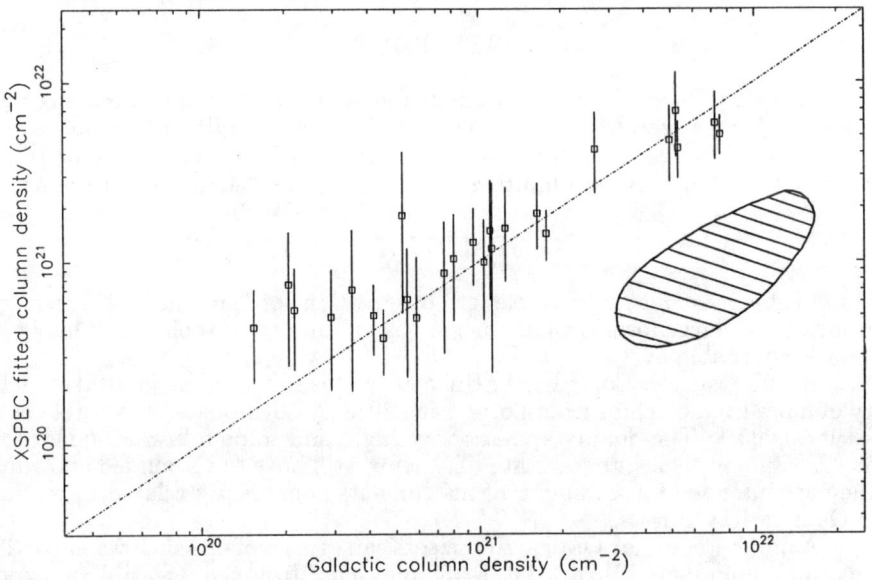

Fig. 3. A simulation of 500 extragalactic bursts randomly distributed on the sky (see text). The best-fit columns for those bursts with a SNR > 50 are plotted as a function of the total galactic column in the appropriate direction.

REFERENCES

1. E.P. Mazets, *et al.*, Nature, **290**, 278 (1981).
2. T. Murakami *et al.*, *Nature*, **335**, 234 (1988).
3. E.P. Mazets, *et al.*, Nature, **282**, 587 (1979).
4. K.S. Wood, *et al.*, Ap. J., **247**, 632 (1981).
5. A. Owens, *et al.*, Ap. J., **352**, 741 (1990).
6. B.J. Teegarden and T.L. Cline, Ap. J. *(Letters)*, **236**, L67 (1980).
7. C.A. Meegan, *et al.*, Nature, **355**, 143 (1992).
8. B.J. Teegarden, *et al.*, Proc. 2nd Compton Symp., AIP, New York, 860 (1993).
9. A. Owens & B.E. Schaefer, *Comments on Astrophysics*, **17**, 119 (1994)..
10. B.E. Schaefer, these proceedings.
11. R. Schafer *et al.*, *XSPEC, An X-ray Spectral Fitting Package*, ESA TM-9, (1991).
12. K. Hurley, in *Gamma-Ray Bursts*, AIP New York, 1 (1986).
13. C. Barber, private communication.
14. F. Marshall *et al.*, Ap. J., **235**, 4 (1980).
15. A.A. Stark, Ap. J. Supp., **79**, 77 (1992).
16. D. Green, *Supernovae and Supernova Remnants*, IAU coll. 145 (1993) in press.

IDEAS FOR A LARGE DETECTOR FOR OPTICAL TRANSIENTS

Holger Pedersen, Michael Andersen
Copenhagen University Observatory
Øster Voldgade 3, DK 1350 Copenhagen K, Denmark

Michel Boër
Centre d'Etude Spatiale des Rayonnements (CNRS/UPS)
BP 4346, F31029 Toulouse Cedex, France

ABSTRACT

Large CCDs are becoming available in significant numbers, making them attractive for a Large Detector for Optical Transients. With the proposed set-up, about 8 BATSE events annually can be imaged. Positive detections will be accurate to 20", otherwise a limiting ratio $L_{optical}/L_\gamma < 4 \; 10^{-4}$ can be reached.

BASIC IDEA

In the quest for multiwavelength observations of Gamma-Ray Bursts, the application of large format CCDs mosaics is an important tool, which lately has become affordable.

As an example, Copenhagen University Observatory has just taken delivery of more than 60 chips from Loral, each 2048 by 2048 pixels, at a total cost of about 90,000 \$. The quality appears very high, and some full-wafer quadruples may function in their integral state. The chips still have to be bonded and wired. They are intended for a number of instruments not directly related to research on Gamma Ray Bursts.

A unit detector for Optical Transients can consist of one such wafer (or 2 by 2 mosaic), illuminated by a f/2.0, F=60 mm lens. It appears feasible to expose for 10 s and read in 3 s or less. The fast reading implies a trivial modification of the detector design, to include more reading registers. No penalty in terms of 1/f noise is foreseen, because of the rather high background level. In view of the short integration time, the camera can be operated at -20 °C. The dewar may thus be of the Krypton-filled, Peltier-cooled type, whereby daily service is limited.

From a given site, four cameras can survey one quarter of the celestial sphere to 50" resolution. Assuming 8 hours night time, 2/3 clear sky, and other losses 50 percent, the efficiency becomes 1/36, implying that about 8 BATSE events annually are within the FOV, while exposing. Two sites are required, separated by several kilometers, for rejection of near-Earth events. We propose to place the LDOT in the southern hemisphere (La Silla, Paranal), so as to complement other initiatives, in particular the ETC[1].

SENSITIVITY

In order to make the event detection routine most sensitive (no pixel-to-pixel motion of field stars) the cameras should be mounted equatorially. The platform will reposition each half hour and the computer will have access to the corresponding reference frames.

For an integration time of 10 s, the sky signal will be about 100 photo-electrons, permitting detection of 7σ optical transients containing as little as 70 e, or n = 10 e cm^{-2}, corresponding to a fluence of

$$L_{optical} = n\, h\, \nu = 4\ 10^{-11}\ \text{erg cm}^{-2}$$

This can be compared to the maximum brightness expected among the the annually occurring BATSE/LDOT events, $L_\gamma \sim 10^{-6}$ erg cm^{-2}. Thus, in the absence of an optical flash, we will have

$$L_{optical}/L_\gamma < 4\ 10^{-5}$$

Even though image aberrations may degrade the performance by an factor 10, this is significantly better than the best upper limit 10^{-2} deduced from photographic observations of classical (non-repeating) GRBs[2]. It is also interesting in terms of recent theoretical predictions[3]. We have assumed dark sky (no Moon) and operation in the 0.42 to 0.65 nm bandpass. During bright of the Moon, the bandpass can be 0.60 to 0.90 nm. This will impose a ten-fold higher optical detection threshold.

DATA REDUCTION

The data flow rate from each CCD will be 0.8 MB s^{-1}, or in total 25.6 MB s^{-1}. Clearly, a most effective procedure has to be utilized, in order to digest this quantity of data, while not producing more than a manageable frecuency of candidate events. For on-line data reduction we suggest to use one IBM Power PC per chip, with 64 MB RAM. At up to 100 MHz this computer will have sufficient speed to perform a pretty advanced image analysis. The limiting factor may be I/O operations.

The data analysis will take different time scales into account: single frames, average of 5 frames, average of 25 frames, etc. It is envisioned to store only tiny parts of the images, around points of candidate events. Files may be kept with other conspicuous data (variable stars, asteroids, meteors, satellite glints, etc). Average frames, covering longer time intervals can be kept on EXABYTE tapes, at full spatial resolution.

A data link between the two sites shall enable strong candidate events to be defined in real time. The results can be transmitted over Internet, for comparison with BATSE and HETE data and for possible ground-based follow-up. Conversely, a special mode of data recording can be initiated upon receipt of a BACODINE signal[4].

REFERENCES

1. Vanderspek, R., Krimm, H., and Ricker, G., these proceedings , (1993).
2. Grindlay, J.E., Wright, E.L., and McCrosky, R.E., Ap.J.L. **192**, L113 (1974).
3. Mészáros, P., and Rees, M.J., Ap.J.L. , preprint (1993).
4. Barthelmy, S., et al., these proceedings , (1993).

THE KAPTEYN (LOGNORMAL) DISTRIBUTION AS A MODEL OF GRB SPECTRA AND PULSE PROFILES

M. Brock, C. Meegan, R. Wilson
Marshall Space Flight Center, AL 35812

C. Kouveliotou
Universities Space Research Association, Huntsville, AL 35812

W. Paciesas, G. Pendleton, M. Briggs, R. Preece
University of Alabama in Huntsville, Huntsville, AL 35899

ABSTRACT

We propose a simple model of the spectrum and time profile of a gamma-ray burst with little emphasis on physical interpretation, and we suggest the usefulness of a standard model. We adopt as our model a probability density function attributable to the astronomer J. C. Kapteyn. We suggest a simple interpretation of the distribution in the context of gamma-ray astrophysics, avoiding any discussion of progenitor objects.

Below we present the time profile of a gamma-ray burst. The burst is archetypical, not a random sample. The burst exemplifies characteristics common to many gamma-ray bursts.

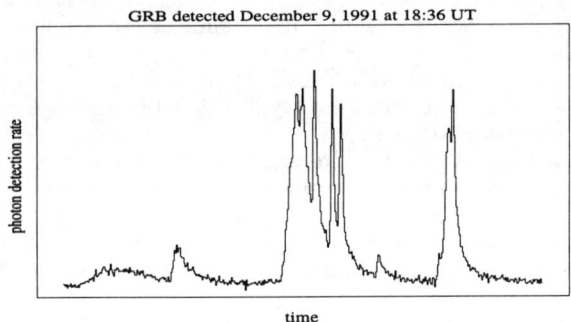

We summarize our observations of the burst and many similar bursts as follows.

1 The time profile of many gamma-ray bursts is well modeled by the superposition of pulses, each characterized by four parameters, a translation, an amplitude and two parameters describing characteristics of the pulse shape other than the amplitude. The profile of a single pulse is well modeled by the Kapteyn distribution.

2 The Kapteyn distribution also well represents the distribution of photon energies in many gamma-ray bursts.

Our application of the Kapteyn distribution to modeling gamma-ray burst spectra preceded our application of the distribution to modeling time profiles.

We emphasize the latter here, believing time profile analysis to be a more fruitful and less harvested field. BATSE data analysis software available from the GRO Science Support Center permits spectral modeling with the Kapteyn distribution.

THE KAPTEYN DISTRIBUTION

Because the Kapteyn distribution enjoys such wide application, the distribution is sometimes called by a less sectarian name, the Lognormal distribution. We here honor the contributions of Kapteyn. The probability density function describing the Kapteyn distribution is commonly written as follows.

$$K(y; \mu, \sigma^2) = \frac{1}{\sqrt{2\pi}\sigma y} \exp(-\frac{1}{2\sigma^2}(\ln y - \mu)^2)$$

Note by substituting $y = \exp(x)$ that the function describes the distribution of a variate the logarithm of which has a Gaussian (Normal) distribution with mean μ and variance σ^2. The distribution may be derived from the following corollary of the central limit theorem of mathematical statistics.

Theorem: Let $\{y_i, i = 1, \ldots, n\}$ be a sequence of positive, independent, identically distributed random variables with finite mean and variance, μ and σ^2. Then the product, $y = y_1 y_2 \cdots y_n$, has a distribution which is approximately Lognormal, and the mean and the variance of $\log y$ are $n\mu$ and $n\sigma^2$, respectively.

The proof of the theorem follows easily from the central limit theorem with the substitution $y_i = \exp(x_i)$. Aitchison and Brown prove more general theorems within the theory of proportionate effect.

We describe gamma-ray burst observations with two parameterizations of the Kapteyn distribution, one a model of the time profile of photon detection rates within a single burst pulsation and the other a model of the distribution of photon energies within a pulsation.

PHOTON DETECTION RATE

We choose the following parameterization of the Kapteyn distribution to describe pulsations in gamma-ray bursts.

$$t^\uparrow = \exp(\mu - \sigma^2)$$

$$t^{\frac{1}{2}\downarrow} = t^\uparrow(\exp(\sigma\sqrt{\ln 4}) - 1).$$

$$A = \frac{f}{\sqrt{2\pi}\sigma} \exp(\frac{1}{2}\sigma^2 - \mu)$$

where $t^\uparrow$ is the time required for the pulse to reach its peak intensity, $t^{\frac{1}{2}\downarrow}$ is the time required for the pulse to decay to half its peak intensity and A is the pulse amplitude. f is the integrated photon flux over the duration of a pulse. We write $A \cdot P(t - t^0; t^\uparrow, t^{\frac{1}{2}\downarrow})$ for the pulse profile, where t^0 is the pulse translation parameter and $P(y; t^\uparrow, t^{\frac{1}{2}\downarrow}) = 0$ for $y \leq 0$. Noting the following relationship, we spare you the complete expression for P and instead present an illustration.

$$A \cdot P(y; t^\uparrow, t^{\frac{1}{2}\downarrow}) = f \cdot K(y; \mu, \sigma^2)$$

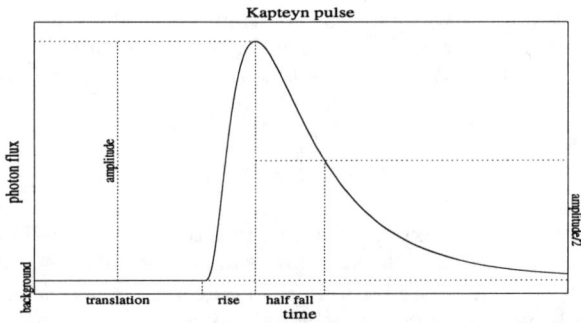

The following figures present a model of a portion of the illustrated burst time profile without and with Poisson noise. We chose the model parameters by eye. The parameters do not represent a best fit of the burst to the Kapteyn model.

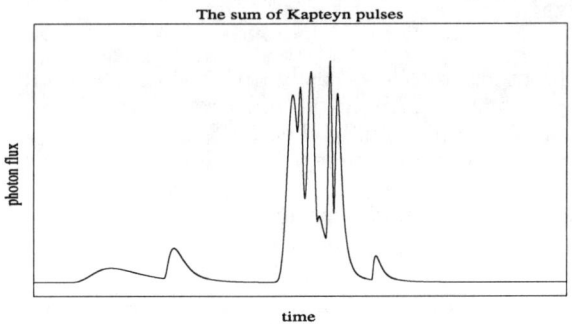

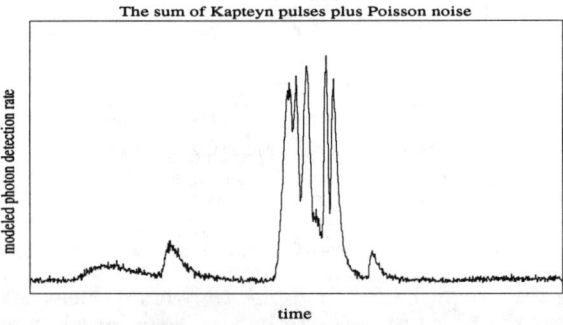

PHOTON ENERGY DISTRIBUTION

After replacing y with E in $K(y; \mu, \sigma^2)$, algebraic manipulation reveals the following form which we choose as our model of the photon energy distribution in a gamma-ray burst, photon flux per unit of energy.

$$\frac{dN}{dE} \propto E^{-\alpha - \beta \ln E}$$

$$\alpha = 1 - \frac{\mu}{\sigma^2}$$

$$\beta = \frac{1}{2\sigma^2}$$

The function may be normalized for all α and $\beta > 0$.

$$\int_0^\infty E^{-\alpha - \beta \ln E} = \sqrt{\pi \beta} \exp\left(\frac{(1-\alpha)^2}{4\beta}\right)$$

Recalling the theorem, note that α is independent of n while β is inversely proportional to n. Therefore, the Kapteyn distribution resembles a power-law distribution, $E^{-\alpha}$, when n is large. Gamma-ray burst spectra often resemble power-laws. Note also that the logarithm of $E^{-\alpha - \beta \ln E}$ is a quadratic equation in $\ln E$ just as the logarithm of a power-law is a linear equation. Paczyński proposed the exponential of a quadratic equation as a vehicle with which to search for signs of cosmological redshift.

Letting γ represent the total photon flux at all energies, we have the following expression for the spectral model.

$$\frac{dN}{dE} = \frac{\gamma}{\sqrt{\pi \beta}} \exp\left(-\frac{(1-\alpha)^2}{4\beta}\right) E^{-\alpha - \beta \ln E}$$

We hesitate to speculate on the nature of physical processes generating gamma-ray bursts; however, a simple scattering process aids our intuition.

Recalling the theorem, let y_i be the fractional change in the energy of a particle during each of a sequence of interactions, and assume the particle's energy is converted into a gamma-ray at the end of the sequence. The interactions randomize the particle's energy. Assuming the fractional change in a particle's energy during each interaction is not much affected by the particle's energy before each interaction (energy obeys the law of proportionate effect), the distribution of particle energies after the interactions resembles the Kapteyn distribution. Elsewhere in these proceedings, Pendleton describes features in the spectrum of a gamma-ray burst with the superposition of smooth spectra describing separate pulses in the burst.

APPLICATIONS OF A STANDARD MODEL

Our observation of burst time profiles leads us to expect a clustering of pulse shapes and a correlation between pulse shape and spectral shape. Standard models of pulse and spectral shape may simplify both comparing the work of cooperating researchers and developing standard analytical tools. In some cases, theorists may fruitfully observe the regions of a standard parameter space filled by experimental measurements before analyzing the often large and complex data sets collected by experiments.

Standards are best adopted by consensus. We offer the model presented here for the reader's consideration. We propose the Kapteyn model because it is simple, well-studied and not associated with a concrete theory of burst genesis. We believe Kapteyn's work in statistics merits greater attention, particularly in light of current interest in the similarity of natural phenomena on different geometric scales and other scales. The Kapteyn distribution has application in economic theories which were the subject of early work by the mathematician credited with generating much of that interest, Benoit Mandelbrot. A recent paper by Korsan on *Fractals and Time Series Analysis* may interest some readers.

We acknowledge the influence of presentations by Jay Norris, Jonathan Katz and Dawn Meredith on the development of this work.

REFERENCES

1. Aitchison and Brown, The Lognormal Distribution (Cambridge University Press, 1969).
2. Craig and Brown, Inverse Problems in Astronomy (Adam Hilger Ltd, 1986).
3. Daubechies, Ingrid, Ten Lectures on Wavelets (SIAM, 1992).
4. Gurbatov, Saichev, Chaos **3**, 333 (1993).
5. Kapteyn, J. C., Skew Frequency Curves in Biology and Statistics (Astronomical Library at Groningen, 1903).
6. Korsan, Robert J., The Mathematica Journal **3**, 39 (1993).
7. Paczýnski, Bohden, Nature **355**, 521 (1992).

NEW TECHNIQUES IN THE FITTING OF GAMMA-RAY BURST CYCLOTRON LINES

P. E. Freeman, C. Graziani, and D. Q. Lamb
Dept. of Astronomy and Astrophysics, University of Chicago, IL, 60637
T. J. Loredo
Dept. of Astronomy, Cornell University, NY, 14853

ABSTRACT

We provide an approximate prescription for determining the Bayesian odds favoring spectral models with lines over models without lines, using a new line parametrization which has several advantages over previous parametrizations. We use an exponentiated Gaussian line model, parameterized in terms of the equivalent width, W_E, and the full width at half maximum, $W_{1/2}$, of the line itself (*not* the full width of the Gaussian). Unlike other, equivalent, parametrizations, this parametrization yields Bayesian posterior probability distributions (and frequentist χ^2 surfaces) which are approximately Gaussian near the maximum likelihood parameter values, and allows the formulation of prior probability distributions for the continuum and line parameters which are independent of each other. It has the additional advantage that it easily treats "saturated" lines, in which $W_E = W_{1/2}$. We use Bayesian inference and this parameterization to determine whether the Burst and Transient Source Experiment (BATSE) Spectroscopy Detector (SD) can detect the single and harmonically-spaced spectral lines observed by *Ginga* in the time periods S1 and S2 of GB870303. We find that a single SD can marginally detect the single, strong, line in S1 at ≈ 20 keV at burst incidence angles $\lesssim 45°$, and the harmonically-spaced, weaker, lines in S2 at ≈ 20 and 40 keV at incidence angles $\lesssim 60°$.

INTRODUCTION

The single and harmonically-spaced low-energy absorption-like lines seen by *Ginga* in the spectra of a few bursts[1,2,3] and the success of the cyclotron scattering model in explaining them[4] implies that some γ-ray bursts come from strongly magnetic neutron stars. Given the importance of this implication, it is imperative to develop a rigorous method one can use to establish the existence of spectral lines. Here we describe a new parameterization of the exponentiated Gaussian line model which allows us to do this within Bayesian inference. We then use this method to determine whether a single SD can detect lines like those seen by *Ginga*. Our approach follows that outlined in Freeman *et al.*,[5] except that we use Bayesian odds, rather than frequentist significance, to determine the sensitivity of the BATSE SD.

$W_E - W_{1/2}$ PARAMETRIZATION

When fitting a dip in the spectrum of a γ-ray burst, one must choose a line model and its parametrization. In frequentist statistics, the signficance of the line is often evaluated using the difference $\Delta\chi^2$ in the minimum χ^2 between the model with lines and the model without.[6] In frequentist statistics, ease of computation is a basis for deciding upon the parametrization, since there is no difference in the minimum χ^2 for different parametrizations of the same model. The behavior of χ^2 as a function of the model parameters is of no importance, since it plays no role in the $\Delta\chi^2$-test.

In Bayesian inference,[7,8] the existence of a line is established by calculating the odds favoring the model with a line over that without. In this approach an "Ockham factor" naturally arises which penalizes the more complicated model. The more complicated model is favored only if its greater ability to describe the data (represented by the likelihood ratio) outweighs its additional complexity (represented by the Ockham factor). In order to calculate the Ockham factor, a parametrization of the line model in which the posterior probability distribution of the model parameters is approximately Gaussian is highly desirable.

Before we can determine which parameterization is the best to use, we must choose a line model. Astrophysicists frequently use either a additive Gaussian model, $F(E) = C(E) - \beta G(E)$; or an exponentiated Gaussian model, $F(E) = C(E)e^{-\beta G(E)}$. Here $F(E)$ is the total spectral flux, $C(E)$ is the continuum flux, and

$$G(E) = \exp(-\frac{(E-E_c)^2}{2\sigma^2}), \qquad (1)$$

where E_c is the line centroid energy, and β and σ are the unnormalized strength and width of the Gaussian.

We do not use the additive Gaussian model because $F(E)$ can be negative, which is unphysical. Instead, we use the exponentiated Gaussian model, which does not have this problem. There are many ways to parametrize the exponentiated Gaussian model. We use two criteria to choose among these possible parametrizations:

• We choose parameterizations that yield elliptical contours when we take two-dimensional slices of the parameter space and plot the contours of constant likelihood corresponding to a substantial amount of integrated probability (or of constant χ^2 corresponding to 2 or 3σ).

• If two or more parameterizations fulfill the first criterion, we prefer the one with more easily integrable prior probability distributions.

While many parametrizations fulfill these criteria when the signal-to-noise is large and/or the line is strong (but not "saturated"), we find that only the parametrization in terms of equivalent width W_E and full width at half maximum $W_{1/2}$ always meets these criteria.[9] We therefore adopt the W_E-$W_{1/2}$ parametrization. In this parametrization,

$$W_E = \int dE \frac{C(E) - F(E)}{C(E)} = \sqrt{2}\sigma \Phi(\beta), \qquad (2)$$

where

$$\Phi(\beta) = \int_{-\infty}^{\infty} dx [1 - \exp(-\beta e^{-x^2})]. \qquad (3)$$

We define the full width at half maximum, $W_{1/2}$, by the relation

$$\frac{1 - \exp[-G(E_0 + F/2)]}{1 - \exp[-G(E_0)]} = \frac{1}{2}. \qquad (4)$$

We have shown[9] that

$$\lim_{\beta \to 0} \frac{W_E}{W_{1/2}} = 0 \quad \text{and} \quad \lim_{\beta \to \infty} \frac{W_E}{W_{1/2}} = 1. \qquad (5)$$

The ratio $W_E/W_{1/2}$ is nearly, but not quite, a monotonic function of β: it rises sharply from 0, reaching $W_E/W_{1/2} = 1$ when $\beta \approx 4.75$ and peaking at ≈ 1.015 when $\beta \approx 20$, before tapering off to 1 as $\beta \to \infty$. Thus the line begins to saturate (i.e., to exhibit a flat bottom) when $\beta = 4.75$, and becomes completely saturated (i.e., approaches a square well) as $\beta \to \infty$.

An observed line has a value of $W_E/W_{1/2}$ which falls in the range $0 \leq W_E/W_{1/2} \leq 1$ (and therefore β in the range $0 \leq \beta \leq 4.75$), unless it is highly saturated. Even then, the likelihood value for a highly saturated line model differs only slightly from that for a line model with $W_E/W_{1/2} = 1$ and $\beta = 4.75$, unless the signal-to-noise of the observed line is very large. In the latter case, the existence of the observed line is easily established using either a line model with $\beta = 4.75$ or with $\beta \to \infty$. We therefore constrain $W_E/W_{1/2}$ to lie in the range $0 \leq W_E/W_{1/2} \leq 1$ with little loss in generality.

A particular advantage of this parametrization is that the ranges for the line parameters are particularly simple; e.g., $0 \leq W_E \leq W_{1/2}$, as discussed above; $0 \leq W_{1/2} \leq 2E_1$ (otherwise the observed spectrum exhibits a low-energy turnover rather than a line); and $E_{\min} \leq E_1 \leq E_{\max}/m$. Here E_1 is the line centroid energy, $E_{\min} = 4$ keV and $E_{\max} = 100$ keV are the minimum and maximum energies at which we seek a line, and m is the number of lines in the fit.

SENSITIVITY OF THE BATSE SD

To investigate the sensitivity of the BATSE SD to lines like those seen by *Ginga*, we simulate bursts using the best-fit spectral parameters derived from the *Ginga* data for period S1 of GB870303 (which extends for 4 s and has a fluence of 2.5×10^{-6} erg cm^{-2}), and period S2 of GB870303 (which extends for 9 s and has a fluence of 9×10^{-6} erg cm^{-2}). This is the weakest burst with lines seen by *Ginga*.

We consider a single SD. We assume a gain setting of 7X nominal and truncate the incident photon energy spectra at 1479 keV, which is the highest photon energy in the *Ginga* response matrix; both assumptions maximize the sensitivity of the SD to lines in the energy range $\approx 20 - 40$ keV.

In our simulations, we assume that bursts strike the SD at four incident angles θ: 30°, 45°, 60°, and 75° (the last for S2 only). We simulate 25 bursts at each angle for both S1 and S2. We fold the incident photon spectrum of each burst through the appropriate response matrix and take into account detector dead time[10] in order to derive the expected burst counts spectrum. We add to this spectrum a typical SD background spectrum that the BATSE team has provided to us. We then apply Poisson statistics to the expected counts spectrum to generate a simulated spectrum.

Assuming that the likelihood functions are Gaussian near the maximum likelihood values of the parameters, evaluating the prior probability distribution at these best-fit values, and assuming a uniform prior, the odds favoring the continuum plus exponentiated Gaussian line model over the continuum (no line) model are

$$O = e^{\Delta L}(2\pi)^{N/2}\sqrt{\frac{\det C_{C+L}}{\det C_C}}\frac{V_C}{V_{C+L}}, \qquad (6)$$

where ΔL is the difference in log likelihood between the two models, C is the covariance matrix, and V is the prior parameter volume. In the W_E-$W_{1/2}$ parametrization, $V_C/V_{C+L} = 1/V_L$. We use a likelihood function derived from Poisson statistics.

TABLE 1
GB870303: BAYESIAN VS. FREQUENTIST RESULTS

Spectrum	θ	$\log O_{\text{median}}$	$F_{O>100}$	$\log Q_{\text{median}}$	$F_{Q<10^{-5}}$
S1	30°	4.31	0.80	-7.14	0.88
	45°	2.15	0.56	-5.55	0.64
	60°	0.25	0.16	-3.46	0.16
S2	30°	8.10	0.96	-13.19	1.00
	45°	5.28	1.00	-9.95	0.96
	60°	3.02	0.67	-7.63	0.84
	75°	-0.47	0.06	-3.77	0.20

We evaluate the odds favoring the existence of a single line like that seen in GB870303 S1 using a two-parameter saturated line model, parametrized by the centroid energy E_1 and equivalent width $W_E(\equiv W_{1/2})$. In this case, the prior parameter volume is

$$V_2 = \int_{E_{\min}}^{E_{\max}} dE_1 \int_0^{2E_1} dW_{1/2} = E_{\max}^2. \tag{7}$$

We compare frequentist and Bayesian results in Table 1. To derive the frequentist result, we take the value of $\Delta\chi^2$ ($=2\Delta L$) which results from each pair of fits and, from the χ^2-distribution for the number of degrees of freedom equal to the number of line parameters N, determine the probability Q that $\chi^2 > \Delta\chi^2$.[6] Thus, the smaller Q, the more significant the line. Within frequentist statistics, we adopt a rough significance criterion for detection of $Q < 10^{-5}$; within Bayesian inference, we adopt a rough odds criterion for detection of $O_{21} = 100$. With these criteria, we find that the BATSE SD easily detects a line like that in GB870303 S1 at incidence angles $\lesssim 30°$, and marginally at angles $\sim 45°$ (see Fig. 1).

Elsewhere we show that $N = 4$, the "constrained harmonic" model, with $E_2 = 2E_1$ and $W_{1/2,2} = 2W_{1/2,1}$, is the best model to use for the harmonically-spaced lines seen in GB870303 S2.[9] The prior parameter volume for this model is:

$$V_4 = \int_{E_{\min}}^{E_{\max}/2} dE_1 \int_0^{2E_1/3} dW_{1/2} \int_0^{W_{1/2}} dW_{E,1} \int_0^{2W_{1/2}} dW_{E,2} \approx \frac{1}{324} E_{\max}^4, \tag{8}$$

where the upper limit on $W_{1/2}$ is due to the fact that the lines cannot overlap, and the upper limit on E_1 results from the fact that we search for two lines.

We compare frequentist and Bayesian results in Table 1. Calculation of the covariance matrix failed for one simulation at each incident angle. We therefore determine the median odds by interpolating between the two middle odds of the 24 successful simulations. We find that the BATSE SD easily detects harmonically-spaced lines like those in GB870303 S2 at incidence angles $\lesssim 60°$ (see Fig. 2).

This research has been supported in part by NASA grants NAGW-666, NAGW-830, NAG5-1454, NAG5-1758, NASW-4690 and NGT-50778.

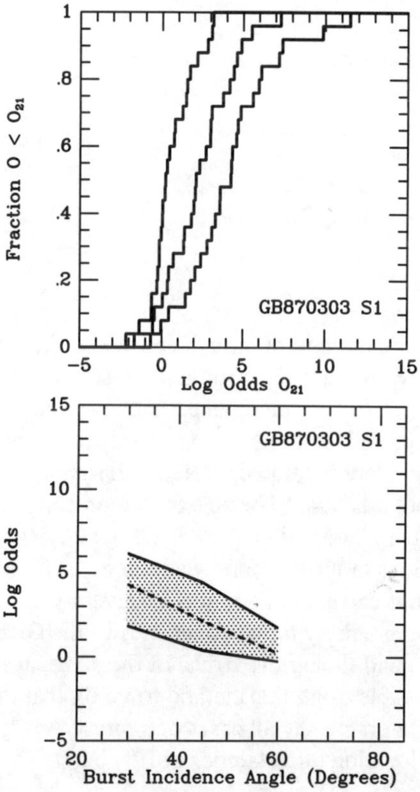

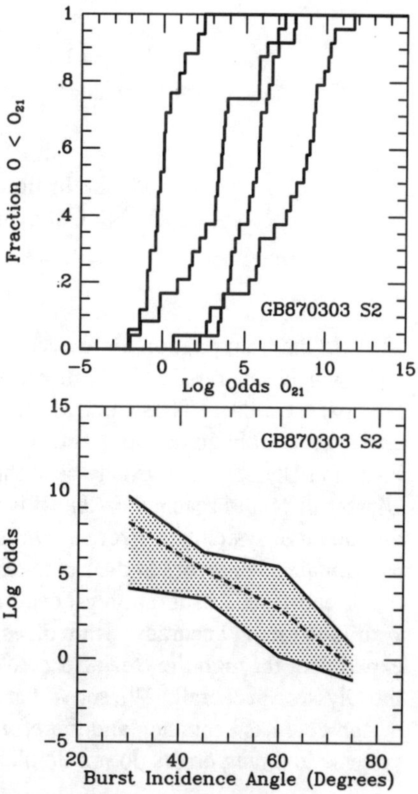

Figure 1. Results for the 25 simulations, at each angle, of GB870303 S1. *Top:* Cumulative distributions for, from left to right, 60°, 45°, and 30°. *Bottom:* The central 68% of odds values, as a function of detector incidence angle.

Figure 2. Same as Figure 1, except for GB870303 S2.

REFERENCES

1. T. Murakami, et al., Nature **335**, 234 (1988).
2. E. E. Fenimore, et al., Ap. J. **335**, L71 (1988).
3. C. Graziani, et al., Gamma-Ray Bursts: Observations, Analyses, and Theories, eds. C. Ho, R. I. Epstein, and E. E. Fenimore (Cambridge: Cambridge University Press, 1992), p. 407.
4. J. C. L. Wang, et al., Phys. Rev. Letters **63**, 1550 (1989).
5. P. E. Freeman, et al., Compton Gamma-Ray Observatory, eds. M. Freidlander, N. Gehrels, and D. J. Macomb (New York: AIP, 1993), p. 922.
6. W. T. Eadie, D. Drijard, F. E. James, M. Roos, and B. Sadoulet, Statistical Methods (Amsterdam: North Holland, 1971).
7. T. J. Loredo, & D. Q. Lamb, Gamma-Ray Bursts, eds. W. S. Paciesas and G. J. Fishman (New York: AIP, 1992), p. 414.
8. T. J. Loredo, Statistical Challenges in Modern Astrophysics, eds. E. Feigelson and G. Babu (New York: Springer-Verlag, 1992), p. 275.
9. C. Graziani, et al., Compton Gamma-Ray Observatory, eds. M. Friedlander, N. Gehrels, and D. J. Macomb (New York: AIP, 1992), p. 897.
10. B. Schaefer, BSAS User's Guide (NASA/GSFC, 1991).

TIMING ACCURACY OF THE ULYSSES GRB EXPERIMENT

K. Hurley
University of California
Space Sciences Laboratory, Berkeley, CA 94720

M. Sommer
Max-Planck Institut für Extraterrestrische Physik
D8046 Garching-bei-München, Germany

Abstract

In the triangulation, or arrival time analysis method of burst localization, the most widely separated spacecraft pair generally provides the strongest constraints on the final error box. Thus a knowledge of the systematic timing uncertainties in these two spacecraft is quite important. In the 3rd Interplanetary Network, spacecraft pairs involving Ulysses have usually been the most widely separated, since the Earth-Ulysses distances reached 6 AU. It is therefore essential to be able to demonstrate that no large systematic errors are present in the Ulysses timing. The design goal was to maintain such errors in the range of several ms at most. Here we describe an in-flight, end-to-end calibration procedure which is carried out about twice every year, to verify the timing accuracy. It involves sending precisely timed commands to the GRB experiment through the Deep Space Network, and timing the arrival of the signal at the Ulysses spacecraft. We show that it is possible using this method to verify that the design goal has been met, and therefore that, in practically all cases of interest, systematic timing errors do not dominate the location uncertainties of IPN bursts.

Introduction

When a gamma-ray burst arrives at two spacecraft with a time delay δT, it may be localized to an annulus whose half-angle Θ with respect to the vector joining the two spacecraft is given by $\cos\Theta = c\delta T/D$, where D is the inter-spacecraft distance and c is the speed of light. Three spacecraft give two annuli which intersect at two positions, and provided that at least one experiment has even coarse directional information (e.g., BATSE), one of the two may be discarded. The width of an annulus $d\Theta$, and thus one dimension of the resulting error box, is $c\sigma(\delta T)/D\sin\Theta$, where $\sigma(\delta T)$ is the uncertainty associated with the time delay. (As shown below, the error associated with the spacecraft coordinates is negligible in practice.) If $\sigma(\delta T)/\sin\Theta$ is approximately equal for two spacecraft pairs, then the most widely separated pair generally will provide the narrowest annulus. The term $\sigma(\delta T)$ in the above formulas has two components, one statistical, and the other systematic. In an ideal network consisting of four or more equidistant spacecraft, systematic uncertainties would not play an important role. Burst locations would be derived from three or more independent annuli of approximately equal width. If the annuli, derived assuming statistical errors only, intersected at or near a single point, this would be good evidence that systematic errors were negligible. Such networks have existed in the past, and will exist again following the launch of the Russian Mars 94 mission. To date, however, the 3rd IPN has consisted of a number of near-Earth spacecraft, Ulysses, and PVO or Mars Observer. Thus provided that the two independent location annuli intersect (a very unrestrictive assumption), systematic timing errors could in principle exist and go undetected. This paper treats the question of Ulysses systematics. The general question of statistical errors is discussed in another paper[1].

General Principles

The Ulysses GRB experiment [2] can be triggered by command from the ground. Upon receipt of this command, the instrument reads the value of the spacecraft clock and inserts it into the telemetry stream, along with count rate and spectral data. Therefore, if a trigger command can be transmitted at a precisely known time, a knowledge of the light-travel time to the spacecraft and of all delays in the relevant spacecraft subsystems allows one to predict the arrival time of the command at the experiment, and check it against the actual time recorded in the telemetry. We review the timing accuracies and delays associated with the elements of these timing tests.

1. The Deep Space Network (DSN). Three stations in the DSN maintain uplink and downlink contact with the Ulysses spacecraft (Madrid, Canberra, and Goldstone), and at each station, a number of dishes may be used for contacts, which generally last 8 hours/day. In normal operations, the DSN does not precisely time the radiation of commands to the spacecraft. For GRB timing tests, however, the DSN is configured to measure the radiation time of the first bit of a command to an accuracy of about 1 ms. (The precise time of a command may not be chosen *a priori*, however; only the time between successive commands may be specified.) The length of the command word is 96 bits, and the transmission bit rate is nominally 15.625 b/s. Doppler corrections require that this frequency be adjusted by up to 0.002 b/s, but the effect on the command transmission time is <1 ms.

2. The One-Way Light Time (OWLT). During tracking passes, spacecraft ranging is accurate to better than 5 km (1σ). The topocentric OWLT is therefore known to an accuracy of better than 0.02 ms, and is calculated for each command, since changes of up to hundreds of ms may occur between successive commands. The errors in the spacecraft coordinates (α,δ) are 500 nrad (~0.00003°, 1σ), and are verifiable using the technique of VLBI tracking, which involves comparing the spacecraft position with the positions of a grid of distant radio sources. The errors are thus negligible for all practical purposes.

3. Spacecraft Subsystem Delays. The trigger command is processed by several subsystems aboard the spacecraft; all but one result in fixed delays in routing the command to the GRB experiment. We list them below.

a) Command decoder   1.7 ms
b) Transfer from decoder to central terminal unit: 0.48 ms
c) Waiting time for onboard data handling bus   1.95 ms
d) Waiting time for remote terminal unit (RTU) command execution 0 - 125 ms
e) Tranfer from remote terminal unit to GRB experiment   0.05 ms

Total: 4.18 - 129.18 ms

4. GRB Experiment Delay. A small amount of time elapses between the receipt of the command at the GRB experiment and the actual issuing of a trigger. To measure this in-flight, we have used a slightly different command procedure. The spacecraft can be programmed in advance with *time-tagged* commands, which are executed when the spacecraft clock reaches a certain value. Since the GRB experiment reads the spacecraft clock into the telemetry with an accuracy of 0.4 ms when it triggers, a comparison of the uploaded time tag and the spacecraft clock value give the GRB time delay. It is approximately 1 ms. Once the trigger has been issued, the experiment requires about 40 m to dump its data to the telemetry; during this time, it cannot be re-triggered. Thus during a given tracking pass, a maximum of about 12 trigger commands can be sent.

5. Spacecraft clock/UT calibration. During all tracking passes, the spacecraft clock is monitored to establish its true frequency (the ratio of the true to nominal frequency is $C \approx 1.000015$). This frequency depends primarily on the clock temperature, and is therefore related to spacecraft activities, which dissipate power and raise the clock temperature slightly. For this reason, the clock frequency may drift slightly (<1%) during a timing test. A clock correction file is thus maintained for the entire mission, and the times of telemetry frames are known to much better than 1 ms.

Taking fixed and variable delays into account, the total delay in executing a trigger command lies between 5.18 and 130.18 ms.

Typical Results

To illustrate the method, we review the results of a timing test which took place from the Madrid station on January 9-10, 1992, during which eight trigger commands were sent. Figure 1 (crosses, solid line, left hand axis) shows the observed minus the predicted command execution times as a function of UT. The observed minus predicted times display a characteristic sawtooth pattern whose minimum value in this case is 10 ms, and whose maximum is 110 ms. Since the total delay must lie between 5.18 and 130.18 ms, the minimum and maximum values in the sawtooth verify that the systematic timing error is in the range -20.18 to +4.82 ms, i.e. consistent with zero.

The sawtooth pattern may be understood as follows. The commands arrive at the spacecraft in a RTU timing cycle whose nominal duration is .125 s, but whose actual duration is 0.125C, where C is the clock correction factor. Suppose that a command is sent from earth at a bit 1 radiation time T_1, and takes a one-way light time $OWLT_1$ to reach the spacecraft. Let this command arrive a time t_1 after the start of the timing cycle; it must wait $.125C-t_1$ to be executed.

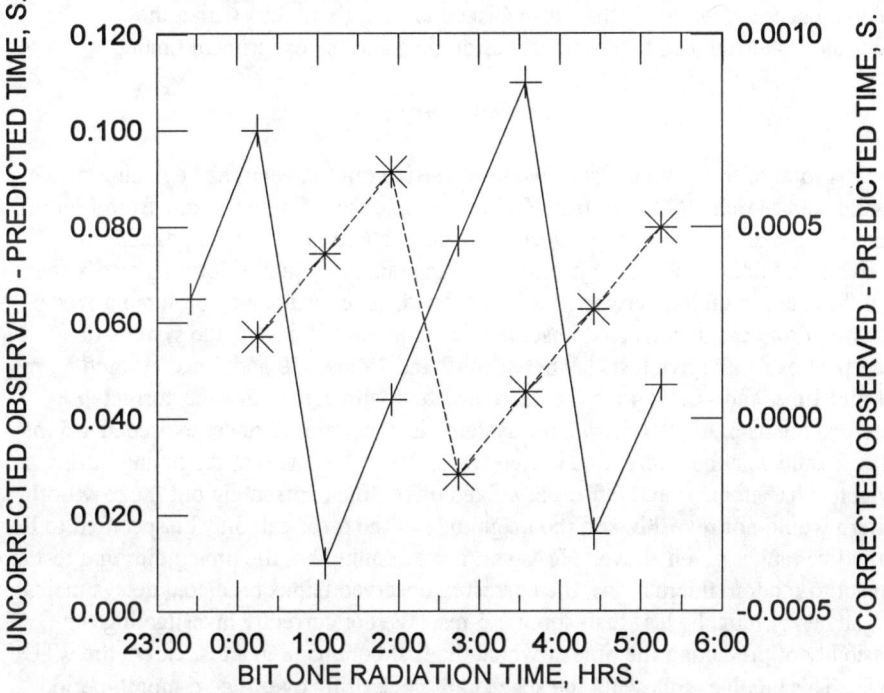

Figure 1. Uncorrected (solid line, left hand axis) and corrected (dashed line, right hand axis) time delays during a typical timing test.

Now suppose that the next command is sent from earth at a bit 1 radiation time T_2, and takes a time $OWLT_2$ to reach the spacecraft. Since the arrival of the first command, $(T_2+OWLT_2-T_1-OWLT_1)/.125C$ timing cycles will have elapsed. The cycle in which this second command arrives will have started a time

$$t_2 = t_1 + \{(T_2+OWLT_2-T_1-OWLT_1)/.125C - INT[(T_2+OWLT_2-T_1-OWLT_1)/.125C]\}.125C$$

seconds ago, where INT is the integer value of the number. The waiting time for this command is $.125C-t_2$. The sawtooth is therefore the result of a time difference between successive commands which is not an integer multiple of the RTU's cycle time. Since the transmission time of the first command in a test cannot be predicted, the RTU cycle is entered at a random time, and the *first* observed minus predicted command time may lie anywhere in the range 5.18-130.18 ms. However, using the above formula, the execution time of all following commands may be calculated accurately (i.e. with the random delay removed). This leads to a second estimate of the systematic errors. Assuming that there are no errors in the timing of the first command, the delays due to the sawtooth pattern may be removed for all the following commands. (This is somewhat analogous to proof by induction, except that at present, there is no proof that the timing of the first command is error-free). The dashed curve in Figure 1 (right hand axis) shows the result of this procedure. The residual uncertainties are in the sub-millisecond range, and thus within the uncertainties introduced by the transmission time and the experiment timing.

Summary

A total of five timing tests have been carried out between the beginning of the mission and March 1993, two from Goldstone, one from Canberra, one from Madrid, and one from all three stations. During the multi-station test, it was possible to receive telemetry from two DSN antennas simultaneously, and rigorously verify that no differences in timing between stations existed. The results may be summarized as follows. From the uncorrected observed minus predicted delays, the systematic uncertainties for the five tests lay between -7 and 29 ms, -20 and 4 ms, -15 and 55 ms, -31 and 1 ms, and -7 and 4 ms (i.e., all consistent with zero). For the corrected observed minus predicted delays, the systematic uncertainties never exceeded 1.5 ms. These results may be interpreted in two ways. If we assume that the timing during a given test has an error in the form of a fixed offset, then correcting out the sawtooth pattern would not reveal it, and the magnitude of the error can only be specified to lie in the five ranges given above. However, if we assume that the timing during a test is subject to random fluctuations, the corrected observed minus predicted delays indicate that any error must be less than about 1.5 ms. We are currently investigating the possibility of predicting the time at which the first command in a test enters the RTU cycle. This involves following the spacecraft clock drifts over the ~6 month period between successive tests. If sufficient accuracy can be achieved, it might be possible to reduce the limits on fixed offset errors below their current values.

References

1. K. Hurley, these proceedings (1993)
2. K. Hurley et al., Astron. Astrophys. Suppl. Ser. **92**, 401, 1992

Acknowledgements

We are grateful to N. Angold, R. Garcia, E. Page, J. Schmidling, and S. Standley for insuring the success of the complex operations involved. We acknowledge the invaluable assistance of Günther Hampel of Dornier Systems in analyzing the spacecraft electronics. Work on these data was supported by JPL Contract 958056 and NASA Grant NAG-1560. The Ulysses GRB experiment was built in France with support from CNES, and in Germany with support by FRG Contracts 01 ON 088 ZA/WRK 275/4-7.12 and 01 ON 88014.

CROSS-CORRELATING GAMMA-RAY BURST TIME HISTORIES

K. Hurley
University of California at Berkeley Space Sciences Laboratory
Berkeley, CA 94720

Abstract

The triangulation, or arrival time analysis method of burst localization depends upon cross-correlating the time histories of gamma-ray bursts observed at different spacecraft. The statistical uncertainty in the burst position is directly related to the uncertainties in the cross correlation, specifically the statistical error in the lag between the time histories of spacecraft pairs. Here I describe one method for determining the most probable lag and its associated statistical error, which is based on a χ^2 minimization technique. It is simple to understand (although not always to implement) and also has the advantage of indicating when two time histories of a given burst are dissimilar (e.g. because of a difference in the energy ranges of the experiments observing it). The results of Monte-Carlo calculations will be presented to illustrate how the uncertainty in the lag scales with parameters such as the total number of counts detected, and the first derivative of the time history (a measure of the "spikiness"). Applications to real data will also be discussed.

Introduction

When a gamma-ray burst arrives at two spacecraft with a time delay δT, it may be localized to an annulus whose half-angle Θ with respect to the vector joining the two spacecraft is given by $\cos\Theta = c\delta T/D$, where D is the inter-spacecraft distance and c is the speed of light. The annulus width $d\Theta$, and thus one dimension of the resulting error box, is $c\sigma(\delta T)/D\sin\Theta$, where $\sigma(\delta T)$ is the uncertainty associated with the time delay. This term has two components, one statistical, and the other systematic. The systematic errors have been discussed in another paper[1]; here, we treat the statistical errors associated with the uncertainty in determining the time lag between a pair of spacecraft.

Suppose that a gamma-ray burst with a simple time history is incident upon two detectors at the same instant, which we define to be lag zero. For the sake of brevity, the following simplifying assumptions are made; they will be relaxed in a more complete version of this paper[2]:
  1. the time history has a triangular shape,
  2. the detectors are identical in size, shape, and time resolution, and have equal backgrounds,
  3. the background is a Poisson random variable, but its variance is zero (i.e., it has been perfectly determined by measurement)

How can the most probable lag between the two detectors be found, along with its 1, 2, and 3σ confidence intervals? Denote the two background-subtracted time histories by X_i, Y_i, i=1,...m. Assume that the X_i, Y_i are normal random variables (rv's) with variances X_i, Y_i. Then at lag zero, $R_i=(X_i-Y_i)/(X_i+Y_i)$ is an rv with mean zero and variance 1. Thus

$$\sum_{i=1}^{m} R_i^2$$

is distributed as χ^2 with m degrees of freedom. Now allow the lag to vary slightly about zero, and form the rv $R_{ij}=(X_i-Y_j)/(X_i+Y_j)$; then

$$R_j^2 = \sum_{i=1}^{m} R_{ij}^2$$

is distributed as χ^2 with m-1 degrees of freedom. As the lag between the two time histories varies, the best correlation between them will occur for some $R_j^2{}_{min}$. The n-σ confidence lag, where n=1,2,3..., will be defined by

$$R_j^2 = R_j^2{}_{min} + \chi^2{}_1(\alpha)$$

where $\chi^2{}_1(\alpha)$ is the value of χ^2 corresponding to significance α and for one degree of freedom[3] (e.g. α=.32 for n=1).

Simulations

The above algorithm has been tested by means of a Monte-Carlo code. A triangular gamma-ray burst time history with 1/64 or 1/128 s binning is generated which is characterized by the number of counts C_{max} at the peak, and the full width at zero maximum (FWZM), T_{max}. (Thus the number of degrees of freedom is 64*T_{max}-1 or 128*T_{max} -1) This time history is randomized to simulate the measured time histories at two identical detectors (see Figure 1). The lag between the two is varied and the following quantities are calculated:
  R_{ij}, for lag zero,
  R_j^2, as a function of lag j, and
  the lag for which R_j^2 is a minimum.
The original time history is then randomized again, and the calculations are repeated. A typical run involves 300 trials. When it is finished, the following distributions are calculated:
  R_{ij} for lag zero (should be distributed with mean zero, variance 1),
  R_j^2 for lag zero (should be distributed as χ^2 with m-1 degrees of freedom),
  the lags for which R_j^2 was a minimum, and
  R_j^2 as a function of lag, averaged over trials.
Gaussians are fit to the first and third distributions, a χ^2 function to the second, and a parabola to the last. C_{max} and T_{max} are then changed to simulate a burst with a different fluence and/or FWZM and a new set of trials is begun. The results are shown

in Figure 2. For concreteness, counts have been converted to fluence in erg cm^{-2} assuming an E^{-2} photon spectrum in the 25-150 keV range incident on Ulysses-size

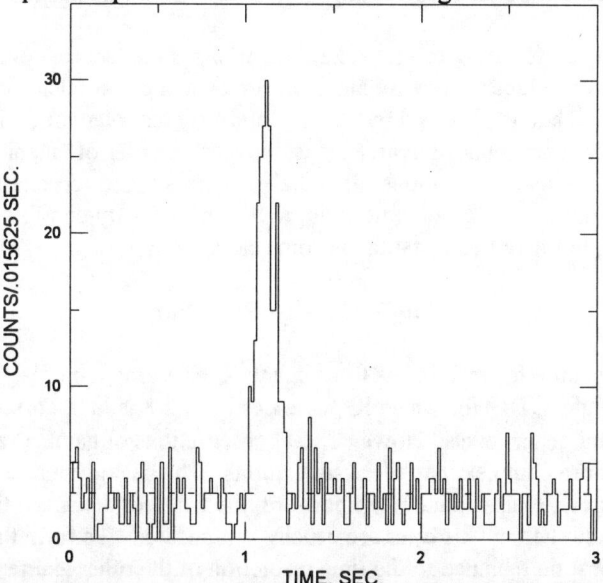

Figure 1. Simulated triangular time history. The background (dashed line) is 3 counts/interval, C_{max}=25 counts, and T_{max}=0.25 s. This burst has fluence 8×10^{-7} erg cm^{-2}, and can be localized to about 0.2 arcminutes.

detectors (20 cm^2),i.e. F=6.65×10^{-9}N erg cm^{-2}, where N is the total number of counts in the time history. To convert the 3σ confidence cross-correlation uncertainty to an error box dimension, a spacecraft pair separation of 4 AU has been assumed, with a burst incident at a 45° degree angle with respect to the line joining the pair, i.e. ΔΘ (arcmin)=2.44σ(δT).

Each curve in Figure 2 has a dependence of the cross-correlation accuracy on N which scales as N$^{-1.5}$. A simple way to understand this is to imagine that, in finding the best lag, we are trying to find the value of $\langle t_i - t_j \rangle$, where t_i and t_j are the weighted average values of the indices of the two (statistically independent) time histories X_i and Y_j. Then $\langle t_i - t_j \rangle = \langle t_i \rangle - \langle t_j \rangle$, and

$$\langle t_i \rangle = \frac{\sum_i t_i X_i}{\sum_i X_i} \propto \frac{1}{N}$$

and similarly for $\langle t_j \rangle$. The variance of a quantity which scales as N^{-1} is N^{-3}, and the standard deviation therefore scales as N$^{-1.5}$. The rise time, or "spikiness" of the time history is proportional to the rise time, $T_{max}/2C_{max}$;. from Figure 2, the cross-correlation accuracy scales approximately as the first power of T_{max}.

Perhaps the most interesting result of the simulations is the fact that even relatively weak bursts may be localized to arcminute accuracy and better. Thus the application of this method to bursts observed by the 3rd Interplanetary Network will give final error boxes which are reduced in area by as much as an order of magnitude with respect to the preliminary locations. It will also be possible to reprocess the older data from previous networks; the amount of improvement will vary considerably from event to event, but may be substantial in some cases.

Application To Real Data

The procedure for real data is to calculate R^2 as a function of lag, find the minimum R^2 and the lags for which R^2 increases by 1, 2.8, and 9, corresponding to 1,2, and 3σ confidence levels. However, real observations of gamma-ray bursts are considerably more complex than these simulations. The backgrounds are different for different detectors, their variances are nonzero, the detector areas and time resolutions are different, and the time histories are usually complicated. To form the rv R_{ij}, one time history must be rebinned to the time resolution of the other and renormalized so that $\langle X_i - Y_i' \rangle = 0$, where Y_i' is the rebinned, renormalized time history. The variance, which must include the renormalization factor, becomes a lengthy expression. However, in all cases of interest, the detector sizes are never less than those of Ulysses; for example, a BATSE LAD is 90 times larger. Thus the count rate statistics are better than those simulated here. These effects have been studied and simulated in the Monte Carlo program, and the complete results, which are not substantially different from those given here, will be presented elsewhere[2]. The energy ranges of different detectors may also differ, resulting in slightly different time histories. The results of this algorithm will indicate when this is the case, since $R_j^2{}_{min}/df$, where df is the number of degrees of freedom, will exceed unity. Many aspects of this method can be checked using the data on real gamma ray bursts from the Venera 11, 12, 13, and 14 spacecraft. Each mission carried two identical gamma-ray burst detectors, and time histories were recorded synchronously from each when one triggered. There are perhaps 100 bursts in this database which were observed with sufficient statistics on both detectors to utilize this algorithm. Often, one detector recorded the burst after absorption in the body of the spacecraft, or with a different energy threshold, leading to different time histories. The application of this algorithm to these data is in progress and will be reported elsewhere[2].

References

1. K. Hurley and M. Sommer, these proceedings (1993)
2. K. Hurley, in preparation, to be submitted to ApJ (1993)

3. M. Lampton, B. Margon, and S. Bowyer, ApJ 208, 177 (1976)

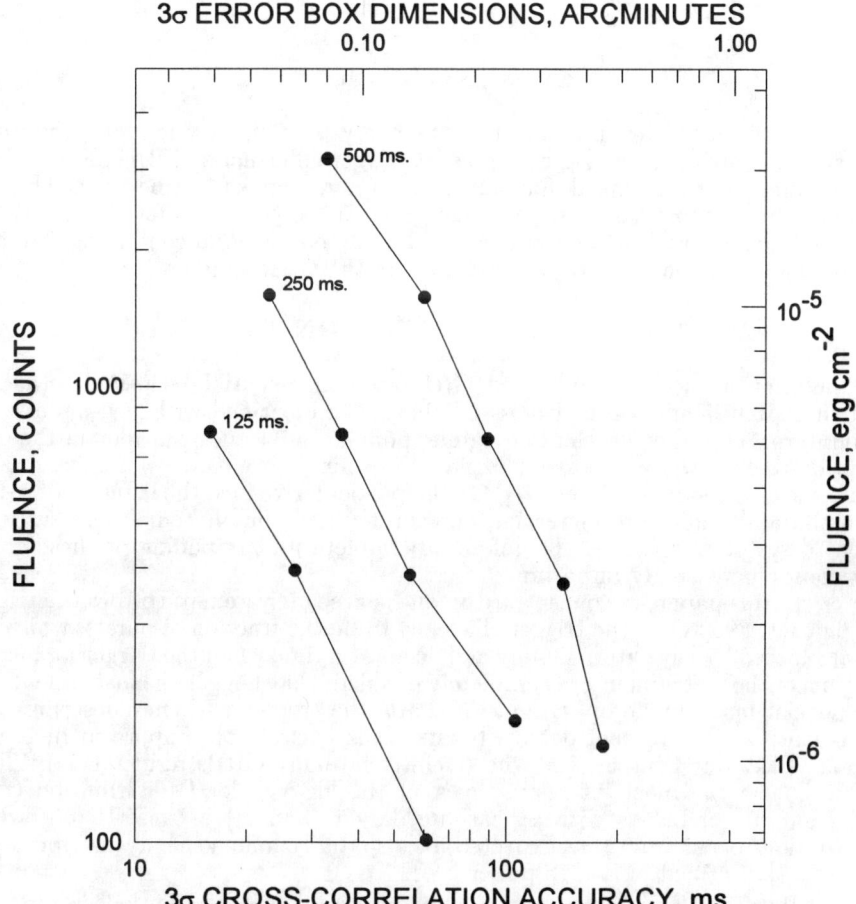

Figure 2. Results of twelve Monte Carlo simulations of gamma-ray burst time histories. The 99.87% confidence cross-correlation accuracy is plotted as a function of the fluence in counts. These have been converted to arcminutes and erg cm^{-2} as explained in the text. Each Monte Carlo run was done for 300 randomizations of the time history. The parameter $T_{max}/2$ (the time to maximum) is indicated for each curve.

Acknowledgments

Work on this project was supported by NASA Grant NAG 5-1560 and by JPL Contract 958056. I am grateful to B. Schaefer for comments on this paper.

TRIGGER EFFICIENCIES OF BATSE AND PVO

J.J.M. in 't Zand & E.E. Fenimore
Los Alamos National Laboratory, SST-9, MS D436, Los Alamos, NM 87545

ABSTRACT

Trigger efficiencies are important in the analysis of the low end of γ-ray burst peak flux distributions. We here present trigger efficiencies of BATSE and PVO as a function of burst peak flux and explain how they should be used. The efficiencies take into account most trigger conditions, typical burst characteristics (including spectra and time profiles) and γ-ray background conditions. We find a strong dependence of trigger efficiency on the GRB time profiles.

INTRODUCTION

Population studies of γ-ray bursts (GRBs) remain essential to the understanding of the GRB phenomenon. Unless GRBs can be identified with well-understood counterparts in any wavelength regime, population studies will remain the only option to narrow down possible explanations for their nature. With the increasing size of the sample of detected GRBs, particularly since the launch of CGRO, population studies are increasing in accuracy and thus become more susceptible to systematic errors. Therefore, a complete understanding of these errors becomes increasingly important.

In this paper we concentrate on one source of systematic error: the trigger efficiency. We define the trigger efficiency to be the fraction of bursts, which are not blocked by any external body and occur at a time when the triggering mechanisms of the instrument are completely enabled, that trigger the instrument into a special burst mode operation. In short, the trigger efficiency describes how efficiently an instrument detects bursts. It is particularly important in the region of low burst intensities, where relatively many GRBs reside. For BATSE, this is also the most interesting part of the $\log N - \log P$ distribution (N is the number of bursts with a peak flux larger than P) because that is where deviations occur from that expected for a spatially homogeneous distribution of bursts (prescribed by a -3/2 power law).

Recently, Fenimore et al.[4] combined $\log N - \log P$ distributions as observed by BATSE and PVO to extend the dynamic range of the peak flux in order to evaluate the possibility of a cosmic origin for the GRBs. Due to the uncertainty in the trigger efficiency, Fenimore et al. excluded bursts in that range of the peak flux where the trigger efficiency was estimated to be deviating from 1 (amounting to $\sim 60\%$ of all BATSE and $\sim 45\%$ of all PVO bursts). With improved values for the trigger efficiencies of both instruments, more data may be utilized in such an analysis and, thus, the BATSE and PVO observations may discriminate even better between different models or may evaluate more accurately model parameter values.

We have studied the trigger efficiencies, taking into account a model for the GRB population which is more complete than used elsewhere (e.g. Fishman et al.[6]), and present in this paper some results for BATSE as well as PVO.

SIMULATION MODEL AND METHOD

In order to calculate the trigger efficiencies of both BATSE and PVO as a function of peak flux, models need to be defined for the instruments as well as the burst population. We employed models that include the following nontrivial details: 1) The distribution of GRB spectral parameters is based on the results as obtained by Band et al.[1] on 54 GRBs with the BATSE spectroscopy detectors. Specifically, the parameters α and β of their 'GRB model' are homogeneously distributed between $\alpha = -1.5$ and 0.0 and $\beta = -2.25$ and -1.75; E_0 is exponentially distributed with an e-folding fall-off constant of 400 keV; 2) As time profiles we used 262 actual time profiles from BATSE and PVO databases (at a 64 ms time resolution). These bursts had observed peak intensities at least ten times the background noise standard deviation, as observed by the BATSE large area detectors during the first 10 months of the mission (up to burst trigger number 1463) or by PVO during its total mission. In each time profile, we filtered out most of the noise; 3) We assumed the GRB sky position distribution to be isotropic and defined the peak flux distribution to be that found by Fenimore et al.[4] (1993); 4) All GRB parameters were chosen to be independent of each other; 5) For BATSE, the probability distribution for the background was modeled by two components: 94% of the time the distribution is Gaussian, with a mean of 2500 c/s and a standard deviation of 116 c/s, and 6% of the time it follows a box-shaped function between 3000 and 3850 c/s. This model for BATSE was checked using the trigger efficiencies as published by Fishman et al.[6]; 6) For the PVO instrument[2], the probability distribution for the *logarithm* of the background B (in c/s) was modeled by two components: 97% of the time the distribution follows a Gaussian, with a mean of $\log B = 2.12$ and a standard deviation of 0.10, and 3% of the time it follows a box-shaped function up to $\log B = 3.1$. This model was determined from the actual PVO data set.

We developed a computer code which simulates the response of both BATSE and PVO to any given GRB population as follows. For each simulated GRB, the spectrum, time profile, sky position and peak flux are picked randomly from the appropriate probability distributions. Each GRB is subsequently convolved with the response of both instruments. This is done for photon energy ranges of 50 to 300 keV for BATSE and 100 to 2000 keV for PVO. Subsequently, the resulting lightcurves (one for each of the 8 BATSE detectors and one for the sum of both PVO detectors) are randomized according to the available count rate statistics (on a 64 ms time resolution). The data thus obtained are analyzed for trigger conditions similar to how it is done by the instruments in reality. For BATSE[5,6] this means checking whether on time scales of either 64, 256 or 1024 ms, count rates of at least two detectors rise at least 5.5 σ above the 17 s averaged background level. For PVO[2,3] the trigger analysis involves checking whether on time scales of 1/4, 1 or 4 s, the total countrate of both detectors rises at least 11.3 σ above the 16 s averaged background level.

TRIGGER EFFICIENCIES

The above code gives the opportunity to compare models with observations of the peak flux distribution, in a fashion common to X-ray spectral analyses: a model is defined, this is folded through the instrument, and the observed distribution is tested against the folded model via for example a χ^2 test. Acceptable results indicate consistent models and parameter values. Alternatively, one might apply a more direct but statistically less consistent method of dividing each burst through the fraction of bursts with the appropriate peak flux that are anticipated to trigger the instrument, thus acquiring a peak flux distribution that is corrected for detection efficiencies. Such distributions may be compared directly to models of intrinsic distributions.

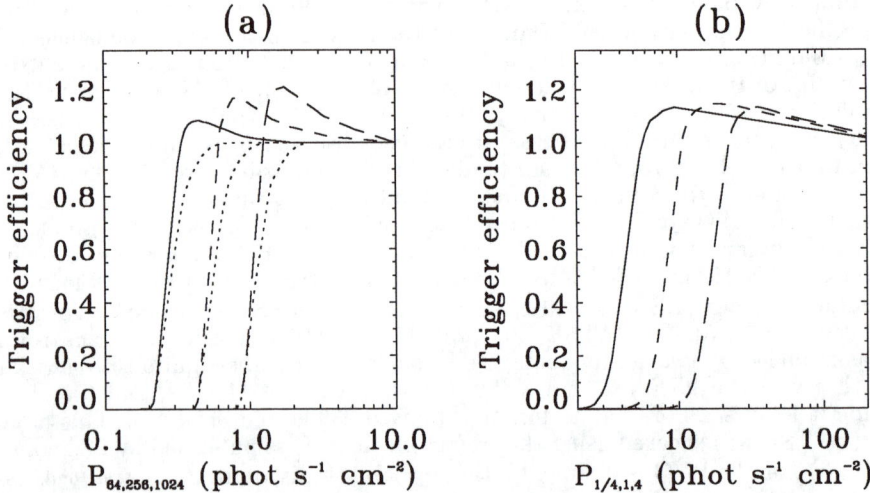

Figure 1. Trigger efficiencies for BATSE (a) and PVO (b) as a function of observed peak flux, for the three time scales appropriate to the respective instruments: the solid curve for the longest, the dashed curve for the middle and the long-dashed curve for the shortest time scale of each instrument. For comparison, the dotted curves in (a) refer to calculations performed by Fishman et al.[6] which did not take into account the peak count rate bias.

These efficiencies may be used on differential observed peak flux distributions, separately for each trigger time scale and excluding overwriting events for BATSE. For example, assume one is interested in a BATSE burst with an *observed* P_{256} (such as listed in the catalog[6]) of 0.5 phot s^{-1}cm^{-2}. The P_{256} curve in (a) above reads a trigger efficiency of 0.4. This means that if 40 such bursts are detected, then on average actually 100 bursts with an *expected* P_{256} of 0.5 phot s^{-1}cm^{-2} will have occurred. Assume now that the same burst has an *observed* P_{64} of 2.0 phot s^{-1}cm^{-2}. Reading the P_{64} curve above (i.e. the efficiency is 1.2) indicates that for 120 such bursts actually 100 bursts with an *expected* P_{64} of 2.0 phot s^{-1}cm^{-2} will have occurred.

In Fig. 1, we present the trigger efficiencies of BATSE and PVO for the three trigger time scales as a function of the peak flux *as observed* on each time scale (this may mean that for one burst three different peak fluxes are measured on the three time scales). We did not include overwriting events. These efficiencies were calculated as follows: the simulation as explained above was performed resolving in a list of bursts which did or did not trigger the instrument, and labeled with the true peak flux, the expected peak flux as measured on each time scale (by this we mean the peak flux expected from sampling the time profile on each time scale, without taking account of statistical effects) and the actual observed peak fluxes per time scale (only for triggering bursts). The trigger efficiency was calculated by dividing the number of triggering bursts in a certain observed peak flux interval through the number of bursts generated in that same interval of expected peak flux. Therefore, by using this efficiency on a differential distribution one will acquire an estimate of the intrinsic peak flux distribution (for each time scale).

Apart from the sensitivity of the instruments, there are a couple of effects that determine the form of the trigger efficiency curves: first, upward statistical fluctuations within the time profile may trigger detection of an otherwise undetectable event ('peak count rate bias'). The extent of this effects depends on the signal-to-noise ratio and the peak duration with respect to the time scale of the trigger (the higher this ratio is, the larger the effect; therefore, the effect is stronger for shorter trigger time scales). Second, for trigger time scales longer than the peak width, the observed peak fluxes will be systematically smaller than the true peak values.

Due to the peak count rate bias, the weaker a burst is, the more the observed peak flux deviates on average from the expected peak flux. The mappings from expected to observed peak flux are illustrated in Fig. 2 for BATSE and PVO. Clearly, this is the reason for the humps in our trigger efficiency curves: one value for the observed peak flux maps to a broad range of expected peak fluxes, particularly for low values. One should be alert that such humps will be apparent in measured peak flux distributions and will give a distorted impression of power-law indices at low peak fluxes. The magnitude of the humps depends on the intrinsic distribution: if the ratio between the number of faint bursts to that of bright bursts is small, so will the hump be, and vice versa. Furthermore, this magnitude depends on the time profiles of the bursts. For example, a time profile with duration equal to that of the shortest trigger time scale will result in no peak count rate bias and, thus, no humps. In this case, the efficiency curves as published by Fishman et al.[6] are reproduced.

Applying the efficiencies in fact is a first order correction. The reason may be illustrated with the example of a delta-function peak flux distribution: this function will be 'smeared' by the above effects and applying the efficiencies to estimate the intrinsic distribution obviously does not result in a delta-function again. However, real distributions are continuous and the error in applying directly trigger efficiencies is not extensive: our calculations show that for power-law distributions with indices between -2.5 and 0.0, the maximum error in the result is 10%. This is less than the accuracy of the burst count statistics as long as the number of burst observed in a peak flux interval is less than 100.

The efficiency curves for PVO show similar behavior as those for BATSE in the above-mentioned sense. However, the ranges in peak flux where the efficiency is larger than 1 are broader because the relation between observed and expected peak flux is less strict due to the spectral diversity of the bursts (see Fig. 2).

CONCLUSION

We have performed calculations of the trigger efficiencies of BATSE and PVO as part of a program to combine observations from BATSE and PVO, excluding overwriting events for BATSE. Since these efficiencies are strongly dependent on time profiles of the GRBs (due to the peak count rate bias), we employed an empirical model for these profiles to take these effect into account. However, our results are not complete. To arrive at more accurate trigger efficiencies, the following uncertainties need to be resolved. First, the signature of

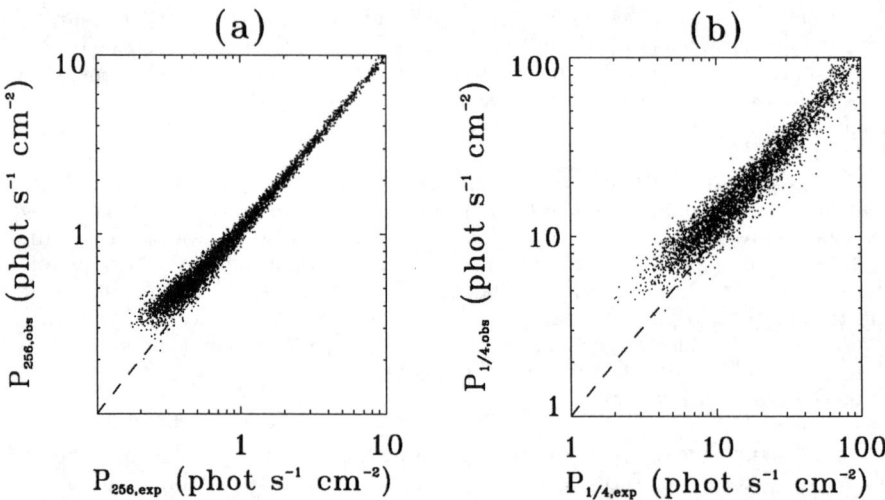

Figure 2. Observed versus expected peak fluxes for 5000 simulated bursts, on trigger time scales of 256 ms and 1/4 s for BATSE (a) and PVO (b) respectively. Note the stronger deviations from the $P_{obs}=P_{exp}$ relation (dashed lines) at lower peak fluxes due to the peak count rate bias. Note also the scatter at high peak flux values, due to the fact that observed peak fluxes are reconstructed with the assumption of a -3/2 power-law spectrum, while the bursts display much more spectral diversity. The difference in scatter at high peak fluxes between both instruments results from the difference in photon energy passband in both instruments trigger criteria.

the parent time profile distribution below peak widths of 64 ms is uncertain (we limited ourselves to a sample of time profiles from BATSE and PVO databases with a time resolution of 64 ms). Second, as noted before[6], atmospheric scattering will affect the trigger efficiencies. We did not evaluate this effect.

Acknowledgements – This research has made use of data obtained through the CGRO Science Support Center Online Service provided by the NASA GSFC, and was supported by the United States Department of Energy.

REFERENCES

1. D. Band, *et al.*, Ap.J. **413**, 281 (1993).
2. W.D. Evans, J.P. Glore, R.W. Klebesadel, J.G. Laros, E.R. Tech, Science **205**, 119 (1979).
3. E.E. Fenimore, R.W. Klebesadel. J. Laros, C. Lacey, C. Madras, M. Meier, G. Schwarz, Compton Gamma-Ray Observatory (eds. M. Friedlander, N. Gehrels and D.J. Macomb, AIP Conference Proceedings 280, 1992), p. 744.
4. E. Fenimore *al.*, Nature **366**, 40 (1993).
5. G.J. Fishman *et al.*, Proc. of the GRO Science Workshop (10-12 April 1989, Greenbelt, ed. W.N. Johnson, 1992), p. 2-39.
6. G.J. Fishman *et al.*, Ap. J. Sup. Ser. , in press (1993).

AN EVALUATION OF BATSE BURST LOCATIONS COMPUTED WITH THE MAXBC DATATYPE

T. M. Koshut, W. S. Paciesas, G. N. Pendleton
Dept. of Physics, University of Alabama in Huntsville, Huntsville, AL 35899

M. N. Brock, G. J. Fishman, C. A. Meegan, R. B. Wilson
Marshall Space Flight Center, ES 64, Huntsville, AL 35812

ABSTRACT

One of the primary objectives of BATSE is to determine the direction to each observed gamma-ray burst, allowing the angular distribution of the burst sources to be studied. Most of the BATSE burst data types are not suitable for this purpose, since they are produced by combining rates from several detectors (details of the BATSE experiment and the various datatypes can be found elsewhere[1,2]). After the loss of the CGRO tape recorders in March, 1992, the availability of data with rates from individual detectors has often been limited by telemetry gaps. Consequently, many burst directions are derived using the MAXBC data, which are almost always available, but have significant limitations. MAXBC data contain the maximum background-subracted count rate (counts per 64 ms) in the energy range of $\sim$ 50–300 keV, averaged over one second, for each of the eight Large Area Detectors. It is important to evaluate the locations computed using MAXBC data because these locations will appear in the BATSE 2B Catalog when no other data are available. We report on the comparison of burst locations computed using MAXBC data with locations derived from the Third Interplanetary Network[3], hereafter known as IPN[3]. We report on the comparison of MAXBC locations of hard solar flares with the known position of the Sun. We also report on the comparison of burst locations computed using MAXBC data with those locations given in the first BATSE catalog, for which the burst data are essentially complete.[4]

METHODOLOGY

The 11 gamma-ray bursts used in these comparisons were selected because each has been well-localized by the IPN[3]. The 22 solar flares were chosen because their spectra are harder than an average flare, thus closer to a typical burst spectrum. Moreover, the flares extend the intensity range of our sample of well-localized events further down than possible with only the IPN[3] bursts. These 33 events are the same events used in the analysis shown in Figure 2 of the BATSE 1B Catalog.

Three locations were calculated for each event. Details concerning the method of calculating locations, and the systematic and statistical errors involved, will not be discussed here but can be found elsewhere.[5,6] One of the three locations was determined using a power-law spectral index of 2.0. This value is used by the BATSE team when calculating locations of events for which there is no other suitable LAD data. Additionally, a value of 2.0 is used by the BATSE team when determining the preliminary location for a strong burst, with the intention of quickly alerting investigators observing at other wavelengths. The second location computed for each event employs a spectral index obtained from a hardness ratio HR look-up table, with

$$HR = \frac{10000 \times Ch2}{Ch2 + Ch3} \quad (1)$$

where $Ch2$ and $Ch3$ are the background-subtracted count rates in the LAD discriminator channel 2 (spanning $\sim$ 50–100 keV) and discriminator channel 3 (spanning $\sim$ 100–300 keV) respectively. This method is used by the BATSE team when the preferred 16-channel CONTINUOUS data and 4-channel DISCLA data are not available due to telemetry gaps. The third location computed for each event utilizes the best-fit spectral index used in calculating the locations in the BATSE 1B Catalog.

The offsets (in degrees) of each location from the well-known locations were calculated. The offsets of each location from the published 1B Catalog location (for which the burst data are essentially complete) are also calculated. The values used for the 1B Catalog solar flare locations are the same values used to make Figure 2 of the 1B Catalog. In addition to locations, the signal-to-noise ratio in the 2 brightest detectors was calculated for each event. If one tries to compare Figure 2 of the 2B Catalog with the results presented here, it must be noted that the value of the signal-to-noise ratio for each event will not necessarily be the same. In the 1B Catalog we were able to use various data types, most of which provided better energy resolution than MAXBC, as well as more flexibility in choosing the burst time interval (recall that MAXBC are the background-subtracted rates only at the peak time of the burst in the 50–300 keV energy range).

RESULTS

Table I. provides a brief summary of the various location comparisons. The two samples being computed are given in the first two columns. The third column specifies the method used to obtain the power-law sepctral indices. The fourth column gives the mean total (systematic + statistical) error $\langle\sigma\rangle$, where

$$\langle\sigma\rangle = \frac{1}{N}\sum_{i=1}^{N}\sigma_i \quad (2)$$

and σ_i is calculated as the difference (in degrees) between the locations from each of the two samples, for the ith event. N is the total number of events in the sample. The fifth column informs the reader which figure contains the corresponding comparison data.

Figure 1 shows the offset of the MAXBC locations from the well-localized locations plotted as a function of the signal-to-noise ratio in the two brightest detectors. The distinguishing characteristic for each of the three plots is the method used to obtain the power-law spectral indices. The dashed line in each figure is a representation of the statistical errors as calculated in the BATSE 1B Catalog.

Figure 2 shows the offset of the MAXBC locations from the locations given in the 1B Catalog plotted as a function of the signal-to-noise ratio in the brightest detectors. Again, the distinguishing factor between these three plots is the method used to obtain the power-law sepctral indices. The dashed line again represents the BATSE 1B Catalog statistical errors.

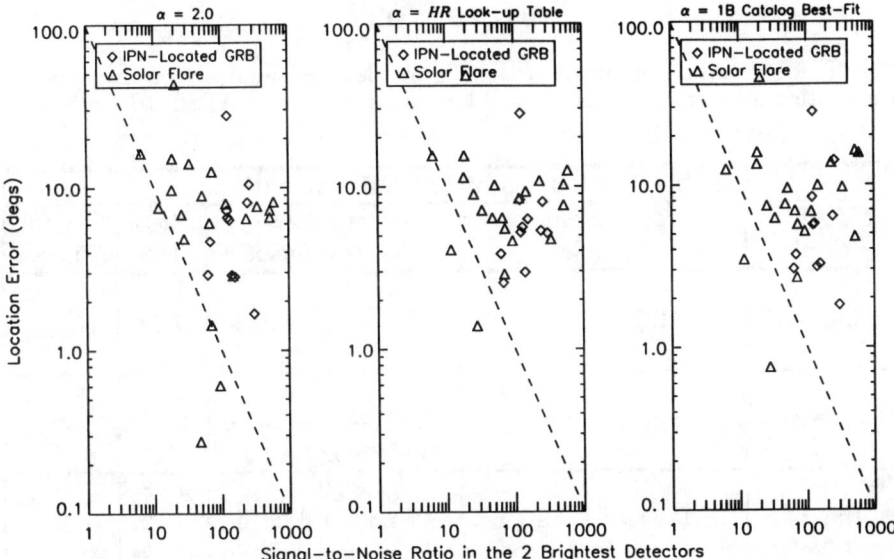

Figure 1.

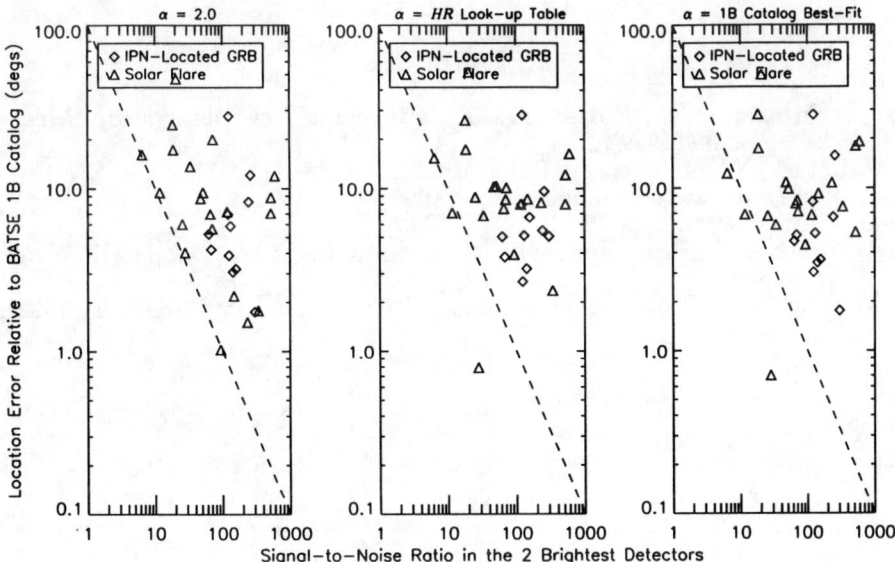

Figure 2.

FUTURE WORK

These results, combined with a more detailed analysis of the seperate systematic and statistical errors, will be included in the BATSE 2B Catalog.

Table I. Analysis of MAXBC Locations					
MAXBC Sample	Comparison Sample	Power-Law Spectral Index	$\langle\sigma\rangle$ (degs)	Figure Number
Bursts	IPN Bursts	2.0	7.41	1
		HR Look-Up Table	7.39	
		1B Catalog Best-Fit	7.57	
Solar Flares	Sun Location	2.0	9.10	1
		HR Look-Up Table	9.80	
		1B Catalog Best-Fit	10.31	
Bursts	1B Catalog Locations	2.0	7.48	2
		HR Look-Up Table	7.52	
		1B Catalog Best-Fit	7.79	
Solar Flares	Figure 2 in 1B Catalog	2.0	10.73	2
		HR Look-Up Table	11.46	
		1B Catalog Best-Fit	11.83	

REFERENCES

1. G. Fishman et al., Proceedings of the Gamma Ray Observatory Science Workshop , 2-39 (1989).
2. J. Horack, NASA Reference Publication No. 1268 , (1991).
3. K. Hurley, private communication , (1993).
4. G. Fishman et al., Ap. JS , in press (1993).
5. M. Brock et al., Proc. Huntsville Gamma-Ray Burst Workshop (AIP, N. Y., 1991), p. 383.
6. M. Brock et al., Proc. Compton Gamma-Ray Observatory (AIP, N. Y., 1992), p. 709.

WAVELET ANALYSIS OF GAMMA RAY BURSTS

Dawn C. Meredith, James M. Ryan, C. Alex Young
University of New Hampshire, Durham, N. H. 03824

John Patrick Lestrade
Dept. of Physics, Mississippi State University, MS 39762

ABSTRACT

The temporal information contained in a GRB may carry significant clues on the nature of the burst object. Various forms of Fourier Transform have been used historically to extract this information. However, non-stationary events such as GRB do not lend themselves easily to such analyses. Wavelet analysis is appropriate for the examination of non-stationary data since Wavelets are localized in both "frequency" and "time" domains. We report here on Wavelet analysis of two cosmic GRB and show that Wavelet analysis is a useful tool in understanding the time evolution of GRB.

WAVELET OVERVIEW

For many decades, the predominance of periodic and quasi-periodic behavior has dominated our interpretation and analysis of time series data and Fourier analysis has provided the basis for this quantative analysis. While this approach has provided much insight in many systems, it is not universally applicable. Even when the time series has clear periodicities when judged by eye, small phase shifts can result in unclear Fourier analysis results. Most Gamma Ray Bursts (GRB), on the other hand, are clearly not periodic and in fact are highly nonstationary events. Therefore, we need another tool to characterize and quantify the information in these time series.

A Wavelet Transform is just such a tool. Wavelet Transforms are similar to the better known coherent states (e.g. a Gaussian wavepacket) which share the delocalization in "time" and "frequency", and are unlike Fourier transforms which are completely localized in frequency and completely delocalized in time. Windowed Fourier transforms are also localized in time and frequency, but not quite in the same way as Wavelets. Windowed Fourier Transforms have the same size window (i.e. same time localization) regardless of the frequency; therefore the lowest frequency component forces the choice of time window, which may be incompatible with the very localized transients. Wavelets, however, have a better time resolution the higher the frequency, thus allowing the analysis to "zoom in" on higher frequencies.

For those interested in learning more about Wavelets, there are many references available. Meyer[1] provides a general overview and historical perspective; Daubechies[2] and Chui[3,4] provide mathematical presentations of the underlying principles; Strang's[5] approach is based on linear algebra, and gives a readable and practical introduction for physicists; Press et al.[6] give an overview of the algorithm for Fast Wavelet Transform, as well as a specific implementation.

There are many kinds of Wavelets[1], however we will concentrate on the orthonormal Wavelet bases which were first understood within the framework of multiresolution analysis. We present here only the essential Wavelet equations.

Each different Wavelet is defined by a set of μ coefficients c_k which obey the following constraints:

$$\sum_k c_{2k} = \sum_k c_{2k+1} = 1 \qquad \sum_k c_k c_{k+2j} = \delta_{0j} \quad . \tag{1}$$

The first constraint ensures accuracy, the last one ensures orthogonality. Once the coefficients have been chosen, we define a scale function by the dilation equation:

$$\phi(t) = \sum_{k=0}^{\mu-1} c_k \phi(2t - k) \quad . \tag{2}$$

At first glance, it appears as though this is a useless definition since $\phi(t)$ appears on both sides of the equation. However, Strang[5] provides many methods for finding $\phi(t)$ once the c_k's are given. One simple method is to pick any initial shape for $\phi(t)$ (e.g. the box function which is only non-zero for $0 \le t \le 1$) and iterate Eq. 2 until $\phi(t)$ no longer changes between iterations. Finally, the Wavelet is given by

$$\psi(t) = \sum_{k=0}^{\mu-1} (-1)^k c_{\mu-k} \phi(2t - k) \quad . \tag{3}$$

In the framework of multiresolution analysis, the scale function is the blurring function which averages the signal on a given scale, whereas the Wavelet function gives differences the signal on a given scale.

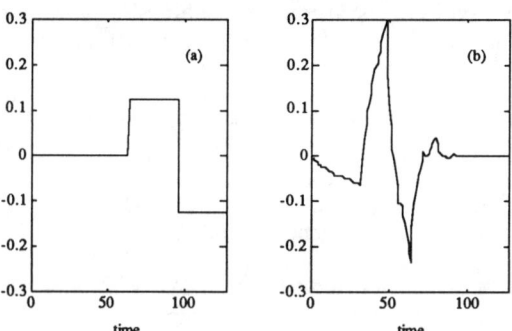

Fig. 1 Plots of the Haar and Daubechies 4 Wavelets.

To give an specific example of a Wavelet, the Haar basis has $\mu = 2$, $c_0 = 1$ and $c_1 = 1$. The scale function is the box function, the Wavelet is shown in Fig. 1a. There is also a family of Wavelets which go by the name Daubechies-μ; the larger the number of coefficients, the smoother the Wavelet (it has approximately $\mu/2$ continuous derivatives) but the greater the delocalization in time. For example, Haar (also known as Daubechies 2) is discontinuous, but very compact. Fig. 1b shows Daubechies 4. As you might guess from the figure, there is no analytical expression for the Daubechies Wavelet; this particular plot was obtained by an inverse Wavelet transform of a unit vector.

It can be shown that we may expand any signal in terms of Wavelets:

$$f(t) = \sum_{n=-\infty}^{\infty} \sum_{k=-\infty}^{\infty} b_{nk} 2^{n/2} \Psi(2^n t - k) \quad , \qquad (4)$$

where the expansion coefficients are

$$b_{nk} = \int_{-\infty}^{\infty} f(t) 2^{n/2} \Psi(2^n t - k) \quad , \qquad (5)$$

i.e. the Wavelets provide a complete orthonormal basis, just as the Fourier Transforms do. We show in Fig. 2 that the Daubechies 8 Wavelet can indeed reconstruct a GRB exactly.

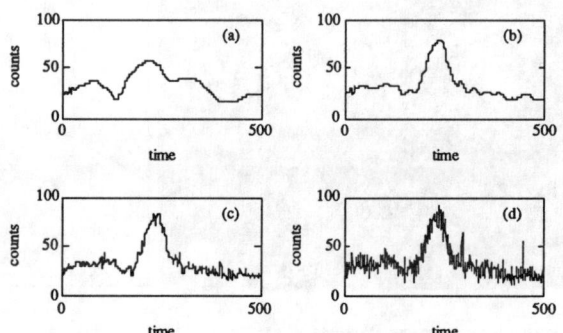

Fig. 2 Reconstruction of Burst 105 using the Daubechies 8 Wavelet. (a) Frequencies 0 to 2, (b) frequencies 0 to 4, (c) frequencies 0 to 6, and (d) all frequencies (0 to 8) which agrees with the original signal.

The algorithm for calculating the coefficients b_{nk} is quite simple and fast; Eq. 5 is misleading in that no integration must be done. The fast Wavelet transform merely finds differences and averages at each scale to calculate the b_{nk}'s[5]; this follows directly from multiresolution analysis. Hence the time needed to calculate the transfrom scales as the number of data points[5], which makes it competitive with Fast Fourier Transforms.

There are, of course, practical problems. The first is that the theory relies on $f(t)$ being a function defined for all t, whereas data are always given for a set of values t_i. In one way this helps, in that the sums in Eq. 4 are now finite. However, another problem appears, known as wrap-around: data at the beginning and end of the data set are averaged and differenced as though they were contiguous in time. One simple fix is to mirror the data about $t = 0$ so that the signal is artificially periodic. This has its own problems if the signal is not flat at the beginning an end (i.e. if the slope is not periodic as well). A final problem for which we know of no fixes, is that Wavelets can only handle uniformly sampled data.

704 Wavelet Analysis of Gamma Ray

WAVELET ANALYSIS OF BURSTS

For this initial analysis, burst catalog numbers 219 (May 22, 1991) and 249 (June 1, 1991) were chosen (we show only the results for 219). They were chosen using visual criteria: both had a well defined and localized structure and were not too spikey. We analyzed the MER data summed over all energy channels, and 2048 time bins were used.

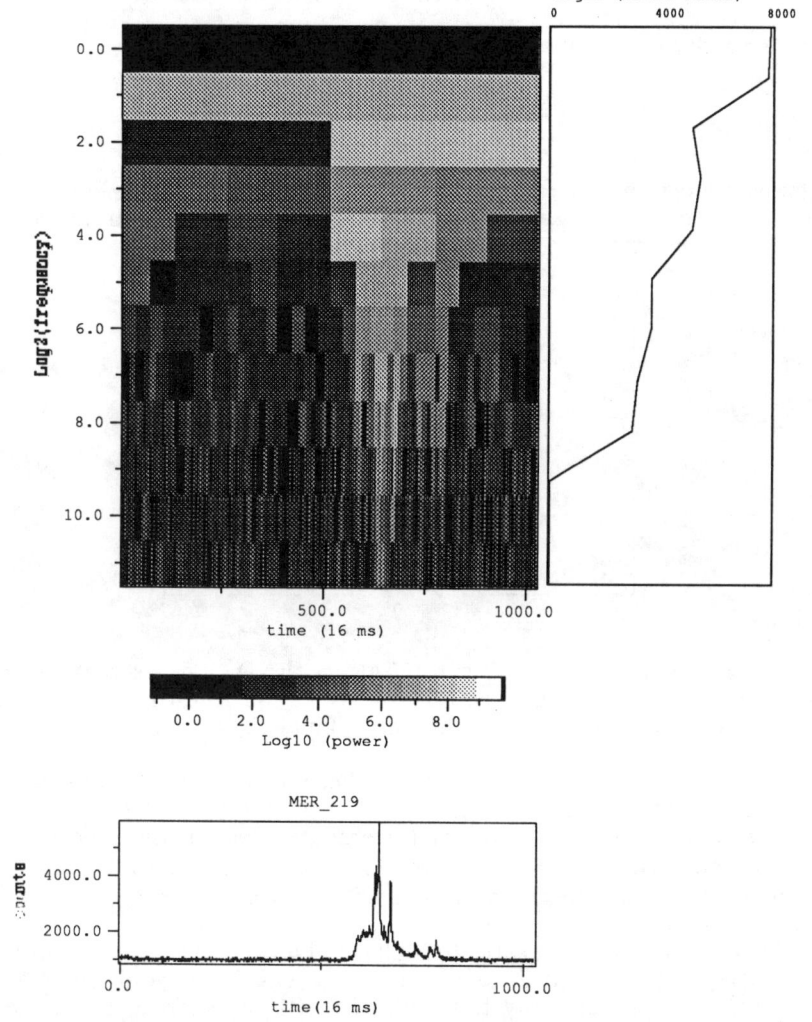

Fig. 3 Dynamic spectrum, time series, and total spectrum for Burst 219

In performing the Wavelet analysis, the MER signals were mirrored, that is, they were reflected about the origin, in order to avoid wrap-around error discussed in the last section. The analysis was performed using the Haar (Daubechies 2) Wavelet.

In order to have a complete view of the spectrum of the burst we have plotted the dynamic spectrum (i.e. the power = $(b_{n,k})^2$) in Figure 3. The horizontal axis is time (labeled by k), the vertical axis is frequency (labeled by n). The higher the frequency, the more time coefficients there are; this is a reflection of the ability of the Wavelets to zoom in (i.e. look at smaller time scales) for higher frequency components. There are $N/2 = 1028$ values of n for the largest value of $k = 11$, and $\mu/2 = 1$ values of n for the smallest two values of $k = 0, 1$. (The division by two in $\mu/2$ arises because the data was mirrored, and half of the $b_{n,k}$, which have redundant information, were thrown away.) The total number of coefficients is $N = 2048$.

Frequency components 0, 1 and 2 represent the burst normalization and global features. The details of the burst become apparent in frequency components 3 and beyond.

This analysis can tell us about the distribution of power within a burst. In Burst 219 we see that the power spectrum is not constant in time nor frequency, reflecting the fact that impulsive spikes are occurring on top of a broader feature. Significant power (although very localized in time) is present down to the resolution of MER (16 ms). This is not evident in the integrated power spectrum. The main pulse in the burst is evident beginning in spectral component 5.

DISCUSSION

There is no doubt that a Wavelet analysis of GRB can be done, but what can we learn from the coefficients? First, Wavelet analysis literally adds a new dimension in the frequency analysis, since we may now look at frequency composition as a function of time. By focusing on times when the burst is strongest, we will not dilute this information with frequency information about the background. With this finer microscope, we have hopes of refining classifications schemes for GRB which rely on analysis of the dynamic spectrum. This may also allow for a new method of searching for gravitationally lensed burst pairs in that the dynamic spectrum may be considered a more detailed fingerprint of the burst. Second, Wavelet Transforms allow us to smooth out the signal by looking at only the lower scale (lower frequency) components, allowing us to reproduce the bursts without noise. Finally, we have a method to reassemble bursts: i.e. by reordering the coeffiecients within certain time intervals we can create similar but distinct bursts for comparison to help guide classification schemes.

We gratefully acknowledge the help of David Meeker who wrote the Wavelet Transform code.

REFERENCES

1. Y. Meyer, Wavelets: Algorithms and Applications (SIAM, 1993).
2. Ingrid Daubechies, Ten Lectures on Wavelets (SIAM, 1992).
3. Chui, An Introduction to Wavelets (Academic Press, 1992).
4. Chui, Wavelets: A Tutorial in Theory and Applications (Academic Press, 1992).
5. Gilbert Strang, SIAM Reives **31**, 614 (1989).
6. W. H. Press, B.P. Flannery, S. A. Teukolsky, W. T.Vetterling, Numerical Recipes: The Art of Scientific Computing, Second Edition (Cambridge University Press, 1992).

A SEARCH TECHNIQUE FOR WEAK AND LONG-DURATION GAMMA-RAY BURSTS FROM BACKGROUND MODEL RESIDUALS

R. T. Skelton* and W. A. Mahoney
Jet Propulsion Laboratory
California Institute of Technology, Pasadena, CA 91109

ABSTRACT

We report a planned search technique for Gamma-Ray Bursts too weak to trigger the on-board threshold. The technique is to search residuals from a physically based background model used for analysis of point sources by the Earth occultation method. Searching residuals (as opposed to raw data) minimizes false triggers from occultation edges and many other effects which lead to a rapid variation in the raw count rate. The background model is based on physical parameters, such as charged particle count rates and atmospheric secondaries. This allows fitting to long periods (e.g., several orbits), which in turn increases search effectiveness for bursts of longer duration. Initial results and expectations are presented.

INTRODUCTION

The Burst and Transient Source Experiment (BATSE) has provided one of the Compton Gamma Ray Observatory's (CGRO) most important scientific results: Gamma Ray Bursts (GRB) are isotropic but distributed inhomogeneously in space, with a deficit of weaker (more distant) bursts.[1,2] The most recent results[3] essentially exclude galactic and galactic halo distributions, thereby essentially eliminating the pre-CGRO favored hypotheses of origin, viz., events associated with galactic neutron stars.

This paper describes a technique for an offline search of BATSE data which should extend BATSE's sensitivity to substantially weaker bursts, particularly to bursts of longer duration (say, $\gtrsim 10\,\text{s}$). In this technique, the count rate is fit to a physically based model; in this manner most variations in count rate are modeled, so that bursts having a duration on the order of the orbital variations can nevertheless be detected. The value of this search will be in providing at least some insight into the spatial distribution of burst sources, and hence into the origin of GRB, provided that a sufficient number of bursts can be detected to, in effect, extend BATSE's $\log N$-$\log S$ curve substantially below its onboard threshold.

METHOD

The search will be conducted on residuals from the background model used in the JPL Enhanced BATSE Occultation Package (EBOP).[4] Use of this model

* National Academy of Sciences/National Research Council Senior Resident Research Associate

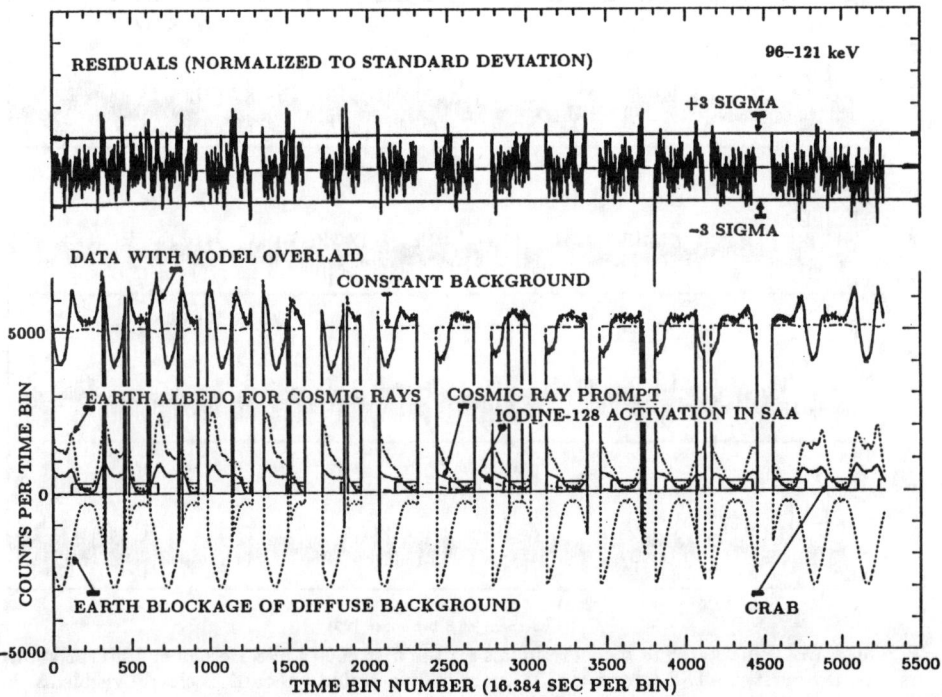

Figure 1. Buildup of components in the background model used in the EBOP.

is a key element of the strategy. The background model incorporates the following terms, which are illustrated in figure 1 for the 96–121 keV energy range:

- Constant Background: This term simply accounts for the diffuse isotropic cosmic background and slowly varying effects, such as long-term cosmic-ray activation.

- Earth Blockage of Cosmic Diffuse Background: This term is expected to be negative by virtue of its definition as a deficit in the diffuse background. When the BATSE Large Area Detector (LAD) is pointed away from the Earth, it remains negative instead of going to zero, probably because of LAD response at angles greater than 90 degrees. At higher energies, the magnitude of this term decreases, and the mathematical fit value can become positive; this is because the mathematical model automatically incorporates some contributions from non-local atmospheric cosmic-ray secondaries into this term. (The Earth Albedo for Cosmic Rays term, described below, accounts for cosmic-ray secondaries proportional to the local cosmic-ray flux.)

- Cosmic Ray Prompt: This term primarily reflects prompt secondaries from cosmic rays striking the instrument or spacecraft but not being vetoed. The cosmic-ray rate is obtained from ULD rates of the 8 BATSE Spectroscopy Detectors.

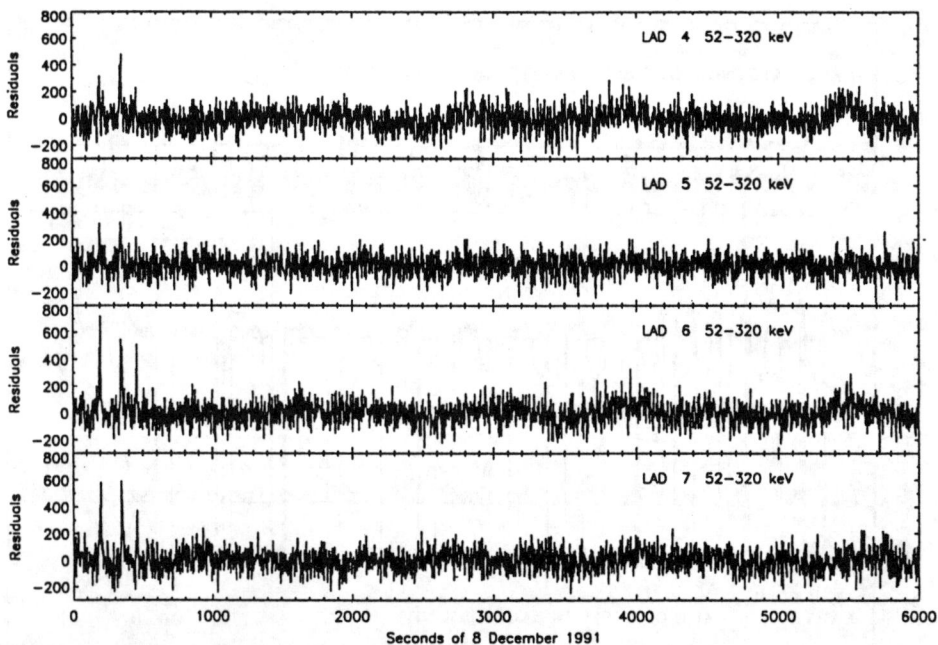

Figure 2. EBOP background model residuals for the first 6000 s of 8 December 1991. A known gamma-ray burst, BATSE trigger #1152, marginally detected onboard, is clearly visible as the series of three pulses within the first 500 s.

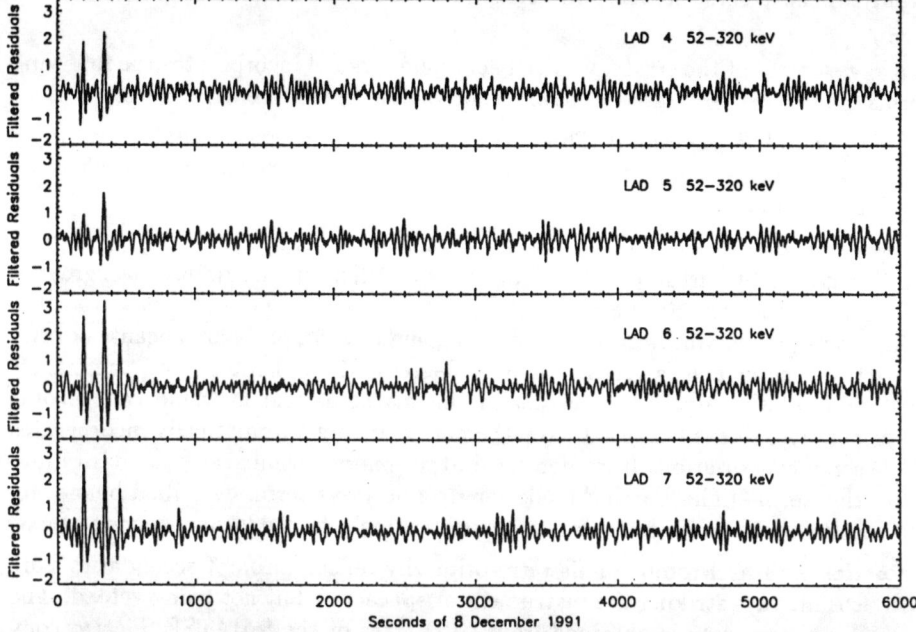

Figure 3. The residuals from figure 2, above, after convolution with a boxcar filter of 16 s width. This value closely matches the width of the pulses in this particular burst.

- Earth Albedo for Cosmic Rays: This term includes factors for both the cosmic-ray rate and the Earth exposure; it reflects the major contribution of atmospheric secondaries, which is proportional to the local cosmic-ray rate.
- ^{128}I Activation in SAA: This term represents contributions to the count rate which decay with the 25-min half-life of ^{128}I. A separate term is added following each South Atlantic Anomaly passage.
- Crab: This rectangular wave represents the one point source in the example of figure 1, namely the Crab nebula and pulsar. The height of the step corresponds to the signal being extracted from the count rate.

In the upper portion residuals for each time bin (normalized to the formal statistical error) are plotted. Use of the residuals from this model, as opposed to raw data, is a crucial element in this search; this is because the raw data contain occultation edges and other periods of rapid variation which would lead to a higher false trigger rate for the same sensitivity.

Once residuals are obtained, the burst search will consist in applying filters to them and looking for positive transients. The baseline filter shape is a boxcar with negative wings; various widths will be used. This "two-sided" filter offers a significant advantage compared to the "one-sided" rising edge exceedance of a floating threshold, as implemented onboard. In order to become a burst candidate, a transient would need to be confirmed by at least 2 LADs, would need to come from unocculted space, would not correlate with known solar flares, and would display consistent temporal behavior among the LADs detecting it. The onboard trigger is somewhat anisotropic[5] owing to the fact that the required second LAD may present a cosine ranging from 0.333 to 0.816 to the burst direction. In cases when the second LAD has a smaller cosine, the third (and even fourth) LAD can have a cosine near that of the second. Consideration will be given to mitigating the anisotropy by allowing multiplicity (third or fourth LAD) to generate a candidate with less threshold exceedance in the second LAD. Since false alarms in the onboard trigger impact the instrument, it is obvious that an offline search can tolerate a higher false alarm rate.

EXAMPLE

Figure 2 shows the potential of this method. Shown are the residuals for LADs 4 through 7 for the first 6000 seconds of 8 December 1991. This period includes a known burst, BATSE trigger #1152. This burst came in three pulses, at 200 s, 340 s, and 460 s; the BATSE trigger was on the middle one, with a threshold exceedance factor of 1.007, which is to say that it was barely detected.

Figure 3 shows the residuals after being filtered by a 16-s square-wave filter of zero net area. Such a filter accentuates features matching its width. This burst is detectable at 9.3 σ in the unfiltered residuals and 13 σ after filtering, where the σ-levels refer to the second-strongest LAD. It is clear that this burst stands out clearly above any other features in the period shown.

SCOPE OF CURRENT AND FOLLOW-ON EFFORT

The current effort is scoped to adapt the existing background model to the burst search and to search 100 days of BATSE CONT (2-second, 16 energy channels) data. These data are already at JPL in connection with the Earth Occultation program. Tasks include development of the software to interface between the Earth occultation model to the burst search, development of the burst search software to examine the data on various time scales, and examine energy band selection options, timing options, and threshold selection. Should the pilot effort achieve a significant increase in sensitivity, a follow-on effort would involve continued analysis and possible extension of the technique in cooperation with the BATSE PI team.

ACKNOWLEDGEMENTS

The assistance of G. J. Fishman and the rest of the BATSE PI team is gratefully acknowledged. The research described in this paper was carried out by the Jet Propulsion Laboratory, California Institute of Technology, under contract to the National Aeronautics and Space Administration.

REFERENCES

1. G. J. Fishman, C. A. Meegan, R. B. Wilson, W. S. Paciesas, and G. N. Pendleton, The BATSE Experiment on the Compton Gamma Ray Observatory: Status and Some Early Results. In: C. R. Shrader, N. Gehrels, and B. Dennis (eds.) Proc. Compton Observatory Science Workshop (NASA Conf. Pub. #3137). NASA, Greenbelt, MD, 26 (1992).
2. C. A. Meegan, G. J. Fishman, R. B. Wilson, et al., Nature **355**, 143 (1992).
3. G. J. Fishman, Gamma-Ray Bursts: Observational Overview. In: Proc. Second CGRO Symp. (College Park, MD, 1993), in press (1993).
4. R. T. Skelton, J. C. Ling, N. F. Ling, R. Radocinski, and Wm. A. Wheaton, Status of the BATSE Enhanced Earth Occultation Analysis Package for Studying Point Sources. In: M. Friedlander, N. Gehrels, and D. Macomb (eds.) Proc. CGRO Symp. AIP, New York, 1189 (1993).
5. M. N. Brock, C. A. Meegan, G. J. Fishman, et al., BATSE's Sky Sensitivity Map. In: W. S. Paciesas and G. J. Fishman (eds.) Gamma-Ray Bursts. AIP, New York, 399 (1992).

USING CONTOUR MAPS TO SEARCH FOR RED-SHIFTED 511 keV FEATURES IN BATSE GRB SPECTRA

P.G. Varmette, J. P. Lestrade
Dept. of Physics, Mississippi State University, MS 39762

G. J. Fishman, C.A. Meegan, M.N. Brock, R.D. Preece
ES 64, Marshall Space Flight Center, Huntsville, AL 35812

M.S. Briggs, G.N. Pendleton, W.S. Paciesas
Univ. Alabama at Huntsville, AL

ABSTRACT

Previous experiments have reported emission features in gamma-ray burst spectra in the region of 400 keV to 500 keV. These lines have been interpreted as gravitationally red-shifted 511 keV annihilation lines produced in the vicinity of neutron stars (Barat 1984, Mazets 1980, Mitrofanov 1984, Nolan 1984). We describe a method to aid in the search for transient emission and absorption features in gamma-ray burst spectra. The process involves fitting a 3-dimensional continuum model to spectra in time-energy space. Significant spectral features form easily recognizable islands in the contour plot of residuals versus energy and time.

INTRODUCTION

Since their discovery twenty years ago, the origin of gamma-ray bursts (GRB's) has remained an intriguing mystery. The quest to understand these objects has given rise to a plethora of competing theories. Several theories suggest that GRB's are galactic in origin while others suggest that GRB's are cosmological (Harding 1993).

One piece of evidence that might provide scientists with a key to understanding the origin of GRB's may be whether or not spectral emission and absorption features exist in burst spectra. If the features exist and can be attributed to either cyclotron lines or to red-shifted 511 keV annihilation lines then credence would be given to those theories that support a galactic origin, i.e. near neutron stars (Barat 1984, Mazets 1980, Mitrofanov 1984, Nolan 1984).

THE SEARCHING METHOD

A method of searching for spectral features in burst spectra (BATSE HER data) will be outlined in this paper. The method was used to investigate the energy range between approximately 350 keV to 600 keV. This energy range was chosen because previous experiments have reported emission features in gamma-ray bursts around 400 keV to 500 keV. These features have been interpreted as gravitationally red-shifted 511 keV annihilation radiation produced near a neutron star (Barat 1984, Mazets 1980, Mitrofanov 1984, Nolan 1984).

The first step was to calculate a background model representing the ambient background radiation. The model was used to separate the burst spectrum from that of the background. Next, we construct the incident "photon" spectrum from the recorded "count" spectrum. To do this involves convolution with matrices that contain information on the detector's efficiency as a function of energy, as a function of angle of incidence of radiation, and also the detector's sensitivity to that fraction of the incident radiation caused by scattering off the Earth's atmosphere. The combination of all of these is called the detector response matrix (DRM) shown in Figure 1.

The BATSE HER data for a single burst can be binned into different time intervals and each interval forms a spectrum. Burst 1B 911221 was binned into 8 spectra each lasting approximately 9 secs. A fit of the spectrum that ranged in time from 9.7 secs to 18.2 secs produced the best fit results. Figure 2 shows the fit that was made to this spectrum using a Broken Power Law, the form of which can be seen in Equation 1.

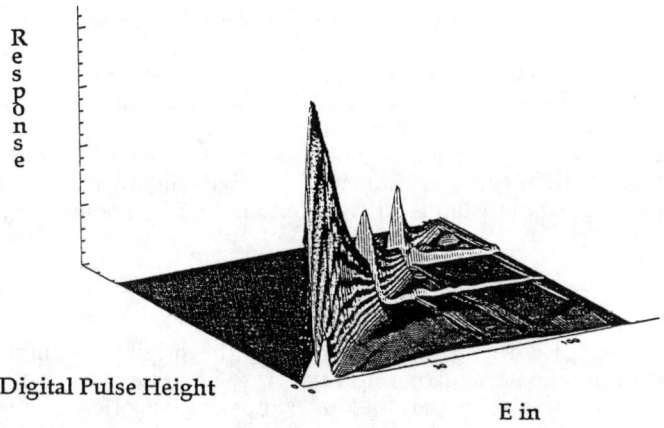

Figure 1. A detector response matrix.

$$\begin{cases} A(E/E_{\text{pivot}})^{\lambda_1}, & \text{for } E \leq E_{\text{break}}; \\ A(E_{\text{break}}/E_{\text{pivot}})^{\lambda_1} + A(E/E_{\text{break}})^{\lambda_2}, & \text{for } E \geq E_{\text{break}} \end{cases} \quad (1)$$

The fit shown in Figure 2 produced a χ^2 of 23.4 with 22 degrees of freedom. After the initial fit was made to this spectrum, a batch fit was made to the other 7 spectra by adjusting the parameters of the first fit to find the best fit for each of the others.

The batch fits form the basis of a continuum model which was then subtracted from the data. These residuals were then divided by the standard deviation, σ, that was associated with each energy value. Contour maps of the residuals plotted against energy and time were then generated. Figure 3 shows the contour map that was generated for burst 1B-911221.

The contour lines are displayed for values of 2σ, 3σ, 4σ, and 5σ. When examining the structure in contour plots the resolution of the detector at the particular energy must be considered in order to determine whether the structure is real or not. Equation 2 gives the resolution of the detector as a function energy.

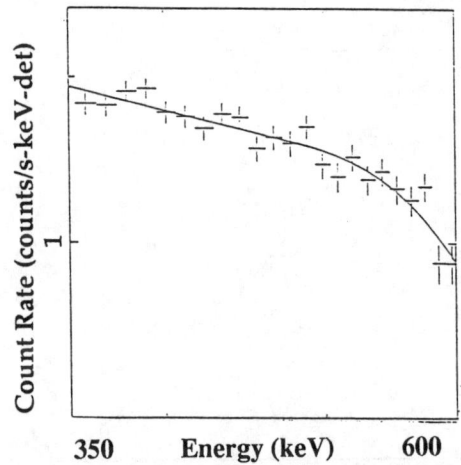

Figure 2. A fit using a Broken Power Law.

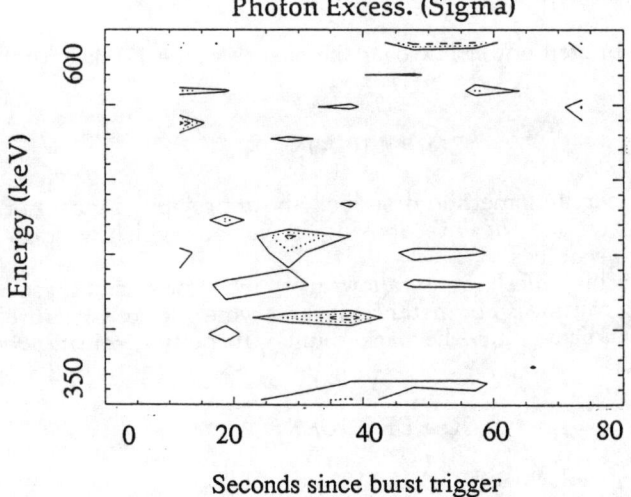

Figure 3. Contour map generated for burst 1B 911221.

$$R = (0.079E)\left(\frac{E}{511}\right)^{-0.42} \quad (2)$$

At 545 keV the resolution is 42 keV. Therefore, the structure seen at 545 keV between 36 secs and 57 secs is probably a detector anomaly. The detector resolution at 490 keV is 39 keV. The observed structure ranges from 480 keV to 510 keV, so the feature is probably not real but further investigation is warranted. Figure 4 shows a plot over a larger energy range chosen to show the features at 490 keV in the context of a larger continuum. The figure shows that, in the energy range of 480 keV to 510 keV, there are no significant features.

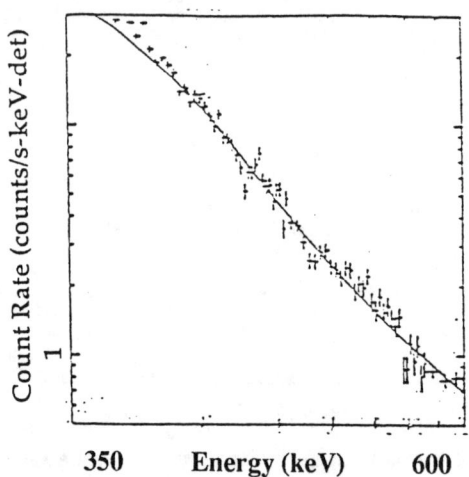

Figure 4. Fit of a broken power law over the energy range 170 keV to 900 keV.

CONCLUSION

The feature searching method described above provides a means of searching through a vast amount of data, looking for regions which warrant further and more thorough searches.

The new searching method also allows us to evaluate our background subtracting and fitting routines. For instance, if there were a lot of structure around 511 keV it might indicate that the background subtraction routines were not working properly.

REFERENCES

1. C. Barat, et al, Ap. J. **286**, L11-L13 (1984).
2. A. K. Harding, Gamma-Ray Burst Theory: Back to the Drawing Board, Ap. J. Supp, Submitted, (1993).
3. E. P. Mazets, et al, Pis'ma Astron. Zh. **6**, 706-711 (1980).
4. I. G. Mitrofanov, et al, Pis'ma Astron. Zh. **61**, 939-943 (1984).
5. P. L. Nolan, et al, Nature **311**, 360-362 (1984).

MISCELLANEOUS

BURSTERS AND THE QUEST FOR "COSMOLOGICAL EFFECTS"

Virginia Trimble

Astronomy Department, University of Maryland, College Park, MD 20742
and
Physics Department, University of California, Irvine, CA 92717 USA

ABSTRACT

From Hubble's time to the present, astronomers have sought observational discriminants for and between Friedmann-Robinson-Walker models of the universe. None has been unambiguously successful, so that confirmation of the constellation of redshift effects (low count rates, long duration, soft spectrum) in gamma ray burst events would have philosophical importance beyond its role in constraining the nature of burst sources.

1. INTRODUCTION: CONFIRMED EFFECTS

An approximation to the $1/R^2$ law for received flux was presumably known to whichever paleolithic tribe that first carried its campfires from place to place. That we look back in time when we look out in space was discovered in 1675 by the Dane Ole Romer (who also built the first meridian transit telescope). He explained irregularities in the timing of eclipses of Jovian satellites by finite light travel speed, putting the earth 11 light minutes from the sun. Since he assumed Cassini's 1672 value of 9.5" for solar parallax, this corresponds to a velocity of about 210,000 km/s, though Romer, working in pre-Revolutionary Paris, must have used different units. Finally, the first third of this century saw the gradual realization that redshifts of spiral nebulae are correlated with their distances, either quadratically (according to Lundmark, 1925; Wirtz, 1925; Stromberg, 1925; and a very few modern disciples like I. Segal) or linearly (according to Hubble [1929] and many others).

And there we come to a screeching halt. In no other case is what we observe about any astronomical phenomenon or object dominated by the large scale structure of the universe, as opposed to astrophysical evolution of intrinsic properties of the objects and/or difficulties with the observations themselves. Direct evidence telling us that we live in some kind of Friedmann-Robertson-Walker universe (that is, that the redshifts are indeed due to the expansion of space-time rather than to tired light, the de Sitter [1933] effect, or something else) or telling us which of the FRW metrics is the best description of the real cosmic four-geometry remains sparse, ambiguous, or worse.

2. THE CLASSICAL TESTS

The redshift-apparent-magnitude diagram at large z would be a relatively pure test for q_0, the deceleration parameter, if galaxies were standard candles (see Sandage, 1988 for thorough discussion of all the classical tests). Unfortunately they are not. Though the light of brightest cluster member elliptical galaxies is dominated by red giants, whose increasing number partly compensates for smaller individual brightness as the main sequence erodes away, evolution nevertheless wins. An important implication of B.M. Tinsley's thesis and later work (Tinsley 1972) is that the luminosity of galaxies with different possible histories diverge from each other with increasing z as fast or faster than the luminosities of galaxies with any one history but in universes with q differing by 0.5.

Counting objects as a function of flux received also goes back to Hubble, who, not surprisingly, found no evidence for deviations from Euclidean space. The application to radio sources (e.g. Ryle 1961) was more disconcerting. Redshift should cause any plot of $\log N$ vs. $\log S$ to droop below the Euclidean -1.5 power law line, by different amounts for different geometries, including steady state. But the data clearly traced out a slope of -1.8 for moderately bright sources. In any evolutionary universe, you can explain this by saying that there were more sources ("density evolution") or brighter sources ("luminosity evolution") in the past. Steady state was allowed no such out, and $\log N$-$\log S$ was its death knell in many minds. The two forms of evolution are not distinguishable with only a single power-law slope to go by, but currently popular cosmological models of gamma bursters would make density evolution the more likely.

Counting galaxies as a function of redshift sounds, at first hearing, less likely to be done in by evolution. Loh and Spillar (1986) made the attempt, claiming $\Omega \approx 1$ (though the test is really primarily sensitive to geometry, K/a^2, not to density. In any case, further analysis (Bahcall & Tremaine 1988) revealed that evolution strikes again and models from empty to closed could not be ruled out.

The observed angular diameter of an object of known physical size and redshift also probes geometry or radius of curvature, K/a^2, of the universe. The optical case has long been recognized as fairly hopeless (e.g. Djorgovski & Spinrad 1981), but radio hope springs eternal. The trouble is that, when you plot authentic raw data in $\log \theta$ vs. $\log z$, you always seem to trace out $\theta \propto 1/z$ (flat, static space), even steady state's $\theta \propto \ln(1+z)^{-1}$ turning up too sharply (e.g. Milsson et al., 1993). Ah, you conclude, the sources were smaller in the past, and, given the redshifts, we can correct for that. Yes, but you must assume a model to do it. And (Milsson et al.), if you assume $q = \frac{1}{2}$ (and $\Lambda = 0$), your corrected data happily trace out the $q = \frac{1}{2}$ theoretical curve; while if you assume $q = 0$, the differently corrected data equally happily trace out its curve. Once again, evolution one, cosmology zero. Kellermann (1993) has tried using more compact radio sources which might be less affected by secular changes in environment. He too finds $q = \frac{1}{2}$, but ....

The test using surface brightness of galaxies vs. redshift straddles the classical and modern periods. Surface brightness within a given metric diameter differs from one FRW model to another. But surface brightness within a given isophotal diameter, which is what can usually be measured, scales as $(1+z)^{-4}$ in any FRW universe. That this should be so feels slightly mysterious (though

Sandage 1988 explains it clearly as the ratio of well-known formulae for received flux and apparent angular diameter. Why then try the test at all? Because non-expansion causes of redshift lead to different exponents on $(1+z)$. The "4" is made up of one power from energy per photon, one from photon arrival rate, and two from an aberration effect on (θ^2) or surface area. Tired light has only the first of these, so SB $\propto (1+z)$. Newtonian expansion has the first and second effects and SB $\propto (1+z)^2$. Surface brightness vs. redshift can, therefore, tell us whether we actually live in a FRW universe.

The most recent application is that of Sandage and Perelmuter (1991). Using their own data out to $z = 0.1$, they find (not surprisingly) that scatter, both observational and real, strews the observed points across all possible SB vs. $(1+z)$ curves, even when the data are normalized to represent galaxies of a standard total brightness ($M_v = -24$) and size. In order to get to larger redshifts where cosmological effects will outweigh the scatter, they apply similar normalization to the data of Djorgovski and Spinrad (1981), extending out to $z = 0.58$ (but ignoring a couple of galaxies at still larger z). They conclude that the normalized data (while not distinguishing one value of q_0 from another) clearly point to a $(1+z)^{-4}$ relation, ruling out tired light and other non-expanding causes of redshift. The original observers, however, feel that the data are not sufficiently robust to support the weight they are being asked to carry (G. Djorgovski, 1993, pr. comm., phrased somewhat more forcefully). The possibility of additional confounding from evolutionary changes in SB remains to be explored.

3. MORE RECENT TESTS

A universe with a positive cosmological constant goes through a "coasting phase" in which a good deal of time is spent at nearly constant expansion parameter. Thus, looking backward, we expect to see many objects with redshift corresponding to that coasting era. This was invoked some years ago as a potential explanation for the apparent pile-up of qso's at redshifts near 1.95. More recently, the numbers of gravitationally-lensed qso's and the redshift distributions of the sources and lenses have been analyzed to look for evidence of such a coasting phase. Turner (1990) concluded that the dimensionless value of λ could not be more than 0.9. Maoz and Rix (1993) have pushed this further down to 0.7, at which point the coasting phase would not last long enough to increase the age of an Ω (total) $= 1$ universe above that of the oldest globular clusters (the main purpose of non-zero Λ in most people's minds). If the limit is correct then $\Lambda \neq 0$ universes cease to be interesting. Krauss and Schramm (1993) point out, however, that the current samples are essentially magnitude-limited ones, so that evolutional changes both in the intrinsic brightness of the sources and in the masses and mass distributions of the lenses may permit a wider range of λ, including interesting values for the age problem. Once again, cosmological effects probably do not dominate what we observe.

Finally, an attempt has been made to look directly for time dilation (that is, expansion vs. tired light or de Sitter effect) in the light curves of Type Ia supernovae. The raw data for the one at largest redshift (SN 1992bi, $z = 0.45$, Pennypacker et al., 1992) have not yet been published. But Nørgaard-Nielsen et al. (1988) show their measured magnitudes and dates for SN 1988U at $z = 0.31$. The points fit well onto a standard SN Ia light curve when time dilation is allowed for. Unfortunately, they did not catch the event at peak light, and the distance modulus to the host galaxy has all the usual uncertainty from $H = 50$ or 85 to 100. Thus the points can, to a certain extent, be slid both vertically and horizontally, permitting at least plausible fits also without time dilation. More events are needed!

The SN Ia case is the one most closely resembling that of the gamma-ray bursters. A light curve from $z \approx 0$ is used as a template, and the high-z light curve undilated to fit it. If Ia supernovae at $z = 0.3$ (or more) were physically different from those here and now, then we will be, at least, misled. The burst case is slightly more difficult, because the template itself must be extracted from the data base that is to be tested for time dilation effects, and because there is no direct, independent measurement of redshifts.

If the very persuasive correlations shown by J. Norris and others at this meeting are indeed the results of large distances and expansion acting on the count rates, time scales, and spectral hardness of otherwise very similar photon bursts, then we have not only a piece of the solution to the gamma-ray burst problem, but also the first astronomical entities in which cosmological effects win out over evolution and observational difficulties. Which of these you think is the more important probably depends on whether the first astronomy book you ever read was written by Fred Hoyle!

4. ACKNOWLEDGEMENTS

The one small idea contained herein is surely not original, but I am not sure from whom I first heard it (the culprit, if he or she confesses, is entitled to a free drink at the next CGRO symposium). The truism that evolution always wins dawned on me very gradually beginning with Caltech graduate lectures by Maarten Schmidt and Peter Scheuer, and continuing with a few months in the early 70's of sharing an office at University of Maryland with the late Beatrice M. Tinsley.

REFERENCES

Bahcall, S.R., & Tremaine, S. 1988. ApJ 326, L1
Djorgovski, S., & Spinrad H. 1981. ApJ 251, 417
Hubble, E. 1929. Proc. NAS 15, 168
Kellermann, K. 1993. Nature 361, 134
Krauss, L.M., & Schramm D. N. 1993. ApJ 405, L43
Loh, E.D., & Spillar, E. J. 1986. ApJ 307, L1
Lundmark, K. 1925. MNRAS 85, 865
Maoz, D., & Rix, H.-W. 1993. ApJ 416, 425
Milsson, K., et al. 1993. ApJ 413, 453

Nørgaard-Nielson, H.M., et al. 1989. Nature 339, 523
Pennypacker, C., et al. 1993. IAUC 5652
Ryle, M. 1961. IAU Symp. 15, 337
Sandage, A. 1988. ARA&A 26, 561
Sandage, A., & Perelmuter, J.-M. 1991. ApJ 370, 455
de Sitter, W. 1933. The Astronomical Aspect of the Theory of Relativity. (Berkeley: U. California Press) p. 196
Stromberg, G. 1925. ApJ 61, 353
Tinsley, B.M. 1972. ApJ 178, 331
Turner, E. L. 1990. ApJ 365, L43
Wirtz, C. 1925. Scientia 38, 303

THE ANGULAR SIZE OF COSMIC GAMMA-RAY BURSTS SOURCES

Igor G. Mitrofanov

Space Research Institute, Profsojuznaya str.84/32, 117810 Moscow, Russia

ABSTRACT

Sources of cosmic gamma-ray bursts could be the objects with the smallest angular size in the observable universe. Both physical and astrophysical issues are discussed connected with this fact.

INTRODUCTION

The old problem of the origin of cosmic gamma-ray bursts (GRBs) becames much more confused now when more observational data had been obtained. In 1990 the dominated majority of scientists have believed that sources of GRBs are very similar to another well-known galactic objects associated with neutron stars (X-ray bursters, or X-ray pulsars, or gamma-ray pulsars), and main discussions were focused on the concrete physical mechanisms of outbursting activity. In 1991-92, when the first data have been obtained from the Compton Observatory, all models has been voted down connected with galactic disk populations, and forgotten exotic models were revived. One of them is associated with neutron stars in the extended galactic corona (e.g. see ref. 1 and 2). Another one has interpreted GRBs as the cosmological phenomenon.[3] In 1992 those models provided the principal alternative, and all GRBs community has been practically equally divided in supporting either the first model or the second one.

However, today, in 1003, one might see even the next stage of confusing. The model of the extended galactic corona has met the additional constraints from the modern BATSE observations,[4] while the cosmological model also has the difficulty connected with the statistical analysis of averaged time histories of "strong" and "weak" events.[5] Moreover, the Workshop'93 discussions have been frequently focused to the question, what one does observe when detect GRBs. As well as in the very beginning of the history of GRBs, the phenomenology becomes again the key issue.

Below questions are considered associated with the fact that sources of GRBs are probably absolutely unique objects with are observed the smallest angular size.

ANGULAR SIZE OF SOURCES OF GRBs

For the first model of GRBs associated with the extended galactic corona, the size of radiating region could be estimated, as

$$\theta_{EGC} = \frac{c \cdot t_{var}}{D} = 10^{-19} \cdot \left(\frac{t_{var}}{1~\mu s}\right), \qquad (1)$$

provided the time of observed variability t_{var} is measured in microseconds.[6] On the other hand, for the case of cosmological model, the angular size is about[6]

$$\theta_{cosm} = \frac{t_{var} H_o}{(1+z)\Phi(z)} = 3 \cdot 10^{-24} \left(\frac{t_{var}}{1\ \mu s}\right) \left(\frac{H_o}{100\ \text{km/s} \cdot \text{Mpc}}\right) (1+z)^{-1} \Phi(z)^{-1}, \qquad (2)$$

where the Hubble constant H_o equals to $100 kms^{-1} Mpc^{-1}$, the density of the universe is assumed to be equal to the critical value, and

$$\Phi(z) = 2 \cdot (z + 1 - \sqrt{(z+1)}) \cdot (z+1)^{-2} \qquad (3)$$

Independently on the model which could be accepted for GRBs, the estimated angular sizes either (1) or (2) correspond to the smallest angle between world lines of detectable photons, which could be expected in the observable universe. Indeed, according to simple estimations,[6] outbursting sources of GRB have the highest density of the radiated electromagnetic energy, which permit them to be observable at very large distances.

PHYSICAL EFFECTS ASSOCIATED WITH POSSIBLE QUANTUM LIMIT FOR ANGLE BETWEEN WORLD LINES OF GRB PHOTONS

The scale of a quantum fluctuation of the electromagnetic field, or of the classical radius of an electron, and the scale of the Planck length correspond to the well-known dimensionless constant

$$l_e / l_{Pl} \sim \cdot 10^{20}, \qquad (4)$$

which have the fundamental physical meaning. It determines the nature of coexistence of the electromagnetic vacuum and the fundamental quantum elements of space. Nobody knows the structure of the space at the Planck scale as well as the physics of photons propagation through this structure.[7] If world lines of free photons could not have arbitrary directions in respect to the Planck-scale structure of space, the inversed constant (3) could be the best estimation for the smallest elementary quantum angle $\theta_{Pl} \sim 5 \cdot 10^{-21}$ (ref. 6). If the assumption of the smallest angle is accepted, the finite number of principal directions $\sim 2 \cdot 10^{41}$ corresponds to the totality of world lines of free photons.[6]

The elementary quantum angle is comparable or even larger, then the angular size of sources of GRBs (1) or (2). Therefore, the actual distances to sources of GRBs could be larger then the critical distance

$$D_* = \frac{L_{em}}{\theta_{Pl}} \sim 30 \left(\frac{L_{em}}{1\ \text{cm}}\right)\ \text{pc}. \qquad (5)$$

where the deviation between neighbor "bright" beams becomes equal to the transversal sizes of beams.

At $D \ll D_*$, there is no observable effects connected with the hypotheses of the quantum smallest angle. At $D > D_*$ there are observable effects, which could be found for GRBs. For simplicity, a GRBs is considered below as a short impulses of gamma-rays, about t_{burst} long, which propagate along directions of "bright" beams. Neighbor "bright" beams of photons are separated by "dark" segments of space between them. Due to the source/detector relative motion, GRB might be recorded while detector is crossing the "bright" segments of space. In the case of transversal relative motion, detector records a burst as several bright pulses at crosses of "bright" beams with the total duration about t_{burst}. In the second case (B) of smaller angle α between line of sight and the relative velocity, a burst could be detected as a one-pulse event with duration $< t_{burst}$. And, in the third case (C) of very small angle α, a burst could be missed at all. So, the following observational effects could be predicted for subsets of events detected with large and small angles α between the relative velocity and the line of sight:

(i) For two equal sectors on the sky, for sector ($\parallel$) with angles α in two segments (0-60) and (120-180), and for another sector ($\perp$) with angles α in the segment (60-120), one might detect different number of events, or larger total durations of those events, in the second one. One might see also different time histories of GRBs, which are detected in sector ($\parallel$) and sector ($\perp$). In the second case GRBs might generally have broader peaks then in the first one.

(ii) Along a "bright" beam there is no dilution of radiation from a source. Therefore, sources at difference distances with similar radiated flux would have the same observed peak fluxes which do not depend on the actual distances to them. This peculiarity strongly influences on the "number/magnitude" statistics of GRBs, where there is the well-known discrepancy between the cases of peak flux or total intensity used as a "magnitude".

POSSIBLE ASTROPHYSICAL EFFECTS OF THE QUANTUM LIMIT FOR THE ANGLE

In the frame of reference of the cosmological objects there is one principal direction, which is the direction of the motion of the Sun in respect to the microwave background, which corresponds to velocity V_o about 400 km/s to the direction of the Leo constellation.[8] There are several different components of the peculiar motion of the Sun (230 km/s of the motion in the Galaxy, 115 km/s of the motion of the Galaxy within the Local Group, etc.), but this one might be considered as the general peculiar "cosmological" motion of the detector.

The transversal component of the relative velocity between a source of GRB and the detector equals to the sum of two components: the transversal component of the "cosmological" velocity of the Sun and transversal component of a peculiar velocity V_s of a source. One could not know the second one, and for selected subset of GRBs the mean transversal velocity could be found by the averaging over the angle α between V_o and line of sight and over the isotropic distribution of transversal components of peculiar velocities of sources V_s. For subsets ($\perp$) and ($\parallel$), the difference between the averaged squared relative velocities $< V_\perp^2 >$ and $< V_\parallel^2 >$ equals to $1/4 \cdot V_o^2$.

The difference between $< V_\perp^2 >$ and $< V_\parallel^2 >$ permits to expect that some observational effects, like (i) and (ii), could be studied to check the hypotheses of the smallest quantum angle. Indeed, during a mean GRB, which is about 30 s long, detector with velocity V_o covers a distance about 10^9 cm. Provided the radiation of GRB would have a structure of "bright/dark" segments along the line of sight, the differences predicted in (i) and (ii) could be found for subsets ($\parallel$) and ($\perp$). This structure could be associated either with the reality of the finite minimal angle, as assumed above, or with another physical reasons for the ultra-narrow angular diagram of radiating objects.

Using the vector of the "cosmological" velocity, as the principal axes, all totality of GRBs from the First Catalog of BATSE[9] has been divided into "parallel" ($\|$) and "perpendicular" subsets ($\perp$). All events with t_{90} shorter 2 s have been excluded from the comparison at all. The following differences have been found between them:

1) There are 83 and 81 events with durations $t_{90} > 2$ s in subsets ($\perp$) and ($\|$), respectively. The averaged durations of events from ($\perp$) and ($\|$)-subset are equal to 20.0 ± 0.13 and 23.6 ± 0.10, respectively. The difference between them is 3.6 ± 0.16 s. This kind of the difference does agree with the prediction (i) of the pervious Section.

2) The difference is known between statistics of "number/magnitude", when for the magnitude the maximal fluxes are used during 64 ms, 256 ms and 1024 ms.[9] There is no commonly accepted explanation, why this difference is observed. If the structure of 'bright/dark" segments really exists, this difference could be naturally explained (see (ii)).

GENERAL CONCLUSIONS

So, GRBs probably provide the absolutely unique possibility to study the geometry of the world, when the scale of angle between gamma-rays is as small as the ratio of the Planck scale (1) and the scale of electromagnetic field (2). So, all facts and predictions should be studied carefully which could be associated with this issue.

The assumption on the smallest quantum angle is one of such predictions, and GRBs permit to check it rather easy. Some tendency in GRBs data is evident compatible with the "bright/dark" segments of space, and larger statistics of BATSE GRBs would certainly check observational predictions (i)-(ii) with much higher significance.

On the other hand, any difference between subsets of ($\|$) and ($\perp$) associated with the "cosmological" velocity of the Sun points out only on the spatial variations of the flux of gamma-rays along the front of propagation. The quantum limit of angle hardly could be the only possible explanation for such peculiarity. Another physical effects, such as radiation self collimation within a source, could be discussed also provided observational indications on the inhomogeneity, would become significant.

REFERENCES

1. M. C. Jennings, High Energy Transients in Astrophysics (Academic Press, N. Y., 1984), p. 412.
2. I. S. Shklovskii & I. G. Mitrofanov, MNRAS **212**, 545 (1985).
3. B. Paczynski, Huntsville Workshop on Gamma-Ray Bursts (Academic Press, N.Y., 1992), p. 144.
4. J. Hakkila, this proceeding (AIP: New York, 1994).
5. G. J. Fishman, this proceedings (AIP: New York, 1994).
6. I. G. Mitrofanov, Ap. J., accepted for publication, (1994).
7. M. Rees, R. Ruffini, & J. A. Wheeler, Black Holes, Gravitational Waves and Cosmology: An Introduction to Current Research (Gordon and Breach Science Publishers, N. Y., London, Paris, 1974).
8. I. A.. Strukov, Russian Astron. Letters **13**, 163 (1987).
9. G. J. Fishman et al., Ap. J. Supp. in press, (1994).

A GAMMA-RAY BURST BIBLIOGRAPHY, 1973 - 1993

K. Hurley
University of California
Space Sciences Laboratory
Berkeley, CA 94720

Abstract

Since the first article was published on cosmic gamma-ray bursts in 1973, approximately 2000 publications, abstracts, and theses have appeared. I have catalogued them chronologically, with key words for subject searches. Here I present this machine-readable bibliography of 20 years of gamma-ray burst literature. Individual copies may be requested in various formats.

Introduction

I have been tracking the gamma-ray burst literature for about the past fifteen years, keeping the authors, titles, and references in a machine-readable form. (To my knowledge, the last major compilation of gamma-ray burst publications dates from 1983[1]). In its current form, this information is in a Microsoft Word for Windows "doc" format. My purpose in doing this was twofold. First, I wanted to be able to retrieve rapidly any articles on a given topic, and second, I wanted to be able to cut and paste references into manuscripts in preparation. The following journals have been scanned on a more or less regular basis starting with the 1973 issues:
  Astronomical Journal
  Astronomy and Astrophysics
  Astrophysical Journal (letters, main journal, and supplements)
  Astrophysical Letters and Communications
  Astrophysics and Space Science
  Monthly Notices of the Royal Astronomical Society
  Nature
  Physical Review
  Scientific American
  Soviet Astronomy (main journal and letters)
In addition, the following journals have been scanned, but in many cases less regularly:
  Annals of Geophysics
  Astrofizika
  Bulletin of the American Astronomical Society
  Bulletin of the American Physical Society
  Chinese Astronomy
  Cosmic Research
  IEEE Transactions on Nuclear Science
  JETP

Journal of Atmospheric and Terrestrial Physics
Journal of the British Interplanetary Society
Journal of the Royal Astronomical Society of Canada
Nuclear Instruments and Methods
Publications of the Astronomical Society of the Pacific
Publications of the Astronomical Society of Japan
Progress in Theoretical Physics
Solar Physics
Soviet Physics

The above lists are not exhaustive. Where theses or internal reports have come to my attention, I have included them, too. To be included, an article had to have something to do with gamma-ray burst theory, observation, or instrumentation, or be closely related to one of these topics (e.g., neutron stars), and must have been published. With only a few exceptions, preprints which were never published have not been included.

Organization of the Bibliography

The overall organization is chronological by year. Within a given year, articles published in journals are listed first, in alphabetical order by first author. Then come theses and conference proceedings articles. The latter are listed in the order in which they appear in the proceedings. An example follows:

289. Taam, R. And Picklum, R., Thermonuclear Runaways On Neutron Stars, Ap. J.,**233**, 327, 1979
Key Words: nuclear reactions, accretion, neutron stars, thermonuclear model

The entries are numbered consecutively, so that paper copies which are kept on file can be retrieved quickly. However, to avoid having to renumber this entire file when a new article is added, numbers are skipped at the end of each year. The complete author list follows, as it appears in the journal, along with the title, volume number, page number, and year. A line containing key words follows this. These are generally not the same key words as the ones listed in the journal, nor are they taken from the title or any particular list. Rather, they are meant to reflect the true content of the article, and provide a list of machine-searchable topics. In general, key words have not been included for conference proceedings articles.

Discussion

The number of articles published each year since 1973 is shown in figure 1. Starting with a modest article per month in 1973, it has already exceeded one every two days in 1993. (But note that there is still about one paper being published per burst). With the sheer number of articles published to date, it is not surprising that many articles cover the same topic. But it is interesting to note that many "new" ideas were in fact proposed long ago. See, for example, the 1974 article by Zwicky[2] in which optical transients were predicted, the 1975 article by Usov and Chibisov[3] with a cosmological interpretation of the log N-log S curve, the unpublished 1980 article by Teller and Johnson[4] and the 1984 article by Blinnikov et al.[5] about extragalactic binaries, or the 1984 article by Usov[6] on superstrong magnetic fields. One gets the impression that the right answer must certainly have appeared in the literature somewhere!

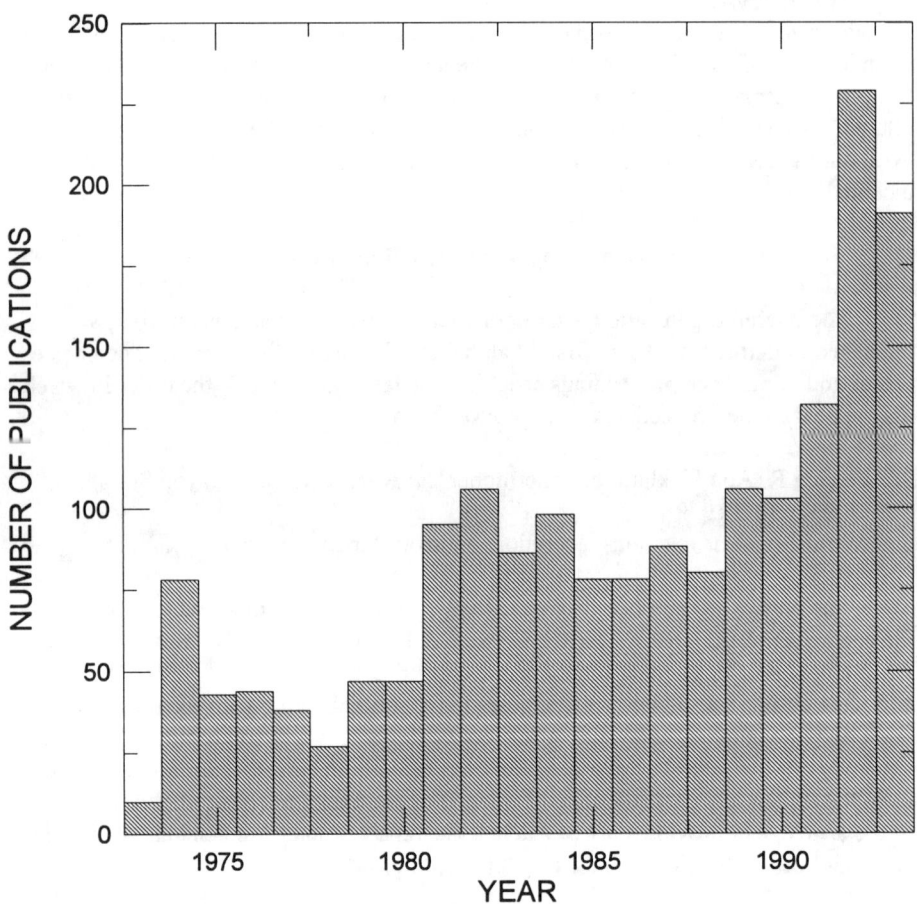

Figure 1. Number of gamma-ray burst publications as a function of year. The cutoff date is October 1993.

Distribution of the Bibliography

Paper and electronic versions of the bibliography will be made available to interested parties as time and money permit. Electronic versions can be either in Word for Windows "doc" format, or plain ASCII. Many word processing programs have conversion utilities which will allow one of these electronic versions to be read. Please contact me by regular or electronic mail (khurley@sunspot.ssl.berkeley.edu) to request copies, and indicate your preference for the format. I update this bibliography frequently, so I would appreciate it if users would communicate errors and omissions to me.

References

1. W. Baity, G. Hueter, and R. Lingenfelter, in High Energy Transients in Astrophysics, Ed. S. Woosley, AIP Conference Proceedings 115 (AIP - New York), 434 (1983)
2. F. Zwicky, Astrophys. Space Sci., 28, 111 (1974)
3. V. Usov and G. Chibisov, Soviet Astron. 19(1), 192 (1975)
4. E. Teller and M. Johnson, Unpublished (1980)
5. S. Blinnikov, I. Novikov, T. Perevodchikova, and A. Polnarev, Sov. Astron. Lett. 10(3), 177 (1984)
6. V. Usov, Astrophys. Space Sci. 107, 191 (1984)

A CENTURY OF GAMMA-RAY BURST MODELS

Robert J. Nemiroff
George Mason University, CSI Institute, Fairfax, VA 22030
NASA Goddard Space Flight Center, Greenbelt, MD 20771

ABSTRACT

More than 100 gamma-ray burst progenitor models have now been published in refereed journals. A list of the models appearing before the beginning of October 1993 is presented and briefly discussed. A previous version of this list and a more complete discussion will appear in Comments on Astrophysics.

ABOUT THE LIST

A list of the more than 100 papers suggesting GRB progenitor models, or variants thereof, is given in Table 1. Brad Schaefer provided a preliminary list on which Table 1 is based. A paper will appear on this list if it was published in a refereed journal, appeared before the 1 October 1993, and proposed a new or revised model for the origin of a GRB. A reasonable effort was made to make Table 1 complete, however it is probable that several papers were missed. A conscience subjective decision *not* to included a paper might have been made if it was deemed that the paper did not create a significantly new progenitor model or did not add significantly to an existing progenitor model (the paper might still be an excellent scientific paper, however). A paper might also have been excluded if it focussed specifically on the physics of a potential mechanism rather than suggesting a significantly different type of mechanism. Papers appearing in journals less circulated in the United States might also have been missed. Please note that the numbers of references in Table 1 are different from the reference numbers of papers cited in the reference section at the end of this paper.

Table 1 is divided into 8 columns. The first column gives a model reference number. Models are listed in chronological order of the date they were received by the journals.

Column 2 lists the lead author. If there were two or more authors, an "et al." is given following the first author. Column 3 lists the year the article was published. Column 4 lists the reference in a compact form. Most of these conform to the accepted modern format, however several non-standard abbreviations were made, primarily due to space limitations. Specifically,
"CJPhys" refers to the Canadian Journal of Physics,
"CosRes" refers to Cosmic Research,
"PRL" refers to Physical Review Letters, and
"SovAstron" refers to Soviet Astronomy.

Column 5 lists the major progenitor body involved in the GRB model. Many abbreviations are straightforward with "NS" meaning neutron star, "WD" meaning white dwarf, "BH" meaning black hole and "AGN" meaning active galactic nucleus. Less standard abbreviations are:
"CS" meaning cosmic strings,
"DG" meaning dust grain,
"GAL" meaning external galaxy,
"MG" meaning magnetic reconnection,

"RE" meaning relativistic elections,
"SS" meaning strange star,
"ST" meaning normal star, and
"WH" meaning white hole.

If a second body is involved in the GRB model, it is listed in column 6, even if it cannot be clearly labelled as a body. The following additional abbreviations were used:
"AGN" meaning active galactic nucleus,
"AST" meaning asteroid,
"COM" meaning comet,
"ISM" meaning interstellar medium,
"MBR" meaning microwave background,
"PLAN" meaning planet, and
"SN" meaning supernova shock.

Column 7 lists the location of the GRB explosion. Here
"COS " refers to a cosmological setting,
"DISK" refers to the disk of our Milky Way Galaxy,
"HALO" refers to the halo of our Galaxy, and
"SOL" refers to the outer solar system. In cases where the GRB location was not well specified between the Galactic disk and the Galactic halo, the later location was typically chosen if energy constraints allowed.

Column 8 gives a brief description of the model (or refinement) proposed. I apologize for the gross generalizations made here and for any inaccuracies. Several times terms and abbreviations are used in the description that need explanation, and I must ask the reader to consult the papers cited for this explanation. Happily, I was not shaken from my belief that once terms are defined, the gist of any good scientific paper can be summarized in five words or less. (Admittedly this makes a better parlor game than a truism.)

The first entry in Table 1 requires explanation. The prime reason the GRB discovery paper[1] gives for the initial search was to test the prediction of GRB existence made by Colgate in Reference 1 of Table 1. GRBs were discovered in this search but found *not* to be coincident with observed supernovae in local galaxies, as this Colgate model predicts. However, since this Colgate paper gave a model for GRBs which fostered GRB detection, it is arguably the first GRB model, even though it predates their detection.

Inspection of Table 1 shows several interesting trends. First of all, most of the models are based in the Galactic disk, and most are based on neutron stars. The most diverse group of models was published immediately after the discovery of GRBs, but over the years a wide variety of distinctly different models have been published. Based on the publication record, it appears that the community had generally settled on the idea of Galactic disk neutron star progenitors in the 1980s, as the majority of papers published then were refinements of this idea (with a few very notable exceptions). Any settling that might have occurred then became unsettled with the announcement of the first BATSE results.

The result of this announcement was a dramatic shift in GRB modeling. Papers submitted in 1992 were, for the first time, predominantly cosmologically placed. There was also a slight shift away from neutron star progenitor models, although NS models still outnumbered all others models combined.

The GRB progenitor problem is now arguably the most prolific in astronomical history, easily surpassing the pulsar problem in this area.[2] (For a list of potential pulsar progenitor models, which numbered 20 about 3 years after their discovery, see Table 2 of Ref. 5. Even at this early date, though, there was a very strong community sentiment toward rotating neutron stars.) Reasons for this include the uncertainty of several important data features, the relatively long period of speculation, the relatively large amount of data needed to solve the dilemma compared to the amount of data taken, the relatively large numbers of astronomers and astronomical journals in the world today, the relative ease which word processing makes published speculation possible, and the pressure to publish in today's academic environment, to name a few. Probably the best reason for the proliferation of GRB models, though, is that new data entering the field has not bolstered any specific model. The BATSE results, in fact, have made the majority of previously published models more tenuous. Therefore, in light of this new data, it is possible that even this list of over 100 models lacks diversity.

DISCUSSION

I apologize if I have omitted or badly described any models in Table 1. I will try to update Table 1 on a regular basis, however, and continually honor requests for a photocopy of it. Therefore, I welcome any comments or corrections anyone may have on this table.

There are, of course many excellent papers about GRBs and GRB models that do not appear in Table 1. There are also some not-so-excellent papers that DO appear. The full criteria for inclusion is discussed above, but includes a) publication in a refereed journal, and b) creating a distinct GRB progenitor model. If the reader knows of a paper that properly belongs on the list but does not appear, please inform me. My address and e-mail are listed in the AAS membership directory.

It will take more than speculation to solve the current GRB model dilemma - it will certainly take more observations. Clearly, several observational uncertainties need to be resolved for theorists to know which data subsets to believe. Do GRB show cyclotron lines, annihilation lines, or repetition? These questions should be answered by the current *Compton* mission. Do GRBs show extra X-ray absorption in the galactic plane? Are GRB positions, when known more accurately, correlated with any known object? These are examples of questions which may be answered with the next generation of GRB measuring instruments. If, when these data arrive, they don't bolster an existing model, we may well be in for yet another era of GRB model speculation!

I thank Brad Schaefer for providing a preliminary list on which Table 1 is based, and for many helpful comments and criticisms. I also thank the following people for humoring me through some unusual discussions: Jay Norris, Thulsi Wickramasinghe, and Jerry Bonnell.

REFERENCES

1. R. Klebesadel, I. B. Strong, R. A. Olson, Ap. J. **182**, L85 (1973).
2. A. Hewish, Ann. Rev. Astron. Astrophysics **265**, (1970).

Nemiroff, R. J., 1993, Comments on Astrophysics, in press.

Table 1

Model #	Author	Year Pub	Reference	Main Body	2nd Body	Place	Description
1.	Colgate	1968	CJPhys, 46, S476	ST		COS	SN shocks stellar surface in distant galaxy
2.	Colgate	1974	ApJ, 187, 333	ST		COS	Type II SN shock brem, inv Comp scat at stellar surface
3.	Stecker et al.	1973	Nature, 245, PS70	ST		DISK	Stellar superflare from nearby star
4.	Stecker et al.	1973	Nature, 245, PS70	WD		DISK	Superflare from nearby WD
5.	Harwit et al.	1973	ApJ, 186, L37	NS	COM	DISK	Relic comet perturbed to collide with old galactic NS
6.	Lamb et al.	1973	Nature, 246, PS52	WD	ST	DISK	Accretion onto WD from flare in companion
7.	Lamb et al.	1973	Nature, 246, PS52	NS	ST	DISK	Accretion onto NS from flare in companion
8.	Lamb et al.	1973	Nature, 246, PS52	BH	ST	DISK	Accretion onto BH from flare in companion
9.	Zwicky	1974	Ap&SS, 28, 111	NS		HALO	NS chunk contained by external pressure escapes, explodes
10.	Grindlay et al.	1974	ApJ, 187, L93	DG		SOL	Relativistic iron dust grain up-scatters solar radiation
11.	Brecher et al.	1974	ApJ, 187, L97	ST		DISK	Directed stellar flares on nearby stars
12.	Schlovskii	1974	SovAstron, 18, 390	WD	COM	DISK	Comet from system's cloud strikes WD
13.	Schlovskii	1974	SovAstron, 18, 390	NS	COM	DISK	Comet from system's cloud strikes NS
14.	Bisnovatyi- et al.	1975	Ap&SS, 35, 23	ST		COS	Absorption of neutrino emission from SN in stellar envelope
15.	Bisnovatyi- et al.	1975	Ap&SS, 35, 23	ST	SN	COS	Thermal emission when small star heated by SN shock wave
16.	Bisnovatyi- et al.	1975	Ap&SS, 35, 23	NS		COS	Ejected matter from NS explodes
17.	Pacini et al.	1974	Nature, 251, 399	NS		DISK	NS crustal starquake glitch; should time coincide with GRB
18.	Narlikar et al.	1974	Nature, 251, 590	WH		COS	White hole emits spectrum that softens with time
19.	Tsygan	1975	A&A, 44, 21	NS		HALO	NS corequake excites vibrations, changing E & B fields
20.	Chanmugam	1974	ApJ, 193, L75	WD		DISK	Convection inside WD with high B field produces flare
21.	Prilutski et al.	1975	Ap&SS, 34, 395	AGN	ST	COS	Collapse of supermassive body in nucleus of active galaxy
22.	Narlikar et al.	1975	Ap&SS, 35, 321	WH		COS	WH excites synchrotron emission, inverse Compton scattering
23.	Piran et al.	1975	Nature, 256, 112	BH		DISK	Inv Comp scat deep in ergosphere of fast rotating, accreting BH
24.	Fabian et al.	1976	Ap&SS, 42, 77	NS		DISK	NS crustquake shocks NS surface
25.	Chanmugam	1976	Ap&SS, 42, 83	WD		DISK	Magnetic WD suffers MHD instabilities, flares
26.	Mullan	1976	ApJ, 208, 199	WD		DISK	Thermal radiation from flare near magnetic WD
27.	Woosley et al.	1976	Nature, 263, 101	NS		DISK	Carbon detonation from accreted matter onto NS
28.	Lamb et al.	1977	ApJ, 217, 197	NS		DISK	Mag gating of accret disk around NS causes sudden accretion
29.	Piran et al.	1977	ApJ, 214, 268	BH		DISK	Instability in accretion onto rapidly rotating BH
30.	Dasgupta	1979	Ap&SS, 63, 517	DG		SOL	Charged intergal rel dust grain enters sol sys, breaks up
31.	Tsygan	1980	A&A, 87, 224	WD		DISK	WD surface nuclear burst causes chromospheric flares
32.	Tsygan	1980	A&A, 87, 224	NS		DISK	NS surface nuclear burst causes chromospheric flares
33.	Ramaty et al.	1981	Ap&SS, 75, 193	NS		DISK	NS vibrations heat atm to pair produce, annihilate, synch cool
34.	Newman et al.	1980	ApJ, 242, 319	NS	AST	HALO	Asteroid from interstellar medium hits NS
35.	Ramaty et al.	1980	Nature, 287, 122	NS		HALO	NS core quake caused by phase transition, vibrations
36.	Howard et al.	1981	ApJ, 249, 302	NS	AST	DISK	Asteroid hits NS, B-field confines mass, creates high temp
37.	Mitrofanov et al.	1981	Ap&SS, 77, 469	NS		DISK	Helium flash cooled by MHD waves in NS outer layers
38.	Colgate et al.	1981	ApJ, 248, 771	NS	AST	DISK	Asteroid hits NS, tidally disrupts, heated, expelled along B lines
39.	van Buren	1981	ApJ, 249, 297	NS	AST	DISK	Asteroid enters NS B field, dragged to surface collision
40.	Kuznetsov	1982	CosRes, 20, 72	MG		SOL	Magnetic reconnection at heliopause
41.	Katz	1982	ApJ, 260, 371	NS		DISK	NS flares from pair plasma confined in NS magnetosphere
42.	Woosley et al.	1982	ApJ, 258, 716	NS		DISK	Magnetic reconnection after NS surface He flash
43.	Fryxell et al.	1982	ApJ, 258, 733	NS		DISK	He fusion runaway on NS B-pole helium lake
44.	Hameury et al.	1982	A&A, 111, 242	NS		DISK	e- capture triggers H flash triggers He flash on NS surface
45.	Mitrofanov et al.	1982	MNRAS, 200, 1033	NS		DISK	B induced cyclo res in rad absorp giving rel e-s, inv C scat
46.	Fenimore et al.	1982	Nature, 297, 665	NS		DISK	BB X-rays inv Comp scat by hotter overlying plasma
47.	Lipunov et al.	1982	Ap&SS, 85, 459	NS	ISM	DISK	ISM matter accum at NS magnetopause then suddenly accretes
48.	Baan	1982	ApJ, 261, L71	WD		HALO	Nonexplosive collapse of WD into rotating, cooling NS
49.	Ventura et al.	1983	Nature, 301, 491	NS	ST	DISK	NS accretion from low mass binary companion
50.	Bisnovatyi- et al.	1983	Ap&SS, 89, 447	NS		DISK	Neutron rich elements to NS surface with quake, undergo fission
51.	Bisnovatyi- et al.	1984	SovAstron, 28, 62	NS		DISK	Thermonuclear explosion beneath NS surface
52.	Ellison et al.	1983	A&A, 128, 102	NS		HALO	NS corequake + uneven heating might yield SGR pulsations
53.	Hameury et al.	1983	A&A, 128, 369	NS		DISK	B field contains matter on NS cap allowing fusion
54.	Bonazzola et al.	1984	A&A, 136, 89	NS		DISK	NS surface nuc explosion causes small scale B reconnection
55.	Michel	1985	ApJ, 290, 721	NS		DISK	Remnant disk ionization instability causes sudden accretion
56.	Liang	1984	ApJ, 283, L21	NS		DISK	Resonant EM absorp during magnetic flare gives hot synch e-s
57.	Liang et al.	1984	Nature, 310, 121	NS		DISK	NS magnetic fields get twisted, recombine, create flare
58.	Mitrofanov	1984	Ap&SS, 105, 245	NS		DISK	NS magnetosphere excited by starquake
59.	Epstein	1985	ApJ, 291, 822	NS		DISK	Accretion instability between NS and disk
60.	Schlovskii et al.	1985	MNRAS, 212, 545	NS		HALO	Old NS in Galactic halo undergoes starquake
61.	Tsygan	1984	Ap&SS, 106, 199	NS		DISK	Weak B field NS spherically accretes, Comptonizes X-rays
62.	Usov	1984	Ap&SS, 107, 191	NS		DISK	NS flares result of magnetic convective-oscillation instability
63.	Hameury et al.	1985	ApJ, 293, 56	NS		DISK	High Landau e-s beamed along B lines in cold atm. of NS
64.	Rappaport et al.	1985	Nature, 314, 242	NS		DISK	NS + low mass binary companion gives GRB + optical flash
65.	Tremaine et al.	1986	ApJ, 301, 155	NS	COM	DISK	NS tides disrupt comet, debris hits NS next pass
66.	Muslimov et al.	1986	Ap&SS, 120, 27	NS		HALO	Radially oscillating NS
67.	Sturrock	1986	Nature, 321, 47	NS		DISK	Flare in the magnetosphere of NS accelerates e-s along B-field
68.	Paczynski	1986	ApJ, 308, L43	NS		COS	Cosmo GRBs: rel e+/- opt thk plasma outflow indicated
69.	Bisnovatyi- et al.	1986	SovAstron, 30, 582	NS		COS	Chain fission of superheavy nuclei below NS surface during SN
70.	Alcock et al.	1986	PRL, 57, 2088	SS	SS	DISK	SN ejects strange mat lump craters rotating SS companion
71.	Babul et al.	1987	ApJ, 316, L49	CS		COS	GRB result of energy released from cusp of cosmic string
72.	Livio et al.	1987	Nature, 327, 398	NS	COM	DISK	Oort cloud around NS can explain soft gamma-repeaters
73.	McBreen et al.	1988	Nature, 332, 234	GAL	AGN	COS	G-wave bkgrd makes BL Lac wiggle across galaxy lens caustic
74.	Curtis	1988	ApJ, 327, L81	WD		COS	WD collapses, burns to form new class of stable particles
75.	Melia	1988	ApJ, 335, 965	NS		DISK	Be/X-ray binary sys evolves to NS accretion with recurrence
76.	Ruderman et al.	1988	ApJ, 335, 306	NS		DISK	e+/- cascades by aligned pulsar outer-mag-sphere reignition
77.	Paczynski	1988	ApJ, 335, 525	CS		COS	Energy released from cusp of cosmic string (revised)
78.	Murikami et al.	1988	Nature, 335, 234	NS		DISK	Absorption features suggest separate colder region near NS
79.	Melia	1988	Nature, 336, 658	NS		DISK	NS + accretion disk reflection explains GRB spectra
80.	Blaes et al.	1989	ApJ, 343, 839	NS		DISK	NS seismic waves couple to magnetospheric Alfen waves

#	Author	Year	Reference				Description
81.	Trofimenko et al.	1989	Ap&SS, 152, 105	WH		COS	Kerr-Newman white holes
82.	Sturrock et al.	1989	ApJ, 346, 950	NS		DISK	NS E- field accelerates electrons which then pair cascade
83.	Fenimore et al.	1988	ApJ, 335, L71	NS		DISK	Narrow absorption features indicate small cold area on NS
84.	Rodrigues	1989	AJ, 98, 2280	WD	WD	DISK	Binary member loses part of crust, through L1, hits primary
85.	Pineault et al.	1989	AJ, 347, 1141	NS	COM	DISK	Fast NS though Oort clouds, fast WD bursts only optical
86.	Melia et al.	1989	ApJ, 346, 378	NS		DISK	Episodic electrostatic accel and Comp scat from rot high-B NSs
87.	Trofimenko	1989	Ap&SS, 159, 301	WH		COS	Different types of white, "grey" holes can emit GRB
88.	Eichler et al.	1989	Nature, 340, 126	NS	NS	COS	NS - NS binary members collide, coalesce
89.	Wang et al.	1989	PRL, 63, 1550	NS		DISK	Cyclo res & Raman scat fits 20, 40 keV dips, magnetized NS
90.	Alexander et al.	1989	ApJ, 344, L1	NS		DISK	QED mag resonant opacity in NS atmosphere
91.	Melia	1990	ApJ, 351, 601	NS		DISK	NS magnetospheric plasma oscillations
92.	Ho et al.	1990	ApJ, 348, L25	NS		DISK	Beaming of radiation necessary from magnetized neutron stars
93.	Mitrofanov et al.	1990	Ap&SS, 165, 137	NS	COM	DISK	Interstellar comets pass through dead pulsar's magnetosphere
94.	Dermer	1990	ApJ, 360, 197	NS		DISK	Compton scattering in strong NS magnetic field
95.	Blaes et al.	1990	ApJ, 363, 612	NS	ISM	DISK	Old NS accretes from ISM, surface goes nuclear
96.	Paczynski	1990	ApJ, 363, 218	NS	NS	COS	NS-NS collision causes v collisions to drive super-Ed wind
97.	Zdziarski et al.	1991	ApJ, 366, 343	RE	MBR	COS	Scattering of microwave background photons by rel e-s
98.	Pineault	1990	Nature, 345, 233	NS	COM	DISK	Young NS drifts through its own Oort cloud
99.	Trofimenko et al.	1991	Ap&SS, 178, 217	WH		HALO	White hole supernova gave simul burst of g-waves from 1987A
100.	Melia et al.	1991	ApJ, 373, 198	NS		DISK	NS B- field undergoes resistive tearing, accelerates plasma
101.	Holcomb et al.	1991	ApJ, 378, 682	NS		DISK	Alfen waves in non-uniform NS atmosphere accelerate particles
102.	Haensel et al.	1991	ApJ, 375, 209	SS	SS	DISK	Strange stars emit binding energy in grav. rad. and collide
103.	Blaes et al.	1991	ApJ, 381, 210	NS	ISM	DISK	Slow interstellar accretion onto NS, e- capture starquakes result
104.	Frank et al.	1992	ApJ, 385, L45	NS		DISK	Low mass X-ray binary evolves into GRB sites
105.	Woosley et al.	1992	ApJ, 391, 228	NS		HALO	Accreting WD collapses to NS
106.	Hojman et al.	1993	ApJ, 411, 541	NS		HALO	NS popul at MW halo boundary expected by hydro density jump
107.	Dar et al.	1992	ApJ, 388, 164	WD		COS	WD accretes to form naked NS, GRBs, cosmic rays
108.	Thompson et al.	1993	ApJ, 408, 194	NS		COS	Sudden NS convection with high B drives e- pairs, gammas.
109.	Hanami	1992	ApJ, 389, L71	NS	PLAN	COS	NS - planet magnetospheric interaction unstable
110.	Meszaros et al.	1992	ApJ, 397, 570	NS	NS	COS	NS - NS collision produces anisotropic fireball
111.	Eichler et al.	1992	Science, 257, 937	NS		HALO	High vel halo pulsars accrete after being kicked from disk
112.	Eichler et al.	1992	Science, 257, 937	WD	WD	HALO	WD merger yields GRB
113.	Carter	1992	ApJ, 391, L67	BH	ST	COS	Normal stars tidally disrupted by galactic nucleus BH
114.	Usov	1992	Nature, 357, 472	NS		COS	WD collapses to form NS, B-field breaks NS rotation instantly
115.	Blaes et al.	1992	ApJ, 399, 634	NS		GAL	Old NS accretes from mol cloud, R-T instab at crust
116.	Narayan et al.	1992	ApJ, 395, L83	NS	NS	COS	NS - NS merger gives optically thick fireball
117.	Narayan et al.	1992,	ApJ, 395, L83	BH	NS	COS	BH-NS merger gives optically thick fireball
118.	Brainerd	1992	ApJ, 394, L33	AGN	JET	COS	Synchrotron emission from AGN jets
119.	Smith et al.	1993	ApJ, 410, 315	NS		DISK	e- beams accel by E-fields near NS with high B
120.	Meszaros et al.	1992	MNRAS, 257, 29P	BH	NS	COS	BH-NS have vs collide to γs in clean fireball
121.	Meszaros et al.	1992	MNRAS, 257, 29P	NS	NS	COS	NS-NS have vs collide to γs in clean fireball
122.	Fatuzzo et al.	1993	ApJ, 407, 680	NS		COS	Alfen waves accel particles which upscatter soft photons
123.	Bisnovatyi-Kogan	1993	A&A Sup, 97, 65	NS		GAL	Absorption by cloud of heavy elements around NS
124.	McBreen et al.	1993	A&A Sup, 97, 81	AGN		COS	Relativistic jets from cocooned AGN
125.	Cline et al.	1992	ApJ, 401, L57	BH		DISK	Primordial BHs evaporating could account for short hard GRBs
126.	Woosley	1993	ApJ, 405, 273	BH		COS	Spinning Wolf-Ray star collapses, failed SN, emits beamed fireball
127.	Melia et al.	1992	ApJ, 398, L85	NS		COS	Crustal adjustments by extragal radio pulsars
128.	Rees et al.	1992	MNRAS, 258, 41P	NS	ISM	COS	Relativistic fireball reconverted to radiation when hits ISM
129.	Kundt et al.	1993	Ap&SS, 200, 151	NS	BH	COS	Spasmodic NS accretion causes beamed cooling `sparks'
130.	Meszaros et al.	1993	ApJ, 405, 278	NS		GAL	Compact binary coalesces, fireball hits external medium
131.	Cheng et al.	1993	MNRAS, 262, 1037	NS		GAL	NS glitch reignites magnetosphere of dead pulsar
132.	Melia et al.	1993	ApJ, 408, L9	NS		COS	NS structural readjustments explain both SGRs and GRBs
133.	Piran et al.	1993	ApJ, 403, L67	NS		GAL	Galactic fireball requires rel ejecta, low T, possible but unlikely
134.	Fabian et al.	1993	MNRAS, 263, 49	NS		LMC	NS accretes after ejected from Mag Cloud by companion SN
135.	Fatuzzo et al.	1993	ApJ, 414, L89	NS		COS	Sheared Alfen waves in NS magsphere dissipate focused power

Gamma-Ray Burst Workshop Attendees
Huntsville, Alabama USA
October 20–22, 1993

Akerlof, Carl W.	University of Michigan
Albergmini, Federico	Instituto TESRE, CNR, Italy
Aptekar, Raphail L.	A.F. Ioffe Phyico-Technical Institute, Russia
Araya, Rafael	Johns Hopkins University
Atteia, Jean Luc	CESR, France
Band, David L.	University of California-San Diego
Baring, Mathew G.	NASA/GSFC
Barthelmy, Scott	NASA/GSFC
Belli, Bianca	Instituto di Astrofisica Spaziale, CNR, Italy
Bionta, Richard M.	Lawrence Livermore National Laboratory
Boer, Michel	CESR, France
Bonnell, Jerry	NASA/GSFC
Boynton, William V.	University of Arizona
Brainerd, Jerome J.	University of Alabama in Huntsville
Briggs, Michael S.	University of Alabama in Huntsville
Brock, Martin	NASA/MSFC
Brown, Lawrence E.	Clemson University
Buccheri, Lino	IFCAI-CNR, Italy
Bulik, Thomasz	University of Chicago
Bunner, Alan	NASA Headquarters
Castro-Tirado, Alberto J.	Danish Space Research Institute, Denmark
Cebral, Juan R.	George Mason University
Chaganti, V. S.	University of Alabama in Huntsville
Chang, Tom	Massachusetts Institute of Technology
Cheeseman, Peter	NASA/Ames
Chernenko, Anton	Space Research Institute, Russia
Chipman, Eric	NASA/GSFC
Chuang, Kuan Wen	University of California-Riverside
Cline, David	University of California-Los Angeles
Cline, Thomas	NASA/GSFC
Cole, Stayce	NASA/MSFC
Colgate, Stirling A.	Los Alamos National Laboratory
Connaughton, Valerie	Smithsonian Institution Fred Lawrence Whipple Observatory
Connors, Alanna	University of New Hampshire
Coppi, Paolo	University of Chicago
Coyne, Donald	University of California-Santa Cruz
Davis, Stanley P.	NASA/GSFC
De Paolis, Francesco	Universita di Lecce, Italy

Workshop Attendees

DeMuth, David	University of Minnesota
Dermer, Charles	Naval Research Laboratory
Dezalay, Jean Pascal	CESR, France
Dingus, Brenda L.	NASA/GSFC
Duncan, Robert C.	University of Texas
Elsner, Ronald	NASA/MSFC
Epstein, Richard	Los Alamos National Laboratory
Fatuzzo, Marco	University of Michigan
Fenimore, Edward	Los Alamos National Laboratory
Fichtel, Carl E.	NASA/GSFC
Finger, Mark	Computer Sciences Corporation/MSFC
Fishman, Gerald	NASA/MSFC
Ford, Lyle	University of California-San Diego
Forrest, David J.	University of New Hampshire
Frail, Dale A.	NRAO
Freeman, Peter E.	University of Chicago
Fry, Jack	University of Wisconsin
Gehrels, Neil	NASA/GSFC
Gilfanov, Marat	Space Research Institute, Russia
Graziani, Carlo	NASA/GSFC
Greiner, Jochen	Max-Plank-Institut für Extraterrestrische Physik, Germany
Guarnieri, A.	Universita de Bologna, Italy
Gursky, Herbert	Naval Research Laboratory
Hack, Felix	University of California-Berkeley
Hagedon, Kathy	Universities Space Research Association/MSFC
Hakkila, Jon	Mankato State University
Hanlon, Lorraine	ESA, ESTEC, The Netherlands
Harding, Alice K.	NASA/GSFC
Harmon, Alan	NASA/MSFC
Hartmann, Dieter	Clemson University
Henze, William	Teledyne Brown/MSFC
Horack, John M.	NASA/MSFC
Horvath, Jorge E.	Rice University
Hudec, Rene	Astronomical Institute, Czech Republic
Hurley, Kevin	University of California-Berkeley
in't Zand, J. J. M.	Los Alamos National Laboratory
Iping, Rosina C.	University of Guam
Isern, Jordi	Centre D'Estudis Avancats De Blanes, Spain
Kargatis, Vincent	Rice University
Katz, Jonathan	Washington University
Kerr, Frank	Universities Space Research Association/GSFC
Kieda, David	University of Utah
Kippen, R. Marc	University of New Hampshire

Kluzniak, Wlodzimierz — University of Wisconsin
Koshut, Thomas — University of Alabama in Huntsville
Kouveliotou, Chryssa — Universities Space Research Association/MSFC
Krimm, Hans A. — Massachusetts Institute of Technology
Lamb, Don Q. — University of Chicago
Laros, John — Los Alamos National Laboratory
Lestrade, John Patrick — Mississippi State University
Lewin, Walter H. G. — Massachusetts Institute of Technology
Li, Hui — Rice University
Li, Peng — University of California-Berkeley
Liang, Edison P. — Rice University
Lingenfelter, Richard E. — University of California-San Diego
Loeb, Abraham — Harvard University
Loredo, Thomas J. — Cornell University
Luginbuhl, Christian B. — U.S. Naval Observatory
MacCallum, Crawford — University of New Mexico
Maccarone, Maria C. — CNR, Italy
Mahoney, William — Jet Propulsion Laboratory
Mallozzi, Robert S. — University of Alabama in Huntsville
Mao, Shude — Harvard-Smithsonian Center For Astrophysics
Marani, Gabriela F. — George Mason University
Matheis, Volker — Max-Plank-Institut für Kernphysik, Germany
Mathews, Grant — Lawrence Livermore National Laboratory
Matteson, James — University of California-San Diego
Matz, Steven M. — Northwestern University
McCollough, Michael — Hughes STX/MSFC
McConnell, Mark L. — University of New Hampshire
McNamara, Bernard — New Mexico State University
Meegan, Charles — NASA/MSFC
Meier, Mike — Los Alamos National Laboratory
Meredith, Dawn — University of New Hampshire
Meszaros, Peter — Pennsylvania State University
Miller, M. Coleman — University of Chicago
Mitrofanov, Igor G. — Space Research Institute, Russia
Miyaji, Shigeki — Chiba University, Japan
Mochkovitch, Robert — Institut D'Astrophysique De Paris, France
Moore, Philip — Boeing/MSFC
Murakami, Toshio — Institute of Space and Astronautical Science, Japan
Narayan, Remesh — Harvard-Smithsonian Center for Astrophysics
Nelson, Robert W. — University of Toronto, Canada
Nemiroff, Robert — NASA/GSFC
Norris, Jay P. — NASA/GSFC
Owens, Alan — Leicester University, United Kingdom
Paciesas, William — University of Alabama in Huntsville

Paczynski, Bohdan	Princeton University
Palmer, David	NASA/GSFC
Park, Hye-Sook	Lawrence Livermore National Laboratory
Pedersen, Holger	Copenhagen University Observatory, Denmark
Pelaez, Francois	University of Chicago
Pendleton, Geoffrey N.	University of Alabama in Huntsville
Peterson, Burl	Universities Space Research Association/MSFC
Petrosian, Vahe	Stanford University
Pilla, Ravi P.	Columbia University
Piran, Tsvi	Hebrew University, Israel
Pizzichini, Graziella	TESRE, CNR, Italy
Pozanenko, Alexi	Space Research Institute, Russia
Preece, Robert D.	National Research Council/MSFC
Quashnock, Jean M.	University of Chicago
Rees, Martin S.	Institute of Astronomy, United Kingdom
Ricker, George R.	Massachusetts Institute of Technology
Rubin, Bradley	Universities Space Research Association/MSFC
Rutledge, Robert	Massachusetts Institute of Technology
Ryan, James M.	University of New Hampshire
Schaefer, Bradley E.	Universities Space Research Association/GSFC
Schmidt, Maarten	California Institute of Technology
Schnee, Richard	University of California-Santa Cruz
Schneid, Edward J.	Grumman Corporation Research Center
Share, Gerald	Naval Research Laboratory
Shemi, Amotz	Tel Aviv University, Israel
Sina, Ramin	NASA/GSFC
Skelton, R. Thomas	Jet Propulsion Laboratory
Slawinski, Raphael	University of Chicago
Smith, Ian	Rice University
Stanek, Krzysztof Z.	Princeton Observatory
Stollberg, Mark T.	University of Alabama in Huntsville
Strohmayer, Tod	Los Alamos National Laboratory
Sutherland, Peter G.	McMaster University, Canada
Takahashi, Y.	University of Alabama in Huntsville
Tanberg-Hanssen, Einar	NASA/MSFC
Tavani, Marco	Princeton University
Teegarden, Bonnard	NASA/GSFC
Terekhov, Oleg	Space Research Institute, Russia
Terrell, James	Los Alamos National Laboratory
Titarchuk, Lev	NASA/GSFC
Trimble, Virginia L.	University of Maryland
Van Paradijs, Jan	University of Amsterdam, The Netherlands, University of Alabama in Huntsville
Vander Velde, Jack	University of Michigan

Vanderspek, Roland	Massachusetts Institute of Technology
Varmette, Peter	Mississippi State University
Vilhu, Osmi	University of Helsinki, Finland
Vo, Van	Mankato State University
Vrba, Frederick J.	U.S. Naval Observatory
Wang, John C. L.	Canadian Institute for Theoretical Astrophysics, Canada
Wang, Virginia C.	University of California, San Diego
Weekes, Trevor C.	Smithsonian Institution Fred Lawrence Whipple Observatory
White, R. Stephen	University of California-Santa Barbara
Wickramasinghe, W. A. D. T.	University of Pennsylvania
Wilson, Robert B.	NASA/MSFC
Wilson-Hodge, Colleen	NASA/MSFC
Winkler, Christoph	ESA, ESTEC, The Netherlands
Woods, Eric	Harvard University
Woosley, Stanford	University of California-Santa Cruz
Yi, Insu	Harvard-Smithsonian Center for Astrophysics
Yoshida, Atsumasa	Institute of Physical and Chemical Research, Japan
Zhang, Shuang-Nan	Universities Space Research Association/MSFC

Author Index

A

Akerlof, C. W., 470, **633**
Alberghini, F., **638**
Alexandreas, D. E., 481
Allen, G. E., 481
Andersen, M., 670
Aoki, T., 489
Atteia, J.-L., 359
Azzam, W. J., 84, 88, 93

B

Bade, N., 408
Band, D. L., **39**, 247, **256**, 261, 266, 271, 280, 283, 298
Barat, C., 207, 359
Baring, M. G., 520, **572**,
Barthelmy, S. D., **392**, 401, 633, **643**
Begelman, M. C., 605
Belli, B. M., **192**
Bennett, K., 275, 418
Berley, D., 481
Bertsch, D. L., 22
Bhat, P. N., 167, **197**, **288**
Bialas, T. G., 643
Biller, S., 481
Bionta, R., 633
Birkle, K., 408
Bade, N.,
Blumenthal, G. R., **117**, 127
Boër, M., 27, 364, 369, 373, **396**, **453**, **458**, 670
Bonnell, J. T., **401**
Boynton, W., 32
Brainerd, J. J., **122**, **346**
Brandt, S., **13**, 17, 364, 404
Briggs, M. S., 3, **44**, 59, 127, 167, 212, 247, 256, 261, 266, 271, 303, 308, 313, 318, 672, 711
Brock, M. N., 3, 44, 59, 127, 266, 318, 648, **672**, 697, 711

*Note: Boldface page numbers denote papers of which an individual is the first author.

Brown, L. E., **49**
Buccheri, R., **377**
Burman, R. L., 481

C

Camerini, U., 377
Castro-Tirado, A. J., 13, **17**, 364, **404**
Cavalli-Sforza, M., 481
Cebral, J. R., 137
Chaffee, F. H., 448
Chang, H.-K., 596
Chang, C. Y., 481
Chantell, M., 470
Chen, M. L., 481
Chernenko, A. M., 187, **293**
Chernych, N. S., 408
Chipman, E., **202**
Chuang, K. W., **145**
Chumney, P., 481
Cline, D. B., **577**
Cline, T. L., 27, 247, 271, 280, 359, 364, 369, **373**, 401, 458, 633, 643, 653
Colgate, S. A., **581**
Collmar, W., 275, 283, 418
Connaughton, V., **470**
Connors, A., 275, 283, 418
Coyne, D., 481
CYGNUS Collaboration, 481

D

Dal Fiume, D., 638
Davis, S. P., 172, **182**
De Paolis, F., **323**
DeMuth, D. M., **475**
den Herder, J. W., 275
Dennis, B., 359
Dermer, C. D., **510**
Diehl, R., 275
Dingus, B. L., **22**, 280, 283
Dion, C., 481
Dion, G. M., 481
Dorfan, D., 481

Duncan, R. C., 79, 600, **625**

E

Eichler, D., **54**
Ellsworth, R. W., 481
Emslie, A. G., **64**, 69

F

Fatuzzo, M., **515**, 633
Fegan, D. J., 470
Fenimore, E. E., 155, 692
Fennell, S., 470
Fichtel, C. E., 22, 283
Fishman, G. J., 3, 27, 44, 59, 122, 127, 150, 167, 172, 182, 187, 197, 212, 237, 247, 280, 283, 288, 313, 364, 369, 373, 396, 401, 408, 418, 428, 458, 643, **648**, 697, 711
Foltz, C. B., 448
Ford, L. A., 247, 256, **261**, 266, 271, **298**
Frail, D. A., **486**
Freeman, P. E., **677**
Frontera, F., 638
Fry, W. F., 377

G

Gaidos, J., 470
Gehrels, N., 401, 633, 643
Getman, V. S., 408
Goodman, J. A., 481
Graziani, C., 227, 677
Greiner, J., 275, **408**, 418, 453, 458, 665
Griffee, J. W., 34
Guziy, S. S., 404

H

Hack, F., 359
Hagan, J., 470
Haines, T. J., 481
Hakkila, J., 3, **59**, 127

Hanlon, L. O., **275**, 418
Harding, A. K., 22, **520**
Harmon, M., 481
Harrison, T., 418, 433
Hartman, R. C., 22
Hartmann, D. H., 49, 117, **127**, 443, 448, **562**
Hermsen, W., 275, 280, 418
Hernanz, M., 525, 537
Higdon, J. C., 222, **586**
Hillas, A. M., 470
Hoffman, C. M., 481
Hong, W., 577
Horack, J. M., 3, 59, **64**, **69**, 127, 648
Horvath, J. E., **591**
Hudec, R., 408, **413**, 448
Hunter, S. D., 22
Hurley, K. C., 22, **27**, 88, 127, 207, 359, **364**, 369, 373, 396, 428, 448, 458, **653**, **682**, **687**, **726**

I

in 't Zand, J. J. M., **692**
Isern, J., **525**, 537

J

Jennings, M. C., 443
Johnson, W. N., 283
Jung, G. V., 283

K

Kahabka, P., 453, 458
Kahn, A., **463**
Kanbach, G., 22, 377
Kargatis, V. E., **207**, 217
Katz, J. I., **529**
Kawai, N., **658**
Kelley, L., 481
Kerrick, A. D., 470
Kippen, R. M., 275, 280, **418**
Klebesadel, R. W., 34, 359, 364, 373, 458

Author Index

Klein, S., 481
Kniffen, D. A., 22
Koshut, T. M., 3, 69, 167, **303**, 308, 313, 648, **697**
Kouveliotou, C., 3, 27, 44, 127, 150, **167**, 172, 182, 187, 232, 237, 288, 313, 318, 328, 359, 364, 369, 373, 396, 401, 408, 418, 428, 458, 643, 648, 672
Krimm, H. A., **423**, 438
Kuiper, L., 275, 283, 418
Kulkarni, S. R., 486, 489
Kundt, W., **596**
Kurfess, J. D., 283
Kurt, V. G., 141, 359
Kuyper, J. R., 643
Kuznetsov, A., 359
Kwok, P. W., 280, 283, 470

L

Lamb, D. Q., **74**, 107, 227, 677
Lamb, R. C., 470
Lapshov, I. Y., 17, 364
Laros, J. G., **32**, 359, 364, 373, 458
Ledebuhr, A., 633
Lee, B., 633
Lee, P., 34
Lee, T. T., 93
Leonard, P. J. T., 581
Lestrade, J. P., **212**, 247, 701, 711
Lewin, W. H. G., 98, 242
Lewis, D. A., 470
Li, H., **79**, 207, **217**, **600**
Li, P., **428**
Liang, E. P., 207, 217, **351**, **662**
Lin, Y. C., 22
Linder, E. V., 117, 127
Lingenfelter, R. E., 160, **222**, 586
Loeb, A., **533**
Loiseau, S., 537,
Loredo, T. J., 677
Luginbuhl, C. B., 448
Lund, N., 13, 17, 364, 404

M

Maccarone, M. C., 377
Macri, J., 418
Mahoney, W. A., 706
Mallozzi, R. S., 69, 303, **308**, 313, 648
Marani, G. F., 137
Marshak, M. L., 475
Martin, X., 525
Matteson, J. L., 3, 247, 256, 261, 266, 271, 283, 298
Mattox, J. R., 22, 377
Matz, S. M., 283
Mayer-Hasselwander, H. A., 22
McCloskey, R., 32
McConnell, M., 275, 418
McNamara, B., 418, **433**
Meegan, C. A., **3**, 27, 44, 59, 84, 122, 127, 150, 167, 172, 182, 187, 197, 212, **232**, 237, 247, 280, 288, 313, 364, 369, 373, 401, 408, 418, 643, 648, 672, 697, 711
Melia, F., 515
Meredith, D. C., **701**
Mészáros, P., **505**, 605
Metlov, V., 408
Metzger, A., 32
Meyer, D. I., 470
Michelson, P. F., 22
Mineo, T., 665
Miralles, J. A., 155
Mitrofanov, I. G., **187**, 293, 722
Mitruka, S., 271
Mochkovitch, R., 525, **537**
Mohanty, G., 470
Moskalenko, E. I., 408
Motch, C., 453, 458
Murakami, T., 333, 466, **489**
Murphy, R. J., 283

N

Nagle, D. E., 481
Narayan, R., **132**
Nelson, R. W., 615
Nemiroff, R. J., **137**, **150**, 172, **237**, 328, 730

Niel, M., 27, 207, 359, 364, 369, 373, 458
Nolan, P. L., 22, 283
Norris, J. P., 137, 150, **172**, **177**, 182, 237, 328, 401

O

Ogasaka, Y., 466, 489
Ögelman, H., 463
Owens, A., **336**, 665

P

Paciesas, W. S., 3, 44, 59, 122, 127, 150, 167, 172, 182, 187, 197, 212, 237, 247, 256, 261, 266, 271, 280, 288, 298, 303, 308, 313, 318, 373, 408, 648, 672, 697, 711
Paczyński, B., **542**
Palmer, D. M., **247**, 256, 261, 266, 271, 392
Park, H.-S., 633
Pedersen, H., 396, **670**
Pendleton, G. N., 3, 44, 59, 127, 167, 247, 256, 261, 271, 280, 288, 303, 308, **313**, 318, 648, 672, 697, 711
Petrosian, V., **84**, **88**, **93**
Pezzuto, S., 323
Piran, T., 132, **495**, **543**
Pizzichini, G., 638
Porter, N. A., 470
Pounds, K., 665
Pozananko, A. S., 187
Preece, R. D., 247, 256, 261, **266**, 271, **318**, 672, 711
Punch, M., 470

Q

Quashnock, J. M., 74, **107**

R

Rees, M. J., 505, **605**, 665
Rhoads, J., 542
Ricker, G. R., 423, 438
Robbins, M. A., 643
Rose, J., 470
Rovero, A. C., 470
Rutledge, R., **98**, **242**
Ryan, J. M., 275, 418, 701

S

Sacco, B., 665
Sagdeev, R. Z., 187
Scarsi, L., 665
Schaefer, B. E., 247, 256, 261, 266, **271**, **280**, 283, 288, **341**, **382**, 392, 665
Schaller, S. C., 481
Schlickeiser, R., 510
Schmidt, D. M., 481
Schnee, R., **481**
Schneid, E. J., 22, 280, 283
Schönfelder, V., 275, 280, 283, 418
Schubnell, M. S., 470
Sembay, S., 665
Sembroski, G., 470
Share, G. H., 88, **283**
Shemi, A., **547**, **548**
Shoup, A, 481
Sinnis, C., 481
Skelton, R. T., 706
Smette, A., 396
Smith, I. A., 79, **103**, 207, 217, **610**
Soldán, J., 413
Sommer, M., 22, 27, 364, 369, 373, 458, 682
Sonobe, T., 489
Sreekumar, R., 22
Stark, M. J., 481
Starr, R., 32
Steinle, H., 275
Storey, S. D., 69
Strohmayer, T. E., **155**
Strong, A., 275
Sunyaev, R. A., 17

T

Tavani, M., 323
Teegarden, B. J., 3, 247, 256, 261, 266, 271, 280, 283, 298
Terekhov, O., 17, 665
Terrell, J., **34**
The, L.-S., 49
Thompson, C., 600, 625
Thompson, D. J., 22, 377
Trimble, V., 717
Tritton, S. B., 408
Trombka, J., 32

U

Usov, V. V., **522**

V

Vanderspek, R. K., 423, 438
van Dijk, R., 275
Varendorff, M., 275, 418
Varmette, P. G., 711
Voges, W., 453, 458
von Montigny, C., 22
Vrba, F. J., **443**, **448**

W

Wagner, G. L., 475
Wang, J. C. L., 615
Wang, V. C., **160**, 222
Weekes, T. C., 470
Weeks, D. D., 481

Wells, A., 665
Wenzel, W., 408
West, M., 470
Whitaker, T., 470
White, R. S., 620
Wickramasinghe, W. A. D. T., 150, **328**
Williams, C., 433
Williams, D. A., 481
Williams, O. R., 275
Wilson, C., 470
Wilson, R. B., 3, 44, 59, 127, 197, 212, 247, 280, 288, 313, 373, 648, 672, 697
Winkler, C., 275, 280, 283, 418
Wu, J.-P., 481

X

Xu, G., 542

Y

Yang, T., 481
Yi, I., **557**
Yodh, G. B., 481
Yoshida, A., **333**, **466**, 489
Young, C. A., 701

Z

Zenchenko, V., 359
Zhang, W., 481
Zharkov, G. F., **141**
Zharkov, V. G., 141
Ziener, R., 408

AIP Conference Proceedings

		L.C. Number	ISBN
No. 261	Rare and Exclusive B&K Decays and Novel Flavor Factories (Santa Monica, CA, 1991)	92-71873	1-56396-055-9
No. 262	Molecular Electronics—Science and Technology (St. Thomas, Virgin Islands, 1991)	92-72210	1-56396-041-9
No. 263	Stress-Induced Phenomena in Metallization: First International Workshop (Ithaca, NY, 1991)	92-72292	1-56396-082-6
No. 264	Particle Acceleration in Cosmic Plasmas (Newark, DE, 1991)	92-73316	0-88318-948-8
No. 265	Gamma-Ray Bursts (Huntsville, AL, 1991)	92-73456	1-56396-018-4
No. 266	Group Theory in Physics (Cocoyoc, Morelos, Mexico, 1991)	92-73457	1-56396-101-6
No. 267	Electromechanical Coupling of the Solar Atmosphere (Capri, Italy, 1991)	92-82717	1-56396-110-5
No. 268	Photovoltaic Advanced Research & Development Project (Denver, CO, 1992)	92-74159	1-56396-056-7
No. 269	CEBAF 1992 Summer Workshop (Newport News, VA, 1992)	92-75403	1-56396-067-2
No. 270	Time Reversal—The Arthur Rich Memorial Symposium (Ann Arbor, MI, 1991)	92-83852	1-56396-105-9
No. 271	Tenth Symposium Space Nuclear Power and Propulsion (Vols. I–III) (Albuquerque, NM, 1993)	92-75162	1-56396-137-7 (set)
No. 272	Proceedings of the XXVI International Conference on High Energy Physics (Vols. I and II) (Dallas, TX, 1992)	93-70412	1-56396-127-X (set)
No. 273	Superconductivity and Its Applications (Buffalo, NY, 1992)	93-70502	1-56396-189-X
No. 274	VIth International Conference on the Physics of Highly Charged Ions (Manhattan, KS, 1992)	93-70577	1-56396-102-4
No. 275	Atomic Physics 13 (Munich, Germany, 1992)	93-70826	1-56396-057-5

No.	Title		
No. 276	Very High Energy Cosmic-Ray Interactions: VIIth International Symposium (Ann Arbor, MI, 1992)	93-71342	1-56396-038-9
No. 277	The World at Risk: Natural Hazards and Climate Change (Cambridge, MA, 1992)	93-71333	1-56396-066-4
No. 278	Back to the Galaxy (College Park, MD, 1992)	93-71543	1-56396-227-6
No. 279	Advanced Accelerator Concepts (Port Jefferson, NY, 1992)	93-71773	1-56396-191-1
No. 280	Compton Gamma-Ray Observatory (St. Louis, MO, 1992)	93-71830	1-56396-104-0
No. 281	Accelerator Instrumentation Fourth Annual Workshop (Berkeley, CA, 1992)	93-072110	1-56396-190-3
No. 282	Quantum 1/f Noise & Other Low Frequency Fluctuations in Electronic Devices (St. Louis, MO, 1992)	93-072366	1-56396-252-7
No. 283	Earth and Space Science Information Systems (Pasadena, CA, 1992)	93-072360	1-56396-094-X
No. 284	US-Japan Workshop on Ion Temperature Gradient-Driven Turbulent Transport (Austin, TX, 1993)	93-72460	1-56396-221-7
No. 285	Noise in Physical Systems and 1/f Fluctuations (St. Louis, MO, 1993)	93-72575	1-56396-270-5
No. 286	Ordering Disorder: Prospect and Retrospect in Condensed Matter Physics: Proceedings of the Indo-U.S. Workshop (Hyderabad, India, 1993)	93-072549	1-56396-255-1
No. 287	Production and Neutralization of Negative Ions and Beams: Sixth International Symposium (Upton, NY, 1992)	93-72821	1-56396-103-2
No. 288	Laser Ablation: Mechanismas and Applications-II: Second International Conference (Knoxville, TN, 1993)	93-73040	1-56396-226-8
No. 289	Radio Frequency Power in Plasmas: Tenth Topical Conference (Boston, MA, 1993)	93-72964	1-56396-264-0
No. 290	Laser Spectroscopy: XIth International Conference (Hot Springs, VA, 1993)	93-73050	1-56396-262-4

No. 291	Prairie View Summer Science Academy (Prairie View, TX, 1992)	93-73081	1-56396-133-4
No. 292	Stability of Particle Motion in Storage Rings (Upton, NY, 1992)	93-73534	1-56396-225-X
No. 293	Polarized Ion Sources and Polarized Gas Targets (Madison, WI, 1993)	93-74102	1-56396-220-9
No. 294	High-Energy Solar Phenomena A New Era of Spacecraft Measurements (Waterville Valley, NH, 1993)	93-74147	1-56396-291-8
No. 295	The Physics of Electronic and Atomic Collisions: XVIII International Conference (Aarhus, Denmark, 1993)	93-74103	1-56396-290-X
No. 296	The Chaos Paradigm: Developments an Applications in Engineering and Science (Mystic, CT, 1993)	93-74146	1-56396-254-3
No. 297	Computational Accelerator Physics (Los Alamos, NM, 1993)	93-74205	1-56396-222-5
No. 298	Ultrafast Reaction Dynamics and Solvent Effects (Royaumont, France, 1993)	93-074354	1-56396-280-2
No. 299	Dense Z-Pinches: Third International Conference (London, 1993)	93-074569	1-56396-297-7
No. 300	Discovery of Weak Neutral Currents: The Weak Interaction Before and After (Santa Monica, CA, 1993)	94-70515	1-56396-306-X
No. 301	Eleventh Symposium Space Nuclear Power and Propulsion (3 Vols.) (Albuquerque, NM, 1994)	92-75162	1-56396-305-1 (Set) 156396-301-9 (pbk. set)
No. 302	Lepton and Photon Interactions/ XVI International Symposium (Ithaca, NY, 1993)	94-70079	1-56396-106-7
No. 304	The Second Compton Symposium (College Park, MD, 1993)	94-70742	1-56396-261-6
No. 305	Stress-Induced Phenomena in Metallization Second International Workshop (Austin, TX, 1993)	94-70650	1-56396-251-9
No. 306	12th NREL Photovoltaic Program Review (Denver, CO, 1993)	94-70748	1-56396-315-9